Quantum Mechanics

A Graduate Course

Written for a two-semester graduate course in quantum mechanics, this comprehensive text helps the reader to develop the tools and formalism of quantum mechanics and its applications to physical systems. It will suit students who have taken some introductory quantum mechanics and modern physics courses at undergraduate level, but it is self-contained and does not assume any specific background knowledge beyond appropriate fluency in mathematics. The text takes a modern logical approach rather than a historical one and it covers standard material, such as the hydrogen atom and the harmonic oscillator, the WKB approximations, and Bohr–Sommerfeld quantization. Important modern topics and examples are also described, including Berry phase, quantum information, complexity and chaos, decoherence and thermalization, and nonstandard statistics, as well as more advanced material such as path integrals, scattering theory, multiparticles, and Fock space. Readers will gain a broad overview of quantum mechanics as a solid preparation for further study or research.

Horaţiu Năstase is a Researcher at the Institute for Theoretical Physics, State University of São Paulo. He completed his PhD at Stony Brook with Peter van Nieuwenhuizen, co-discoverer of supergravity. While in Princeton as a postdoctoral fellow, in a 2002 paper with David Berenstein and Juan Maldacena he started the pp-wave correspondence, a sub-area of the AdS/CFT correspondence. He has written more than 100 scientific articles and five other books, including *Introduction to the AdS/CFT Correspondence* (2015), *String Theory Methods for Condensed Matter Physics* (2017), *Classical Field Theory* (2019), *Introduction to Quantum Field Theory* (2019), and *Cosmology and String Theory* (2019).

"There are a few dozen books on quantum mechanics. Who needs another one? Graduate students. Most of the existing books aim at the undergraduate level. Often, students of graduate courses of QM are all familiar with the motivation and its historical development but have a very varied technical background. This book skips the history of the field and enables all the students to reach a common needed level after going over the first chapters.

This book spans a very wide manifold of QM topics. It provides the tools for further researching any quantum system. An important further value of the book is that it covers the forefront aspects of QM including the Berry phase, anyons, entanglement, quantum information, quantum complexity and chaos, and quantum thermalization. Readers understand a topic only if they are able to solve problems associated with it; the book includes at least 7 exercises for each of its 59 chapters."

— **Professor Jacob Sonnenschein, Tel Aviv University**

"Năstase's *Quantum Mechanics* is another marvellous addition to his encyclopaedic collection of graduate-level courses that take both the dedicated student and the hardened researcher on a grand tour of contemporary theoretical physics. It will serve as an ideal bridge between a comprehensive undergraduate course in quantum mechanics and the frontiers of research in quantum systems."

— **Professor Jeff Murugan, University of Cape Town**

Quantum Mechanics

A Graduate Course

HORAŢIU NĂSTASE
Universidade Estadual Paulista, São Paulo

CAMBRIDGE
UNIVERSITY PRESS

University Printing House, Cambridge CB2 8BS, United Kingdom

One Liberty Plaza, 20th Floor, New York, NY 10006, USA

477 Williamstown Road, Port Melbourne, VIC 3207, Australia

314–321, 3rd Floor, Plot 3, Splendor Forum, Jasola District Centre,
New Delhi – 110025, India

103 Penang Road, #05–06/07, Visioncrest Commercial, Singapore 238467

Cambridge University Press is part of the University of Cambridge.

It furthers the University's mission by disseminating knowledge in the pursuit of education, learning, and research at the highest international levels of excellence.

www.cambridge.org
Information on this title: www.cambridge.org/highereducation/isbn/9781108838733
DOI: 10.1017/9781108976299

First published 2023

A catalogue record for this publication is available from the British Library.

Library of Congress Cataloging-in-Publication Data
Names: Năstase, Horaţiu, 1972– author.
Title: Quantum mechanics : a graduate course / by Horatiu Nastase,
Instituto de Fisica Teorica, UNESP, Saõ Paulo, Brazil.
Description: Cambridge, United Kingdom ; New York, NY : Cambridge University Press, 2022. |
Includes bibliographical references.
Identifiers: LCCN 2022010290 | ISBN 9781108838733 (hardback)
Subjects: LCSH: Quantum theory. | BISAC: SCIENCE / Physics / Quantum Theory
Classification: LCC QC174.12 .N38 2022 | DDC 530.12–dc23/eng20220517
LC record available at https://lccn.loc.gov/2022010290

ISBN 978-1-108-83873-3 Hardback

Additional resources for this publication at www.cambridge.org/Nastase

To the memory of my mother,
who inspired me to become a physicist

Contents

Preface

There are so many books in quantum mechanics, that one has to ask: why write another one? When teaching graduate quantum mechanics in either the US or Brazil (countries I am familiar with), as well as in other places, one is faced with a conundrum: how to address all possible backgrounds, while keeping the course both interesting and also comprehensive enough to offer the graduate student a chance to follow competitively in any area? Indeed, while graduate students have usually very different backgrounds, there is usually a compulsory graduate quantum mechanics course, usually a two-semester one. The students will certainly have some introductory undergraduate quantum mechanics and some modern physics, but some have studied these topics in detail, while others less so.

To that end, I believe there is little need for a long historical introduction, or a detailed explanation of why we need to define quantum mechanics in the way we do. In order to challenge students, we cannot simply repeat what they heard in the undergraduate course. So one needs a tighter presentation, with more emphasis on the building up of the formalism of quantum mechanics rather than its motivation, as well as a presentation that contains more interesting new developments besides the standard advanced concepts. On the other hand, I personally found that many (even very bright) students are caught up between the two systems and miss on important information: they have had an introductory quantum mechanics course that did not treat all the standard classical material (examples: the hydrogen atom in all detail, the harmonic oscillator in various treatments, the WKB and Bohr–Sommerfeld formalisms), yet the graduate course assumes that students have covered all these topics and so they struggle in their subsequent research as a result.

Since I have found no graduate book that adheres to these conditions of comprehensiveness to my satisfaction, I decided to write one, and this is the result. The book is intended as a two-semester course, corresponding to the two parts, each chapter corresponding to one two-hour lecture, though sometimes an extended version of a lecture.

Acknowledgements

I should first thank all the people who have helped and guided me on my journey as a physicist, starting with my mother Ligia, who was the first example I had of a physicist, and from whom I first learned to love physics. My high school physics teacher, Iosif Sever Georgescu, helped to start me on the career of a physicist, and made me realize that it was something that I could do very well. My student exchange advisor at the Niels Bohr Institute, Poul Olesen, first showed me what string theory is, and started me in a career in this area, and my PhD advisor, Peter van Nieuwenhuizen, taught me the beauty of theoretical physics, of rigor and of long calculations, and an approach to teaching graduate physics that I still try to follow.

With respect to teaching quantum mechanics, most of the credit goes to my teachers in the Physics Department of Bucharest University, since during my four undergraduate years there (1990–1994), many of the courses there taught me about various aspects of quantum mechanics. Some recent developments described here I learned from my research, so I thank all my collaborators, students, and postdocs for their help with understanding them. My wife Antonia I thank for her patience and understanding while I wrote this book, as well as for help with an accident during the writing. I would like to thank my present editor at Cambridge University Press, Vince Higgs, who helped me get this book published, as well as my outgoing (now retired) editor Simon Capelin, for his encouragement in starting me on the path to publishing books. To all the staff at CUP, thanks for making sure this book, as well as my previous ones, is as good as it can be.

Introduction

As described in the Preface, this book is intended for graduate students, and so I assume that readers will have had both an introductory undergraduate course in quantum mechanics, which familiarized them with the historical motivation as well as the basic formalism. For completeness, however, I have included a brief historical background as an optional introductory chapter. I do assume a certain level of mathematical proficiency, as might be expected from a graduate student. Nevertheless, the first two chapters are dedicated to the mathematics of quantum mechanics; they are divided into finite- and infinite-dimensional Hilbert spaces, since these concepts are necessary to a smooth introduction.

I start the discussion of quantum mechanics with its postulates and the Schrödinger equation, after which I start developing the formalism. I mostly use the bra-ket notation for abstract states, as it is often cleaner, though for most of the standard systems I use wave functions. I introduce early on the notion of the path integral, since it is an important alternative way to describe quantization, through the sum over all paths, classical or otherwise, though I mostly use the operatorial formalism. I also introduce early on the notion of pictures, including the Heisenberg and interaction pictures, as alternatives for the description of time evolution. The important notion of angular momentum theory is presented within the larger context of symmetries in quantum mechanics, since this is the modern viewpoint, especially within the relativistic quantum field theory that extends quantum mechanics.

With respect to topics, in Part I, on basic concepts, I consider both the older, but still very useful, topics such as WKB and Bohr–Sommerfeld quantization as well as more recent ones such as the Berry phase and Dirac quantization condition. In Part II I consider scattering theory, variational methods, occupation number space, quantum entanglement and information, quantum complexity, quantum chaos, and thermalization; these are the more advanced topics. The advanced foundations of Quantum Mechanics (Part II_a) are discussed in Part II since they are newer and harder to understand even though they could be said to belong to Part I for being foundational issues.

After each chapter, I summarize a set of "Important Concepts to Remember", and present seven exercises whose solution is meant to clarify the concepts in the chapter.

PART I

FORMALISM AND BASIC PROBLEMS

Introduction: Historical Background

In this book, we will describe the formalism and applications of quantum mechanics, starting from the first principles and postulates, and expanding them logically into the fully developed system that we currently have. That means that we will start with those principles and postulates, assuming that the reader has had an undergraduate course in modern physics, as well as an undergraduate course in quantum mechanics. This implies a first interaction with the ideas of quantum mechanics, and the historical development and experiments that led to it. However, for the sake of consistency and completeness, in this chapter we will quickly review these historical developments and how we were led to the current formalism for quantum mechanics.

In classical mechanics and optics, there were deterministic laws, involving two types of objects: particles of matter, following well-defined classical paths, with their evolution defined by their Hamiltonian and the initial conditions, and waves for light, defined by wave functions (for the electric and magnetic fields $\vec{E}$ and $\vec{B}$), and the observable intensities I defined by them, showing interference patterns in the sum of waves. But there was no overlap between the two descriptions, in terms of particles and waves. Notably, Newton had a rudimentary particle description for light (in his *Optics*), but with the development of the classical wave picture of Huyghens for optics, that description was forgotten for a long time.

0.1 Experiments Point towards Quantum Mechanics

Then, around the turn of the nineteenth century into the twentieth, a string of developments changed the classical picture, allowing for a complementary particle description for waves, and for a complementary wave description for particles. One of the first such developments was the discovery of radioactivity by Henri Becquerel in 1896, who, after Wilhelm Roentgen's discovery of X-rays in 1895, showed that radioactive elements produce rays similar to X-rays in that they can go through matter and then leave a pointlike mark on a photographic plate, thus involving emission of highly energetic particles. Now we know that these emitted particles can be α (nucleons), β (electrons and positrons), but also γ (high-energy photons, so lightlike).

Next came what is believed to be the start of the modern quantum era, with the first theoretical idea of a quantum, specifically of light, used to describe blackbody radiation. In 1900, Planck managed to explain the blackbody emission spectrum with a simple but revolutionary idea. Assuming that the energy exchange between matter and the emitted radiation occurs not continuously (as assumed in classical physics), but only in given quanta of energy, proportional to the frequency ν of the radiation,

$$\Delta E = h\nu, \tag{0.1}$$

where the constant h is called the Planck constant, Planck managed to derive a formula for the spectral radiance:

$$B_\nu(\nu,T) = \frac{2h\nu^3}{c^2}\frac{1}{e^{h\nu/k_BT}-1}, \tag{0.2}$$

which matches the emission spectrum of the radiation of a perfect black body, which was then known experimentally. Note that by integrating, we obtain the power per unit of emission area,

$$P = \int_0^\infty d\nu \int d\Omega B_\nu \cos\theta = \sigma T^4, \tag{0.3}$$

where we have used the differential of solid angle $d\Omega = \sin\theta\, d\theta d\phi$ and have defined the Stefan–Boltzmann constant

$$\sigma = \frac{2k_B^4\pi^5}{15c^2h^3}, \tag{0.4}$$

obtaining the (experimentally discovered) Stefan–Boltzmann law. Planck himself didn't quite believe in the physical reality of the quantum of energy (as existing independently of the phenomenon of emission of radiation), but thought of it as a useful trick that manages to provide the correct result.

It was necessary to wait until Einstein's explanation of the photoelectric effect in 1905 for the true start of the quantum era. Indeed, Einstein took Planck's idea to its logical conclusion, and postulated that there is a quantum of energy, that he called a *photon* and that can interact with matter, such that its energy is $E = h\nu$. Such an energy quantum is *absorbed* in the photoelectric effect by the bound electrons with binding energy W, such that they are released with kinetic energy

$$E_{\text{kin}} = \frac{mv^2}{2} = h\nu - W, \tag{0.5}$$

providing an electrical current; see Fig. 0.1a.

Continuing the idea of the interaction of electrons with quanta of light, i.e., photons, we also note the *Compton effect*, observed experimentally by Arthur Compton in 1923, in which a photon scatters off a free electron and changes its wavelength according to the law

$$\Delta\lambda = \frac{2h}{mc}\sin^2\frac{\theta}{2}. \tag{0.6}$$

The law is easily explained by assuming that the photon has a relativistic relation between the energy and momentum, $E = pc$, leading to a momentum

$$p = \frac{h\nu}{c} = \frac{h}{\lambda}. \tag{0.7}$$

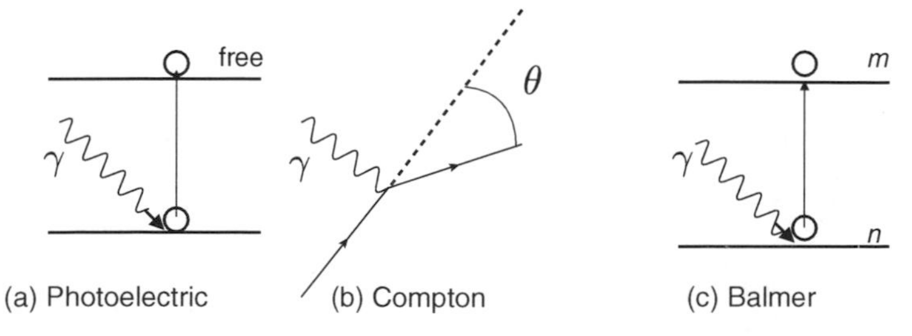

Figure 0.1 Interactions of photons with electrons: (a) photoelectric effect; (b) Compton effect; (c) Balmer relation (transition to excited state).

Then relativistic energy and momentum conservation for the process $e^- + \gamma \to e^- + \gamma$ (see Fig. 0.1b), where the initial e^- is (almost) at rest,

$$\begin{aligned} \vec{p}_\gamma &= \vec{p}'_\gamma + \vec{p}_e \\ mc^2 + p_\gamma c &= \sqrt{p_e^2 c^2 + m^2 c^4} + p'_\gamma c, \end{aligned} \tag{0.8}$$

leads to the Compton law (0.6).

Next, we can also consider the case where the absorbed or emitted energy quantum leaves the electron bound to an atom, just changing its state inside the atom; see Fig. 0.1c. Since not all possible energies for the photon can be absorbed in this way, this means that the allowed energy states for the electron inside the atom are also quantized, i.e., they can only take well-defined values. In fact, experimentally we have the *Balmer relation*, discovered by Johann Balmer in 1885, for the possible frequencies of the absorbed or emitted radiation for the hydrogen (H) atom, given by

$$\nu = R\left(\frac{1}{n^2} - \frac{1}{m^2}\right), \tag{0.9}$$

where R is called the Rydberg constant. Together with Planck's and Einstein's relation $\Delta E = h\nu$, we obtain the law that the possible energies of the electron inside the H atom are given by the *energy levels*

$$E_n = -\frac{hR}{n^2}, \tag{0.10}$$

where n is a natural nonzero number, i.e. $n = 1, 2, 3, \ldots$

The fact that the states of the electron in the H atom (and, in fact, of any atoms or molecules, as we now know) are quantized suggests that it is possible to have more general quantization relations for energy states. That this is true, and that we can turn it into a concrete description for states in quantum mechanics, was proved in another important experiment, the *Stern–Gerlach experiment* of 1922 (by Otto Stern and Walter Gerlach). In this experiment, horizontal paramagnetic atomic beams are passed through a magnetic field varying as a function of the vertical position, $B = B(z)$, created by two magnets oriented on a vertical line, as in Fig. 0.2. The electrons are understood as "spinning", like magnets with a magnetic moment μ, so the energy of the interaction of the "spinning" electrons with the magnetic field is given by

$$\Delta E = -\vec{\mu} \cdot \vec{B}. \tag{0.11}$$

But the magnetic moment is proportional to the "angular momentum" $\vec{l}$, $\vec{\mu} = m\vec{l}$. Classically, we expect any possible value for $\vec{l}$ and, so any possible value for both the magnetic moment $\vec{\mu}$, and its projection μ_z on the z direction. Since the force in the z direction is

$$F_z = -\partial_z(\vec{\mu} \cdot \vec{B}) \simeq \mu_z \partial_z B_z, \tag{0.12}$$

we expect the arbitrary value of μ_z to translate into an arbitrary deviation of the beam, experimentally detected on a screen transverse to the original beam. But in fact one observes only two possible deviations, implying only two possible values of μ_z, symmetric about zero, which in fact can only be $+\mu$ and $-\mu$, with a fixed μ. That in turn means that l_z has only the possible values $+l$ and $-l$, with l fixed.

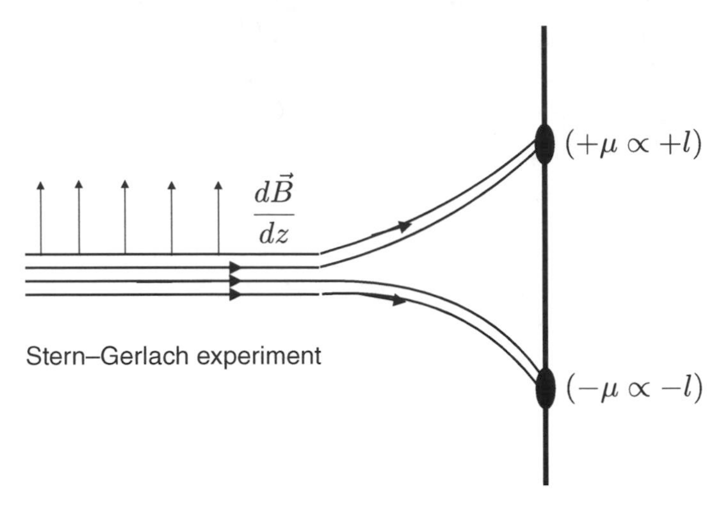

Figure 0.2 The Stern–Gerlach experiment.

0.2 Quantized States: Matrix Mechanics, and Waves for Particles: Correspondence Principle

The conclusion is that the "spinning" states of the electron have a projection onto any axis (since z is a randomly chosen direction) that can have only two given values, which is another example of the quantization of states, involving the simplest possibility: two possible choices only, states "1" and "2". In general, we expect more quantized values for the energy, therefore more, but usually countable, states. Observables in these states can be "diagonal", meaning they map the state to itself (for instance, the energy of a state), or "off-diagonal", meaning they map one state to another (like those related to the transition between matter states due to interaction with light). In total, we obtain a "matrix mechanics", for matrix observables M_{ij} acting on states i. This was defined by Werner Heisenberg, Max Born, and Pascual Jordan in 1925, in three papers (first by Heisenberg, then generalized and formalized by Born and Jordan, then by all three).

But one still had to understand what is the physical meaning of the states, and obtain an alternative to the classical idea of the path of a matter-particle. This came with the idea of a wave associated with any particle, and conversely, a particle associated with any wave, or *particle–wave complementarity*; see Fig. 0.3a. This is illustrated best in the classic double-slit experiment, which today is a table top experiment covered in undergraduate physics. Consider a plane wave, or a beam of particles, moving perpendicularly towards a screen with two slits in it, and observe the result on a detecting parallel screen behind it, as in Fig. 0.3b.

If we have classical waves, as in classical electromagnetism (the description of the macroscopic quantities of light), a *traveling plane wave* is described by a function ψ depending only on the propagation direction x and the time t, as

$$\psi(x,t) = A \exp\left[i\left(\frac{2\pi}{\lambda}x - \frac{2\pi}{T}t\right)\right], \tag{0.13}$$

where ψ is made up from $\vec{E}$ and $\vec{B}$. The general (spherical) form of the traveling wave is

$$\psi(r,t) = \frac{A}{r}\exp\left[i(kr - \omega t)\right], \tag{0.14}$$

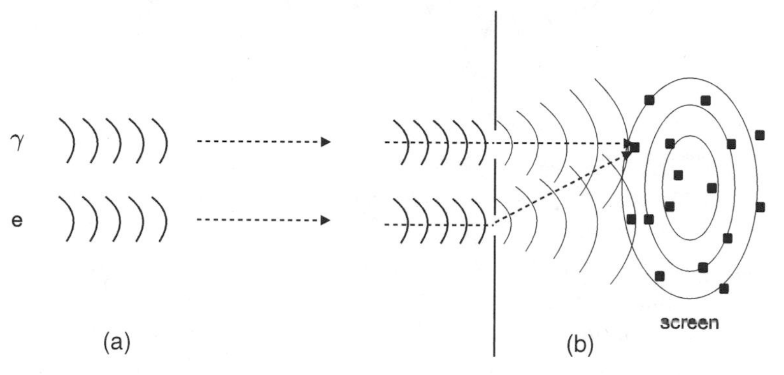

Figure 0.3 (a) Particle–wave complementarity (correspondence principle). (b) Two-slit interference of waves ψ_1, ψ_2 giving a screen image made up of individual points (particles), forming an interference pattern at large times.

and is the form that is relevant for the wave that comes out of a slit. One can however measure only the intensity, given by

$$I = |\psi|^2. \tag{0.15}$$

The role of the double slits is to split the plane wave into two waves, 1 and 2, with origin in each slit, but converging on the same point on the detecting screen. Then one detects the total $I = I_{1+2}$, but what adds up is not I, but rather ψ, so $\psi_{1+2} = \psi_1 + \psi_2$, and

$$I_{1+2} = |\psi_1 + \psi_2|^2. \tag{0.16}$$

This leads to the interference pattern observed on the screen, with many local maxima of decaying magnitude.

On the other hand, for particles, the particle beam divides at the slits in two beams. The "intensity" of the beam is proportional to the number of particles per unit area (with the total number of particles classically a constant, since they cannot be created or destroyed), so for them one finds instead

$$I_{1+2} = I_1 + I_2. \tag{0.17}$$

This would lead to a pattern with a single maximum, situated at the midpoint of the screen, but this is not what it is observed.

In fact, as one knows from the experiment, if one decreases the intensity of the beam, such that, according to the previous description, we have only individual photons (quanta of light) coming through the slits, one can see the individual photons hitting the screen. However, their locations are such that, when sufficiently many of them have hit, they still create the same interference pattern of the waves. That means that, in a sense, a photon can "interfere with itself".

It also implies a *correspondence principle*, that classical physics corresponds to macroscopic effects, whereas microscopic effects are described by quantum mechanics. As long as something becomes macroscopic (through many iterations of the microscopic, like for instance the large number of photons in the double slit experiment), it becomes classical.

So the question arises, is there a wave associated with the photon itself? And what does it correspond to? And if this is true for a photon, could it be true for any particle, even a particle of matter? In fact, historically, this was first proposed by (the Marquis) Louis de Broglie in 1924, who said that there should be a wave number k associated with any particle of momentum p, given by

$$k = \frac{p}{\hbar}, \tag{0.18}$$

where $\hbar = h/(2\pi)$. Then it follows that the de Broglie wavelength of the particle is

$$\lambda = \frac{2\pi}{k} = \frac{h}{p}. \tag{0.19}$$

This gives the correct formula for a photon, since for a photon, $p = E/c$ and $\lambda = c/\nu$. But the formula is supposed to also apply for any matter particle, such as for instance an electron. For electrons, the formula was confirmed experimentally by Davisson and Germer in experiments performed between 1923 and 1927. Today, the principle implied in these experiments is used in the electron microscope, which uses electrons instead of photons in order to "see". This allows it to have a better resolution, since the resolution is of the order of λ and the de Broglie λ for an electron is much smaller than that for a photon of light (since the momentum p is much larger).

0.3 Wave Functions Governing Probability, and the Schrödinger Equation

Because the intensity of a beam of particles is $I = |\psi|^2$, if we allow for the possibility that particles have only *probabilities* of behaving in some way, we arrive at the conclusion that the intensity I must be proportional to this probability, and that as such the probability is given by $|\psi|^2$, where ψ is a *wave function* associated with any particle. This was the interpretation given by Erwin Schrödinger, who then went on to write an equation, now known as *Schrödinger's equation*, for the wave function. The equation for ψ is in general

$$i\hbar\partial_t\psi = \hat{H}\psi, \tag{0.20}$$

where $\hat{H}$ is an operator associated with the classical Hamiltonian H, now acting on the wave function ψ.

This interpretation of quantum mechanics, of the Schrödinger equation for a wave function associated with probability, was complementary to the matrix mechanics of Heisenberg, Born, and Jordan. The two however are united into a single one by the interpretation of Dirac in terms of bra $\langle\psi|$ and ket $|\psi\rangle$ states, which we will develop. In it, thinking of various states $|\psi_i\rangle$ as the states i of matrix mechanics, operators like $\hat{H}$ appear as matrices H_{ij}.

0.4 Bohr–Sommerfeld Quantization and the Hydrogen Atom

Quantization of the electrons in the H atom on the other hand is obtained from the Bohr–Sommerfeld quantization rules, which state that the total *action* over the domain of a state, such as a cycle of the motion of an electron around the H atom, is quantized in units of h,

$$\oint_{H=E} p \cdot dq = nh, \tag{0.21}$$

where E is the energy of the electron. More generally, this should be true for any variable q and its canonically conjugate momentum p, i.e.,

$$\oint p_i dq_i = nh. \tag{0.22}$$

We will study this in more detail later, but in this review we will just recall the principal details.

For the H atom, we have three simple quantizations. For p_ϕ (the momentum associated with the angular variable ϕ),

$$\oint p_\phi d\phi = lh \tag{0.23}$$

leads to quantization of the total angular momentum,

$$L = l\hbar. \tag{0.24}$$

For p_r (the momentum associated with the radial variable r),

$$\oint p_r dr = kh \tag{0.25}$$

leads to the relation ($e_0^2 \equiv e^2/(4\pi\epsilon_0)$)

$$\sqrt{\frac{2\pi^2 m e_0^4}{-E}} - 2\pi L = kh. \tag{0.26}$$

The sum of the quantum numbers k and l is called the principal quantum number,

$$n = l + k. \tag{0.27}$$

This then leads to the derivation of the energy levels obtained from the Balmer law,

$$E_n = -\frac{m e_0^4}{2\hbar^2 n^2}. \tag{0.28}$$

Physically, one describes the original Bohr–Sommerfeld quantization as the particle trajectory being an integral number of de Broglie wavelengths, as in Fig. 0.4.

On the other hand, the projection of the angular momentum L on any direction z must also be quantized (as we deduced from the Stern–Gerlach experiment), so

$$\oint L_z d\phi = mh, \tag{0.29}$$

leading to

$$L_z = m\hbar. \tag{0.30}$$

Thus the H atom is described by the three quantum numbers (n, l, m).

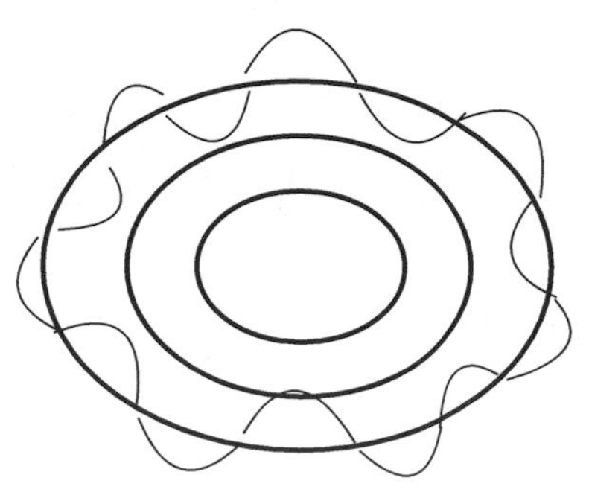

Figure 0.4 Bohr–Sommerfeld quantization, physical interpretation.

0.5 Review: States of Quantum Mechanical Systems

To summarize, in the case of quantum mechanical systems we have the following possibilities for the states.

- We can have a finite number of discrete states, as in the case of spin systems. *Quantization* in this case means exactly that: instead of a continuum of possible states (for a classical angular momentum, in this case), now we have a discrete set of states. In particular, for spin $s = 1/2$ (as in the case of electrons, proven by the Stern–Gerlach experiment), we have a two-state system, the simplest possible, which will be analyzed first.
- We can have an infinite number of discrete states, as in the case of the hydrogen atom, considered above. In this case, *quantization* means the same: instead of a continuum of states, we have a discrete set, with an allowed set of energies and states, as well as other observables.
- We can also have a continuum of states, as in the case of free particles. In this case, quantization refers only to the types of states allowed; there is no restriction on the allowed number of energies and states associated with them.

Important Concepts to Remember

- Some states for some systems are quantized (quantized energy states, for instance), leading to matrix mechanics.
- There is always a wave associated with a particle and a particle with a wave, leading to a complementarity.
- There is always a (complex-valued) wave function ψ describing probabilities for a measurement via $P = |\psi|^2$. It satisfies the Schrödinger equation.
- We can always have a discrete infinite number of states, or even a continuous infinite number, but both are still described by matrix mechanics. In any case, there is a wave function.

Further Reading

See, for instance, Messiah's book [2]. It has a more thorough treatment of the historical background.

Exercises

(1) Derive the Stefan–Boltzmann law from the Planck law for the spectral radiance.

(2) In the photoelectric effect, if the incoming flux of light has frequency ν and the released electrons get thermalized through interaction, what is the resulting temperature? If ν is within the visible spectrum and W is comparable with the energy of the photon, what kind of temperature range do you expect?

(3) Derive the Compton effect $\Delta\lambda$ from the relativistic collision of the photon from a free electron.

(4) Considering a *relativistic* traveling wave of frequency ν, calculate the distance between the local maxima (or local minima) in the interference pattern on a screen.

(5) Calculate the de Broglie wavelength for an electron at "room temperature". How does it compare with a corresponding wavelength in the radiation spectrum? What does this imply for the feasibility of microscopes using electrons versus those using radiation?

(6) Can we apply Bohr–Sommerfeld quantization to a circular pendulum? Why, and would it be useful to do so?

(7) Do you think that the Schrödinger equation is applicable for light as well? Explain why.

1 The Mathematics of Quantum Mechanics 1: Finite-Dimensional Hilbert Spaces

Before we define quantum mechanics through its postulates, we will review relevant issues about the mathematics of vector spaces, formulating it in a way more easily applicable to the physical case in which we are interested, and in particular the vector spaces we will call Hilbert spaces, relevant for quantum mechanics. In this chapter we will consider the case of finite-dimensional spaces, which are easier to define and analyze, and in the next chapter we will consider the more complicated, but more physically relevant, case of infinite-dimensional Hilbert spaces, which are more involved.

1.1 (Linear) Vector Spaces $\mathbb{V}$

We start with the general definition of a vector space. A vector space is a generalization of the space of vectors familiar from classical mechanics. As such, it is a collection $\{|v\rangle\}$ of elements denoted by $|v\rangle$ (in the so-called "ket" notation defined by Dirac) together with an addition rule "+" and a multiplication rule "·", and a set of axioms about them, as follows.

- There is an addition operation "+" that respects the space (a group property): $\forall |v\rangle, |w\rangle \in \mathbb{V}$, $|v\rangle + |w\rangle \in \mathbb{V}$.
- There is a multiplication by a scalar operation "·" that also respects the space: $\forall \alpha \in \mathbb{C}$ and $|v\rangle \in \mathbb{V}$, $\alpha|v\rangle \in \mathbb{V}$.
- The addition of scalars is distributive: $\forall \alpha, \beta \in \mathbb{C}$ and $|v\rangle \in \mathbb{V}$, $(\alpha + \beta)|v\rangle = \alpha|v\rangle + \beta|v\rangle$.
- The addition of vectors is also distributive: $\forall \alpha \in \mathbb{C}$ and $|v\rangle, |w\rangle \in \mathbb{V}$, $\alpha(|v\rangle + |w\rangle) = \alpha|v\rangle + \alpha|w\rangle$.
- The addition of vectors "+" is commutative and associative, and so is the multiplication by scalars "·".
- There is a unique zero vector $|0\rangle$, such that $|v\rangle + |0\rangle = |v\rangle$.
- There is a unique inverse under vector addition: $\forall |v\rangle \; \exists |-v\rangle$ such that $|v\rangle + |-v\rangle = |0\rangle$.

Given a set of vectors $|v_1\rangle, |v_2\rangle, \ldots, |v_n\rangle$, we can form linear combinations of them. We say that the vectors are *linearly dependent* if there is a linear combination of them that vanishes, i.e., if $\exists$ $\alpha_i \neq 0$ (at least two of them nonzero), complex numbers such that

$$\sum_{i=1}^{n} \alpha_i |v_i\rangle = |0\rangle. \tag{1.1}$$

In this case, we can write one of the vectors in terms of the others. For instance, if $\alpha_n \neq 0$, we write

$$|v_n\rangle = -\sum_{i=1}^{n-1} \frac{\alpha_i}{\alpha_n} |v_i\rangle = \sum_{i=1}^{n-1} \alpha_i' |v_i\rangle. \tag{1.2}$$

We will then call a vector space $\mathbb{V}$ n-dimensional if there are n linearly independent vectors $|v_1\rangle, |v_2\rangle, \dots, |v_n\rangle$, but $\forall$ sets of $n+1$ vectors $|w_1\rangle, |w_2\rangle, \dots, |w_{n+1}\rangle$, they are linearly dependent. We call $\{|v_1\rangle, |v_2\rangle, \dots, |v_n\rangle\}$ a *basis* for the space.

We can then decompose any vector $|v\rangle \in \mathbb{V}$ in this basis, i.e., there is a unique set of coefficients α_i, such that

$$|v\rangle = \sum_{i=1}^{n} \alpha_i |v_i\rangle. \tag{1.3}$$

The fact that the set of coefficients is unique follows from the fact that if there were a second set $\tilde{\alpha}_i$ satisfying the same condition,

$$|v\rangle = \sum_{i=1}^{n} \tilde{\alpha}_i |v_i\rangle, \tag{1.4}$$

then by taking the difference of the two decompositions, we obtain

$$\sum_{i=1}^{n} (\alpha_i - \tilde{\alpha}_i)|v_i\rangle = 0, \tag{1.5}$$

which would mean that the $|v_i\rangle$ would be linearly dependent, which is a contradiction.

Subspaces

If there is a subset $\mathbb{V}' \subset \mathbb{V}$ such that, $\forall |v\rangle \in \mathbb{V}'$ and $\forall \alpha \in \mathbb{C}$, then $\alpha|v\rangle \in \mathbb{V}'$ as well, and also $\forall |v\rangle, |w\rangle \in \mathbb{V}'$, $|v\rangle + |w\rangle \in \mathbb{V}'$, we call $\mathbb{V}'$ a subspace of the vector space.

In this case, the subspace will have dimension $k \leq n$ (n being the dimension of $\mathbb{V}$), and there will be a basis $(|v_1\rangle, |v_2\rangle, \dots, |v_k\rangle)$ for $\mathbb{V}'$.

Scalar Product and Orthonormality

On a vector space we can define a notion of a scalar product, which associates with any two vectors $|v\rangle$ and $|w\rangle$ a complex number $\langle w|v\rangle$ such that:

(1) $\langle w|v\rangle = \langle v|w\rangle^*$.

(2) The product is distributive in the second term, i.e., $\langle w|(\alpha|v\rangle + \beta|u\rangle) = \langle w|\alpha v + \beta u\rangle = \alpha\langle w|v\rangle + \beta\langle w|u\rangle$.

(3) The product of a vector with itself is nonnegative, $\langle v|v\rangle \geq 0$, and is zero only for the null vector,

$$\langle v|v\rangle = 0 \quad \Leftrightarrow \quad |v\rangle = |0\rangle. \tag{1.6}$$

Note that the first and second axioms imply also that the product is distributive in the first term,

$$\langle \alpha v + \beta u|w\rangle = (\alpha^*\langle v| + \beta^*\langle u|)|w\rangle = \alpha^*\langle v|w\rangle + \beta^*\langle u|w\rangle. \tag{1.7}$$

Given this scalar product, we can define *orthogonality*, namely when

$$\langle v|w\rangle = 0. \tag{1.8}$$

We also deduce, using the second axiom, that the product of any vector with the null vector gives zero, since

$$\langle v|0\rangle = \langle v|w\rangle - \langle v|w\rangle = 0. \tag{1.9}$$

We define the norm as the scalar product of a vector with itself, as

$$||v|| \equiv \sqrt{\langle v|v\rangle}. \tag{1.10}$$

Next, we can define orthonormality as orthogonality with a unit norm for each vector, $||v|| = 1, \forall |v\rangle$. If a basis vector does not have unit norm, we can always find a complex number c such that $||cv|| = 1$, since

$$\langle cv|cv\rangle = cc^*\langle v|v\rangle \quad \Rightarrow \quad |c|^2 = \frac{1}{||v||^2}. \tag{1.11}$$

We can consider a basis of orthonormal vectors $|i\rangle$, and decompose an arbitrary vector $|v\rangle$ using it, as

$$|v\rangle = \sum_i v_i |i\rangle. \tag{1.12}$$

Considering also another vector decomposed in this way,

$$|w\rangle = \sum_i w_i |i\rangle, \tag{1.13}$$

we obtain for the scalar product

$$\langle v|w\rangle = \sum_{i,j} v_i^* w_j \langle i|j\rangle. \tag{1.14}$$

If the basis is orthonormal, we have $\langle i|j\rangle = \delta_{ij}$, so

$$\langle v|w\rangle = \sum_i v_i^* w_i. \tag{1.15}$$

Then the norm is given by

$$||v||^2 = \langle v|v\rangle = \sum_i |v_i|^2. \tag{1.16}$$

By multiplying the decomposition of $|v\rangle$ by $\langle k|$ (considering the scalar product with $|k\rangle$), we obtain (using $\langle k|i\rangle = \delta_{ik}$)

$$\langle k|v\rangle = \sum_i v_i \langle k|i\rangle = v_k. \tag{1.17}$$

Since $v_i = \langle i|v\rangle$, the decomposition becomes

$$|v\rangle = \sum_i |i\rangle\langle i|v\rangle. \tag{1.18}$$

The Gram–Schmidt theorem (orthonormalization)

Given a linearly independent basis, we can always find an orthonormal basis by making linear combinations of the basis vectors $|v\rangle$.

1.2 Operators on a Vector Space

An operator $\hat{A}$ is an action on the vector space that takes us back into the vector space, i.e., if $|v\rangle \in \mathbb{V}$ then also $|w\rangle = \hat{A}|v\rangle \in \mathbb{V}$. In other words, it associates with any vector another vector,

$$\hat{A} : |v\rangle \mapsto |w\rangle. \tag{1.19}$$

In other words, it is the generalization of the notion of a function for an action on a vector space.

We will mostly be interested in *linear operators*, which means that

$$\hat{A}(\alpha|v\rangle + \beta|w\rangle) = \alpha\hat{A}|v\rangle + \beta\hat{A}|w\rangle. \tag{1.20}$$

Note however that there are also useful *antilinear operators*, relevant for the time-reversal invariance (T) operator,

$$\hat{A}(\alpha|v\rangle + \beta|w\rangle) = \alpha^*\hat{A}|v\rangle + \beta^*\hat{A}|w\rangle. \tag{1.21}$$

Nevertheless, at present we will consider only linear operators (except later on in the book, when we will consider the T symmetry).

In the space of operators (or "functions" on the vector space) there exists the notion of $\hat{0}$ (a neutral element under addition) and $\hat{1}$ (a neutral element under multiplication), namely defined by

$$\hat{0}|v\rangle = |0\rangle, \quad \hat{1}|v\rangle = |v\rangle. \tag{1.22}$$

We also have the same definition of the product with a scalar (in general a complex number), and of addition, as in the vector space itself,

$$\begin{aligned}(\alpha\hat{A})|v\rangle &= \alpha(\hat{A}|v\rangle),\\ (\alpha\hat{A} + \beta\hat{B})|v\rangle &= \alpha(\hat{A}|v\rangle) + \beta(\hat{A}|v\rangle).\end{aligned} \tag{1.23}$$

Since the operator is the generalization of the notion of a function for a vector space, we can also define the product of operators as the generalization of the composition of functions, then $\hat{A} \cdot \hat{B} \to A \circ B$, namely

$$\hat{A} \cdot \hat{B}|v\rangle \equiv \hat{A}(\hat{B}|v\rangle). \tag{1.24}$$

On the space of operators, obtained from the vector space, we can define a *completeness relation*, saying that an orthonormal basis of vectors $|i\rangle$ is also complete in the space, i.e., that we can expand any vector in the vector space in terms of them. As we saw, this expansion formula was

$$|v\rangle = \sum_i |i\rangle\langle i|v\rangle, \tag{1.25}$$

and by identifying this with $\hat{1}|v\rangle$ (by the definition of $\hat{1}$), we obtain the completeness relation

$$\sum_i |i\rangle\langle i| = \hat{1}. \tag{1.26}$$

This "ket-bra" notation for operators is used more generally than just for the completeness relation, and it can be shown that we can write an operator $\hat{A}$ in the form

$$\hat{A} = \sum_{a\in A, b\in B} |a\rangle\langle b|, \tag{1.27}$$

where A, B are some sets of vectors, since by acting on an arbitrary vector $|v\rangle$ we obtain

$$\hat{A}|v\rangle = \sum_{a\in A, b\in B} |a\rangle\langle b|v\rangle = \sum_{a\in A, b\in B} v_b|a\rangle \equiv |w\rangle, \tag{1.28}$$

where $v_b \equiv \langle b|v\rangle$.

Matrix Representation of Operators

We have talked until now in an abstract way of vector spaces and operators acting on them, but we now show that we can represent operators as matrices acting on column vectors. Then in the case of quantum mechanics one obtains the representation of the "matrix mechanics" of Heisenberg.

Consider an (orthonormal) basis $|1\rangle, |2\rangle, \ldots, |n\rangle$ for the vector space. Then an arbitrary vector space $|v\rangle$, decomposed as

$$|v\rangle = \sum_i |i\rangle v_i = \sum_i |i\rangle\langle i|v\rangle, \tag{1.29}$$

can be represented by the $\langle i|v\rangle$ numbers represented as a column vector,

$$|v\rangle = \begin{pmatrix} v_1 \\ v_2 \\ \vdots \\ v_n \end{pmatrix} = \begin{pmatrix} \langle 1|v\rangle \\ \langle 2|v\rangle \\ \vdots \\ \langle n|v\rangle \end{pmatrix}, \tag{1.30}$$

where the first form is general (in terms of coefficients v_i of the basis), and the second is for an orthonormal basis.

In this representation, the ith basis vector corresponds to the column vector with only a 1 in the ith position,

$$|1\rangle = \begin{pmatrix} 1 \\ 0 \\ \vdots \\ 0 \end{pmatrix}, \ldots, \; |i\rangle = \begin{pmatrix} 0 \\ \vdots \\ 1_i \\ \vdots \\ 0 \end{pmatrix}. \tag{1.31}$$

Under this representation of the vectors, the operators $\hat{A}$ associating a vector with another vector are associated with a matrix. The matrix element of the operator in the basis $\{|i\rangle\}$ is

$$A_{ij} = \langle i|\hat{A}|j\rangle. \tag{1.32}$$

If moreover the set $\{|i\rangle\}$ is complete then the product of operators $\hat{A}\cdot\hat{B}$ is associated with the matrix product of matrices. Indeed, inserting the identity written as the completeness relation, we obtain

$$(\hat{A}\cdot\hat{B})_{ij} = \langle i|\hat{A}\cdot\hat{B}|j\rangle = \sum_k \langle i|\hat{A}|k\rangle\langle k|\hat{B}|j\rangle = \sum_k A_{ik}B_{kj}. \tag{1.33}$$

Then it follows also that the operator inverse $\hat{A}^{-1}$, defined by

$$\hat{A} \cdot \hat{A}^{-1} = \hat{A}^{-1} \cdot \hat{A} = \hat{1}, \tag{1.34}$$

is associated with the matrix inverse $(\hat{A})^{-1}_{ij}$.

1.3 Dual Space, Adjoint Operation, and Dirac Notation

We represented the vectors $|v\rangle$ as column vectors (1.30), where the basis vectors $|i\rangle$ correspond to each component of the column vector. But it follows that we can define the adjoint matrix of this column vector, i.e., the transposed and complex conjugate matrix, a row vector, defined as the adjoint of the column vector:

$$(|v\rangle)^\dagger = (v_1^*, v_2^*, \dots, v_n^*). \tag{1.35}$$

Then we can also write the scalar product of two vectors in an orthonormal basis, which we saw was written as $\sum_i v_i^* u_i$, as the matrix product of row and column vectors,

$$\langle v|u\rangle = \sum_i v_i^* u_i = (v_1^* \; v_2^* \; \cdots \; v_n) \begin{pmatrix} u_1 \\ u_2 \\ \vdots \\ u_n \end{pmatrix}. \tag{1.36}$$

But that means that we can represent the adjoint of the vector as the formal "bra" vector,

$$\langle v| = (|v\rangle)^\dagger \equiv (v_1^*, v_2^*, \dots, v_n^*). \tag{1.37}$$

The "bra" and "ket" notation is due to Dirac, and it is a splitting of the word "bracket", which refers to the scalar product $\langle v|u\rangle$. Then $|u\rangle$, the vectors in the space $\mathbb{V}$, are called ket vectors and $\langle v|$, the vectors in the "dual" (adjoint) space, denoted by $\mathbb{V}^*$, are called bra vectors.

We can define the adjoint operation on vectors in a vector space a bit more generally, in order to better define $\mathbb{V}^*$, by defining the adjoint operation under multiplication by scalars $a \in \mathbb{C}$ and under a sum of the basis vectors,

$$\begin{aligned} \langle av| &= (|av\rangle)^\dagger = (a|v\rangle)^\dagger = a^* \langle v| \\ |v\rangle = \sum_i v_i |i\rangle &\Rightarrow \langle v| = (|v\rangle)^\dagger = \sum_i v_i^* \langle i|, \end{aligned} \tag{1.38}$$

where we have used that $|i\rangle^\dagger = \langle i|$ for the basis vectors.

Having defined the adjoint operation for the vectors in the space, we can also define the same for the operators relating two vectors, also via the adjoint matrix, i.e., the transpose and complex conjugate, for the basis vectors $|i\rangle$, so

$$(\hat{A}^\dagger)_{ij} \equiv (\hat{A}_{ij})^\dagger = \left(\langle i|\hat{A}|j\rangle\right)^\dagger = \langle j|\hat{A}^\dagger|i\rangle = A_{ji}^*. \tag{1.39}$$

We can write this adjoint operation formally without reference to a basis of vectors. Since an operator associates a vector with another vector,

$$\hat{A}|v\rangle = |\hat{A}v\rangle, \tag{1.40}$$

it follows that

$$\langle \hat{A}v| = \left(\hat{A}|v\rangle\right)^\dagger = \langle v|\hat{A}^\dagger. \tag{1.41}$$

The matrix definition, or this abstract definition, also implies that the dagger operation acts in reverse on a product, as follows:

$$(\hat{A}\cdot\hat{B})^\dagger = \hat{B}^\dagger\cdot\hat{A}^\dagger. \tag{1.42}$$

Change of Basis

Consider a change of basis, from the orthonormal basis $\{|i\rangle\}$ to $\{|i'\rangle\}$. The fact that both are bases means that we can expand a vector in one basis in the other basis, as the linear combination

$$|i'\rangle = \sum_j U_{ji}|j\rangle. \tag{1.43}$$

Multiplying by the bra vector $\langle k|$ and using orthonormality, $\langle k|i\rangle = \delta_{ik}$, we obtain

$$U_{ki} = \langle k|i'\rangle. \tag{1.44}$$

We can consider the reverse mapping, and represent this matrix as an abstract operator $\hat{U}$, with matrix elements U_{ki}, i.e.,

$$U_{ki} = \langle k|\hat{U}|i\rangle = \langle k|\hat{U}i\rangle, \tag{1.45}$$

which implies that formally we have

$$|i'\rangle = \hat{U}|i\rangle = |\hat{U}i\rangle. \tag{1.46}$$

For the inverse relation, we can expand the vector $|i\rangle$ in the basis $|i'\rangle$, giving the linear combination,

$$|i\rangle = \sum_j V_{ji}|j'\rangle. \tag{1.47}$$

Multiplying by the vector $\langle k'|$ and using the orthonormality relation $\langle k'|j'\rangle = \delta_{jk}$, we obtain

$$V_{ki} = \langle k'|i\rangle = (\langle i|k'\rangle)^* = U^*_{ik} = (\hat{U}^\dagger)_{ki}, \tag{1.48}$$

implying that formally we have $\hat{V} = \hat{U}^\dagger$. Then, applying the transformation twice, from $|i\rangle$ to $|i'\rangle$ and back to $|i\rangle$, we obtain

$$|i\rangle = \hat{U}^\dagger\hat{U}|i\rangle \Rightarrow \hat{U}^\dagger\hat{U} = \hat{1}, \tag{1.49}$$

and similarly $\hat{U}\hat{U}^\dagger = \hat{1}$, implying that

$$\hat{U}^{-1} = \hat{U}^\dagger. \tag{1.50}$$

Such an operator is called a *unitary operator*. Thus, operators that change basis in a vector space are unitary.

Moreover, for such an operator, taking the determinant of the relation (mapped to matrices) $\hat{U}\hat{U}^\dagger = \hat{1}$, and using the facts that the determinant of the matrix product is the product of the matrix determinants and that $\det U^T = \det U$, so that $\det U^\dagger = (\det U)^*$, we obtain

$$1 = \det(\hat{U} \cdot \hat{U}^\dagger) = \det U \cdot \det U^\dagger = |\det U|^2. \tag{1.51}$$

These unitary matrices U form a group G, which means that their multiplication is such that:

(a) $\forall \hat{A}, \hat{B} \in G, \hat{A} \cdot \hat{B} \in G$.
(b) $\exists$ a unit operator (matrix) $\hat{1}$, such that $\hat{A} \cdot \hat{1} = \hat{1} \cdot \hat{A} = \hat{A}, \forall \hat{A} \in G$.
(c) $\forall \hat{A}, \exists \hat{A}^{-1}$ such that $\hat{A}^{-1} \cdot \hat{A} = \hat{A} \cdot \hat{A}^{-1} = \hat{1}$.

The group of such unitary $n \times n$ matrices is called the unitary group $U(n)$. Such a unitary transformation of basis $\hat{U}$ preserves the scalar product, meaning that if

$$|v'\rangle = U|v\rangle, \quad |w'\rangle = U|w\rangle, \tag{1.52}$$

then the scalar product is invariant,

$$\langle w'|v'\rangle = \langle Uw|Uv\rangle = \langle w|U^\dagger U|v\rangle = \langle w|v\rangle. \tag{1.53}$$

The transformation of the basis is called an "active" transformation, $|v\rangle \to U|v\rangle$. Then the matrix elements of an operator $\hat{A}$ change as follows:

$$\langle i|\hat{A}|j\rangle \to \langle i'|\hat{A}|j'\rangle = \langle Ui|\hat{A}|Uj\rangle = \langle i|U^\dagger \hat{A} U|j\rangle. \tag{1.54}$$

Equivalently, describing this as a "passive" transformation, we can think of the transformation acting on the operators instead as;

$$\hat{A} \to U^\dagger \hat{A} U = U^{-1} \hat{A} U. \tag{1.55}$$

If, however, we consider the matrix element as invariant, the change of basis $|i\rangle \to U|i\rangle$ is accompanied by a compensating inverse transformation on operators $\hat{A}$ given by $\hat{A} \to U\hat{A}U^{-1}$.

1.4 Hermitian (Self-Adjoint) Operators and the Eigenvalue Problem

We call an operator Hermitian or self-adjoint if

$$\hat{A}^\dagger = \hat{A}, \tag{1.56}$$

and anti-Hermitian if

$$\hat{A}^\dagger = -\hat{A}. \tag{1.57}$$

The notion of Hermitian and anti-Hermitian operators corresponds to the notion of real or imaginary complex numbers. As for complex numbers (where we can decompose any complex number into a real and an imaginary part), we can decompose any operator into a Hermitian and an anti-Hermitian part,

$$\hat{A} = \frac{\hat{A} + \hat{A}^\dagger}{2} + \frac{\hat{A} - \hat{A}^\dagger}{2} \equiv \hat{A}_H + \hat{A}_{A\text{-}H}. \tag{1.58}$$

Indeed, $\hat{A}_H = (\hat{A} + \hat{A}^\dagger)/2$ is Hermitian and $\hat{A}_{A\text{-}H} = (\hat{A} - \hat{A}^\dagger)/2$ is anti-Hermitian.

In matrix notation, a Hermitian (self-adjoint) operator satisfies

$$A_{ij} = \langle i|\hat{A}j\rangle = \langle i|\hat{A}|j\rangle = \left(\langle j|\hat{A}^\dagger|i\rangle\right)^* = \left(\langle j|\hat{A}|i\rangle\right)^* = A^*_{ji}. \tag{1.59}$$

We will be interested in eigenvalue problems for operators, especially Hermitian operators.

An eigenvalue problem amounts to finding a vector $|v\rangle$, called an eigenvector, for which the operator acts on it simply as multiplication by a complex number λ, called the eigenvalue. Thus we have

$$\hat{A}|v\rangle = \lambda|v\rangle = \lambda\hat{1}|v\rangle, \tag{1.60}$$

or equivalently

$$(\hat{A} - \lambda\hat{1})|v\rangle = 0. \tag{1.61}$$

Multiplying this equation with a bra vector, we obtain an equation describable as a matrix equation (since there is a basis in which $|v\rangle$ is a basis vector), for which we can find the determinant, and obtain the eigenvalue condition

$$\det(\hat{A} - \lambda\hat{1}) = 0. \tag{1.62}$$

As mentioned previously, we are mostly interested in Hermitian operators. For them, we have the following:

Theorem The eigenvalues of a Hermitian operator are all real.

Proof We have

$$\hat{A}|v\rangle = \lambda|v\rangle \Rightarrow \langle v|\hat{A}|v\rangle = \lambda||v||^2. \tag{1.63}$$

On the other hand, for a Hermitian operator we also have

$$\left(\hat{A}|v\rangle\right)^\dagger = \langle v|\hat{A}^\dagger = \langle v|\hat{A}. \tag{1.64}$$

It then follows that we can rewrite the diagonal element of $\hat{A}^\dagger$ in two ways:

$$\begin{aligned} \langle v|\hat{A}^\dagger|v\rangle &= \langle v|\hat{A}|v\rangle = \lambda||v||^2 \\ &= \left(\langle v|\hat{A}^\dagger\right)|v\rangle = \left(\langle v|\lambda^*\right)|v\rangle = \lambda^*||v||^2. \end{aligned} \tag{1.65}$$

Therefore $\lambda = \lambda^*$, so λ is real. *q.e.d.*

We can use the eigenvectors of a Hermitian operator to build an orthonormal basis, according to the following:

Theorem For a Hermitian operator, $\hat{A}^\dagger = \hat{A}$, there is an orthonormal basis of eigenvectors for $\mathbb{V}$. In it, if the λ_i are all different,

$$\hat{A} = \begin{pmatrix} \lambda_1 & 0 & \cdots & \cdots & 0 \\ 0 & \lambda_2 & 0 & \cdots & 0 \\ 0 & & \cdots & & 0 \\ 0 & \cdots & \cdots & 0 & \lambda_n \end{pmatrix}. \tag{1.66}$$

Then, the eigenvalue condition (1.62) corresponds to an nth-order polynomial in λ, $P_n(\lambda) = 0$, with n solutions $\lambda_1, \lambda_2, \ldots, \lambda_n$.

That means that we can diagonalize a Hermitian operator by a unitary transformation, and on the diagonal we have the eigenvalues of the operator.

1.5 Traces and Tensor Products

We can define the trace of an operator formally, as corresponding to the trace of the matrix in its matrix representation,

$$\mathrm{Tr}\,\hat{A} = \sum_i \langle i|\hat{A}|i\rangle = \sum_i A_{ii}. \tag{1.67}$$

We can also define tensor products of vector spaces, and operator actions on them, in a way also obvious from the behavior of matrices.

Consider the vector spaces $\mathbb{V}$ and $\mathbb{V}'$. Then define the tensor product $\mathbb{V} \otimes \mathbb{V}'$ as follows. If $|i\rangle \in \mathbb{V}$ is a basis for $\mathbb{V}$ and $|a\rangle \in \mathbb{V}'$ is a basis for $\mathbb{V}'$, we denote the basis for $\mathbb{V} \otimes \mathbb{V}'$ as the tensor product of the basis elements, i.e.,

$$|ia\rangle \equiv |i\rangle \otimes |a\rangle, \tag{1.68}$$

and any vector $|v\rangle \in \mathbb{V} \otimes \mathbb{V}'$ is expandable in this basis,

$$|v\rangle = \sum_{i,a} v_{ia}|ia\rangle. \tag{1.69}$$

Then operators on the tensor product space are also tensor products $\hat{A} = \hat{A}_V \otimes \hat{A}_{V'}$, acting as tensor products of the corresponding matrices in the basis, i.e., acting independently in each space,

$$\hat{A}|v\rangle = \left(\hat{A}_V \otimes \hat{A}_{V'}\right) \sum_{i,a} v_{ia}|i\rangle \otimes |a\rangle = \sum_{i,a} v_{ia} \left(\hat{A}_V|i\rangle\right) \otimes \left(\hat{A}_{V'}|a\rangle\right). \tag{1.70}$$

1.6 Hilbert Spaces

Finally, we come to the definition of Hilbert spaces, which are the spaces needed for states in quantum mechanics. We will forgo a more precise and rigorous definition until the next chapter, where we will also generalize Hilbert spaces to infinite-dimensional spaces.

Briefly, Hilbert spaces are vector spaces, with a notion of a scalar product and the norm derived from it, which contain only *proper vectors*, which can be normalized to unity, $||v|| = 1$, and for which we have a notion of continuity and completeness of the space with respect to a metric.

Physically, as mentioned above the Hilbert space is the state space for quantum mechanics. In the finite-dimensional case, a Hilbert space can be thought of as a regular topological vector space, associated with the metric distance defined from the scalar product (and its norm). In the infinite-dimensional case more care is needed, since the space of functions will be included in it, so in the next chapter we will define the Hilbert space more carefully.

Note that for vectors $|v\rangle$ in a Hilbert space, if they are normalizable, which means we can write $\langle v|v\rangle = 1$ and expand the vectors in an orthonormal basis $|i\rangle$, $\langle j|i\rangle = \delta_{ij}$, we have

$$|v\rangle = \sum_i v_i|i\rangle = \sum_i \langle i|v\rangle|i\rangle, \tag{1.71}$$

so that

$$1 = \langle v|v\rangle = \sum_{i,j} \langle j|v_j^* v_i|i\rangle = \sum_i |v_i|^2 = \sum_i |\langle i|v\rangle|^2. \tag{1.72}$$

Important Concepts to Remember

- Linear vector spaces are a collection of elements together with the addition of vectors and multiplication by scalars, and some axioms.
- Vector spaces have bases. A basis contains a maximal number of linearly independent vectors, such that any element in the space can be decomposed into it.
- A subspace is a subset of elements of the space, such that the addition and multiplication by scalars defined for the space apply also to its subspaces.
- The scalar product associates a complex scalar with two vectors. In Dirac's bra-ket notation, the scalar product is $\langle v|w\rangle$.
- The norm is the square root of the scalar product of a vector with itself.
- An orthonormal basis is a basis of vectors of unit norm that are orthogonal to each other (the scalar product of two different basis vectors vanishes), i.e., $\langle v_i|v_j\rangle = \delta_{ij}$.
- Operators $\hat{A}$ on a vector space associates any vector with another. A linear operator respects linearity.
- Representing vectors as column vectors whose elements are the coefficients of the expansion into a basis, an operator is represented as a matrix acting on these column vectors.
- When changing the basis between two orthonormal bases, the corresponding operator $\hat{U}$ is unitary, $\hat{U} \in U(N)$, and the operators are changed as $\hat{A} \to \hat{U}\hat{A}\hat{U}^{-1}$.
- Hermitian operators, $\hat{A}^\dagger = \hat{A}$, admit an eigenvalue–eigenstate problem, $\hat{A}|v\rangle = \lambda|v\rangle$, with real eigenvalues $\lambda \in \mathbb{R}$.
- Traces of operators are associated with the traces of their associated matrices, while tensor products of vector spaces mean that for any element of one space, we associate with it a full copy of the other vector space.
- Hilbert spaces are vector spaces with scalar products and norms and vectors that can be normalized, and for which we have a notion of continuity and completeness of the space with respect to a metric. Physically, they are the state space for quantum mechanics.

Further Reading

See, for instance, Shankar's book [3] for more details on vector spaces, from the point of view of a physicist.

Exercises

(1) Write formally for a general vector space the triangle inequality for three vectors, generalized from vectors in a three-dimensional space, with $\vec{c} = \vec{a} + \vec{b}$.

(2) Considering three linearly independent vectors in a three-dimensional Euclidean space, $\vec{a}, \vec{b}, \vec{c}$, construct an orthonormal basis out of them.

(3) Using the bra-ket Dirac notation for operators, show that we can put the product $\hat{A} \cdot \hat{B}$ into the same form and that the product is associated with the matrix product of the associated matrices.

(4) Show that the trace of a product of Hermitian operators is real.
(5) What is the trace of a tensor product of two operators?
(6) Show that, modulo some discrete symmetries, $U(n)$ can be split up into $SU(n)$ times a group $U(1)$ of complex phases $e^{i\alpha}$ (and show that these phases do form a group).
(7) Find the two eigenvalues of an operator associated with a 2×2 matrix with arbitrary elements.

2 The Mathematics of Quantum Mechanics 2: Infinite-Dimensional Hilbert Spaces

In this chapter we move to the more complicated case of an infinite-dimensional Hilbert space, in particular the vector space of functions.

Before that, however, we will take a second, more rigorous, look at the definition of a Hilbert space and various related definitions.

2.1 Hilbert Spaces and Related Notions

We define a Hilbert space as a complex vector space with a scalar product, with which we associate a distance function ("metric") for any two vectors x and y in the space,

$$d(x,y) = ||x-y|| = \sqrt{\langle x-y|x-y\rangle}, \tag{2.1}$$

and which is a *complete metric space* with respect to this distance function.

A *metric space*, sometimes also called a pre-Hilbert space, is a vector space with a metric or distance $d(x,y)$ defined on it. The distance $d(x,y)$ must obey the triangle inequality:

$$d(x,z) \leq d(x,y) + d(y,z). \tag{2.2}$$

In the case (as for a Hilbert space) where this comes from a scalar product, the triangle inequality follows from the Cauchy–Schwarz inequality

$$|\langle x|y\rangle| \leq ||x||\ ||y||. \tag{2.3}$$

A metric space M is called *complete* (or a "Cauchy space") if a so-called Cauchy sequence of points in M, that is, a sequence v_n of points that become arbitrarily close to each other (in the sense of the metric or distance) as $n \to \infty$, has a limit, as $n \to \infty$, that is also in M. A counterexample to this, i.e., an incomplete metric space, is the space of rational numbers $\mathbb{Q}$, since there are irrationals such as $\sqrt{2}$ that are the limit of rational numbers (in a Cauchy sequence), yet they are not part of the space itself.

A metric space (pre-Hilbert) that is complete is called Hilbert.

Related to the Hilbert space is the notion of a *Banach space*, which means a vector space with an associated notion of a norm (but that does not come from a scalar product), that is also complete. That means that any Hilbert space is Banach, but the converse is not true: there are Banach spaces that are not Hilbert.

Any Hilbert space is also a *topological vector space*, which means a vector space that is also a topological space. A topological space is a space admitting a notion of continuity and having a uniform topological structure, i.e., a notion of convergence (for the definition of Cauchy sequences). Thus any Banach space is also a topological vector space.

2.2 Functions as Limits of Discrete Sets of Vectors

We are interested in the relevant and nontrivial case of infinite-dimensional Hilbert spaces. Indeed, quantum mechanical systems are usually of this type, so we need to generalize to this case. Moreover, in the quantum mechanical case we have functions $\psi(x)$ as vectors, and operators acting on them, so we need to generalize further to the case of infinite-dimensional vectors. Thus we need to find a way to think of $\psi(x)$ as a vector space the same as we do for the x's themselves, and to define operators acting on them, $\hat{A}\psi(x)$. That is, we need to replace a set of vectors $|f_k\rangle$ with a function $f(x)$.

A way to do that is to understand $f(x)$ as a limit of a function defined on discrete points, $f(x_k)$. Though the details are a bit different. The first step is to define a basis $|x_i\rangle$ that is orthonormal,

$$\langle x_i | x_j \rangle = \delta_{ij}, \tag{2.4}$$

and complete in the space of n vectors,

$$\sum_{i=1}^{n} |x_i\rangle\langle x_i| = \mathbb{1}, \tag{2.5}$$

where $\mathbb{1}$ is the unit operator, which means that any vector $|f\rangle$ can be expanded in this basis as

$$|f\rangle = \sum_{i=1}^{n} f(x_i)|x_i\rangle. \tag{2.6}$$

If we understand the components $f(x_i)$ as, by definition,

$$f(x_i) \equiv \langle x_i | f \rangle, \tag{2.7}$$

then, in view of the completeness relation, we obtain an identity,

$$|f\rangle = \sum_{i=1}^{n} |x_i\rangle\langle x_i | f\rangle. \tag{2.8}$$

Note that here we define $|x_i\rangle$ as the column vector with a 1 only in the ith position and zeroes in the other positions:

$$|x_i\rangle = \begin{pmatrix} 0 \\ \vdots \\ \vdots \\ 1_i \\ \vdots \\ \vdots \\ 0 \end{pmatrix}. \tag{2.9}$$

The decomposition of $|f\rangle$ in $|x_i\rangle$ and the fact that $f(x_i) = \langle x_i | f\rangle$ suggest that the scalar product can be understood in a similar way, as

$$\langle f | g\rangle = \sum_{i=1}^{n} \langle f | x_i\rangle\langle x_i | g\rangle = \sum_{i=1}^{n} f^*(x_i) g(x_i), \tag{2.10}$$

since

$$\langle f|x_i\rangle = (\langle x_i|f\rangle)^* = f^*(x_i). \tag{2.11}$$

Then we also obtain that the norm of a vector $|f\rangle$ is

$$||f||^2 = \langle f|f\rangle = \sum_{i=1}^{n} |f(x_i)|^2. \tag{2.12}$$

2.3 Integrals as Limits of Sums

In both the scalar product and the norm defined above we saw sums appearing, which we need to generalize to the continuum case. It seems obvious that sums should generalize to integrals. Indeed, the Riemann integral is defined as the limit of a Riemann sum.

A very small (infinitesimal) constant step of integration $\Delta x_i = \epsilon \to 0$ becomes the differential dx, where $\epsilon = (b - a)/n$, n is the number of points and a, b the limits of integration. Then the sum tends to an integral in the limit $n \to \infty$,

$$\sum_{i=1}^{n} f(x_i) \to \int_a^b dx\, f(x). \tag{2.13}$$

However, it turns out that a better limit of the sum is the more general Lebesgue integral, which allows for integration in the presence of distributions such as Dirac's delta function, treated next. The limiting Lebesgue integral is

$$\int f\, d\mu = \int_0^{\infty} f^*(t)dt, \tag{2.14}$$

where

$$f^*(t) = \mu(\{x\}|f(x) > t) \tag{2.15}$$

is the measure (the values of x for which $f(x)$ is greater than a given t). With respect to the Lebesgue integral, we can define $L^2(\mathbb{X}, \mu)$, the space of square integrable functions over the domain, which is a space $\mathbb{X}$, with measure μ. We say that f belongs to $L^2(\mathbb{X}, \mu)$ if

$$\int_{\mathbb{X}} |f|^2 d\mu < \infty. \tag{2.16}$$

This means that the scalar product can be generalized from a sum to a Riemann integral, as

$$\langle f|g\rangle = \sum_{i=1}^{n} f^*(x_i)g(x_i) \to \int_a^b f^*(x)g(x)dx, \tag{2.17}$$

or more generally to a Lebesgue integral, as

$$\langle f|g\rangle = \int f^*(t)g(t)w(t)dt, \tag{2.18}$$

where the Lebesgue measure is

$$\mu(A) = \int_A w(t)dt. \tag{2.19}$$

Thus the norm of a function can be written as a Riemann integral or as a Lebesgue integral,

$$||f||^2 = \int_a^b |f(x)|^2 dx \to \int_A |f|^2 d\mu. \tag{2.20}$$

2.4 Distributions and the Delta Function

Since the so-called Dirac delta function, which famously is not a function, but rather a *distribution*, will also make an appearance in infinite-dimensional Hilbert spaces, we need to define the theory of distributions, and specialize to the delta function.

In short, a distribution T is a linear and continuous application (not a function), defined on the space of test functions of compact support, and with complex (or real) values; thus $T : D \to \mathbb{C}$.

The *support* means the *effective* domain of a function, which is smaller than the full domain, and it is supposed that the function vanishes outside the support. The space of test functions for distributions is then defined as the set of infinitely differentiable functions with compact support,

$$D \equiv \{\phi : \mathbb{R}^n \to \mathbb{C} | \text{supp}(\phi) \text{ compact}, \quad \phi \in \mathbb{C}^\infty(\mathbb{R}^n)\}. \tag{2.21}$$

This space is a vector space, with convergence structure (so that it is a topological space) $\phi_n \to \phi$ in D.

Distributions are in some sense generalizations of functions that allow us to make sense of quantities like the Dirac delta function. Indeed, for any locally integrable function, $f(x) \in L^1_{\text{loc}}(A)$, we can associate a distribution $\tilde{f}$ with it, $\tilde{f} : D \to \mathbb{C}$, such that, for any test function ϕ,

$$\langle \tilde{f} | \phi \rangle = \int_A f^*(x)\phi(x)dx. \tag{2.22}$$

The space of distributions D^* is called the dual of the space of test functions D,

$$D^* = \{T : D \to \mathbb{C} | T \text{ a distribution}\}. \tag{2.23}$$

It is also a vector space, and has a convergence structure (so it is a topological space): if the sequence $\{T_n\}_n \in D^*$, then $T_n \to T$ (weak convergence) $\Leftrightarrow$ for any $\phi \in D$ we have $\langle T_n|\phi\rangle \to \langle T|\phi\rangle$.

On this space, we can define multiplication of the distribution $T \in D^*$ by any infinitely differentiable function $a(x) \in C^\infty(\mathbb{R}^n)$, by

$$\langle aT|\phi\rangle \equiv \langle T|a^*(x)\phi(x)\rangle. \tag{2.24}$$

If there exists a function f such that a distribution has a function associated with it, $T = \tilde{f}$, we call the distribution T a regular distribution. If there is no such function, we call the distribution singular; for example, the Dirac delta function is a singular distribution:

$$\delta_{x_0} : D \to \mathbb{R}, \quad \langle \delta_{x_0}|\phi\rangle = \phi(x_0). \tag{2.25}$$

On the other hand the Heaviside (step) distribution is regular. It is defined by

$$\langle H_{x_0}|\phi\rangle = \int_{x_0}^\infty \phi(x)dx. \tag{2.26}$$

Then the Heaviside function associated with it is defined as

$$H_{x_0}(x) = \begin{cases} 1, & x > x_0 \\ 0, & x \leq x_0. \end{cases} \tag{2.27}$$

The Heaviside function is continuous, but not differentiable at x_0. It is only differentiable as a distribution.

We can define the derivative of a distribution $T \in D^*$ by a bracket with a test function $\phi \in D$, as

$$\left\langle \frac{\partial}{\partial x_j} T \middle| \phi \right\rangle \equiv -\left\langle T \middle| \frac{\partial \phi}{\partial x_j} \right\rangle. \tag{2.28}$$

It is a good definition, since it is self-consistent for regular distributions (those associated with functions). For $T = \tilde{f}$, the relation is written as

$$\left\langle \frac{\partial}{\partial x_j} \tilde{f} \middle| \phi \right\rangle \equiv -\left\langle \tilde{f} \middle| \frac{\partial \phi}{\partial x_j} \right\rangle. \tag{2.29}$$

But the right-hand side can be written as follows:

$$-\int_{\mathbb{R}^n} f(x)^* \frac{\partial \phi}{\partial x_j} dx = -\int_{\mathbb{R}^n} dx \frac{\partial}{\partial x_j} (f^* \cdot \phi(x)) + \int_{\mathbb{R}^n} \frac{\partial f^*}{\partial x_j} \phi(x) dx = \left\langle \left(\widetilde{\frac{\partial f}{\partial x_j}} \right) \middle| \phi \right\rangle, \tag{2.30}$$

where in the first equality we have used partial integration, and in the second we used the fact that the test functions have compact support, so the boundary integral on $\mathbb{R}^n$ vanishes and the remaining term is the definition of the scalar product.

Furthermore, *as a distribution*, we have that the delta function is the derivative of the Heaviside function,

$$H'_{x_0} = \delta_{x_0}. \tag{2.31}$$

To prove this, note that

$$\langle H'_{x_0} | \phi \rangle = -\langle H_{x_0} | \phi' \rangle = -\int_{x_0}^{\infty} \phi'(x) dx = \phi(x_0) - \phi(\infty) = \phi(x_0) = \langle \delta_{x_0} | \phi \rangle, \tag{2.32}$$

where we have used the fact that $\phi(\infty) = 0$ because of the compact support of ϕ.

We can also define a notion of the support of a distribution $T \in D^*$,

$$\text{supp}(T) \equiv \{x \in \mathbb{R}^n | T \neq 0 \text{ in } X\}. \tag{2.33}$$

Then the support of the delta function is a single point,

$$\text{supp}(\delta_{x_0}) = \{x_0\}. \tag{2.34}$$

Moreover, there is a theorem that, for any function $f \in L^1_{\text{loc}}(A)$, we have $\text{supp}(\tilde{f}) \subset \text{supp}(f)$ and for any continuous function f, $\text{supp}(\tilde{f}) = \text{supp}(f)$.

2.5 Spaces of Functions

We can now go back to defining the notion of a space of functions. As we saw, the scalar product of two functions f, g on an interval (a, b) is defined as

$$\langle f|g\rangle = \int_a^b f^*(x)g(x)dx. \tag{2.35}$$

We can consider the case when one of the function vectors is replaced by the $|x\rangle$ vector itself. It follows that we need to define the product of two such vectors, giving something like $\delta_{x,y}$. Of course, now we know that the correct thing to do is to use the delta function, or rather, distribution, $\delta(x-y)$, so

$$\langle x|y\rangle = \delta(x-y), \tag{2.36}$$

which is zero outside the support x_0 and gives 1 on integration,

$$\int_a^b \delta(x-y)dy = 1, \quad a < x < b. \tag{2.37}$$

In fact, we can restrict the domain of integration to an infinitesimal region around the zero of the argument of δ, and write

$$\begin{aligned} \int_a^b \delta(x-y)f(y)dy &= f(x) \\ &= \int_{x-\epsilon}^{x+\epsilon} \delta(x-y)f(y)dy. \end{aligned} \tag{2.38}$$

On the other hand, the completeness relation for $|x\rangle$, defined as a (Riemann) sum, is now generalized to an integral:

$$\sum_{i=1}^{n} |x_i\rangle\langle x_i| = \mathbb{1} \rightarrow \int_a^b dx|x\rangle\langle x| = \mathbb{1}. \tag{2.39}$$

Then we can also generalize $\langle x_i|f\rangle \equiv f(x_i)$ to a continuous function $\langle x|f\rangle = f(x)$, since

$$f(x) = \langle x|f\rangle = \langle x|\mathbb{1}|f\rangle = \int_a^b \langle x|x'\rangle\langle x'|f\rangle = \int_a^b dx' f(x')\langle x|x'\rangle \tag{2.40}$$

only makes sense if $\langle x|x'\rangle = \delta(x-x')$.

Note that the delta function is not a function (it is a distribution), but can be obtained as the limit when $\Delta \rightarrow 0$ of a Gaussian,

$$f_\Delta(x-y) = \frac{1}{\sqrt{\pi\Delta^2}} e^{-(x-y)^2/\Delta^2}. \tag{2.41}$$

The delta function $\delta(x-y)$ is even, owing to its definition as a distribution, and since it is the limit of f_Δ, and also because

$$\delta(x-y) = \langle x|y\rangle = (\langle y|x\rangle)^* = \delta^*(y-x) = \delta(y-x). \tag{2.42}$$

From the fact that $\delta(x-y)$ is a distribution, its derivative with respect to its argument,

$$\delta'(x-y) = \frac{d}{dx}\delta(x-y) = -\frac{d}{dy}\delta(x-y), \tag{2.43}$$

acts in reality as

$$\delta(x-y)\frac{d}{dy}. \tag{2.44}$$

(the proof is by partial integration, as we saw from the general derivative of a distribution). *Note that here we have assumed that the space of integration for the test functions is in y, $f(y)dy$, which is the opposite convention to the usual one, but it makes sense if we want to think of the integration as a generalization of summation coming from the matrix product.*

This fact in particular means that $\delta'(x-y)$ is a kind of matrix in (x, y) space (since it has indices).

2.6 Operators in Infinite Dimensions

We saw that the space of functions $|f\rangle$ such that $\langle x|f\rangle = f(x)$ is a Hilbert space, which means from the general theory that we can define operators $\hat{A}$ on the functions, acting as $\hat{A}|f\rangle$.

Among such operators we have the trivial ones, corresponding to the usual multiplication by a number. One nontrivial case is the differentiation operator, which can be thought as a matrix too. It is defined as

$$\frac{d}{dx} : f(x) \to f'(x) = \frac{d}{dx}f(x), \tag{2.45}$$

but we should extend this definition to the abstract space $|f\rangle$ (without the basis $|x\rangle$) as D, where

$$D|f\rangle \equiv \left|\frac{df}{dx}\right\rangle. \tag{2.46}$$

Then we can also multiply by $\langle x|$, in order to obtain a matrix element,

$$\langle x|D|f\rangle = \left\langle x \,\middle|\, \frac{df}{dx}\right\rangle = \frac{df}{dx}(x). \tag{2.47}$$

On the other hand, introducing the completeness relation inside this expression, we obtain

$$\frac{df}{dx}(x) = \int dx'\langle x|D|x'\rangle\langle x'|f\rangle \equiv \int dx' D_{xx'} f(x'), \tag{2.48}$$

which means that the matrix element of D is (note the integration over x' in the delta function)

$$D_{xx'} = \delta'(x-x') = \delta(x-x')\frac{d}{dx}. \tag{2.49}$$

2.7 Hermitian Operators and Eigenvalue Problems

Consider a linear operator $\hat{A} : H \to H$ on a Hilbert space H. Then its adjoint $\hat{A}^\dagger$ is defined as before (in the finite-dimensional case), by

$$\langle \hat{A}^\dagger f|g\rangle \equiv \langle f|\hat{A}g\rangle. \tag{2.50}$$

Then a Hermitian (or self-adjoint) operator is one for which

$$\hat{A} = \hat{A}^\dagger. \tag{2.51}$$

One thing we have to be careful about is that in principle f and g live in different Hilbert spaces (the vector space of functions and its dual), but $\hat{A} = \hat{A}^\dagger$ is only meaningful if the domain (more precisely, the support) of the two operators is the same. This is a subtle issue, since there are operators that are formally self-adjoint, but act a priori on different spaces, and requiring that the spaces are identical imposes constraints.

We have several important theorems about Hermitian operators:

Theorem For $\hat{A} : H \to H$ Hermitian, i.e., $\langle \hat{A} f | g \rangle = \langle f | \hat{A} g \rangle$ for all $f, g \in H$, it follows that $\hat{A}$ is bounded.

Theorem For $\hat{A} : H \to H$ ($\hat{A} \in L(H)$), there is a unique adjoint $\hat{A}^\dagger \in L(H)$, and it has the same norm, $||\hat{A}^\dagger|| = ||\hat{A}||$.

Theorem For $A \in L(H)$ Hermitian, $\hat{A}^\dagger = \hat{A}$, the diagonal elements are real, $\langle \hat{A} f | f \rangle \in \mathbb{R}$.

Theorem For a Hermitian operator $\hat{A} \in L(H)$, $\hat{A}^\dagger = \hat{A}$, orthogonal eigenvectors correspond to different eigenvalues. This leads to the same Gram–Schmidt diagonalization algorithm as that considered in the finite-dimensional case.

We next consider the *eigenvalue problem* (or spectral problem) $\hat{A} f = \lambda f$. It follows that

$$(\hat{A} - \lambda \mathbb{1}) f = 0. \tag{2.52}$$

Then we have another theorem:

Theorem (Hilbert–Schmidt) For a Hermitian operator $\hat{A} = \hat{A}^\dagger$, with a set of eigenvalues λ_i and eigenvectors f_i, $\{\lambda_i, f_i\}_{i \in \mathbb{N}}$, the set of eigenvectors forms a complete basis, so

$$\hat{A} | f \rangle = \sum_{i \in \mathbb{N}} \lambda_i \langle f_i | f \rangle | f_i \rangle. \tag{2.53}$$

Theorem (Fredholm) For an operator $\hat{A} : H \to H$, consider the homogenous and inhomogenous eigenvalue equations,

$$\begin{aligned} (1)&: \quad (\hat{A} - \lambda \mathbb{1}) x = z \\ (2)&: \quad (\hat{A} - \lambda \mathbb{1}) x = 0. \end{aligned} \tag{2.54}$$

Then we have two statements:

(a) (1) admits a unique solution $\Leftrightarrow$ λ is not an eigenvalue.

(b) If λ is an eigenvalue, then (1) admits a solution $\Leftrightarrow$ (2) admits a nontrivial solution, and $z \in H^\perp$ (the transverse subspace).

2.8 The Operator $D_{xx'}$

We can ask, is the operator $D_{xx'}$ from (2.49) Hermitian? No, but from it we can make a potentially Hermitian operator, $-iD_{xx'}$, since

$$-iD_{xx'} = (-iD_{xx'})^* = +iD_{x'x}, \tag{2.55}$$

which can be rewritten as

$$-i\delta'(x-x') = +i\delta'(x'-x). \tag{2.56}$$

But if the domain of definition of the functions is actually finite (as opposed to just having a compact support), there can be an issue with boundary terms. Indeed, for a Hermitian operator $\hat{A}$, we have the equality

$$\langle g|\hat{A}|f\rangle = \langle g|\hat{A}f\rangle = \left(\langle \hat{A}f|g\rangle\right)^* = \langle f|\hat{A}|g\rangle^*, \tag{2.57}$$

and by introducing two completeness relations we can write the relations in terms of the xx' matrix components of the operators:

$$\int_a^b dx \int_a^b dx' \langle g|x\rangle\langle x|\hat{A}|x'\rangle\langle x'|f\rangle = \left(\int_a^b dx \int_a^b dx' \langle f|x\rangle\langle x|\hat{A}|x'\rangle\langle x'|g\rangle\right)^*. \tag{2.58}$$

We want to see whether this can be satisfied by $A_{xx'} = -iD_{xx'}$. The left-hand side becomes

$$\int_a^b dx \int_a^b dx' g^*(x)\left(-i\delta(x-x')\frac{d}{dx'}\right) f(x') = \int_a^b dx g^*(x)\left(-i\frac{d}{dx}\right) f(x), \tag{2.59}$$

whereas the right-hand side gives, by partial integration,

$$\left[\int_a^b dx \int_a^b dx' f^*(x)\left(-i\delta(x-x')\frac{d}{dx'}\right) g(x')\right]^* = -i\int_a^b g^*(x)\frac{d}{dx}f(x) + ig^*(x)f(x)|_a^b. \tag{2.60}$$

That means that we get equality only if the boundary term

$$+\, ig^*(x)f(x)|_a^b \tag{2.61}$$

vanishes.

The eigenvalue problem for $\hat{A} = -iD_{xx'}$, defined as

$$\hat{A}|k\rangle = k|k\rangle \tag{2.62}$$

gives

$$k\langle x|k\rangle = \langle x|\hat{A}|k\rangle = \int dx' \langle x|\hat{A}|x'\rangle\langle x'|k\rangle, \tag{2.63}$$

which in turn becomes (since $A_{xx'} = -iD_{xx'} = -i\delta(x-x')d/dx'$ and can be integrated, and defining $\langle x|k\rangle \equiv \psi_k(x)$)

$$-i\frac{d}{dx}\psi_k(x) = k\psi_k(x), \tag{2.64}$$

which is solved by

$$\psi_k(x) = Ae^{ikx} = \frac{1}{\sqrt{2\pi}}e^{ikx}. \tag{2.65}$$

The normalization constant A has been chosen so that $\langle k|k'\rangle = \delta(k-k')$. Note then that in this new $|k\rangle$ basis we have that

$$-iD_{kk'} = \langle k| - iD|k'\rangle = k'\langle k|k'\rangle = k'\delta(k-k') \tag{2.66}$$

is diagonal.

Important Concepts to Remember

- The distance on a metric or pre-Hilbert space is $d(x, y) = ||x - y||$, and a Hilbert space is a metric space that is also complete, meaning that the limit of a series in the space also belongs to the space, and the norm comes from a scalar product.
- A Banach space is like a Hilbert space, except that the norm does not come from a scalar product.
- For quantum mechanics, we need infinite-dimensional Hilbert spaces in which the vectors are functions. To discretize, $f(x_i) = \langle x_i | f \rangle$ but, in the continuous case, $|x_i\rangle$ is replaced by the continuous variable $|x\rangle$.
- The scalar product on the Hilbert space of functions is $\langle f|g\rangle = \int_a^b f^*(x)g(x)dx$, but the completeness relation for the $|x\rangle$ basis vectors is $\int_a^b dx|x\rangle\langle x| = \mathbb{1}$, reducing the previous definition of a scalar product to a trivial identity.
- Distributions are linear and continuous applications that act on the space of some test functions and give a complex number result. For instance, the distribution $T = \tilde{f}$ is defined such that $\langle \tilde{f}|\phi\rangle = \int_A f^*(x)\phi(x)dx$.
- The "delta function" distribution δ_{x_0} is defined by $\langle \delta_{x_0}|\phi\rangle = \phi(x_0)$, or, more commonly, by $\phi(x_0) = \int \delta(x - x_0)\phi(x)dx$.
- The derivative of a distribution is defined by $\langle \partial T/\partial x_j|\phi\rangle = -\langle T|\partial\phi/\partial x_j\rangle$.
- The Heaviside function is a function, but it can also be associated with a distribution. As a distribution, $H'_{x_0} = \delta_{x_0}$.
- The orthonormality relation of the $|x\rangle$ vectors is $\langle x|y\rangle = \delta(x - y)$.
- Nontrivial operators on the Hilbert space of functions are those that involve derivatives, in particular the operator D, with $D|f\rangle = |df/dx\rangle$ and with matrix element $D_{xx'} = \delta'(x - x') = \delta(x - x')d/dx$.
- For Hermitian (or self-adjoint) operators $\hat{A}^\dagger = \hat{A}$ there are many solutions to the eigenvalue-eigenstate problem (or spectral problem) $(\hat{A} - \lambda\mathbb{1})f = 0$. In fact, the set of (linearly independent) eigenvectors forms a complete basis for the Hilbert space.
- The operator $-iD_{xx'}$, thought of as a distribution or as an operator on the space of functions is Hermitian, but only if the Hilbert space (the "space of test functions for the distribution") involves functions with compact support or otherwise if the boundary terms match for the matrix element with two functions.

Further Reading

See, for instance, Shankar's book [3] for more details about spaces of functions, from the point of view of a physicist.

Exercises

(1) Discretize the product of two functions, as compared to discretizing each function independently, and describe what that means in the language of kets.

(2) For a tensor product of kets, describe what the norm is in the abstract sense, and then in the function form (with integrals).

(3) Is the square of the "delta function" a distribution? If so, prove it using Dirac's bra-ket notation.
(4) Show how $\delta''(x-y)$ (the second derivative with respect to x) acts as a distribution on functions.
(5) Is a real function of a Hermitian operator $\hat{A}$, $f(\hat{A})$, also Hermitian? Give examples.
(6) Consider the unitary operator $e^{i\hat{A}}$, with $\hat{A}$ Hermitian and acting on function space. Can it be diagonalized? If so, write an expression for its diagonal elements.
(7) Diagonalize e^{σ_1}, where σ_1 is the first Pauli matrix $\begin{pmatrix} 0 & 1 \\ 1 & 0 \end{pmatrix}$.

3 The Postulates of Quantum Mechanics and the Schrödinger Equation

After developing a somewhat lengthy mathematical background, we are now ready to define quantum mechanics from some first principles, or postulates. The word "postulates" implies some mathematical-type axiomatic system, but the situation is not quite so clear cut. There isn't a mathematically rigorous collection of assumptions, only a set of rules that can be presented in many ways. The number, order, and content of the postulates varies, though the total physical content of the system is the same. Here I will present my own viewpoint on the postulates. We will start with the postulates themselves, and then we will explain them and add comments.

The crucial difference from classical mechanics is that we don't any longer have classical paths defined by a Hamiltonian $H(p_i, q_i)$ and initial conditions for the phase-space variables (p_i, q_i) (giving a "state" at time t). Instead of that, we have *probabilities* for everything we can observe, including analogs of classical variables like (p_i, q_i); but there are also new variables that have no classical counterpart. We thus have to define states, observables, probabilities for them, time evolution, and the postulates dealing with these concepts.

3.1 The Postulates

First postulate, P1

At every time t, the state of a physical system is defined by a ket $|\psi\rangle$ in a Hilbert space.

Second postulate, P2

For every observable described classically by a quantity A, there is a Hermitian operator $\hat{A}$ acting on the physical states $|\psi\rangle$ of the system. The fact that the operator is Hermitian means that its eigenvalues are real, which as we will shortly see means that the observables are real, as they should be.

Third postulate, P3

The only possible result of a measurement of an observable is an *eigenvalue* λ of the operator corresponding to it:

$$\hat{A}|\psi_\lambda\rangle = \lambda|\psi_\lambda\rangle. \tag{3.1}$$

Note that if the spectrum of the operator $\hat{A}$ is discrete, there is a discrete number of values λ_n, explaining the "quantum" in "quantum mechanics". Since the operators $\hat{A}$ corresponding to observables are Hermitian, it follows that the Hilbert space has a basis $|n\rangle$ (corresponding to λ_n) formed by eigenvectors of the operator $\hat{A}$, so every state can be expanded into it,

$$|\psi\rangle = \sum_n c_n |n\rangle. \tag{3.2}$$

Fourth postulate, P4

The probability of obtaining an eigenvalue λ_n corresponding to the eigenstate $|v_n\rangle$ in a measurement of the observable $\hat{A}$ in a normalized state $|\psi\rangle$ is

$$P_n = |\langle v_n|\psi\rangle|^2, \tag{3.3}$$

where, besides $|\psi\rangle$, the $|v_n\rangle$ are also normalized states. Here we have to assume that the sum of all probabilities is one, $\sum_n P_n = 1$.

Fifth postulate, P5

After a measurement of a variable corresponding to an operator $\hat{A}$, giving as a result the eigenvalue λ_n corresponding to the eigenvector $|v_n\rangle$, the state of the system has changed from the original $|\psi\rangle$ to $|v_n\rangle$.

This is the strangest of the postulates, which clashes most with our intuition since it implies a discontinuous, nonlinear, change. But it is experimentally found to be true. If we re-measure the same quantity right after the first measurement, we always find the same result λ_n.

Sixth postulate, P6

The time evolution of a state is given by

$$|\psi(t)\rangle = \hat{U}(t, t_0)|\psi(t_0)\rangle, \tag{3.4}$$

where the operator $\hat{U}$ is unitary, $\hat{U}^\dagger \hat{U} = \hat{1}$, in order to preserve norms, and thus the total probability (which equals 1), i.e.,

$$\langle\psi(t)|\psi(t)\rangle = \langle\psi(t_0)|\psi(t_0)\rangle. \tag{3.5}$$

This evolution operator (also known as a "propagator") is found by imposing the requirement that the time dependence of the state satisfies the *Schrödinger equation*,

$$i\hbar \frac{d}{dt}|\psi(t)\rangle = \hat{H}|\psi(t)\rangle. \tag{3.6}$$

Note that here we have assumed that states change with time but operators corresponding to observables do not, which is known as the "Schrödinger picture", but there are other pictures, as we will see later on in the book.

We next start to comment on and expand on the various postulates.

3.2 The First Postulate

As we have just seen, the state of a system is defined at a given time (in the Schrödinger picture), but it depends on time, with the time variation given by the Schrödinger equation. We described abstract states $|\psi\rangle$, but for states described by quantities that depend on spatial coordinates, such as, for instance states corresponding to single particles with a classical position, we can define the function

$$\langle x|\psi\rangle - \psi(x), \tag{3.7}$$

known as the wave function. Since $|x\rangle$ is an eigenvector of the position operator $\hat{X}$, from postulate 4 it follows that if we measure the position through $\hat{X}$ then $|v_n\rangle \to |x\rangle$, so the probability of obtaining the value x is

$$|\langle v_n|\psi\rangle|^2 \to |\langle x|\psi\rangle|^2 = |\psi(x)|^2. \tag{3.8}$$

Thus it is this function that defines position probabilities through its modulus squared, justifying the name wave function.

3.3 The Second Postulate

One observation to make is that there is some ambiguity about the operator $\hat{A}$ corresponding to a classical mechanics quantity A: there is an issue concerning the order of the operators (such as $\hat{X}, \hat{P}$) involved in the expression for a composite operator. For instance, consider a Hamiltonian $H(p,q)$ that classically contains a term pq^2. Quantum mechanically, should we replace that with $\hat{P}\hat{Q}^2$, $\hat{Q}\hat{P}\hat{Q}$, $\hat{Q}^2\hat{P}$, or some linear combination of the three possibilities?

The other observation is that for a Hermitian operator $\hat{A}$, the eigenvalues λ_n in $\hat{A}|v_n\rangle = \lambda_n|v_n\rangle$ are real, and the $|v_n\rangle$ eigenstates form a complete basis for the Hilbert space, so any state $|\psi\rangle$ can be expanded in them:

$$|\psi\rangle = \sum_n c_n|v_n\rangle. \tag{3.9}$$

It could be, however, (this is more generic) that there are other observables that can also be measured, which could mean that there are several states of given eigenvalue λ_n for $\hat{A}$, which are distinguished by some other index b (corresponding perhaps to eigenvalues of some other observables), i.e.,

$$\hat{A}|v_n, b\rangle = \lambda_n|v_n, b\rangle. \tag{3.10}$$

But we can also have a linear combination of the states with index b, so

$$\hat{A}\left(\sum_b c_b|v_n, b\rangle\right) = \lambda_n\left(\sum_b c_b|v_n, b\rangle\right). \tag{3.11}$$

Then an arbitrary state can still be expanded in this set, but now we write

$$|\psi\rangle = \sum_{n,b} c_{n,b}|v_n, b\rangle. \tag{3.12}$$

3.4 The Third Postulate

As we saw, we measure only eigenvalues of operators, which can be discrete, i.e., indexed by a natural number, λ_n. This is the reason we talk about *quantum* mechanics.

(1) The eigenvalues, and the corresponding eigenstates associated with them, can be not only discrete but also finite in number, in which case we have a finite-dimensional Hilbert space. The standard example is the system with only two states that arises from a spin 1/2 particle: spin up $\uparrow$ or spin down $\downarrow$, or more precisely

$$|s_z = +1/2\rangle \text{ and } |s_z = -1/2\rangle. \tag{3.13}$$

This system has no classical counterpart.

(2) Another possibility is a discrete spectrum, for an operator that corresponds to a classical continuous variable, but still to have a countable (indexable by $\mathbb{N}$) set of eigenvalues. As we said, in this case we talk about *quantization* of the observable. The standard example in this case is the energy of an electron in a hydrogen atom, E_n. It could also be, as in this case, that the energy spectrum is *degenerate*, meaning that there are other observables besides $\hat{H}$ (giving the energy) that add an index b to the state,

$$\hat{H}|\psi_n, b\rangle = E_n|\psi_n, b\rangle. \tag{3.14}$$

3.5 The Fourth Postulate

Since the probability of finding the eigenvalue λ_i corresponding to eigenstate $|v_i\rangle$ of $\hat{A}$ is

$$P_i = |\langle v_i|\psi\rangle|^2, \tag{3.15}$$

and we can expand the state in the complete and orthonormal ($\langle v_i|v_j\rangle = \delta_{ij}$) set of the eigenstates of $\hat{A}$, we write

$$|\psi\rangle = \sum_i c_i|v_i\rangle, \tag{3.16}$$

so that the probability of measuring λ_i is

$$P_i = |c_i|^2. \tag{3.17}$$

Since the sum of all the probabilities is one, we obtain

$$\sum_i P_i = \sum_i |c_i|^2 = 1, \tag{3.18}$$

which amounts (since the $|v_i\rangle$ are normalized states) to normalization of the state,

$$\langle\psi|\psi\rangle = \sum_{i,j} c_i^* c_j \langle v_i|v_j\rangle = \sum_i |c_i|^2 = 1. \tag{3.19}$$

This also means that only normalizable states $|\psi\rangle$ can be physical (hence our definition for the Hilbert space), since normalizable states imply that we can construct a probability set that sums to one.

In the case of degenerate states, $\hat{A}|v_n, b\rangle = \lambda_n|v_n, b\rangle$, instead of projecting onto a single state $\langle v_i|$, we project onto the set of states with the same eigenvalue λ_n, using the *projector* $\mathbb{P}_{V_n}$, so onto the state

$$(\mathbb{P}_{V_n}|\psi\rangle)^\dagger. \tag{3.20}$$

A projector must satisfy the relations

$$\mathbb{P}_{V_n}^2 = \mathbb{P}_{V_n}, \quad \mathbb{P}_{V_n}^\dagger = \mathbb{P}_{V_n}, \tag{3.21}$$

the first stating that projecting an already projected state changes nothing, so $\mathbb{P}^2 = \mathbb{P}$, and the second that the projections onto the bra and ket are identical.

It is easy to realize that the projector must be the part of the completeness relation involving only the states in question, so in our case

$$\mathbb{P}_{V_n} - \sum_a |v_n, a\rangle\langle v_n, a|. \tag{3.22}$$

Acting on the general state

$$|\psi\rangle = \sum_{m,b} c_{m,b}|v_m, b\rangle \tag{3.23}$$

gives

$$\mathbb{P}_{V_n}|\psi\rangle = \sum_{a,m,b} |v_n, a\rangle\langle v_n, a|v_m, b\rangle c_{m,b}. \tag{3.24}$$

Then we must replace $P_n = |\langle v_n|\psi\rangle|^2 = |c_n|^2$ by the more general

$$P_n = \langle\psi|\mathbb{P}_{V_n}|\psi\rangle = \langle\mathbb{P}_{V_n}\psi|\mathbb{P}_{V_n}\psi\rangle = \sum_a c_{n,a}^* c_{n,a}. \tag{3.25}$$

Finally, we come to the possibility of having a continuous spectrum for an operator. As we said, this is the case for the position operator, with matrix elements

$$\langle x|\hat{X}|x'\rangle = x\delta(x - x'), \tag{3.26}$$

and the momentum operator, with matrix elements

$$\langle x|\hat{P}|x'\rangle = -i\hbar\delta'(x - x'). \tag{3.27}$$

The position operator has eigenstates $|x\rangle$,

$$\hat{X}|x\rangle = x|x\rangle, \tag{3.28}$$

which means we can define the wave function $\langle x|\psi\rangle \equiv \psi(x)$. In terms of it, the abstract relation for eigenvectors of operators,

$$\hat{A}|\psi\rangle = \lambda|\psi\rangle \tag{3.29}$$

becomes, by inserting the identity as a completeness relation using the states $|x\rangle$,

$$\int dx'\langle x|\hat{A}|x'\rangle\langle x'|\psi\rangle = \lambda\langle x|\psi\rangle. \tag{3.30}$$

Writing this in terms of wave functions, we obtain

$$\int dx' A_{xx'}\psi(x') = \lambda\psi(x). \tag{3.31}$$

In the case of the momentum operator, $\hat{A} = \hat{P}$, we obtain

$$\int dx' \, (-i\hbar\delta'(x - x')) \, \psi(x') = p\psi(x), \tag{3.32}$$

and, since $\delta'(x - x')$ acts as $\delta(x - x')d/dx'$, we finally obtain

$$-i\hbar \frac{d\psi(x)}{dx} = p\psi(x). \tag{3.33}$$

On the other hand, for the position operator $\hat{A} = \hat{X}$, since $\hat{X}_{xx'} = x\delta(x - x')$ we obtain just an identity, $x\psi(x) = x\psi(x)$.

We have presented three cases, of finite dimensional Hilbert spaces (such as for electron spin), of discrete countable states (such as the energy states of the hydrogen atom), and of continuous states (such as position and momentum). But we can have a system that has eigenvalues corresponding to more than one observable, for instance both spin and position like the electron. In such a case, the total Hilbert space is a tensor product of the individual Hilbert spaces,

$$\mathcal{H} = \mathcal{H}_1 \otimes \mathcal{H}_2. \tag{3.34}$$

Example For the electron, with spin 1/2, consider that $\mathcal{H}_1$ corresponds to spin 1/2, with basis $|s_z = +1/2\rangle, |s_z = -1/2\rangle$ (or $|+\rangle, |-\rangle$), and $\mathcal{H}_2$ corresponds to position, with $\langle x|\psi\rangle = \psi(x)$. Then we consider states of the type $|s\rangle \otimes |\psi\rangle$ and basis states of the type $\langle +| \otimes \langle x|$.

If we have two independent observables that can be diagonalized simultaneously (so we can measure their eigenvalues simultaneously), corresponding to operators $\hat{A}_1, \hat{A}_2$, so that

$$\hat{A}_1|\lambda_1, \lambda_2\rangle = \lambda_1|\lambda_1, \lambda_2\rangle, \quad \hat{A}_2|\lambda_1, \lambda_2\rangle = \lambda_2|\lambda_1, \lambda_2\rangle, \tag{3.35}$$

we can take the commutator of the two operators, and obtain

$$(\hat{A}_1\hat{A}_2 - \hat{A}_2\hat{A}_1)|\lambda_1, \lambda_2\rangle = 0. \tag{3.36}$$

Then, more generally (since the vectors $|\lambda_1, \lambda_2\rangle$ form a basis for the Hilbert space), we can write a commutation condition for operators:

$$[\hat{A}_1, \hat{A}_2] = 0. \tag{3.37}$$

Thus operators corresponding to independent observables commute.

From the fourth postulate we also obtain a result for the experimentally measured average value of an observable, which is sometimes included as part of the postulates:

$$\langle A\rangle \equiv \langle\psi|\hat{A}|\psi\rangle = \sum_{n,m} c_n^*\langle v_n|\hat{A}c_m|v_m\rangle = \sum_n |c_n|^2 a_n = \sum_n P_n a_n, \tag{3.38}$$

where we have used $\hat{A}|v_n\rangle = a_n|v_n\rangle$ and the orthonormality of the eigenstates. Since the right-hand side ($\sum_n P_n a_n$) is the experimentally measured average value, it follows that the average value of $\hat{A}$ in the state $|\psi\rangle$ is consistently defined as

$$\langle A\rangle \equiv \langle\psi|\hat{A}|\psi\rangle. \tag{3.39}$$

We also saw that the sum of probabilities must be 1, which means that states must be normalized, $\sum_n P_n = \sum_n |c_n|^2 = 1$. But this relation, the conservation of probability, must be preserved by any transformation, such as a change of basis by an operator $\hat{U}$, or time evolution through the operator $\hat{U}(t, t_0)$. That means in both cases that $\hat{U}$ must be unitary, since

$$|\psi\rangle \to \hat{U}|\psi\rangle \quad \Rightarrow \quad \langle\psi|\psi\rangle \to \langle\psi|\hat{U}^\dagger\hat{U}|\psi\rangle = \langle\psi|\psi\rangle \tag{3.40}$$

implies $\hat{U}^\dagger\hat{U} = \hat{1}$.

3.6 The Fifth Postulate

As we said, this is the weirdest postulate, the one that most contradicts our common intuition but is experimentally verified. On the one hand it looks very singular, since the state changes abruptly by *measurement*, which is an interaction with a *classical* apparatus, and on the other it seems to happen instantaneously. Moreover, it is a nonlinear process. Other quantum processes, as we just saw, are obtained by unitary evolution (in either time, or change of basis), but this one process is nonunitary as well as nonlinear. Since classical processes are supposed to appear in the limit of a large number of components (as a macroscopic object is formed of a very large number of atoms), they should also be related somehow to quantum mechanical processes. Yet measurements, interactions with a classical system, somehow give a nonlinear and nonunitary process. This is very puzzling, and its interpretation is still the subject of some debate.

3.7 The Sixth Postulate

We now consider the Schrödinger equation, and how it leads to the time evolution operator. The eigenstates $|E\rangle$, corresponding to eigenvalues, or energies E, of the Hamiltonian $\hat{H}$ form a complete set, so we introduce their completeness relation in the definition of a state $|\psi(t)\rangle$, to find

$$|\psi(t)\rangle = \sum_E |E\rangle\langle E|\psi(t)\rangle \equiv \sum_E \psi_E(t)|E\rangle, \tag{3.41}$$

and substitute in the *time-dependent* Schrödinger equation

$$i\hbar\frac{\partial}{\partial t}|\psi(t)\rangle = \hat{H}|\psi(t)\rangle, \tag{3.42}$$

while imposing the *time-independent* Schrödinger equation, or eigenvalue problem for the Hamiltonian,

$$\hat{H}|E\rangle = E|E\rangle. \tag{3.43}$$

We obtain

$$\left(i\hbar\frac{\partial}{\partial t} - \hat{H}\right)|\psi(t)\rangle = \sum_E \left(i\hbar\frac{d}{dt}\psi_E(t) - E\psi_E(t)\right)|E\rangle = 0, \tag{3.44}$$

with solution

$$\psi_E(t) = e^{-iE(t-t_0)/\hbar}\psi_E(t_0), \tag{3.45}$$

leading to the full time-dependent state

$$|\psi(t)\rangle = \sum_E e^{-iE(t-t_0)/\hbar}\psi_E(t_0)|E\rangle. \tag{3.46}$$

Then the time evolution operator is

$$\hat{U}(t,t_0) = \sum_E |E\rangle\langle E|e^{-iE(t-t_0)/\hbar}. \tag{3.47}$$

For a degenerate energy spectrum, we will have a sum over states $|E,a\rangle$:

$$\begin{aligned}|\psi(t)\rangle &= \sum_{E,a} e^{-iE(t-t_0)/\hbar}\psi_{E,a}(t_0)|E,a\rangle,\\ \hat{U}(t,t_0) &= \sum_{E,a} |E,a\rangle\langle E,a|e^{-iE(t-t_0)/\hbar}.\end{aligned} \tag{3.48}$$

3.8 Generalization of States to Ensembles: the Density Matrix

Until now we have considered the "pure quantum mechanical" case, where the state of the system is defined by a unique state $|\psi\rangle$ in a Hilbert space, also known as a "pure state". However, in practice, many important cases correspond to a combination of quantum mechanical and classical systems. Specifically, instead of having a pure quantum state, we have a "collection" or *ensemble* of states in which the system may be found each one with its own *classical* probability p_i. We characterize this situation by a "density matrix" (though it is really an operator, and it should be thought of this way),

$$\hat{\rho} = \sum_i p_i|i\rangle\langle i|. \tag{3.49}$$

Note that $|i\rangle$ can be any kind of state, not necessarily an eigenstate of some observable.

Then we say that the average value of an observable associated with an operator $\hat{A}$ is the result of a double averaging: first a classical averaging (denoted with a line over the operator) giving probabilities p_i, then the standard quantum averaging over states $|i\rangle$,

$$\langle\bar{A}\rangle = \sum_i p_i\langle i|\hat{A}|i\rangle. \tag{3.50}$$

This average is obtained from the quantity

$$\mathrm{Tr}(\hat{\rho}\hat{A}) = \sum_n\sum_i p_i\langle n|i\rangle\langle i|\hat{A}|n\rangle = \sum_i p_i\sum_n|\langle i|n\rangle|^2\lambda_n = \sum_i p_i\langle A\rangle_i = \langle\bar{A}\rangle, \tag{3.51}$$

where in the first equality we have used the fact that there is an orthonormal basis $|n\rangle$ of eigenstates of $\hat{A}$, and we have taken the trace of the corresponding matrix, and in the second we have used the eigenvalue equation, $\hat{A}|n\rangle = \lambda_n|n\rangle$, and finally we have used that the quantum probability of finding the state $|i\rangle$ in the state $|n\rangle$ is $P_{i,n} = |\langle i|n\rangle|^2$.

Note that the density matrix is normalized, i.e., we assume that the sum of all the probabilities is one, which amounts to (setting formally $\hat{A} = \mathbb{1}$, or $\lambda_n = 1$)

$$\mathrm{Tr}\,\hat{\rho} = \sum_i p_i\sum_n|\langle i|n\rangle|^2 = \sum_i p_i\sum_n P_{i,n} = 1. \tag{3.52}$$

To come back to the original case, of a "pure state" (as opposed to a "mixed state", i.e., an ensemble with a density matrix), or, more properly stated, a "pure ensemble", we now take into account that $p_i = \delta_{i,\psi}$, thus the density matrix is

$$\hat{\rho} = |\psi\rangle\langle\psi|. \tag{3.53}$$

In this case, the average reduces to the quantum average, since

$$\langle\bar{A}\rangle = \mathrm{Tr}(\hat{\rho}\hat{A}) = \sum_n \langle n|\psi\rangle\langle\psi|\hat{A}|n\rangle = \sum_n \lambda_n |\langle n|\psi\rangle|^2 = \langle A\rangle_\psi, \tag{3.54}$$

the same as for a pure state.

The formalism of density matrices is relevant for the transition from quantum to classical regimes which, as we alluded to already, is a tricky issue. As we said, we have a combination of classical and quantum aspects, and in the presence of temperature we obtain something of this type. Temperature means a degree of "thermalization", or interaction with a classical temperature bath, that means classicalization to some degree. These are however issues that will be discussed more at length later on, in Part $\mathrm{II_a}$ of this book.

Important Concepts to Remember

- The postulates of quantum mechanics are just a convenient way to package the physical content of quantum mechanics, not mathematical axioms.
- The first postulate: at every time, the state of a physical system is a ket $|\psi\rangle$ in a Hilbert space.
- The second postulate: observables correspond to a Hermitian operator $\hat{A}$ acting on $|\psi\rangle$. The eigenvalues of $\hat{A}$ are real, as they must be in order to correspond to observables.
- The third postulate: the result of a measurement of an observable A is an eigenvalue λ_n on $\hat{A}$. If the λ_n are discrete, we have quantization. We can expand a state $|\psi\rangle$ in terms of a basis consisting of the eigenstates $|v_n\rangle$ of $\hat{A}$.
- The fourth postulate: the probability of obtaining the eigenvalue λ_n associated with $|v_n\rangle$ is $P_n = |\langle v_n|\psi\rangle|^2$. Since $|\psi\rangle$ and $|v_n\rangle$ are normalized, $\sum_n P_n = 1$.
- The fifth postulate: after measuring the variable A and obtaining λ_n, the state of the system has changed from $|\psi\rangle$ to $|v_n\rangle$.
- The sixth postulate: the evolution in time of a state is governed by the Schrödinger equation $i\hbar d/dt|\psi(t)\rangle = \hat{H}|\psi(t)\rangle$.
- If we consider the position of a particle in a state $|\psi\rangle$, with vector $|x\rangle$, then the probability of finding the particle at position x is $|\psi(x)|^2$.
- Independent observables can be diagonalized simultaneously and have commuting operators: $[\hat{A}, \hat{B}] = 0$.
- For an eigenvalue E of the Hamiltonian of a system, we have the time-independent Schrödinger equation $\hat{H}|\psi_E\rangle = E|\psi_E\rangle$; then $|\psi_E(t)\rangle = e^{-iE(t-t_0)/\hbar}|\psi_E(t_0)\rangle$.
- The evolution operator $\hat{U}(t, t_0)$ is defined by $|\psi(t)\rangle = \hat{U}(t, t_0)|\psi(t_0)\rangle$.
- A classical ensemble of quantum states corresponds to a density matrix $\hat{\rho} = \sum_i |i\rangle\langle i|$, normalized as $\mathrm{Tr}\,\hat{\rho} = 1$ and with average observable value $\langle\bar{A}\rangle = \mathrm{Tr}(\rho\hat{A})$.

Further Reading

See, for instance, Messiah's book [2] or Shankar's book [3] for an alternative approach.

Exercises

(1) Suppose that the probability of finding some particle 1 at x_1 is a Gaussian around x_1, with standard deviation σ_1, and the probability of finding another particle 2 at x_2 is a Gaussian around x_2 with standard deviation σ_2. What is the condition on the wave functions of the two particles such that the probability to find either particle at the position mentioned is the sum of the two Gaussians?

(2) Consider the classical Hamiltonian

$$H = \alpha p^2 + \beta q^2 + p f(q). \tag{3.55}$$

Write a quantum Hamiltonian that is ordering-symmetric with respect to $\hat{Q}$ and $\hat{P}$, taking into account that $[\hat{Q}, \hat{P}]$ is a constant.

(3) Consider infinite and discrete energy spectra E_n. If the spectrum extends by a finite amount, between $E_{\rm min}$ and $E_{\rm max}$, what conditions can you impose on the E_n very close to either of these values? What if $E_{\rm max} = +\infty$ or $E_{\rm min} = -\infty$?

(4) Consider a (spinless) particle in a three-dimensional potential. How many commuting operators associated with observables are there?

(5) Consider the Hamiltonian

$$H = x^2. \tag{3.56}$$

Solve the Schrödinger equation and find the time evolution operator.

(6) Consider the density matrix

$$\hat{\rho} = \frac{1}{3}(|1\rangle\langle 1| + |2\rangle, \langle 2| + |3\rangle\langle 3|), \tag{3.57}$$

where $|1\rangle$ and $|2\rangle \pm |3\rangle$ are eigenstates of some operator $\hat{A}$. Calculate the average value of A.

(7) In exercise 6, if the states $|1\rangle, |2\rangle, |3\rangle$ are eigenstates of the Hamiltonian $\hat{H}$, write the evolution in time of the density matrix $\hat{\rho}$.

4 Two-Level Systems and Spin-1/2, Entanglement, and Computation

In this chapter we consider the simplest possible situation, that of a Hilbert space with dimension 2, i.e., a system with only two orthonormal states leading to operators represented as 2×2 matrices, also known as a two-level system. We will only analyze the time-independent case, leaving the time-dependent case for later. Then, we will consider the case of several such two-level systems coupled together, with a tensor product Hilbert space (as we described previously). The relevant new phenomenon that we will observe is called entanglement, related to the quantum coupling of the Hilbert spaces. Moreover, this will allow us to define a quantum version of classical computation, which here we will only touch upon; both entanglement and quantum computation will be addressed in more detail later on in the book.

4.1 Two-Level Systems and Time Dependence

We will motivate the study of general two-level systems by first considering the simplest such system, of a fermion (such as an electron) with spin $|\vec{S}| = 1/2$ (more precisely, the spin is 1/2 times the quantum unit of angular momentum, which is $\hbar$, so $|\vec{S}| = \hbar/2$, but we will suppress the $\hbar$ for the moment). In this case, as we know experimentally, for instance from the Stern–Gerlach experiment (described in Chapter 0), the projection of the spin onto any direction, here denoted by z, S_z, can take only the values $+1/2$ and $-1/2$. In the Stern–Gerlach experiment, we put a magnetic field in the z direction, splitting an electron beam passing through it transversally.

Therefore, the independent states in this simplest system are $|S = 1/2, S_z = +1/2\rangle$, also called $|+\rangle$ or $|\uparrow\rangle$, and $|S = 1/2, S_z = -1/2\rangle$, also called $|-\rangle$ or $|\downarrow\rangle$. In a matrix notation for the Hilbert space and the operators on it, the states are (see Fig. 4.1)

$$|\uparrow\rangle = |+\rangle = \begin{pmatrix} 1 \\ 0 \end{pmatrix}, \quad |\downarrow\rangle = |-\rangle = \begin{pmatrix} 0 \\ 1 \end{pmatrix}. \tag{4.1}$$

Operators acting on this space are then represented by 2×2 matrices acting on the above states.

The most important operators are the spin operators themselves, or more precisely the spin projections onto the three axes S_x, S_y, S_z. We will describe the general theory of spin and angular momentum later, but for the moment we note that, since the spin projections are $\pm 1/2$, we want the matrices representing S_x, S_y, S_z to satisfy

$$S_x^2 = S_y^2 = S_z^2 = \frac{1}{4}. \tag{4.2}$$

Moreover, we want to have (we will explain this condition later)

$$(S_x + iS_y)^2 = (S_x - iS_y)^2 = 0, \tag{4.3}$$

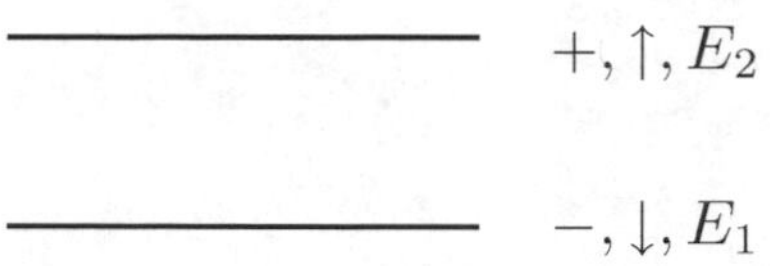

$$+, \uparrow, E_2$$

$$-, \downarrow, E_1$$

Figure 4.1 A two-level system, with a common notation.

which leads to

$$S_x S_y + S_y S_x = 0, \tag{4.4}$$

i.e., these matrices anticommute. The above conditions, together with the anticommutativity of S_x and S_y with S_z, are actually sufficient to define the matrices as $S_i = \frac{1}{2}\sigma_i$, with σ_i the Pauli matrices,

$$\sigma_1 = \sigma_x = \begin{pmatrix} 0 & 1 \\ 1 & 0 \end{pmatrix}, \quad \sigma_2 = \sigma_y = \begin{pmatrix} 0 & -i \\ i & 0 \end{pmatrix}, \quad \sigma_3 = \sigma_z = \begin{pmatrix} 1 & 0 \\ 0 & -1 \end{pmatrix}. \tag{4.5}$$

More precisely, the Pauli matrices satisfy

$$\sigma_i \sigma_j = \delta_{ij} \mathbb{1}_{2\times 2} + i\epsilon_{ijk}\sigma_k, \tag{4.6}$$

which we can check explicitly; we have used implicit Einstein summation over the index k, and the Levi–Civita tensor is $\epsilon_{ijk} = +1$ for $(ijk) = 123$ and for cyclic permutations thereof, and -1 otherwise.

The energy due to the interaction of the spin of the electron, manifested through its magnetic moment, and its interaction with the magnetic field $\vec{B}$ is

$$\Delta H_{\text{spin}} = -\vec{\mu} \cdot \vec{B}, \tag{4.7}$$

where $\vec{\mu}$ is the magnetic moment of the electron. Classically,

$$\vec{\mu} = \frac{e}{2m}\vec{L}. \tag{4.8}$$

Quantum mechanically, as we will see, the quantum unit of $|\vec{L}|$ is $\hbar$. Then the quantum of $|\vec{\mu}|$ is called the Bohr magneton, and is

$$\mu_B = \frac{e\hbar}{2m}. \tag{4.9}$$

On the other hand the electron, as a fermion, has half-integer spin (i.e., intrinsic angular momentum), and more precisely

$$|\vec{S}| = \frac{\hbar}{2}. \tag{4.10}$$

Moreover, the quantum magnetic moment of the electron due to its spin has an extra factor, the Landé g-factor, so

$$\vec{\mu}_S = g\frac{e}{2m}\vec{S}, \tag{4.11}$$

where $g \simeq 2$. The magnetic moment as an operator, or more precisely a matrix, acting on the two spin states, is then

$$\vec{\mu}_S = \frac{ge\hbar}{4m}\vec{\sigma}, \tag{4.12}$$

so the spin energy, leading to spin splitting of the electron's energy levels, is

$$\Delta H_{\text{spin}} = -\vec{\mu}_S \cdot \vec{B} = -\mu\vec{\sigma} \cdot \vec{B}. \tag{4.13}$$

If $\vec{\mu} = \mu_z \vec{e}_z$ and B is time independent, then we finally obtain

$$\Delta H_{\text{spin}} = -\mu B_z \begin{pmatrix} 1 & 0 \\ 0 & -1 \end{pmatrix}. \tag{4.14}$$

Its eigenvectors are the basis vectors,

$$|\uparrow\rangle = |+\rangle = \begin{pmatrix} 1 \\ 0 \end{pmatrix}, \quad |\downarrow\rangle = |-\rangle = \begin{pmatrix} 0 \\ 1 \end{pmatrix}. \tag{4.15}$$

If we are measuring this energy, for instance as in the Stern–Gerlach experiment, a prototype of a quantum state that does not have a well-defined energy but rather *probabilities* for energies, is the normalized linear combination

$$|\psi_\pm\rangle \equiv \frac{1}{\sqrt{2}}(|\uparrow\rangle \pm |\downarrow\rangle) = \frac{1}{\sqrt{2}}\begin{pmatrix} 1 \\ \pm 1 \end{pmatrix}. \tag{4.16}$$

In this state, the probabilities for the occurrence of each eigenstate $|i\rangle = |\pm\rangle$ are equal, and given by

$$p_i = |c_i|^2 = \frac{1}{2}, \tag{4.17}$$

for both values of the measured energy,

$$E_+ = E + \mu B_z \quad \text{and} \quad E_- = E - \mu B_z. \tag{4.18}$$

We could in principle also add a time-dependent magnetic field with circular polarization in the (x, y) directions,

$$B_x = -b\cos\omega t, \quad B_y = b\sin\omega t. \tag{4.19}$$

Then its interaction with the magnetic moment of the electron gives

$$-(B_x\sigma_x + B_y\sigma_y) = b\begin{pmatrix} 0 & \cos\omega t + i\sin\omega t \\ \cos\omega t - i\sin\omega t & 0 \end{pmatrix} = b\begin{pmatrix} 0 & e^{i\omega t} \\ e^{-i\omega t} & 0 \end{pmatrix}, \tag{4.20}$$

leading to a total spin Hamiltonian

$$\Delta H_{\text{spin}} = -\mu B_z \begin{pmatrix} 1 & 0 \\ 0 & -1 \end{pmatrix} + \mu b \begin{pmatrix} 0 & e^{i\omega t} \\ e^{-i\omega t} & 0 \end{pmatrix}. \tag{4.21}$$

However, this case is more complicated and will not be analyzed here, but later in the book.

4.2 General Stationary Two-State System

Consider a complete basis $|\phi_1\rangle, |\phi_2\rangle$ corresponding to some part of the Hamiltonian $\hat{H}_0$. In coordinate space, this would correspond to wave functions $\langle x|\phi_1\rangle = \phi_1(x)$ and $\langle x|\phi_2\rangle = \phi_2(x)$. As before, representing $|\phi_1\rangle$ by $|+\rangle = \begin{pmatrix} 1 \\ 0 \end{pmatrix}$ and $|\phi_2\rangle$ by $|-\rangle = \begin{pmatrix} 0 \\ 1 \end{pmatrix}$, a general wave function can be decomposed as

$$\psi(x,t) = c_1(t)\phi_1(x) + c_2(t)\phi_2(x), \tag{4.22}$$

which is written formally as

$$|\psi(t)\rangle = c_1(t)|\phi_1\rangle + c_2(t)|\phi_2\rangle = \begin{pmatrix} c_1(t) \\ c_2(t) \end{pmatrix}. \tag{4.23}$$

Then the Schrödinger equation for the wave function in coordinate space is

$$i\hbar \frac{d}{dt}\psi(x,t) = \hat{H}\psi(x,t), \tag{4.24}$$

and in matrix form for the formal case it is

$$i\hbar \frac{d}{dt}\begin{pmatrix} c_1(t) \\ c_2(t) \end{pmatrix} = \hat{H}\begin{pmatrix} c_1(t) \\ c_2(t) \end{pmatrix} = \begin{pmatrix} H_{11} & H_{12} \\ H_{21} & H_{22} \end{pmatrix}\begin{pmatrix} c_1(t) \\ c_2(t) \end{pmatrix}. \tag{4.25}$$

Here the matrix elements of the Hamiltonian are, by definition,

$$H_{ij} = \langle \phi_i|\hat{H}|\phi_j\rangle = \int dx \int dy \langle \phi_i|x\rangle\langle x|\hat{H}|y\rangle\langle y|\phi_j\rangle = \int dx \int dy \phi_i^*(x)\hat{H}_{xy}\phi_j(y), \tag{4.26}$$

where we have introduced two identities written using the completeness relation in order to express the Hamiltonian matrix elements in terms of wave functions.

As we saw in the general theory, in order to solve the Schrödinger equation we write an eigenvalue problem for the Hamiltonian, and in terms of it we write an ansatz for the time dependence as

$$\begin{pmatrix} c_1(t) \\ c_2(t) \end{pmatrix} = e^{-iEt/\hbar}\begin{pmatrix} c_1 \\ c_2 \end{pmatrix}. \tag{4.27}$$

Here c_1, c_2 are constants and E is an eigenvalue for the eigenvalue–eigenvector problem for the Hamiltonian, satisfying

$$\begin{pmatrix} H_{11} & H_{12} \\ H_{21} & H_{22} \end{pmatrix}\begin{pmatrix} c_1 \\ c_2 \end{pmatrix} = E\begin{pmatrix} c_1 \\ c_2 \end{pmatrix}. \tag{4.28}$$

As we know, the existence of the eigenvalue E is equivalent to the vanishing of $\det(\hat{H} - E\mathbb{1})$,

$$\left|\begin{pmatrix} H_{11} - E & H_{12} \\ H_{21} & H_{22} - E \end{pmatrix}\right| = 0. \tag{4.29}$$

Moreover, remembering that the Hamiltonian (like all observables) must be a Hermitian operator, $\hat{H} = \hat{H}^\dagger$, in matrix form it means that we must have

$$H_{21} = H_{12}^*. \tag{4.30}$$

The above equation for the eigenvalues then becomes

$$(E - H_{11})(E - H_{22}) - |H_{12}|^2 = E^2 - E(H_{11} + H_{22}) + H_{11}H_{22} - |H_{12}|^2 = 0, \tag{4.31}$$

with solutions

$$E_\pm = \frac{H_{11} + H_{22}}{2} \pm \sqrt{\frac{(H_{11} - H_{22})^2}{4} + |H_{12}|^2}. \tag{4.32}$$

Given these eigenvalues, the eigenstate equations are

$$\begin{pmatrix} H_{11} & H_{12} \\ H_{21} & H_{22} \end{pmatrix}\begin{pmatrix} c_1^\pm \\ c_2^\pm \end{pmatrix} = E_\pm\begin{pmatrix} c_1^\pm \\ c_2^\pm \end{pmatrix}. \tag{4.33}$$

Because of the eigenvalue condition, the two equations are degenerate (the determinant of the linear equation for the variables $c_1^\pm, c_2^\pm$ vanishes), so only one of the two equations is independent, say, the second,

$$H_{21}c_1^\pm + (H_{22} - E_\pm)c_2^\pm = 0, \tag{4.34}$$

while the second condition for the pair of variables is the normalization condition

$$|c_1^\pm|^2 + |c_2^\pm|^2 = 1, \tag{4.35}$$

obtained from

$$\langle\psi_+|\psi_+\rangle = \langle\psi_-|\psi_-\rangle = 1. \tag{4.36}$$

The solution of the two equations is

$$\begin{pmatrix} c_1^\pm \\ c_2^\pm \end{pmatrix} = \frac{\eta^\pm}{\sqrt{1 + \left|\frac{H_{21}}{E_\pm - H_{22}}\right|^2}} \begin{pmatrix} 1 \\ \frac{H_{21}}{E_\pm - H_{22}} \end{pmatrix}, \tag{4.37}$$

where $|\eta^\pm|^2 = 1$, so $\eta^\pm$ is a phase, which we will choose to be 1. We can simplify the equations using the definitions

$$\frac{H_{11} + H_{22}}{2} \equiv \bar{E}, \quad \frac{H_{22} - H_{11}}{2} \equiv \Delta, \quad H_{12} = H_{21}^* \equiv V, \quad 2\sqrt{\Delta^2 + |V|^2} \equiv \hbar\Omega. \tag{4.38}$$

Then the eigenenergies are

$$E_\pm = \bar{E} \pm \sqrt{\Delta^2 + |V|^2} = \bar{E} \pm \frac{\hbar\Omega}{2}, \tag{4.39}$$

the Hamiltonian is

$$\hat{H} = \begin{pmatrix} \bar{E} - \Delta & V \\ V^* & \bar{E} + \Delta \end{pmatrix}, \tag{4.40}$$

and the eigenstates are

$$\begin{aligned}
\begin{pmatrix} c_1^\pm \\ c_2^\pm \end{pmatrix} &= \frac{\eta^\pm}{\sqrt{1 + \frac{|V|^2}{(\Delta \pm \sqrt{\Delta^2 + |V|^2})^2}}} \begin{pmatrix} 1 \\ \frac{V^*}{\Delta \pm \sqrt{\Delta^2 + |V|^2}} \end{pmatrix} \\
&= \frac{\eta^\pm(\pm\Delta + \sqrt{\Delta^2 + |V|^2})}{\sqrt{|V|^2 + (\pm\Delta + \sqrt{\Delta^2 + |V|^2})^2}} \begin{pmatrix} 1 \\ \frac{V^*}{\Delta \pm \sqrt{\Delta^2 + |V|^2}} \end{pmatrix} \\
&= \frac{\eta^\pm}{\sqrt{|V|^2 + (\pm\Delta + \sqrt{\Delta^2 + |V|^2})^2}} \begin{pmatrix} \pm\Delta + \sqrt{\Delta^2 + |V|^2} \\ \pm V^* \end{pmatrix}.
\end{aligned} \tag{4.41}$$

We choose $\eta^\pm = 1$, a real V ($V = V^* \in \mathbb{R}$), and then we can define

$$\begin{pmatrix} c_1^+ \\ c_2^+ \end{pmatrix} = \begin{pmatrix} -\sin\theta \\ \cos\theta \end{pmatrix}, \quad \begin{pmatrix} c_1^- \\ c_2^- \end{pmatrix} = \begin{pmatrix} \cos\theta \\ \sin\theta \end{pmatrix}. \tag{4.42}$$

Note that this definition is self-consistent since $c_2^- = -c_1^+$, which gives

$$\frac{V}{\sqrt{V^2 + (-\Delta + \sqrt{\Delta^2 + V^2})^2}} = \frac{\Delta + \sqrt{\Delta^2 + V^2}}{\sqrt{V^2 + (\Delta + \sqrt{\Delta^2 + V^2})^2}} \equiv \sin\theta, \tag{4.43}$$

as we can check by squaring the equation and multiplying with the denominators.

Moreover, we obtain that

$$\sin 2\theta = 2\sin\theta\cos\theta = -\frac{2V(\Delta + \sqrt{\Delta^2 + V^2})}{V^2 + (\Delta + \sqrt{\Delta^2 + V^2})^2} = -\frac{V}{\sqrt{\Delta^2 + V^2}}, \tag{4.44}$$

which also implies that

$$\cos 2\theta = \frac{\Delta}{\sqrt{\Delta^2 + V^2}}. \tag{4.45}$$

In conclusion, the eigenstates corresponding to $E_\pm = \bar{E} \pm \hbar\Omega/2$ are

$$\begin{aligned} |\psi_-\rangle &= \cos\theta|\phi_1\rangle + \sin\theta|\phi_2\rangle \\ |\psi_+\rangle &= -\sin\theta|\phi_1\rangle + \cos\theta|\phi_2\rangle, \end{aligned} \tag{4.46}$$

so the general time-dependent state is given by

$$|\psi(t)\rangle = c_- e^{-iE_- t/\hbar}|\psi_-\rangle + c_+ e^{-iE_+ t/\hbar}|\psi_+\rangle, \tag{4.47}$$

where, since $|\psi_\pm\rangle$ are orthonormal states, by multiplying with $\langle\psi_\pm|$ we obtain that the coefficients $c_\pm$ are given by

$$c_\pm = \langle\psi_\pm|\psi(t=0)\rangle. \tag{4.48}$$

4.3 Oscillations of States

Choosing as initial state an eigenstate of the diagonal part of the Hamiltonian, for instance $|\phi_1\rangle = \begin{pmatrix}1\\0\end{pmatrix}$, but not of the full Hamiltonian, we will see that the system actually oscillates between the state $|\phi_1\rangle$ and the state $|\phi_2\rangle$. This phenomenon is actually the one responsible for neutrino oscillations, and what we have described here applies almost directly.

Consider then the state $|\psi(t=0)\rangle = |\phi_1\rangle$ as the initial state. Inverting (4.46), we have

$$\begin{aligned} |\phi_1\rangle &= \cos\theta|\psi_-\rangle - \sin\theta|\psi_+\rangle \\ |\phi_2\rangle &= \sin\theta|\psi_-\rangle + \cos\theta|\psi_+\rangle, \end{aligned} \tag{4.49}$$

which means that from $c_\pm = \langle\psi_\pm|\psi(t=0)\rangle$ we get

$$c_- = \cos\theta, \quad c_+ = -\sin\theta. \tag{4.50}$$

Substituting into the general state, we obtain

$$|\psi(t)\rangle = \left(\cos^2\theta\, e^{-iE_- t/\hbar} + \sin^2\theta\, e^{-iE_+ t/\hbar}\right)|\phi_1\rangle + \sin\theta\cos\theta\left(e^{-iE_- t/\hbar} - e^{-iE_+ t/\hbar}\right)|\phi_2\rangle \equiv c_1|\phi_1\rangle + c_2|\phi_2\rangle, \tag{4.51}$$

which means that, from the point of view of the $|\phi_1\rangle, |\phi_2\rangle$ basis, the state oscillates between the basis states. This is indeed the phenomenon of neutrino oscillations, if we consider the neutrino states as $|\nu_1\rangle = |\phi_1\rangle$ and $|\nu_2\rangle = |\phi_2\rangle$ and that they are actually not eigenstates of the Hamiltonian.

The probability of oscillation, that is, of finding the system in the basis state $|\phi_2\rangle$, is

$$\begin{aligned} p_2(t) = |c_2|^2 &= \sin^2\theta\cos^2\theta\left|e^{-iE_- t/\hbar} - e^{-iE_+ t/\hbar}\right|^2 = (\sin 2\theta)^2\frac{1-\cos(E_+ - E_-)t/\hbar}{2} \\ &= \frac{V^2}{\Delta^2 + V^2}\sin^2\frac{\sqrt{\Delta^2+V^2}t}{\hbar} \\ &= \frac{V^2}{2(\Delta^2+V^2)}(1-\cos\Omega t), \end{aligned} \tag{4.52}$$

where we have used $E_+ - E_- = 2\sqrt{\Delta^2 + V^2} = \hbar\Omega$. Then the probability of remaining in the original state is

$$p_1(t) = |c_1(t)|^2 = 1 - p_2(t) = \frac{2\Delta^2 + V^2}{2(\Delta^2+V^2)} + \frac{V^2}{2(\Delta^2+V^2)}\cos\Omega t, \tag{4.53}$$

as we can check explicitly. These probabilities oscillate with frequency Ω around an average, as can be seen.

Consider next two special cases:

- First, $\Delta = 0$, in which case $\hbar\Omega = 2|V|$ and

$$\begin{aligned} p_1(t) &= \frac{1}{2}\left(1 + \cos\frac{2|V|t}{\hbar}\right) \\ p_2(t) &= \frac{1}{2}\left(1 - \cos\frac{2|V|t}{\hbar}\right). \end{aligned} \tag{4.54}$$

  In this case, the oscillations happen around the average value of 1/2, but there are times where the system finds itself fully in the state $|\phi_2\rangle$, since $p_1 = 0, p_2 = 1$, and others where it is fully back at $|\phi_1\rangle$, since $p_1 = 1, p_2 = 0$.

- Second, consider $V = 0$, in which case we are back at the spin 1/2 case with $B = B_z$, and $E_- = H_{11} = E_1, E_+ = H_{22} = E_2$. Then there is no oscillation, since we get $\sin 2\theta = 0$, and $p_2(t) = 0$. Moreover, in this case $|\psi_\pm\rangle = |\phi_{1,2}\rangle$ are the basis states of the matrix. But for a general initial state $|\psi(t=0)\rangle = c_-|\phi_1\rangle + c_+|\phi_2\rangle$, with $c_\pm$ arbitrary, we find

$$|\psi(t)\rangle = e^{-iE_1 t/\hbar}c_-|\phi_1\rangle + e^{-iE_2 t/\hbar}c_+|\phi_2\rangle. \tag{4.55}$$

  That means, however, that the probability of the system being in state 1 is $p_1(t) = |c_-|^2$ and of its being in state 2 is $p_2(t) = |c_+|^2$; these probabilities are time independent, so we have no oscillations.

4.4 Unitary Evolution Operator

Consider the general initial state

$$|\psi(t=0)\rangle = b|\phi_1\rangle + a|\phi_2\rangle = \begin{pmatrix} b \\ a \end{pmatrix} = a(\sin\theta|\psi_-\rangle + \cos\theta|\psi_+\rangle) + b(\cos\theta|\psi_-\rangle - \sin\theta|\psi_+\rangle), \tag{4.56}$$

that is, with

$$c_- = a\sin\theta + b\cos\theta, \quad c_+ = a\cos\theta - b\sin\theta. \tag{4.57}$$

Then substituting into the general form of the time-dependent state (4.56), we obtain

$$\begin{aligned} |\psi(t)\rangle &= c_-|\psi_-\rangle e^{-iE_-t/\hbar} + c_+|\psi_+\rangle e^{-iE_+t/\hbar} \\ &= \left[(a\sin\theta\cos\theta + b\cos^2\theta)e^{-iE_-t/\hbar} + (-a\sin\theta\cos\theta + b\sin^2\theta)e^{-iE_+t/\hbar}\right]|\phi_1\rangle \\ &+ \left[(a\sin^2\theta + b\sin\theta\cos\theta)e^{-iE_-t/\hbar} + (a\cos^2\theta - b\sin\theta\cos\theta)e^{-iE_+t/\hbar}\right]|\phi_2\rangle. \end{aligned} \tag{4.58}$$

This can be written in the form of a matrix $U(t)$ acting on the initial state,

$$|\psi(t)\rangle = U(t)\begin{pmatrix} b \\ a \end{pmatrix} = U(t)|\psi(t=0)\rangle, \tag{4.59}$$

where the $U(t)$ is given by

$$U(t) = e^{-i\bar{E}t/\hbar}\begin{pmatrix} \cos^2\theta\, e^{i\Omega t/2} + \sin^2\theta\, e^{-i\Omega t/2} & 2i\sin\theta\cos\theta\sin\Omega t/2 \\ 2i\sin\theta\cos\theta\sin\Omega t/2 & \sin^2\theta\, e^{i\Omega t/2} + \cos^2\theta\, e^{-i\Omega t/2} \end{pmatrix}. \tag{4.60}$$

We can check explicitly that the matrix is unitary,

$$U^\dagger(t)U(t) = \hat{1}. \tag{4.61}$$

This agrees with the general theory explained before, stating that the time evolution of a state, as well as any change of basis, must be a unitary operation in order for the sum of all probabilities to be equal to 1 (the conservation of probability). This also means that the time evolution is a reversible process: we can act with $U^{-1} = U^\dagger$ on the final state, and obtain the initial state.

The only possible nonunitary process, in quantum theory, is the process of measurement, which is not reversible.

4.5 Entanglement

We have described a single two-level system, for instance a spin 1/2 system, but we can also consider several such systems interacting with each other. In the simplest case, consider just two such systems, A and B. In this case, as we explained in the general theory, the total Hilbert space is a tensor product of the individual Hilbert spaces, so $\mathcal{H}_{AB} = \mathcal{H}_A \otimes \mathcal{H}_B$.

But there is one possibility of interest that we will consider here, namely that the interaction between the systems is such that the state of one system influences the state of the other, in such a way that the state of the total system is not describable as the product of states in each system, i.e.,

$$|\psi\rangle_{AB} \neq |\psi_A\rangle \otimes |\psi_B\rangle. \tag{4.62}$$

We call such a situation an *entangled state*, and the phenomenon is called *entanglement*.

The quintessential example is one that is called "maximally entangled", where the state of one system completely determines the state of the other. For instance, consider a system, such as an atom or nucleus, that has a total spin equal to zero but decays into two systems (decay products) that each have spin 1/2 (such as an electron and a "hole", or ionized atom in the case of an atom). Then spin conservation means that the total spin of the decaying products must still add up to zero, meaning that if system A has $S_z = +1/2$ then system B has $S_z = -1/2$, and vice versa (if A has $S_z = -1/2$ then B has $S_z = +1/2$); thus the state of system A completely determines the state of system B. In this case, the probabilities of the two spin situations are each equal to 1/2, so the total state of the system is

$$|\psi_\pm\rangle_{AB} = \frac{1}{\sqrt{2}}\left(|\uparrow\rangle_A \otimes |\downarrow\rangle_B \pm |\downarrow\rangle_A \otimes |\uparrow\rangle_B\right) \equiv \frac{1}{\sqrt{2}}(|10\rangle \pm |01\rangle), \tag{4.63}$$

where in the last equality we have introduced a new notation for the two states of the two-level system: $|0\rangle$ and $|1\rangle$. We note that, as desired, the probabilities of the two cases are $p_i = |c_i|^2 = 1/2$ and that the state of the system is normalized,

$${}_{AB}\langle\psi_\pm|\psi_\pm\rangle_{AB} = 1, \tag{4.64}$$

as well as having orthogonal basis states, ${}_{AB}\langle\psi_+|\psi_-\rangle_{AB} = 0$.

Another way to analyze entangled states is to consider the density matrix associated with the state. Indeed, in a more general situation we could also consider the entanglement associated with a nontrivial density matrix, though we will not do it here. The density matrix of the above entangled state is

$$\rho_{AB} = |\psi_\pm\rangle\langle\psi_\pm| = \frac{1}{2}\left(|\uparrow\downarrow\rangle \pm |\downarrow\uparrow\rangle\right)\left(\langle\uparrow\downarrow| \pm \langle\downarrow\uparrow|\right). \tag{4.65}$$

If we consider the trace of the density matrix operator over the Hilbert space of system B, corresponding to a *sum over all the possibilities for the system B* (for instance, assuming that we don't know which possibility is correct), we obtain

$$\begin{aligned}\mathrm{Tr}_B\,\rho_{AB} &= \mathrm{Tr}_B\,|\psi_\pm\rangle_{AB\,AB}\langle\psi_\pm| = {}_B\langle\uparrow|\psi_\pm\rangle_{AB\,AB}\langle\psi_\pm|\uparrow\rangle_B + {}_B\langle\downarrow|\psi_\pm\rangle_{AB\,AB}\langle\psi_\pm|\downarrow\rangle_B \\ &= \frac{1}{2}(|\downarrow\rangle_{AA}\langle\downarrow| + |\uparrow\rangle_{AA}\langle\uparrow|) = \frac{1}{2}\hat{\mathbb{1}}_A,\end{aligned} \tag{4.66}$$

where we have used the orthonormality of the eigenstates for the B system.

The result is that after taking the trace over the B system's density matrix, i.e., summing over its possibilities, we obtain a completely random system for state A: on measuring the spin we get the probability 1/2 of obtaining spin up, and probability 1/2 of obtaining spin down. This is another way to determine that we have maximum entanglement in the system.

There is much more that can be said about entanglement, but we will postpone this until the second part of the book.

4.6 Quantum Computation

We have introduced another notation for two-level states, of $|0\rangle$ and $|1\rangle$, which recalls the bits of classical computation, also denoted by states 0 and 1. This is not incidental: we will call a two-level system a "qubit", or quantum version of the computer bit.

In the entangled system, the two interacting spins (the two qubits) are now denoted by $|a\rangle \otimes |b\rangle$, where $a, b = 0$ or 1.

On individual two-level systems, i.e., qubits, as well as on the two-qubit states, i.e., tensor product states, in the absence of measurements the only possible operations allowed by quantum mechanics, time evolution and changes of basis, are unitary operations, as we said earlier.

This allows one to define a quantum version of classical computation in a computer. In a classical computer, we encode the data in classical states of 0s and 1s, and act on them with "gates", which are operators with a well-defined action on the products of two-state systems (taking a given combination of 0s and 1s to be well defined). Any classical computation can be reduced to a product of gates.

In a quantum computer, similarly, we would encode the data in sets, or tensor products, of qubits. All qubits are then evolved in time with unitary evolution operations (coming from Hamiltonians), corresponding to any computation. Thus a quantum computation is a unitary evolution, which can also be reduced to a product of gates. Each gate will be a quantum unitary operator acting on a tensor product of qubits, usually taken to be an operator acting on the two-qubit state $|a\rangle \otimes |b\rangle$.

Unlike a classical computation however, now we can have a *wave function* containing components in many different individual qubit states, and the unitary evolution will act on all of them at once. This allows for a version of parallel computing, which however is usually more efficient than anything classical, thus sometimes allowing an exponentially faster solution than in the classical case. The quintessential case is an initial state of the type

$$|\psi_\pm\rangle = \frac{1}{\sqrt{2}}(|\uparrow\rangle \pm |\downarrow\rangle) = \frac{1}{\sqrt{2}}(|0\rangle \pm |1\rangle), \tag{4.67}$$

or in the case of the two-qubit state, an entangled state like $|\psi_\pm\rangle_{AB}$.

The downside of such a calculation is that, when we "collapse the wave function" by making a measurement, which is the only nonunitary operation (and one which breaks the otherwise reversibility of the quantum calculation), we obtain only probabilities for individual states; so we need to make sure that (1) the result of the calculation can be extracted from a particular state and that (2) the probability of obtaining such a state is high enough that we can do this with a sufficiently small number of trials.

There is much more information to be given about quantum computation, but we will expand upon it in the second part of the book.

Important Concepts to Remember

- The simplest quantum mechanical Hilbert space is one with two basis elements, i.e., that is "two-level", such as a spin $s = \pm 1/2$ fermion for instance.
- For the spin $s = \pm 1/2$ case, the spin operators are $S_i = \frac{1}{2}\sigma_i$, where σ_i are the Pauli matrices, satisfying $\sigma_i \sigma_j = \delta_{ij} + i\epsilon_{ijk}\sigma_k$.

- Since $\Delta H_{\text{spin}} = -\vec{\mu} \cdot \vec{B}$, and classically the magnetic moment $\vec{\mu}_L = (e/2m)\vec{L}$, quantum mechanically $\vec{\mu}_S = g(e/2m)\vec{S}$, with $g \simeq 2$.
- Since $|\vec{S}| = \hbar/2$, $\Delta H_{\text{spin}} = -\mu\vec{\sigma} \cdot \vec{B}$, with $\mu = g(e\hbar/2m)$.
- Then the eigenvalues (for a magnetic field on the z direction) are $\pm\mu B_z$, and the eigenstates are $|\psi_\pm\rangle = \frac{1}{\sqrt{2}}(|\uparrow\rangle \pm |\downarrow\rangle)$.
- For a general two-state system Hamiltonian, with elements H_{11}, H_{22}, H_{12} and $H_{21} = H_{12}^*$, the eigenenergies are $E_\pm = (H_{11} + H_{22})/2 \pm \hbar\Omega/2$, with $\hbar\Omega/2 = \sqrt{\Delta^2 + |V|^2}$ where $\Delta = (H_{22} - H_{11})/2$ and $V = H_{12}$.
- If the initial state is not one of the eigenstates $|\psi_+\rangle$, then the time-dependent state oscillates between $|\psi_+\rangle$ and $|\psi_-\rangle$ with frequency Ω.
- If the original state is $|\uparrow\rangle$, the probability of switching spin is $p_2(t) = \dfrac{V^2}{2(\Delta^2 + V^2)}(1 - \cos\Omega t)$. This corresponds to the phenomenon of neutrino oscillation.
- For several two-level systems, for instance two such systems, we can have states of the coupled system that are not separable, $|\psi\rangle_{AB} \neq |\psi_A\rangle \otimes |\psi_B\rangle$, which are called entangled. The system is defined by a density matrix ρ_{AB}.
- The two-level system is the basis for the bits 0, 1 of the quantum version of computation.
- In quantum computation the wave function is evolved in time during a calculation, so we have to extract from it the correct result.

Further Reading

See Preskill's Caltech Notes on Quantum Entanglement and Computation [12]. For more on quantum computation, see [13]–[15].

Exercises

(1) Using the Pauli matrices σ_i: $(\sigma_1, \sigma_2, \sigma_3)$, show that we can construct four matrices γ_a, $a = 1, 2, 3, 4$, as tensor products of Pauli matrices, $\gamma_a = \sigma_i \otimes \sigma_j$, such that $\gamma_a\gamma_b + \gamma_b\gamma_a = 2\delta_{ab}$. Find explicitly an example of such tensor products.

(2) Consider a system with two spin 1/2 electrons in a constant magnetic field parallel to the z direction, B_z. Assume there is no other degree of freedom (not even momentum). Solve the Schrödinger equation and find the eigenstates of the system.

(3) Consider the Hamiltonian for a two-level system

$$H = a_0\mathbb{1} + \sum_{i=1}^{3} a_i\sigma_i. \tag{4.68}$$

Calculate its eigenstates and the associated energies.

(4) (Neutrino oscillations). Consider a two-level system with eigenstates of the Hamiltonian $|\psi_1\rangle$ and $|\psi_2\rangle$, of energies E_1 and E_2, respectively ($E_2 > E_1$), corresponding to the mass (and, of course, momentum) eigenstates of two massive neutrinos. Consider also flavor eigenstates $|\phi_1\rangle = |\nu_\mu\rangle$ and $|\phi_2\rangle = |\nu_\tau\rangle$, rotated by a mixing angle θ with respect to the mass eigenstates, and an initial muon neutrino eigenstate,

$|\psi(t = 0)\rangle = |\nu_\mu\rangle$, of energy E. If the neutrinos are ultra-relativistic ($E_1 \gg m_1$, $E_2 \gg m_2$), find the formula for the oscillation probability (the probability of finding the system in the tau neutrino flavor eigenstate) as a function of θ, time, E and $\Delta m^2 \equiv m_2^2 - m_1^2$.

(5) Find the unitary evolution operator corresponding to the previous case, that of neutrino oscillations.

(6) Consider the two-qubit state

$$|\phi\rangle = C\left[|1\rangle \otimes |1\rangle + |0\rangle \otimes |1\rangle + a|1\rangle \otimes 0\rangle + |0\rangle \otimes |0\rangle\right], \tag{4.69}$$

where C is a normalization constant. Find C as a function of a. When is the state entangled (at what values of a), and when is it not entangled?

(7) Calculate the density matrix of system A for the two-qubit state above (in exercise 6), when the trace is taken over system B, as a function of a. Find the maximum and minimum probabilities that system A is in state $|1\rangle$, independently of system B.

5 Position and Momentum and Their Bases; Canonical Quantization, and Free Particles

After analyzing the simplest discrete system, with just two states, in this chapter we move on to the simplest continuum systems, with an infinite dimensional and continuum (not countable) number of dimensions.

A classical particle is usually described in phase space, in terms of variables x, p which are continuous. In quantum theory, these classical observables are promoted to quantum operators $\hat{X}, \hat{P}$, which will have continuous eigenvalues. Each operator will have, according to general theory, a complete set of eigenstates $|x\rangle$, $|p\rangle$ such that

$$\hat{X}|x\rangle = x|x\rangle, \quad \hat{P}|p\rangle = p|p\rangle. \tag{5.1}$$

Both the basis $\{|x\rangle\}$ and the basis $\{|p\rangle\}$ are orthonormal. But the question is, how does the momentum operator $\hat{P}$ act on the $|x\rangle$ basis?

5.1 Translation Operator

To answer the above question, we have to first define the operator corresponding to translation by an infinitesimal amount dx, written as $\hat{T}(dx)$. It is thus defined as an operator that translates the eigenvalue x by dx, so

$$\hat{T}(dx)|x\rangle = |x + dx\rangle. \tag{5.2}$$

Since this translation operator preserves the scalar product, we have

$$\langle x|x'\rangle = \langle x + dx|x' + dx\rangle = \delta(x - x'), \tag{5.3}$$

but on the other hand we obtain

$$\langle x|\hat{T}^\dagger(dx)\hat{T}(dx)|x'\rangle = \langle x|x'\rangle, \tag{5.4}$$

which implies that the translation operator is unitary,

$$\hat{T}^\dagger(dx)\hat{T}(dx) = \hat{\mathbb{1}}. \tag{5.5}$$

Moreover, the translation operation obeys composition, so that

$$\hat{T}(dx)\hat{T}(dx') = \hat{T}(dx + dx'), \tag{5.6}$$

and the inverse operator must be

$$\hat{T}^{-1}(dx) = \hat{T}(-dx). \tag{5.7}$$

Finally, translation by zero must give the identity,

$$\hat{T}(dx \to 0) \to \mathbb{1}. \tag{5.8}$$

In fact, we can construct the action of the translation operator even for a finite translation by a in the coordinate (x) representation, that is, on wave functions we have

$$\hat{T}^{\dagger}_{xx'}(a) = \hat{T}^{-1}_{xx'}(a) = \delta(x - x')e^{ad/dx}, \tag{5.9}$$

since

$$\begin{aligned}\langle x|\hat{T}^{\dagger}(a)|\psi\rangle &= \langle x + a|\psi\rangle = \psi(x + a)\\ &= \hat{T}(a)_{xx}\langle x|\psi\rangle = e^{ad/dx}\psi(x).\end{aligned} \tag{5.10}$$

But on the other hand, by the Taylor expansion,

$$\psi(x + a) = \psi(x) + a\frac{d}{dx}\psi(x) + \frac{a^2}{2}\frac{d^2}{dx^2}\psi(x) + \cdots = \sum_{n\geq 0}\frac{a^n}{n!}\frac{d^n}{dx^n}\psi(x) \equiv e^{ad/dx}\psi(x), \tag{5.11}$$

which proves the above identity. Then, for an infinitesimal translation, we have

$$\hat{T}_{xx}(dx) \simeq \hat{\mathbb{1}} - dx\frac{d}{dx}. \tag{5.12}$$

And, in abstract terms, the unitarity property means that

$$\hat{T}^{\dagger}(dx) = \hat{T}^{-1}(dx) \Rightarrow \hat{\mathbb{1}} - idx\hat{K}, \tag{5.13}$$

where $\hat{K}$ is Hermitian ($\hat{K}^{\dagger} = \hat{K}$). Indeed, in the x representation, we find that

$$\hat{K}_{xx'} = \left(-i\frac{d}{dx}\right)_{xx'}, \tag{5.14}$$

which is indeed Hermitian, as we have proven earlier.

Again in abstract terms, we find that

$$\begin{aligned}\hat{X}\hat{T}(dx')|x\rangle &= \hat{X}|x + dx'\rangle = (x + dx')|x + dx'\rangle,\\ \hat{T}(dx')\hat{X}|x\rangle &= \hat{T}(dx')x|x\rangle = x|x + dx'\rangle,\end{aligned} \tag{5.15}$$

meaning that in general, as a condition on operators (since it is valid on an orthonormal basis of the Hilbert space) we have

$$[\hat{X}, T(dx')] = dx'\hat{\mathbb{1}} \Rightarrow [\hat{X}, \hat{K}] = i. \tag{5.16}$$

This is of course satisfied by (5.14), its form in the coordinate representation.

To continue, and to see that in fact $\hat{K}$ is $\hat{P}$ up to a real constant, we review a bit of classical mechanics.

5.2 Momentum in Classical Mechanics as a Generator of Translations

The formal definition of dynamics in classical mechanics is given in terms of the Hamilton equations on the phase space (q_i, p_i). The Hamilton equations become more formal in terms of Poisson brackets $\{,\}_{P.B.}$, defined for functions f and g on the phase space as

$$\{f(q,p), g(q,p)\}_{P.B.} = \sum_i \left(\frac{\partial f}{\partial q_i}\frac{\partial g}{\partial p_i} - \frac{\partial f}{\partial p_i}\frac{\partial g}{\partial q_i}\right). \tag{5.17}$$

This is an antisymmetric bracket,

$$\{f, g\}_{P.B.} = -\{g, f\}_{P.B.}, \tag{5.18}$$

that anticommutes with (complex) numbers, which are not functions of phase space,

$$\{f(q, p), c\}_{P.B.} = 0, \tag{5.19}$$

and satisfies the Jacobi identity

$$\{\{A, B\}, C\} + \{\{B, C\}, A\} + \{\{C, A\}, B\} = 0. \tag{5.20}$$

In terms of (5.18), the Hamilton equations

$$\frac{\partial H}{\partial p_i} = \dot{q}_i, \qquad \frac{\partial H}{\partial q_i} = -\dot{p}_i \tag{5.21}$$

become

$$\dot{q}_i = \{q_i, H\}_{P.B.}, \qquad \dot{p}_i = \{p_i, H\}_{P.B.}. \tag{5.22}$$

For an arbitrary function of phase space and time $f(q, p; t)$, we have the time evolution

$$\frac{df(q, p; t)}{dt} = \frac{\partial f}{\partial q_i}\dot{q}_i + \frac{\partial f}{\partial p_i}\dot{p}_i + \frac{\partial f}{\partial t} = \{f, H\}_{P.B.} + \frac{\partial f}{\partial t}. \tag{5.23}$$

This means that the Hamiltonian, through its Poisson brackets, is the generator of the time translations (the motion of the system in time).

Next, consider an infinitesimal canonical transformation on phase space, from (q, p) to (Q, P), and take the *active* point of view, which is that this is a transformation that changes points in phase space from (q, p) to $(Q, P) = (q + \delta q, p + \delta p)$. Then we have the *generating function* of the canonical transformation:

$$F(q, P, t) = qP + \epsilon G(q, P, t), \tag{5.24}$$

where ϵ is an infinitesimal parameter. It generates the canonical transformation through the equations

$$\begin{aligned} p &\equiv \frac{\partial F}{\partial q} = P + \epsilon\frac{\partial G}{\partial q} \\ Q &\equiv \frac{\partial F}{\partial P} = q + \epsilon\frac{\partial G}{\partial P} \simeq q + \epsilon\frac{\partial G}{\partial p}, \end{aligned} \tag{5.25}$$

where in the last equation we used $P \simeq p$, and so the variations are

$$\begin{aligned} \delta p &= -\epsilon\frac{\partial G}{\partial q} = \epsilon\{p, G\}_{P.B.} \\ \delta q &= \epsilon\frac{\partial G}{\partial p} = \epsilon\{q, G\}_{P.B.}. \end{aligned} \tag{5.26}$$

Doing the same thing for the Hamiltonian $H(p, q)$, we obtain

$$\delta H(q, p) = \epsilon\{H(q, p), G\}_{P.B.} - \epsilon\frac{\partial G}{\partial t} = -\epsilon\frac{dG}{dt}. \tag{5.27}$$

That means that if $dG/dt = 0$, we have $\delta H = 0$. This implies that a constant of motion (a function $G(q, p)$ that is independent of time) is a generating function of canonical transformations that leaves the Hamiltonian H invariant.

In particular, considering a cyclic coordinate q_i i.e., $\partial H/\partial q_i = 0$, it follows from the Hamilton equations that the corresponding momentum is a constant of motion, $dp_i/dt = 0$. In particular, take this constant of motion as the generating function G, i.e.,

$$G(q, p) = p_i. \tag{5.28}$$

Then the canonical transformation gives

$$\begin{aligned} \delta q_j &= \epsilon \delta_{ij} \\ \delta p_j &= 0, \end{aligned} \tag{5.29}$$

which means that the momentum p_i is the generator of translations in the coordinate q_i canonically conjugate to it ($\delta_{\text{can},p_i} q_i = \epsilon$).

5.3 Canonical Quantization

It follows that in the quantum case, where classical observables correspond to operators, we can equate the momentum operator $\hat{P}$ with the translation generator $\hat{K}$, at least up to a multiplicative real constant. In fact, the dimensions of $\hat{K}$ (inverse length) and $\hat{P}$ (momentum) are different, so that linking them there must be at least a constant with dimensions of momentum times length, or in other words the dimensions of action. One such constant is $\hbar = h/(2\pi)$, so at least up to a number, we have

$$\hat{K} = \frac{\hat{P}}{\hbar}. \tag{5.30}$$

In fact, there is no further multiplicative number. A way to see this is to consider that, using de Broglie's assumption on the relation between the wavelength λ and the momentum p of a particle,

$$p = \frac{h}{\lambda} \Rightarrow \frac{p}{\hbar} = k = \frac{2\pi}{\lambda}, \tag{5.31}$$

so $p/\hbar$ is the wave number, and it is natural to associate it with the translation operator (for instance on a lattice).

Finally, then, when acting on a wave function $\psi(x) = \langle x|\psi\rangle$, the momentum operator becomes

$$\hat{P}_{xx'} = \left(-i\hbar\frac{d}{dx}\right)_{xx'}. \tag{5.32}$$

In this x representation, we have the commutator

$$[\hat{X}, \hat{P}] = \left[x, -i\hbar\frac{d}{dx}\right] = i\hbar, \tag{5.33}$$

which replaces the classical canonical Poisson bracket formula

$$\{q, p\}_{P.B.} = 1. \tag{5.34}$$

Moreover, the previous analysis of the Hamiltonian applies to any coordinates q_i and the momenta canonically conjugate to them p_i, so we obtain the general *canonical quantization conditions*

$$[\hat{X}_i, \hat{P}_j] = i\hbar\delta_{ij}, \tag{5.35}$$

replacing the classical canonical commutation relations

$$\{q_i, p_j\}_{P.B.} = \delta_{ij}. \tag{5.36}$$

In other words, in the general canonical quantization prescription, the Poisson brackets are replaced with the commutator divided by $i\hbar$,

$$\{,\}_{P.B.} \to \frac{1}{i\hbar}[,]. \tag{5.37}$$

The other canonical Poisson brackets are

$$\{q_i, q_j\}_{P.B.} = \{p_i, p_j\}_{P.B.} = 0, \tag{5.38}$$

and under canonical quantization, they would become

$$[\hat{X}_i, \hat{X}_j] = [\hat{P}_i, \hat{P}_j] = 0. \tag{5.39}$$

These relations are indeed satisfied: for $i = j$ they are trivial and for $i \neq j$ the variables are independent (do not influence each other), so the corresponding operators commute.

5.4 Operators in Coordinate and Momentum Spaces

Operators in Coordinate (x) Space

The eigenstates $|x\rangle$ of the coordinate operator $\hat{X}$ are orthonormal, $\langle x|x'\rangle = \delta(x - x')$, and in the coordinate representation we use the following identity in terms of them, $\hat{\mathbb{1}} = \int dx|x\rangle\langle x|$.

Then the action of the matrix element of the operator $\hat{A}$ between the states $|\psi\rangle$ and $|\chi\rangle$ is given by

$$\langle\chi|\hat{A}|\psi\rangle = \int dx \int dx'\langle\chi|x\rangle\langle x|\hat{A}|x'\rangle\langle x'|\psi\rangle = \int dx \int dx'\chi^*(x)A_{xx'}\psi(x'). \tag{5.40}$$

For an observable corresponding to a function of position, $\hat{A} = \hat{f}(\hat{X})$, we have

$$A_{xx'} = \langle x|\hat{f}(\hat{X})|x'\rangle = f(x')\langle x|x'\rangle = f(x')\delta(x - x'). \tag{5.41}$$

On the other hand, for $\hat{A} = \hat{P}$, we have

$$A_{xx'} = \left(-i\hbar\frac{d}{dx}\right)_{xx'} = -i\hbar\delta(x - x')\frac{d}{dx'}, \tag{5.42}$$

as we saw in Chapter 2, so

$$\langle\chi|\hat{P}|\psi\rangle = \int dx \int dx'\chi^*(x)(-i\hbar\delta(x - x'))\frac{d}{dx'}\psi(x') = \int dx\chi^*(x)\left(-i\hbar\frac{d}{dx}\psi(x)\right). \tag{5.43}$$

Operators in Momentum (p) Space

In momentum space we use momentum eigenstates $|p\rangle$, which are also orthonormal, $\langle p|p'\rangle = \delta(p - p')$, and the identity written as a completeness relation in terms of them, $\hat{\mathbb{1}} = \int dp|p\rangle\langle p|$.

Then the matrix element (5.40) of an operator $\hat{A}$ is

$$\langle\chi|\hat{A}|\psi\rangle = \int dp \int dp' \langle\chi|p\rangle\langle p|\hat{A}|p'\rangle\langle p'|\psi\rangle = \int dp \int dp' \chi^*(p) A_{pp'} \psi(p), \tag{5.44}$$

where $\langle p|\psi = \psi(p)$ is the wave function in momentum space.

However, to do this integral we have to calculate the transformation of a wave function from coordinate space to momentum space. Consider the matrix element of $\hat{P}$ between the $\langle x|$ and the $|p\rangle$ states, and introduce the completeness relation in coordinate space, giving

$$\begin{aligned}\langle x|\hat{P}|p\rangle &= \int dx' \langle x|\hat{P}|x'\rangle\langle x'|p\rangle = \int dx' P_{xx'}\langle x'|p\rangle = \int dx' \left(-i\hbar\delta(x-x')\frac{d}{dx'}\right)\langle x'|p\rangle \\ &= -i\hbar\frac{d}{dx}\langle x|p\rangle.\end{aligned} \tag{5.45}$$

On the other hand, by using $\hat{P}|p\rangle = p|p\rangle$, we also obtain $p\langle x|p\rangle$, so

$$-i\hbar\frac{d}{dx}\langle x|p\rangle = p\langle x|p\rangle, \tag{5.46}$$

with solution

$$\langle x|p\rangle = Ne^{ipx/\hbar}. \tag{5.47}$$

The normalization constant N is found from the normalization condition by inserting a completeness relation for $|x\rangle$,

$$\delta(x-x') = \langle x|x'\rangle = \int dp\langle x|p\rangle\langle p|x'\rangle. \tag{5.48}$$

Substituting the solution above into this equation, we find

$$\delta(x-x') = N^2 \int dp e^{ipx/\hbar} e^{-ipx'/\hbar} = N^2 2\pi\delta\left(\frac{x-x'}{\hbar}\right) = 2\pi\hbar N^2\delta(x-x'), \tag{5.49}$$

meaning that $N = 1/\sqrt{2\pi\hbar}$, and so

$$\langle x|p\rangle = \frac{1}{\sqrt{2\pi\hbar}} e^{ipx/\hbar}. \tag{5.50}$$

That means that the transformation between the x and the $p/\hbar$ basis (which has the same dimensions) is just the Fourier transform,

$$\begin{aligned}\langle x|\psi\rangle &= \int dp\langle x|p\rangle\langle p|\psi\rangle \Rightarrow \\ \psi(x) &= \frac{1}{\sqrt{2\pi\hbar}} \int dp e^{ipx/\hbar}\psi(p),\end{aligned} \tag{5.51}$$

and the inverse relation (the inverse Fourier transform) is

$$\begin{aligned}\langle p|\psi\rangle &= \int dx\langle p|x\rangle\langle x|\psi\rangle \Rightarrow \\ \psi(p) &= \frac{1}{\sqrt{2\pi\hbar}} \int dx e^{-ipx/\hbar}\psi(x).\end{aligned} \tag{5.52}$$

5.5 The Free Nonrelativistic Particle

The Schrödinger equation for an arbitrary time-dependent state is

$$i\hbar\frac{d}{dt}|\psi\rangle = \hat{H}|\psi\rangle. \tag{5.53}$$

But for a free classical particle, $H(q,p) = p^2/2m + 0$, leading to the quantum Hamiltonian

$$\hat{H} = \frac{\hat{p}^2}{2m}. \tag{5.54}$$

We will solve the Schrödinger equation by the general procedure we described: first we consider the eigenvalue problem for the Hamiltonian,

$$\hat{H}|\psi_E\rangle = E|\psi_E\rangle, \tag{5.55}$$

and for the eigenstate we can solve the time evolution exactly,

$$|\psi_E(t)\rangle = e^{-iEt/\hbar}|\psi_E(t=0)\rangle = e^{-iEt/\hbar}|\psi_E\rangle. \tag{5.56}$$

In our case, the Hamiltonian eigenvalue problem is

$$\hat{H}|\psi_E\rangle = \frac{\hat{p}^2}{2m}|\psi_E\rangle = E|\psi_E\rangle \;\Rightarrow\; \left(\frac{\hat{p}^2}{2m} - E\right)|\psi_E\rangle = 0. \tag{5.57}$$

This means that $|\psi_E\rangle = |p\rangle$ is a momentum eigenstate, with $p = \pm\sqrt{2mE}$. That in turn means that we have a degeneracy, with states

$$|E,+\rangle \equiv |p = +\sqrt{2mE}\rangle, \quad |E,-\rangle \equiv |p = -\sqrt{2mE}\rangle. \tag{5.58}$$

Since the system is degenerate with respect to energy, we have that, in general,

$$|\psi_E(t)\rangle = e^{-iEt/\hbar}\left(\alpha|p = +\sqrt{2mE}\rangle + \beta|p = -\sqrt{2mE}\rangle\right). \tag{5.59}$$

Then the coordinate (x) space wave function for a given energy is

$$\langle x|\psi_E(t)\rangle = \psi_E(x,t) = \frac{e^{-iEt/\hbar}}{\sqrt{2\pi\hbar}}\left(\alpha e^{+ix\sqrt{2mE}/\hbar} + \beta e^{-ix\sqrt{2mE}/\hbar}\right). \tag{5.60}$$

Choosing a sign for the momentum (and so a state of given momentum), say $p > 0$, so that $\alpha = 1, \beta = 0$, means that the probability density of finding the free particle in this state at position x is

$$\frac{dP(x)}{dx} = |\langle x|\psi_E(t)\rangle|^2 = \frac{1}{2\pi\hbar}, \tag{5.61}$$

which is constant. This means that the position x is completely arbitrary once we fix the momentum p. We will see the implications of this fact in more detail in the next chapter.

Note that the above relation is consistent with the normalization condition

$$1 = \int dx\frac{dP(x)}{dx} = \int dx|\langle x|\psi\rangle|^2 = \int dx\langle\psi|x\rangle\langle x|\psi\rangle = \langle\psi|\psi\rangle, \tag{5.62}$$

which is therefore satisfied.

We should comment on the fact that we want the variables x, p to be approximately classical, that is, defined up to some error. That means that the fact that the position is arbitrary in the above discussion arises because having a perfectly defined momentum p is an idealization. In reality, we need to consider so-called *wave packets*, a linear combination (superposition) of various momentum states with varying p. In this case, we can approximately localize the position x around an average value, perhaps with a moving (time-dependent) average value. We will see in the next chapter that if we center the state of given momentum p on some x by adding a multiplicative Gaussian profile at time $t = 0$ then, as t progresses, this *wave packet* will spread out in space, tending towards the same constant wave function with respect to x.

Important Concepts to Remember

- A constant of motion is a generating function of canonical transformations that leaves the Hamiltonian invariant, and, in particular, the momentum p_i is the generator of translations in the coordinate q_i canonically conjugate to it.
- The generator of translations is $\hat{K} = \hat{P}/\hbar$, so $\hat{P}_{xx'} = (-i\hbar d/dx)_{xx'}$.
- The canonical quantization conditions are $[\hat{X}_i, \hat{P}_j] = i\hbar\delta_{ij}$ (and $[\hat{X}_i, \hat{X}_j] = [\hat{P}_i, \hat{P}_j] = 0$), replacing $\{q_i, p_j\}_{P.B.} = \delta_{ij}$ (and $\{q_i, q_j\}_{P.B.} = \{p_i, p_j\}_{P.B.} = 0$) in classical mechanics; so canonical quantization means that we have the replacement $\{,\}_{P.B.} \to (1/i\hbar)[,]$.
- In x space, $\langle x|f(\hat{X})|x'\rangle = f(x')\delta(x - x')$ and $\langle\chi|\hat{P}|\psi\rangle = \int dx\, \chi^*(x)\,(-i\hbar d/dx)\,\psi(x)$.
- The product of coordinate and momentum space basis elements is $\langle x|p\rangle = e^{ipx/\hbar}/\sqrt{2\pi\hbar}$, so the wave function in momentum space is the Fourier transform (in units with $\hbar = 1$) $\psi(p) = (1/2\pi\hbar)\int dx e^{-ipx/\hbar}\psi(x)$.
- For a free particle, the time evolution is $|\psi_E(t)\rangle = e^{-iEt/\hbar}|\psi_E\rangle$, $E|\psi_E\rangle = (\hat{P}^2/2m)|\psi_E\rangle$, and there is a degeneracy with respect to momentum, $|E, +\rangle = |p = +\sqrt{2mE}\rangle$, $|E, -\rangle = |p = -\sqrt{2mE}\rangle$.
- The state of perfectly defined momentum p is an idealization. In reality we have wave packets, states given as linear combinations of momentum states, peaked around a given classical momentum.

Further Reading

See any other fairly advanced quantum mechanics book, e.g., [2, 1, 4, 3].

Exercises

(1) Consider possible terms in the (quantum) Hamiltonian for a one-dimensional system,

$$H_1 = \int dx\, \psi^*(x) e^{a\frac{d}{dx}} \psi(x), \quad H_2 = \int dx\, \psi^*(x) \frac{1}{d^2/dx^2} \psi(x), \tag{5.63}$$

where $\psi(x)$ is a continuous complex function of the spatial variable x, and $1/(d^2/dx^2)$ is understood as the inverse of an operator, acting on the function. Can these terms be understood to be coming from *local* interactions in x (interactions defined at a single point x), and if so, why? Use a calculation to explain your reasoning.

(2) Consider the angular momentum of a classical particle, $\vec{L} = \vec{r} \times \vec{p}$, and a system for which it is invariant. Use it as a generating function to define the relevant canonical transformations.

(3) Consider a system with Lagrangian

$$L = \frac{\dot{q}_1^2}{2} + \dot{q}_1 \dot{q}_2 + \frac{\dot{q}_2^2}{2} - k(q_1 + q_2). \tag{5.64}$$

Write down the Hamiltonian and Poisson brackets, and canonically quantize the system.

(4) Consider a system of N free particles in one spatial dimension x. Write down the general eigenstate and eigenenergy, its degeneracy and time evolution.

(5) Consider a system with Hamiltonian

$$H = \frac{p^2}{2m} + \alpha p + \beta x^2 + \gamma x^3. \tag{5.65}$$

Write down its Schrödinger equation for wave functions in coordinate space. If $\beta = \gamma = 0$, write down its general eigenstate wave function and time evolution, for a given energy E.

(6) Consider a free particle in three spatial dimensions, and a system of coordinates relative to a point O not on the path of the particle. Write down the integrals of motion conjugate to Cartesian, polar, and spherical coordinates respectively, and the canonical quantization for each set of coordinates.

(7) Considering the free particle in three spatial dimensions from exercise 6, what is the probability density of finding a particle at a point $\vec{r}$, given a momentum $\vec{p}$?

6 The Heisenberg Uncertainty Principle and Relations, and Gaussian Wave Packets

In this chapter, following the explicit analysis of Gaussian wave packets, we will see that products in the variances of canonically conjugate variables have a nonzero value. Then we will prove a theorem saying that such products in general obey a lower bound, called the Heisenberg uncertainty relation(s). Moreover, we will then show that the minimum is attained in the case of the same Gaussian wave packets.

6.1 Gaussian Wave Packets

As we said in the previous chapter, free particles of a given momentum are an abstraction, and moreover imply a coordinate that is completely arbitrary. In a realistic situation, a superposition of waves, or wave packet, represents a real particle, as in Fig. 6.1. The quintessential example is the Gaussian wave packet at $t = 0$, where we multiply a free wave of given momentum by a Gaussian in position space, centered on $x = 0$ and with variance $\sigma = d$, so that

$$\langle x|\psi_p(t=0)\rangle = Ne^{ipx/\hbar}e^{-x^2/2\sigma^2}. \tag{6.1}$$

The probability density is a pure Gaussian (since the wave multiplying the Gaussian has a constant probability density):

$$\frac{dP}{dx} = |\langle x|\psi\rangle|^2 = N^2 e^{-x^2/d^2}. \tag{6.2}$$

Imposing normalization of the probability, $\int dx dP/dx = 1$, we obtain

$$N^2 \int_{-\infty}^{+\infty} dx e^{-x^2/d^2} = N^2 d\sqrt{\pi} = 1 \;\Rightarrow\; N = \frac{1}{(\pi d^2)^{1/4}}, \tag{6.3}$$

so the Gaussian wave packet at $t = 0$ is

$$\langle x|\psi_p(t=0)\rangle = \langle x|p\rangle e^{-x^2/2d^2}\frac{\sqrt{2\pi\hbar}}{(\pi d^2)^{1/4}}. \tag{6.4}$$

Now we calculate the averages of the x, x^2, p, p^2 operators in this Gaussian wave packet state. First,

$$\langle x\rangle \equiv \langle\psi|\hat{X}|\psi\rangle = \int dx\langle\psi|x\rangle\langle x|\hat{X}|\psi\rangle = \int_{-\infty}^{+\infty} dx\; x|\psi(x)|^2 = 0, \tag{6.5}$$

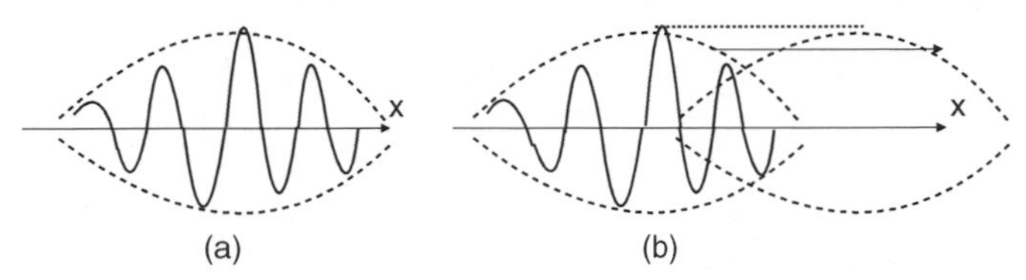

Figure 6.1 A wave packet, representing a particle. (a) A static wave packet, centered on a position in x space. (b) A time-dependent wave packet, where the central position in x moves with time.

by the symmetry in x of the integral. The average of x^2 is nonzero, however:

$$
\begin{aligned}
\langle x^2 \rangle \equiv \langle \psi | \hat{X}^2 | \psi \rangle &= \int_{-\infty}^{+\infty} dx\, x^2 |\psi(x)|^2 = \int_{-\infty}^{+\infty} dx \frac{x^2}{d\sqrt{\pi}} e^{-x^2/d^2} \\
&= \frac{d^2}{\sqrt{\pi}} \Gamma(3/2) = \frac{d^2}{2}.
\end{aligned}
\tag{6.6}
$$

Next, the average of p is also nonzero,

$$
\begin{aligned}
\langle p \rangle \equiv \langle \psi | \hat{P} | \psi \rangle &= \int dx \int dx' \langle \psi | x \rangle \langle x | \hat{P} | x' \rangle \langle x' | \psi \rangle \\
&= \int dx \int dx' \psi^*(x) \left(-i\hbar \delta(x - x') \frac{d}{dx'} \psi(x') \right) \\
&= -i\hbar \int_{-\infty}^{+\infty} dx \psi^*(x) \frac{d}{dx} \psi(x) = \int_{-\infty}^{+\infty} dx |\psi(x)|^2 \left(p + \frac{i\hbar x}{d^2} \right) = p.
\end{aligned}
\tag{6.7}
$$

And, finally, the average of p^2 is

$$
\begin{aligned}
\langle p^2 \rangle \equiv \langle \psi | \hat{P}^2 | \psi \rangle &= (-i\hbar)^2 \int dx \psi^*(x) \frac{d^2}{dx^2} \psi(x) \\
&= (-i\hbar)^2 \int dx |\psi(x)|^2 \left[-\frac{1}{d^2} + \frac{1}{(-i\hbar)^2} \left(p + \frac{i\hbar x}{d^2} \right)^2 \right] \\
&= \left(p^2 + \frac{\hbar^2}{d^2} \right) - \frac{\hbar^2}{d^4} \langle x^2 \rangle = p^2 + \frac{\hbar^2}{2d^2}.
\end{aligned}
\tag{6.8}
$$

Now we can calculate the deviations in x and p,

$$
\begin{aligned}
\langle (\Delta x)^2 \rangle &= \langle (x - \langle x \rangle)^2 \rangle = \langle x^2 \rangle - \langle x \rangle^2 = \frac{d^2}{2} \\
\langle (\Delta p)^2 \rangle &= \langle (p - \langle p \rangle)^2 \rangle = \langle p^2 \rangle - \langle p \rangle^2 = \frac{\hbar^2}{2d^2}.
\end{aligned}
\tag{6.9}
$$

Multiplying them, we obtain

$$
\langle (\Delta x)^2 \rangle \langle (\Delta p)^2 \rangle = \frac{\hbar^2}{4}.
\tag{6.10}
$$

We will see that this actually yields the lower bound coming from the general Heisenberg uncertainty relations.

For completeness, we will write the Gaussian wave packet with momentum p' in momentum space (we set $a = i(p' - p)d^2/\hbar$):

$$\begin{aligned}
\langle p|\psi_{p'}\rangle = \psi_{p'}(p) &= \int dx \langle p|x\rangle\langle x|\psi_{p'}\rangle \\
&= \frac{1}{\sqrt{2\pi\hbar}} \frac{1}{(\pi d^2)^{1/4}} \int_{-\infty}^{+\infty} dx\, e^{-ipx/\hbar} \exp\left(+i\frac{p'x}{\hbar} - \frac{x^2}{2d^2}\right) \\
&= \frac{1}{\sqrt{2\pi\hbar}} \frac{e^{+a^2/2d^2}}{(\pi d^2)^{1/4}} \int_{-\infty}^{+\infty} dx\, e^{-(x-a)^2/2d^2} \\
&= \frac{\sqrt{d}}{(\pi\hbar^2)^{1/4}} e^{-(p-p')^2 d^2/2\hbar^2}.
\end{aligned} \tag{6.11}$$

6.2 Time Evolution of Gaussian Wave Packet

We have defined the Gaussian wave packet with momentum p_0 at $t = 0$, but in order to find its time evolution, we need to remember the time evolution of states of definite momentum,

$$|\psi_E(t)\rangle = e^{-iEt/\hbar}|p = \sqrt{2mE}\rangle, \tag{6.12}$$

which we could also write in the coordinate representation.

But we need to calculate the time evolution operator for the more general wave function that we have. In order to do this, we use the evolution operator formalism, where

$$|\psi(t)\rangle = \hat{U}(t)|\psi(t = 0)\rangle \tag{6.13}$$

is seen to imply (given the relation (6.12), which defines the time evolution for $|p\rangle$ states, with $p = \sqrt{2mE}$) that

$$\hat{U}(t) = \int_{-\infty}^{+\infty} dp\, |p\rangle\langle p| e^{-iE(p)t/\hbar} = \int_{-\infty}^{+\infty} dp\, |p\rangle\langle p| e^{-ip^2 t/2m\hbar}. \tag{6.14}$$

The matrix elements of this operator in the coordinate basis $|x\rangle$ are

$$\begin{aligned}
U(x, x'; t) \equiv \langle x|\hat{U}(t)|x'\rangle &= \int_{-\infty}^{+\infty} dp \langle x|p\rangle \langle p|x'\rangle e^{-ip^2 t/2m\hbar} \\
&= \int dp \frac{e^{ip(x-x')/\hbar}}{2\pi\hbar} e^{-ip^2 t/2m\hbar} = \sqrt{\frac{m}{2\pi\hbar i t}} e^{im(x-x')^2/2\hbar t}.
\end{aligned} \tag{6.15}$$

Taking the relation (6.13) in the coordinate representation, i.e., multiplying with $\langle x|$ and then introducing a completeness relation $\int dx'|x'\rangle\langle x'|$ on the right of the evolution operator, we get

$$\psi(x, t) = \int dx' U(x, x'; t)\psi(x', t = 0), \tag{6.16}$$

with $U(x, x'; t)$ as above and $\psi(x', t = 0)$ the Gaussian wave packet from the previous section:

$$\psi(x', t = 0) = \frac{e^{ip_0 x'/\hbar} e^{-x'^2/2d^2}}{\sqrt{d\sqrt{\pi}}}. \tag{6.17}$$

On performing the Gaussian x' integral, the result is

$$\langle x|\psi(t)\rangle = \frac{1}{\sqrt{\sqrt{\pi}\,(d + i\hbar t/md)}} \exp\left[-\frac{(x - p_0 t/m)^2}{2d^2(1 + i\hbar t/md^2)}\right] \exp\left[-\frac{ip_0}{\hbar}\left(x - \frac{p_0 t}{2m}\right)\right], \tag{6.18}$$

which implies the probability density

$$\begin{aligned} \rho(x,t) \equiv \frac{dP(x,t)}{dx} &= |\psi(x,t)|^2 \\ &= \frac{1}{\sqrt{\pi}\sqrt{d^2 + \hbar^2 t^2/m^2 d^2}} \exp\left[-\frac{(x - p_0 t/m)^2}{d^2 + (\hbar^2 t^2/m^2 d^2)}\right]. \end{aligned} \tag{6.19}$$

We observe two properties of this result:

- the mean value moves with momentum p_0, as it is a function of $x - p_0 t/m$;
- the width $\Delta x(t)$ of the Gaussian grows in time as

$$\Delta x = \frac{d}{\sqrt{2}}\sqrt{\left(1 + \frac{\hbar^2 t^2}{m^2 d^4}\right)}. \tag{6.20}$$

That is, the initial wave function spreads out until it becomes more like the wave associated with a free particle of given momentum.

6.3 Heisenberg Uncertainty Relations

We have seen that $\langle \Delta x \rangle$ is correlated with $\langle \Delta p \rangle$, and also that $[\hat{X}, \hat{P}] = i\hbar$. But how general is this correlation, in the case where two operators do not commute, $[\hat{A}, \hat{B}] \neq 0$?

In the case where operators do commute, which means that the related observables are *compatible*, the states have simultaneous eigenvalues,

$$\hat{A}\hat{B}|a,b\rangle = \hat{A}b|a,b\rangle = ab|a,b\rangle = \hat{B}\hat{A}|a,b\rangle = \hat{B}a|a,b\rangle. \tag{6.21}$$

In the case where they do not commute, this would be impossible (we would obtain a contradiction from the above relation). For instance, $\hat{X}$ and $\hat{P}$ are incompatible and, as we saw, if we localize the momentum (p is given), then space is completely delocalized (x is arbitrary), and moreover if $\Delta x \neq 0$, we also have $\Delta p \neq 0$.

In fact this is part of a general theory of noncommuting operators. We have the following:

Theorem If $[\hat{A}, \hat{B}] \neq 0$ and $\hat{A}, \hat{B}$ are associated with observables, so that they are Hermitian operators, then we have

$$\langle (\Delta A)^2 \rangle \langle (\Delta B)^2 \rangle \geq \frac{1}{4} |\langle [\hat{A}, \hat{B}] \rangle|^2. \tag{6.22}$$

These relation are known as *Heisenberg's uncertainty relation*.

Proof First, note that if $[\hat{A}, \hat{B}] = i\hat{C}$ then $\hat{C}$ is Hermitian, $\hat{C} = \hat{C}^\dagger$. Indeed,

$$(i\hat{C})^\dagger = -i\hat{C}^\dagger = ([\hat{A}, \hat{B}])^\dagger = [\hat{B}^\dagger, \hat{A}^\dagger] = [\hat{B}, \hat{A}] = -[\hat{A}, \hat{B}] = -i\hat{C}. \tag{6.23}$$

Also, if $\{\hat{A},\hat{B}\} = \hat{D}$ then $\hat{D}$ is Hermitian, $\hat{D}^\dagger = \hat{D}$:

$$D^\dagger = (\{\hat{A},\hat{B}\})^\dagger = \{\hat{B}^\dagger,\hat{A}^\dagger\} = \{\hat{B},\hat{A}\} = \{\hat{A},\hat{B}\} = \hat{D}. \tag{6.24}$$

The quantities in the Heisenberg uncertainty relations are defined as

$$\begin{aligned} \langle[\hat{A},\hat{B}]\rangle &\equiv \langle\psi|(\hat{A}\hat{B}-\hat{B}\hat{A})|\psi\rangle \\ \langle(\Delta A)^2\rangle &\equiv \langle\psi|(\hat{A}-\langle A\rangle)^2|\psi\rangle \\ \langle(\Delta B)^2\rangle &\equiv \langle\psi|(\hat{B}-\langle B\rangle)^2|\psi\rangle. \end{aligned} \tag{6.25}$$

Then note that if $\hat{A},\hat{B}$ are Hermitian, $\hat{A}-\langle A\rangle$ and $\hat{B}-\langle B\rangle$ are also Hermitian.

Since $\hat{A},\hat{B}$ are operators acting on a Hilbert space, consider the states $|\phi\rangle, |\chi\rangle$ in this Hilbert space, which, as we saw in Chapters 1 and 2, obey the Cauchy–Schwarz inequality,

$$|\langle\phi|\chi\rangle|^2 \leq ||\phi||^2||\chi||^2 = \langle\phi|\phi\rangle\langle\chi|\chi\rangle, \tag{6.26}$$

with equality only if $|\chi\rangle \propto |\phi\rangle$.

Now consider the states

$$|\phi\rangle = (\hat{A}-\langle A\rangle)|\psi\rangle, \qquad |\chi\rangle = (\hat{B}-\langle B\rangle)|\psi\rangle. \tag{6.27}$$

The norms of these states are (using the fact that $\hat{A}-\langle A\rangle$ and $\hat{B}-\langle B\rangle$ are Hermitian)

$$\begin{aligned} ||\phi||^2 &= \langle\psi|(\hat{A}-\langle A\rangle)^2|\psi\rangle = \langle(\Delta A)^2\rangle \\ ||\chi||^2 &= \langle\psi|(\hat{B}-\langle A\rangle)^2|\psi\rangle = \langle(\Delta B)^2\rangle. \end{aligned} \tag{6.28}$$

Moreover, also using the Hermitian nature of ΔA and ΔB, we find

$$|\langle\phi|\chi\rangle|^2 = |\langle\psi|(\hat{A}-\langle A\rangle)(\hat{B}-\langle B\rangle)|\psi\rangle|^2. \tag{6.29}$$

But then

$$\begin{aligned} (\hat{A}-\langle A\rangle)(\hat{B}-\langle B\rangle) &= \frac{1}{2}[\hat{A}-\langle A\rangle,\hat{B}-\langle B\rangle] + \frac{1}{2}\{\hat{A}-\langle A\rangle,\hat{B}-\langle B\rangle\} \\ &= \frac{1}{2}[\hat{A},\hat{B}] + \frac{1}{2}\{\hat{A}-\langle A\rangle,\hat{B}-\langle B\rangle\}, \end{aligned} \tag{6.30}$$

and, as we saw, $[\hat{A},\hat{B}] = i\hat{C}$ with $\hat{C}^\dagger = \hat{C}$, and so the anticommutator above is a Hermitian operator. These two operators then act as real and imaginary numbers (remembering that the eigenvalues of a Hermitian operator are real, this is actually what we get in an eigenstate basis), so the modulus squared equals the real (Hermitian) part squared plus the imaginary (anti-Hermitian) part squared.

In conclusion, the Cauchy–Schwarz inequality becomes

$$\begin{aligned} \langle(\Delta A)^2\rangle\langle(\Delta B)^2\rangle &\geq |\langle\psi|\Delta\hat{A}\Delta\hat{B}|\psi\rangle|^2 \geq \left|\frac{1}{2}\langle\psi|[\hat{A},\hat{B}]|\psi\rangle\right|^2 + \left|\frac{1}{2}\langle\psi|\{\Delta\hat{A},\Delta\hat{B}\}|\psi\rangle\right|^2 \\ &\geq \frac{1}{4}\left|\langle\psi|[\hat{A},\hat{B}]|\psi\rangle\right|^2. \end{aligned} \tag{6.31}$$

q.e.d.

We have thus proven the Heisenberg uncertainty relations in the general case.

Specializing to the case of canonically conjugate operators, $\hat{A} = \hat{q}_i$ and $\hat{B} = \hat{p}_i$, we obtain first

$$[\hat{A},\hat{B}] = [\hat{q}_i,\hat{p}_i] = i\hbar, \tag{6.32}$$

and, substituting in the Heisenberg uncertainty relations,

$$\langle(\Delta q_i)^2\rangle\langle(\Delta p_i)^2\rangle \geq \frac{\hbar^2}{4}, \tag{6.33}$$

or, more simply,

$$\Delta q_i \Delta p_i \geq \frac{\hbar}{2}, \tag{6.34}$$

in terms of the standard deviations of q_i, p_i. This is the original form of Heisenberg's uncertainty relations, in terms of the x, p observables themselves (rather than for general canonically conjugate variables).

We also note the situation when we have equality in the uncertainty relations (i.e., when we "saturate" the inequality). There were two inequalities used, and each comes with its own condition for equality, so both must hold:

(1) For the first inequality, equality arises when $|\chi\rangle \propto |\phi\rangle$, so that

$$\hat{A}|\psi\rangle = c\hat{B}|\psi\rangle. \tag{6.35}$$

(2) For the second inequality, equality arises when the (average of the) anticommutator vanishes, so that

$$\langle\psi|\{\hat{A}, \hat{B}\}|\psi\rangle = 0. \tag{6.36}$$

6.4 Minimum Uncertainty Wave Packet

We now find the wave packet for the free particle that saturates the minimum of the standard inequality.

Apply the Heisenberg uncertainty relations for $\hat{A} = \hat{X}, \hat{B} = \hat{P}$. Then equality arises when

(1)

$$(\hat{P} - \langle p\rangle)|\psi\rangle = c(\hat{X} - \langle x\rangle)|\psi\rangle, \tag{6.37}$$

(2)

$$\langle\psi|\left[(\hat{P} - \langle p\rangle)(\hat{X} - \langle x\rangle) + (\hat{X} - \langle x\rangle)(\hat{P} - \langle p\rangle)\right]|\psi\rangle = 0. \tag{6.38}$$

Applying these conditions in coordinate space, i.e., by acting with $\langle x|$ from the left on the conditions, we find from (1) that

$$\begin{aligned}\left(-i\hbar\frac{d}{dx} - \langle p\rangle\right)\psi(x) = c(x - \langle x\rangle)\psi(x) \quad \Rightarrow \\ \frac{d}{dx}\ln\psi(x) = \frac{i}{\hbar}[\langle p\rangle + c(x - \langle x\rangle)],\end{aligned} \tag{6.39}$$

with the solution

$$\psi(x) = \psi(x=0)e^{i\frac{\langle p\rangle x}{\hbar}}\exp\left[i\frac{c(x - \langle x\rangle)^2}{2\hbar}\right]. \tag{6.40}$$

Then, multiplying condition (1) with $\langle\psi|(\hat{X} - \langle x\rangle)$, considering the Hermitian conjugate of the resulting relation, summing the equation and its conjugate, and then using condition (2), we finally obtain

$$(c + c^*)\langle\psi|(\hat{X} - \langle x\rangle)^2|\psi\rangle = 0. \tag{6.41}$$

Since the expectation value cannot be zero independently of $|\psi\rangle$, we must have a purely imaginary coefficient,

$$c = i|c|, \tag{6.42}$$

which means that finally the wave function is

$$\psi(x) = \psi(x = 0)e^{i\frac{\langle p\rangle x}{\hbar}} \exp\left[-\frac{|c|}{2\hbar}(x - \langle x\rangle)^2\right]. \tag{6.43}$$

This is the Gaussian wave packet we used before (in which case, indeed, we found saturation of the Heisenberg uncertainty relations), with Gaussian variance

$$d^2 = \frac{\hbar}{|c|}. \tag{6.44}$$

6.5 Energy–Time Uncertainty Relation

We have found an (x, p) uncertainty relation, but we also can have an (E, t) (energy versus time) uncertainty relation,

$$\Delta E \Delta t \geq \frac{\hbar}{2}. \tag{6.45}$$

Since time t does not correspond to an operator in quantum theory, this relation cannot strictly speaking be derived in the same way as the (x, p) relation. However, we can argue for it because:

- relativistically (i.e., in a relativistic theory), E and t are the zero components of the 4-vectors p^μ and x^μ, so $\Delta E \Delta t$ is $\Delta p^0 \Delta x^0$, meaning that by relativistic invariance we must have (6.45).
- We also know that E is the eigenvalue (observable value) for the quantum Hamiltonian $\hat{H}$, which as we saw generates evolution in time t. But we also saw that the canonically conjugate momentum $\hat{P}$ generates translations in x, so the same Heisenberg uncertainty relation should be valid for the pair (E, t).

The meaning of the uncertainty relation is however the same: if for instance we know the energy of a system with infinite precision, such as in the case when a system is confined to a discrete energy state (without transitions: say the state is a ground state), then the time it spends there is infinite, so $\Delta t = \infty$. Conversely, at a given time the quantum energy is arbitrary (cannot be determined). Another way to apply this relation is to say that, for an unstable particle, the lifetime Δt of the particle and its uncertainty in energy ΔE are related by the uncertainty relation. This is what happens for instance for virtual particles (particles created from the vacuum for a short time) in a quantum theory: if the particles exist for a time Δt, then their energy is not well defined but has uncertainty ΔE.

Important Concepts to Remember

- In a Gaussian wave packet in x space, at $t = 0$, $\psi(x, t = 0) = N e^{i\frac{px}{\hbar}} e^{-\frac{x^2}{2\sigma^2}}$ and we have $\Delta x^2 \Delta p^2 = \hbar^2/4$.
- A Gaussian wave packet travels with momentum p_0, so the x dependence of the probability density is encoded in a function of $x - tp_0/m$ (but there is an extra t dependence), and the wave packet spreads out in time, with $\Delta x = d/\sqrt{2}(1 + \hbar^2 t^2/m^2 d^4)$.
- Heisenberg's uncertainty relation for incompatible operators ($[\hat{A}, \hat{B}] \neq 0$) comes from $\langle(\Delta A)^2\rangle\langle(\Delta B)^2\rangle \geq \frac{1}{4}|\langle[\hat{A}, \hat{B}]\rangle|^2$.
- The standard form of Heisenberg's uncertainty relation is $\Delta q_i \Delta p_i \geq \hbar/2$.
- A Gaussian wave packet minimizes Heisenberg's uncertainty relation.
- The energy–time uncertainty relation, $\Delta E \Delta t \geq \hbar/2$, is not derived as above, and it applies either to errors or to an energy difference and time of decay.

Further Reading

See any other book on quantum mechanics, for instance [2] or [1].

Exercises

(1) Consider a Gaussian wave packet in momentum space, equation (6.11). Calculate $(\Delta x)^2$ and $(\Delta p)^2$ in this p space, and check again the saturation of Heisenberg's uncertainty relations.

(2) Do the integral for the Gaussian wave packet with evolution operator, to prove (6.18), and calculate $\Delta x(t)$ for it, to prove (6.20).

(3) Can we measure simultaneously the momentum and the angular momentum in three spatial dimensions and, if not, what are the Heisenberg uncertainty relations corresponding to them?

(4) Consider a system with Hamiltonian

$$H = \frac{p^2}{2m} + \alpha p x. \tag{6.46}$$

Can we measure simultaneously the energy and the momentum of the system?

(5) Consider two energy levels of an atomic system, $E_1 = E_*$ and $E_2 = E_* + 0.5$ eV. What is the minimum possible decay time from E_2 to E_1?

(6) Consider a superposition of two Gaussian wave packets with the same momentum p_0, initially (at time t_0) at the same position x_0, but with different variances, $\sigma_1 = d_1$ and $\sigma_2 = d_2$. Calculate $\langle(\Delta x)^2\rangle$ and $\langle(\Delta p)^2\rangle$, and check the Heisenberg uncertainty relation, at time t_0.

(7) For the situation in exercise 6, calculate the time dependence of the uncertainty in position, $\Delta x(t)$.

7 One-Dimensional Problems in a Potential $V(x)$

After analyzing two-state systems and free particles and wave packets, which are the simplest systems in the discrete-Hilbert-space and continuous-Hilbert-space cases, we consider the next simplest systems: particles in one dimension, with a standard nonrelativistic kinetic term and with a potential $V(x)$.

7.1 Set-Up of the Problem

We want to solve the Schrödinger equation in position space, so we multiply by $\langle x|$ the equation

$$i\hbar\partial_t|\psi(t)\rangle = \hat{H}|\psi(t)\rangle, \tag{7.1}$$

where the quantum Hamiltonian is

$$\hat{H} = \frac{\hat{P}^2}{2m} + \hat{V}(\hat{x}). \tag{7.2}$$

We note that $\hat{P}$ and $\hat{x}$ are Hermitian operators, $\hat{P}^\dagger = \hat{P}$, $\hat{x}^\dagger = \hat{x}$, and so is $\hat{V}(\hat{x})$, and therefore the Hamiltonian $\hat{H}$. However, note that this is actually not enough: one also needs to have the same domain for $\hat{H}$ and $\hat{H}^\dagger$, not just for them to act in the same way (which they do, since $\hat{P}$ and $\hat{x}$ do). This is a subtlety which is irrelevant in most cases, but implies some interesting counterexamples.

The kinetic and potential terms act as follows (note that $\hat{P}^2$ is the matrix product of two $\hat{P}$s, and involves a sum or integral over the middle indices, turning two delta functions into a single one):

$$\begin{aligned}\left\langle x\left|\frac{\hat{P}^2}{2m}\right|x'\right\rangle &= \delta(x-x')\left(-i\hbar\frac{\partial}{\partial x}\right)^2\\ \langle x|\hat{V}(\hat{x})|x'\rangle &= \delta(x-x')V(x).\end{aligned} \tag{7.3}$$

Then the Schrödinger equation in coordinate space becomes (introducing a completeness relation via $\hat{1} = \int dx'|x'\rangle\langle x'|$)

$$\begin{aligned}i\hbar\partial_t\langle x|\psi(t)\rangle &= \int dx'\langle x|\hat{H}|x'\rangle\langle x'|\psi(t)\rangle \;\Rightarrow\\ i\hbar\partial_t\psi(x,t) &= \int dx' H_{xx'}\psi(x',t).\end{aligned} \tag{7.4}$$

Substituting the matrix elements of the kinetic and potential operators, and doing the x' integral, we obtain

$$i\hbar\partial_t\psi(x,t) = \left[-\frac{\hbar^2}{2m}\frac{\partial^2}{\partial x^2} + V(x)\right]\psi(x,t). \tag{7.5}$$

This is solved, as earlier, by the separation of variables. We consider a wave function $\psi_E(x,t)$ that is an eigenfunction of the Hamiltonian, so that

$$\begin{aligned} i\hbar\partial_t\psi_E(x,t) &= E\psi_E(x,t) \\ \hat{H}\psi_E(x,t) &= E\psi_E(x,t). \end{aligned} \tag{7.6}$$

Then we have the *stationary* solution

$$\psi_E(x,t) = e^{-iEt/\hbar}\psi_E(x,t=0) = e^{-iEt/\hbar}\psi_E(x). \tag{7.7}$$

Note that for a stationary solution the probability density is independent of time,

$$\rho(x,t) \equiv \frac{dP(x,t)}{dx} = |\psi(x,t)|^2 = |\psi(x)|^2. \tag{7.8}$$

The eigenfunction problem for the Hamiltonian, also called the stationary (time-independent) Schrödinger equation, in our case is

$$\hat{H}\psi_E(x) = \left[-\frac{\hbar^2}{2m}\frac{\partial^2}{\partial x^2} + V(x)\right]\psi_E(x) = E\psi_E(x). \tag{7.9}$$

Defining the rescaled variables

$$\frac{2mE}{\hbar^2} \equiv \epsilon, \qquad \frac{2mV(x)}{\hbar^2} = U(x), \tag{7.10}$$

the Schrödinger equation becomes

$$\psi_E''(x) = -(\epsilon - U(x))\psi_E(x). \tag{7.11}$$

This is the equation we will study in this chapter. It is a real equation (with real coefficients), so it is enough to consider its real eigenfunctions, since complex eigenfunctions can be obtained from them. However, for simplicity sometimes we will consider complex eigenfunctions directly.

7.2 General Properties of the Solutions

Next we make a general description of the solutions, without considering specific potentials $U(x)$.

- The one-dimensional Schrödinger equation discussed above is a second-order linear (in the variable $\psi(x)$) differential equation, similar to the classical equation for the position of a particle $x(t)$ in a potential. As in that case, we could give the initial conditions at some point, $\psi(x_0)$ and $\psi'(x_0)$, like the values for x and $p = m\dot{x}$ in the initial condition for the classical particle, and find the evolution in x (corresponding to evolution in time for the particle). This would amount to "integrating" the differential equation, and it would clearly lead (via integration) to a continuous solution $\psi(x)$, so $\psi(x) \in C^0(\mathbb{R})$ (the space of continuous functions).
- Besides being continuous, $\psi(x)$ is also bounded as $x \to \pm\infty$, since otherwise the probability density $|\psi(x,t)|^2$ would not be normalizable to 1, as needed.
- If $U(x)$ doesn't have infinite discontinuities (jumps), or in another words if it doesn't have delta functions, then on integrating $\psi'' = (\epsilon - U(x))\psi$ we obtain that $\psi'(x)$ is continuous, so $\psi \in C^1(\mathbb{R})$ (the space of once-differentiable functions with continuous derivative). Indeed, if $U(x)$ has a finite jump then ψ'' also has a finite jump, which means that ψ' is still continuous.

- If we are in a classically allowed region, with energy greater than the potential, $\epsilon > U(x)$, so that $\epsilon - U(x) > 0$, then the equation $\psi'' = -(\epsilon - U(x))\psi$ in the $U(x) \simeq$ constant case has sinusoidal/cosinusoidal solutions or, taking a complex basis, e^{ikx} and e^{-ikx} solutions, with

$$k = \sqrt{\epsilon - U}. \tag{7.12}$$

- If on the other hand we are in a classically forbidden region, with energy smaller than the potential, $\epsilon - U(x) < 0$, the equation $\psi'' = +(U(x) - \epsilon)\psi$ in the $U(x) \simeq$ constant case has exponentially increasing and decreasing solutions, $\psi \sim e^{\kappa x}$ and $\sim e^{-\kappa x}$, with

$$\kappa = \sqrt{U - \epsilon}. \tag{7.13}$$

- The energy spectrum can be continuous (as for a free particle) or discrete (as for a H atom or the two-level system studied earlier). It can also be degenerate (with two or more states with the same energy) or nondegenerate.
- We will define $\lim_{x\to+\infty} U(x) \equiv U_+$ and $\lim_{x\to-\infty} \equiv U_-$ and will assume that $U_+ > U_-$; if not, we can set $x \to -x$, and retrieve this situation. Then as a function of the energy ϵ, we have three possible cases:

  (I) $\epsilon > U_+ > U_-$. In this case, there are two independent solutions (sin and cos, or e^{ikx} and e^{-ikx}) at each end of the real domain. That means that we can define two solutions that are bounded everywhere, including at both ends of the domain, for any such energy ϵ, which means that the spectrum is continuous, i.e., there is no restriction on ϵ, and degenerate with degeneracy 2, since for each energy there are two solutions. These states are *unbound* states, like free particles, with kinetic energy $\epsilon - U$.

  (II) $U_- < \epsilon < U_+$. In this case $\epsilon - U(x)$ is negative at $+\infty$ and positive at $-\infty$. The solution at $+\infty$ is exponentially increasing or decaying. Since the exponentially increasing solution is non-normalizable, we must choose the exponentially decreasing solution. But then this solution, when continued to $-\infty$, will correspond to only a given linear combination of the two (sin and cos) solutions there. That means that there is a unique solution for every energy ϵ in this region, so the spectrum is still continuous but now nondegenerate, and we also have unbound states.

  (III) $\epsilon < U_- < U_+$. In this case $\epsilon - U(x)$ is negative at both $+\infty$ and $-\infty$, which means there is a unique solution (exponentially decaying) at both ends. Consider the unique solution at $-\infty$, f_1, and continue it to $+\infty$, where it will become a linear combination of the exponentially decaying (g_1) and increasing (g_2) solutions, $f_1 = \alpha g_1 + \beta g_2$. But the coefficients α, β are functions of the energy ϵ, so the condition that $\beta(\epsilon) = 0$ (so that we have a bounded condition at $+\infty$ also), gives a constraint on the possible energies, having as solutions a discrete (and also nondegenerate) spectrum. Moreover, because of the exponentially decaying function at both $\pm\infty$, the eigenfunction is bounded in space to a finite region, and we say we have *bound* states. The existence or not of eigenvalues for the energy depends on the details of the potential $U(x)$ in the finite region in the middle, as does the total number of eigenvalues, which can be anything from 0 to infinity.

- **The Wronskian theorem**. This is a theorem for general linear second-order equations. Define the Wronskian of two functions y_1, y_2 as

$$W(y_1, y_2) = y_1 y_2' - y_2 y_1', \tag{7.14}$$

where y_1 and y_2 are solutions to the equations

$$\begin{aligned} y_1'' + f_1(x)y_1 &= 0 \qquad (1) \\ y_2'' + f_2(x)y_2 &= 0 \qquad (2). \end{aligned} \tag{7.15}$$

Here $f_1(x), f_2(x)$ are real functions that are piecewise continuous in the interval $x \in (a, b)$, meaning they can have jumps at a finite number of points but are otherwise continuous. Then the Wronskian theorem states that

$$W(y_1, y_2)|_a^b = \int_a^b dx(f_1(x) - f_2(x))y_1y_2. \tag{7.16}$$

Proof Multiplying equation (1) above by y_2 and equation (2) by y_1, and subtracting the two, we get

$$y_2y_1'' - y_1y_2'' + (f_1(x) - f_2(x))y_1y_2 = 0 \;\Rightarrow\; \frac{d}{dx}(y_2y_1' - y_1y_2') + (f_1(x) - f_2(x))y_1y_2 = 0, \tag{7.17}$$

and by integration over x we get the result of the theorem. *q.e.d.*

- Applying the Wronskian theorem in our case, with $f(x) = \epsilon - U(x)$, for two energies ϵ_1, ϵ_2, we get $f_1(x) - f_2(x) = \epsilon_1 - \epsilon_2$. Then the theorem says that

$$(y_1y_2' - y_2y_1')|_a^b = W(y_1, y_2)|_a^b = (\epsilon_1 - \epsilon_2)\int_a^b dxy_1y_2, \quad \forall a, b. \tag{7.18}$$

  In particular, for $\epsilon_1 = \epsilon_2$, we obtain that $W(y_1, y_2)$ is independent of x.
- The Wronskian theorem in our case can be used to prove *rigorously* that indeed if $\epsilon > U(x)$ for $x > x_0$ we have two oscillatory solutions, and if $\epsilon < U(x)$ and $U(x) - \epsilon \geq M^2 > 0$ for $x > x_0$, there is a unique solution going faster than or as fast as e^{-Mx} to zero at $+\infty$, and other solutions increase exponentially. It also follows that in a classically forbidden region ($\epsilon < U(x)$), the eigenfunction $\psi(x)$ can have *at most one zero*, since it is either increasing or decreasing exponentially.
- If $\epsilon < U(x)$ *everywhere*, there is no solution since it means that we have exponential solutions everywhere, but this would mean that the solution would have to increase at least on one side ($+\infty$ or $-\infty$), since the continuity of the derivative at possible jump points means that the increasing or decreasing property is conserved at jumps of $U(x)$ also. Therefore this solution will not be normalizable, so there is no solution to the eigenvalue problem.
- This also means that if $U(x)$ has a minimum $U_{\min}$ somewhere then the eigenvalues of E are necessarily larger than $U_{\min}$.
- The *number of nodes* (zeroes of the wave function) can be analyzed as follows. Consider two different energies $\epsilon_2 > \epsilon_1$. Then the Wronskian theorem for the case where a, b are two *consecutive* zeroes (or nodes) of y_1, so that $y_1(a) = y_1(b) = 0$, means that

$$y_2y_1'|_a^b = (\epsilon_2 - \epsilon_1)\int_a^b dxy_1y_2. \tag{7.19}$$

  But if a, b are consecutive zeroes, it means that between them, in the interval (a, b), y_1 has the same sign, say $y_1 > 0$, which implies $y_1'(a) > 0, y_1'(b) < 0$. Thus y_2 must change sign in the interval (a, b), since otherwise the right-hand side of the Wronskian theorem has the same sign as y_2 (as $\epsilon_2 - \epsilon_1 > 0$), whereas the left-hand side has the opposite sign (since both $y_1'(b)$ and $-y_1'(a)$ are negative).

That in turn means that y_2 has at least one zero in between the two consecutive zeroes of y_1, a, and b. But since both y_1 and y_2 also vanish asymptotically at $+\infty$ and $-\infty$, besides the finite zeroes (or nodes) it follows that if y_1 has n nodes (so that there are $n+1$ intervals (a,b), including $-\infty$ and $+\infty$ as boundaries) then y_2 has $n+1$ nodes. In turn, that means that there are at least $n-1$ nodes for the nth eigenfunction of the system.

- Finally, if we have a discrete spectrum, which implies that the eigenstates satisfy $y(\pm\infty) = 0$, then two eigenstates y_1, y_2 for two different eigenenergies ($\epsilon_1 \neq \epsilon_2$) are orthonormal, since the Wronskian theorem for $a = -\infty, b = +\infty$ implies that (dividing by the nonzero $\epsilon_1 - \epsilon_2$)

$$\int_{-\infty}^{+\infty} dx\, y_1 y_2 = 0. \tag{7.20}$$

7.3 Infinitely Deep Square Well (Particle in a Box)

For an infinitely high potential barrier, i.e., if $U(x) = \infty$ at some $x = x_0$, it is clear that we have $\psi(x_0) = 0$. Consider then the case where the potential is an infinitely deep square well (or box; see Fig. 7.1a),

$$\begin{aligned} U(x) &= 0, \qquad |x| < L/2 \\ &= \infty, \qquad |x| > L/2. \end{aligned} \tag{7.21}$$

In this case, it is clear that $\psi(x) = 0$ for $|x| \geq L/2$. Note that if $U(x) = U_0$ for $|x| < L/2$, we can make the rescaling $\epsilon - U_0 \to \epsilon$, and so get back to this case.

The differential equation in the only relevant region, $|x| \leq L/2$, is

$$\psi'' + \epsilon\psi = 0, \tag{7.22}$$

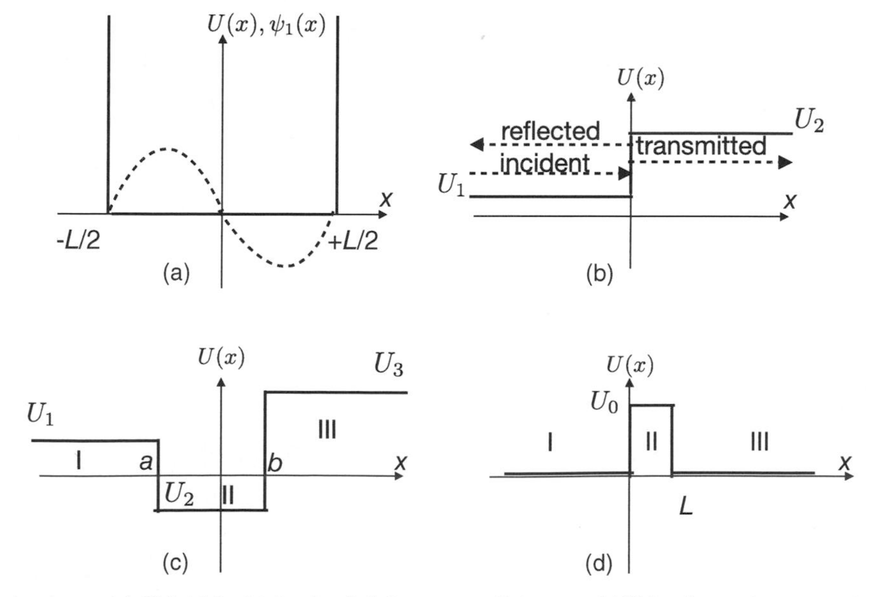

Figure 7.1 One-dimensional potentials $U(x)$. (a) Particle in a box (infinite square well): its potential $U(x)$ and ground state wave function $\psi_1(x)$. (b) Potential step and wave components. (c) Finite square well. (d) Potential barrier and tunneling effect.

with the boundary conditions that $\psi(x) = 0$ at $x = \pm L/2$. This is basically a stationary violin string set-up, with discrete eigenmodes (harmonics).

In the relevant region II, $|x| \leq L/2$, the solution is

$$\psi(x) = Ae^{ikx} + Be^{-ikx}, \tag{7.23}$$

where

$$k = \sqrt{\epsilon} = \sqrt{\frac{2mE}{\hbar^2}}. \tag{7.24}$$

Imposing continuity of the wave function at both ends, i.e.,

$$\psi(-L/2) = \psi(+L/2) = 0, \tag{7.25}$$

implies the system of equations

$$\begin{aligned} &Ae^{+ikL/2} + Be^{-ikL/2} = 0 \\ &Ae^{-ikL/2} + Be^{+ikL/2} = 0, \end{aligned} \tag{7.26}$$

which has nontrivial solutions only if the determinant of coefficients vanishes,

$$\left| \begin{pmatrix} e^{ikL/2} & e^{-ikL/2} \\ e^{-ikL/2} & e^{ikL/2} \end{pmatrix} \right| = 0 \Rightarrow e^{ikL} = e^{-ikL}, \tag{7.27}$$

which means that the wave numbers must be quantized,

$$k = k_n = \frac{n\pi}{L} \quad \Rightarrow \quad e^{ik_nL} = (-1)^n. \tag{7.28}$$

Consequently the eigenenergies are

$$E_n = \frac{\hbar^2}{2m}\frac{n^2\pi^2}{L^2}. \tag{7.29}$$

Here, naively, we would have $n \in \mathbb{Z}$, so that $n = 0, \pm 1, \pm 2, \pm 3, \ldots$, but in fact $n = 0$ is excluded, since that would imply that $\psi(x)$ is constant, which would mean we don't have $\psi(\pm L/2) = 0$, or that the wave function vanishes.

On the eigenenergies, we have $B = -Ae^{ik_nL}$, which means that the eigenfunctions are

$$\psi_n(x) = A(e^{ik_nx} + (-1)^n e^{-ik_nx}) = \begin{cases} 2iA\sin\frac{n\pi x}{L}, & n \text{ even} \\ 2A\cos\frac{n\pi x}{L}, & n \text{ odd}. \end{cases} \tag{7.30}$$

The remaining constant, A, is determined from the normalization condition for probability,

$$\int_{-L/2}^{+L/2} dx \frac{dP}{dx} = \int_{-L/2}^{L/2} |\psi(x)|^2 = 1. \tag{7.31}$$

This gives

$$1 = 4|A|^2 \int_{-L/2}^{+L/2} dx \sin^2\frac{n\pi x}{L} = 2|A|^2 L \Rightarrow |A| = \frac{1}{\sqrt{2L}}, \tag{7.32}$$

so the eigenfunctions are

$$\psi_n(x) = \begin{cases} \sqrt{\frac{2}{L}}\sin\frac{n\pi x}{L}, & n \text{ even} \\ \sqrt{\frac{2}{L}}\cos\frac{n\pi x}{L}, & n \text{ odd}. \end{cases} \tag{7.33}$$

We observe that the ground state of the quantum system is not the classical value $E = 0$, but is now

$$E_1 = \frac{\pi^2\hbar^2}{2mL^2}, \tag{7.34}$$

and the remaining states satisfy

$$E_n = n^2 E_1. \tag{7.35}$$

The fact that the energy of the ground state is nonzero, and its order of magnitude, can be understood from Heisenberg's uncertainty relations. Since the particle is confined to be in a box of size L, for $|x| \leq L/2$, it means that the variance of x (the uncertainty in the position) is $\Delta x = L/2$. From the Heisenberg uncertainty relation we obtain

$$\Delta p \geq \frac{\hbar/2}{L/2}. \tag{7.36}$$

Assuming equality,

$$p_{\min} = \Delta p = \frac{\hbar}{L}, \tag{7.37}$$

we find

$$E \geq \frac{p_{\min}^2}{2m} = \frac{\hbar^2}{2mL^2}. \tag{7.38}$$

This gives the right result except for a missing factor of π^2 (so the inequality is valid, just the possibility of equality is not).

The next observation is that in a bound state $\langle p \rangle = 0$. Having a stationary state means that $\langle p \rangle$ is time independent, and if the result were nonzero, it would mean that the particle translates on average and would eventually escape from the box, but that cannot happen, since the box has infinite sides.

The average position can be calculated explicitly,

$$\langle x \rangle = \int_{-L/2}^{+L/2} dx\, x|\psi(x)|^2 = \frac{2}{L}\int_{-L/2}^{+L/2} dx\, x \sin^2\frac{n\pi x}{L} = 0 \tag{7.39}$$

by the $x \to -x$ symmetry.

To calculate Δx, we find that, given $\langle x \rangle = 0$, we have

$$\begin{aligned}\langle(\Delta x)^2\rangle_n = \langle x^2\rangle_n &= \int_{-L/2}^{+L/2} dx\, x^2|\psi(x)|^2 = \frac{2}{L}\int_{-L/2}^{+L/2} dx\, x^2\frac{1-\cos 2\pi n x/L}{2}\\ &= \frac{L^2}{12} - \frac{L^2}{\pi^3 n^3}\int_{-\pi n/2}^{+\pi n/2} d\theta\, \theta^2\cos 2\theta = L^2\left(\frac{1}{12} - \frac{1}{2n^2\pi^2}\right),\end{aligned} \tag{7.40}$$

which gives

$$\Delta x = L\sqrt{\frac{1}{12} - \frac{1}{2n^2\pi^2}}. \tag{7.41}$$

This tends, as $n \to \infty$, to $L/\sqrt{12}$, which is the classical result: indeed, the classical result for the average of x^2 is $(1/L)\int_{-L/2}^{+L/2} dx\, x^2 = L^2/12$.

We can also calculate Δp in a similar way, considering that $\langle p \rangle = 0$:

$$\langle p^2 \rangle_n = \int_{-L/2}^{+L/2} dx |\hat{P}\psi(x)|^2 = \hbar^2 \int_{-L/2}^{+L/2} dx |\psi'(x)|^2 = \frac{2n^2\pi^2\hbar^2}{L^3} \int_{-L/2}^{+L/2} dx \frac{1 + \cos 2\pi n x/L}{2}$$
$$= \frac{\hbar^2 n^2 \pi^2}{L^2} = 2mE_n. \tag{7.42}$$

It follows that we have the expected classical-like relation

$$\langle (\Delta p)^2 \rangle_n = \langle p^2 \rangle_n = 2mE_n. \tag{7.43}$$

7.4 Potential Step and Reflection and Transmission of Modes

After considering the simplest stationary case, the particle in a box, we now move on to a different type of potential, one that allows for states at infinity, states propagating in time and reflecting and being transmitted through a "wall". The simplest example in this class is the potential step, with two different values of the potential, changing at $x = 0$, as in Fig. 7.1b,

$$U(x) = \begin{cases} U_1, & x < 0 \\ U_2, & x > 0. \end{cases} \tag{7.44}$$

Consider for concreteness that $U_2 > U_1$, so that the step is an increase, looking a bit like a wall to an incoming wave.

(a) The case $U_2 > \epsilon > U_1$

We treat first the case where the energy of the particle – or particles? is in between the two values of the potential. According to the general theory described at the beginning of the chapter, we expect a continuous and nondegenerate spectrum. For $x < 0$ we have $\epsilon > U_1$, so we have a sinusoidal solution, whereas for $x > 0$ we have $\epsilon < U_2$, so we have an exponentially decaying solution. Thus the solution can be parametrized as

$$\phi(x) = \begin{cases} A \sin(k_1 x + \phi), & x < 0 \\ B e^{-\kappa_2 x}, & x > 0, \end{cases} \tag{7.45}$$

where $k_1 = \sqrt{\epsilon - U_1}$ and $\kappa_2 = \sqrt{U_2 - \epsilon}$.

We first impose the condition of continuity of $\phi(x)$ at $x = 0$,

$$A \sin \phi = B, \tag{7.46}$$

and thus A is a normalization constant. Next, we impose the continuity of the derivative $\psi'(x)$ at $x = 0$, or better, of the logarithmic derivative (under the log, the derivative gets rid of otherwise arbitrary multiplicative constants), i.e. $\psi'(x)/\psi(x)$ at $x = 0$,

$$\frac{\psi'(x - \tilde{\epsilon})}{\psi(x - \tilde{\epsilon})} = \frac{\psi'(x + \tilde{\epsilon})}{\psi(x + \tilde{\epsilon})}, \tag{7.47}$$

implying a solution for ϕ:

$$-\kappa_2 = k_1 \cot\phi \Rightarrow \phi = \arctan\frac{-k_1}{\kappa_2}. \tag{7.48}$$

Substituting back, we can find value for B/A from

$$B = A\sin\phi = -A\frac{k_1}{\sqrt{k_1^2+\kappa_2^2}} = -A\sqrt{\frac{\epsilon - U_1}{U_2 - U_1}}. \tag{7.49}$$

Then A is found from the normalization condition

$$\int dx \phi_E^*(x)\phi_{E'}(x) = \delta(E - E'). \tag{7.50}$$

The full time-dependent solution is

$$A\sin(k_1 x + \phi)e^{-iEt} = \tilde{A}\left[e^{i(k_1 x - Et)+\phi} - e^{-i(k_1 x + Et)-\phi}\right], \tag{7.51}$$

which is the sum of an *incident* (original) traveling wave and a *reflected* wave, traveling in the opposite direction. The interference of the two waves gives the resulting solution.

(b) The case $\epsilon > U_2$

The second case is for an energy above both values of the potential. Again, according to the general theory described earlier, in this case we have a continuous and degenerate spectrum, with degeneracy 2. Both for $x > 0$ and for $x < 0$ we have sinusoidal solutions and, besides the incident wave for $x < 0$, we must have both a reflected wave (for $x < 0$) and a transmitted wave (for $x > 0$). Thus the wave function (where we have taken out an overall normalization constant A and replaced it with 1) is

$$\begin{aligned}\psi(x) &= e^{ik_1 x} + Re^{-ik_1 x}, \quad x < 0\\ &= Se^{ik_2 x}, \quad x > 0,\end{aligned} \tag{7.52}$$

where the wave with factor R is the reflected wave and the wave with factor S is the transmitted wave, and $k_1 = \sqrt{\epsilon - U_1} > k_2 = \sqrt{\epsilon - U_2}$. Continuity of the wave function at $x = 0$ gives

$$1 + R = S, \tag{7.53}$$

whereas continuity of the logarithmic derivative (of $\psi'(x)/\psi(x)$), also at $x = 0$, gives

$$ik_1\frac{1-R}{1+R} = ik_2 \Rightarrow R = \frac{k_1 - k_2}{k_1 + k_2} \Rightarrow S = 1 + R = \frac{2k_1}{k_1 + k_2} > 1. \tag{7.54}$$

Then, introducing the time dependence, we have an incident wave $\sim e^{ik_1 x - i\omega t}$, where $\omega = E/\hbar$, a reflected wave $\sim e^{-ik_1 x - i\omega t}$, and a transmitted wave $\sim e^{ik_2 x - i\omega t}$.

7.5 Continuity Equation for Probabilities

We can define the probability current density for continuous systems from

$$\langle\psi(t)|\psi(t)\rangle = \int \rho dV, \tag{7.55}$$

where the probability $\rho = dP/dV$ (written for the more general case of more than one dimension, with dV replacing dx), so we obtain in the coordinate representation

$$\rho = \frac{dP}{dV} = \left|\psi(\vec{r},t)\right|^2. \tag{7.56}$$

Then writing the Schrödinger equation and its complex conjugate in the coordinate representation,

$$\begin{aligned} i\hbar\frac{\partial\psi}{\partial t} &= -\frac{\hbar^2}{2m}\vec{\nabla}^2\psi + V\psi \\ -i\hbar\frac{\partial\psi^*}{\partial t} &= -\frac{\hbar^2}{2m}\vec{\nabla}^2\psi^* + V\psi^*, \end{aligned} \tag{7.57}$$

and then multiplying the first equation by ψ and the second by ψ^* and subtracting the two, we obtain

$$\begin{aligned} i\hbar(\psi^*\partial_t\psi + \psi\partial_t\psi^*) &= -\frac{\hbar^2}{2m}(\psi^*\vec{\nabla}^2\psi - \psi\vec{\nabla}^2\psi^*) \Rightarrow \\ i\hbar\partial_t(|\psi|^2) = i\hbar\partial_t\rho &= -\frac{\hbar^2}{2m}\vec{\nabla}(\psi^*\vec{\nabla}\psi - \psi\vec{\nabla}\psi^*), \end{aligned} \tag{7.58}$$

which can be written in terms of a continuity equation for probability flow,

$$\partial_t\rho = -\vec{\nabla}\cdot\vec{j}, \tag{7.59}$$

where the probability current density is

$$\vec{j} \equiv \frac{\hbar}{2mi}(\psi^*\vec{\nabla}\psi - \psi\vec{\nabla}\psi^*). \tag{7.60}$$

Applying this formalism for a one-dimensional problem with an incident wave $\psi = Ae^{ikx}$, we find the current density

$$\vec{j} = \frac{\hbar k}{m}|A|^2\vec{e}_x. \tag{7.61}$$

Even in one dimension, the vector $\vec{e}_x$ makes sense, taking the values ± 1 depending on the direction of propagation.

Consider then the conservation of probability at a point of reflection and refraction (i.e., transmission),

$$\vec{j}_{\text{inc}} + \vec{j}_{\text{refl}} = \vec{j}_{\text{transmitted}}. \tag{7.62}$$

In our case, these currents become

$$\begin{aligned} \vec{j}_{\text{inc}} &= \frac{\hbar k_1}{m}\vec{e}_x \\ \vec{j}_{\text{refl}} &= \frac{\hbar k_1}{m}|R|^2(-\vec{e}_x) \\ \vec{j}_{\text{trans}} &= \frac{\hbar k_2}{m}|S|^2\vec{e}_x, \end{aligned} \tag{7.63}$$

which means that the conservation of probability becomes

$$k_1(1-|R|^2)=k_2|S|^2 \Rightarrow 1-|R|^2=\frac{k_2}{k_1}|S|^2\equiv T, \tag{7.64}$$

where the right-hand side, T, is the *transmission coefficient* (since $|R|^2$ is the probability of reflection and T is the probability of transmission, so that the two sum up to 1). We find specifically

$$T=\frac{4k_1k_2}{(k_1+k_2)^2},\quad |R|^2=\frac{(k_1-k_2)^2}{(k_1+k_2)^2}. \tag{7.65}$$

7.6 Finite Square Well Potential

We next consider a square well potential, where the potential "dips" for a short while; for generality we consider the two sides to have different potentials, so that

$$U(x)=\begin{cases} U_1, & x<a, & \text{(I)}\\ U_2, & a<x<b, & \text{(II)}\\ U_3, & x>b, & \text{(III)},\end{cases} \tag{7.66}$$

where $U_2<U_1<U_3$, as in Fig. 7.1c.

If $\epsilon<U_2$, there is no solution, since the solution must be exponentially decaying or increasing, and this implies that at least one side of the solution will increase exponentially, making it impossible to normalize the probability to one.

(a) The case $U_1>\epsilon>U_2$

In this case, according to the general analysis, the spectrum is discrete, and represents bound states. The solution is exponentially decreasing to zero on both sides of the well, and sinusoidal inside the well, so we write

$$\psi(x)=\begin{cases} Ce^{-\kappa_3 x}, & x>b\\ B\sin(k_2x+\phi)=B_1\sin k_2x+B_2\cos k_2x, & a>x>b\\ Ae^{+\kappa_1 x}, & x<a,\end{cases} \tag{7.67}$$

where

$$\kappa_1=\sqrt{U_1-\epsilon},\quad k_2=\sqrt{\epsilon-U_2},\quad \kappa_3=\sqrt{U_3-\epsilon}. \tag{7.68}$$

Next, we impose continuity at $x=a$ and $x=b$ for $\psi(x)$, obtaining the equations

$$\begin{aligned} \psi(a+)=\psi(a-) &\Rightarrow Ae^{+\kappa_1 a}=B\sin(k_2a+\phi)\\ \psi(b+)=\psi(b-) &\Rightarrow B\sin(k_2b+\phi)=Ce^{-\kappa_3 b}. \end{aligned} \tag{7.69}$$

Finally, we impose the continuity of $\psi'(x)$ at the above points or, better still, considering that we have already imposed the continuity of $\psi(x)$, the continuity of the logarithmic derivative $\psi'(x)/\psi(x)$, giving

$$\frac{\psi'}{\psi}(a+) = \frac{\psi'}{\psi}(a-) \Rightarrow +\kappa_1 = k_2 \cot(k_2 a + \phi)$$
$$\frac{\psi'}{\psi}(b+) = \frac{\psi'}{\psi}(b-) \Rightarrow k_2 \cot(k_2 b + \phi) = -\kappa_3. \tag{7.70}$$

The equations giving the continuity of ψ equations solve for B, C as a function of A (then A is found from the normalization condition), whereas one of the equations for the continuity of ψ'/ψ can be solved for ϕ, and then the other gives an equation for the eigenenergies ϵ_n. We write these last two equations as two equations for the same ϕ, whose consistency will give ϵ_n:

$$\phi = \arctan\left(\frac{k_2}{\kappa_1}\right) - k_2 a$$
$$= -\arctan\left(\frac{k_2}{\kappa_3}\right) - k_2 b + n\pi. \tag{7.71}$$

Equating the two values, we find

$$\arctan\frac{k_2}{\kappa_3} + \arctan\frac{k_2}{\kappa_1} + k_2(b-a) - n\pi = 0, \tag{7.72}$$

which is a transcendental equation for ϵ, indexed by n (one solution per value of n) giving a discrete and finite sequence for ϵ_n. Defining the quantities

$$\cos\gamma \equiv \sqrt{\frac{U_1 - U_2}{U_3 - U_2}}, \quad \xi \equiv \sqrt{\frac{\epsilon - U_2}{U_1 - U_2}}, \quad K = \sqrt{U_1 - U_2}, \quad L = b - a, \tag{7.73}$$

we find the equation

$$\arcsin\xi + \arcsin(\xi\cos\gamma) = n\pi - \xi KL; \tag{7.74}$$

to find a solution for it, we need the condition

$$KL \geq (n-1)\pi + \gamma \tag{7.75}$$

which implies that there is a maximum value for n, given by

$$n \leq n_{\max} = 1 + \frac{KL - \gamma}{\pi}. \tag{7.76}$$

This means that, indeed, we have a finite spectrum, as promised. Moreover, n is the number of nodes or zeroes of ψ (see our previous general discussion, equating this number with the index of the eigenenergy), since we have a factor $\sin(k_2 x + \phi)$ in the wave function, vanishing for the $n\pi$ term.

If $U_1 = U_3 = U$ and $U_2 = U_0$ (for a symmetric well), we find $\kappa_1 = \kappa_3$, which means $\cos\gamma = 1$ and thus $\gamma = 0$, and if moreover $L\sqrt{U - U_0} = KL \ll 1$ then there is a unique solution: the condition $KL \geq (n-1)\pi + 0$ can only have $n = 1$ as a solution, given by the transcendental equation

$$2\arctan\frac{k_2}{\kappa} = \pi - k_2 L, \tag{7.77}$$

from which we find $k_2/\kappa \to \infty$, and the unique eigenenergy is

$$E = V - \frac{1}{2}KL(V - V_0), \tag{7.78}$$

and the angle $\phi \simeq \pi/2$.

(b) The case $U_1 < \epsilon < U_3$

In this case, when the energy is in between the two potentials at $-\infty$ and $+\infty$, according to the general analysis we must find a continuous, nondegenerate spectrum. As for the finite step potential, in region I ($x < a$), where $\epsilon > U(x)$, we write the wave function as an incident wave and a reflected wave, while in region III ($x > b$), where $\epsilon < U(x)$, we find an exponentially decaying solution. All in all,

$$\psi(x) = \begin{cases} e^{ik_1x} + e^{-ik_1x}e^{2i\phi_1}, & x < a \\ 2Ae^{i\phi_1}\sin(k_2x + \phi_2), & a < x < b \\ 2Be^{i\phi_1}e^{-\kappa_3 x}, & x > b. \end{cases} \tag{7.79}$$

Note that the equation is basically the previous set-up (with $\cos(k_1x - \phi_1)$ in region I and $\sin(k_2x+\phi_2)$ in region II), multiplied by $e^{i\phi_1}$, and by redefining $e^{2i\phi_1} \equiv R$ and $\tilde{A} = Ae^{i\phi_1}$, $\tilde{B} = Be^{i\phi_1}$, we get the set-up with incident and reflected waves.

As before, A and B are obtained from the continuity equations for $\psi(x)$ at $x = a$ and $x = b$. On the other hand, the angles ϕ_1 and ϕ_2 are found from the continuity of the logarithmic derivative, $\psi'(x)/\psi(x)$ at $x = a$ and $x = b$, giving

$$\begin{aligned} ik_1\frac{1 - e^{2i(k_1a+\phi_1)}}{1 + e^{2i(k_1a+\phi_1)}} &= k_1\tan(k_1a + \phi_1) = k_2\cot(k_2a + \phi_2) \\ k_2\cot(k_2b + \phi_2) = -\kappa_3 &\Rightarrow \phi_2 = -k_2b - \tan\frac{k_2}{\kappa_3}. \end{aligned} \tag{7.80}$$

Substituting ϕ_2 in the first equation, we find ϕ_1 as well. That means that all the parameters are fixed, but we have no new equations for the energy, meaning the energy spectrum is indeed continuous.

(c) The case $\epsilon > U_3$

Finally, consider the case when $\epsilon > U(x)$ everywhere, meaning that the solution is sinusoidal everywhere, and we have two good solutions in every region, I, II, and III. This means that the spectrum is continuous and degenerate with degeneracy 2.

As in the similar case with a single step, we write an incident and a reflected wave in region I, and a transmitted wave in region III. In region II, we also write a combination of a forward moving and a backward moving, wave, so

$$\psi(x) = \begin{cases} e^{ik_1x} + Re^{-ik_1x}, & x < a \\ Pe^{ik_2x} + Qe^{-ik_2x}, & a < x < b \\ Se^{ik_3x}, & x > b. \end{cases} \tag{7.81}$$

Continuity of the function $\psi(x)$ and of its logarithmic derivative $\psi'(x)/\psi(x)$ at $x = a$ and $x = b$ allows us to fix the four constants P, Q, R, S. The four equations are

$$\begin{aligned} e^{ik_1a} + Re^{-ik_1a} &= Pe^{ik_2a} + Qe^{-ik_2a} \\ Pe^{ik_2b} + Qe^{-ik_2b} &= Se^{i\kappa_3b} \\ ik_1\frac{1 - Re^{-2ik_1a}}{1 + Re^{-2ik_1a}} &= ik_2\frac{1 - (Q/P)e^{-2ik_2a}}{1 + (Q/P)e^{-2ik_2a}} \\ ik_2\frac{1 - (Q/P)e^{-2ik_2b}}{1 + (Q/P)e^{-2ik_2b}} &= i\kappa_3. \end{aligned} \tag{7.82}$$

The last equation can be solved for Q/P, then the third for R, then the first for P, and finally the second for S. The one interesting coefficient is the transmission coefficient,

$$T = \frac{\kappa_3}{\kappa_1}|S|^2 = \frac{4\eta\zeta\xi^2}{\xi^2(\eta+\zeta)^2\cos^2\xi KL + (\xi^2+\eta\zeta)^2\sin^2\xi KL}, \tag{7.83}$$

where we have defined

$$K = \sqrt{U_1 - U_2}, \quad \xi = \frac{k_2}{K}, \quad \eta = \frac{\kappa_1}{K}, \quad \zeta = \frac{\kappa_3}{K}. \tag{7.84}$$

The reflection coefficient is

$$|R|^2 = 1 - T, \tag{7.85}$$

an equation that can be explicitly checked.

7.7 Penetration of a Potential Barrier and the Tunneling Effect

We end the analysis of one-dimensional problems with the problem of a potential barrier consisting of a potential step of length L, which would be classically impenetrable, if $\epsilon < U$ inside this middle region.

Consider then the potential in Fig. 7.1d,

$$U(x) = \begin{cases} 0, & x < 0 \\ U_0 > 0, & 0 < x < L \\ 0, & x > L, \end{cases} \tag{7.86}$$

This means that, in the absence of any potential barrier (for $L = 0$), we would have just a free particle propagating. But, for $L \neq 0$, classically the particle starting in region I ($x < 0$) would be unable to reach region II ($x > L$) for $\epsilon < U_0$, though quantum mechanically there is a nonzero probability for the particle to "tunnel" through the barrier.

In regions I and III we write respectively the incident plus reflected waves and the transmitted wave,

$$\psi(x) = \begin{cases} e^{i\sqrt{\epsilon}x} + Re^{-i\sqrt{\epsilon}x}, & x < 0 \\ Se^{i\sqrt{\epsilon}x}, & x > L. \end{cases} \tag{7.87}$$

In region II ($0 < x < L$), we write a sum of exponentials, which are either real or imaginary, depending on the energy:

$$\psi(x) = \begin{cases} Ae^{\kappa x} + Be^{-\kappa x}, & \epsilon < U_0 \\ Ce^{ikx} + De^{-ikx}, & \epsilon > U_0, \end{cases} \tag{7.88}$$

where $\kappa = \sqrt{U_0 - \epsilon}$ and $k = \sqrt{\epsilon - U_0}$.

(a) The case $\epsilon < U_0$

In this case, the continuity conditions at $x = 0$ and $x = L$ for $\psi(x)$ and $\psi'(x)/\psi(x)$ give

$$\begin{aligned}
1 + R &= A + B \\
Ae^{\kappa L} + Be^{-\kappa L} &= Se^{i\sqrt{\epsilon}L} \\
i\sqrt{\epsilon}\frac{1 - R}{1 + R} &= \kappa\frac{A - B}{A + B} \\
\kappa\frac{Ae^{\kappa L} - Be^{-\kappa L}}{Ae^{\kappa L} + Be^{-\kappa L}} &= i\sqrt{\epsilon}.
\end{aligned} \tag{7.89}$$

Solving the equations as before (the first for R in terms of A and B, the second for S in terms of the same, and then the last two for A and B), we find the transmission coefficient (since $k_2 = k_1 = \sqrt{\epsilon}$)

$$T = |S|^2 = \frac{1}{1 + \dfrac{U_0^2 \sinh^2 \kappa L}{4\epsilon(U_0 - \epsilon)}} < 1, \tag{7.90}$$

and the reflection coefficient is $|R|^2 = 1 - T$ as before. This is the *tunneling probability*, namely the probability of passing through a classically impenetrable barrier. This is relevant for radioactivity, for instance, where an α particle can "tunnel out" of a potential barrier binding it to a radioactive nucleus. We see that if $\kappa L \gg 1$, we have an exponentially small probability of tunneling, as expected:

$$T \simeq \frac{16\epsilon(U_0 - \epsilon)}{U_0^2} e^{-2\kappa L}. \tag{7.91}$$

(b) The case $\epsilon > U_0$

In this case, the particle can go "over" the barrier classically, as if it wasn't there, but quantum mechanically there will be reflection as well as transmission.

The continuity conditions at $x = 0$ and $x = L$ for $\psi(x)$ and $\psi'(x)/\psi(x)$ give

$$\begin{aligned}
1 + R &= C + D \\
Ce^{ikL} + De^{-ikL} &= Se^{i\sqrt{\epsilon}L} \\
i\sqrt{\epsilon}\frac{1 - R}{1 + R} &= ik\frac{C - D}{C + D} \\
ik\frac{Ce^{ikL} - De^{-ikL}}{Ce^{ikL} + De^{-ikL}} &= i\sqrt{\epsilon},
\end{aligned} \tag{7.92}$$

which similarly lead to a transmission coefficient

$$T = |S|^2 = \frac{1}{1 + \dfrac{U_0^2}{4\epsilon(\epsilon - U_0)} \sin^2 kL} < 1, \tag{7.93}$$

(and a reflection coefficient $|R|^2 = 1 - T$); note that now the transmission coefficient oscillates between 1 and $1/(1 + U_0^2/4\epsilon(\epsilon - U_0)) < 1$ as a function of k (classically, we expect only the value $T = 1$).

Important Concepts to Remember

- For a one-dimensional particle in a potential, the wave function is $\psi(x) \in C^0(\mathbb{R})$ and, if the potential has no delta functions, $\psi(x) \in C^1(\mathbb{R})$ and is bounded at $x = \pm\infty$.
- The wave function is of sin/cos type in a classically allowed region, and of exponentially decaying/growing type in a classically forbidden region (being exactly sin/cos or exp only if the potential is constant in the region).
- For energies above the values of the potential at $+\infty$ and $-\infty$, we have a continuous spectrum, of unbound states of degeneracy 2, as for free particles.
- For energies between the values of the potential at $+\infty$ and $-\infty$, we have a continuous spectrum, of unbound states but nondegenerate.
- For energies below the values of the potential at $+\infty$ and $-\infty$, we have a discrete and nondegenerate spectrum of bound states, i.e., they are constrained to a finite region (exponentially decaying outside it).
- There are at least $n - 1$ nodes (zeroes) for the nth eigenfunction of the system.
- The infinitely deep square well has eigenenergies $E_n = \frac{\hbar^2\pi^2}{2mL^2}n^2$, with $n = \pm1, \pm2, \pm3$, etc., and eigenfunctions $\sin n\pi x/L$ for n even and $\cos n\pi x/L$ for n odd.
- In the case of a step-type wall, we can define a transmitted wave and a reflected wave, $\psi(x) = e^{ik_1x} + Re^{-ik_2x}$ for $x < 0$ and $\psi(x) = Se^{-ik_2x}$ for $x > 0$.
- The continuity equation for probabilities is $\partial_t\rho + \vec{\nabla}\vec{j} = 0$, where $\rho = |\psi|^2$ is the probability density and $\vec{j} = \frac{\hbar}{2m}(\psi^*\vec{\nabla}\psi - \psi\vec{\nabla}\psi^*)$ is the probability current.
- For both the step-function case and the case of a finite barrier, we can define the transmission coefficient T and the reflection coefficient R for the probability, with $T + R = 1$.
- For a potential barrier of length L and height U_0, for $L\sqrt{U_0 - E}L \gg 1$, the tunneling probability is $T \propto e^{-2\sqrt{U_0-E}L}$.

Further Reading

See [2] for more details.

Exercises

(1) Consider a potential that depends only on the radial direction r in a three-dimensional space, $V(r)$, with $V(r$ such that $\to \infty) = 0$ while $V(0) = -V_0 < 0$ and $V_{\max} = \max_r V(r) = V_1 > 0$, reached at $r = r_1$, and there is a unique solution r_0 to the equation $V(r_0) = 0$. Describe the spectrum of the system in the various energy regimes.

(2) Repeat the previous exercise in the case $V(r \to 0) \to -\infty$.

(3) Consider the delta function potential in one dimension,

$$V = -V_0\delta(x). \tag{7.94}$$

Calculate its bound state spectrum.

(4) Calculate $\langle x^m\rangle_n$ and $\langle p^m\rangle_n$ for a particle in a box (an infinitely deep square well) for arbitrary integers $n, m > 0$. In which limits do we have a classical result?

(5) Consider a periodic cosinusoidal potential,

$$V(x) = V_0 \cos(ax). \tag{7.95}$$

Calculate the bound state spectrum, and the number of nodes for the corresponding eigenstates.

(6) For the finite square well potential, prove the formula for the transmission coefficient (7.83), calculate R, and prove that $|R|^2 + T = 1$.

(7) Consider an asymmetric potential barrier, with

$$U(x) = \begin{cases} 0, & x < 0 \\ U_0 > U_1, & 0 < x < L \\ U_1 > 0, & x > L. \end{cases} \tag{7.96}$$

Calculate the tunneling probability T for energy $U_1 < E < U_0$, and for energy $E > U_0$.

8 The Harmonic Oscillator

We next move to the simplest one-dimensional system with a truly space-dependent potential $V(x)$ (as opposed to one that is piecewise constant), the harmonic oscillator. It is an excellent tool, both for learning quantum mechanics, and for applying it. Indeed, we can say that:

- This is the most basic quantum mechanical system, the prototype that we can use for a quantum mechanical problem. It has all the important features that allow one to understand the basics.
- If understood *in detail*, it can be generalized to any problem. Famously, Sidney Coleman stated it thus: "The career of a young theoretical physicist consists of treating the harmonic oscillator in ever-increasing levels of abstraction". As such, while the basic treatment is described here, we will return to the harmonic oscillator from time to time, using it as an example and application.
- Most systems can be *approximated* by one, or several, harmonic oscillators. Thus the harmonic oscillator can be understood as a first approximation in a perturbation theory for a general system.

8.1 Classical Set-Up and Generalizations

The basic system is a spring of constant k, so that we have Hooke's law $F = -kx$ and the potential is (since $\vec{F} = -\vec{\nabla}V$)

$$V = \frac{1}{2}kx^2, \tag{8.1}$$

writing $k = m\omega^2$, where ω is the classical angular frequency of oscillation. Then the classical Hamiltonian is

$$H = T + V = \frac{p^2}{2m} + \frac{m\omega^2}{2}x^2, \tag{8.2}$$

so that the quantum Hamiltonian operator is

$$\hat{H} = \frac{\hat{P}^2}{2m} + \frac{m\omega^2}{2}\hat{X}^2. \tag{8.3}$$

However, rather than this exact and basic system we can consider a more general set-up: a particle in a general potential $V(x)$ that has a stable minimum $x = x_0$ (so that $V''(x_0) > 0$). Around this minimum, we can do a Taylor expansion:

$$\begin{aligned} V(x) &\simeq V(x_0) + V'(x_0)(x - x_0) + \frac{V''(x_0)}{2}(x - x_0)^2 + \cdots \\ &= V(x_0) + \frac{V''(x_0)}{2}(x - x_0)^2. \end{aligned} \tag{8.4}$$

Thus, with $x - x_0 \equiv y$ and $V''(x_0) \equiv m\omega^2$, we obtain approximately the above harmonic oscillator.

For a generic multiparticle coupled system with variables $x_1, \ldots, x_n$ and potential $V(x_1, \ldots, x_n)$, and a *stable* minimum (so the determinant of the Hessian matrix is positive) written as $\vec{x}_0 = (x_{1,0}, \ldots, x_{n,0})$, we obtain similarly

$$V(x_1, \ldots, x_n) \simeq V(\vec{x}_0) + \frac{1}{2} \sum_{i,j=1}^{n} \partial_i \partial_j V(\vec{x}_0)(x_i - x_{i,0})(x_j - x_{j,0}). \tag{8.5}$$

By writing $\delta x_i \equiv x_i - x_{i,0}$ and

$$\partial_i \partial_j V(\vec{x}_0) \equiv V_{ij} = V_{ji}, \tag{8.6}$$

we obtain the Hamiltonian

$$H = \sum_{i,j=1}^{n} \frac{1}{2} p_i \frac{\delta_{ij}}{m} p_j + \frac{1}{2} \sum_{i,j=1}^{n} \delta x_i V_{ij} \delta x_j. \tag{8.7}$$

This corresponds to a matrix in (i, j) space, which can be diagonalized to find the eigenmodes, i.e., the eigenvectors and eigenvalues for H_{ij}. In the diagonalized form, the system reduces to a system of decoupled harmonic oscillators.

Finally, similarly considering a field $\phi(x, t)$ in Fourier space, it can be decomposed in an infinite set of approximately harmonic oscillators, as in the above finite dimensional case. We will not do this here, however, since field theory is beyond the scope of this book.

The classical equations of motion of the harmonic oscillator Hamiltonian (Hamilton's equations) are:

$$\dot{x} = \frac{\partial H}{\partial p} = \frac{p}{m}, \quad \dot{p} = -\frac{\partial H}{\partial x} = -m\omega^2 x, \tag{8.8}$$

leading to a single equation for $x(t)$,

$$\ddot{x} + \omega^2 x = 0, \tag{8.9}$$

with independent solutions $e^{i\omega t}$ and $e^{-i\omega t}$, or $\sin \omega t$ and $\cos \omega t$, with general real solution

$$x(t) = A \sin(\omega t + \phi). \tag{8.10}$$

8.2 Quantization in the Creation and Annihilation Operator Formalism

We want to solve the Schrödinger equation for the quantum harmonic oscillator,

$$i\hbar\partial_t |\psi\rangle = \hat{H}|\psi\rangle, \tag{8.11}$$

where the quantum Hamiltonian is, as we saw above,

$$\hat{H} = \frac{\hat{P}^2}{2m} + \frac{m\omega^2}{2}\hat{X}^2. \tag{8.12}$$

We do the usual separation of variables, writing the eigenvalue problem for $\hat{H}$ (the time-independent Schrödinger equation),

$$\hat{H}|\psi\rangle = E|\psi\rangle. \tag{8.13}$$

In order to do that, and to diagonalize the Hamiltonian, i.e., find its eigenstates, we consider a change of quantum operators from $(\hat{X}, \hat{P})$ to complex operators $(\hat{a}, \hat{a}^\dagger)$, defined by

$$\begin{aligned} \hat{a} &= \frac{1}{\sqrt{2}}\left(\sqrt{\frac{m\omega}{\hbar}}\hat{X} + i\frac{1}{\sqrt{m\omega\hbar}}\hat{P}\right) \equiv \frac{\hat{\mathcal{Q}} + i\hat{\mathcal{P}}}{\sqrt{2}} \\ \hat{a}^\dagger &= \frac{1}{\sqrt{2}}\left(\sqrt{\frac{m\omega}{\hbar}}\hat{X} - i\frac{1}{\sqrt{m\omega\hbar}}\hat{P}\right) \equiv \frac{\hat{\mathcal{Q}} - i\hat{\mathcal{P}}}{\sqrt{2}}, \end{aligned} \tag{8.14}$$

with the inverse

$$\hat{\mathcal{Q}} = \frac{\hat{a} + \hat{a}^\dagger}{\sqrt{2}}, \quad \hat{\mathcal{P}} = \frac{\hat{a} - \hat{a}^\dagger}{i\sqrt{2}}. \tag{8.15}$$

Under this change, the canonical commutation relation $[\hat{X}, \hat{P}] = i\hbar\mathbb{1}$ becomes

$$[\hat{a}, \hat{a}^\dagger] = \frac{1}{2}[\hat{\mathcal{Q}} + i\hat{\mathcal{P}}, \hat{\mathcal{Q}} - i\hat{\mathcal{P}}] = +i[\hat{\mathcal{P}}, \hat{\mathcal{Q}}] = \hat{\mathbb{1}}, \tag{8.16}$$

and of course we also have $[\hat{a}, \hat{a}] = [\hat{a}^\dagger, \hat{a}^\dagger] = 0$.

The Hamiltonian then becomes

$$\begin{aligned} \hat{H} &= \frac{m\omega\hbar}{2m}\frac{(\hat{a} - \hat{a}^\dagger)^2}{(-2)} + \frac{m\omega^2}{2}\frac{\hbar}{m\omega}\frac{(\hat{a} + \hat{a}^\dagger)^2}{2} = \frac{\hbar\omega}{2}(\hat{a}\hat{a}^\dagger + \hat{a}^\dagger\hat{a}) \\ &= \hbar\omega\left(\hat{a}^\dagger\hat{a} + \frac{1}{2}\right) \equiv \hbar\omega\left(\hat{N} + \frac{1}{2}\right), \end{aligned} \tag{8.17}$$

where in the second line we have used the commutation relation $[\hat{a}, \hat{a}^\dagger] = 1$ and have defined the "number operator" (we will see shortly why it is called this)

$$\hat{N} = \hat{a}^\dagger\hat{a}, \tag{8.18}$$

which we can easily see is Hermitian, $\hat{N}^\dagger = \hat{a}^\dagger\hat{a} = \hat{N}$.

This means that the eigenstates of $\hat{H}$ are also eigenstates of $\hat{N}$, so we will consider the eigenvalue problem for the latter, with the eigenvalue called n:

$$\hat{N}|n\rangle = n|n\rangle. \tag{8.19}$$

Since we have a Hilbert space, the norms of the states $|n\rangle$ must be positive and nonzero, i.e., nonnegative, $\langle n|n\rangle > 0$. But now consider the expectation value

$$\begin{aligned} \langle n|\hat{N}|n\rangle &= \langle n|\hat{a}^\dagger\hat{a}|n\rangle = ||\hat{a}|n\rangle||^2 \geq 0 \\ &= n\langle n|n\rangle \geq 0, \end{aligned} \tag{8.20}$$

leading to the fact that the eigenvalue n is positive, $n \geq 0$, and moreover

$$n = 0 \quad \Leftrightarrow \quad a|n\rangle = 0. \tag{8.21}$$

That means that the *ground state* of the harmonic oscillator, the state of minimum energy and therefore also the minimum eigenvalue for $\hat{N}$, has $n = 0$ and so is denoted $|0\rangle$ and is called the "vacuum" state. It is defined by

$$\hat{a}|0\rangle = 0. \tag{8.22}$$

Next we note that n (the eigenvalue for $\hat{N}$) is increased by the action of $a^\dagger$ and decreased by the action of a, since

$$\begin{aligned}[\hat{N}, \hat{a}] &= \hat{a}^\dagger \hat{a}\hat{a} - \hat{a}\hat{a}^\dagger \hat{a} = [\hat{a}^\dagger, \hat{a}]\hat{a} = -\hat{a} \\ [\hat{N}, \hat{a}^\dagger] &= \hat{a}^\dagger \hat{a}\hat{a}^\dagger - \hat{a}^\dagger \hat{a}^\dagger \hat{a} = \hat{a}^\dagger[\hat{a}, \hat{a}^\dagger] = \hat{a}^\dagger.\end{aligned} \tag{8.23}$$

Indeed, then $\hat{N}|n\rangle = n|n\rangle$ implies that

$$\begin{aligned}\hat{N}(\hat{a}|n\rangle) &= (\hat{a}\hat{N} - \hat{a})|n\rangle = (n-1)(\hat{a}|n\rangle) \quad \Rightarrow \quad |n-1\rangle \propto \hat{a}|n\rangle \\ \hat{N}(\hat{a}^\dagger|n\rangle) &= (\hat{a}^\dagger \hat{N} + \hat{a}^\dagger)|n\rangle = (n+1)(\hat{a}^\dagger|n\rangle) \quad \Rightarrow \quad \hat{a}^\dagger|n\rangle \propto |n+1\rangle.\end{aligned} \tag{8.24}$$

This means that $\hat{a}^\dagger$ creates a quantum with energy $\hbar\omega$, since when acting on a state it increases the energy as $E \to E + \hbar\omega$, and $\hat{a}$ annihilates a quantum of energy $\hbar\omega$, decreasing the energy of a state as $E \to E - \hbar\omega$. Thus $\hat{a}$ is called an annihilation or lowering operator, and $\hat{a}^\dagger$ is called a creation or raising operator.

Since n can jump by only one, up or down, and we have $n = 0$ for the state defined by $a|0\rangle = 0$, it follows that $n \in \mathbb{N}$ is a natural number, and the *vacuum* state $|0\rangle$ is defined by $a|0\rangle = 0$. The states are then indexed by this natural number n,

$$|0\rangle, \quad |1\rangle \propto a^\dagger|0\rangle, \quad |2\rangle \propto a^\dagger|1\rangle \propto (a^\dagger)^2|0\rangle, \quad \ldots \tag{8.25}$$

and the eigenvalue for the state $|n\rangle$ is

$$E_n = \left(n + \frac{1}{2}\right)\hbar\omega. \tag{8.26}$$

We require orthonormal states, $\langle n|n\rangle = 1$, and, more generally,

$$\langle n|m\rangle = \delta_{nm}, \tag{8.27}$$

which means that the proportionality constants for the action of $\hat{a}$ and $\hat{a}^\dagger$, given by

$$\begin{aligned}\hat{a}|n\rangle &= \alpha_n|n-1\rangle \\ \hat{a}^\dagger|n\rangle &= \beta_n|n+1\rangle,\end{aligned} \tag{8.28}$$

can be calculated. Indeed, from the relations found previously,

$$n||\,|n\rangle||^2 = ||a|n\rangle||^2 = |\alpha_n|^2||\,|n-1\rangle||^2 \quad \Rightarrow \quad |\alpha_n| = \sqrt{n}. \tag{8.29}$$

Choosing the phase of α_n to be unity (the trivial choice), we obtain $\alpha_n = \sqrt{n}$. Similarly, from the previous relations,

$$\begin{aligned}\langle n|\hat{a}\hat{a}^\dagger|n\rangle = ||\hat{a}^\dagger|n\rangle||^2 &= \langle n|(\hat{a}^\dagger\hat{a} + 1)|n\rangle = (n+1)||\,|n\rangle||^2 \\ &= |\beta_n|^2||\,|n\rangle||^2 \Rightarrow |\beta_n| = \sqrt{n+1}.\end{aligned} \tag{8.30}$$

Then, choosing the phase of β_n to also be unity, we obtain $\beta_n = \sqrt{n+1}$. All in all,

$$\hat{a}|n\rangle = \sqrt{n}|n-1\rangle, \quad \hat{a}^\dagger|n\rangle = \sqrt{n+1}|n+1\rangle. \tag{8.31}$$

Since $\langle n|m\rangle = \delta_{mn}$, it means that the matrix elements of $\hat{a}$ and $\hat{a}^\dagger$ are

$$\begin{aligned}\langle m|\hat{a}|n\rangle &= \sqrt{n}\delta_{m,n-1}\\ \langle m|\hat{a}^\dagger|n\rangle &= \sqrt{n+1}\delta_{m,n+1},\end{aligned} \tag{8.32}$$

which means the opearators a and $a^\dagger$ are represented in the basis with $n = 0, 1, 2, \dots$ by the matrices

$$a = \begin{pmatrix} 0 & \sqrt{1} & 0 & & & \\ 0 & 0 & \sqrt{2} & 0 & & \\ & 0 & 0 & \sqrt{3} & & \\ & & 0 & 0 & \sqrt{4} & \\ & & & & & \dots \end{pmatrix}, \quad a^\dagger = \begin{pmatrix} 0 & 0 & 0 & & & \\ \sqrt{1} & 0 & 0 & & & \\ 0 & \sqrt{2} & 0 & & & \\ & & \sqrt{3} & 0 & 0 & \\ & & & \sqrt{4} & 0 & \\ & & & & & \dots \end{pmatrix}. \tag{8.33}$$

Moreover, we obtain the general state as a repeated action with $\hat{a}^\dagger$, using the recursion relation n times,

$$|n\rangle = \frac{\hat{a}^\dagger}{\sqrt{n}}|n-1\rangle = \frac{\hat{a}^\dagger\hat{a}^\dagger}{\sqrt{n(n-1)}}|n-2\rangle = \dots = \frac{(\hat{a}^\dagger)^n}{\sqrt{n!}}|0\rangle. \tag{8.34}$$

8.3 Generalization

We can easily generalize the single harmonic oscillator to the case of several oscillators, coming either from several particles with harmonic oscillator potentials, or from diagonalizing a total interacting Hamiltonian expanded around a stable minimum. Then we have

$$H = \sum_q H_q, \tag{8.35}$$

with each H_q being an independent harmonic oscillator of $\omega = \omega_q$,

$$H_q = \hbar\omega_q\left(\hat{a}^\dagger\hat{a} + \frac{1}{2}\right). \tag{8.36}$$

The fact that the harmonic oscillators are independent means that we have

$$[\hat{a}_q, \hat{a}^\dagger_{q'}] = \delta_{q,q'}\hat{\mathbb{1}} \tag{8.37}$$

(and the rest of the commutators vanish). Then the states of the total system are, as in the general theory described previously, tensor products of the states for each Hamiltonian H_q, meaning that the general state, with occupation numbers $n_1, n_2, \dots, n_q, \dots$ in each mode, is

$$|n_1, n_2, \dots, n_q, \dots\rangle = \frac{(\hat{a}^\dagger_1)^{n_1}}{\sqrt{n_1!}} \dots \frac{(\hat{a}^\dagger_q)^{n_q}}{\sqrt{n_q!}} \dots |0\rangle, \tag{8.38}$$

where the total vacuum state is the tensor product of the vacuum state of each oscillator,

$$|0\rangle \equiv |0\rangle_1 \otimes |0\rangle_2 \otimes \cdots \tag{8.39}$$

8.4 Coherent States

We can define other bases than $\{|n\rangle\}$ (eigenstates of the Hamiltonian) in the Hilbert space. One such basis is the basis of *coherent states* $|\alpha\rangle$, defined as eigenstates of the annihilation operator $\hat{a}$,

$$a|\alpha\rangle = \alpha|\alpha\rangle. \tag{8.40}$$

We find that the coherent states can be written in terms of the basis of $|n\rangle$ states, as

$$|\alpha\rangle = e^{\alpha\hat{a}^\dagger}|0\rangle \equiv \sum_{n\geq 0}\frac{\alpha^n}{n!}(\hat{a}^\dagger)^n|0\rangle. \tag{8.41}$$

Indeed, then

$$\hat{a}|\alpha\rangle = [\hat{a}, e^{\alpha\hat{a}^\dagger}]|0\rangle = \alpha e^{\alpha\hat{a}^\dagger}|0\rangle = \alpha|\alpha\rangle, \tag{8.42}$$

where in the first equality we have used $\hat{a}|0\rangle = 0$ and in the second we have used

$$\begin{aligned}[]
[\hat{a}, e^{\alpha\hat{a}^\dagger}] &= \sum_{n\geq 0}\frac{\alpha^n}{n!}\left([\hat{a},\hat{a}^\dagger](\hat{a}^\dagger)^{n-1} + \hat{a}^\dagger[\hat{a},\hat{a}^\dagger](\hat{a}^\dagger)^{n-2} + \cdots + (\hat{a}^\dagger)^{n-1}[\hat{a},\hat{a}^\dagger]\right)\\
&= \alpha\sum_{n\geq 0}\frac{\alpha^{n-1}}{(n-1)!}(\hat{a}^\dagger)^{n-1} = \alpha e^{\alpha a^\dagger}.
\end{aligned} \tag{8.43}$$

Similarly, we define the complex conjugate state,

$$\langle\alpha^*| \equiv \langle 0|e^{\alpha^* a}, \tag{8.44}$$

obeying the complex conjugate relation

$$\langle\alpha^*|\hat{a}^\dagger = \langle\alpha^*|\alpha^*. \tag{8.45}$$

The inner product of the bra and ket states is found to be

$$\langle\alpha^*|\alpha\rangle = \langle 0|e^{\alpha^* a}|\alpha\rangle = e^{\alpha\alpha^*}\langle 0|\alpha\rangle = e^{\alpha^*\alpha}, \tag{8.46}$$

since we have (using $\langle 0|a^\dagger = 0$)

$$\langle 0|\alpha\rangle = \langle 0|e^{\alpha a^\dagger}|0\rangle = \langle 0|0\rangle = 1. \tag{8.47}$$

We also have a completeness relation for these bra and ket states,

$$\hat{\mathbb{1}} = \int\frac{d\alpha\, d\alpha^*}{2\pi i}e^{-\alpha\alpha^*}|\alpha\rangle\langle\alpha^*|, \tag{8.48}$$

but we will leave its proof as an exercise.

8.5 Solution in the Coordinate, $|x\rangle$, Representation (Basis)

Until now, we have worked abstractly with ket states, without choosing a representation. But we now want to obtain the probabilities of finding the harmonic oscillator in coordinate (x) space, so we need to calculate the wave functions $\psi_n(x) \equiv \langle x|n\rangle$.

According to the general theory defined in the previous chapter, the Schrödinger equation $\hat{H}|n\rangle = E_n|n\rangle$ becomes

$$\left(-\frac{\hbar^2}{2m}\frac{d^2}{dx^2} + \frac{1}{2}m\omega^2 x^2\right)\psi_n(x) = E_n\psi_n(x), \tag{8.49}$$

by inserting the completeness relation $\int dx|x\rangle\langle x| = \hat{1}$ and multiplying with $\langle x|$. Rewriting it as

$$\psi_n''(x) + \left[\epsilon_n - \left(\frac{m\omega x}{\hbar}\right)^2\right]\psi_n(x) = 0, \tag{8.50}$$

where we have defined as before

$$\epsilon = \frac{2m}{\hbar^2}E, \tag{8.51}$$

or, better still, defining the coordinate $y = x\sqrt{m\omega/\hbar}$ and

$$\tilde{\epsilon} = \frac{E}{\hbar\omega}, \tag{8.52}$$

we have

$$\psi_n''(y) + (2\tilde{\epsilon}_n - y^2)\psi_n(y) = 0. \tag{8.53}$$

In order to solve this differential equation and find its eigenvalues ϵ_n, we use a method, called the Sommerfeld method, that can be used in more general situations. We first find the behavior of the equation, and the corresponding solutions, at the extremes of the region, $y \to \infty$ and $y \to 0$, and then we factor them out of $\psi(y)$, and write an equation for the reduced function.

(a) The limit $y \to \infty$

In this limit, equation (8.53) becomes

$$\psi'' - y^2\psi = 0, \tag{8.54}$$

on neglecting the constant in the second term. The solution is of the general form

$$\psi(y) = (Ay^m)e^{\pm y^2/2}, \tag{8.55}$$

where for completeness we have considered also the subleading polynomial factor (Ay^m), but in fact this is not needed, since all we need to do is to factor out the leading behavior. Indeed, then

$$\psi''(y) = Ay^{m+2}e^{\pm y^2/2}\left[1 \pm \frac{2m+1}{y^2} + \frac{m(m-1)}{y^4}\right] \to y^2\psi(y), \tag{8.56}$$

independently of the polynomial in the $y \to \infty$ limit. From the condition that $\psi(y)$ must be a normalizable function (since its modulus squared gives probabilities), we can only have the negative sign in the exponent, $e^{-y^2/2}$.

(b) Limit $y \to 0$

In this case we can ignore the y^2 term with respect to the constant in the second term, obtaining

$$\psi''(y) + 2\tilde{\epsilon}\psi(y) = 0, \tag{8.57}$$

with the general solution

$$\psi(y) = \tilde{A}\sin\sqrt{2\tilde{\epsilon}}y + \tilde{B}\cos\sqrt{2\tilde{\epsilon}}y. \tag{8.58}$$

However, since we are considering the $y \to 0$ limit, we can ignore order-y^2 terms in the solution (as in the equation), and write only

$$\psi(y) \to \tilde{A}\sqrt{2\epsilon}y + \tilde{B}. \tag{8.59}$$

Putting together the two limits, we factor out the leading behaviors at infinity ($e^{-y^2/2}$) and at zero (just 1, a constant), redefining

$$\psi(y) = e^{-y^2/2}H(y), \tag{8.60}$$

where from the *general* (including subleading) behavior at infinity, and the behavior at zero, we know that $H(y)$ should be a polynomial ($\sim Ay^m$ at $y \to \infty$ and $\sim \tilde{C}y + \tilde{B}$ at $y \to 0$). We calculate first

$$\psi''(y) = e^{-y^2/2}[H''(y) - 2yH'(y) + (y^2 - 1)H(y)], \tag{8.61}$$

which means that the equation for $H(y)$ is

$$H''(y) - 2yH'(y) + (2\tilde{\epsilon} - 1)H(y) = 0. \tag{8.62}$$

But from the fact that the good (i.e., normalizable) solution at $y \to \infty$ is $\sim e^{-y^2/2}$ and not $\sim e^{+y^2/2}$, which in general is a linear combination of the two possible solutions at zero ($\sin\sqrt{2\epsilon}y \sim \sqrt{2\epsilon}y$ and $\cos\sqrt{2\epsilon}y \sim 1$), while we want a nonzero solution at $y = 0$, we get a constraint on $\epsilon = \epsilon_n$, that is, a *quantization condition for the energy*. Indeed, this good behavior both at $y = 0$ and at $y = \infty$ of the solution will lead to the condition

$$2\tilde{\epsilon} = 2\frac{E_n}{\hbar\omega} = 2n + 1, \quad n \in \mathbb{N}, \tag{8.63}$$

which will mean that the $H_n(x)$ are *Hermite polynomials*. To see this, we first write $H(y)$ as an infinite Taylor series,

$$H(y) = \sum_{n\geq 0} C_n y^n. \tag{8.64}$$

Then, substituting in the equation for $H(y)$, equation (8.62), we find

$$\sum_{n\geq 0} C_n[n(n-1)y^{n-2} - 2ny^n + (2\tilde{\epsilon} - 1)y^n] = 0. \tag{8.65}$$

But we can redefine the sums above to have the same y^n factor (and verify that in doing so we don't change the $n = 0$ and $n = 1$ terms):

$$\sum_{n\geq 0} y^n[(n+2)(n+1)C_{n+2} - C_n(2\epsilon - 1 - 2n)] = 0. \tag{8.66}$$

Since the y^n are linearly independent functions, we can set all their coefficients to zero, obtaining a recurrence relation for C_n,

$$C_{n+2} = C_n\frac{2\tilde{\epsilon} - 1 - 2n}{(n+2)(n+1)}. \tag{8.67}$$

Here we use the fact that equation (8.62) is a rewriting (by redefining $y(x)$ as $-y^2/2$) of (8.53), which we know has two possible solutions at infinity, $e^{\pm y^2/2}$; so (8.62) also has two possible solutions at infinity, one polynomial, and one $\sim e^{+y^2}$. Generically, then, the series $H(y) = \sum_{n\geq 0} C_n y^n$

would lead, if all C_ns were nonzero all the way to infinity, to $\sim e^{+y^2}$ behavior. To avoid that possibility, we require the Taylor series to terminate at a finite n, so that $H(y)$ is a Hermite polynomial $H_n(x)$. Given the above recurrence relation, this is only possible if $C_{n+2}/C_n = 0$ for some n, implying indeed that

$$\tilde{\epsilon}_n = n + \frac{1}{2}, \quad n \in \mathbb{N}. \tag{8.68}$$

For such an n, we obtain the eigenfunction

$$\psi_n(x) = A_n e^{-m\omega x^2/2\hbar} H_n\left(\sqrt{\frac{m\omega}{\hbar}} x\right), \tag{8.69}$$

where A_n is a normalization constant, which can be found from the condition

$$\int_{-\infty}^{+\infty} dx |\psi_n(x)|^2 = 1, \tag{8.70}$$

leading to

$$A_n = \left[\frac{m\omega}{\pi\hbar 2^{2n}(n!)^2}\right]^{1/4}. \tag{8.71}$$

8.6 Alternative to $|x\rangle$ Representation: Basis Change from $|n\rangle$ Representation

As an alternative to the above derivation, we could go from the $|n\rangle$ to the $|x\rangle$ basis directly. First, we consider the ground state, defined by $\hat{a}|0\rangle = 0$. Introducing the completeness relation $\hat{\mathbb{1}} = \int dx'|x'\rangle\langle x'|$ on the right-hand side of $\hat{a}$, multiplying with $\langle x|$, and defining the ground state wave function

$$\langle x|0\rangle \equiv \psi_0(x), \tag{8.72}$$

we obtain

$$0 = \int dx' \langle x|\hat{a}|x'\rangle \psi_0(x'). \tag{8.73}$$

Using the matrix element

$$\begin{aligned} \langle x|\hat{a}|x'\rangle &= \frac{1}{\sqrt{2}}\langle x|\left(\sqrt{\frac{m\omega}{\hbar}}\hat{X} + \frac{1}{\sqrt{m\omega\hbar}} i\hat{P}\right)|x'\rangle = \frac{1}{\sqrt{2}}\delta(x-x')\left(\sqrt{\frac{m\omega}{\hbar}}x' + \frac{1}{\sqrt{m\omega\hbar}}\hbar\frac{d}{dx'}\right) \\ &= \frac{\delta(x-x')}{\sqrt{2}}\left(y' + \frac{d}{dy'}\right), \end{aligned} \tag{8.74}$$

where we have defined $y \equiv \sqrt{m\omega/\hbar}x$ as before, we find the equation

$$0 = \int dx' \frac{\delta(x-x')}{\sqrt{2}}\left(y' + \frac{d}{dy'}\right)\psi_0(x') = \frac{1}{\sqrt{2}}\left(y + \frac{d}{dy}\right)\psi_0(y). \tag{8.75}$$

Its solution is

$$\psi_0(x) = A_0 e^{-m\omega x^2/2\hbar}. \tag{8.76}$$

To proceed to the other states, we act with $\langle x|$ on

$$|n\rangle = \frac{(\hat{a}^\dagger)^n}{\sqrt{n!}}|0\rangle, \tag{8.77}$$

and again by inserting a completeness relation on the left of $|0\rangle$, we find

$$\begin{aligned}\psi_n(x) &= \frac{1}{\sqrt{n!}}\int dx'\langle x'|(\hat{a}^\dagger)^n|x'\rangle\psi_0(x') = \frac{A_0}{\sqrt{n!}}\left(\frac{y-d/dy}{\sqrt{2}}\right)^n e^{-y^2/2}\\ &= \frac{A_0}{\sqrt{n!}2^{n/2}}e^{-y^2/2}H_n(x),\end{aligned} \tag{8.78}$$

since the Hermite polynomials can be defined by

$$H_n(y) \equiv e^{y^2/2}\left(y-\frac{d}{dy}\right)^n e^{-y^2/2}. \tag{8.79}$$

This result matches what we found before, including the normalization constant.

8.7 Properties of Hermite Polynomials

The Hermite polynomials form an orthonormal set with respect to the weight e^{y^2}, and so have scalar product

$$\int_{-\infty}^{+\infty} H_n(y)H_m(y)e^{-y^2}dy = \delta_{nm}(2^n n!\sqrt{\pi}). \tag{8.80}$$

They can be defined by

$$H_n(x) = (-1)^n e^{x^2}\left(\frac{d}{dx}\right)^n e^{-x^2}, \tag{8.81}$$

have parity change given by the order n,

$$H_n(-x) = (-1)^n H_n(x), \tag{8.82}$$

and satisfy the recurrence relations

$$\begin{aligned}H_{n+1}(x) - 2xH_n(x) + 2nH_{n-1}(x) &= 0\\ H_n'(x) &= 2nH_{n-1}(x).\end{aligned} \tag{8.83}$$

They also form a complete set on the space of (doubly differentiable) functions on the interval $(-\infty,+\infty)$ with weight e^{x^2}, $L^2_{e^{x^2}}(\mathbb{R})$, since they satisfy

$$\sum_{n\geq 0}\psi_n(x)\psi_n(x) = \delta(x'-x). \tag{8.84}$$

They can be obtained by Taylor expansion from the generating function

$$e^{-s^2+2sz} = \sum_{n\geq 0}\frac{s^n}{n!}H_n(z). \tag{8.85}$$

8.8 Mathematical Digression (Appendix): Classical Orthogonal Polynomials

A set of polynomials $\{P_n(x)\}$ constitutes a set of classical orthogonal polynomials with weight ρ on the interval (a, b) if

$$\int_a^b \rho(x) P_n(x) P_m(x) = \delta_{nm} \tag{8.86}$$

and ρ satisfies

$$[\sigma(x)\rho(x)]' = \tau(x)\rho(x), \tag{8.87}$$

where

$$\sigma(x) = \begin{cases} (x-a)(x-b), & a, b \text{ finite} \\ x - a, & b = \infty,\ a \text{ finite} \\ b - x, & a = -\infty,\ b \text{ finite}, \end{cases} \tag{8.88}$$

$\tau(x)$ is a linear function and

$$\rho(x) = \begin{cases} (x-a)^\beta (b-x)^\alpha, \quad \alpha = -\dfrac{\tau(b)}{b-a} - 1, \quad \beta = \dfrac{\tau(a)}{b-a} - 1, & a, b \text{ finite} \\ (x-a)^\beta e^{x\tau'(x)}, \quad \beta = \tau(a) - 1, & b = \infty, a \text{ finite} \\ (b-x)^\alpha e^{-x\tau'(x)}, \quad \alpha = -\tau(b) - 1, & a = -\infty, b \text{ finite} \\ \exp\left[\int \tau(x) dx\right], & (a, b) = (-\infty, +\infty). \end{cases} \tag{8.89}$$

Then, one can prove the limits

$$\begin{aligned} \lim_{x \to a} x^m \sigma(x) \rho(x) &= 0 \\ \lim_{x \to b} x^m \sigma(x) \rho(x) &= 0. \end{aligned} \tag{8.90}$$

One can take the general classical polynomials to their canonical forms, by making changes of variables for x and P_n.

(1) The case where (a, b) are finite

In this case, we can change variables to set $(a, b) \to (-1, 1)$ and obtain

$$\begin{aligned} \rho(t) &= (1-t)^\alpha (1+t)^\beta \\ \sigma(t) &= 1 - t^2 \\ \tau(t) &= -(\alpha + \beta + 2)t + \beta - \alpha. \end{aligned} \tag{8.91}$$

The resulting classical orthogonal polynomials are the *Jacobi polynomials*.

In the particular case of $\alpha = \beta = 0$ (so that $\rho(t) = 1$ is trivial), we obtain for $P_n(x)$ the Legendre polynomials.

In the particular case $\alpha = \beta = +1/2$ and $\alpha = \beta = -1/2$, we obtain the Chebyshev polynomials of the first kind, $T_n(x)$, and second kind, $U_n(x)$. In the cases $\alpha = \beta = \lambda - 1/2$, we obtain the Gegenbauer polynomials $G_n(x)$.

(2) The cases $(-\infty, b)$ and $(a, +\infty)$

In these cases, we change variables (a, b) to $(0, +\infty)$. We obtain the Laguerre polynomials $L_n^{\alpha}(x)$, with

$$\begin{aligned} \rho(t) &= t^{\alpha} e^{-t} \\ \sigma(t) &= t \\ \tau(t) &= -t + \alpha + 1. \end{aligned} \tag{8.92}$$

(3) The case $(a, b) = (-\infty, +\infty)$

In this case, we can put

$$\begin{aligned} \rho(t) &= e^{-t^2} \\ \sigma(t) &= 1 \\ \tau(t) &= -2t, \end{aligned} \tag{8.93}$$

and the resulting polynomials are the Hermite polynomials.

Properties of the Classical Orthogonal Polynomials

The only orthogonal polynomials whose derivatives are also orthogonal polynomials are the classical orthogonal polynomials, which we are describing in this appendix here.

They satisfy the eigenvalue equation

$$\frac{d}{dx}\left(\sigma(x)\rho(x)\frac{d}{dx}P_n(x)\right) + \lambda_n \rho(x) P_n(x) = 0, \tag{8.94}$$

where the eigenvalue is

$$\lambda_n = -n\left[\tau'(x) + \frac{n-1}{2}\sigma''(x)\right]. \tag{8.95}$$

They are also given by Rodrigues' formula,

$$P_n(x) = A_n \frac{1}{\rho(x)} \frac{d^n}{dx^n}\left(\sigma^n(x)\rho(x)\right), \tag{8.96}$$

and they form a basis for the Hilbert space in $L^2_\rho(a, b)$, so that

$$\sum_{n\geq 0} P_n(x) P_n(x') = \delta(x - x'). \tag{8.97}$$

They satisfy the recurrence formulas

$$xP_n(x) = \alpha_n P_{n+1}(x) + \beta_n P_n(x) + \gamma_n P_{n-1}(x), \tag{8.98}$$

valid for a general orthogonal polynomial, and

$$\begin{aligned}\sigma(x)P'_n(x) &= \alpha_n^1 P_{n+1}(x) + (\beta_n^1 + \gamma_n^1 x)P_n(x)\\ P_n(x) &= \alpha_n^2 P'_{n+1}(x) + (\beta_n^2 + \gamma_n^2 x)P'_n(x),\end{aligned} \tag{8.99}$$

valid only for classical orthogonal polynomials.

The generating function $K(x,t)$ can be Taylor expanded to give the classical orthogonal polynomials,

$$K(x,t) = \sum_{n\geq 0} \frac{\tilde{P}_n(x)}{n!} t^n, \tag{8.100}$$

and it can be calculated from the formula

$$K(x,t) = \frac{1}{\rho(x)} \frac{\rho(z_1)}{1 - t\sigma'(z_1)}, \tag{8.101}$$

where z_1 is the closest zero to x of the function

$$f(z) = z - x - \sigma(z)t. \tag{8.102}$$

For the Legendre polynomials $P_n(x)$, we have

$$\begin{aligned}P_n(x) &= \frac{1}{2^n n!}\frac{d^n}{dx^n}[(x^2-1)^n]\\ \frac{1}{\sqrt{1-2tx+t^2}} &= \sum_{n\geq 0} P_n(x)t^n.\end{aligned} \tag{8.103}$$

For the Hermite polynomials $H_n(x)$, we have

$$\begin{aligned}H_n(x) &= (-1)^n e^{x^2}\frac{d^n}{dx^n}(e^{-x^2})\\ e^{2tx-x^2} &= \sum_{n\geq 0}\frac{t^n}{n!}H_n(x).\end{aligned} \tag{8.104}$$

For the Laguerre polynomials $L_n^\alpha(x)$, we have

$$K(x,t) = \frac{e^{tx/t-1}}{(1-t)^{\alpha+1}}. \tag{8.105}$$

Important Concepts to Remember

- The harmonic oscillator is an important prototype for a quantum mechanical system, which can be used to test formalisms and to approximate systems to the quadratic order.
- One can define its quantum mechanics in terms of creation and annihilation operators $\hat{a}^\dagger, \hat{a}$, satisfying $[\hat{a}, \hat{a}^\dagger] = 1$, which create and annihilate quanta of frequency ω. This is also called the occupation number basis.
- The states of n quanta are created by acting with $\hat{a}^\dagger$ n times on the vacuum $|0\rangle$, satisfying $\hat{a}|0\rangle = 0$, i.e., $|n\rangle = [(\hat{a}^\dagger)^n/\sqrt{n!}]|0\rangle$; their number is measured by $N = \hat{a}^\dagger\hat{a}$, $N|n\rangle = n|n\rangle$ and their energy is $E_n = (n+1/2)\hbar\omega$.

- Coherent states $|\alpha\rangle$ are eigenstates of the annihilation operator $\hat{a}$, $\hat{a}|\alpha\rangle = \alpha|\alpha\rangle$, and are written as $|\alpha\rangle = e^{\alpha\hat{a}^\dagger}|0\rangle$.
- The solution of the harmonic oscillator in the $|x\rangle$ (coordinate) representation is obtained by the Sommerfeld method, by finding the behavior of the harmonic oscillator equation, and of its solution, at the limits of the domain (here $x \to 0$ and $x \to \infty$), and factoring this out of the general solution, which usually reduces the solution to a polynomial.
- In the Sommerfeld method, one usually obtains a quantization condition by imposing that the unique normalizable solution at one end of the domain matches the unique normalizable solution at the other end, so that an infinite series reduces to a polynomial, usually a classical orthogonal one.
- For the harmonic oscillator, the Sommerfeld method leads to the Hermite polynomials and so the solution $\psi(y) = e^{-y^2/2}H_n(y)$, with $y = x\sqrt{m\omega/\hbar}$.
- Classical orthogonal polynomials are polynomials that are orthogonal on an interval (a, b), with a weight $\rho(x)$, whose derivatives are also orthogonal polynomials.
- In the case of a, b finite, redefining the domain to $(-1, 1)$ we have the Jacobi polynomials; in the case of $(a, +\infty)$ or $(-\infty, b)$ we redefine the domain to $(0, +\infty)$, obtaining the Laguerre polynomials, and in the case $(-\infty, +\infty)$ we have the Hermite polynomials.

Further Reading

See [2] for more details.

Exercises

(1) Consider the Lagrangian

$$L = \frac{\dot{q}_1^2}{2} + \frac{\dot{q}_2^2}{2} - \frac{\alpha_1}{2}\sin^2(\beta_1 q_1) - \frac{\alpha_2}{2}\sin^2(\beta_2 q_2) - \frac{\alpha_{12}}{2}\sin^2(\beta_{12}(q_1 + q_2)). \tag{8.106}$$

Approximate the system by two harmonic oscillators, and quantize it in the creation and annihilation operator (occupation number) representation.

(2) Consider the Hamiltonian for a (large) number N of oscillators $\hat{a}_i, \hat{a}_i^\dagger$ (with $[\hat{a}_i, \hat{a}_j^\dagger] = \delta_{ij}$):

$$H = \sum_{i=1}^{N}\left[\hat{a}_i^\dagger\hat{a}_i + \left(\alpha\hat{a}_{i+1}^\dagger\hat{a}_i + \text{h.c.}\right)\right], \tag{8.107}$$

where $\hat{a}_{N+1} \equiv \hat{a}_1, \hat{a}_{N+1}^\dagger \equiv \hat{a}_1$. Diagonalize it and find its eigenstates.

(3) Show that the completeness relation for coherent states is (8.48).

(4) Calculate (in terms of a single ket state and no operators)

$$\exp\left\{i[a^\dagger a + \beta(\hat{a} + \hat{a}^\dagger)^3]\right\}|\alpha\rangle. \tag{8.108}$$

(5) Calculate $\langle x^2\rangle_n$ and $\langle x^3\rangle_n$ in the state $|n\rangle$ of the harmonic oscillator.

(6) Use the Sommerfeld method for a particle in one dimension with a quartic potential, instead of a quadratic one, $V(x) = \lambda x^4$. What is the resulting reduced equation, and can you describe the quantization condition for bound states?

(7) Consider two harmonic oscillators, one with frequency ω and one with frequency 2ω. Calculate the *symmetric* wave function (in x space) corresponding to the energy $E = (15/2)\hbar\omega$.

9 The Heisenberg Picture and General Picture; Evolution Operator

Until now, we have considered the Schrödinger equation as acting on states $|\psi_S(t)\rangle$ which are time dependent, with operators that are time independent. This is known as the *Schrödinger picture*. However, we can consider other "pictures" with which to describe quantum mechanics, for instance one in which the *operators* are time dependent but states are time independent; this is known as the *Heisenberg picture*.

9.1 The Evolution Operator

In order to define the Heisenberg picture, we remember that in the Schrödinger picture the time evolution was packaged in a single operator, the "time evolution" operator $\hat{U}(t, t_0)$, defined in the Schrödinger picture by

$$|\psi_S(t)\rangle = \hat{U}(t, t_0)|\psi_S(t_0)\rangle. \tag{9.1}$$

Indeed, in this usual, Schrödinger, picture we can define the evolution operator as follows. We can note that the evolution from $|\psi(t_0)\rangle$ to $|\psi(t)\rangle$ is a linear operation, preserving in time the linear superposition of states, so we can define this time evolution as the result of an operator $\hat{U}(t, t_0)$ as in the relation (9.1) above. The relation must be a unitary transformation, $\hat{U}(t, t_0)^\dagger = \hat{U}^{-1}(t, t_0)$, since it must preserve the norm of the states in time,

$$\langle\psi(t)|\psi(t)\rangle = \langle\psi(t_0)|\psi(t_0)\rangle, \tag{9.2}$$

because the norm is associated with probabilities, and conservation of the total probability must be imposed.

We also saw that, for a *conservative* system, for which $\hat{H} \neq \hat{H}(t)$, on eigenstates $|\psi_E(t)\rangle$ of the Hamiltonian,

$$\hat{H}|\psi_E(t)\rangle = E|\psi_E(t)\rangle, \tag{9.3}$$

the time evolution of states is obtained as

$$|\psi_E(t)\rangle = e^{-iE(t-t_0)/\hbar}|\psi_E(t_0)\rangle = e^{-i\hat{H}(t-t_0)/\hbar}|\psi_E(t_0)\rangle, \tag{9.4}$$

so the time evolution operator is

$$\hat{U}(t, t_0) = e^{-i\hat{H}(t-t_0)/\hbar}. \tag{9.5}$$

In the energy eigenstates basis, the completeness relation is written as $\hat{\mathbb{1}} = \sum_{n,a} |E_{n,a}\rangle\langle E_{n,a}|$, so, inserting it in front of the above relation, the time evolution operator can be written also as

$$\hat{U}(t, t_0) = \sum_{n,a} |E_{n,a}\rangle\langle E_{n,a}| e^{-iE(t-t_0)/\hbar}. \tag{9.6}$$

Moreover, we can write a differential equation for this time evolution operator by acting on (9.5) with d/dt, obtaining

$$i\hbar\frac{d}{dt}\hat{U}(t,t_0) = \hat{H}\hat{U}(t,t_0). \tag{9.7}$$

The difference is that now we can *define* the time evolution operator to be that found from the above differential equation, together with the boundary condition $\hat{U}(t_0,t_0) = \hat{\mathbb{1}}$, even in the case of a nonconservative system, with $\hat{H} = \hat{H}(t)$. In this latter case, we can "integrate" the equation over a very small interval $t - t_0$, and write

$$\hat{U}(t,t_0) = \hat{\mathbb{1}} - \frac{i}{\hbar}\int_{t_0}^{t} \hat{H}(t')\hat{U}(t',t_0)dt'. \tag{9.8}$$

This is in fact equivalent to the action of the Schrödinger equation acting on states,

$$i\hbar\frac{d}{dt}|\psi(t)\rangle = \hat{H}|\psi(t)\rangle, \tag{9.9}$$

since by integrating it we obtain the same time evolution operator:

$$|\psi(t+dt)\rangle = \left(1 - \frac{i}{\hbar}\hat{H}dt\right)|\psi(t)\rangle = \hat{U}(t+dt,t)|\psi(t)\rangle \simeq e^{-i\hat{H}(t)dt/\hbar}|\psi(t)\rangle. \tag{9.10}$$

We note that in this expression, since we have a Hermitian Hamiltonian, $\hat{H}^\dagger = \hat{H}$, the evolution operator is indeed unitary, $\hat{U}^\dagger = \hat{U}^{-1}$, so that finite time evolution amounts to an infinite sequence of unitary infinitesimal operations, leading to a total unitary operation. The final state, after this sequence of operations, each one centered on a t_n, is

$$|\psi(t)\rangle = \prod_{i=1}^{n} \exp\left[-\frac{i}{\hbar}\hat{H}(t_n)dt_n\right]|\psi(t_0)\rangle. \tag{9.11}$$

We need to use this complicated formula since, in general, the Hamiltonians at different times need not commute,

$$[\hat{H}(t_1),\hat{H}(t_2)] \neq 0, \tag{9.12}$$

so if we were to write an integral $\int H(t')dt'$ in the exponent instead of the product of infinitesimal exponentials, this would be in general different from the above product of exponentials. Therefore we have to define the integral in the exponential only with a *time ordering operator*, which puts products of operators in time order: $T(A(t_1)\cdots A(t_n)) = A(t_1)\cdots A(t_n)$ if $t_1 > t_2 > \cdots t_n$. Then

$$|\psi(t)\rangle = T\left\{\exp\left[-\frac{i}{\hbar}\int_{t_0}^{t}\hat{H}(t')dt'\right]\right\}|\psi(t_0)\rangle, \tag{9.13}$$

where the *time-ordered exponentials* are defined as

$$T\left\{\exp\left[-\frac{i}{\hbar}\int_{t_0}^{t}\hat{H}(t')dt'\right]\right\} = \lim_{N\to\infty}\prod_{n=0}^{N-1}\exp\left[-i\hat{H}(t'_n)dt'_n\right]. \tag{9.14}$$

Noncommuting Exponentials

For such noncommuting exponentials, we have the following theorem:

$$e^{\hat{A}+\hat{B}} = e^{\hat{A}}e^{\hat{B}}\exp\left(-\frac{[\hat{A},\hat{B}]}{2}\right), \tag{9.15}$$

which can be proven as follows. Consider the function

$$f(x) = e^{\hat{A}x} e^{\hat{B}x}, \tag{9.16}$$

and consider its derivative,

$$\begin{aligned} \frac{df}{dx} &= \hat{A} e^{\hat{A}x} e^{\hat{B}x} + e^{\hat{A}x} \hat{B} e^{\hat{B}x} \\ &= \left(\hat{A} + e^{\hat{A}x} \hat{B} e^{-\hat{A}x}\right) f(x). \end{aligned} \tag{9.17}$$

On the other hand, we also find that

$$[B, e^{-Ax}] = \sum_{n \geq 0} \frac{(-x)^n}{n!} [B, A^n] = -x \sum_{n \geq 0} \frac{(-x)^{n-1} n}{n!} A^{n-1} [B, A] = -x e^{-Ax} [B, A], \tag{9.18}$$

allowing us to write

$$\frac{df}{dx} = (\hat{A} + \hat{B} + [\hat{A}, \hat{B}]x) f(x). \tag{9.19}$$

This differential equation can be integrated, with boundary condition $f(0) = 1$, to obtain

$$f(x) = e^{(\hat{A}+\hat{B})x} e^{x^2/2[\hat{A},\hat{B}]}. \tag{9.20}$$

Putting $x = 1$, we arrive at the stated equation.

9.2 The Heisenberg Picture

Having defined the evolution operator, it follows that we can define time-independent states, which by definition will be the states in the Heisenberg picture, by reversing the evolution operator, so these states, denoted by $|\psi_H\rangle$, will be

$$|\psi_H\rangle = \hat{U}^{-1}(t, t_0) |\psi_S(t)\rangle = |\psi_S(t_0)\rangle. \tag{9.21}$$

This change in the space of states can be thought of as a "unitary transformation", or change of basis in the Hilbert space, with the unitary operator $U^{-1}(t, t_0) = U^\dagger(t, t_0)$. Through it, the time dependence encoded in $\hat{U}(t, t_0)$ can be put onto the operators, since the change of basis implies that the operators are also changed from the Schrödinger picture operators $\hat{A}_S$ to the Heisenberg picture operators $\hat{A}_H$, as

$$\hat{A}_S \to \hat{A}_H(t) = \hat{U}^\dagger(t, t_0) \hat{A}_S \hat{U}(t, t_0), \tag{9.22}$$

in such a way that the matrix elements (or expectation values) are unchanged,

$$\langle \psi_S | \hat{A}_S | \chi_S \rangle = \langle \psi_H | \hat{A}_H | \chi_H \rangle. \tag{9.23}$$

Note therefore that when we write $\hat{U}(t, t_0)$, we mean the evolution operator in the Schrödinger picture, $\hat{U}_S(t, t_0)$.

The time evolution of the new operators is found by using the differential equation for $\hat{U}(t, t_0)$ in equation (9.7), and its complex conjugate, in the time derivative of the definition of $\hat{A}_H$ (acting on three terms), obtaining

$$
\begin{aligned}
i\hbar\frac{d}{dt}\hat{A}_H &= -\hat{U}_S^\dagger(t,t_0)\hat{H}_S\hat{A}_S\hat{U}(t,t_0) + i\hbar\hat{U}^\dagger\frac{\partial\hat{A}_S}{\partial t}\hat{U} + \hat{U}^\dagger\hat{A}_S\hat{H}_S\hat{U} \\
&= \hat{U}^\dagger[\hat{A}_S,\hat{H}]\hat{U} + i\hbar\hat{U}^\dagger\frac{\partial\hat{A}_S}{\partial t}\hat{U}.
\end{aligned}
\tag{9.24}
$$

Defining the Hamiltonian in the Heisenberg representation,

$$
\hat{H}_H = \hat{U}^\dagger\hat{H}_S\hat{U}, \tag{9.25}
$$

in such a way that

$$
\hat{U}^\dagger[\hat{A}_S,\hat{H}_S]\hat{U} = [\hat{A}_H,\hat{H}_H], \tag{9.26}
$$

and if the operators have *explicit* time dependence $\partial/\partial t\hat{A}_S \neq 0$, such that

$$
\frac{\partial\hat{A}_H}{\partial t} = \hat{U}^\dagger\frac{\partial\hat{A}_S}{\partial t}\hat{U}, \tag{9.27}
$$

we obtain the equation for the time evolution of the Heisenberg operators,

$$
i\hbar\frac{d\hat{A}_H}{dt} = [\hat{A}_H,\hat{H}_H] + i\hbar\frac{\partial}{\partial t}\hat{A}_H. \tag{9.28}
$$

This *Heisenberg equation of motion* replaces the Schrödinger equation, since now operators have time dependence, not the states (now $\partial_t|\psi_H\rangle = 0$ always).

This equation of motion is the quantum mechanical counterpart of the classical equation in the Hamiltonian formalism, in terms of Poisson brackets,

$$
\frac{dA_{\rm cl}}{dt} = \{A_{\rm cl}, H_{\rm cl}\}_{P.B.} + \frac{\partial A_{\rm cl}}{\partial t}. \tag{9.29}
$$

This is the generalization of the Hamilton equations written with Poisson brackets,

$$
\begin{aligned}
\frac{dq_i}{dt} &= \{q_i, H\}_{P.B.} = \frac{\partial H}{\partial p_i} \\
\frac{dp_i}{dt} &= \{p_i, H\}_{P.B.} = -\frac{\partial H}{\partial q_i}.
\end{aligned}
\tag{9.30}
$$

For the quantum mechanical case, the above Hamilton equations become the Heisenberg equations of motion,

$$
\begin{aligned}
\frac{d\hat{q}_i(t)}{dt} &= -\frac{i}{\hbar}[\hat{q}_i,\hat{H}] \\
\frac{d\hat{p}_i(t)}{dt} &= -\frac{i}{\hbar}[\hat{p}_i,\hat{H}].
\end{aligned}
\tag{9.31}
$$

9.3 Application to the Harmonic Oscillator

We will use as the simplest example our standard toy model, the harmonic oscillator.

The quantum mechanical Heisenberg equations for $\hat{a}, \hat{a}^\dagger$, taking the role of the Hamilton equations, are

$$i\hbar\frac{d\hat{a}}{dt} = [\hat{a}, \hat{H}] = \hbar\omega\hat{a}$$
$$i\hbar\frac{d\hat{a}^\dagger}{dt} = [\hat{a}^\dagger, \hat{H}] = -\hbar\omega\hat{a}^\dagger, \tag{9.32}$$

where we have used $\hat{H} = \hbar\omega(\hat{a}^\dagger\hat{a} + 1/2)$ and $[\hat{a}, \hat{a}^\dagger] = 1$.

The solution of these equations is

$$\hat{a}(t) = \hat{a}_0 e^{-i\omega t}$$
$$\hat{a}^\dagger(t) = \hat{a}_0^\dagger e^{+i\omega t}, \tag{9.33}$$

which implies for the phase space variables $\hat{q}(t), \hat{p}(t)$ that

$$\hat{q}(t) = \sqrt{\frac{\hbar}{2m\omega}}\left(\hat{a}_0^\dagger e^{i\omega t} + \hat{a}_0 e^{-i\omega t}\right)$$
$$\hat{p}(t) = i\sqrt{\frac{m\hbar\omega}{2}}\left(\hat{a}_0^\dagger e^{i\omega t} - \hat{a}_0 e^{-i\omega t}\right). \tag{9.34}$$

9.4 General Quantum Mechanical Pictures

The Schrödinger and Heisenberg pictures are the best known ones, but there are others. Here we write a general analysis for such pictures.

To gain an understanding, we look to classical mechanics in the Hamiltonian and Poisson bracket formalism, as we did above for the Heisenberg picture case.

In classical mechanics, there are *canonical transformations* on phase space, changing (p, q) to (p', q'), under which functions of phase space $F(q, p)$ and $G(q, p)$, with Poisson brackets

$$\{F(q, p), G(q, p)\}_{P.B.} = K(q, p) \tag{9.35}$$

transform to functions $F'(q', p')$ and $G'(q', p')$ that obey the same relation,

$$\{F'(q', p'), G'(q', p')\}_{P.B.} = K'(q', p'). \tag{9.36}$$

In quantum mechanics, the role of such canonical transformations on phase space is taken by unitary transformations on the Hilbert space,

$$|\psi'\rangle = \hat{U}|\psi\rangle, \tag{9.37}$$

with inverse $|\psi\rangle = \hat{U}^{-1}|\psi'\rangle$, and under which the operators also transform,

$$\hat{A}' = \hat{U}\hat{A}\hat{U}^{-1}. \tag{9.38}$$

Since quantization involves the replacement of $\{,\}_{P.B.}$ with $1/i\hbar[,]$, we must obtain for the commutator a transformation law similar to that for Poisson brackets in classical mechanics. Indeed, we find

$$[\hat{F}, \hat{G}] \to [\hat{F}', \hat{G}'] = \hat{U}[\hat{F}, \hat{G}]\hat{U}^{-1} = \hat{U}\hat{K}\hat{U}^{-1} = \hat{K}'. \tag{9.39}$$

Consider a unitary transformation by an operator $\hat{W}(t)$ that can depend explicitly on time,

$$|\psi_W\rangle = \hat{W}(t)|\psi\rangle, \quad \hat{A}_W = \hat{W}(t)\hat{A}\hat{W}^{-1}(t). \tag{9.40}$$

The Hamiltonian generating the new time dependence is found from the Schrödinger equation:

$$\begin{aligned} i\hbar\partial_t|\psi\rangle &= \hat{H}|\psi\rangle \Rightarrow \\ i\hbar\partial_t|\psi_W\rangle &= i\hbar\partial_t(\hat{W}|\psi\rangle) = i\hbar\hat{W}\partial_t|\psi\rangle + i\hbar(\partial_t\hat{W})|\psi\rangle \\ &= (\hat{W}\hat{H}\hat{W}^{-1})|\psi_W\rangle + i\hbar(\partial_t\hat{W})\hat{W}^\dagger|\psi_W\rangle, \end{aligned} \tag{9.41}$$

where in the last line we have used the Schrödinger equation for $|\psi\rangle$. Equating the final result to $H'|\psi_W\rangle$ allows us to write $\hat{H}'$, the Hamiltonian generating the new time evolution, as

$$\hat{H}' = \hat{H}_W + i\hbar(\partial_t\hat{W})\hat{W}^\dagger. \tag{9.42}$$

The time evolution of the transformed operators $\hat{A}_W = \hat{W}\hat{A}\hat{W}^\dagger$ is found as before, by acting on each term:

$$\begin{aligned} i\hbar\frac{\partial}{\partial t}\hat{A}_W &= i\hbar\hat{W}\left(\frac{\partial\hat{A}}{\partial t}\right)\hat{W}^\dagger + i\hbar(\partial_t\hat{W})\hat{A}\hat{W}^{-1} + i\hbar\hat{W}\hat{A}\partial_t\hat{W}^{-1} \\ &= i\hbar\left(\frac{\partial}{\partial t}\hat{A}\right)_W + i\hbar(\partial_t\hat{W})\hat{W}^\dagger\hat{A}_W - i\hbar\hat{A}_W(\partial_t\hat{W})\hat{W}^{-1}, \end{aligned} \tag{9.43}$$

which finally becomes

$$i\hbar\frac{\partial}{\partial t}\hat{A}_W = i\hbar(\partial_t\hat{A})_W + [i\hbar(\partial_t\hat{W})\hat{W}^\dagger, \hat{A}_W]. \tag{9.44}$$

From the above, we can also find the new time evolution operator. Since in the Schrödinger picture we had $|\psi(t)\rangle = \hat{U}(t,t_0)|\psi(t_0)\rangle$, in the new W picture we find

$$|\psi_W(t)\rangle = \hat{W}(t)|\psi(t)\rangle = \hat{W}(t)\hat{U}(t,t_0)\hat{W}^\dagger(t_0)|\psi_W(t_0)\rangle, \tag{9.45}$$

from which we deduce that the time evolution operator in the W picture is

$$\hat{U}'(t,t_0) = \hat{W}(t)\hat{U}(t,t_0)\hat{W}^\dagger(t_0) \neq \hat{U}_W. \tag{9.46}$$

Indeed, note that the new evolution operator is not simply the Schrödinger operator transformed into the W picture, because we have $\hat{W}(t)$ to the left but $\hat{W}^\dagger(t_0)$ to the right (at a different time t_0).

Application to the Heisenberg Picture

Note that we can define the Heisenberg picture by the condition that states are time independent, $\partial_t|\psi_H\rangle = 0$, so the new Hamiltonian must be trivial, $H' = 0$. But that means that the Schrödinger Hamiltonian in the new picture is

$$\hat{H}_H = \hat{H}_S = -i\hbar(\partial_t\hat{W})\hat{W}^\dagger. \tag{9.47}$$

By choosing the two pictures, the W picture and the Schrödinger picture, to be equal at time t_0, we finally obtain

$$\hat{W}(t) = \hat{U}_S^{-1}(t,t_0) = \hat{U}_S(t_0,t). \tag{9.48}$$

9.5 The Dirac (Interaction) Picture

The most interesting new example of a picture that remains is the Dirac, or interaction, picture. This is defined in the case where the Hamiltonian H can be split into a free particle part $\hat{H}_0$ and an interaction part $\hat{H}_1$,

$$\hat{H} = \hat{H}_0 + \hat{H}_1. \tag{9.49}$$

We can then use just the free part, $\hat{H}_0$, in order to go to a sort of Heisenberg picture for $\hat{H}_0$, i.e., using a unitary transformation $\hat{W}(t)$ defined by

$$\hat{W}(t) = \hat{U}_{S,0}^{-1}(t,t_0) = \left\{\exp\left[-i\frac{\hat{H}_0(t-t_0)}{\hbar}\right]\right\}^{-1} = \exp\left[+i\frac{\hat{H}_0(t-t_0)}{\hbar}\right]. \tag{9.50}$$

In this case, we obtain that the interaction picture states $|\psi_I(t)\rangle$ and operators $\hat{A}(t)$ *both* depend on time, and their time dependences are given by

$$\begin{aligned} i\hbar\frac{\partial}{\partial t}|\psi_I(t)\rangle &= H_{1,I}|\psi_I(t)\rangle \\ i\hbar\frac{\partial}{\partial t}\hat{A}_I(t) &= [\hat{A}_I(t), \hat{H}_0]\left(+i\hbar(\partial_t\hat{A}_S)_I\right). \end{aligned} \tag{9.51}$$

The evolution operator in the interaction picture is, according to the general theory,

$$\hat{U}_I(t,t_0) = \hat{W}(t)\hat{U}(t,t_0)\hat{W}^{-1}(t_0), \tag{9.52}$$

for the canonical transformation

$$\hat{W}(t) = e^{i\hat{H}_0(t-t_0)/\hbar}, \tag{9.53}$$

so that $\hat{W}(t_0) = \mathbb{1}$. In the conservative case, of a time-independent Hamiltonian, $\hat{U}(t,t_0) = e^{-i\hat{H}(t-t_0)/\hbar}$, so that

$$\hat{U}_I(t,t_0) = e^{i\hat{H}_0(t-t_0)/\hbar}e^{-i\hat{H}(t-t_0)/\hbar}. \tag{9.54}$$

We can write a differential equation for it, using the Schrödinger picture equation derived before, $i\hbar\frac{d}{dt}\hat{U}(t,t_0) = \hat{H}\hat{U}(t,t_0)$. We find

$$\begin{aligned} i\hbar\frac{d}{dt}\hat{U}_I(t,t_0) &= i\hbar\frac{d}{dt}[\hat{W}(t)\hat{U}(t,t_0)] = i\hbar\left(\frac{d}{dt}\hat{W}(t)\right)\hat{W}^{-1}(t)\hat{U}_I(t,t_0) + \hat{W}(t)i\hbar\frac{d}{dt}\hat{U}(t,t_0) \\ &= [i\hbar(\partial_t\hat{W}(t))\hat{W}^\dagger(t)]\hat{U}_I(t,t_0) + \hat{W}(t)\hat{H}\hat{W}^{-1}(t)U_I(t,t_0) \\ &= \hat{H}_I\hat{U}_I(t,t_0), \end{aligned} \tag{9.55}$$

where $\hat{H}_I$ is the Hamiltonian giving the time evolution in the interaction picture.

Moreover, since

$$\hat{U}_I(t,t_0) = e^{i\hat{H}_0(t-t_0)/\hbar} e^{-i\hat{H}(t-t_0)/\hbar}, \tag{9.56}$$

it follows that by differentiation we get

$$i\hbar\partial_t \hat{U}_I(t,t_0) = e^{i\hat{H}_0(t-t_0)/\hbar}(\hat{H} - \hat{H}_0)e^{-i\hat{H}(t-t_0)/\hbar}, \tag{9.57}$$

where, when differentiating both factors, we have chosen to write the resulting Hamiltonian in the middle of the two exponentials. Then, using the fact that $\hat{H} - \hat{H}_0 = \hat{H}_1$ in the Schrödinger picture, and that in the interaction picture we have

$$H_{1,I} = e^{i\hat{H}_0(t-t_0)} H_{1,S} e^{-i\hat{H}_0(t-t_0)}, \tag{9.58}$$

where $H_{1,I}$ is H_1 in the interaction picture, we derive that

$$i\hbar\partial_t \hat{U}_I(t,t_0) = H_{1,I} e^{i\hat{H}_0(t-t_0)} e^{-i\hat{H}(t-t_0)} = \hat{H}_{1,I}\hat{U}_I(t,t_0). \tag{9.59}$$

Comparing the two differential equations, we see that the Hamiltonian in the interaction picture equals $\hat{H}_1$ in the interaction picture,

$$\hat{H}_I = \hat{H}_{1,I}. \tag{9.60}$$

We can solve the differential equation (9.59) for something like $U(t,0) \sim e^{-iH_I t/\hbar}$, by analogy. More precisely, we see that we can write the expansion of the exponential, taking care to integrate only over the *time-ordered* product of Hamiltonians:

$$\hat{U}_I(t,t_0) = 1 + (-i/\hbar)\int_{t_0}^{t} dt_1 \hat{H}_{1,I}(t_1) + (-i/\hbar)^2 \int_{t_0}^{t} dt_1 \int_{t_0}^{t_1} dt_2 H_{1,I}(t_1)H_{1,I}(t_2) + \cdots \tag{9.61}$$

Calling the three terms zeroth-order (1), first-order, and second-order, we see that in the second-order term we are integrating only over a triangle, satisfying the condition of time ordering, $t_1 > t_2$ in $H_{1,I}(t_1)H_{1,I}(t_2)$, instead of dividing by 2, as would happen from the expansion of an exponential. When doing this, we find that, indeed,

$$\begin{aligned} i\hbar\partial_t(\text{first-order}) &= \hat{H}_{1,I}(t)(\text{zeroth-order}) \\ i\hbar\partial_t(\text{second-order} &= \hat{H}_{1,I}(t)(\text{first-order}). \end{aligned} \tag{9.62}$$

However, as we said, the integration over the triangle actually equals half the integration over the whole rectangle $\int_{t_0}^{t} dt_1 \int_{t_0}^{t} dt_2$. But, since in general the interaction Hamiltonians at different times don't commute,

$$[\hat{H}_{1,I}(t), \hat{H}_{1,I}(t')] \neq 0, \tag{9.63}$$

in order to get the correct result we must include a time ordering as well, so

$$\hat{U}_I(t,t_0) = 1 + (-i/\hbar)\int_{t_0}^{t} dt_1 \hat{H}_{1,I}(t_1) + \frac{(-i/\hbar)^2}{2!}\int_{t_0}^{t} dt_1 \int_{t_0}^{t} dt_2 T\{H_{1,I}(t_1)H_{1,I}(t_2)\} + \cdots \tag{9.64}$$

This pattern continues at higher orders, and we find that we must in fact put the time-ordering operator in front of all the terms coming from the exponential, so

$$\hat{U}_I(t,t_0) = T\left\{\exp\left[-i\int_{t_0}^{t} dt' H_{1,I}(t')\right]\right\}. \tag{9.65}$$

This is the same formula as we obtained in the Schrödinger picture in (9.13) but now defined rigorously (and in the interaction picture).

Important Concepts to Remember

- The evolution operator $\hat{U}(t,t_0)$ can be written as the solution of a Schrödinger-like equation, $i\hbar\frac{d}{dt}\hat{U}(t,t_0) = \hat{H}\hat{U}(t,t_0)$, with boundary condition $\hat{U}(t_0,t_0) = \mathbb{1}$, leading to a solution as a time-ordered exponential, $T\left\{\exp\left[-\frac{i}{\hbar}\int_{t_0}^{t} dt'\ H(t')\right]\right\}$.
- Quantum mechanics can be described in different pictures, the usual Schrödinger picture being the one where operators are time independent and states are time dependent (and obey the Schrödinger equation).
- In the Heisenberg picture, states are time independent, $|\psi_H\rangle = \hat{U}^{-1}(t,t_0)|\psi(t)\rangle = |\psi_S(t_0)\rangle$, and operators are time dependent, $\hat{A}_H = \hat{U}^{-1}(t,t_0)\hat{A}_S\hat{U}(t,t_0)$, evolving with the Hamiltonian via the Heisenberg equation of motion, $i\hbar d\hat{A}_H/dt = [\hat{A}_H,\hat{H}_H] + i\hbar\partial_t\hat{A}_H$.
- A general picture is related to the Schrödinger picture by a canonical transformation, $|\psi_W\rangle = \hat{W}(t)|\psi\rangle$, $\hat{A}_W = \hat{W}(t)A\hat{W}^{-1}(t)$, resulting in a new Hamiltonian for the time evolution of states $i\hbar\partial_t|\psi_W\rangle = i\hbar\hat{H}'|\psi_W\rangle$, and a new time evolution of operators, $i\hbar\partial_t\hat{A}_W = i\hbar(\partial_t\hat{A})_W + [i\hbar(\partial_t\hat{W})\hat{W}^{-1},\hat{A}_W]$.
- In the Dirac, or interaction, picture, $\hat{H} = \hat{H}_0 + \hat{H}_1$ and $\hat{W}(t) = \hat{U}_{S,0}^{-1}(t,t_0)$, so it is a sort of Heisenberg picture for $\hat{H}_0$ only, in which states evolve with $\hat{H}_{1,I}$, $i\hbar\partial_t|\psi_I(t)\rangle = \hat{H}_{1,I}|\psi_I(t)\rangle$ (and $\hat{H}_I = \hat{H}_{1,I}$), but operators evolve with $\hat{H}_0$, $i\hbar\partial_t\hat{A}_I(t) = [\hat{A}_I(t),\hat{H}_0]$.
- The evolution operator in the interaction picture is also a time-ordered exponential, $\hat{U}_I(t,t_0) = T\left\{\exp\left[-\frac{i}{\hbar}\int_{t_0}^{t} dt'\ \hat{H}_{1,I}(t')\right]\right\}$.

Further Reading

See [2] and [1] for more details.

Exercises

(1) Write down the time evolution operator and the differential equation that it satisfies for a one-dimensional particle in a potential $V(x)$.

(2) Write down the time evolution equation for Heisenberg operators, calculating explicitly the evolution Hamiltonian, in the case of a free particle, and then calculate explicit expressions for $\hat{x}(t)$ and $\hat{p}(t)$.

(3) For a harmonic oscillator of frequency ω, calculate the explicit time-dependent $a(t), a^\dagger(t)$ operators in the picture with $\hat{W}(t) = \exp\left[2i\omega\hat{a}^\dagger\hat{a}(t-t_0)\right]$, and the explicit differential equations for a general operator and state.

(4) Consider a harmonic oscillator perturbed by $H_1 = \lambda(a+a^\dagger)^3$. Write down the explicit evolution equations for states (the first two terms in the expansion in ω) and the $a, a^\dagger$ operators in the interaction picture.

(5) Calculate the evolution operator for the interaction picture in the case in exercise 4.

(6) Consider a Hamiltonian $\hat{H}(p)$ only (no x dependence). What are the operators in the theory that evolve nontrivially in the Heisenberg picture? What about in a possible interaction picture, if $\hat{H}(p)$ can be separated into two parts?

(7) Can one have a picture in which operators don't evolve in time other than the Schrödinger picture?

10 The Feynman Path Integral and Propagators

Having defined the Heisenberg picture and evolution operators, we can now attack quantum mechanics in a different way, proposed by Feynman: instead of states and operators acting on them, we talk about something closer to the classical concept of a particle moving on a path. Instead of having a single path, Feynman thought about the fact that in quantum mechanics, we should be able to sum over *all* possible paths between two points, even discontinuous ones, with the weight given by the phase e^{iS}, S being the classical action for the particle. This resulting *Feynman path integral* formulation of quantum mechanics is actually equivalent to the Schrödinger formulation (or the Heisenberg, or Dirac, formulations for that matter). The path integral for a generalized coordinate $q(t)$ is

$$\int \mathcal{D}q(t)e^{iS[q]/\hbar}, \tag{10.1}$$

and, in the case where it is possible to have a nearly classical motion, we will find that the result is approximated by

$$\sim e^{iS_{\rm cl}[q_{\rm cl}]/\hbar} \times (\text{corrections}), \tag{10.2}$$

where $q_{\rm cl}(t)$ is the classical path, and $S_{\rm cl}$ is the classical on-shell action (in the classical solution). We will derive this next.

10.1 Path Integral in Phase Space

The path integral formalism is applied to the *propagator*, that is, the evolution operator in the coordinate representation, defined as the product of two Heisenberg (bra and ket) states,

$${}_H\langle x', t'|x, t\rangle_H \equiv U(x', t'; x, t) = \langle x'|\hat{U}(t', t)|x\rangle, \tag{10.3}$$

where as we showed, in the case of a conservative system, $\hat{U}(t', t) = e^{-i\hat{H}(t'-t)/\hbar}$ and represents the transition amplitude between the initial and final points.

The term "propagator" is related to the fact that we can write the evolution of a state in the coordinate representation (by multiplying with $\langle x'|$), and insert a completeness relation for $|x\rangle$, to get

$$\begin{aligned} |\psi(t)\rangle &= \hat{U}(t, t_0)|\psi(t_0)\rangle \quad \Rightarrow \\ \psi(x', t) &= \langle x'|\hat{U}(t, t_0)|\psi(t_0)\rangle = \int dx\, U(x', t'; x, t)\psi(x, t_0). \end{aligned} \tag{10.4}$$

We saw earlier that the evolution operator can be written in terms of the energy eigen-basis as

$$\hat{U}(t, t_0) = \sum_{n,a} |E_n, a\rangle\langle E_n, a|e^{-iE_n(t-t_0)/\hbar}. \tag{10.5}$$

Then the propagator at equal times becomes

$$\lim_{t'\to t} U(x',t';x,t) = \sum_{n,a}\langle x'|E_n,a\rangle\langle E_n,a|x\rangle = \langle x'|\hat{\mathbb{1}}|x\rangle = \delta(x-x'). \tag{10.6}$$

Moreover, the relation (10.4) implies that, since $\psi(x,t)$ satisfies the Schrödinger equation,

$$\left(-\frac{\hbar^2}{2m}\frac{d^2}{dx^2} + V(x) - i\hbar\frac{\partial}{\partial t}\right)\psi(x,t) = 0, \tag{10.7}$$

the propagator $U(x',t';x,t)$ also satisfies the Schrödinger equation for $t' > t$. Then, adding the definition that

$$U(x',t';x,t) = 0, \quad \text{for} \quad t' < t, \tag{10.8}$$

we find that $U(x',t';x,t)$ becomes a Green's function for the Schrödinger operator. Indeed, because of the discontinuity at $t' \to t$, we find that

$$\left(-\frac{\hbar^2}{2m}\frac{d^2}{dx^2} + V(x) - i\hbar\frac{\partial}{\partial t}\right)U(x,t;x_0,t_0) = \delta(x-x_0)\delta(t-t_0). \tag{10.9}$$

The Heisenberg state $|x,t\rangle_H$ is an eigenstate of the Heisenberg operator $\hat{x}_H(T)$ at time $T=t$, i.e.,

$$\hat{x}_H(T=t)|x,t\rangle_H = x(t)|x,t\rangle_H, \tag{10.10}$$

and is not an eigenstate at $T \neq t$. The relation to the Schrödinger states is given by

$$|x,t\rangle_H = e^{i\hat{H}t/\hbar}|x\rangle \Rightarrow |x\rangle = e^{-i\hat{H}t/\hbar}|x,t\rangle_H. \tag{10.11}$$

The Heisenberg operator is related to the Schrödinger operator by

$$\hat{X}(t) = e^{i\hat{H}t/\hbar}\hat{X}_S e^{-i\hat{H}t/\hbar}, \tag{10.12}$$

where $\hat{X}_S|x\rangle = x|x\rangle$.

To write a path integral formulation for the propagator $U(x',t';x,t)$, we first divide the interval (t,t') into $n+1$ intervals of length ϵ,

$$\epsilon = \frac{t'-t}{n+1} \quad \Rightarrow \quad t_0 = t,\ t_1 = t+\epsilon, \ldots, t_{n+1} = t'. \tag{10.13}$$

At any fixed time t_i, the set of all possible Heisenberg states $\{|x_i,t_i\rangle | x_i \in \mathbb{R}\}$ is a complete set. The completeness relation is then

$$\int dx_i |x_i,t_i\rangle\langle x_i,t_i| = \hat{\mathbb{1}}. \tag{10.14}$$

Because of this, we can insert in $U(x',t';x,t)$ at each t_i, $i=1,\ldots,n$, an operator $\hat{\mathbb{1}}$ written as the completeness relation for time t_i, obtaining

$$U(x',t';x,t) = \int dx_1 \cdots dx_n \langle x',t'|x_n,t_n\rangle\langle x_n,t_n|x_{n-1},t_{n-1}\rangle \cdots |x_1,t_1\rangle\langle x_1,t_1|x,t\rangle. \tag{10.15}$$

In the above equation, $x_i = x(t_i)$ is a discretized path but it is not a classical path, in that we are integrating arbitrarily and independently over x_i and x_{i+1}, which means that generically the distance between x_i and x_{i+1} is large, and does not go to zero (i.e., we do not have $x_{i+1} \to x_i$) if $\epsilon \to 0$. Instead of a classical path, we get a discontinuous quantum path, as in Fig. 10.1.

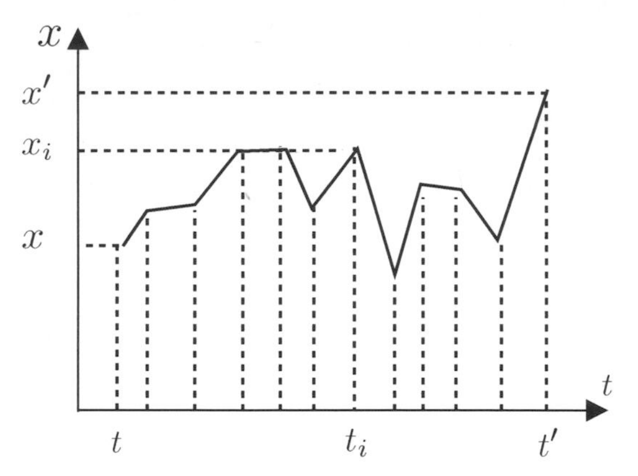

Figure 10.1 For the definition of the quantum mechanical "path integral", we integrate over all possible discretized paths, not just the smooth ones (as would be suggested by classical paths). Indeed, we divide the path into a large number of discrete points, after which we integrate over the positions of all these points, *independently* of the positions of their nearby points.

The product of the integrations at each intermediate point between x and x' is called the *path integral*, or integral over all possible quantum paths, and is denoted by

$$\int \mathcal{D}x(t) = \lim_{n\to\infty} \prod_{i=1}^{n} \int dx(t_i). \tag{10.16}$$

Since we can expand $|x\rangle$ in the $|p\rangle$ basis, and vice versa,

$$|x\rangle = \int dp|p\rangle\langle p|x\rangle, \quad |p\rangle = \int dx|x\rangle\langle x|p\rangle, \tag{10.17}$$

and since we have (as we saw earlier)

$$\langle x|p\rangle = \frac{e^{ipx/\hbar}}{\sqrt{2\pi\hbar}} \Rightarrow \int dx\langle p'|x\rangle\langle x|p\rangle = \int dx \frac{e^{ix(p-p')/\hbar}}{2\pi\hbar} = \delta(p-p'), \tag{10.18}$$

we can write the generic scalar product of two Heisenberg states appearing in the path integral as

$$\begin{aligned} {}_H\langle x(t_i), t_i|x(t_{i-1}), t_{i-1}\rangle_H &= \langle x(t_i)|e^{-i\epsilon\hat{H}/\hbar}|x(t_{i-1})\rangle \\ &= \int dp(t_i)\langle x(t_i)|p(t_i)\rangle\langle p(t_i)|e^{-i\epsilon\hat{H}/\hbar}|x(t_{i-1})\rangle. \end{aligned} \tag{10.19}$$

As a technical point, we now impose that $\hat{H}$ *is to be ordered so that momenta* $\hat{p}$ *are to the left of coordinates* $\hat{x}$. Then, to order ϵ, we can write

$$\begin{aligned} \langle p(t_i)|\exp\left(-\frac{i}{\hbar}\epsilon\hat{H}\right)|x(t_{i-1})\rangle &\simeq \exp\left[-\frac{i}{\hbar}\epsilon H(p(t_i), x(t_{i-1}))\right]\langle p(t_i)|x(t_{i-1})\rangle \\ &= \exp\left[-\frac{i}{\hbar}\epsilon H(p(t_i), x(t_{i-1}))\right], \end{aligned} \tag{10.20}$$

where in the first relation, we acted with $\hat{P}$ on the left and with $\hat{X}$ on the right, and neglected commutator corrections, which are of order ϵ^2: for instance, in $\hat{H}(\hat{P},\hat{X})\hat{H}(\hat{P},\hat{X})$ we have $\hat{P}\hat{X}\hat{P}\hat{X}$-type terms, which would be more complicated.

All in all, using the formulas derived, we obtain

$$
\begin{aligned}
U(x',t';x,t) &= \int \prod_{i=1}^{n} dp(t_i) \prod_{j=1}^{n} dx(t_j) \langle x(t_{n+1})|p(t_{n+1})\rangle\langle p(t_{n+1})|e^{-i\epsilon\hat{H}/\hbar}|x(t_n)\rangle \\
&\quad \cdots \langle x(t_1)|p(t_1)\rangle\langle p(t_1)|e^{-i\epsilon\hat{H}/\hbar}|x(t_0)\rangle \\
&= \int \mathcal{D}p(t) \int \mathcal{D}x(t) \exp\left\{\frac{i}{\hbar}\left[p(t_{n+1})(x(t_{n+1}) - x(t_n)) + \cdots \right.\right. \\
&\quad \left.\left. + p(t_1)(x(t_1) - x(t_0)) - \epsilon\left(H(p(t_{n+1}), x(t_n)) + \cdots + H(p(t_1), x(t_0))\right)\right]\right\} \\
&= \int \mathcal{D}p(t) \int \mathcal{D}x(t) \exp\left\{\frac{i}{\hbar}\int_{t_0=t}^{t_{n+1}=t'} dt[p(t)\dot{x}(t) - H(p(t), x(t))]\right\},
\end{aligned}
\tag{10.21}
$$

where we have made the replacement $x(t_{i+1}) - x(t_i) \to dt\dot{x}(t)$.

This is the *path integral in phase space* (in the Hamiltonian formulation). Note that this is a physicist's "rigorous" derivation (mathematically, of course, even the notion of path integration is not well defined, but once that is accepted, everything is rigorous). However, as we said at the beginning, we would be more interested in the path integral in configuration space, in terms of only $x(t)$.

In order to obtain this, we must do a Gaussian integration over $p(t)$. For that, we need one more technical requirement on the Hamiltonian: it must be quadratic in the momenta p: $H(p,x) = p^2/2 + V(x)$. However, before we do the integration over $p(t)$ we must calculate some Gaussian integration formulas.

10.2 Gaussian Integration

We will have a mathematical interlude at this point, in order to find out how to do the integration we must perform. The basic Gaussian integral is

$$
I = \int_{-\infty}^{+\infty} dx\, e^{-\alpha x^2} = \sqrt{\frac{\pi}{\alpha}}. \tag{10.22}
$$

Squaring it, we obtain

$$
I^2 = \int dx \int dy\, e^{-(x^2+y^2)} = \int_0^{2\pi} d\phi \int_0^{\infty} r dr\, e^{-r^2} = \pi, \tag{10.23}
$$

where in the second equality we transformed to polar coordinates in the (x, y) plane.

Generalizing this Gaussian integration formula to an n-dimensional space, we can say that

$$
\int d^n x\, e^{-x_i A_{ij} x_j/2} = (2\pi)^{n/2} \frac{1}{\sqrt{\det A}}. \tag{10.24}
$$

This formula can be proven by diagonalizing the (constant) matrix A_{ij}, in which case $\det A = \prod_i \alpha_i$, where α_i are the eigenvalues of the matrix A_{ij}.

Then, defining a scalar that is identified with a quadratic "action" in a discretized form,

$$
S = \frac{1}{2} x^T \cdot A \cdot x + b^T \cdot x, \tag{10.25}
$$

and defining its "classical solution",

$$\frac{\partial S}{\partial x_i} = 0 \Rightarrow x_{\text{cl}} = -A^{-1} \cdot b, \tag{10.26}$$

we write the on-shell (classical) action as

$$S(x_{\text{cl}}) = -\frac{1}{2} b^T \cdot A^{-1} \cdot b, \tag{10.27}$$

and the quantum action (including fluctuations) is rewritten as

$$S = \frac{1}{2}(x - x_{\text{cl}})^T \cdot A \cdot (x - x_{\text{cl}}) - \frac{1}{2} b^T \cdot A^{-1} \cdot b. \tag{10.28}$$

Then the Gaussian integration over the vector x, shifted to $x - x_{\text{cl}}$, is

$$\int d^n x e^{-S(x)} = (2\pi)^{n/2} (\det A)^{-1/2} e^{-S(x_{\text{cl}})} = (2\pi)^{n/2} \frac{1}{\sqrt{\det A}} \exp\left(+\frac{1}{2} b^T \cdot A^{-1} \cdot b\right). \tag{10.29}$$

10.3 Path Integral in Configuration Space

We are now ready to do the path integral over $\int \mathcal{D}p(t)$, which explicitly (in the discrete version) is

$$\begin{aligned} &\int \mathcal{D}p(t) \exp\left\{\frac{i}{\hbar} \int_t^{t'} d\tau \left[p(\tau)\dot{x}(\tau) - \frac{1}{2} p^2(\tau)\right]\right\} \\ &= \prod_{i=1}^{n} \int \frac{dp(\tau_i)}{2\pi} \exp\left[\frac{i}{\hbar} \Delta\tau \left(p(\tau_i)\dot{x}(\tau_i) - \frac{1}{2} p^2(\tau_i)\right)\right]. \end{aligned} \tag{10.30}$$

This means that, with respect to the above formula for Gaussian integration, we have the identifications

$$x_i \to p(t_i), \quad A_{ij} \to i\Delta\tau\delta_{ij}, \quad b = -i\Delta\tau \dot{x}(\tau), \tag{10.31}$$

so that the result of the path integration is

$$\int \mathcal{D}p(t) \exp\left\{\frac{i}{\hbar} \int_t^{t'} d\tau \left[p(\tau)\dot{x}(\tau) - \frac{1}{2} p^2(\tau)\right]\right\} = \mathcal{N} \exp\left[\frac{i}{\hbar} \int_t^{t'} d\tau \frac{\dot{x}^2(\tau)}{2}\right], \tag{10.32}$$

where the *normalization constant* $\mathcal{N}$ contains factors of $i, 2, \pi, \Delta\tau$, all constant and so irrelevant once we normalize the probability to one.

Finally, this means that the path integral for the propagator becomes a path integral only in configuration space,

$$\begin{aligned} U(x', t'; x, t) &= \mathcal{N} \int \mathcal{D}x(t) \exp\left\{\frac{i}{\hbar} \int_t^{t'} d\tau \left[\frac{\dot{x}^2(\tau)}{2} - V(x)\right]\right\} \\ &= \mathcal{N} \int \mathcal{D}x(t) \exp\left[\frac{i}{\hbar} \int_t^{t'} d\tau L(x(\tau), \dot{x}(\tau))\right] \\ &= \mathcal{N} \int \mathcal{D}x(t) \exp\left[\frac{i}{\hbar} S(x)\right]. \end{aligned} \tag{10.33}$$

This is indeed the formula we argued for at the beginning of the chapter, now derived rigorously from the phase space path integral, under the following two technical requirements: the Hamiltonian is to be ordered with ps on the left and xs on the right, and moreover the Hamiltonian is only quadratic in momenta.

This formula, besides having a simple physical interpretation that generalizes the classical path (the amplitude for the transition probability (see equation (10.3)) is the sum over all paths weighted by the phase $e^{iS[x]}$), also implies a simple *classical limit.*

Indeed, the classical path is the extremum of the action $S[x]$, so we expand any action around its value on the classical solution (the "on-shell value"), up to at least quadratic order (and maybe higher),

$$S[x] \simeq S_{\rm cl}[x_{\rm cl}] + \frac{1}{2}(x - x_{\rm cl}) \cdot A \cdot (x - x_{\rm cl}), \tag{10.34}$$

as before. Then the first approximation to the propagator is

$$U(x', t'; x, t) \simeq e^{i\frac{S_{\rm cl}[x_{\rm cl}]}{\hbar}}, \tag{10.35}$$

while the second comes with a factor of $1/\sqrt{\det A}$, as we saw from the Gaussian integration formula.

We can calculate exactly the propagator for free particles, though we will not do it here but will leave it for later.

Also, we can show that the propagator $U(x', t'; x, t)$, calculated as a Feynman path integral, does also satisfy the Schrödinger equation, so it is indeed a Green's function for the Schrödinger operator, meaning that any wave function evolved with it will also satisfy the Schrödinger equation. Thus the Feynman path integral representation of quantum mechanics is actually equivalent to the original Schrödinger representation.

10.4 Path Integral over Coherent States (in "Harmonic Phase Space")

Our prototype (toy model) for a simple quantum system is, as we have seen, the harmonic oscillator. A more complicated system can be described as a perturbed harmonic oscillator or a system of harmonic oscillators. Therefore it is of interest to see if there is any other way to describe the path integral for it. In fact, there is, using the so-called "harmonic phase space" instead of configuration space or phase space.

This space is defined using the notion of coherent states of the harmonic oscillator. For a harmonic oscillator, the Hamiltonian can be written in terms of creation and annihilation operators as

$$\hat{H} = \hbar\omega\left(\hat{a}^\dagger \hat{a} + \tfrac{1}{2}\right). \tag{10.36}$$

But, while the original intention in introducing $\hat{a}$ and $\hat{a}^\dagger$ was to deal with eigenenergy states $|n\rangle$, whose index n is changed by $\hat{a}$ and $\hat{a}^\dagger$ (hence the name creation and annihilation), we can also introduce eigenstates *for these creation and annihilation operators*, called *coherent states*.

We can define $|\alpha\rangle$, as we showed in Chapter 8, as an eigenstate of $\hat{a}$,

$$|\alpha\rangle \equiv e^{\alpha\hat{a}^\dagger}|0\rangle \quad \Rightarrow \quad \hat{a}|\alpha\rangle = \alpha|\alpha\rangle \tag{10.37}$$

and similarly $\langle\alpha^*|$ as an eigenstate of $\hat{a}^\dagger$,

$$\langle\alpha^*| \equiv \langle 0|e^{\alpha^*\hat{a}} \quad \Rightarrow \quad \langle\alpha|\hat{a}^\dagger = \langle\alpha^*|\alpha^*. \tag{10.38}$$

Moreover, they satisfy a completeness relation

$$\int \frac{d\alpha d\alpha^*}{2\pi i} e^{-\alpha\alpha^*}|\alpha\rangle\langle\alpha^*| = \hat{\mathbb{1}}. \tag{10.39}$$

We can compute the propagator for these states, i.e., the scalar product of Heisenberg states constructed out of them,

$$U(\alpha^*, t'; \alpha, t) \equiv {}_H\langle\alpha^*, t'|\alpha, t\rangle_H = \langle\alpha^*|\hat{U}(t', t)|\alpha\rangle, \tag{10.40}$$

where as always $\hat{U}(t', t) = e^{-i\hat{H}(t'-t)}$.

On the other hand, we find that the expectation value of $\hat{H}$ in the coherent state basis is

$$\langle\alpha^*|\hat{H}(\hat{a}^\dagger, \hat{a})|\beta\rangle = H(\alpha^*, \beta)\langle\alpha^*|\beta\rangle = H(\alpha^*, \beta)e^{\alpha^*\beta}. \tag{10.41}$$

We define the same discretization of a classical path from α to α^* in $n+1$ intervals, with

$$\epsilon = \frac{t'-t}{n+1}, \quad t_0 = t,\ t_1, \ldots, t_n,\ t_{n+1} = t'. \tag{10.42}$$

We follow the same steps as in phase space, inserting a completeness relation at each intermediate point, obtaining

$$\begin{aligned} U(\alpha^*, t'; \alpha, t) = \prod_i \int \left[\frac{d\alpha(t_i)d\alpha^*(t_i)}{2\pi i} e^{-\alpha^*(t_i)\alpha(t_i)}\right] \langle\alpha^*(t')|e^{-i\epsilon\hat{H}/\hbar}|\alpha(t_n)\rangle \\ \times \langle\alpha^*(t_n)|e^{-i\epsilon\hat{H}/\hbar}|\alpha(t_{n-1})\rangle\langle\alpha^*(t_{n-1})|\cdots\langle\alpha^*(t_1)|e^{-i\epsilon\hat{H}/\hbar}|\alpha(t)\rangle. \end{aligned} \tag{10.43}$$

But the generic scalar expectation value is

$$\langle\alpha^*(t_{i+1})|\exp\left(-\frac{i}{\hbar}\epsilon\hat{H}\right)|\alpha(t_i)\rangle = \exp\left[-\frac{i}{\hbar}\epsilon H(\alpha^*(t_{i+1}), \alpha(t_i))\right]\exp\left[\alpha^*(t_{i+1})\alpha(t_i)\right], \tag{10.44}$$

if $\hat{H}$ is normally ordered (with $\hat{a}^\dagger$ to the left and $\hat{a}$ to the right).

Then we obtain for the propagator

$$\begin{aligned} &U(\alpha^*, t'; \alpha, t) \\ &= \prod_i \int \left[\frac{d\alpha(t_i)d\alpha^*(t_i)}{2\pi i}\right] \exp\Bigg[\alpha^*(t')\alpha(t_n) - \alpha^*(t_n)\alpha(t_n) \\ &\quad + \alpha^*(t_n)\alpha(t_{n-1}) - \alpha^*(t_{n-1})\alpha(t_{n-1}) + \cdots + \alpha^*(t_1)\alpha(t) - \frac{i}{\hbar}\int_t^{t'} d\tau\ H(\alpha^*(\tau), \alpha(\tau))\Bigg], \\ &= \prod_i \int \left[\frac{d\alpha(t_i)d\alpha^*(t_i)}{2\pi i}\right] \exp\left\{\int_t^{t'} d\tau\left[\dot{\alpha}^*(\tau)\alpha(\tau) - \frac{i}{\hbar}H(\alpha^*(\tau), \alpha(\tau)) + \alpha^*(t)\alpha(t)\right]\right\}, \end{aligned} \tag{10.45}$$

so that finally we can write it as

$$U(\alpha^*, t'; \alpha, t) = \int \mathcal{D}(\alpha(\tau)) \int \mathcal{D}(\alpha^*(\tau)) \exp\left\{\frac{i}{\hbar}\int_t^{t'} d\tau\left[\frac{\hbar}{i}\dot{\alpha}^*(\tau)\alpha(\tau) - H\right] + \alpha^*(t)\alpha(t)\right\}. \tag{10.46}$$

10.5 Correlation Functions and Their Generating Functional

We can consider a more general observable than a transition probability, namely one where we insert an operator $\hat{X}(t_a)$ in between Heisenberg states,

$$ {}_H\langle x', t'|\hat{X}(t_a)|x, t\rangle_H, \tag{10.47} $$

obtaining what is known as a *correlation function*, specifically a *one-point function*. This observable is harder to understand, but is of great theoretical interest, especially in that it can be generalized.

To calculate it, we follow the same steps as before and discretize the path. The only constraint in doing so is to choose t_a as one of the intermediate t_is in the path integral (we can choose the division into $n+1$ steps in such a way that this is true).

Then, if $t_a = t_i$, so that $x(t_a) = x(t_i)$, we find

$$ \langle x_{i+1}, t_{i+1}|\hat{X}(t_a)|x_i, t_i\rangle = x(t_a)\langle x_{i+1}, t_{i+1}|x_i, t_i\rangle; \tag{10.48} $$

now following the same steps as before, we find the path integral

$$ \langle x', t'|\hat{X}(t_a)|x, t\rangle = \int \mathcal{D}x(t) e^{\frac{i}{\hbar}S[x]} x(t_a). \tag{10.49} $$

Next, we can define the two-point function (which is also a correlation function)

$$ \langle x', t'|\hat{X}(t_b)\hat{X}(t_a)|x, t\rangle. \tag{10.50} $$

If $t_a < t_b$, we can work as before: choose $t_a = t_i$ and $t_b = t_j$, where $j > i$, and find

$$ \begin{aligned} &\langle x_{j+1}, t_{j+1}|\hat{X}(t_b)|x_j, t_j\rangle \cdots \langle x_{i+1}, t_{i+1}|\hat{X}(t_a)|x_i, t_i\rangle \\ &\quad = x(t_b)\langle x_{j+1}, t_{j+1}|x_j, t_j\rangle \cdots x(t_a)\langle x_{i+1}, t_{i+1}|x_i, t_i\rangle, \end{aligned} \tag{10.51} $$

so we obtain

$$ \langle x', t'|\hat{X}(t_b)\hat{X}(t_a)|x, t\rangle = \int \mathcal{D}x(t) e^{iS[x]/\hbar} x(t_b) x(t_a). \tag{10.52} $$

Conversely, the path integral gives the *time-ordered product*

$$ \int \mathcal{D}x(t) e^{iS[x]/\hbar} x(t_a) x(t_b) = \langle x', t'|T\{\hat{X}(t_a)\hat{X}(t_b)\}|x, t\rangle, \tag{10.53} $$

where the time ordering is defined as usual,

$$ T\{\hat{X}(t_a)\hat{X}(t_b)\} = \begin{cases} \hat{X}(t_a)\hat{X}(t_b), & t_a > t_b \\ \hat{X}(t_b)\hat{X}(t_a), & t_a < t_b, \end{cases} \tag{10.54} $$

and generalized to

$$ T\{\hat{X}(t_{a_1}) \cdots \hat{X}(t_{a_n})\} = \hat{X}(t_{a_1}) \cdots \hat{X}(t_{a_n}), \quad \text{if} \quad t_{a_1} > t_{a_2} > \cdots > t_{a_n}. \tag{10.55} $$

Then we can also calculate *n-point functions* (general correlation functions) as a path integral,

$$ G_n(t_{a_1}, \ldots, t_{a_n}) = \langle x', t'|T\{\hat{X}(t_{a_1}) \cdots \hat{X}(t_{a_n})\}x, t\rangle = \int \mathcal{D}x(t) e^{iS[x]/\hbar} x(t_{a_1}) \cdots x(t_{a_n}). \tag{10.56} $$

Finally, we can write a generating function for all the n-point correlation functions. Indeed, for numbers a_n we can find a *generating function*

$$f(x) = \sum_{n\geq 0} \frac{a_n x^n}{n!}, \tag{10.57}$$

and the coefficient a_n is found from the nth order derivatives at zero,

$$a_n = \frac{d^n}{dx^n} f(x)\bigg|_{x=0}. \tag{10.58}$$

In the present case, we can define a *generating functional* $Z[J]$ for all the Green's functions G_n, defined as

$$Z[J] = \sum_{n\geq 0} \int dt_1 \cdots \int dt_n \frac{(i/\hbar)^n}{n!} G_n(t_1, \ldots, t_n) J(t_1) \cdots J(t_n). \tag{10.59}$$

Substituting the Green's functions as path integrals, we obtain

$$Z[J] = \int \mathcal{D}x(t) e^{iS[x]/\hbar} \sum_{n\geq 0} \frac{1}{n!} \left[\int dt \frac{i}{\hbar} x(t) J(t) \right]^n, \tag{10.60}$$

which is seen to easily sum to

$$Z[J] = \int \mathcal{D}x(t) \exp\left[\frac{i}{\hbar} S(x, J)\right] \equiv \int \mathcal{D}x(t) \exp\left[\frac{i}{\hbar} S(x) + \frac{i}{\hbar} \int dt J(t) x(t)\right]; \tag{10.61}$$

then the Green's functions are obtained from it via

$$\frac{\delta^n}{(i/\hbar)\delta J(t_1) \ldots (i/\hbar)\delta J(t_n)} Z[J]\bigg|_{J=0} = \int \mathcal{D}x(t) e^{iS[x]/\hbar} x(t_1) \cdots x(t_n) = G_n(t_1, \ldots, t_n). \tag{10.62}$$

Important Concepts to Remember

- The Feynman path integral is another way to define quantum mechanics, as a sum over all possible paths, not necessarily continuous, between initial points and endpoints.
- The path integral is $\int \mathcal{D}q(t) \exp\left\{\frac{i}{\hbar} S[q(t)]\right\}$, and is approximated (at a saddle point) by $\exp\left\{\frac{i}{\hbar} S_{\text{cl}}[q_{\text{cl}}(t)]\right\}$, with corrections coming from the Gaussian around the saddle point, as $1/\sqrt{\det A}$, where $S = S_{\text{cl}}[q_{\text{cl}}] + \frac{1}{2}(q - q_{\text{cl}}) \cdot A \cdot (q - q_{\text{cl}})$.
- The path integral is derived in phase space as $U(x', t'; x, t) = \int \mathcal{D}x(t) \int \mathcal{D}p(t) \exp\left\{\frac{i}{\hbar} \int_t^{t'} dt(p\dot{q} - H)\right\}$, under the technical requirement to have p to the left of x in H.
- From the path integral in phase space, we derive the path integral in coordinate space, $U(x', t'; x, t) = \mathcal{N} \int \mathcal{D}x(t) \exp\left\{\frac{i}{\hbar} S[x(t)]\right\}$, under the extra technical requirement that H is quadratic in p, $H = p^2/2 + V(x)$.
- The best definition of the path integral is in harmonic phase space, related to coherent states of the harmonic oscillator, α, α^*, where

$U(\alpha^*, t'; \alpha, t) = \int \mathcal{D}\alpha^*(\tau) \int \mathcal{D}\alpha(\tau) \exp\left\{ i/\hbar \int_t^{t'} d\tau\, [\hbar/i\dot{\alpha}^*(\tau)\alpha(\tau) - H] + \alpha^*(t)\alpha(t) \right\}$, and is valid for normal-ordered Hamiltonians.

- The correlation functions or n-point functions are expectation values of time-ordered products of position operators, rewritten as path integrals, $G_n(t_{a_1}, \dots, t_{a_n}) = \langle x', t' | T(\hat{X}(t_{a_1}) \cdots \hat{X}(t_{a_n})) | x, t \rangle = \int \mathcal{D}x(t) \exp\left\{ \frac{i}{\hbar} S[x(t)] \right\} x(t_{a_1}) \cdots x(t_{a_n})$.
- The generating functional for all the correlations functions is $Z[J] = \int \mathcal{D}x(t) \exp\left[\frac{i}{\hbar} S[x] + i \int dt\, J(t)x(t) \right]$, from which we obtain the correlation functions by taking derivatives.

Further Reading

See [3] and [5] for more details (the latter for details on correlation functions).

Exercises

(1) We have used the idea of Gaussian integration around the classical solution for the action to argue that the first-order result for the path integral is $\exp\left\{ \frac{i}{\hbar} S_{\text{cl}}[x_{\text{cl}}(t)] \right\}$. However, is Gaussian integration, and more generally the path integral itself, well defined for this particular exponential? How would you expect to modify the exponential in order to make sense of the Gaussian integration?

(2) For the path integral in phase space, we have assumed that the Ps are always on the left of the Xs in the Hamiltonian. Consider a case where this is not true, for instance having an extra term in the Hamiltonian of the type $\alpha(\hat{P}\hat{X}^2 + \hat{X}\hat{P}\hat{X} + \hat{X}^2 P)$. Redo the calculation, and see what you obtain for the path integral in phase space.

(3) Consider the case when $H = \frac{3}{4}p^{4/3} + V(x)$ and the path integral is in phase space. Is the resulting path integral in coordinate space approximated in any way by the usual expression $\int \mathcal{D}x(t) e^{\frac{i}{\hbar} S[x(t)]}$?

(4) Consider the Hamiltonian, in terms of $a, a^\dagger$ (with $[a, a^\dagger] = 1$),

$$H = \hbar\omega \left(a^\dagger a + \tfrac{1}{2} \right) + \lambda (a + a^\dagger)^3. \tag{10.63}$$

Derive the harmonic path integral in phase space for it (without calculating the path integral, which is not Gaussian).

(5) Consider a generating functional

$$Z[J(t)] = \mathcal{N} \exp\left[-\frac{1}{2} \int dt \int dt' J(t) \Delta(t, t') J(t') \right]. \tag{10.64}$$

Calculate the two-point, three-point, and four-point functions.

(6) Consider a harmonic oscillator, with Lagrangian $L = (\dot{q}^2 - \omega^2 q^2)/2$. Using a *naive* generalization of the Gaussian integration, show that the generating functional is of the type given in exercise 5. Write a formal expression for $\Delta(t, t')$.

(7) Consider the generating functional

$$Z[J(t)] = \mathcal{N} \exp\left[-\frac{1}{2}\int dt \int dt' J(t)\Delta(t,t')J(t') + \lambda \int dt_1 \int dt_2 \int dt_3 \int dt_4 J(t_1)J(t_2)J(t_3)J(t_4)\right]. \tag{10.65}$$

Calculate the four-point function.

11 The Classical Limit and Hamilton–Jacobi (WKB Method), the Ehrenfest Theorem

Until now, we have developed the formalism of quantum mechanics from postulates, and we have shown how to match with experiments. We have not used classical mechanics at all, except sometimes as a guide. But we know that the macroscopic world is approximately classical, and most systems have a classical limit. We must therefore consider a way to obtain a classical limit, plus quantum corrections, from the quantum formalism we have developed so far.

First we will show that the classical equations of motion play a role in the quantum theory itself, in the form of the Ehrenfest theorem, which however does not in itself suggest a classical limit. The classical limit is suggested by the Feynman path integral formalism, developed in the previous chapter, for transition amplitudes, perhaps with the insertion of an operator. The action $S[x]$ appearing in the path integral has as a minimum the classical action, the action on the classical path, for which $\delta S = 0 \Rightarrow x = x_{\rm cl}(t)$; this allows us to expand the action around the classical on-shell action,

$$S \simeq S_{\rm cl}[x_{\rm cl}] + (\delta x)^2(\dots) = S_{\rm cl}[x_{\rm cl}] + \text{fluctuations}, \tag{11.1}$$

and integrate over the fluctuations:

$$\begin{aligned}\left\langle x', t \left| T\left(\hat{A}(\{t_i\})\right)\right| x, t\right\rangle &= \int \mathcal{D}x(t) e^{(iS[x]/\hbar)} \mathcal{A}(\{t_i\}) \simeq e^{iS_{\rm cl}[x_{\rm cl}]/\hbar} \mathcal{A}_{\rm cl}(\{t_i\}) \times \text{fluctuations} \\ &\equiv e^{iS_{\rm cl}[x_{\rm cl}]/\hbar} \mathcal{A}_{\rm cl}(\{t_i\}) e^{iS_{\rm qu}/\hbar}.\end{aligned} \tag{11.2}$$

Thus the classical action has some relevance to the classical limit of quantum mechanics and, as we will see, it is related to the classical mechanics Hamilton–Jacobi equation and formalism. But before analyzing that, we will prove the Ehrenfest theorem.

11.1 Ehrenfest Theorem

A statement of the Ehrenfest theorem is: The classical mechanics equations of motion are valid for the quantum average over quantum states.

Proof Consider a quantum operator $\hat{A}$ and its expectation value in the state $|\psi\rangle$,

$$\langle A\rangle_\psi \equiv \langle\psi|\hat{A}|\psi\rangle. \tag{11.3}$$

Its time evolution is calculated as

$$\frac{d}{dt}\langle A\rangle_\psi = \left(\frac{d\langle\psi|}{dt}\right)\hat{A}|\psi\rangle + \left\langle\psi\left|\hat{A}\frac{d}{dt}\right|\psi\right\rangle + \left\langle\psi\left|\frac{\partial\hat{A}}{\partial t}\right|\psi\right\rangle. \tag{11.4}$$

But we can use the Schrödinger equation acting on $|\psi\rangle$, namely

$$\frac{d}{dt}|\psi\rangle = \frac{1}{i\hbar}\hat{H}|\psi\rangle \;\Rightarrow\; \frac{d}{dt}\langle\psi| = -\frac{1}{i\hbar}\langle\psi|\hat{H}, \tag{11.5}$$

on the time evolution above, and obtain

$$\begin{aligned}\frac{d}{dt}\langle A\rangle_\psi &= \frac{1}{i\hbar}\langle\psi|(\hat{A}\hat{H}-\hat{H}\hat{A})|\psi\rangle + \left\langle\psi\left|\frac{\partial\hat{A}}{\partial t}\right|\psi\right\rangle\\ &= \frac{1}{i\hbar}\langle[\hat{A},\hat{H}]\rangle_\psi + \left\langle\frac{\partial\hat{A}}{\partial t}\right\rangle_\psi.\end{aligned} \tag{11.6}$$

This is the first version of the Ehrenfest theorem, which is the quantum mechanical equivalent of the classical mechanics evolution equation

$$\frac{d}{dt}A = \{A, H\}_{P.B.} + \frac{\partial A}{\partial t}. \tag{11.7}$$

Consider next a Hamiltonian on a general phase space, $H = H(q_i, p_i)$, where p_i is canonically conjugate to the variable q_i. In this case, $\partial(\hat{q}_i)/\partial t = \partial(\hat{p}_i)/\partial t = 0$, so we obtain

$$\begin{aligned}\frac{d}{dt}\langle q_i\rangle_\psi &= \frac{1}{i\hbar}\langle[\hat{q}_i,\hat{H}]\rangle_\psi\\ \frac{d}{dt}\langle p_i\rangle_\psi &= \frac{1}{i\hbar}\langle[\hat{p}_i,\hat{H}]\rangle_\psi.\end{aligned} \tag{11.8}$$

On the other hand, the commutations come from Poisson brackets, in particular the canonical commutators from the canonical Poisson brackets,

$$\frac{1}{i\hbar}[\hat{q}_i,\hat{p}_i] = 1 \quad\leftarrow\quad \{q_i, p_i\}_{P.B.} = 1, \tag{11.9}$$

which means that we can represent the coordinates by derivatives with respect to momenta, and momenta as derivatives with respect to coordinates, as

$$\frac{1}{i\hbar}\hat{q}_i = \frac{\partial}{\partial p_i}, \qquad \frac{1}{i\hbar}\hat{p}_i = -\frac{\partial}{\partial q_i}, \tag{11.10}$$

which allows us to replace them in the commutators on the right-hand side of the Ehrenfest theorem (11.6).

Better still, we have that the specific commutators in the Ehrenfest theorem come from Poisson brackets that amount to derivatives of the Hamiltonian, which can be extended to operators as

$$\begin{aligned}\frac{1}{i\hbar}[\hat{p}_i,\hat{H}] &\leftarrow \{p_i, H\}_{P.B.} = -\frac{\partial H}{\partial q_i} \to -\frac{\partial\hat{H}}{\partial\hat{q}_i}\\ \frac{1}{i\hbar}[\hat{q}_i,\hat{H}] &\leftarrow \{q_i, H\}_{P.B.} = +\frac{\partial H}{\partial p_i} \to +\frac{\partial\hat{H}}{\partial\hat{p}_i}.\end{aligned} \tag{11.11}$$

Finally, then, we obtain the Hamiltonian equations of motion as operator equations quantum averaged over a state $|\psi\rangle$,

$$\begin{aligned}\frac{d}{dt}\langle q_i\rangle_\psi &= \left\langle\frac{\partial H}{\partial p_i}\right\rangle_\psi\\ \frac{d}{dt}\langle p_i\rangle_\psi &= -\left\langle\frac{\partial H}{\partial q_i}\right\rangle_\psi,\end{aligned} \tag{11.12}$$

which is the more common way to describe the Ehrenfest theorem. *q.e.d.*

As an example, consider the case of a coordinate $\vec{q}$, and a Hamiltonian with a kinetic term and a potential term,

$$\hat{H} = \frac{\hat{p}^2}{2m} + V(\hat{q}). \tag{11.13}$$

In this case, the two Hamiltonian equations of motion can be turned into a single (Newtonian) equation of motion, both for the quantum average:

$$\begin{gathered}\frac{d}{dt}\langle\vec{q}\rangle_\psi = \frac{1}{m}\langle\vec{p}\rangle_\psi, \quad \frac{d}{dt}\langle\vec{p}\rangle_\psi = -\langle\vec{\nabla}V\rangle_\psi \quad \Rightarrow \\ m\frac{d^2}{dt^2}\langle\vec{q}\rangle_\psi = -\langle\vec{\nabla}V\rangle_\psi = \langle\vec{F}\rangle_\psi.\end{gathered} \tag{11.14}$$

11.2 Continuity Equation for Probability

In Chapter 7, we saw that the norm of a time-dependent state amounts to an integral of a probability density,

$$\langle\psi(t)|\psi(t)\rangle = \int \rho dV = \int \frac{dP}{dV}dV, \tag{11.15}$$

and in the coordinate representation this implies that

$$\rho = \frac{dP}{dV} = |\psi(\vec{r},t)|^2. \tag{11.16}$$

Moreover, the Schrödinger equation implies the continuity condition for probability,

$$\partial_t \rho = -\vec{\nabla}\cdot\vec{j}, \tag{11.17}$$

where $\vec{j}$ is the probability current density (when we think of probability as a fluid),

$$\vec{j} = \frac{\hbar}{2mi}(\psi^*\vec{\nabla}\psi - \psi\vec{\nabla}\psi^*). \tag{11.18}$$

But we can expand on this formalism, and write the wave function in coordinate space as a real normalization constant times a phase,

$$\psi(\vec{r},t) = Ae^{iS(\vec{r},t)/\hbar}, \tag{11.19}$$

where $A \in \mathbb{R}$ and, from the above, we have

$$A = \sqrt{\rho(\vec{r},t)}. \tag{11.20}$$

Moreover, the probability current density is

$$\vec{j} = \frac{\hbar}{2mi}A^2\frac{2i}{\hbar}\vec{\nabla}S = \frac{\rho}{m}\vec{\nabla}S. \tag{11.21}$$

Thinking of probability as a fluid, we can define its "velocity" $\vec{v}$ by writing the fluid equation $\vec{j} = \rho\vec{v}$, obtaining

$$\vec{v} = \frac{1}{m}\vec{\nabla}S. \tag{11.22}$$

Since a wave function is written in terms of the propagator (the time evolution operator in the coordinate basis), which itself is a path integral that can be expanded in terms of the classical action plus a quantum correction, as in (11.2), we obtain

$$\psi(\vec{r},t) = U(\vec{r},t;\vec{r}',0)\psi(\vec{r}',0) \simeq \exp\left[\frac{i}{\hbar}(S_{\rm cl}(\vec{r}_{\rm cl},t) + S_{\rm qu}(\vec{r},t))\right]\psi(\vec{r}',0), \tag{11.23}$$

which is consistent with the previous ansatz for the wave function (11.19).

Part of the time evolution of the ansatz (11.19), namely the factor $A = \sqrt{\rho}$, was defined from the continuity equation, which can now be written more explicitly,

$$\begin{aligned}\partial_t A^2 = \partial_t \rho = -\vec{\nabla}(\rho\vec{v}) = -\frac{1}{m}\vec{\nabla}(A^2\vec{\nabla}S) \quad \Rightarrow \\ \partial_t A = -\frac{1}{m}\left(\vec{\nabla}A\cdot\vec{\nabla}S + \frac{1}{2}A\vec{\nabla}^2 S\right).\end{aligned} \tag{11.24}$$

This continuity equation was derived from the Schrödinger equation, but we have used only part of it; the other part must be the equation for S. Indeed, from the Schrödinger equation in coordinate space,

$$\left[-\frac{\hbar^2}{2m}\vec{\nabla}^2 + V(\vec{r})\right]\psi(\vec{r},t) = i\hbar\partial_t\psi(\vec{r},t), \tag{11.25}$$

which for the ansatz becomes

$$\begin{aligned}-\frac{\hbar^2}{2m}e^{\frac{i}{\hbar}S}\left[\vec{\nabla}^2 A + \frac{2i}{\hbar}\vec{\nabla}A\cdot\vec{\nabla}S + \frac{i}{\hbar}A\vec{\nabla}^2 S - \frac{1}{\hbar^2}A(\vec{\nabla}S)^2\right] + V(\vec{r})Ae^{iS/\hbar} \\ = i\hbar e^{\frac{i}{\hbar}S}\left(\partial_t A + \frac{i}{\hbar}A\partial_t S\right),\end{aligned} \tag{11.26}$$

we obtain, using the continuity equation (11.24) to cancel terms involving A between the left- and right-hand sides,

$$\frac{1}{2m}(\vec{\nabla}S)^2 + V(\vec{r}) + \partial_t S - \frac{\hbar^2}{2m}\frac{\vec{\nabla}^2 A}{A} = 0. \tag{11.27}$$

We see that the quantum value $\hbar$ appears only in the last term, the other three terms being of the same, classical, order.

That means that *if we define the classical limit as the limit* $\hbar \to 0$, in which for instance the canonical quantization condition $[\hat{q}_i, \hat{p}_j] = i\hbar\delta_{ij}$ becomes just the commutation of classical variables, $[q_i, p_j] = 0$, we obtain an equation involving just the first three terms, which turns out to be the *classical Hamilton–Jacobi equation*.

11.3 Review of the Hamilton–Jacobi Formalism

In order to continue, we quickly review the Hamilton–Jacobi formalism of classical mechanics. There is a more complete and rigorous derivation for it, but here we will show a somewhat shorter version, focusing on the physical interpretation.

If we consider a *canonical transformation* between two representations of classical mechanics for the same system, the classical action S must change by a total derivative, given that the endpoints of the action are fixed:

$$
\begin{aligned}
S \;\rightarrow\; S' = S - \int dt \frac{d}{dt}\mathcal{S}(\vec{x},t) \;\Rightarrow \\
\int_{t_1}^{t_2} L\, dt \;\rightarrow\; \int_{t_1}^{t_2} (L\, dt - d\mathcal{S}) = \int_{t_1}^{t_2} L\, dt - \mathcal{S}(\vec{x}(t_2),t_2) + \mathcal{S}(\vec{x}(t_1),t_1),
\end{aligned}
\tag{11.28}
$$

where $\mathcal{S}(\vec{x}(t_{1,2}),t_{1,2})$ are constants and so do not change the physics.

That means that a Lagrangian depending on general variables q_i changes as follows:

$$
\begin{aligned}
L \;\rightarrow\; L' = L - \frac{d\mathcal{S}}{dt} &= \sum_i m_i \frac{\dot{q}_i^2}{2} - V - \frac{\partial \mathcal{S}}{\partial t} - \sum_i \frac{\partial \mathcal{S}}{\partial q_i}\dot{q}_i \\
&= \sum_i \frac{1}{2m_i}\left(p_i - \frac{\partial \mathcal{S}}{\partial q_i}\right)^2 - V - \frac{\partial \mathcal{S}}{\partial t} - \sum_i \frac{1}{2m_i}\left(\frac{\partial \mathcal{S}}{\partial q_i}\right)^2.
\end{aligned}
\tag{11.29}
$$

The extremum of the new Lagrangian is obtained when we have both

$$
p_i = \frac{\partial}{\partial q_i}\mathcal{S} \tag{11.30}
$$

and

$$
\frac{\partial}{\partial t}\mathcal{S} + \sum_i \frac{1}{2m_i}\left(\frac{\partial \mathcal{S}}{\partial q_i}\right)^2 + V(q,t) = 0. \tag{11.31}
$$

This latter equation (11.31) is the Hamilton–Jacobi equation, and $\mathcal{S}(\vec{q},t)$ is Hamilton's principal function.

The usefulness of the Hamilton–Jacobi equation comes when, after the canonical transformation, $H' = 0$, which means that $L' = P_i\dot{Q}_i$, and this is actually zero, since P_i and Q_i are constant, as we will see shortly. Thus $\mathcal{S}$ is the action of the theory.

Since $H' = 0$, one finds that the new momenta P_i are constants of motion, which will be called α_i, and the principal function will depend on these constants as well: $\mathcal{S} = \mathcal{S}(t, q_i, \alpha_i)$. Then, moreover, the derivatives with respect to these constants α_i give the new coordinates Q_i and are new constants of motion, called β_i,

$$
\frac{\partial \mathcal{S}}{\partial \alpha_i} = \beta_i = Q_i. \tag{11.32}
$$

This, together with $P_i = \alpha_i$, defines the constants of motion, also known as integrals of motion.

There is no algorithmic solution to the Hamilton–Jacobi equation, but often one can solve it by the separation of variables. Indeed, if the potential $V(\vec{r})$ is (explicitly) time independent, the equation admits a solution that separates the variables,

$$
\mathcal{S}(t, q_i) = f(t) + s(q_i). \tag{11.33}
$$

Then, substituting into (11.31), we find

$$
\dot{f}(t) + \sum_i \frac{1}{2m}\left(\frac{\partial s}{\partial q_i}\right)^2 + V(q_i) = 0, \tag{11.34}
$$

which means that time dependence equals q_i dependence, allowing us to set to zero both independently,

$$
\dot{f}(t) = \text{constant} \equiv -E \;\Rightarrow\; f(t) = -Et + \text{constant}, \tag{11.35}
$$

and substituting into (11.34) we find the *time-independent Hamilton–Jacobi equation*

$$\sum_i \frac{1}{2m_i}\left(\frac{\partial s}{\partial q_i}\right)^2 + V(q_i) = E. \tag{11.36}$$

For Hamilton's principal function we obtain

$$\mathcal{S}(t, q_i, E, \alpha_i) = s(q_i, E, \alpha_i) - Et, \tag{11.37}$$

and the integrals of motion satisfy

$$\frac{\partial \mathcal{S}}{\partial \alpha_i} = \frac{\partial s}{\partial \alpha_i} = \beta_i, \qquad \frac{\partial \mathcal{S}}{\partial E} = \frac{\partial s(q_i, E, \alpha_i)}{\partial E} - t = t_0. \tag{11.38}$$

We can also separate more variables, in the case where the Hamiltonian is independent of other variables, but we will not describe that here.

11.4 The Classical Limit and the Geometrical Optics Approximation

We see then that, in the $\hbar \to 0$ limit, the quantum version of the Hamilton–Jacobi equation, (11.27), reduces to the classical Hamilton–Jacobi equation, (11.31). That means that $S(\vec{x}, t)$ appearing in $\psi = Ae^{iS/\hbar}$ is identified with Hamilton's principal function $\mathcal{S}(\vec{x}, t)$, which in turn equals the action if $H' = 0$, and thus $L' = P\dot{Q} = 0$. Indeed, then, from the path integral formalism, we have

$$\psi(\vec{r}, t) \simeq e^{iS_{\rm cl}(\vec{r}, t)/\hbar} \times (\cdots). \tag{11.39}$$

Considering then the time-independent (stationary) case for $S(\vec{x}, t)$ as in the Hamilton–Jacobi equation, we write

$$S(\vec{x}, t) = W(\vec{r}) - Et, \tag{11.40}$$

which leads to the following equation for W:

$$\frac{1}{2m}(\vec{\nabla} W)^2 + V(\vec{r}) - E - \frac{\hbar^2}{2m}\frac{\vec{\nabla}^2 A}{A} = 0. \tag{11.41}$$

But we still haven't defined the classical limit physically or defined rigorously the smallness of the extra term in (11.27). Physically, we see that the limit is a generalization of the *geometrical optics approximation* of classical wave mechanics. In it, a wave is replaced by "paths of light rays", corresponding now to replacing the integral over all paths with just the motion of classical particles.

Using the particle–wave duality defined by de Broglie, the wave associated with a particle of momentum p has wavelength $\lambda = h/p$. In the presence of a potential $V(\vec{r})$, $p^2/(2m) = E - V(\vec{r})$, so the position-dependent de Broglie wavelength is

$$\lambda = \frac{h}{p} = \frac{h}{\sqrt{2m(E - V(\vec{r}))}}. \tag{11.42}$$

Replacing this in (11.41), we can rewrite it as

$$(\vec{\nabla} W)^2 = \frac{h^2}{\lambda^2}\left[1 + \frac{\lambda^2}{(2\pi)^2}\frac{\vec{\nabla}^2 A}{A}\right]. \tag{11.43}$$

Then it is indeed clear that neglecting the term with $(\lambda/(2\pi))^2$ as being small, we obtain the geometrical optics approximation. Indeed, the resulting equation is

$$(\vec{\nabla} W)^2 = \frac{h^2}{\lambda^2}, \tag{11.44}$$

which is the equation of a wave front in geometrical optics. The planes of constant W are planes of constant phase, i.e., wave fronts. If we have $V(\vec{r}) = 0$, λ is constant, so the solution of the equation is

$$W = \vec{p} \cdot \vec{r} + \text{constant}, \tag{11.45}$$

where $|\vec{p}| = h/\lambda$, and so wave fronts are perpendicular to the direction of $\vec{p}$, whereas light rays are in a direction parallel to $\vec{p}$.

11.5 The WKB Method

A related way to expand the wave function semiclassically is the WKB method, found by Wentzel, Kramers, and Brillouin; it is a more general method, but roughly speaking can be understood here as an expansion in $\hbar$ and so as a "semiclassical" expansion of the wave function.

Specifically, we saw that

$$\psi(\vec{r},t) = e^{iS(\vec{r},t)/\hbar} = \exp\left[\frac{i}{\hbar}(W(\vec{r}) - Et)\right] \Rightarrow \psi(\vec{r}) = e^{iW(\vec{r})/\hbar}, \tag{11.46}$$

and moreover we write

$$W(\vec{r}) = s(\vec{r}) + \frac{\hbar}{i}T(\vec{r}), \tag{11.47}$$

so that

$$\psi(\vec{r}) = e^{T(\vec{r})}e^{is(\vec{r})/\hbar} \equiv Ae^{is(\vec{r})/\hbar}, \tag{11.48}$$

where we can *uniquely* define $s(\vec{r})$ and $T(\vec{r})$ by saying that they are both even in $\hbar$. We can moreover expand them in $\hbar$, and keep only the lowest order in the expansion.

The Schrödinger equation for $\psi(\vec{r},t)$ becomes an equation for $S(\vec{r},t)$,

$$\frac{\partial S}{\partial t} + \sum_i \frac{1}{2m_i}\left[\left(\frac{\partial S}{\partial q_i}\right)^2 + \frac{\hbar}{i}\frac{\partial^2 S}{\partial q_i^2}\right] + V(q_i) = 0. \tag{11.49}$$

Neglecting the term in $\hbar$, specifically because

$$\hbar\left|\frac{\partial^2 S}{\partial q_i^2}\right| \ll \left(\frac{\partial S}{\partial q_i}\right)^2, \tag{11.50}$$

we obtain again the classical Hamilton–Jacobi equation.

As we saw, this limit is a geometrical optics approximation, and, remembering that $\partial S/\partial q_i = p_i$, the above condition becomes

$$\hbar\left|\frac{\partial p_i}{\partial q_i}\right| \ll p_i \Rightarrow \left|\frac{\partial(\hbar/p_i)}{\partial q_i}\right| \ll 1. \tag{11.51}$$

Remembering that $\hbar/p_i = \lambda_i$, we write the condition as

$$|\vec{\nabla}\lambda| \ll 1, \tag{11.52}$$

or, considering the variation of λ over a distance of order λ, so $\delta\lambda \simeq (\vec{\nabla}\lambda)\delta x$ for $\delta x = \lambda$, we obtain the condition

$$\left|\frac{\delta_\lambda \lambda}{\lambda}\right| \ll 1, \tag{11.53}$$

which is the precise form of the geometrical optics approximation.

WKB in One Dimension

The simplest application of the WKB method is for a one-dimensional system. Putting $\hbar \to 0$ in (11.41), and restricting to one dimension, we obtain

$$\frac{W'^2}{2m} \simeq E - V(x), \tag{11.54}$$

with approximate solution

$$W(x) \simeq \pm \int^x dx' \sqrt{2m(E - V(x'))} \equiv \pm \int^x dx' p(x'), \tag{11.55}$$

where $p(x) = \sqrt{2m(E - V(x))}$ is the (space-dependent) momentum. Then the wave function is

$$\psi(x,t) = A \exp\left[\pm \frac{i}{\hbar} \int_{x_0}^x dx' p(x') - \frac{iEt}{\hbar}\right]. \tag{11.56}$$

The equation for A is, as we have seen already, the continuity equation (coming from $\rho = A^2$), now written as

$$\frac{\partial \rho}{\partial t} + \frac{\partial}{\partial x}\left(\frac{\rho}{m}\frac{\partial S}{\partial x}\right) = 0. \tag{11.57}$$

Since $\partial\rho/\partial t = 0$ (the stationary case) and $\partial S/\partial x = dW/dx$, we obtain that $\rho W'(x)$ is constant. But on the other hand, from (11.55), $\rho W' = \pm\rho\sqrt{2m(E - V(x))}$, meaning we obtain

$$A = \sqrt{\rho} = \frac{\text{constant}}{[E - V(x)]^{1/4}}. \tag{11.58}$$

Finally, then the *WKB solution to the one-dimensional problem* is

$$\psi(x,t) = \frac{\text{constant}}{[E - V(x)]^{1/4}} \exp\left[\pm \frac{i}{\hbar} \int_{x_0}^x dx' p(x') - \frac{iEt}{\hbar}\right]. \tag{11.59}$$

Important Concepts to Remember

- The statement of the Ehrenfest theorem is that the classical equations of motion in the Hamiltonian formulation are valid for the quantum average, for quantum states.

- The wave function can be written as $\psi(\vec{r},t) = \sqrt{\rho(\vec{r},t)}e^{iS(\vec{r},t)/\hbar}$, so that the "velocity of the probability fluid" is $\vec{v} = \frac{1}{m}\vec{\nabla}S$.
- Then, from the Schrödinger equation (and the continuity equation for probability), the function $S(\vec{r},t)$ satisfies a quantum-corrected Hamilton–Jacobi equation, with a term of order $\hbar^2$.
- In the Hamilton–Jacobi formalism, $S(\vec{q},t)$ is Hamilton's principal function, and its derivatives with respect to the constants of motion α_i, the values of P_i, are other constants of motion β_i, the values of Q_i.
- There is no algorithmic way to solve the Hamilton–Jacobi equation; one generally uses some separation of variables, at least for time, leading to the time-independent Hamilton–Jacobi equation.
- The classical limit, ignoring the $\hbar^2$ term in the Hamilton–Jacobi equation, is a generalization of the geometrical optics approximation, for the case where the de Broglie wavelength satisfies, $|\delta_\lambda \lambda/\lambda| \ll 1$.
- An equivalent expression for the wave function is $\psi(\vec{r},t) = \exp\left[\frac{i}{\hbar}(W(\vec{r}) - Et)\right]$, where $W(\vec{r}) = s(\vec{r}) - \frac{\hbar}{i}T(\vec{r})$.
- The WKB method in one dimension is a first-order correction to the classical solution $Ae^{is_{\rm cl}/\hbar}$, in which $W(x) = \pm\int^x dx'\sqrt{2m(E - V(x'))} - \frac{\hbar}{i}\ln A$ and A is no longer constant anymore but equals const $/[E - V(x)]^{1/4}$.

Further Reading

See [2] and [3] for more details.

Exercises

(1) Consider the Hamiltonian

$$H = \frac{p^2}{2} + \lambda x^4 \tag{11.60}$$

and the observable

$$A = x^3 + p^3. \tag{11.61}$$

Find the equation of motion for A in the Hamiltonian formalism and the corresponding quantum version of this equation of motion.

(2) Consider a radial (central) potential for motion in three dimensions, $V(r)$. Write the quantum version of the Hamilton–Jacobi equation, and reduce it to a single equation for the radial motion (in r).

(3) (Review from classical mechanics) Consider the classical Hamilton–Jacobi formalism, for the case of the motion of a particle in three spatial dimensions, in a central potential $V(r) = -B/r$. Solve the Hamilton–Jacobi equation for the motion of a particle coming in from infinity and being deflected, and find the deflection angle.

(4) In the parametrization $\psi = A \exp\left[\frac{i}{\hbar}W(\vec{r}) - \frac{i}{\hbar}Et\right]$, what is the equation of motion for A? What happens to it in the geometrical optics approximation? If also $V(\vec{r}) = 0$, solve the equation for A.

(5) Consider a one-dimensional harmonic oscillator. Can one apply the WKB approximation to it? If so, why, and when?

(6) Write down the WKB approximation for the one-dimensional potential $V = -B/x$, $B > 0$, and find the domain of validity for it.

(7) For the case in exercise 6, write down the explicit equations for the wave function outside the geometrical optics approximation, and find a way to introduce the next-order corrections to the WKB approximation that are based on the geometrical optics approximation.

12 Symmetries in Quantum Mechanics I: Continuous Symmetries

We now begin the analysis of symmetries in quantum mechanics. First, in this chapter, we will analyze simple continuous symmetries, such as translation invariance. But before that, we will review symmetries in classical mechanics. Then we will generalize the results to quantum mechanics, and show that the symmetries of quantum mechanics form groups. In the next chapter, we will analyze discrete symmetries, such as parity invariance, and "internal" symmetries. After that, in the following chapters, we will consider rotational invariance and the theory of angular momentum.

12.1 Symmetries in Classical Mechanics

Consider an infinitesimal transformation of the variables q_i of a system with Lagrangian $L(q_i, \dot{q}_i, t)$, $q_i \rightarrow q_i + \delta q_i$, where δq_i is a *specific* change. Then we have a symmetry of the system if the Lagrangian is invariant, $\delta L/\delta q_i = 0$, or more precisely, if the action is invariant,

$$\frac{\delta S}{\delta q_i} = 0. \tag{12.1}$$

Note that therefore the action $S = \int dt\, L$ is invariant if L varies by at most the total derivative of a function f, i.e., df/dt.

We can consider two simple cases:

- We can have the invariance of L under *any* translation by δq_i, so that $\partial L/\partial q_i = 0$. Then the Lagrange equations of motion,

$$\frac{\partial L}{\partial q_i} - \frac{d}{dt}\frac{\partial L}{\partial \dot{q}_i} = 0, \tag{12.2}$$

where the canonically conjugate momentum is $p_i \equiv \partial L/\partial \dot{q}_i$, imply that this conjugate momentum is conserved, i.e., constant in time,

$$\frac{dp_i}{dt} = 0. \tag{12.3}$$

This is the simplest form of the *Noether theorem*, which states that: for any symmetry of the theory there is a conserved charge (i.e., a charge that is constant in time). Here the conserved charge is the canonically conjugate momentum p_i. Note that usually the Noether theorem is defined within classical field theory, which is outside our scope, but here we present the simple (nonrelativistic) classical-mechanics form.

- We can also have a symmetry corresponding to invariance under a transformation that is *continuous* (so that there exists an infinitesimal form) and *linear* (so that it is proportional to the q_i itself) and is of the type

$$\delta q_i = \sum_a \epsilon^a (iT_a)_i{}^j q_j, \tag{12.4}$$

where the ϵ^a are arbitrary (real or complex) infinitesimal parameters and $(iT_a)_i{}^j$, for $a = 1, \ldots, N$, are constant matrices called the "generators of the symmetry".

In this more general case, consider a Lagrangian that (for simplicity) has no explicit time dependence, so that $L = L(q_i, \dot{q}_i)$. Then the variation of the action under the symmetry is

$$\begin{aligned} \delta S &= \int dt \left(\frac{\partial L}{\partial q_i} \delta q_i + \frac{\partial L}{\partial \dot{q}_i} \delta \dot{q}_i \right) \\ &= \int dt \left[\delta q_i \left(\frac{\partial L}{\partial q_i} - \frac{d}{dt} \left(\frac{\partial L}{\partial \dot{q}_i} \right) \right) + \frac{d}{dt} \left(\frac{\partial L}{\partial \dot{q}_i} \delta q_i \right) \right], \end{aligned} \tag{12.5}$$

where in the second line we have used $\delta \dot{q}_i = \frac{d}{dt}(\delta q_i)$ (since $[\partial_t, \delta] = 0$) and partial integration.

If the transformation above is a symmetry then the action is invariant, $\delta S = 0$. Moreover, using the classical Lagrange equations of motion (so, "on-shell"), we find that

$$0 = \int_{t_1}^{t_2} dt \frac{d}{dt} \left(\frac{\partial L}{\partial \dot{q}_i} \delta q_i \right) = \epsilon^a \left(\frac{\partial \mathcal{L}}{\partial \dot{q}_i} (iT_a)_i{}^j q_j \right) \Bigg|_{t_1}^{t_2}, \tag{12.6}$$

which vanishes for any ϵ^a, allowing us to peel off (factorize out) the ϵ^a, and, for any t_1, t_2, which means that the quantity, a priori time dependent,

$$Q_a \equiv \left(\frac{\partial L}{\partial \dot{q}_i} (iT_a)_i{}^j q_j \right) (t), \tag{12.7}$$

is actually time independent, and is known as a "conserved charge" associated with the symmetry (12.4). This is the more common version of the Noether theorem in classical mechanics.

The Noether theorem allows for one more generalization, for the case where the Lagrangian is not invariant, but only the action is invariant. The Lagrangian can change by a total derivative, so

$$\delta S = \int dt \frac{d}{dt} (\epsilon^a \tilde{Q}_a) = \epsilon^a \tilde{Q}_a \Big|_{t_1}^{t_2}, \tag{12.8}$$

which vanishes only because of the boundary conditions on $\tilde{Q}_a$ at t_1, t_2. Then, actually instead of Q_a, it is

$$(Q_a - \tilde{Q}_a)(t) \tag{12.9}$$

that is independent of time.

We can make a number of observations about this analysis:

- Observation 1. The total new variable, after the transformation, is

$$q_i' = q_i + \delta q_i = \left(\delta_i^j + \epsilon^a (iT_a)_i{}^j \right) q_j \equiv M_i{}^j q_j, \tag{12.10}$$

where the matrix $M_i{}^j$ defines the linear transformation of q_i.

- Observation 2. Consider the time translation, $t \to t + \delta t$; under it q_i varies by $\delta q_i = \dot{q}_i \delta t$. If time translation is an invariance (so that the *action* is invariant under it), the infinitesimal conserved charge and the variation of the Lagrangian are

$$\begin{aligned} \epsilon^a Q_a &= \frac{\partial L}{\partial \dot{q}_i} \dot{q}_i \delta t \\ \delta L &= \partial_t L \delta t, \end{aligned} \tag{12.11}$$

and from the last relation we deduce that $\partial_t L = \partial_t \tilde{Q} \Rightarrow L = \tilde{Q}$. Substituting $p_i = \partial L/\partial \dot{q}_i$, we find the conservation law

$$0 = \frac{d}{dt}\left(\frac{\partial L}{\partial \dot{q}_i}\dot{q}_i - L\right) = \frac{d}{dt}(p_i \dot{q}_i - L) = \frac{d}{dt}H, \tag{12.12}$$

meaning that the Hamiltonian, or energy, is the Noether charge associated with time translations.

- Observation 3. We can write the linear and continuous variation of the q_i as a Poisson bracket for the charge. Indeed, we find that

$$Q_a - \frac{\partial L}{\partial \dot{q}_i}(iT_a)_i{}^j q_j - p_i (iT_a)_i{}^j q_j, \tag{12.13}$$

so the variation of q_i is the Poisson bracket

$$\delta q_k = -\{\epsilon^a Q_a, q_k\}_{P.B.} = \delta^i_k \epsilon^a (iT_a)_i{}^j q_j = \epsilon^a (iT_a)_k{}^j q_j, \tag{12.14}$$

as written before. Here we have used the fundamental Poisson bracket $\{q_k, p^i\} = \delta^i_k$.

- Observation 4. We can generalize the formalism to *discrete* symmetries (meaning that there is no infinitesimal version), by writing the finite transformation

$$q_i \to q_i' = M_i{}^j q_j, \tag{12.15}$$

instead of $M = \mathbf{1} + \mathcal{O}(\epsilon)$. Then, if S changes to $S' = S$ or, more restrictively, if L changes to $L' = L$, the matrix M defines the action of the symmetry on the variables q_i.

12.2 Symmetries in Quantum Mechanics: General Formalism

In quantum mechanics, the classical matrix $M_i{}^j$ becomes the operator $\hat{M}_i^j$ and the charge Q_a becomes the charge operator $\hat{Q}_a$, which as we saw, can be defined to be a function of the phase space variables, now operators, $\hat{q}_i, \hat{p}_i$.

The linear transformation, written as a Poisson bracket, with the quantization rule becomes ($1/(i\hbar)$ times) the commutator,

$$\delta q_k = -\{\epsilon^a Q_a, q_k\}_{P.B.} \to -\frac{1}{i\hbar}[\epsilon^a \hat{Q}_a, \hat{q}_k]. \tag{12.16}$$

This is a natural relation in canonical quantization.

In classical mechanics, we focused on symmetries in the Lagrangian formalism, when the action (or sometimes the Lagrangian) is invariant, $\delta S = 0$. However, in the Hamiltonian formalism, useful for quantum mechanics, the equivalent statement is the invariance of the Hamiltonian, $\delta H(q_i, p_i) = 0$. In the classical mechanics version, this means that

$$\delta H(q_i, p_i) = -\{\epsilon^a Q_a, H\}_{P.B.} = 0, \tag{12.17}$$

which translates in quantum mechanics into

$$\frac{1}{i\hbar}[\epsilon^a \hat{Q}_a, \hat{H}] = 0 \;\Rightarrow\; [\hat{Q}_a, \hat{H}] = 0, \tag{12.18}$$

i.e., that the transformed Hamiltonian operator is the same as the original one,

$$\hat{H}' \equiv \hat{Q}_a \hat{H} \hat{Q}_a^{-1} = \hat{H}. \tag{12.19}$$

More importantly, the Ehrenfest theorem, stating that the quantum averages on states should have the same property as the classical values, must hold.

In the case of symmetry transformations, we can have two points of view.

(1) The *active* point of view, where a transformation changes the state of the quantum system,

$$|\psi\rangle \to |\psi'\rangle = \hat{U}|\psi\rangle, \tag{12.20}$$

but not the operators.

Under this change, the energy, or quantum average of the Hamiltonian, the equivalent of the classical quantity, should be invariant. Thus

$$\begin{aligned}\langle\psi'|\hat{H}|\psi'\rangle &= \langle\psi|\hat{U}^\dagger \hat{H}\hat{U}|\psi\rangle \\ &= \langle\psi|\hat{H}|\psi\rangle,\end{aligned} \tag{12.21}$$

while the norms of the states should also be invariant,

$$\langle\psi'|\psi'\rangle = \langle\psi'|\hat{U}^\dagger\hat{U}|\psi\rangle = \langle\psi|\psi\rangle, \tag{12.22}$$

implying that the operator $\hat{U}$ should be unitary, $\hat{U}^\dagger = \hat{U}^{-1}$. Then the invariance of the quantum average of the Hamiltonian above, for any state $|\psi\rangle$, means that the transformed operator equals the original one,

$$\hat{U}^{-1}\hat{H}\hat{U} = \hat{H} \quad \Rightarrow \quad [\hat{U}, \hat{H}] = 0, \tag{12.23}$$

so that the Hamiltonian commutes with the operator $\hat{U}$, representing the symmetry transformation on states.

(2) The *passive* point of view, where the transformation acts on operators but not on states,

$$\hat{A} \to \hat{A}' = \hat{U}^{-1}\hat{A}\hat{U}, \quad |\psi\rangle \to |\psi\rangle. \tag{12.24}$$

This point of view should lead to the same results as the active one: we can see that the quantum average transforms in the same way.

But, the change from an active to a passive point of view is more general than its application above to invariances of the Hamiltonian. It is in fact valid simply for *transformations* of the states or operators, corresponding to classical transformations. By the Ehrenfest theorem, the quantum version of the classical transformations of an observable A are transformed quantum averages for the quantum operator $\hat{A}$,

$$\langle A'\rangle = \langle\psi|\hat{A}'|\psi\rangle \;\text{ or }\; \langle\psi'|\hat{A}|\psi'\rangle, \tag{12.25}$$

where $\hat{A}' = \hat{U}^{-1}\hat{A}\hat{U}$.

Note that the invariance of the quantum Hamiltonian $\hat{H}$ under a symmetry transformation with operator $\hat{Q}_a$ implies a degeneracy of the eigenstates of the Hamiltonian. Indeed, if $|\psi\rangle$ is an eigenstate of $\hat{H}$ then so is $\hat{Q}_a|\psi\rangle$, since

$$\hat{H}(\hat{Q}_a|\psi\rangle) = \hat{Q}_a(\hat{H}|\psi\rangle) = E_\psi(Q_a|\psi\rangle). \tag{12.26}$$

12.3 Example 1. Translations

A classical translation of the coordinate q_i acts on classical phase space, and the corresponding quantum averages, as

$$\begin{aligned} q_i &\to q_i + a_i \Rightarrow \langle q_i \rangle \to \langle q_i \rangle' = \langle q_i \rangle + a_i \\ p_i &\to p_i \Rightarrow \langle p_i \rangle \to \langle p_i \rangle. \end{aligned} \tag{12.27}$$

But in Chapter 5 we saw that a translation by δq is represented by an operator

$$\hat{T}(\delta q) = \hat{\mathbb{1}} - i\delta q \hat{K}, \tag{12.28}$$

where $\hat{K}$ is Hermitian, $\hat{K}^\dagger = \hat{K}$, and acts on wave functions as

$$\hat{K} \to -i\frac{d}{dq}. \tag{12.29}$$

More precisely, since, as we saw, $\epsilon/(i\hbar)$ appears naturally in the transition from classical mechanics, we write

$$\hat{T}(\epsilon) = \hat{\mathbb{1}} - \frac{i\epsilon}{\hbar}\hat{K}. \tag{12.30}$$

Then, indeed, for a finite transformation by a instead of ϵ,

$$\hat{T}(a) = e^{-i(a/\hbar)\hat{K}} \equiv \sum_{n\geq 0}\left(-\frac{ia}{\hbar}\right)^n \frac{\hat{K}^n}{n!}. \tag{12.31}$$

This is so since for wave functions, when $\hat{K} = -id/dq$, this is just Taylor expansion on $\psi(x)$.

By the general definition,

$$\hat{T}(a)|q\rangle = |q + a\rangle, \tag{12.32}$$

and this means that for the wave function in the coordinate (q) representation,

$$\begin{aligned} \langle q|\hat{T}(a)|\psi\rangle &= \psi(q + a) \\ &= \hat{T}(a)\psi(q), \end{aligned} \tag{12.33}$$

as required.

In the *passive* point of view, the action of translation on the operators $\hat{q}$ is

$$\hat{q}' = \hat{T}^{-1}(\epsilon)\hat{q}\hat{T}(\epsilon) = \hat{q} + \epsilon\hat{\mathbb{1}}, \tag{12.34}$$

since then, indeed, the action on the quantum averages is

$$\langle q \rangle_\psi \to \langle q \rangle'_\psi = \langle q \rangle_\psi + \epsilon. \tag{12.35}$$

12.4 Example 2. Time Translation Invariance

As we saw, at the classical level, time translation invariance leads to a Hamiltonian that is invariant in time (conserved), $dH/dt = 0$, which by the Ehrenfest theorem becomes in quantum mechanics

$$\left\langle \frac{d\hat{H}}{dt} \right\rangle_{\psi} = 0. \tag{12.36}$$

This means that if we start *with a given state* at $t = 0$ or $t = \tau$ and evolve in time by t, we should obtain the same state, up to a phase $e^{i\alpha(t,\tau)}$ at the most, for any state:

$$|\psi(t)\rangle = \hat{U}(t,0)|\psi(0)\rangle = e^{i\alpha(t,\tau)}|\psi'(t+\tau)\rangle = e^{i\alpha(t,\tau)}\hat{U}(t+\tau,\tau)|\psi'(\tau)\rangle, \tag{12.37}$$

where $|\psi'(\tau)\rangle = |\psi(0)\rangle = |\psi_0\rangle$ is the same state.

That in turn implies that in fact we have a relation between evolution operators,

$$\hat{U}(t,0) = e^{i\alpha(t,\tau)}\hat{U}(t+\tau,\tau), \tag{12.38}$$

for any t and τ.

But, since in the infinitesimal form $\hat{U} \simeq \hat{\mathbb{1}} - \frac{i}{\hbar}\hat{H}dt$, we obtain

$$\mathbb{1} - \frac{i}{\hbar}\hat{H}(0)dt = (1 + if(\tau)dt)\left(\mathbb{1} - \frac{i}{\hbar}\hat{H}(\tau)dt\right), \tag{12.39}$$

so that finally

$$\hat{H}(\tau) = \hat{H}(0) + \hbar f(\tau)\hat{\mathbb{1}}. \tag{12.40}$$

However, in any case the function $f(\tau)$ coming from the expansion of the phase $e^{i\alpha(t,\tau)}$ is trivial and can be absorbed in redefinitions, so we can just drop it, resulting in the fact that the Hamiltonian operator is time independent, as expected from the classical analysis.

Equivalently, and dropping the phase $e^{i\alpha}$ from the beginning, we have

$$\begin{aligned} |\psi(t_1 + dt)\rangle &\simeq \left(\mathbb{1} - \frac{idt}{\hbar}\hat{H}(t_1)\right)|\psi_0\rangle \\ |\psi(t_2 + dt)\rangle &\simeq \left(\mathbb{1} - \frac{idt}{\hbar}\hat{H}(t_2)\right)|\psi_0\rangle, \end{aligned} \tag{12.41}$$

so we obtain

$$\frac{d\hat{H}}{dt} = 0, \tag{12.42}$$

which in particular implies also that

$$\left\langle \frac{d\hat{H}}{dt} \right\rangle_{\psi} = 0, \quad \forall|\psi\rangle, \tag{12.43}$$

as needed.

12.5 Mathematical Background: Review of Basics of Group Theory

General Linear and Continuous Symmetry

In general, we can write for the transformation operator,

$$\hat{M} = \hat{\mathbb{1}} - \frac{i\epsilon^a}{\hbar}\hat{Q}_a. \tag{12.44}$$

Then the transformed operator is

$$\hat{q}'_i = \hat{M}^{-1}\hat{q}_i\hat{M} = \left(\mathbb{1} + \frac{i\epsilon^a}{\hbar}\hat{Q}_a\right)\hat{q}_i\left(\mathbb{1} - \frac{i\epsilon^a}{\hbar}\hat{Q}_a\right) \simeq \hat{q}'_i - \frac{1}{i\hbar}[\epsilon^a\hat{Q}_a, \hat{q}_i]. \tag{12.45}$$

Thus we can represent the operator $\hat{Q}_a$ as

$$\hat{Q}_a = \sum_{i,j}\hat{p}^i(iT_a)_i{}^j\hat{q}_j \Rightarrow -\frac{1}{i\hbar}[\epsilon^a\hat{Q}_a, \hat{q}_k] = \epsilon^a(iT_a)_i{}^j\hat{q}_j = \delta\hat{q}_i. \tag{12.46}$$

Groups and Invariance under Groups

Symmetry transformations form a *group*.

We say we have a group G if we have a set of elements, together with a multiplication between them, represented as $G \times G \to G$, such that we have the following properties:

(a) The multiplication respects the group, i.e., $\forall f, g \in G$, $h = f \cdot g \in G$.
(b) The multiplication is associative, i.e., $\forall f, g, h \in G$, $(f \cdot g) \cdot h = f \cdot (g \cdot h)$.
(c) There is an element e called the identity, such that $e \cdot f = f \cdot e = f, \forall f \in G$.
(d) There is an element called the inverse, f^{-1}, associated with any $f \in G$, such that $f \cdot f^{-1} = f^{-1} \cdot f = e$.

Indeed, in the case of symmetries $q_i \to M_i{}^j q_j$, with $M \in G$, $\forall\, M_1, M_2 \in G$, $M_1 \cdot M_2 = M_3 \in G$. Also, matrix multiplication is associative and admits an inverse.

As examples, we can consider both time translation and translation of some q_i by a, so that the operator is $T(a)$. Then indeed

$$T(a_1) \cdot T(a_2) = T(a_1 + a_2). \tag{12.47}$$

Example of a Group: $\mathbb{Z}_2$

The simplest group is the group $\mathbb{Z}_2$, which has just two elements, $a = e$ and b.

It can be represented on $\mathbb{R}$ as the numbers $a = e = +1$ and $b = -1$. This implies that the *multiplication table* for the group is

$$a \cdot a = a, \quad b \cdot b = a, \quad a \cdot b = b \cdot a = b. \tag{12.48}$$

It also means that the group is *Abelian*, so that $g_1 \cdot g_2 = g_2 \cdot g_1$, $\forall\, g_1, g_2 \in G$. Abelian groups are named after the Norwegian mathematician Niels Henrik Abel, whose name is associated with the most important prize in mathematics, the Abel prize.

To have invariance of a system under a group G in classical mechanics means that we need to have invariance under the transformations $q_i \to g q_i$, $\forall g \in \mathbb{G}$. Thus $L(g q_i) = L(q_i)$, or at least $S[g q_i] = S[q_i]$ (invariant action).

Example of an Invariant System under $\mathbb{Z}_2$ Acting on q

As a simple example of invariance under $\mathbb{Z}_2$, consider a system with one real coordinate $q \in \mathbb{R}$, but with a potential depending only on its modulus, $V = V(|q|)$, so that

$$L = m\frac{\dot{q}^2}{2} - V(|q|). \tag{12.49}$$

Then $q' = gq$ for $g = a$ or b. For the case $g = a = +1$, we obtain $q' = q$, so this is trivially a symmetry. For the case $g = b = -1$, we need $L(-q) = L(q)$. But, indeed, $\dot{q}^2 = (-\dot{q})^2$ and $|-q| = |q|$, implying $L(-q) = L(q)$. Then also $S[q] = S[-q]$ (the action is invariant), though the reverse is not true in general (invariance of the action invariant doesn't imply invariance of the Lagrangian).

Example of a System with Invariance of the Action but not of the Lagrangian

We can modify the above example in such a way that the action is still invariant, $S[-q] = S[q]$, but the Lagrangian is not. Consider then the new Lagrangian

$$L = m\frac{\dot{q}^2}{2} - V(|q|) + \alpha\frac{d}{dt}q, \tag{12.50}$$

together with the boundary condition $q(t_2) = q(t_1)$. Then we have explicitly that $L(-q) \neq L(q)$, since the new term is odd under $q \to -q$, but the action is invariant, since the new term gives

$$\alpha \int_{t_1}^{t_2} dt \frac{d}{dt} q = \alpha(q(t_2) - q(t_1)) = 0, \tag{12.51}$$

because of the boundary condition.

Generalization: Cyclic Groups

The next ("cyclic") group in terms of dimension is $\mathbb{Z}_3$, made up of three elements, $\{e, a, b\}$, that can be represented in $\mathbb{C}$ as complex numbers, equal to the third roots of unity, i.e., $x \in \mathbb{C}$ such that $x^3 = 1$. We have then

$$e = 1 = e^0, \quad a = e^{2\pi i/3}, \quad b = e^{4\pi i/3}, \tag{12.52}$$

which implies the multiplication table

$$a^2 = b, \quad b^2 = a, \quad ab = ba = e. \tag{12.53}$$

This is also an Abelian group, as we can easily see.

As an example of a classical Lagrangian that is invariant under the above group, consider a generalization of it. Consider a complex variable $q = q_1 + iq_2$, $q_1, q_2 \in \mathbb{R}$, and

$$L(q) = m\frac{|\dot{q}|^2}{2} - V(q^3). \tag{12.54}$$

Then, under $q \to q' = gq$, we have $|\dot{q}| = |g\dot{q}| = |\dot{q}|$ and

$$q^3 \to q'^3 = (gq)^3 = q^3, \tag{12.55}$$

since $g^3 = 1$ for all $g \in \mathbb{Z}_3$.

Our final generalization is to the N-element cyclic group $\mathbb{Z}_N$, with elements $\{e, a_1, \dots, a_{N-1}\}$ that can be represented in the complex numbers as the Nth roots of unity, g, such that $g^N = 1$, specifically

$$e = 1, a_1 = e^{2\pi i/N}, \dots, a_{N-1} = e^{2\pi i(N-1)/N}. \tag{12.56}$$

An example of a Lagrangian invariant under this group is a further generalization of the same Lagrangian as before, for a complex variable q,

$$L(q, \dot{q}) = m\frac{|\dot{q}|^2}{2} - V(q^N), \tag{12.57}$$

since then indeed

$$q^N \to q'^N = (gq)^N = q^N. \tag{12.58}$$

$\mathbb{Z}_N$-Invariant System in Quantum Mechanics

We can translate the invariance of the above classical system into a quantum mechanical invariance. For that, we define the quantum Hamiltonian

$$\hat{H} = m\frac{|\hat{\dot{q}}|^2}{2} + V(\hat{q}^N). \tag{12.59}$$

Then we can check that the Hamiltonian is invariant under $\hat{g} \in \mathbb{Z}_N$, acting as $\hat{q} \to \hat{g}^{-1}\hat{q}\hat{g}$, since

$$\hat{H}' = \hat{g}^{-1}\hat{H}\hat{g} = m\frac{|\hat{g}^{-1}\hat{\dot{q}}\hat{g}|^2}{2} + V\left((\hat{g}^{-1}\hat{q}\,\hat{g})^N\right) = \hat{H}. \tag{12.60}$$

Representations of Groups

So far, we have defined the groups $\mathbb{Z}_N$ by the complex numbers that define their multiplication table. But it is important to understand that we have different kinds of *representations* of the group, in terms of different kinds of objects. For a representation R, we say that the element g is represented by $D_R(g)$.

In the case of the fundamental (defining) representation of the $\mathbb{Z}_N$ defined above, since all the elements g are complex numbers, we have a one-dimensional complex vector space. But for $\mathbb{Z}_N$ with $N \geq 3$, there is (at least) one other representation, called the *regular* representation.

In the case of $\mathbb{Z}_3$, we define it in terms of 3×3 matrices,

$$D(e) = \begin{pmatrix} 1 & 0 & 0 \\ 0 & 1 & 0 \\ 0 & 0 & 1 \end{pmatrix}, \quad D(a) = \begin{pmatrix} 0 & 0 & 1 \\ 1 & 0 & 0 \\ 0 & 1 & 0 \end{pmatrix}, \quad D(b) = \begin{pmatrix} 0 & 1 & 0 \\ 0 & 0 & 1 \\ 1 & 0 & 0 \end{pmatrix}, \tag{12.61}$$

so it permutes the three elements of the vector space between each other.

This is a three-dimensional representation, meaning that these are matrix operators acting on a three-dimensional vector space $\begin{pmatrix} x \\ y \\ z \end{pmatrix}$. Consider a Cartesian basis for this vector space, and denote its components according to the three elements of the group, as

$$|e\rangle = \begin{pmatrix} 1 \\ 0 \\ 0 \end{pmatrix} \equiv |e_1\rangle, \quad |a\rangle = \begin{pmatrix} 0 \\ 1 \\ 0 \end{pmatrix} \equiv |e_2\rangle, \quad |b\rangle = \begin{pmatrix} 0 \\ 0 \\ 1 \end{pmatrix} \equiv |e_3\rangle. \tag{12.62}$$

The notation has been chosen in such a way that we have

$$D(g_1)|g_2\rangle = |g_1 g_2\rangle \equiv |h\rangle, \tag{12.63}$$

as we can check explicitly. Moreover, since the states are orthonormal, $\langle e_i | e_j \rangle = \delta_{ij}$, we find that the matrix elements of $D(g)$ are given by

$$(D(g))_{ij} = \langle e_i | D(g) | e_j \rangle = \langle e_i | D(g) e_j \rangle. \tag{12.64}$$

This regular representation is for the same group as the group defined by the roots of unity, but it is *inequivalent* to it, since the vector space has a different dimension (one versus three).

Equivalent Representations

Equivalent representations means that, first, the spaces must have the same dimension, and second, there must be a similarity transformation S, defining a change of basis, under which we get from one representation to the other, i.e.,

$$D(g) \to D'(g) = S^{-1} D(g) S. \tag{12.65}$$

Indeed, in this case, the matrix elements are the same,

$$(D'(g))_{ij} \equiv \langle e'_i | D'(g) | e'_j \rangle = \langle e'_i | S^{-1} D(g) S | e'_j \rangle = \langle e_i | D(g) | e_j \rangle = (D(g))_{ij}, \tag{12.66}$$

where in the second equality we have used the change of basis $|e_i\rangle = S|e'_j\rangle$ and the unitarity of S, $S^{-1} = S^\dagger$, which is true for $\mathbb{Z}_N$ and for all the classical Lie groups.

Reducible Representations

If there is an invariant subspace $\mathcal{H} \subset \mathcal{G}$ in the representation vector space, i.e.,

$$\forall |h\rangle \in \mathcal{H}, \quad D(g)|h\rangle \in \mathcal{H}, \quad \forall g \in G, \tag{12.67}$$

we say that we have a reducible representation. If not, we say we have an irreducible one.

A reducible representation is always block-diagonal,

$$D(g) = \begin{pmatrix} D(h) & 0 & 0 & \cdots \\ 0 & D(\tilde{h}) & 0 & \\ 0 & 0 & D(\tilde{\tilde{h}}) & \\ & & & \cdots \end{pmatrix}. \tag{12.68}$$

Indeed, in this case, the vector space splits as

$$\begin{pmatrix} |h\rangle \\ |\tilde{h}\rangle \\ |\tilde{\tilde{h}}\rangle \end{pmatrix}, \tag{12.69}$$

such that then, for each subspace, $D(g)|h\rangle \in \mathcal{H}$, etc.

An irreducible representation is a representation that is not reducible.

Important Concepts to Remember

- A symmetry is a transformation on the variables of the Lagrangian that leaves the action invariant.
- The Noether theorem says that for any symmetry of the theory there is a conserved charge.
- For a continuous linear symmetry, $\delta q_i = \sum_a \epsilon^a (T_a)_i{}^j q_j$, the conserved charge (invariant in time) is $Q_a = \frac{\partial L}{\partial \dot{q}_i}(T_a)_i{}^j q_j$.
- The momenta p^i are the charges associated with invariance under translations in q_i, and the Hamiltonian is the charge associated with time translation invariance.
- Linear transformations can be written as $\delta q_k = -\{\epsilon^a Q_a, q_k\}_{P.B.}$.
- Invariance in quantum mechanics means that $[\hat{Q}_a, \hat{H}] = 0$, corresponding to $\{Q_a, H\}_{P.B.} = 0$.
- Symmetry in quantum mechanics can be understood either as an active transformation, on states, $|\psi\rangle \to \hat{U}|\psi\rangle$, but not on operators, or as a passive transformation on operators, $\hat{A} \to \hat{A}' = \hat{U}^{-1}\hat{A}\hat{U}$, but not on states. Either way, the expectation values transform in the same manner.
- Symmetry transformations form a group.
- The simplest groups are the discrete cyclic groups $\mathbb{Z}_N$; these can be represented by the Nth roots of unity, which are Abelian groups, i.e., $g_1 g_2 = g_2, g_1, \forall g_1, g_2 \in G$.
- Groups have different representations; for $\mathbb{Z}_N$, we have the fundamental representation as Nth roots of unity with a multiplication table, and the regular representation as real matrices acting on the group elements as elements in an N-dimensional vector space, etc.
- Equivalent representations can be mapped to each other; reducible representations have block-diagonal matrices and so can be reduced (split) into lower-dimensional representations; irreducible representations cannot.

Further Reading

See [8] for more on group theory for quantum mechanics.

Exercises

(1) Consider the Lagrangian ($q, \tilde{q} \in \mathbb{C}$)

$$L = \frac{m}{2}(|\dot{q}|^2 + |\dot{\tilde{q}}|^2) - V((q^2 + \tilde{q}^2)^2). \tag{12.70}$$

What are its symmetries? What are the representations of the group in which $q, \tilde{q}$ belong?

(2) Calculate the Noether charge for the continuous symmetry(ies) in exercise 1, and check that the infinitesimal variations of $q, \tilde{q}$ are indeed generated by the Noether charge via the Poisson bracket with $q, \tilde{q}$.

(3) Consider two harmonic oscillators of the same mass and frequency, with Hamiltonian

$$H = \frac{1}{2m}(p_1^2 + p_2^2) + \frac{k}{2}(x_1^2 + x_2^2). \tag{12.71}$$

Find the continuous symmetries of the system and write down the resulting conserved charges as a function of the phase space variables. Then quantize the system, and show that the charges do indeed commute with the quantum Hamiltonian.

(4) For the system in exercise 3, write down the equation of motion for $x_1^2 + x_2^2 \equiv r^2$ and then the corresponding Ehrenfest theorem equation and its transformed version under the continuous symmetries, in both the active and the passive sense.

(5) Consider the Lagrangian for $q \in \mathbb{C}$

$$L = \frac{m}{2}|\dot{q}|^2 - V(q), \tag{12.72}$$

in the cases

$$\begin{aligned} V &= \frac{1}{q^3} + \frac{1}{\bar{q}^3} + \frac{1}{q^5} + \frac{1}{\bar{q}^5} \quad \text{(I)} \\ V &= \frac{1}{|q|^3} + \frac{1}{|q|^5} \quad \text{(II)} \\ V &= \frac{1}{q^3} \quad \text{(III)}. \end{aligned} \tag{12.73}$$

What are the symmetries in each case?

(6) Consider the matrices

$$A = \begin{pmatrix} 1 & 0 & 0 \\ 0 & 1 & 0 \\ 0 & 0 & 1 \end{pmatrix}, \quad B = \begin{pmatrix} 0 & 0 & 1 \\ 0 & 1 & 0 \\ 1 & 0 & 0 \end{pmatrix}. \tag{12.74}$$

(a) Do they form a representation of $\mathbb{Z}_2$? Why?
(b) If so, is the representation reducible?
(c) If so, is this a *regular* representation?
(d) If so, is this equivalent with the roots of unity representation?

(7) Write down the generalization of the regular representation for $\mathbb{Z}_3$ to the $\mathbb{Z}_N$ case, for a cyclic permutation by one step of the basis elements.

13 Symmetries in Quantum Mechanics II: Discrete Symmetries and Internal Symmetries

After analyzing mostly continuous spacetime symmetries (spatial and temporal translations), as well as considering the general theory of symmetries and of discrete groups, we move on to applications of discrete symmetries in quantum mechanics.

13.1 Discrete Symmetries: Symmetries under Discrete Groups

We start with $\mathbb{Z}_2$-type symmetries (symmetries that can be described in terms of the group $\mathbb{Z}_2$). We consider symmetries that are present at the classical level also:

- Parity symmetry P, or spatial inversion (like mirror reflection, but for all spatial coordinates instead of just one). By definition, that means that the space gets a minus sign. But inverting the direction of space also inverts the direction of momentum, so we have

$$\vec{x} \to \vec{x}\,' = -\vec{x}, \quad \vec{p} \to \vec{p}\,' = -\vec{p}. \tag{13.1}$$

  In the presence of spin (understood as rotation around the direction of the momentum), spin does not get inverted; see Fig. 13.1a.
- Time reversal invariance T, inverting the direction of time,

$$t \to -t. \tag{13.2}$$

  Of course, unlike parity, which could be thought of as a simple change of reference frame, inverting the direction of time is not physically possible. Moreover, we know that there are phenomena such as the increase in entropy that define the arrow of time. But in this case, we simply mean "filming" the evolution in time of a system and rolling it backwards. In other words, consider the dynamics of the system and change t into $-t$ in its equations.
- "Internal" symmetries, i.e., symmetries that are not associated with spacetime but rather with some internal degrees of freedom. Examples are "charge conjugation", which changes particles into antiparticles, R-parity in supersymmetric theories, isospin parity in the Standard Model of particle physics, etc.

Another example of a discrete symmetry is lattice translation (translation by a fixed amount a) in condensed matter, but we will not analyze it here.

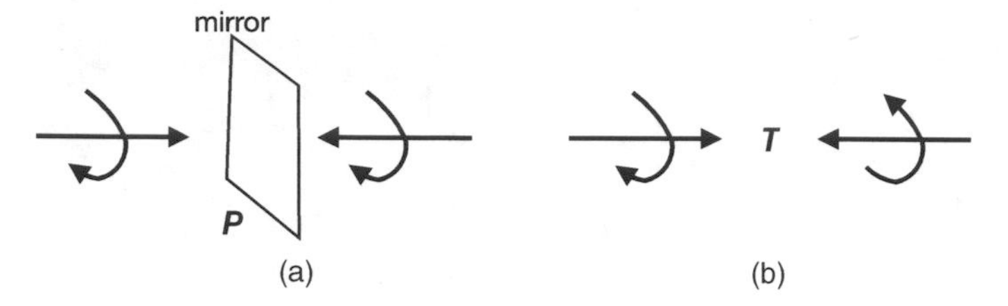

Figure 13.1 (a) Parity is understood as inversion in a mirror: momentum gets inverted, spin (rotation around the direction of the momentum) does not. (b) Time reversal invariance: both momentum and spin get inverted.

13.2 Parity Symmetry

Corresponding to the classical parity symmetry acting on coordinate and momenta by inverting them, we defined a parity operator $\hat{\pi}$ that acts on states and/or operators. Specifically, for eigenstates of $\hat{X}$ and $\hat{P}$, in the *active* point of view,

$$\hat{\pi}|x\rangle = |-x\rangle, \quad \hat{\pi}|p\rangle = |-p\rangle. \tag{13.3}$$

Equivalently, in the *passive* point of view, with the parity operator acting on operators,

$$\begin{aligned} \hat{X}' &= \hat{\pi}^{-1}\hat{X}\hat{\pi} = -\hat{X} \\ \hat{P}' &= \hat{\pi}^{-1}\hat{P}\hat{\pi} = -\hat{P}. \end{aligned} \tag{13.4}$$

This means that $[\hat{\pi}, \hat{X}] \neq 0, [\hat{\pi}, \hat{P}] \neq 0$.

Either way, we find the correct transformation law for the quantum averages of operators, which by the Ehrenfest theorem are the quantum equivalents of classical transformations,

$$\begin{aligned} \langle x\rangle'_\psi &= \langle\psi|\hat{\pi}^{-1}\hat{X}\hat{\pi}|\psi\rangle = -\langle\psi|\hat{X}|\psi\rangle = -\langle x\rangle_\psi \\ \langle p\rangle'_\psi &= \langle\psi|\hat{\pi}^{-1}\hat{P}\hat{\pi}|\psi\rangle = -\langle\psi|\hat{P}|\psi\rangle = -\langle p\rangle_\psi. \end{aligned} \tag{13.5}$$

Moreover, we can find the action of parity on wave functions $\psi(x) = \langle x|\psi\rangle$ by introducing the completeness relation for $|x\rangle$ states in $\langle x|\hat{\pi}|\psi\rangle$,

$$\langle x|\hat{\pi}|\psi\rangle = \int dx'\langle x|\hat{\pi}|x'\rangle\langle x'|\psi\rangle = \int dx'\langle x|-x'\rangle\langle x'|\psi\rangle = \langle -x|\psi\rangle = \psi(-x), \tag{13.6}$$

where we have used the orthonormal relation $\langle x'|x\rangle = \delta(x - x')$.

Some observations about the parity operator: carrying out the classical parity operation twice, we get back to the same situation. Correspondingly, we find that the parity operator applied twice also gives the identity,

$$\hat{\pi}^2|x\rangle = |x\rangle \;\Rightarrow\; \hat{\pi}^2 = \hat{\mathbb{1}} \Rightarrow \hat{\pi}^{-1} = \hat{\pi}. \tag{13.7}$$

Since π is a symmetry transformation, $\hat{\pi}$ is also unitary, according to the general theory, so for all $\hat{\pi}$ we have $\hat{\pi}^\dagger = \hat{\pi}^{-1}$.

Since $\hat{\pi}^2 = \hat{\mathbb{1}}$, the eigenvalues of $\hat{\pi}$ are ± 1, so for *parity eigenstates* we have

$$\hat{\pi}|\psi\rangle = \pm|\psi\rangle. \tag{13.8}$$

Then, for the wave functions of these parity eigenstates we have

$$\psi(-x) = \pm\psi(x), \tag{13.9}$$

which are called *even parity* and *odd parity*, respectively.

But not all states are parity eigenstates. In particular, eigenstates of $\hat{X}, \hat{P}$ are not, since, as we saw, $\hat{\pi}|x\rangle = |-x\rangle$ and $\hat{\pi}|p\rangle = |-p\rangle$. Also, as we saw, $[\hat{\pi}, \hat{X}] \neq 0, [\hat{\pi}, \hat{P}] \neq 0$, so we cannot have simultaneously an eigenstate of $\hat{X}$ or $\hat{P}$ and of $\hat{\pi}$.

However, we can have a Hamiltonian that is invariant under parity,

$$\hat{\pi}\hat{H}\hat{\pi}^{-1} = \hat{H} \Rightarrow [\hat{\pi}, \hat{H}] = 0, \tag{13.10}$$

which means that such a Hamiltonian *can* have an eigenstate simultaneously with π. Then, if ($[\hat{\pi}, \hat{H}] = 0$ and) $|n\rangle$ is an eigenstate of $\hat{H}$, $\hat{H}|n\rangle = E_n|n\rangle$, and $|n\rangle$ is nondegenerate, it follows that $|n\rangle$ is also a $\hat{\pi}$ eigenstate.

Example: Harmonic Oscillator

For the harmonic oscillator, the ground state $|0\rangle$ is *parity even*, since the wave function is symmetric around 0. Since

$$a^\dagger = \alpha\hat{X} + \beta\hat{P}, \quad \alpha, \beta \in \mathbb{C}, \tag{13.11}$$

which is parity odd (since both $\hat{X}$ and $\hat{P}$ are parity odd), it follows that the first excited state is also *parity odd*, since

$$|1\rangle \propto a^\dagger|0\rangle, \tag{13.12}$$

and, more generally,

$$|n\rangle \propto (a^\dagger)^n|0\rangle \Rightarrow \pi|n\rangle = (-1)^n|n\rangle, \tag{13.13}$$

so the parity of the state $|n\rangle$ is $(-1)^n$.

Parity Selection Rules

We can also consider *selection rules*, meaning rules for a matrix element to be nonzero.

If $|\alpha\rangle, |\beta\rangle$ are parity eigenstates, $|\alpha\rangle$ has parity ϵ_α, and $|\beta\rangle$ has parity ϵ_β then (since $\hat{X}, \hat{P}$ are odd under parity)

$$\langle\beta|\hat{X}|\alpha\rangle = 0 = \langle\beta|\hat{P}|\alpha\rangle, \tag{13.14}$$

unless $\epsilon_\alpha = -\epsilon_\beta$ (if not, the states $|\beta\rangle$ and $\hat{X}|\alpha\rangle, \hat{P}|\alpha\rangle$ are orthogonal, corresponding to different eigenvalues for the Hermitian operator $\hat{\pi}$).

13.3 Time Reversal Invariance, *T*

As we mentioned, the time reversal operator corresponds to the reversal of motion, meaning running backwards the dynamics of the theory, which includes inverting the momentum. This leads to the classical action of T,

$$\vec{x} \to \vec{x}, \quad \vec{p} \to -\vec{p}. \tag{13.15}$$

In classical mechanics, if $\vec{x}(t)$ is a solution to the Newtonian equation of motion,

$$m(\ddot{\vec{x}}) = -\vec{\nabla} V, \tag{13.16}$$

then $\vec{x}(-t)$ is also a solution, since making the replacement $t \to -t$ amounts to replacing d^2/dt^2 by $d^2/d(-t)^2$ in the equation, which leaves it invariant. Defining this as a new solution $\vec{x}\,'(t) \equiv \vec{x}(-t)$, we indeed find that

$$\frac{d^2}{dt^2}\vec{x}\,'(t) = \frac{d^2}{dt^2}\vec{x}(t). \tag{13.17}$$

On the other hand, in quantum mechanics, the Schrödinger equation,

$$i\hbar\frac{\partial}{\partial t}\psi = \left(-\frac{\hbar^2}{2m}\vec{\nabla}^2 + V(\vec{r})\right)\psi, \tag{13.18}$$

is not invariant under time inversion, because of the presence of a single time derivative: if $\psi(\vec{x}, t)$ is a solution, then $\psi(\vec{x}, -t)$ is not a solution, since the time reversed equation is

$$-i\hbar\frac{\partial}{\partial t}\psi = \left(-\frac{\hbar^2}{2m}\vec{\nabla}^2 + V(\vec{r})\right)\psi. \tag{13.19}$$

However, $\psi^*(\vec{x}, -t)$ is a solution to (13.9). Indeed, the complex conjugate of the Schrödinger equation is

$$-i\hbar\frac{\partial}{\partial t}\psi^* = \left(-\frac{\hbar^2}{2m}\vec{\nabla}^2 + V(\vec{r})\right)\psi^*, \tag{13.20}$$

which is the same as the time reversed equation, except that ψ is replaced by ψ^*.

Then, for an energy eigenfunction, the wave function is stationary, and we have

$$\psi(\vec{x}, t) = \psi_n(\vec{x})e^{-iE_n t/\hbar} \Rightarrow \psi^*(\vec{x}, -t) = \psi_n^*(\vec{x})e^{-iE_n t/\hbar}. \tag{13.21}$$

Considering a fixed time, such as $t = 0$, we obtain

$$\psi_n(\vec{x}) = \langle \vec{x}|n\rangle \xrightarrow{T} \psi_n^*(\vec{x}) = (\langle \vec{x}|n\rangle)^* = \langle n|\vec{x}\rangle. \tag{13.22}$$

This suggests that the $\hat{T}$ operator in quantum mechanics is an antilinear and unitary operator, or *anti-unitary* operator, so (according to the mathematics reviewed in the first few chapters)

$$\hat{T}(c|\psi\rangle) = c^*\hat{T}(|\psi\rangle), \quad \forall c \in \mathbb{C}. \tag{13.23}$$

By definition, in the passive point of view the action of $\hat{T}$ on the operators $\hat{X}, \hat{P}$ is the equivalent of the classical action on x, p, i.e.,

$$\hat{T}\hat{\vec{X}}\hat{T}^{-1} = \hat{\vec{X}}, \quad \hat{T}\hat{\vec{P}}\hat{T}^{-1} = -\hat{\vec{P}}. \tag{13.24}$$

Therefore

$$\hat{T}^2 = \hat{\mathbb{1}} \Rightarrow \hat{T}^{-1} = \hat{T}, \tag{13.25}$$

though only in the case without spin. As we will see later, on states with spin, the action of $\hat{T}$ is different.

Having T as an anti-unitary operator solves another issue. Indeed, *for a time-reversal invariant system*, evolving a state by an infinitesimal time dt after the action of T should be equivalent to the action of T after a time evolution by $-dt$. Since in the infinitesimal case $\hat{U}(dt, 0) \simeq \hat{\mathbb{1}} - i\hat{H}dt/\hbar$, we obtain

$$\begin{aligned}\left(\hat{\mathbb{1}} - \frac{i}{\hbar}\hat{H}dt\right) T|\psi\rangle = T\left(\hat{\mathbb{1}} - \frac{i}{\hbar}\hat{H}(-dt)\right)|\psi\rangle \Rightarrow \\ -i\hat{H}\left(T|\psi\rangle\right) = T\left(i\hat{H}|\psi\rangle\right).\end{aligned} \tag{13.26}$$

In the above, we have divided by $dt/\hbar$; both quantities in the ratio are real numbers, so the ratio can be taken out of both linear and antilinear operators.

However, we cannot divide by i, as for a linear operator, and conclude that $\hat{H}\hat{T} = -\hat{T}\hat{H}$, which would be nonsensical, since it would imply that

$$\hat{H}\left(\hat{T}|E_n\rangle\right) = -\hat{T}\hat{H}|E_n\rangle = -E_n\left(\hat{T}|E_n\rangle\right), \tag{13.27}$$

which would give negative eigenvalues of the Hamiltonian, and this is physically impossible.

Instead, if we accept that $\hat{T}$ is antilinear, which is consistent with the previously derived action on wave functions $\psi_n(\vec{x}) \xrightarrow{T} \psi_n^*(\vec{x})$, when dividing out i from $\hat{T}$, it becomes $-i$, so we obtain

$$\hat{H}(\hat{T}|\psi\rangle) = \hat{T}(\hat{H}|\psi\rangle), \tag{13.28}$$

which can be stated as an action on the Hamiltonian,

$$\hat{H}\hat{T} = \hat{T}\hat{H} \Rightarrow \hat{T}\hat{H}\hat{T}^{-1} = \hat{H}, \tag{13.29}$$

saying that the Hamiltonian is time-reversal invariant, which is consistent with classical physics and our assumption.

We note that in the above analysis, we didn't need to define the action of $\hat{T}$ on a "bra" state, $\langle\chi|T$, which would be hard to do since the notation of bra and ket states was defined for linear operators.

For a general observable, associated with a Hermitian operator $\hat{A}$, if the observable has a well-defined action under T, we have

$$\hat{T}\hat{A}\hat{T}^{-1} = \pm\hat{A}, \tag{13.30}$$

which means $\hat{A}$ is even or odd under time reversal. Then, for quantum averages, we have

$$\langle\psi|\hat{A}|\chi\rangle = \pm\langle\hat{T}\psi|\hat{A}|\hat{T}\chi\rangle, \tag{13.31}$$

which is consistent with the Ehrenfest theorem.

Coming back to the action of $\hat{T}$ on wave functions, we note that the antilinear property of $\hat{T}$ implies that action as well: since

$$\hat{T}(c|x\rangle) = c^*\hat{T}|x\rangle = c^*|x\rangle, \tag{13.32}$$

inserting the $|x\rangle$ completeness relation in front of a state $|\psi\rangle$,

$$|\psi\rangle = \int dx|x\rangle\langle x|\psi\rangle = \int dx|x\rangle\psi(x), \tag{13.33}$$

implies the action of T on it as

$$T|\psi\rangle = \int dx \psi^*(x)|x\rangle, \tag{13.34}$$

so we re-obtain the previously found action $\psi(x) \xrightarrow{T} \psi^*(x)$.

We note two more properties of $\hat{T}$:

- If $\hat{H}$ is invariant under $\hat{T}$, and the energy eigenstates $|n\rangle$ are nondegenerate, then $\langle x|n\rangle$ is real, since

$$T : \langle x|n\rangle \to \langle x|n\rangle^* = \langle x|n\rangle. \tag{13.35}$$

- If a system has spin $\vec{S}$, the spin is odd under time reversal, $\vec{S} \xrightarrow{T} -\vec{S}$. The reason is that we can think of the spin as a sort of rotation around the axis of the momentum, and running time backwards means the rotation changes direction; see Fig. 13.1b. In quantum theory, this means that

$$\hat{T}\hat{\vec{S}}\hat{T}^{-1} = -\hat{\vec{S}}. \tag{13.36}$$

A more precise notion has to wait for a better description of spin, here we merely note that for a state of spin j, we have

$$\hat{T}^2|\text{spin } j\rangle = (-1)^{2j}|\text{spin } j\rangle. \tag{13.37}$$

For spinless states ($j = 0$), we recover $\hat{T}^2 = \hat{\mathbb{1}}$.

13.4 Internal Symmetries

Having dealt with discrete spacetime symmetries, we turn to discrete internal symmetries, having to do with internal degrees of freedom.

Consider our prototype, the $\mathbb{Z}_N$ group, specifically $\mathbb{Z}_2$, for the case where there is no spacetime interpretation (unlike for $\hat{\pi}, \hat{T}$).

We considered the example of a $\mathbb{Z}_N$-invariant system with complex variable $\hat{q}$ and Hamiltonian

$$\hat{H} = m\frac{|\hat{q}|^2}{2} + V(\hat{q}^N). \tag{13.38}$$

Then, indeed,

$$\hat{g}\hat{H}\hat{g}^{-1} = \hat{H}, \quad \forall g \in \mathbb{Z}_N. \tag{13.39}$$

Since therefore $[\hat{g}, \hat{H}] = 0$, we obtain a degeneracy of states, with $\hat{g}|E_n\rangle$ having the same energy E_n as $|E_n\rangle$ for all $g \in \mathbb{Z}_N$, meaning that there are N related states of the same energy. However, since in this q space the g_k are represented as complex phases, $g_k = e^{2\pi ik/N}$, the states all represent the same physical state.

Another example is better suited to show the relevant degeneracy: consider another representation of $\mathbb{Z}_N$. In the case of $\mathbb{Z}_2$, we can choose the representation in a two-dimensional vector space permuting the group elements,

$$D(e) = \begin{pmatrix} 1 & 0 \\ 0 & 1 \end{pmatrix}, \quad D(b) = \begin{pmatrix} 0 & 1 \\ 1 & 0 \end{pmatrix}, \tag{13.40}$$

where e and b are elements of $\mathbb{Z}_2$. This representation can be used to construct a relevant case of a $\mathbb{Z}_2$-invariant Hamiltonian, with two real variables q_1 and q_2,

$$\hat{H} = \frac{m}{2}(\dot{\hat{q}}_1^2 + \dot{\hat{q}}_2^2) + V(\hat{q}_1 \cdot \hat{q}_2). \tag{13.41}$$

Indeed, the Hamiltonian is invariant under the above representation of $\mathbb{Z}_2$, with the only nontrivial element $D(b)$ acting on the vector space (q_1, q_2) by interchanging them. Interchanging q_1 and q_2 leaves $\hat{H}$ invariant. Then, in this case also, $|E_n\rangle$ and $D(b)|E_n\rangle$ are states degenerate in energy, but now this means something nontrivial, as $D(b)$ interchanges q_1 and q_2, so one obtains a different energy eigenstate.

We have given an example of $\mathbb{Z}_2$ invariance, where there are two group elements and correspondingly two states (degeneracy 2), but for a finite discrete group with N elements we have N states, so the degeneracy is N.

13.5 Continuous Symmetry

Above we have considered only discrete internal symmetries, but we can also have continuous symmetries that can be both internal (considered later on in the book) and spacetime, e.g., rotations, which will be considered in the next chapter.

As an example of continuous symmetry, consider the $SO(2)$ rotation, with matrix element

$$g = g(\alpha) = \begin{pmatrix} \cos\alpha & \sin\alpha \\ -\sin\alpha & \cos\alpha \end{pmatrix}, \quad \forall \alpha \in [0, 2\pi]. \tag{13.42}$$

This matrix element acts on a two-dimensional vector $\begin{pmatrix} x \\ y \end{pmatrix}$ (for a rotational symmetry in the plane defined by Cartesian coordinates x and y) or $\begin{pmatrix} q_1 \\ q_2 \end{pmatrix}$ (for an internal symmetry acting on two real variables as before).

Consider the case of an internal symmetry for a system with two variables q_1 and q_2,

$$\begin{aligned} L &= \frac{m}{2}(\dot{q}_1^2 + \dot{q}_2^2) - V(q_1^2 + q_2^2) = T - V \quad \Rightarrow \\ \hat{H} &= \frac{m}{2}(\dot{\hat{q}}_1^2 + \dot{\hat{q}}_2^2) + V(\hat{q}_1^2 + \hat{q}_2^2) = \hat{T} + \hat{V}. \end{aligned} \tag{13.43}$$

But then, under the $SO(2)$ internal symmetry, the terms in the Hamiltonian transform as

$$\begin{aligned} q_1^2 + q_2^2 &\to (q_1 \cos\alpha + q_2 \sin\alpha)^2 + (-q_1 \sin\alpha + q_2 \cos\alpha)^2 = q_1^2 + q_2^2 \\ \dot{q}_1^2 + \dot{q}_2^2 &\to (\dot{q}_1 \cos\alpha + \dot{q}_2 \sin\alpha)^2 + (-\dot{q}_1 \sin\alpha + \dot{q}_2 \cos\alpha)^2 = \dot{q}_1^2 + \dot{q}_2^2, \end{aligned} \tag{13.44}$$

which means that both T and V are independently invariant.

13.6 Lie Groups and Algebras and Their Representations

The group $SO(2)$ considered above is an example of a *Lie group*, which is a continuous group, depending on continuous parameter(s) α^a, so that $g = g(\alpha)$. It is also Abelian, though in fact it is the only Abelian Lie group, up to equivalences.

For a Lie group, by convention we have $g(\alpha^a = 0) = e$. Then, Taylor expanding in α^a around $\alpha = 0$, we have

$$D(g(\alpha^a)) \simeq \hat{\mathbb{1}} + id\alpha^a X_a + \cdots, \tag{13.45}$$

which means that the constant elements X_a, called the *generators of the Lie group*, are found as

$$X_a \equiv -i \frac{\partial}{\partial \alpha^a} D(g(\alpha^a))\bigg|_{\alpha^a=0}. \tag{13.46}$$

The factor i is conventional (there is another convention without it), but it is chosen because if $D(g)$ is unitary, so that $[D(g)]^\dagger = [D(g)]^{-1}$, then X_a is Hermitian ($X_a^\dagger = X_a$).

Lie groups are named after Sophus Lie, who showed that the generators above can be defined independently of the particular representation and also defined the Lie algebra, which will be described shortly.

For a finite group transformation by a parameter α^a, we can consider infinitesimal pieces $d\alpha^a = \alpha^a/k$, as $k \to \infty$, and then

$$D(g(\alpha)) = \lim_{k\to\infty}\left(1 + \frac{i\alpha^a X_a}{k}\right)^k = \exp\left(i\alpha^a X_a\right). \tag{13.47}$$

The simplest example of a Lie group is the group of unitary numbers, $U(1)$, which is equivalent to the $SO(2)$ group. Indeed, consider

$$D(g(\alpha)) = e^{i\alpha}, \tag{13.48}$$

when acting on a complex number $q = q_1 + iq_2$. Then

$$q \to e^{i\alpha} q = (\cos\alpha\, q_1 - \sin\alpha\, q_2) + i(q_1 \sin\alpha + q_2\cos\alpha), \tag{13.49}$$

meaning that we have an action

$$\begin{pmatrix} q_1 \\ q_2 \end{pmatrix} \to \begin{pmatrix} \cos\alpha & -\sin\alpha \\ \sin\alpha & \cos\alpha \end{pmatrix} \begin{pmatrix} q_1 \\ q_2 \end{pmatrix}, \tag{13.50}$$

equivalent to the $SO(2)$ action above by a permutation of elements.

Lie Algebra

A Lie algebra is the algebra (defined by a commutator) satisfied by the generators X_a, $X_a \in \mathcal{L}(G)$. The commutator acts as a product on the Lie algebra space $\mathcal{L}(G)$, $[,] : \mathcal{L}(G) \times \mathcal{L}(G) \to \mathcal{L}(G)$. Because of the group property, $\forall g, h \in G$,

$$g = e^{i\alpha^a X_a}, \quad h = e^{i\beta^a X_a} \Rightarrow g \cdot h = e^{i\gamma^a X_a}. \tag{13.51}$$

Then, after a somewhat simple analysis, one can show that this implies that we have the "algebra"

$$[X_a, X_b] = f_{ab}{}^c X_c, \tag{13.52}$$

which defines the constants $f_{ab}{}^c$, known as "structure constants". From its definition, we find the antisymmetry property

$$f_{ab}{}^c = -f_{ba}{}^c. \tag{13.53}$$

In fact, the structure constants are completely antisymmetric in a, b, c, which can be shown by considering the Jacobi identity, an identity (of the type $0 = 0$) described as

$$[X_a, [X_b, X_c]] + \text{cyclic permutations of } (a, b, c) = 0, \tag{13.54}$$

which implies, by calculating both the commutators in (13.54) in terms of structure constants,

$$f_{bc}{}^d f_{ad}{}^e + f_{ab}{}^d f_{cd}{}^e + f_{ca}{}^d f_{bd}^e = 0. \tag{13.55}$$

Representations for Lie Groups

Lie groups can be represented as matrices, e.g., the $SO(n)$ can be represented as special (of determinant 1) and orthogonal matrices, the $SU(N)$ as special and unitary matrices, etc. This is called the fundamental representation. But we have other representations denoted by R, in which the generators T_a are $(T_a^{(R)})_{ij}$. Another important representation is the adjoint one, defined by

$$(T_a)_b{}^c = -i f_{ab}{}^c. \tag{13.56}$$

It is indeed a representation, since we have

$$([T_a, T_b])_c{}^e = i f_{ab}{}^d (T_d)_c{}^e, \tag{13.57}$$

which is true owing to the Jacobi identity, as we can easily check.

Important Concepts to Remember

- Discrete symmetries are: parity, time reversal invariance, lattice translation, and internal.
- Parity eigenstates, $\hat{\pi}|\psi\rangle = \pm|\psi\rangle$ have even parity (the + eigenvalue) or odd parity (the − eigenvalue), but not all states are parity eigenstates in a parity-invariant theory.
- For the harmonic oscillator (which is parity invariant), the parity of the state $|n\rangle$ is $(-1)^n$, but $|x\rangle$ and $|p\rangle$ are not parity eigenstates.
- A selection rule for some particular symmetry is a rule for a matrix to be nonzero on the basis of the symmetry.
- The Schrödinger equation is not time-reversal invariant; one needs to take the complex conjugate also of ψ. A related feature is that $\hat{T}$ (the time reversal operator in quantum mechanics) is antilinear and unitary, or anti-unitary.
- In a time-reversal-invariant theory, $\hat{T}\hat{H}\hat{T}^{-1} = \hat{H}$, leading to $\langle x|n\rangle \in \mathbb{R}$.
- The action of time reversal on spin is $\hat{T}\hat{S}\hat{T}^{-1} = -\hat{S}$, and $\hat{T}^2|\text{spin } j\rangle = (-1)^{2j}|\text{spin } j\rangle$.
- Lie groups are continuous groups depending on continuous parameters, with generators T_a that are $(-i)$ times the derivative of the group element with respect to the parameter, at zero parameter, so $D(g(\alpha)) = \exp(i\alpha^a X_a)$.
- The Lie algebra is the algebra of the generators of the Lie group, $[X_a, X_b] = f_{ab}{}^c X_c$, with $f_{ab}{}^c$ the structure constants, satisfying the Jacobi identity, $f_{bc}{}^d f_{ad}{}^e$ + cyclic permutations of $(a, b, c) = 0$.
- The group $SO(2)$, for the rotation of two real objects, is equivalent to the group $U(1)$ with group element $e^{i\alpha}$ rotating a complex element.
- For the classical Lie groups we have the fundamental representation, on which the generators act as matrices on a space, and the adjoint representation, for which $(T_a)_b{}^c = -i f_{ab}{}^c$, and other representations.

Further Reading

See [8] for more about groups and quantum mechanics.

Exercises

(1) Consider a three-dimensional harmonic oscillator (a harmonic oscillator with the same mass and frequency in each direction). Is it parity invariant? If so, what is the parity of a generic state?

(2) Consider the probability density of the nth state of the one-dimensional harmonic oscillator. Is it invariant under time-reversal invariance? How about under parity?

(3) Consider a system with two degrees of freedom in one dimension, q_1 and q_2, and Hamiltonian

$$H = \frac{p_1^2}{2} + \frac{p_2^2}{2} + V(|q_1|) + V(|q_2|) + V_{12}(|q_1 - q_2|). \tag{13.58}$$

Does it have any continuous internal symmetries?

(4) Consider the algebra for three generators A, B, C,

$$[A, B] = C, \quad [B, C] = A, \quad [C, A] = B. \tag{13.59}$$

Is it a Lie algebra? If so, write its adjoint representation in terms of matrices.

(5) Consider a Hamiltonian for two spins,

$$H = \alpha(\vec{S}_1^2 + \vec{S}_2^2) + \beta \vec{S}_1 \cdot \vec{S}_2. \tag{13.60}$$

Is the system time-reversal invariant? What about parity invariant?

(6) What is the dimension of the adjoint representation of $SU(N)$, $N > 2$? Is this adjoint representation equivalent to the fundamental representation?

(7) Consider the following Lagrangian for N degrees of freedom q_i,

$$L = \sum_{i=1}^{N} \frac{|\dot{q}_i|^2}{2} - V\left(\sum_{i=1}^{N} |q_i|^2\right). \tag{13.61}$$

If $q_i \in \mathbb{R}$, what is the internal continuous symmetry group? What about if $q_i \in \mathbb{C}$?

14 Theory of Angular Momentum I: Operators, Algebras, Representations

In this chapter we start the analysis of angular momentum. We first analyze the rotation group $SO(3)$ and its equivalence to the group $SU(2)$. We then define representations of both groups in terms of values for angular momentum (or "spin").

14.1 Rotational Invariance and $SO(n)$

A rotation is a linear transformation on a system: a transformation either of coordinates or of the system, depending on whether one takes a passive or active point of view, that leaves the lengths $|\Delta\vec{r}_{ij}| = |\vec{r}_i - \vec{r}_j|$ of objects (the distances between points) invariant.

Considering a rotation around a point O with vector $\vec{r}_O$, with linear transformation

$$(\vec{r} - \vec{r}_O)'_a = \Lambda_a{}^b(\vec{r} - \vec{r}_O)_b, \tag{14.1}$$

where $a, b = 1, 2, 3$ (these are spatial coordinates), by subtracting the relation for two points i and j we find that indeed

$$(\vec{r}\,'_i - \vec{r}\,'_j)_a = \Lambda_a{}^b(\vec{r}_i - \vec{r}_j)_b. \tag{14.2}$$

Imposing invariance, so that $|\vec{r}\,'_i - \vec{r}\,'_j| = |\vec{r}_i - \vec{r}_j|$, we find that

$$\begin{aligned} |\vec{r}\,'_i - \vec{r}\,'_j|^2 &= \Lambda_a{}^b(\vec{r}_i - \vec{r}_j)_b\Lambda^{ac}(\vec{r}_i - \vec{r}_j)_c = (\vec{r}_i - \vec{r}_j)_b(\Lambda^T\Lambda)^{bc}(\vec{r}_i - \vec{r}_j)_c \\ &= (\vec{r}_i - \vec{r}_j)_b\delta^{bc}(\vec{r}_i - \vec{r}_j)_c, \end{aligned} \tag{14.3}$$

which means that the matrix Λ defining the linear transformation obeys

$$\Lambda^\dagger\Lambda = \mathbb{1} \;\Rightarrow\; \Lambda^T = \Lambda^{-1}, \tag{14.4}$$

and thus is an *orthogonal* matrix. These orthogonal matrices form a group, since if $\Lambda_1^T = \Lambda_1^{-1}$ and $\Lambda_2^T = \Lambda_2^{-1}$ then

$$(\Lambda_1\Lambda_2)^T = \Lambda_2^T\Lambda_1^T = \Lambda_2^{-1}\Lambda_1^{-1} = (\Lambda_1\Lambda_2)^{-1}, \tag{14.5}$$

so their product is also orthogonal. The group of orthogonal matrices is called the orthogonal group and is denoted by $O(3)$ in the case of 3×3 matrices acting on three-dimensional vectors (spatial vectors).

Moreover, using the fact that $\det(A \cdot B) = \det A \cdot \det B$ and $\det A^T = \det A$, we find that

$$\begin{aligned} \det(\Lambda^T\Lambda) &= \det\Lambda \det\Lambda^T = (\det\Lambda)^2 \\ &= \det\mathbb{1} = 1, \end{aligned} \tag{14.6}$$

implying that $\det\Lambda = \pm 1$. But that is easily understood: $O(3)$ contains also the parity operation P, which as we saw acts as $P\vec{r} = \vec{r}\,' = -\vec{r}$, which is not an operation continuously connected with the identity as is needed for a Lie group.

This means that, in order to obtain a Lie group we need to eliminate $P = -\mathbb{1}$ from $O(3)$ by imposing the *special* condition, $\det\Lambda = +1$. This condition respects the group property since if $\det\Lambda_1 = \det\Lambda_2 = +1$ then

$$\det\Lambda_1\Lambda_2 = \det\Lambda_1 \det\Lambda_2 = 1. \tag{14.7}$$

We thus define the *special orthogonal group*, here $SO(3)$, for 3×3 matrices. We have

$$SO(3) = O(3)/\mathbb{Z}_2, \tag{14.8}$$

where $\mathbb{Z}_2 = \{+\mathbb{1}, -\mathbb{1}\}$, so $O(3)$ corresponds to two copies of the Lie group $SO(3)$.

The analysis above trivially generalizes to n dimensions, for $n\times n$ matrices acting on n-dimensional vectors, forming the group $O(n)$ with Lie subgroup $SO(n) = O(n)/\mathbb{Z}_2$. The Lie algebra of $SO(n)$ will be denoted by lower case letters, as $so(n)$.

14.2 The Lie Groups *SO*(2) and *SO*(3)

We have that $SO(2)$, the Abelian Lie group studied before, is a subgroup: $SO(2)\subset SO(3)$. As we saw, the matrix (which we can easily see is orthogonal and special)

$$M = \begin{pmatrix} \cos\theta & -\sin\theta \\ \sin\theta & \cos\theta \end{pmatrix} \tag{14.9}$$

acts on two-dimensional spatial vectors $\begin{pmatrix} x \\ y \end{pmatrix}$ as

$$\begin{pmatrix} x' \\ y' \end{pmatrix} = M\begin{pmatrix} x \\ y \end{pmatrix} = \begin{pmatrix} x\cos\theta - y\sin\theta \\ x\sin\theta + y\cos\theta \end{pmatrix}. \tag{14.10}$$

Then we have rotational invariance, $x'^2 + y'^2 = x^2 + y^2$, and the action of this $SO(2)$ matrix is equivalent to the action of a unitary number (a complex phase) on a complex number, thus to the action of

$$M = e^{i\theta} = \cos\theta + i\sin\theta \tag{14.11}$$

on $z = x + iy$, such that

$$Mz = z' = x' + iy', \tag{14.12}$$

as we can easily check, so $SO(2) \simeq U(1)$ (where $U(1)$ is the group of unitary numbers, i.e., phases).

This $SO(2) \simeq U(1)$ Abelian group (the unique Abelian Lie group) is a subgroup of any compact Lie group, in particular of any $SO(n)$ for $n \geq 2$. The reason is that we can always pick two coordinates out of the n on which $SO(n)$ acts to rotate, hence defining $SO(2) \subset SO(n)$.

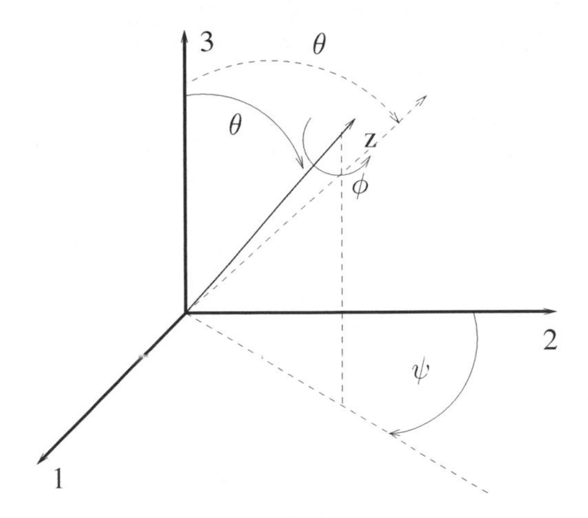

Figure 14.1 Euler angles parametrizing a rotation in three-dimensional space: two angles, θ and ψ, relate to the Cartesian coordinate system, and one, ϕ, corresponds to a rotation around the axis itself.

The Group $SO(3)$

Considering now rotations in three-dimensional space, we can always pick a planar ($SO(2)$) rotation of angle ϕ around a fixed axis $\hat{z}$. Choosing the coordinate system in such a way as to have this axis $\hat{z}$ in the third direction, $e_z = \begin{pmatrix} 0 \\ 0 \\ 1 \end{pmatrix}$, we then have

$$R_3(\phi) = \begin{pmatrix} \cos\phi & -\sin\phi & 0 \\ \sin\phi & \cos\phi & 0 \\ 0 & 0 & 1 \end{pmatrix}. \tag{14.13}$$

However, in general the vector $\hat{z}$, in a fixed coordinate system with Cartesian directions $\hat{1}, \hat{2}, \hat{3}$, is defined by two angles, θ and ψ. Consider θ to be the angle made by $\hat{z}$ with $\hat{3}$, and project $\hat{z}$ onto the plane defined by $(\hat{1}, \hat{2})$. Then $\hat{z}$ makes an angle $\pi/2 - \theta$ with its projection onto the $(\hat{1}, \hat{2})$ plane, and the projection makes an angle ψ with $\hat{2}$, as in Fig. 14.1. The angles ϕ, θ, ψ are known as the Euler angles.

To obtain the general rotation, we first rotate by ψ around $\hat{3}$, thus aligning the projection of $\hat{z}$ with $\hat{2}$, and then rotate around $\hat{1}$ by θ, aligning $\hat{z}$ with $\hat{3}$, and finally rotate by ϕ around it. Then, the general three-dimensional rotation parametrized by the Euler angles is

$$\begin{aligned} g(\phi, \theta, \psi) &= g(\phi)g(\theta)g(\psi) = R_3(\phi)R_1(\theta)R_3(\psi) \\ &= \begin{pmatrix} \cos\phi & -\sin\phi & 0 \\ \sin\phi & \cos\phi & 0 \\ 0 & 0 & 1 \end{pmatrix} \begin{pmatrix} 1 & 0 & 0 \\ 0 & \cos\theta & -\sin\theta \\ 0 & \sin\theta & \cos\theta \end{pmatrix} \begin{pmatrix} \cos\psi & -\sin\psi & 0 \\ \sin\psi & \cos\psi & 0 \\ 0 & 0 & 1 \end{pmatrix}. \end{aligned} \tag{14.14}$$

To obtain the most general rotation axis $\hat{z}$ in three-dimensional space, the rotation range of ψ is $[0, 2\pi)$, and then of θ is $[0, \pi)$. Finally, the rotation by ϕ has the standard $[0, 2\pi)$ range.

There are three $U(1) \simeq SO(2)$ Abelian subgroups in $SO(3)$, corresponding to the three Euler angles, and three ways to pick two coordinates for the rotation plane from the three coordinates.

To find the Lie algebra of $SO(3)$, we define its three generators J_i as being associated with the three Euler angles. Each corresponds to rotation in a plane. For instance, $R_3(\phi)$ corresponds to rotation in the $(1,2)$ plane, and, for an infinitesimal ϕ, we obtain

$$R_3(\phi) = \begin{pmatrix} \cos\phi & -\sin\phi & 0 \\ \sin\phi & \cos\phi & 0 \\ 0 & 0 & 1 \end{pmatrix} \simeq \begin{pmatrix} 1 & -\phi & 0 \\ \phi & 1 & 0 \\ 0 & 0 & 1 \end{pmatrix} + \mathcal{O}(\phi)^2 = \mathbb{1} + \phi \begin{pmatrix} 0 & -1 & 0 \\ 1 & 0 & 0 \\ 0 & 0 & 0 \end{pmatrix} + \mathcal{O}(\phi^2). \quad (14.15)$$

Then the generator for this rotation around the 3-direction is

$$iT_3 = iJ_3 = \begin{pmatrix} 0 & -1 & 0 \\ 1 & 0 & 0 \\ 0 & 0 & 0 \end{pmatrix}, \quad (14.16)$$

and we can do a similar calculation for the other directions. The group elements for finite rotations by angles $\theta^i, i = 1,2,3$, are

$$\exp\left(i\theta^i J_i\right) \equiv \exp\left(i\epsilon^{ijk}\omega_{jk} J_i\right). \quad (14.17)$$

Moreover, a unitary infinitesimal matrix means that

$$(\mathbb{1} + \theta^i iJ_i)^\dagger = \mathbb{1} + \theta(iJ_i)^\dagger = (\mathbb{1} + i\theta^i J_i)^{-1} = \mathbb{1} - \theta^i iJ_i, \quad (14.18)$$

and for real iJ_i we find an antisymmetric matrix, as above for iJ_3.

Rotationally Invariant Systems

For a rotationally ($SO(3)$) invariant Lagrangian L (and thus Hamiltonian $\hat{H}$), for instance

$$L = \sum_i \frac{m_i}{2}\dot{\vec{r}}_i^2 - V(|\vec{r}_i - \vec{r}_j|) = \sum_i \frac{m_i}{2}\left[\frac{d}{dt}(\vec{r}_i - \vec{r}_O)\right]^2 - V\left((\vec{r}_i - \vec{r}_j)^2\right), \quad (14.19)$$

the Lagrangian of the system must be in a representation of the rotation group $SO(3)$. Here, as we can see, we have the fundamental representation, where the rotation matrices act on the three-dimensional position vectors $\vec{r}$.

However, the *state* of the system, $|\psi\rangle$, need not be in the same representation, but can be in an arbitrary representation (since it is not related to $\vec{x}$ as a vector space).

Before discussing general representations, we consider the equivalence of $SO(3)$ and $SU(2)$ in the Lie algebra.

14.3 The Group *SU*(2) and Its Isomorphism with *SO*(3) Mod $\mathbb{Z}_2$

Consider the group of unitary matrices, i.e., the complex matrices U satisfying $U^\dagger = U^{-1}$. Then, indeed, if $U_1^\dagger = U_1^{-1}$ and $U_2^\dagger = U_2^{-1}$,

$$(U_1U_2)^\dagger = U_2^\dagger U_1^\dagger = U_2^{-1}U_1^{-1} = (U_1U_2)^{-1}, \quad (14.20)$$

which means that unitarity respects the group property. The group of unitary $n \times n$ matrices is called $U(n)$ and acts on an n-dimensional complex vector.

For a unitary matrix U,

$$\det(U^\dagger U) = (\det U)^* \det U \\ = \det \mathbb{1} = 1 \quad \Rightarrow \quad |\det U| = 1 \Rightarrow \det U = e^{i\alpha}, \tag{14.21}$$

which means that the determinant of the unitary matrix forms a $U(1)$ group.

Imposing the special condition $\det U = +1$, which respects the group property as we saw, we obtain the *special unitary group*, $SU(n)$. However, now (since the determinant forms a $U(1)$ group), $U(n) \simeq SU(n) \times U(1)$, modulo topological issues. In fact,

$$U(n) \simeq (SU(n) \times U(1))/\mathbb{Z}_n. \tag{14.22}$$

In Lie algebra, $u(n) = su(n) \times u(1)$. In the relevant case of $n = 2$, we have also

$$SO(3) \simeq SU(2)/\mathbb{Z}_2, \tag{14.23}$$

which means that $SU(2)$ winds around $SO(3)$ twice. Indeed, $-\mathbb{1}_{2\times 2} \in SU(2)$, but $-\mathbb{1}_{3\times 3}$ is not in $SO(3)$.

As the $SO(2) \subset SO(3)$ rotation equals $U(1) \subset SU(2)$, we can identify the rotation in the plane $(1, 2)$ (around $\hat{3}$) as an $SU(2)$ matrix with diagonal elements that have opposite phases:

$$R_3(\phi) \to \begin{pmatrix} e^{i\phi/2} & 0 \\ 0 & e^{-i\phi/2} \end{pmatrix} \equiv \tilde{R}_3(\phi), \tag{14.24}$$

acting on the two-dimensional complex vector $\begin{pmatrix} z_1 \\ z_2 \end{pmatrix}$. The matrix is unitary, since

$$\tilde{R}_3^\dagger = \begin{pmatrix} e^{-i\phi/2} & 0 \\ 0 & e^{i\phi/2} \end{pmatrix} = \tilde{R}_3^{-1}, \tag{14.25}$$

and $\det \tilde{R}_3 = 1$.

The rotation around $\hat{1}$ can be represented by a different unitary matrix,

$$R_1(\theta) \to \begin{pmatrix} \cos\theta/2 & i\sin\theta/2 \\ i\sin\theta/2 & \cos\theta/2 \end{pmatrix} \equiv \tilde{R}_1(\theta). \tag{14.26}$$

We can then check that

$$\tilde{R}_1^\dagger = \begin{pmatrix} \cos\theta/2 & -i\sin\theta/2 \\ -i\sin\theta/2 & \cos\theta/2 \end{pmatrix} = \tilde{R}_1^{-1}, \tag{14.27}$$

and $\det \tilde{R}_1 = +1$.

Then the general element of $SU(2)$ is represented via the Euler angles ϕ, θ, ψ as

$$g(\phi,\theta,\psi) = \tilde{R}_3(\phi)\tilde{R}_1(\theta)\tilde{R}_3(\psi) \\ = \begin{pmatrix} \cos\theta/2\, e^{i(\phi+\psi)/2} & i\sin\theta/2\, e^{i(\phi-\psi)/2} \\ i\sin\theta/2\, e^{-i(\phi-\psi)/2} & \cos\theta/2\, e^{-i(\phi+\psi)/2} \end{pmatrix}. \tag{14.28}$$

But, as we said, $-\mathbb{1}_{2\times 2} \in SU(2)$, so we need to obtain $\pm$ times the above generic matrix, which was defined over the original range of the $(SO(3))$ Euler angles. Since $\cos(\pi + \alpha) = -\cos\alpha$ and $\sin(\pi + \alpha) = -\alpha$, we can achieve that by noting that

$$\tilde{R}_3(\phi)\tilde{R}_1(\theta + 2\pi)\tilde{R}_3(\psi) = -\tilde{R}_3(\phi)\tilde{R}_1(\theta)\tilde{R}_3(\psi), \tag{14.29}$$

that is, by doubling the range of θ, thus proving the statement from before, that $SU(2)$ winds twice around $SO(3)$ (it is a double cover of $SO(3)$).

On the other hand, that also means that the Lie algebras of $SU(2)$ and $SO(3)$ are the same, $su(2) = so(3)$, which means that their representations are equivalent.

In particular, it means that we can represent $SO(3)$ via complex 2×2 matrices also! We will see that this is the minimum representation space (of "spin $j = 1/2$") that we can have for $SO(3)$.

Example for *SU*(2)

An example of such a representation, for a system with "internal" $SU(2)$ symmetry, is a system defined by four real coordinates, $x_1, x_2, y_1, y_2 \in \mathbb{R}$, combined into two complex coordinates $q_1 = x_1 + iy_1$ and $q_2 = x_2 + iy_2$, and further into the vector space with two complex dimensions, $q = \begin{pmatrix} q_1 \\ q_2 \end{pmatrix}$, with Lagrangian

$$L = m\frac{|\dot{q}|^2}{2} - V(|q|^2) \equiv \frac{m}{2}(|\dot{q}_1|^2 + |\dot{q}_2|^2) - V(|q_1|^2 + |q_2|^2). \tag{14.30}$$

Then, for $q \to q' = gq$, we obtain

$$\begin{aligned} |q|^2 &\to |q'|^2 = |gq|^2 = q^\dagger g^\dagger g q = q^\dagger q = |q|^2 \\ |\dot{q}|^2 &\to |\dot{q}'|^2 = |g\dot{q}|^2 = \dot{q}^\dagger g^\dagger g \dot{q} = \dot{q}^\dagger \dot{q} = |\dot{q}|^2. \end{aligned} \tag{14.31}$$

Again, just because the (classical or quantum) system is in the two-dimensional representation of $SU(2)$, it doesn't mean that the state is in that representation, since the state is not related to q_1, q_2 as a vector space. States $|\psi\rangle$ can be in a different representation of $SU(2)$.

14.4 Generators and Lie Algebras

From the Noether theorem, we saw in Chapter 12 that, for a symmetry with generators T_a, we have a conserved charge associated with it,

$$Q_a = \frac{\partial L}{\partial \dot{q}_i}(iT_a)_i{}^j q_j = p^i (iT_a)_i{}^j q_j. \tag{14.32}$$

Considering the symmetry to be a rotation, specifically around $z = x_3$, and $q_i = x_i = (x_1, x_2, x_3)$, we found that

$$iT_3 = \begin{pmatrix} 0 & -1 & 0 \\ +1 & 0 & 0 \\ 0 & 0 & 1 \end{pmatrix}, \tag{14.33}$$

and similarly (just permuting the vector space 1, 2, 3) for iT_1, iT_2 we find

$$Q_3 = p \cdot iT_3 \cdot q = x_1 p_2 - x_2 p_1. \tag{14.34}$$

Moreover, for the other two charges, we find

$$\begin{aligned} Q_2 &= x_3 p_1 - x_1 p_3 \\ Q_1 &= x_2 p_3 - x_3 p_2, \end{aligned} \tag{14.35}$$

or, written together,

$$Q_i - \sum_{j,k} \epsilon_{ijk} x_j p_k, \tag{14.36}$$

using the Levi-Civita antisymmetric symbol ϵ_{ijk}, for which $\epsilon_{123} = +1$ and which is totally antisymmetric (so that, for instance, $\epsilon_{132} = -\epsilon_{123} = -1$).

However, ϵ_{ijk} defines the three-dimensional vector product, so we have in fact

$$\vec{Q} = \vec{r} \times \vec{p} = \vec{L}, \tag{14.37}$$

which is the angular momentum of a particle! Therefore the conserved charge associated with rotational invariance is the angular momentum.

Moreover, denoting $iT_a = iJ_a$, we see that $Q_a = L_a$ is the charge associated with the generator J_a, so the generators of $SO(3)$ rotations are (associated with) angular momenta. This is similar to the fact that the generators of translations K_a are (associated with) the momenta P_a.

Classically, we can calculate the Poisson brackets of angular momenta, using the canonical expressions $\{x_i, p_j\}_{P.B.} = \delta_{ij}$ or, equivalently, the definitions of the Poisson brackets:

$$\begin{aligned} \{L_i, L_j\}_{P.B.} &= \epsilon_{ikl}\epsilon_{jmn}\{x_k p_l, x_m p_n\}_{P.B.} = \epsilon_{ilk}\epsilon_{jmk} x_m p_l + \epsilon_{ikl}\epsilon_{jmn} x_k p_n \\ &= -\epsilon_{ijk}\epsilon_{kmn} x_m p_n = -\epsilon_{ijk} L_k. \end{aligned} \tag{14.38}$$

This is the same algebra as the algebra of generators iJ_i for $SO(3)$ (a Lie algebra), as we can directly check:

$$[iJ_1, iJ_2] = \left[\begin{pmatrix} 0 & 0 & 0 \\ 0 & 0 & -1 \\ 0 & 1 & 0 \end{pmatrix}, \begin{pmatrix} 0 & 0 & -1 \\ 0 & 0 & 0 \\ 1 & 0 & 0 \end{pmatrix} \right] = -iJ_3 = - \begin{pmatrix} 0 & -1 & 0 \\ 1 & 0 & 0 \\ 0 & 0 & 0 \end{pmatrix}, \tag{14.39}$$

so we can identify the generators of rotations with the angular momenta.

Finally, we will see in the next chapter that *for wave functions* $\psi(\vec{r})$ the rotation acts according to the operator $\hat{\vec{L}}$:

$$\hat{L}_i \psi(\vec{x}) = \sum_{jk} \epsilon_{ijk} \hat{X}_j \hat{P}_k \psi(\vec{x}) = -i\hbar \sum_{jk} \epsilon_{ijk} x_j \frac{\partial}{\partial x_k} \psi(\vec{x}). \tag{14.40}$$

14.5 Quantum Mechanical Version

In quantum mechanics, the Poisson bracket $\{,\}_{P.B.}$ is replaced by $(1/i\hbar)[,]$, so we obtain the algebra

$$[\hat{L}_i, \hat{L}_j] = i\hbar \sum_k \epsilon_{ijk} \hat{L}_k. \tag{14.41}$$

Equivalently, and more directly, x_i and p_i become the operators $\hat{X}_i$ and $\hat{P}_i$, so we have

$$\hat{L}_1 = \hat{X}_2\hat{P}_3 - \hat{X}_3\hat{P}_2, \text{ etc.} \Rightarrow \hat{L}_i = \sum_k \epsilon_{ijk}\hat{X}_j\hat{P}_k, \tag{14.42}$$

and, using the canonical commutation relations $[\hat{X}_i, \hat{P}_j] = i\hbar\delta_{ij}$, we find again

$$[\hat{L}_i, \hat{L}_j] = i\hbar \sum_k \epsilon_{ijk}\hat{L}_k. \tag{14.43}$$

Moreover, constructing the square of the total angular momentum,

$$\hat{\vec{L}}^2 = \hat{L}_1^2 + \hat{L}_2^2 + \hat{L}_3^2, \tag{14.44}$$

we then obtain

$$[\hat{L}_i, \hat{\vec{L}}^2] = \hat{L}_j[\hat{L}_i, \hat{L}_j] + [\hat{L}_i, \hat{L}_j]\hat{L}_j = i\hbar \sum_k \epsilon_{ijk}(\hat{L}_j\hat{L}_k + \hat{L}_k\hat{L}_j) = 0. \tag{14.45}$$

Thus a general "angular momentum", for either rotation $SO(3)$ generators or generators of an internal $SO(3)$, has the commutation relations

$$[\hat{J}_i, \hat{J}_k] = i\hbar \sum_k \epsilon_{ijk}\hat{J}_k, \quad [\hat{\vec{J}}^2, \hat{J}_i] = 0, \tag{14.46}$$

where the $\hat{J}_i$ are Hermitian.

14.6 Representations

Finally, we can construct the representations of the rotation group $SU(2) \simeq SO(3)$ corresponding to the possible Hilbert spaces for the rotationally invariant systems.

We define the operators

$$J_+ = J_1 + iJ_2, \quad J_- = J_1 - iJ_2, \quad J_z = J_3, \tag{14.47}$$

with $(J_+)^\dagger = J_-$ and $(J_-)^\dagger = J_+$.

Their algebra is

$$\begin{aligned}
&[J_+, J_-] = 2i[J_2, J_1] = 2\hbar J_z \\
&[J_z, J_+] = \hbar J_+ \\
&[J_z, J_-] = -\hbar J_-.
\end{aligned} \tag{14.48}$$

From the fact that $[\vec{J}^2, J_z] = 0$, it follows that there is a complete set of simultaneous eigenstates for both $\vec{J}^2$ and J_z, denoted $|\lambda^2, m, \ldots\rangle$ by their eigenvalues,

$$\vec{J}^2|\lambda^2, m, \ldots\rangle = \lambda^2\hbar^2|\lambda^2, m, \ldots\rangle, \quad J_z|\lambda^2, m, \ldots\rangle = m\hbar|\lambda^2, m, \ldots\rangle. \tag{14.49}$$

But, since J_i is Hermitian, we find that

$$\begin{aligned}
\sum_i \langle\lambda^2, m, \ldots|J_i^2|\lambda^2, m, \ldots\rangle &= \sum_i ||J_i|\lambda^2, m, \ldots\rangle||^2 \geq 0 \\
&= \lambda^2\hbar^2 \Rightarrow \lambda^2 \geq 0,
\end{aligned} \tag{14.50}$$

where we have used the fact that the norm of the eigenstates is one (or in any case, positive and nonzero). Thus $\lambda^2 \geq 0$, so λ is real (which shows that it was a good definition).

In a similar manner, using the Hermiticity of J_1, J_2 we find

$$\begin{aligned}\langle\lambda^2, m, \dots|(J_1^2 + J_2^2)|\lambda^2, m, \dots,\rangle &= \langle\lambda^2, m, \dots|(\vec{J}^2 - J_z^2)|\lambda^2, m, \dots\rangle = (\lambda^2 - m^2)\hbar^2\\ &= \|J_1|\lambda^2, m, \dots\rangle\|^2 + \|J_2|\lambda^2, m, \dots\rangle\|^2 \geq 0\\ &\Rightarrow \lambda^2 - m^2 \geq 0.\end{aligned} \tag{14.51}$$

On the other hand, J_+ acts as a raising operator for m and J_- as a lowering operator, since (using $[J_z, J_+] = \hbar J_+$ and $[J_z, J_-] = \hbar J_-$) we find

$$\begin{aligned}J_z(J_+|\lambda^2, m, \dots\rangle) &= J_+(J_z|\lambda^2, m, \dots\rangle) + \hbar J_+|\lambda^2, m, \dots\rangle = (m+1)\hbar(J_+|\lambda^2, m, \dots\rangle)\\ J_z(J_-|\lambda^2, m, \dots\rangle) &= J_-(J_z|\lambda^2, m, \dots\rangle) - \hbar J_+|\lambda^2, m, \dots\rangle = (m-1)\hbar(J_-|\lambda^2, m, \dots\rangle).\end{aligned} \tag{14.52}$$

Since $m^2 \leq \lambda^2$, it follows that there is a maximum value and a minimum value of m, such that

$$\begin{aligned}J_+|\lambda^2, m_{\text{max}}, \dots\rangle = 0 &\Rightarrow |\lambda^2, m_{\text{max}} + 1, \dots\rangle = 0\\ J_-|\lambda^2, m_{\text{min}}, \dots\rangle = 0 &\Rightarrow |\lambda^2, m_{\text{min}} - 1, \dots\rangle = 0.\end{aligned} \tag{14.53}$$

Also, because the step in m is 1, it follows that a representation of given λ^2 has a finite number of states. Moreover, as we have

$$J_-J_+ = (J_1 - iJ_2)(J_1 + iJ_2) = J_1^2 + J_2^2 + i[J_1, J_2] = \vec{J}^2 - J_z^2 - \hbar J_z \tag{14.54}$$

and $J_- = (J_+)^\dagger$, we find that

$$\begin{aligned}\langle\lambda^2, m, \dots|J_-J_+|\lambda^2, m, \dots\rangle &= \langle\lambda^2, m, \dots|(\vec{J}^2 - J_z^2 - \hbar J_z)|\lambda^2, m, \dots\rangle = (\lambda^2 - m^2 - m)\hbar^2\\ &= \|J_+|\lambda^2, m, \dots\rangle\|^2 \geq 0 \Rightarrow \lambda^2 - m^2 - m \geq 0.\end{aligned} \tag{14.55}$$

Similarly, as we have

$$J_+J_- = (J_1 + iJ_2)(J_1 - iJ_2) = J_1^2 + J_2^2 - i[J_1, J_2] = \vec{J}^2 - J_z^2 + \hbar J_z \tag{14.56}$$

and $J_+ = (J_-)^\dagger$, we obtain

$$\begin{aligned}\langle\lambda^2, m, \dots|J_+J_-|\lambda^2, m, \dots\rangle &= \langle\lambda^2, m, \dots|(\vec{J}^2 - J_z^2 + \hbar J_z)|\lambda^2, m, \dots\rangle = (\lambda^2 - m^2 + m)\hbar^2\\ &= \|J_-|\lambda^2, m, \dots\rangle\|^2 \geq 0 \Rightarrow \lambda^2 - m^2 + m \geq 0.\end{aligned} \tag{14.57}$$

Further, the equality to zero happens, for m_{max}, when the norm of the state is zero, since the state itself vanishes, meaning when

$$J_+|\lambda^2, m, \dots\rangle = 0 \Rightarrow \lambda^2 - m_{\text{max}}^2 - m_{\text{max}} = 0, \tag{14.58}$$

and for m_{min} when similarly

$$J_-|\lambda^2, m, \dots\rangle = 0 \Rightarrow \lambda^2 - m_{\text{min}}^2 + m_{\text{min}} = 0. \tag{14.59}$$

However, subtracting (14.58) from (14.59) we find

$$m_{\text{max}}(m_{\text{max}} + 1) = m_{\text{min}}(m_{\text{min}} - 1), \tag{14.60}$$

whose only solution is

$$m_{\rm max} = -m_{\rm min}. \tag{14.61}$$

Indeed, there is no solution if both $m_{\rm max}$ and $m_{\rm min}$ are positive, or both are negative.

Finally, we have

$$-m_{\rm max} \leq m \leq m_{\rm min}, \tag{14.62}$$

and since m changes by ± 1 only (through $J_\pm$), it follows that

$$2m_{\rm max} = m_{\rm max} - m_{\rm min} = n \in \mathbb{N}, \tag{14.63}$$

so

$$m_{\rm max} = \frac{n}{2} \equiv j. \tag{14.64}$$

This "quantum number" j is thus half-integer.

Finally, (14.58) implies that

$$\lambda^2 = j(j+1). \tag{14.65}$$

Assuming that there are no other quantum numbers (so we have only angular momentum as a variable), the state is $|jm\rangle$ and

$$\begin{aligned} J_z|jm\rangle &= m\hbar|jm\rangle \\ \vec{J}^2|jm\rangle &= j(j+1)\hbar^2|jm\rangle. \end{aligned} \tag{14.66}$$

The interpretation is that the index j describes *possible representations of* $SO(3)$, or of angular momentum (since j defines the total angular momentum), and m describes the states in the representation.

Matrix Elements

Finally, we compute the matrix elements of the generators in the $|jm\rangle$ representation. From (14.55),

$$||J_+|jm\rangle||^2 = \hbar^2[j(j+1) - m^2 - m], \tag{14.67}$$

whereas from (14.52),

$$J_+|jm\rangle \propto |j, m+1\rangle, \tag{14.68}$$

which finally gives

$$J_+|jm\rangle = \sqrt{(j-m)(j+m+1)}\hbar|jm\rangle \equiv \alpha_{j,m+1}\hbar|j, m+1\rangle, \tag{14.69}$$

where

$$\alpha_{jm} \equiv \sqrt{j(j+1) - m(m-1)} = \sqrt{(j+m)(j-m+1)}. \tag{14.70}$$

As this is a matrix element in $|jm\rangle$ space, we have

$$(J_+)_{jm',jm} = \hbar\alpha_{j,m+1}\delta_{m',m+1}. \tag{14.71}$$

Similarly, from (14.57),

$$||J_-|jm\rangle||^2 = \hbar^2[j(j+1) - m^2 + m] \tag{14.72}$$

whereas from (14.52)

$$J_-|jm\rangle \propto |j, m-1\rangle, \tag{14.73}$$

which finally gives

$$J_-|jm\rangle = \sqrt{(j+m)(j-m+1)}\hbar|jm\rangle \equiv \alpha_{jm}\hbar|j, m-1\rangle. \tag{14.74}$$

As a matrix element in $|jm\rangle$ space, this means that

$$(J_-)_{jm',jm} = \alpha_{jm}\hbar\delta_{m',m-1}. \tag{14.75}$$

Finally, since $J_1 = (J_+ + J_-)/2$ and $J_2 = (J_+ - J_-)/(2i)$, we find also

$$\begin{aligned}(J_1)_{jm',jm} &= \frac{\hbar}{2}(\alpha_{j,m+1}\delta_{m',m+1} + \alpha_{jm}\delta_{m',m-1})\\ (J_2)_{jm',jm} &= \frac{\hbar}{2i}(\alpha_{j,m+1}\delta_{m',m+1} - \alpha_{jm}\delta_{m',m-1}).\end{aligned} \tag{14.76}$$

To these, we add the diagonal elements

$$\begin{aligned}(J_3)_{jm',jm} &= m\hbar\delta_{m'm}\\ (\vec{J}^2)_{jm',jm} &= j(j+1)\hbar^2\delta_{m'm}.\end{aligned} \tag{14.77}$$

We now specialize to the $j = 1/2$ case (which will be described in much more detail later on, as it corresponds in particular to the electron spin 1/2 case), and note that the above matrices are now 2×2 matrices, acting on the two-dimensional vector space $|jm\rangle = |+1/2, \pm 1/2\rangle$. Then, as a matrix,

$$J_3 = \hbar\begin{pmatrix}1/2 & 0\\ 0 & -1/2\end{pmatrix}. \tag{14.78}$$

This indeed matches with the generators of $SU(2)$, since for infinitesimal Euler angle parameters, we find

$$\begin{aligned}\tilde{R}_3(\phi) &\simeq \begin{pmatrix}1+i\phi/2 & 0\\ 0 & 1-i\phi/2\end{pmatrix} + \mathcal{O}(\phi^2) = \mathbb{1} + i\phi\begin{pmatrix}1/2 & 0\\ 0 & -1/2\end{pmatrix} + \mathcal{O}(\phi^2) \Rightarrow\\ J_3 &= \begin{pmatrix}1/2 & 0\\ 0 & -1/2\end{pmatrix}\\ \tilde{R}_1(\theta) &\simeq \begin{pmatrix}1 & +i\theta/2\\ +i\theta/2 & 0\end{pmatrix} + \mathcal{O}(\theta^2) = \mathbb{1} + i\theta\begin{pmatrix}0 & 1/2\\ 1/2 & 0\end{pmatrix} + \mathcal{O}(\theta^2) \Rightarrow\\ J_1 &= \begin{pmatrix}0 & 1/2\\ 1/2 & 0\end{pmatrix}.\end{aligned} \tag{14.79}$$

The last formula is consistent with the matrices we found for the spin 1/2 representation, since we find

$$\begin{aligned}J_+|1/2, \pm 1/2\rangle &= \begin{pmatrix}0\\1\end{pmatrix} \times |1/2, \mp 1/2\rangle\\ J_-|1/2, \pm 1/2\rangle &= \begin{pmatrix}1\\0\end{pmatrix} \times |1/2, \mp 1/2\rangle \Rightarrow\\ J_1 &= \frac{J_+ + J_-}{2} = \frac{1}{2}\begin{pmatrix}0 & 1\\ 1 & 0\end{pmatrix}.\end{aligned} \tag{14.80}$$

Important Concepts to Remember

- The special orthogonal group $SO(n)$ is the group of matrices that are orthogonal, $\Lambda^T = \Lambda^{-1}$ and special, $\det \Lambda = 1$, which correspond to rotations in n-dimensional space. Globally, $SO(n) = O(n)/\mathbb{Z}_2$.
- The special unitary group $SU(n)$ is the group of matrices that are unitary, $U^\dagger = U^{-1}$, and special, $\det U = 1$, with $U(n) \simeq (SU(n) \times U(1))/\mathbb{Z}_n$.
- For $n = 2$ we also have $SO(3) \simeq SU(2)/\mathbb{Z}_2$, so $SU(2)$ winds twice around $SO(3)$: $SO(3)$ is generated by three Euler angles for rotations around three axes (in a plane; with the z axis towards the plane; and around the final z axis), while $SU(2)$ has twice the range of the angles, so that $-\mathbb{1} \in SU(2)$.
- The charges associated with $SO(3)$ rotations of coordinates are the angular momenta $L_i = \epsilon_{ijk} x_k p_k$, which classically satisfy the Poisson brackets $\{L_i, L_j\}_{P.B.} = -\epsilon_{ijk} L_k$.
- Quantum mechanically, the $\hat{L}_i$ give the Lie algebra $[\hat{L}_i, \hat{L}_j] = i\hbar\epsilon_{ijk}\hat{L}_k$, with $[\hat{\vec{L}}^2, \hat{L}_k] = 0$, which is the Lie algebra of $SU(2)$, with generators generally denoted by $\hat{J}_i$.
- The representations of $SU(2)$ are indexed by a half-integer $j \in \mathbb{N}/2$ and within a representation, the states are indexed by m, $-j \le m \le j$, and thus by $|jm\rangle$, with $\vec{J}^2|jm\rangle = j(j+1)\hbar^2|jm\rangle$ and $J_z|jm\rangle = m\hbar|jm\rangle$.
- The matrix elements within the representation are $J_+|jm\rangle = \alpha_{j,m+1}\hbar|j, m+1\rangle$, $J_-|jm\rangle = \alpha_{j,m}\hbar|j, m-1\rangle$, and $J_3|jm\rangle = m\hbar|jm\rangle$, with $\alpha_{j,m} = \sqrt{(j+m)(j-m+1)}$.
- The spin 1/2 ($j = 1/2$) generators J_i correspond to the generators of $SU(2)$ via the Euler angles.

Further Reading

See [2], [1], [3] for more information.

Exercises

(1) Write explicitly the matrices of the spin 1 representation and the adjoint representation of $SU(2)$, and relate them.

(2) Write explicitly the 2×2 matrices $g = e^{i\alpha_i\sigma_i/2}$, where σ_i are the Pauli matrices, and compare with $g(\theta, \phi, \psi)$ for $SO(3)$, to find an explicit map between α_i and the Euler angles (θ, ϕ, ψ).

(3) For a general matrix

$$A = \begin{pmatrix} a & b \\ c & d \end{pmatrix} \tag{14.81}$$

belonging to $SU(2)$, i.e., such that $A^\dagger = A^{-1}$ and $\det A = 1$, find the Euler angles in terms of a, b, c, d.

(4) Consider the complex degrees of freedom $q_1, q_2 \in \mathbb{C}$ forming a doublet $q = \begin{pmatrix} q_1 \\ q_2 \end{pmatrix}$, $A = \sum_{i=1}^3 a_i\sigma_i \in \mathbb{C}$, where σ_i are the Pauli matrices, and the Lagrangian

$$L = \dot{q}^\dagger \dot{q} + \det e^{\dot{A}} - q^\dagger A q. \tag{14.82}$$

Show that L is invariant under $SU(2)$.

(5) Write explicitly the matrices for the generators of the group $SU(2)$ in the $j = 3/2$ representation, and check that they satisfy the Lie algebra of $SU(2)$.

(6) In the classical limit for the angular momentum, i.e., for large j, we expect to see quantum averages $\langle J_1 \rangle, \langle J_2 \rangle, \langle J_3 \rangle$ close to the classical values, as well as a small quantum fluctuation for these angular momenta. Show that this is the case.

(7) Consider a three-dimensional rotationally invariant harmonic oscillator. What quantum numbers describe a state of the oscillator, viewed as a system with a potential $V = V(r)$?

15 Theory of Angular Momentum II: Addition of Angular Momenta and Representations; Oscillator Model

In the previous chapter, we saw that the angular momentum of a system acts as a generator of the $SO(3)$ rotation group, and that the values of the angular momentum j define the representation of $SO(3)$ in which the *states* of the system belong. Classically, energy and mass are scalars under rotations, meaning they do not change, while momentum $\vec{p}$ and angular momentum $\vec{J}$ are vectors, and so do transform under the fundamental (or, defining) representation of $SO(3)$. Quantum mechanically, $\vec{p}$ and $\vec{J}$ no longer characterize *states* (which are in a given representation) but rather are *operators*; in particular they are *generators* of the translation and rotation ($SO(3)$) groups, respectively, and as such are in fixed representations of the groups themselves. On the other hand, *states* are in different representations, unrelated to those of the operators: they are in representations defined by an eigenvalue j:

- for $j = 0$, we have the "scalar" representation, with a single state, $|0, 0\rangle$;
- for $j = 1/2$, we have the "spinor" representation, with two states, either $|1/2, -1/2\rangle$ and $|1/2, +1/2\rangle$, or $|1/2\uparrow\rangle$ and $|1/2\downarrow\rangle$.
- for $j = 1$, we have the "vector" representation, with three states, $|1, -1\rangle, |1, 0\rangle$, and $|1, -1\rangle$.
- etc.

15.1 The Spinor Representation, $j = 1/2$

The simplest nontrivial representation, with angular momentum $j = 1/2$, has generators J_i in this representation given by the following matrices (from the general formulas of the last chapter):

$$J_1^{1/2} = \frac{\hbar}{2}\begin{pmatrix} 0 & 1 \\ 1 & 0 \end{pmatrix}, \quad J_2^{1/2} = \frac{\hbar}{2}\begin{pmatrix} 0 & -i \\ i & 0 \end{pmatrix}, \quad J_3^{1/2} = \frac{\hbar}{2}\begin{pmatrix} 1 & 0 \\ 0 & -1 \end{pmatrix}. \tag{15.1}$$

We take out the prefactor

$$J_i^{1/2} \equiv \frac{\hbar}{2}\sigma_i, \tag{15.2}$$

and thus defining the Pauli matrices σ_i,

$$\sigma_1 = \begin{pmatrix} 0 & 1 \\ 1 & 0 \end{pmatrix}, \quad \sigma_2 = \begin{pmatrix} 0 & -i \\ i & 0 \end{pmatrix}, \quad \sigma_3 = \begin{pmatrix} 1 & 0 \\ 0 & -1 \end{pmatrix}. \tag{15.3}$$

We can check that these matrices satisfy a number of properties:

(1) $\sigma_i^2 = \mathbb{1}$ (matrix normalization):
(2) $\{\sigma_i, \sigma_j\} = 0$ for $i \neq j$ (anticommutation):
(3) $\sigma_1\sigma_2 = i\sigma_3$ and cyclic permutations thereof: $\sigma_2\sigma_3 = i\sigma_1$, $\sigma_3\sigma_1 = i\sigma_2$ (commutation); since $\sigma_1\sigma_2 = -\sigma_2\sigma_1$, etc. (anticommutation), we obtain the algebra

$$[\sigma_i, \sigma_j] = 2i\epsilon_{ijk}\sigma_k. \tag{15.4}$$

Putting together properties (1) and (3), we have

$$\sigma_i\sigma_j = \delta_{ij}\mathbb{1} + i\epsilon_{ijk}\sigma_k. \tag{15.5}$$

Further properties of the Pauli matrices are as follows:

(4) $\text{Tr}[\sigma_i] - 0$ (tracelessness):
(5) $\text{Tr}[\sigma_i^2] = 2$ (matrix normalization);
(6) using properties (3)–(5), we find also $\text{Tr}[\sigma_i\sigma_j] = 0$ for $i \neq j$, so we have a (trace orthonormalization) condition

$$\text{Tr}[\sigma_i\sigma_j] = 2\delta_{ij}; \tag{15.6}$$

(7) (completeness of $(\mathbb{1}, \sigma_i)$) in the space of complex 2×2 matrices, we can decompose any matrix M in terms of the four matrices σ_i and $\mathbb{1}$, as

$$M = \alpha_0\mathbb{1} + \sum_{i=1,2,3} \alpha_i\sigma_i; \tag{15.7}$$

indeed, multiplying with 1 or σ_j and taking the trace, and then using the trace orthonormalization condition at property (6), we find

$$\alpha_0 = \frac{1}{2}\text{Tr}[M], \quad \alpha_i = \frac{1}{2}\text{Tr}[M\sigma_i]; \tag{15.8}$$

(8) $\sigma_i^\dagger = \sigma_i$ (Hermiticity);
(9) putting together (1) and (8), we find also that $\sigma_i^\dagger = \sigma_i^{-1}$ (unitarity).

Finally, we obtain that the *spin 1/2 representation is the fundamental representation of $SU(2)$* (which is equal in Lie algebra to the rotation group $SO(3)$, as we saw), since the $\sigma_i = (2/\hbar)J_i^{1/2}$ are 2×2 complex Hermitian matrices and generators, so that $g = e^{i\vec{\alpha}\cdot\vec{J}}$ is unitary: $g^\dagger = g^{-1}$. Moreover, property (4) (the tracelessness of σ_I) means that the matrices g are also special. Indeed, we have

$$\det M = e^{\text{Tr}\ln M}, \tag{15.9}$$

which can be proven as follows. Diagonalize the matrix M, by writing it as $S^{-1}\tilde{M}S$, with $\tilde{M}$ diagonal with diagonal elements λ_i, and using $\det(S^{-1}MS) = \det S^{-1}\det M \det S = \det M$ and $\text{Tr}[S^{-1}MS] = \text{Tr}\, M$, so that

$$\text{Tr}[M] = \sum_i \lambda_i \Rightarrow \det M = \prod_i \lambda_i = e^{\sum_i \ln\lambda_i} = e^{\text{Tr}\ln M}. \tag{15.10}$$

Further, writing $M = e^A$, we obtain

$$\det e^A = e^{\text{Tr}A}, \tag{15.11}$$

which means that if $\text{Tr}\, A = 0$ then $\det e^A = 1$. In our case, $A = \vec{\alpha}\cdot\vec{\sigma}$ and $e^A = g$, so $\det g = 1$.

On the other hand, the adjoint representation of $SO(3)$ is defined by the structure constants of $SO(3)$, $f_{abc} = \epsilon_{abc}$,

$$(T_a)_{bc} = -if_{abc} = -i\epsilon_{abc}. \tag{15.12}$$

This means specifically that

$$T_1 = -i\begin{pmatrix} 0 & 0 & 0 \\ 0 & 0 & 1 \\ 0 & -1 & 0 \end{pmatrix}, \quad T_2 = -i\begin{pmatrix} 0 & 0 & -1 \\ 0 & 0 & 0 \\ 1 & 0 & 0 \end{pmatrix}, \quad T_3 = -i\begin{pmatrix} 0 & 1 & 0 \\ -1 & 0 & 0 \\ 0 & 0 & 0 \end{pmatrix}, \tag{15.13}$$

which is the same set of matrices as $J_i^1/\hbar$ (the generators, or angular momenta in the $j = 1$ representation), as can be checked using the explicit formulas from the last chapter.

With regard to the representations of $j \geq 3/2$, we can obtain them from the previous representations by *composition*, which will be discussed next.

15.2 Composition of Angular Momenta

Consider a composite system, such as for instance an atom, that contains several objects (such as electrons, etc.), each with angular momentum. Thus we have eigenstates of $\vec{J}^2$ and $\vec{J}_z$ for each object, but we must have also eigenstates of the system for the total momentum $\vec{J} = \sum_i \vec{J}_i$.

Two Angular Momenta, $\vec{J} = \vec{J}_1 + \vec{J}_2$

It is in fact enough to consider only two momenta, since we can add up two momenta, then add another to the sum of the first two, etc., until we have done the whole sum.

Since for $\vec{J}^2$, J_z we have eigenstates $|jm\rangle$, with j defining the representation and m an index for the states in the representation, for two such momenta $\vec{J}_1, \vec{J}_2$ we have a tensor product total state, written as

$$|j_1 m_1\rangle \otimes |j_2 m_2\rangle \equiv |j_1 j_2; m_1 m_2\rangle. \tag{15.14}$$

The algebra of the angular momenta is

$$\begin{aligned} [J_{1i}, J_{1j}] &= i\hbar\epsilon_{ijk}J_{1k} \\ [J_{2i}, J_{2j}] &= i\hbar\epsilon_{ijk}J_{2k} \\ [J_{1i}, J_{2j}] &= 0 \\ [J_1^2, J_{1z}] &= 0 \\ [J_2^2, J_{2z}] &= 0. \end{aligned} \tag{15.15}$$

For fixed j_1, j_2, we have that m_1, m_2 must define the system, as a tensor product representation. But this representation is in fact too large, considered as a representation of $SO(3)$. In fact, we already know that *all* the representations of $SO(3)$ are defined by a fixed j and indexed by m values. This means that the basis of states above, while complete in the Hilbert space, is too large, and we must impose an extra condition that $\vec{J} = \vec{J}_1 + \vec{J}_2$ is also an angular momentum,

$$[J_i, J_j] = i\hbar\epsilon_{ijk}J_k, \tag{15.16}$$

with quantum numbers j, m defining the representation of the total system.

We note that on the space of states $|j_1 m_1\rangle \otimes |j_2 m_2\rangle$, we can act with the matrices in the corresponding representations for each of the two Hilbert subspaces, thus with

$$\begin{aligned} D_{1R}(g) \otimes D_{2R}(g) &= D_R\left(e^{-i\vec{\alpha}\cdot\vec{J}_1/\hbar}\right) \otimes D_R\left(e^{-i\vec{\alpha}\cdot\vec{J}_2/\hbar}\right) \\ &\simeq \left(\mathbb{1} - i\frac{\vec{\alpha}\cdot\vec{J}_1}{\hbar}\right) \otimes \left(\mathbb{1} - i\frac{\vec{\alpha}\cdot\vec{J}_2}{\hbar}\right) \\ &\simeq \mathbb{1}\otimes\mathbb{1} - \frac{i}{\hbar}\vec{\alpha}\cdot\left(\vec{J}_1 \otimes \mathbb{1}_2 + \mathbb{1}_1 \otimes \vec{J}_2\right) \\ &\equiv \mathbb{1} - \frac{i}{\hbar}\vec{\alpha}\cdot\vec{J}, \end{aligned} \tag{15.17}$$

meaning that more precisely we have

$$\vec{J} = \vec{J}_1 \otimes \mathbb{1}_2 + \mathbb{1}_1 \otimes \vec{J}_2. \tag{15.18}$$

In the basis $|j_1 j_2; m_1 m_2\rangle$, we have the eigenvalues

$$\begin{aligned} \vec{J}_1^2 |j_1 j_2; m_1 m_2\rangle &= j_1(j_1+1)\hbar^2 |j_1 j_2; m_1 m_2\rangle \\ \vec{J}_2^2 |j_1 j_2; m_1 m_2\rangle &= j_2(j_2+1)\hbar^2 |j_1 j_2; m_1 m_2\rangle \\ J_{1z} |j_1 j_2; m_1 m_2\rangle &= m_1\hbar |j_1 j_2; m_1 m_2\rangle \\ J_{2z} |j_1 j_2; m_1 m_2\rangle &= m_2\hbar |j_1 j_2; m_1 m_2\rangle. \end{aligned} \tag{15.19}$$

But since j_1, j_2 define the *representation*, they must be kept in an alternative basis, while we can replace m_1, m_2 (the eigenvalues of J_{1z}, J_{2z}) with the eigenvalues of $\vec{J}^2$ and J_z, represented by j and m. Indeed, the set $\vec{J}_1^2, \vec{J}_2^2, \vec{J}^2, J_z$ is mutually commuting, so we can have simultaneous eigenvalues, whereas adding J_{1z}, J_{2z} spoils the mutual commutativity. Indeed, we can calculate

$$\begin{aligned} [\vec{J}^2, \vec{J}_1^2] &= [\vec{J}_1^2 + \vec{J}_2^2 + 2\vec{J}_1\cdot\vec{J}_2, \vec{J}_1^2] \\ &= [\vec{J}_1^2 + \vec{J}_2^2 + 2J_{1z}J_{2z} + J_{1+}J_{2-} + J_{1-}J_{2+}, \vec{J}_1^2] = 0 \\ [\vec{J}^2, \vec{J}_2^2] &= [\vec{J}_1^2 + \vec{J}_2^2 + 2\vec{J}_1\cdot\vec{J}_2, \vec{J}_2^2] \\ &= [\vec{J}_1^2 + \vec{J}_2^2 + 2J_{1z}J_{2z} + J_{1+}J_{2-} + J_{1-}J_{2+}, \vec{J}_2^2] = 0 \\ [J_z, \vec{J}_1^2] &= [J_{1z} + J_{2z}, \vec{J}_1^2] = 0 \\ [J_z, \vec{J}_2^2] &= [J_{1z} + J_{2z}, \vec{J}_2^2] = 0, \end{aligned} \tag{15.20}$$

but on the other hand

$$\begin{aligned} &[\vec{J}^2, J_{1z}] \neq 0 \quad (\text{since } [J_{1z}, J_{1\pm}] \neq 0) \\ &[\vec{J}^2, J_{2z}] \neq 0 \quad (\text{since } [J_{2z}, J_{2\pm}] \neq 0). \end{aligned} \tag{15.21}$$

Therefore the new basis of states is $|j_1 j_2; jm\rangle$, with eigenvalues

$$\begin{aligned} \vec{J}_1^2 |j_1 j_2; jm\rangle &= j_1(j_1+1)\hbar^2 |j_1 j_2; jm\rangle \\ \vec{J}_2^2 |j_1 j_2; jm\rangle &= j_2(j_2+1)\hbar^2 |j_1 j_2; jm\rangle \\ \vec{J}^2 |j_1 j_2; jm\rangle &= j(j+1)\hbar^2 |j_1 j_2; jm\rangle \\ J_z |j_1 j_2; jm\rangle &= m\hbar |j_1 j_2; jm\rangle, \end{aligned} \tag{15.22}$$

and this basis is also complete in the Hilbert space of states of the system. Thus we can expand the elements of one basis in terms of the other basis. In particular, by using the completeness relation

$$\mathbb{1} = \sum_{m_1,m_2} |j_1 j_2; m_1 m_2\rangle\langle j_1 j_2; m_1 m_2|, \tag{15.23}$$

we can write

$$|j_1 j_2; jm\rangle = \sum_{m_1,m_2} |j_1 j_2; m_1 m_2\rangle\langle j_1 j_2; m_1 m_2|j_1 j_2; jm\rangle. \tag{15.24}$$

The matrix elements $\langle j_1 j_2; m_1 m_2|j_1 j_2; jm\rangle$ are known as the *Clebsch–Gordan coefficients*.

Properties of the Clebsch–Gordan coefficients:

(1) The first property derives from acting with $J_z = J_{1z} + J_{2z}$ on states, specifically taking matrix elements between the two basis states,

$$0 = \langle j_1 j_2; m_1 m_2|(J_z - J_{1z} - J_{2z})|j_1 j_2; jm\rangle = (m - m_1 - m_2)\langle j_1 j_2; m_1 m_2|j_1 j_2; jm\rangle, \tag{15.25}$$

where we have acted with J_z on the right, and with J_{1z}, J_{2z} on the left. The equation above means that the Clebsch–Gordan coefficients can only be nonzero for $m = m_1 + m_2$ (otherwise they vanish, according to the equation).

(2) The second property gives a range for the total angular momentum j:

$$|j_1 - j_2| \leq j \leq j_1 + j_2. \tag{15.26}$$

As a first check, we note that at large angular momenta j_1, j_2 (in the classical regime), when $j_i^2 \simeq \vec{J}_i^2/\hbar^2$, (15.26) is obviously true:

$$\vec{J}^2 = (\vec{J}_1 + \vec{J}_2)^2 = \vec{J}_1^2 + \vec{J}_2^2 + 2\vec{J}_1 \cdot \vec{J}_2 \Rightarrow (j_1 - j_2)^2 \leq \frac{\vec{J}^2}{\hbar^2} \leq (j_1 + j_2)^2. \tag{15.27}$$

But (15.26) is actually true in quantum mechanics as well. We first note that the maximum projection of J_z is $m_{\max} = m_{1,\max} + m_{2,\max} = j_1 + j_2$, which means that

$$j_{\max} = j_1 + j_2. \tag{15.28}$$

The representation with this $j_{\max}$ has $2j_{\max} + 1$ states, and the whole representation can be obtained from the state with $m_{\max} = j_1 + j_2$ by acting with J_- (which lowers m by one unit). For the next highest value of m, $m_{\max} - 1$, we have more than one state: the state $J_-|j_1 j_2 j_{\max} m_{\max}\rangle$ and another state, with $j = j_{\max} - 1$. From the latter state we generate another representation with $j = j_{\max} - 1$ by acting with J_-, for a total of $2(j_{\max} - 1) + 1$ states. Then, at $m_{\max} - 2$, we have two states coming from the previous two representations, plus one more, with $j = j_{\max} - 2$, etc. In total, assuming that indeed $j_{\min} = |j_1 - j_2|$, and assuming for concreteness that $j_1 \geq j_2$, we find that the total number of states is

$$\begin{aligned}\sum_{j=j_1-j_2}^{j_1+j_2} (2j+1) &= \frac{2}{2}[(j_1 + j_2 + 1)(j_1 + j_2) - (j_1 - j_2 - 1)(j_1 - j_2)] + 2j_2 + 1 \\ &= (2j_1 + 1)(2j_2 + 1),\end{aligned} \tag{15.29}$$

which is indeed the total number of states in the basis with j_1, j_2, m_1, m_2 at fixed j_1, j_2, so we have found all the states in that basis, confirming the correctness of the hypothesis of $j_{\min} = |j_1 - j_2|$.

Decomposition of Product of Representations

We have thus arrived at the following logic: when we take a product of two representations of $SO(3)$, one of dimension j_1 and the other of dimension j_2, we can decompose the resulting states into representations of $SO(3)$ of varying j, between $|j_1 - j_2|$ and $j_1 + j_2$, formally:

$$j_i \otimes j_2 = |j_1 - j_2| \oplus |j_1 - j_2| + 1 \oplus \cdots \oplus j_1 + j_2. \tag{15.30}$$

15.3 Finding the Clebsch–Gordan Coefficients

The Clebsch–Gordan coefficients are the matrix elements for a change of basis in the Hilbert space of the system, from $|j_1 j_2; m_1 m_2\rangle$ to $|j_1 j_2; jm\rangle$, which by the general theory of Hilbert spaces are unitary matrices. In general, a unitary matrix has complex elements, but since we can multiply all the elements of both bases by arbitrary phases without changing the physics of the system, we can *choose* a convention where the Clebsch–Gordan coefficients are all real (the phases turn all complex elements into real ones). Moreover, we can also choose a convention such that $\langle j_1 j_2, m_1 = j_1, m_2 | j_1 j_2; jm = j\rangle \in \mathbb{R}_+$ is not only real, but positive.

Then, since we have unitary transformations on the Hilbert space, these matrices obey orthonormality conditions,

$$\begin{aligned}
&\sum_{j,m} \langle j_1 j_2; m_1 m_2 | j_1 j_2; jm\rangle\langle j_1 j_2; m_1' m_2' | j_1 j_2; jm\rangle = \delta_{m_1 m_1'}\delta_{m_2 m_2'} \\
&\sum_{m_1, m_2} \langle j_1 j_2; m_1 m_2 | j_1 j_2; jm\rangle\langle j_1 j_2; m_1 m_2 | j_1 j_2; j'm'\rangle = \delta_{jj'}\delta_{mm'}.
\end{aligned} \tag{15.31}$$

For $j = j', m = m'$ in the second relation, we obtain a normalization condition for the Clebsch–Gordan coefficients.

Instead of the Clebsch–Gordan coefficients, one sometimes uses a set of symbols related to them by constants, Wigner $3j$ symbols, defined by

$$\langle j_1 j_2; m_1 m_2 | j_1 j_2; jm\rangle = (-1)^{j_1 - j_2 + m}\sqrt{2j+1}\begin{pmatrix} j_1 & j_2 & j \\ m_1 & m_2 & -m \end{pmatrix}. \tag{15.32}$$

Recursion Relations

We can find recursion relation between the Clebsch–Gordan coefficients by acting with $J_\pm = J_{1\pm} + J_{2\pm}$ on the basis decomposition relation, since successive actions of $J_\pm$ construct the j representation, of $J_{1\pm}$ construct the j_1 representation, and of $J_{2\pm}$ construct the j_2 representation:

$$J_\pm |j_1 j_2; jm\rangle = (J_{1\pm} + J_{2\pm}) \sum_{m_1, m_2} |j_1 j_2; m_1 m_2\rangle\langle j_1 j_2; m_1 m_2 | j_1 j_2; jm\rangle. \tag{15.33}$$

Using the formulas from the previous chapter for the actions of $J_\pm$ on $|jm\rangle$, we obtain the following relation between states:

$$\begin{aligned}&\sqrt{(j\mp m)(j\pm m+1)}|j_1j_2;jm\pm 1\rangle\\&\quad=\sum_{m_1,m_2}\Big(\sqrt{(j_1\mp m_1)(j_1\pm m_1+1)}|j_1j_2,m_1\pm 1,m_2\rangle\\&\qquad+\sqrt{(j_2\mp m_2)(j_2\pm m_2+1)}|j_1j_2;m_1,m_2\pm 1\rangle\Big)\langle j_1j_2;m_1m_2|j_1j_2;jm\rangle.\end{aligned} \tag{15.34}$$

However defining

$$(m_1'\equiv m_1\pm 1,m_2'=m_2)\ \text{ and }\ (m_1'=m_1,m_2'=m_2\pm 1), \tag{15.35}$$

for the two states, respectively, and multiplying from the left with $\langle j_1j_2;m_1'm_2'|$, we find that

$$\begin{aligned}&\sqrt{(j\mp m)(j\pm m+1)}\langle j_1j_2;m_1'm_2'|j_1j_2;j,m\pm 1\rangle\\&\quad=\sqrt{(j_1\mp m_1'+1)(j_1\pm m_1')}\langle j_1j_2;m_1'\mp 1,m_2'|j_1j_2;jm\rangle\\&\qquad+\sqrt{(j_2\mp m_2'+1)(j_2\pm m_2')}\langle j_1j_2;m_1',m_2'\mp 1|j_1j_2;jm\rangle.\end{aligned} \tag{15.36}$$

These recursion relations, together with the normalization condition and the phase and sign convention, are enough to specify the Clebsch–Gordan coefficients completely, though we will not give examples here.

15.4 Sums of Three Angular Momenta, $\vec{J}_1+\vec{J}_2+\vec{J}_3$: Racah Coefficients

In the case where three angular momenta are to be added, we have the following starting basis for the Hilbert space:

$$|j_1m_2\rangle\otimes|j_2m_2\rangle\otimes|j_3m_3\rangle\equiv|j_1j_2j_2;m_1m_2m_3\rangle. \tag{15.37}$$

To sum the angular momenta, and decompose these states into representations (of given j) of the rotation group $SO(3)$, we can first add two momenta, and then add the remaining one to the sum of the first two. This can be done in three different ways.

(1) We can first set $\vec{J}_1+\vec{J}_2=\vec{J}_{12}$, and then set $\vec{J}_{12}+\vec{J}_3=\vec{J}$. In the first step, we replace m_1 and m_2 with j_{12} and m_{12}, and in the second, we replace m_3 and m_{12} with j and m (i.e., total angular momentum and its projection onto z), for a total basis state of $|j_1j_2j_{12}j_3;j,m\rangle$. Decomposing this basis state in terms of the original basis, using the two steps above, leads to

$$|j_1j_2j_{12}j_3;jm\rangle=\sum_{m_1,m_2,m_{12},m_3}|j_1j_2j_3;m_1m_2m_3\rangle\langle j_1j_2m_1m_2|j_{12}m_{12}\rangle\langle j_{12}j_3;m_{12}m_3|jm\rangle, \tag{15.38}$$

where

$$\begin{aligned}\langle j_1j_2m_1m_2|j_{12}m_{12}\rangle&\equiv\langle j_1j_2j_3;m_1m_2m_3|j_1j_2j_{12}j_3;m_{12}m_3\rangle\\\langle j_{12}j_3m_{12}m_3|jm\rangle&\equiv\langle j_1j_2j_{12}j_3;m_{12}m_3|j_1j_2j_{12}j_3j;m\rangle.\end{aligned} \tag{15.39}$$

(2) Alternatively, we can first set $\vec{J}_2+\vec{J}_3=\vec{J}_{23}$, and then $\vec{J}_{23}+\vec{J}_1=\vec{J}$. In the first step, we replace m_2 and m_3 with j_{23} and m_{23}, and in the second, we replace m_1 and m_{23} with j and m. The

new decomposition is

$$|j_1j_2j_3j_{23};jm\rangle = \sum_{m_1,m_2,m_3,m_{23}} |j_1j_2j_3;m_1m_2m_3\rangle\langle j_2j_3m_2m_3|j_{23}m_{23}\rangle\langle j_1j_{23};m_1m_{23}|jm\rangle. \quad (15.40)$$

(3) The third possibility is to set $\vec{J}_1 + \vec{J}_3 = \vec{J}_{13}$, and then $\vec{J}_{13} + \vec{J}_2 = \vec{J}$. The corresponding last decomposition is

$$|j_1j_3j_{13}j_2;jm\rangle = \sum_{m_1,m_2,m_3,m_{23}} |j_1j_2j_3;m_1m_2m_3\rangle\langle j_1j_3m_1m_3|j_{13}m_{13}\rangle\langle j_2j_{13};m_2m_{13}|jm\rangle. \quad (15.41)$$

So, all three methods are equivalent, and we should arrive at similar bases. This means that, in particular, the bases obtained in points (1) and (2) are related by a unitary transformation. The matrix elements between them can be written as (by taking out real factors) the Racah W coefficients, or equivalently the Wigner $6j$ symbols,

$$\begin{aligned}\langle j_1j_2j_{12}j_3;jm|j_1j_2j_3j_{23};jm\rangle &\equiv \sqrt{(2j_{12}+1)(2j_{23}+1)}W(j_1,j_2,j_{13};j_{12},j_{23})\\ &\equiv (-1)^{j_1+j_2+j_3+j}\sqrt{(2j_{12}+1)(2j_{23}+1)}\begin{pmatrix} j_1 & j_2 & j_{12}\\ j_3 & j & j_{23}\end{pmatrix}.\end{aligned} \quad (15.42)$$

These matrices are rather complicated and tedious to obtain, though Racah showed that the W coefficients have workable expressions.

15.5 Schwinger's Oscillator Model

Finally, we note a very useful way to obtain the representations of the Lie algebra of $SU(2) \simeq SO(3)$, by building its generators from the algebra of harmonic oscillators. It turns out that we need *two* harmonic oscillators, one with operators $a_+, a_+^\dagger$, one with $a_-, a_-^\dagger$. The two oscillators will be associated with the positive and negative values of $m = J_z$, respectively.

We start with a single harmonic oscillator, with $a, a^\dagger$ and number operator $N = a^\dagger a$. Thus its algebra is

$$\begin{aligned}&[a, a^\dagger] = 1\\ &[N, a^\dagger] = a^\dagger a a^\dagger - a^\dagger a^\dagger a = a^\dagger[a, a^\dagger] = +a^\dagger\\ &[N, a] = a^\dagger a a - a a^\dagger a = -[a, a^\dagger]a = -a.\end{aligned} \quad (15.43)$$

The following commutation relations of N look indeed like the commutation relations

$$\left[\left(2\frac{J_z}{\hbar}\right), \left(\frac{J_\pm}{\hbar}\right)\right] = \pm\left(2\frac{J_z}{\hbar}\right), \quad (15.44)$$

but the remaining one is not quite right, since it is

$$\left[\left(\frac{J_+}{\hbar}\right), \left(\frac{J_-}{\hbar}\right)\right] = \left(2\frac{J_z}{\hbar}\right), \quad (15.45)$$

which doesn't match $[a^\dagger, a] = -1$.

This is why we actually need *two* oscillators in order to construct the Lie algebra, commuting one with the other, so with simultaneous eigenvalues,

$$[a_+, a_-^\dagger] = [a_-, a_+^\dagger] = 0. \quad (15.46)$$

For this system, we can construct simultaneous eigenstates, in the usual manner, as a tensor product of the states of the two oscillators,

$$|n_+, n_-\rangle = |n_+\rangle \otimes |n_-\rangle = \frac{(a_+^\dagger)^{n_+}(a_-^\dagger)^{n_-}}{\sqrt{n_+!\, n_-!}}|0,0\rangle. \tag{15.47}$$

Then we can construct the generators of the Lie algebra in terms of these two oscillators,

$$\begin{aligned}\frac{J_+}{\hbar} &\equiv a_+^\dagger a_-,\\ \frac{J_-}{\hbar} &\equiv a_-^\dagger a_+,\\ \frac{2J_z}{\hbar} &\equiv a_+^\dagger a_+ - a_-^\dagger a_- = N_+ - N_-.\end{aligned} \tag{15.48}$$

We can check that, indeed this has the right commutation relations. For instance, the commutation relation that did not work with a single oscillator now does work:

$$\left[\frac{J_+}{\hbar}, \frac{J_-}{\hbar}\right] = \left[a_+^\dagger a_-, a_-^\dagger a_+\right] = a_+^\dagger[a_-, a_-^\dagger]a_+ - a_-^\dagger[a_+, a_+^\dagger]a_- = N_+ - N_- = \frac{2J_z}{\hbar}. \tag{15.49}$$

Moreover, the action of the angular momentum operators on the basis consisting of occupation number states $|n_+, n_-\rangle$ is

$$\begin{aligned}\frac{J_+}{\hbar}|n_+, n_-\rangle &= a_+^\dagger a_-|n_+, n_-\rangle = \sqrt{n_-(n_+ + 1)}|n_+ + 1, n_- - 1\rangle\\ \frac{J_-}{\hbar}|n_+, n_-\rangle &= a_-^\dagger a_+|n_+, n_-\rangle = \sqrt{n_+(n_- + 1)}|n_+ - 1, n_+ + 1\rangle\\ \frac{2J_z}{\hbar}|n_+, n_-\rangle &= (N_+ - N_-)|n_+, n_-\rangle = (n_+ - n_-)|n_+, n_-\rangle.\end{aligned} \tag{15.50}$$

We see that we can identify the occupation states for the oscillators, $|n_+, n_-\rangle$, with the states in the representation of angular momentum j, by making the identifications

$$n_+ = j + m, \quad n_- = j - m. \tag{15.51}$$

Defining the total occupation number, $N = N_+ + N_-$, we obtain that

$$n_+ + n_- \equiv N = 2j. \tag{15.52}$$

Substituting (15.51) in the general state $|n_+, n_-\rangle$, we obtain the states in the spin j representation of the $SO(3)$ rotation group as oscillator states, with

$$|jm\rangle = \frac{(a_+^\dagger)^{j+m}(a_-^\dagger)^{j-m}}{\sqrt{(j+m)!\,(j-m)!}}|0\rangle. \tag{15.53}$$

In this way, we can construct *all* the states in the representation of angular momentum j: we note that $m_{\max} = j$ (so that $j - m = 0$) and $m_{\min} = -j$ (so that $j + m = 0$), otherwise the above state (15.53) does not make sense. We also note that the state with $m = j$ is

$$|jj\rangle = \frac{(a_+^\dagger)^{2j}}{\sqrt{2j!}}|0\rangle, \tag{15.54}$$

which can be interpreted as being composed of $2j$ elements with angular momentum 1/2 each, all with "spin up", $j_z = +1/2$. Thus the total angular momentum can be thought of as a sum of angular momenta each having the minimum value, 1/2:

$$\vec{j} = \sum_{i=1}^{2j} (\vec{1/2})_i \,. \tag{15.55}$$

Thus, as we mentioned at the beginning, we can indeed construct all the representations of $SO(3)$ from just sums of the simplest, "spinor" representation.

In this Schwinger oscillator model, $a_+^\dagger$ creates a spin 1/2 state with spin up, and $a_-^\dagger$ creates one with spin down, each state having $N = 2j$ oscillators, $j + m$ with spin up and $j - m$ with spin down.

The Schwinger construction presented here for $SO(3) \simeq SU(2)$ is actually much more general. In fact, *any* Lie algebra can be constructed in terms of harmonic oscillators $a_i, a_i^\dagger$, which in turn allows us, as above, to explicitly construct its representations. Indeed, in Lie group theory, one can show that we can reduce the Lie algebra to a kind of product of $SU(2)$ factors, after which we can just follow the above analysis. We will not do this here, but the result is that this quantum mechanical construction is useful for solving the issue of representations of general (Lie) groups.

Thus we can say that quantum mechanics helps not just with physics, but also with mathematics, in terms of representations of Lie groups!

Important Concepts to Remember

- The angular momentum $\vec{J}$ is the generator of the $SO(3)$ group of spatial rotations, and the values j of the angular momentum correspond to representations of $SO(3)$ in which physical $|\psi\rangle$ states belong.
- The spinor representation of $SO(3) \simeq SU(2)/\mathbb{Z}$, $j = 1/2$, has generators given by the Pauli matrices, $J_i = \frac{\hbar}{2}\sigma_i$, and is the fundamental representation of $SU(2)$.
- The Pauli matrices σ_i satisfy $\sigma_i\sigma_j = \delta_{ij}\mathbb{1} + i\epsilon_{ijk}\sigma_k$; they are Hermitian and, since $\sigma_i^2 = \mathbb{1}$, are also unitary and traceless.
- The matrices $(\mathbb{1}, \sigma_i)$ are complete in the space of 2×2 matrices.
- The $j = 1$ representation is the adjoint representation of $SO(3)$.
- We can compose two angular momenta to give $\vec{J} = \vec{J}_1 + \vec{J}_2$, in which case the states can be described either as a tensor product of the original states, $|j_1 m_1\rangle \otimes j_2 m_2\rangle \equiv |j_1 j_2; m_1 m_2\rangle$, or as states of $\vec{J}$, with j_1, j_2 defined also, thus as $|j_1 j_2; jm\rangle$, since we have two mutually commuting sets, $(\vec{J}_1^2, \vec{J}_2^2, J_{1z}, J_{2z})$ and $(\vec{J}_1^2, \vec{J}_2^2, \vec{J}^2, J_z)$.
- The relation between the two possible bases is given as

  $$|j_1 j_2; jm\rangle = \sum_{m_1, m_2} |j_1 j_2; m_1 m_2\rangle\langle j_1 j_2; m_1 m_2 | j_1 j_2; jm\rangle;$$

  the coefficients $\langle j_1 j_2; m_1 m_2 | j_1 j_2; jm\rangle$ are the Clebsch–Gordan coefficients.
- In the Clebsch–Gordan coefficients we have (providing that the coefficient is nonzero) $m = m_1 + m_2$ and $|j_1 - j_2| \leq j \leq j_1 + j_2$, which means that we have the decomposition of the product of representations into a sum of representations, as $j_1 \otimes j_2 = |j_1 - j_2| \oplus |j_1 - j_2| + 1 \oplus \cdots \oplus j_1 + j_2$.
- Up to a rescaling, the Clebsch–Gordan coefficients are the same as the Wigner $3j$ symbols, $\begin{pmatrix} j_1 & j_2 & j \\ m_1 & m_2 & -m \end{pmatrix}$.

- When composing three angular momenta, $\vec{J}_1 + \vec{J}_2 + \vec{J}_3 = \vec{J}$, we can do this by first adding 1 and 2, or first adding 2 and 3 (or first adding 1 and 3), which are equivalent procedures, leading to coefficients for the transition, $\langle j_1 j_2 j_{12} j_3; jm | j_1 j_2 j_3 j_{23}; jm\rangle$, which up to some rescaling are the Racah coefficients $W(j_1, j_2, j_{13}; j_{12}, j_{23})$ or the Wigner $6j$ symbols $\begin{pmatrix} j_1 & j_2 & j_{12} \\ j_3 & j & j_{23} \end{pmatrix}$.
- In Schwinger's oscillator model, we can construct the generators of $SU(2)$ or $SO(3)$, the angular momenta, in terms of two sets of oscillators $a_+, a_+^\dagger, a_-, a_-^\dagger$, as $J_+/\hbar = a_+^\dagger a_-$, $J_-/\hbar = a_-^\dagger a_+$, $2J_z/\hbar = N_+ - N_-$, and the states in a representation in terms of the creation operators,

$$|jm\rangle = \frac{(a_+^\dagger)^{j+m}}{\sqrt{(j+m)!}} \frac{(a_-^\dagger)^{j-m}}{\sqrt{(j-m)!}} |0\rangle,$$

  so $a_+^\dagger$ creates a spin 1/2 up and $a_-^\dagger$ a spin 1/2 down.
- The Schwinger oscillator model is much more general; any Lie algebra can be represented in terms of harmonic oscillators, and the states in a representation in terms of the creation operators acting on a vacuum.

Further Reading

See [2], [1], [3] for more on angular momenta, and [8] for more on mathematical constructions for $SU(2)$ representations and for other Lie algebras.

Exercises

(1) Calculate

$$\mathrm{Tr}\, e^{\alpha_i \sigma_i/2}, \tag{15.56}$$

where σ_i are the Pauli matrices.

(2) Decompose the product of two spin 1 representations into (irreducible) representations of $SU(2)$, and list explicitly the relation between the basis elements for the two bases in terms of Clebsch–Gordan coefficients.

(3) For the case in exercise 2, find all the Clebsch–Gordan coefficients for the lowest value of the total spin j, using the recursion relations and the normalization conditions.

(4) For the sum of three spin 1 representation, in the case where the final spin is 2, write explicitly the three decompositions of the final basis in terms of the tensor product basis (with a product of two Clebsch–Gordan coefficients) and also write explicitly the resulting Racah coefficients.

(5) Rewrite the relation between the two bases in exercise 2 in terms of Schwinger oscillators.

(6) Consider the Lie algebra for $J_i, K_i, i = 1, 2, 3$,

$$[J_i, J_j] = i\epsilon_{ijk} J_k, \quad [K_i, K_j] = i\epsilon_{ijk} J_k, \quad [J_i, K_i] = i\epsilon_{ijk} K_k. \tag{15.57}$$

Write it in terms of Schwinger oscillators, and give the general representation in terms of Schwinger oscillators acting on a vacuum.

(7) Decompose the product of four spin 1 representations of $SU(2)$, $1 \otimes 1 \otimes 1 \otimes 1$, into irreducible $SU(2)$ representations.

16 Applications of Angular Momentum Theory: Tensor Operators, Wave Functions and the Schrödinger Equation, Free Particles

After having described how to construct representations, and how to compose them (and to decompose the products of representations into sums of representations, or $\vec{J}_1 + \vec{J}_2$ into $\vec{J}$), we now apply these notion to physics. First, we learn how to write operators, associated with observables, transforming in a given representation under rotation, i.e., "tensor operators". Then, we find how wave functions transform under rotations, and how the Schrödinger equation decomposes under them. Then, we apply this formalism to the simplest case, that of free particles.

16.1 Tensor Operators

Consider classical measurable vectors, e.g., $\vec{r}, \vec{p}, \vec{L}$, collectively called V_i, that transform according to the rule

$$V_i \to V_i' = \sum_j R_{ij} V_j, \tag{16.1}$$

where R_{ij} is an $SO(3)$ matrix, i.e., a *rotation matrix* in the defining $j = 1$, or vector, representation (with orthogonal 3×3 matrices).

Then, also classically, one considers tensors as objects transforming as products of the basic vector representation,

$$T_{i_1 i_2 \dots i_n} \to T'_{i_1 i_2 \dots i_n} = R_{i_1 i_1'} R_{i_2 i_2'} \cdots R_{i_n i_n'} T_{i_1' i_2' \dots i_n'}. \tag{16.2}$$

However, as we saw in the previous chapter, this representation, the product of vector representations, is not irreducible. It can in fact be reduced, i.e., decomposed into irreducible representations of $SO(3)$, which are all of angular momentum j. From the previous chapter, we can formally say that we have

$$\sum_i \vec{1}_i = \vec{j}, \tag{16.3}$$

where j has several values that can be calculated, or, in a different notation, where we denote the representation by its value of j, and denote the effect of the angular momentum sum on states (tensor product, decomposed as a sum),

$$1 \otimes 1 \otimes \cdots \otimes 1 = \cdots \oplus (n-1) \oplus n. \tag{16.4}$$

Note that we have a tensor product of n quantities on the left and that on the right the possible j's take only *integer* values, since we are summing integer values (of $j = 1$) many times. In yet another notation, we can denote the representation by its multiplicity (the number of states in it), for instance $j = 1$ as 3, $j = 2$ as 5, etc., so that

$$3 \otimes 3 \otimes \cdots \otimes 3 = \cdots \oplus (2(n-1)+1) \oplus 2n+1. \tag{16.5}$$

This relation would be true as numbers also, i.e., $3 \times 3 \times \cdots \times 3 = \cdots + (2(n-1)+1) + (2n+1)$.

In particular, for a sum of only two vectors (giving a tensor product of representations) and thus for the tensor T_{ij}, we have

$$1 \otimes 1 = 0 \oplus 1 \oplus 2, \tag{16.6}$$

meaning that, for $j_1 = j_2 = 1$, we have $0 = |1-1| \leq j \leq 1+1 = 2$, so $j = 0, 1, 2$. Equivalently, denoting the representation by its multiplicity, we have

$$3 \otimes 3 = 1 \oplus 3 \oplus 5, \tag{16.7}$$

which is also true as regular products and sums, $3 \times 3 = 1 + 3 + 5 = 9$, giving the total number of states in the system.

Quantum mechanically, we have operators $\hat{V}_i$ corresponding to the observables V_i, for instance $\hat{r}_i$, $\hat{p}_i, \hat{L}_i$. Then the transformed operators $\hat{V}_i'$ have the same transformation rules as the classical observables. This is so, since if the states $|\psi\rangle$ in the Hilbert space transform in some representation with angular momentum j under a rotation with rotation matrix R,

$$|\psi\rangle \to |\psi'\rangle \equiv D_{(j)}(R)|\psi\rangle, \tag{16.8}$$

where $D_{(j)}(R)$ is the *rotation matrix in the* (j) *representation*, then the classically observable quantum averages of the operators $\hat{V}_i$, classically observable, must transform in the same way as the classical objects, i.e.,

$$\langle\psi|\hat{V}_i|\psi\rangle \to \langle\psi|D^\dagger_{(j)}(R)\hat{V}_i D_{(j)}(R)|\psi\rangle = \sum_j R_{ik}\langle\psi|\hat{V}_k|\psi\rangle, \tag{16.9}$$

implying the transformation for operators

$$\hat{V}_i \to \hat{V}_i' = D^\dagger_{(j)}(R)\hat{V}_i D_{(j)}(R) = \sum_k R_{ik}\hat{V}_k. \tag{16.10}$$

The infinitesimal form of the transformation law can be obtained by remembering the general form of a rotation matrix in a representation (j), in terms of the generators J_i in this representation. In the infinitesimal case, we have

$$D_{(j)}(R) = \exp\left[-\frac{i}{\hbar}\alpha^i (J_i)_{(j)}\right] \simeq 1 - \frac{i}{\hbar}\alpha^i (J_i)_{(j)}, \tag{16.11}$$

and in the defining, or vector representation, which for $SO(3)$ happens to be also the adjoint one, we have

$$R_{ij} = \left[\exp\left(-\frac{i}{\hbar}\alpha^k J_k\right)\right]_{ij} \simeq \mathbb{1}_{ij} - \frac{i}{\hbar}\alpha^k (J_k)_{ij} = \delta_{ij} - \frac{1}{\hbar}\alpha^k \epsilon_{kij}, \tag{16.12}$$

where we have used $(J_k)_{ij} = -if_{ijk} = -i\epsilon_{ijk}$.

Then the transformation law $D^\dagger \hat{V}_i D = \sum_j R_{il}\hat{V}_l$ implies

$$\begin{aligned} -\frac{i\alpha^k}{\hbar}[\hat{V}_i, (J_k)]_{(j)} &= -\frac{i\alpha^k}{\hbar}(-i\epsilon_{kil}\hat{V}_l)_{(j)} \Rightarrow \\ [\hat{V}_i, J_k]_{(j)} &= i\epsilon_{ikl}(\hat{V}_l)_{(j)}, \end{aligned} \tag{16.13}$$

which can be taken as an alternative definition of a vector operator.

Thus a quantum mechanical tensor operator is an object $\hat{T}_m^{(j)}$, in a representation of integer (not half-integer!) angular momentum $j = k \in \mathbb{N}$ where m is the standard index in the representation, $m = -j, \ldots, j$, that therefore transforms according to the rotation matrices in the representation j under a transformation of the states by matrices in a different representation $\tilde{j}$,

$$D^\dagger_{(\tilde{j})}(R)\hat{T}_m^{(j)}D_{(\tilde{j})}(R) = \sum_{m'=-j}^{j} D^{(j)*}_{mm'}(R)\hat{T}_{m'}^{(j)}, \tag{16.14}$$

where we have used $D^T(R^{-1}) = D^*(R)$ to write the complex conjugate of the rotation matrix.

The infinitesimal form of the above transformation law for tensors is similar to the one for vectors, except that now we must use the (complex conjugate of the) matrix of the generator J_k in the j representation,

$$(J_k)^*_{mm'} \equiv \langle jm|J_k|jm'\rangle^* = \langle jm'|J_k|jm\rangle, \tag{16.15}$$

to obtain

$$[\hat{T}_m^{(j)}, \hat{J}_k] = -(J_k)^*_{mm'}\hat{T}_{m'}^{(j)}, \tag{16.16}$$

or

$$[\hat{J}_k, \hat{T}_m^{(j)}] = \hat{T}_{m'}^{(j)}\langle jm'|J_k|jm\rangle. \tag{16.17}$$

16.2 Wigner–Eckhart Theorem

This is a very useful theorem, which states that the matrix elements of the tensor operator $\hat{T}_m^{(j)}$ with respect to $|\alpha\tilde{j}\tilde{m}\rangle$ eigenstates (α stands for all other indices) are given by Clebsch–Gordan coefficients, times a function that is independent of m or, denoting the tensor operator by $\hat{T}_q^{(k)}$ in order to match standard notation and to avoid confusing j and $\tilde{j}$,

$$\langle\alpha', j'm'|\hat{T}_q^{(k)}|\alpha, jm\rangle = \langle jk; mq|jk; j'm'\rangle\frac{\langle\alpha', j'||T^{(k)}||\alpha, j\rangle}{\sqrt{2j+1}}, \tag{16.18}$$

where the double verticals indicate that the matrix element is independent of m.

Proof. To prove the relation, we consider the action of the tensor operator on a state,

$$\hat{T}_q^{(k)}|\alpha, jm\rangle, \tag{16.19}$$

and construct a linear combination with coefficients given by the Clebsch–Gordan coefficients,

$$|\tau, \tilde{j}\tilde{m}\rangle \equiv \sum_{j,m}\hat{T}_q^{(k)}|\alpha, jm\rangle\langle jk; mq|jk\tilde{j}\tilde{m}\rangle. \tag{16.20}$$

Using the orthogonality property of the Clebsch–Gordan coefficients, we reverse the relation and write

$$\hat{T}_q^{(k)}|\alpha, jm\rangle = \sum_{\tilde{j},\tilde{m}}|\tau, \tilde{j}\tilde{m}\rangle\langle jk; mq|\tilde{j}\tilde{m}\rangle. \tag{16.21}$$

Using (16.17) and (16.21), we find that $\hat{J}_k$ acts on $|\tau\tilde{j}\tilde{m}\rangle$ as on $|\tilde{j}\tilde{m}\rangle$. The proof of that is left as an exercise. Thus

$$\langle \alpha jm|\tau\tilde{j}\tilde{m}\rangle \propto \delta_{j\tilde{j}}\delta_{m\tilde{m}}, \tag{16.22}$$

which implies the statement of the theorem. *q.e.d.*

As an application of the theorem, we see that the matrix elements of $\hat{T}_q^{(k)}$ are zero, unless

$$\begin{aligned} q &= m - m' \\ |j - j'| &\leq k \leq j + j'. \end{aligned} \tag{16.23}$$

This selection rule is very useful for quantum processes involving matrix elements for transition between states, as we will see later on in the book.

16.3 Rotations and Wave Functions

For quantum observables, one can measure the eigenvalues of the corresponding operators. But we saw that the classical angular momentum,

$$L_i = \epsilon_{ijk} x_j p_k \quad (\vec{L} = \vec{r} \times \vec{p}), \tag{16.24}$$

corresponds to a quantum operator

$$\hat{L}_i = \epsilon_{ijk} \hat{X}_j \hat{P}_k, \tag{16.25}$$

acting as a generator of the rotation group $SO(3)$; hence its eigenvalues $\hbar^2 j(j+1)$, defined by a given j, are something that one can measure. Therefore measurements will correspond to a given j, and so a given representation of $SO(3)$, a quantization condition for angular momentum that will be reflected in the wave functions.

In the coordinate basis, angular momenta have matrix elements

$$\langle \vec{x}|\hat{L}_i|\vec{x}'\rangle = \epsilon_{ijk}\langle \vec{x}|\hat{X}_j\hat{P}_k|\vec{x}'\rangle = \epsilon_{ijk}x_j\langle \vec{x}|\hat{P}_k|\vec{x}'\rangle = -i\hbar\epsilon_{ijk}x_j\delta(\vec{x}-\vec{x}')\frac{\partial}{\partial x'_k}. \tag{16.26}$$

When acting on a wave function in coordinate, $|\vec{x}\rangle$, space, the usual manipulations lead to

$$\begin{aligned} \langle \vec{x}|\hat{L}_i|\psi\rangle &= \int d\vec{x}'\langle \vec{x}|\hat{L}_i|\vec{x}'\rangle\langle \vec{x}'|\psi\rangle \\ &= -i\hbar\epsilon_{ijk}x_j\frac{\partial}{\partial x_k}\psi(\vec{x}) = -i\hbar\left(\vec{r}\times\vec{\nabla}\psi\right)_i. \end{aligned} \tag{16.27}$$

It is useful to go to spherical coordinates r, θ, ϕ, defined from the Cartesian coordinates x_1, x_2, x_3 by

$$\begin{aligned} x_1 &= r\sin\theta\cos\phi \\ x_2 &= r\sin\theta\sin\phi \\ x_3 &= r\cos\theta, \end{aligned} \tag{16.28}$$

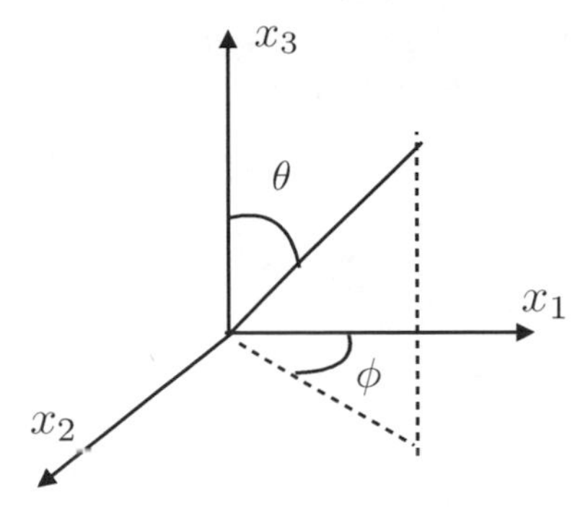

Figure 16.1 Coordinate system.

as in Fig. 16.1. The new system of coordinates has basis vectors $\vec{e}_r, \vec{e}_\theta$, and $\vec{e}_\phi$, which obey the relations

$$\vec{e}_r = \vec{e}_\theta \times \vec{e}_\phi, \quad \vec{e}_\theta = \vec{e}_\phi \times \vec{e}_r, \quad \vec{e}_\phi = \vec{e}_r \times \vec{e}_\theta, \tag{16.29}$$

and then the $\vec{\nabla}$ operator becomes in spherical coordinates

$$\vec{\nabla} = \vec{e}_r \frac{\partial}{\partial r} + \vec{e}_\theta \frac{1}{r}\frac{\partial}{\partial \theta} + \vec{e}_\phi \frac{1}{r \sin\theta}\frac{\partial}{\partial \phi}. \tag{16.30}$$

This allows us to write the form of the angular momentum operator acting on wave functions in spherical coordinates:

$$\vec{L} = -i\hbar \vec{r} \times \vec{\nabla} = -i\hbar \left[\vec{e}_\phi \frac{\partial}{\partial \theta} - \vec{e}_\theta \frac{1}{\sin\theta}\frac{\partial}{\partial \phi} \right]. \tag{16.31}$$

By using the decomposition of the spherical basis vectors in the Cartesian basis,

$$\begin{aligned}
\vec{e}_r &= \sin\theta\cos\phi\,\vec{e}_1 + \sin\theta\sin\phi\,\vec{e}_2 + \cos\theta\,\vec{e}_3 \\
\vec{e}_\theta &= \sin\phi\,\vec{e}_1 + \cos\phi\,\vec{e}_2 \\
\vec{e}_\phi &= \cos\theta\cos\phi\,\vec{e}_1 + \cos\theta\sin\phi\,\vec{e}_2 - \sin\theta\,\vec{e}_3,
\end{aligned} \tag{16.32}$$

we can also decompose $\vec{L}$, written above in spherical coordinates, into L_1, L_2, L_3, as follows:

$$\begin{aligned}
L_1 &= -i\hbar \left(-\sin\phi \frac{\partial}{\partial \theta} - \cos\phi \cot\theta \frac{\partial}{\partial \phi} \right) \\
L_2 &= -i\hbar \left(\cos\phi \frac{\partial}{\partial \theta} - \sin\phi \cot\theta \frac{\partial}{\partial \phi} \right) \\
L_3 &= -i\hbar \frac{\partial}{\partial \phi}.
\end{aligned} \tag{16.33}$$

Then we can also construct operators that change m,

$$\begin{aligned}
\hat{L}_+ &= \hat{L}_1 + i\hat{L}_2 = \hbar e^{i\phi} \left[\frac{\partial}{\partial \theta} + i \cot\theta \frac{\partial}{\partial \phi} \right] \\
\hat{L}_- &= \hat{L}_1 - i\hat{L}_2 = \hbar e^{-i\phi} \left[-\frac{\partial}{\partial \theta} + i \cot\theta \frac{\partial}{\partial \phi} \right].
\end{aligned} \tag{16.34}$$

We want to write wave functions in radial coordinates to match the symmetry of the relevant angular momentum operators, which means that we must consider basis coordinate states of the type

$$|\vec{x}\rangle \equiv |r, \vec{n}\rangle = |r\rangle \otimes |\vec{n}\rangle, \tag{16.35}$$

where $\vec{n} = \vec{n}(\theta, \phi)$ is a unit vector defined by the angles θ and ϕ, and corresponding factorized states

$$|\psi\rangle = |\psi_r\rangle \otimes |\psi_{\vec{n}}\rangle. \tag{16.36}$$

Then the wave functions factorize as well, or in other words, we can consider the *separation of variables* in order to be able to act with $\hat{L}_i$ on them and obtain eigenvalues,

$$\begin{aligned}\langle r, \vec{n}|\hat{L}_i|\psi\rangle &= \langle r|\psi_r\rangle\langle\vec{n}|\hat{L}_i|\psi_{\vec{n}}\rangle \\ &\equiv -i\hbar\psi_r(r)\left[\vec{e}_\phi\frac{\partial}{\partial\theta} - \vec{e}_\theta\frac{1}{\sin\theta}\right]\langle\vec{n}|\psi_{\vec{n}}\rangle.\end{aligned} \tag{16.37}$$

Here we have defined the radial wave function

$$\langle r|\psi_r\rangle \equiv \psi_r(r). \tag{16.38}$$

We also define

$$\langle\vec{n}(\theta, \phi)|\psi_{\vec{n}}\rangle \equiv \psi(\theta, \phi). \tag{16.39}$$

Since, as we said, we are interested in eigenfunctions of $\vec{L}$, it means that $|\psi_{\vec{n}}\rangle$ is a state $|lm\rangle$. Moreover, we will see shortly that in fact we are interested only in $l \in \mathbb{N}$ (so $m \in \mathbb{Z}$), in which case the angular wave function is a *spherical harmonic*,

$$Y_{lm}(\theta, \phi) \equiv \langle\vec{n}(\theta, \phi)|lm\rangle. \tag{16.40}$$

Then, the $Y_{lm}(\theta, \phi)$ are eigenfunctions of $\vec{L}^2$ and L_z, namely

$$\begin{aligned}\vec{L}^2 Y_{lm}(\theta, \phi) &= \hbar^2 l(l+1) Y_{lm}(\theta, \phi) \\ L_z Y_{lm}(\theta, \phi) &= m\hbar Y_{lm}(\theta, \phi).\end{aligned} \tag{16.41}$$

But, since we saw that in spherical coordinates $L_z = -i\hbar\partial/\partial\phi$, we obtain that

$$Y_{lm}(\theta, \phi) = e^{im\phi} y_{lm}(\theta). \tag{16.42}$$

We will give a better definition of these spherical harmonics later; for the moment this is all that we need to say about them.

16.4 Wave Function Transformations under Rotations

Under rotations, since the $Y_{lm}(\theta, \phi)$ describe states $|\psi\rangle \propto |lm\rangle$, they must transform under the l representation of $SO(3)$. Under $\vec{n}' = R\vec{n}$, these states transform under the rotation matrix $D(R)$,

$$|\vec{n}'\rangle = D(R)|\vec{n}\rangle, \tag{16.43}$$

and correspondingly wave functions transform with the rotation matrix in the l representation, since then

$$\begin{aligned}
&\langle \vec{n}'|lm\rangle = \langle \vec{n}|D(R)^\dagger|lm\rangle \Rightarrow \\
&Y_{lm}(\vec{n}') \equiv \langle \vec{n}'|lm\rangle = \langle \vec{n}|D(R)^\dagger|lm\rangle = D^{(l)*}_{mm'}(R)\langle \vec{n}|lm'\rangle \\
&= D^{(l)*}_{mm'}(R)Y_{lm'}(\vec{n}).
\end{aligned} \tag{16.44}$$

We now make an important observation. Since we have found that

$$Y_{lm}(\theta,\phi) \propto e^{im\phi}, \tag{16.45}$$

it means that indeed, for periodicity of $Y_{lm}(\theta,\phi)$ we must have only $m \in \mathbb{Z}$, so that $l \in \mathbb{N}$, as we have assumed before.

However, it also means that if we considered instead $m \in \mathbb{Z}/2$, for half-integer spin, in particular for $j - 1/2$, we would obtain

$$Y_{lm}(\theta,2\pi) = e^{i\pi}Y_{lm}(\theta,0) \Rightarrow \psi(r,\theta,2\pi) = -\psi(r,\theta,0), \tag{16.46}$$

and only a rotation by 4π would bring us back to the original state.

Equivalently, consider the formula from Chapter 14 concerning a rotation with Euler angle ϕ around the third axis, corresponding to the $j = 1$ representation, i.e., the vector or fundamental representation,

$$R_3(\phi) = \begin{pmatrix} \cos\theta & -\sin\theta & 0 \\ \sin\theta & \cos\theta & 0 \\ 0 & 0 & 1 \end{pmatrix}, \tag{16.47}$$

for which $R_3(2\pi) = 1$, as expected for $j = 1$ (integer). On the other hand, the same rotation acting on the space of $j = 1/2$ states, or "spinor" representation, i.e., acting as an $SU(2)$ matrix, gives

$$\tilde{R}_3(\phi) = \begin{pmatrix} e^{i\phi/2} & 0 \\ 0 & e^{-i\phi/2} \end{pmatrix}, \tag{16.48}$$

so that

$$\tilde{R}_3(2\pi) = -\mathbb{1} \Rightarrow \tilde{R}_3(4\pi) = \mathbb{1}. \tag{16.49}$$

This condition, in general gives a good definition of a spinor: it is a representation in which only a rotation by 4π, but not by 2π, gives back the same state. Indeed, in general we find that a rotation by ϕ in a representation of angular momentum j has a group element

$$g \propto \exp\left[\frac{i}{\hbar}\phi J_3^{(j)}\right]. \tag{16.50}$$

Thus, if J_3 has eigenvalue $j\hbar$, and $\phi = 2\pi$, we obtain

$$g(2\pi) = e^{2\pi ij}g(0) = (-1)^{2j}g(0). \tag{16.51}$$

We might ask: is it not contradictory to have a rotation by 2π that changes the state? The answer is that it is not, as long as the change is unobservable, which it is: we will see that a single spinor is unobservable; only two spinors are observable, and, for a state of two spinors, nothing changes.

Finally, we note that rotation matrices *in the j representation* can be found using the Schwinger method of representing generators using oscillators. We could calculate them explicitly and obtain the so-called Wigner formula (so, in this case, the method is also known as the Wigner method), though we will not do this here.

16.5 Free Particle in Spherical Coordinates

Our first application of the formalism that we have developed will be to a free particle in spherical coordinates, as a warm-up for later, more involved, cases. We choose, instead of the $|\vec{r}\rangle = |x, y, z\rangle$ coordinate basis, the spherical coordinate basis $|r, \theta, \phi\rangle$.

Just as we had before ket states $|\vec{p}\rangle = |p_x p_y p_z\rangle$ corresponding to the eigenvalues of three commuting observables $\hat{P}_x, \hat{P}_y, \hat{P}_z$, so now we choose instead three other commuting observables, related to the rotational invariance. That means that $\vec{L}^2$ and L_z are certainly involved, but we must choose a third. The simplest choice is the Hamiltonian $\hat{H}$, giving the energy E.

Indeed, for a free particle, the Hamiltonian is

$$\hat{H} = \frac{\hat{\vec{P}}^2}{2m}, \tag{16.52}$$

with eigenvalue

$$E = \frac{\vec{p}^2}{2m}. \tag{16.53}$$

Since

$$\hat{L}_i = \epsilon_{ijk} \hat{X}_j \hat{P}_k, \tag{16.54}$$

we can check that $[\hat{H}, \hat{L}_z] = 0$ and also $[\hat{H}, \hat{L}^2] = 0$.

The Schrödinger equation in a coordinate basis,

$$\langle \vec{x} | \hat{H} | \psi \rangle = \left[-\frac{\hbar^2}{2m} \vec{\nabla}^2 \right] \psi(\vec{x}) = E\psi(\vec{x}), \tag{16.55}$$

can be rewritten in terms of $\vec{L}^2$, since (using $\sum_i \epsilon_{ijk} \epsilon_{ilm} = \delta_{jl}\delta_{km} - \delta_{jm}\delta_{kl}$, which can be proven by analyzing all possibilities)

$$\begin{aligned} \hat{L}_i \hat{L}_i &= \epsilon_{ijk} \epsilon_{ilm} \hat{X}_j \hat{P}_k \hat{X}_l \hat{P}_m = \hat{X}_j \hat{P}_k \hat{X}_j \hat{P}_k - \hat{X}_j \hat{P}_k \hat{X}_k \hat{P}_j \\ &= \hat{X}_j \hat{X}_j \hat{P}_k \hat{P}_k - \hat{X}_j \hat{P}_j \hat{X}_k \hat{P}_k - i\hbar \hat{X}_j \hat{P}_j, \end{aligned} \tag{16.56}$$

where in the second line we have used the canonical commutation relations. Then, we rewrite (16.56) in spherical coordinates,

$$\hat{L}_i \hat{L}_i = -i\hbar \hat{r} \hat{p}_r - \hat{r} \hat{p}_r \hat{r} \hat{p}_r + \hat{r}^2 \hat{\vec{p}}^2 = \hat{r}^2 (\hat{\vec{p}}^2 - \hat{p}_r^2), \tag{16.57}$$

where we have used $[\hat{r}, \hat{p}_r] = i\hbar$. We see that therefore

$$\hat{\vec{p}}^2 = \hat{p}_r^2 + \frac{1}{r^2} \hat{\vec{L}}^2. \tag{16.58}$$

We can also prove this directly in the coordinate basis, since then

$$p_r = -i\hbar \frac{1}{r} \frac{\partial}{\partial r} r \tag{16.59}$$

and

$$\frac{\vec{L}^2}{\hbar^2} = \Delta_{\theta,\phi} = \frac{1}{\sin\theta} \frac{\partial}{\partial \theta} \left(\sin\theta \frac{\partial}{\partial \theta} \right) + \frac{1}{\sin^2\theta} \frac{\partial^2}{\partial \phi^2}, \tag{16.60}$$

and finally

$$\Delta = \frac{\partial^2}{\partial r^2} + \frac{2}{r}\frac{\partial}{\partial r} + \frac{1}{r^2}\Delta_{\theta,\phi}, \tag{16.61}$$

leading to

$$\vec{p}^2 = -\hbar^2\Delta = -\hbar^2\vec{\nabla}^2 = p_r^2 + \frac{\vec{L}^2}{r^2}, \tag{16.62}$$

for $r \neq 0$.

The Schrödinger equation becomes

$$\frac{\vec{p}^2}{2m}\psi(r,\theta,\phi) = \frac{1}{2m}\left(p_r^2 + \frac{\vec{L}^2}{r^2}\right)\psi(r,\theta,\phi) = E\psi(r,\theta,\phi), \tag{16.63}$$

and can be solved by separation of variables, writing the wave function as a product of a radial function and a spherical harmonic,

$$\psi_{lm}(r,\theta,\phi) = f_l(r)Y_{lm}(\theta,\phi), \tag{16.64}$$

where the spherical harmonics are eigenfunctions of $\vec{L}^2$ and L_z,

$$\begin{aligned} \vec{L}^2 Y_{lm}(\theta,\phi) &= l(l+1)\hbar^2 Y_{lm}(\theta,\phi) \\ L_z Y_{lm}(\theta,\phi) &= m\hbar Y_{lm}(\theta,\phi). \end{aligned} \tag{16.65}$$

The Schrödinger equation then reduces to an eigenvalue equation for the radial wave function $f_l(r)$,

$$\begin{aligned} \left[\frac{p_r^2}{2m} + \frac{l(l+1)\hbar^2}{2mr^2}\right] f_l(r) = E f_l(r) \Rightarrow \\ \frac{\hbar^2}{2m}\left[-\left(\frac{1}{r}\frac{\partial}{\partial r}r\right)^2 + \frac{l(l+1)}{r^2} - \frac{2mE}{\hbar^2}\right] f_l(r) = 0. \end{aligned} \tag{16.66}$$

Defining

$$k^2 \equiv \frac{2mE}{\hbar^2} \tag{16.67}$$

and $\rho = kr$, we find the equation

$$\left[\frac{\partial^2}{\partial\rho^2} + \frac{2}{\rho}\frac{\partial}{\partial\rho} + \left(1 - \frac{l(l+1)}{\rho^2}\right)\right] f_l(\rho) = 0. \tag{16.68}$$

This is in fact the equation for the spherical Bessel functions $j_l(\rho)$, meaning that the solution to the Schrödinger equation in spherical coordinates is

$$\langle r,\theta,\phi|lm\rangle = \psi_{lm}(r,\theta,\phi) = Y_{lm}(\theta,\phi)j_l(kr). \tag{16.69}$$

In fact, the solution to the Schrödinger equation in Cartesian coordinates,

$$\langle\vec{x}|\vec{k}\rangle = e^{i\vec{k}\cdot\vec{r}}, \tag{16.70}$$

can be expanded in the basis of the solutions in spherical coordinates as

$$e^{i\vec{k}\cdot\vec{r}} = \sum_{l=0}^{\infty}\sum_{m=-l}^{l} a_{lm}(\vec{k})Y_{lm}(\theta,\phi)j_l(kr). \tag{16.71}$$

We will continue this analysis in Chapter 19; for the moment we just note that the final result can be rewritten as

$$
\begin{aligned}
e^{ikr\cos\theta} &= \sum_{l=0}^{\infty}(2l+1)i^l j_l(kr)P_l(\cos\theta)\\
&= 4\pi\sum_{l=0}^{\infty}\sum_{m=-l}^{l}(2l+1)i^l j_l(kr)Y^*_{lm}(\vec{n}_1)Y_{lm}(\vec{n}_2),
\end{aligned}
\tag{16.72}
$$

where in the last form we used two directions $\vec{n}_1, \vec{n}_2$, for the spherical harmonics, that have an angle θ between them.

Important Concepts to Remember

- The decomposition of products of spin j representations, in particular spin 1 representations, into irreducible representations is denoted equivalently as $\sum_i \vec{1}_i = \vec{j}$, as $1\otimes 1\otimes\cdots\otimes 1 = \cdots\oplus(n-1)\oplus n$, or (using the dimension of the representation) as $3\otimes 3\otimes\cdots\otimes 3 = \cdots\oplus(2(n-1)+1)\oplus 2n+1$, which is also true for products of numbers.
- The classical tensor transformation law under a rotation with matrix R_{ij}, $T_{i_1 i_2\ldots i_n} \to T'_{i_1 i_2\ldots i_n} = R_{i_1 i'_1}R_{i_2 i'_2}\ldots R_{i_n i'_n}T_{i'_1 i'_2\ldots i'_n}$, becomes the transformation law for states $|\psi\rangle \to |\psi'\rangle \equiv D_{(j)}(R)|\psi\rangle$ and, for operators $\hat{T}^{(j)}_m$ in a representation $j\in\mathbb{N}$, $\hat{T}^{(j)}_m \to D^\dagger_{(j)}(R)\hat{T}^{(j)}_m D_{(j)}(R) = \sum_{m'=-j}^{j} D^{(j)*}_{mm'}(R)\hat{T}^{(j)}_{m'}$ or $[\hat{J}_k, \hat{T}^{(j)}_m] = \hat{T}^{(j)}_{m'}\langle jm'|J_k|jm\rangle$ (for vectors, $[\hat{V}_i, J_k]_{(j)} = i\epsilon_{ikj}(\hat{V}_j)_{(j)}$).
- The Wigner–Eckhart theorem states that the matrix elements (in $|\alpha jm\rangle$) of tensor operators $\hat{T}^{(k)}_q$ are given by Clebsch–Gordan coefficients for summing $\vec{j}+\vec{k}=\vec{j}'$, times a function independent of m: $\langle\alpha', j'm'|\hat{T}^{(k)}_q|\alpha, jm\rangle = \langle jk; mq|jk; j'm'\rangle\frac{\langle\alpha', j'\|T^{(k)}\|\alpha, j\rangle}{\sqrt{2j+1}}$.
- For a given representation of the $SO(3)$ rotation group, i.e., a given *integer* angular momentum $\vec{L}^2$, we have $|\psi\rangle = |\psi_r\rangle\otimes|\psi_{\vec{n}}\rangle$, or, for wave functions, $\psi(r,\theta,\phi) = \psi_r(r)Y_{lm}(\theta,\phi)$, where $Y_{lm}(\theta,\phi)$ is the spherical harmonic $e^{im\phi}y_{lm}(\theta)$.
- We see that single-valued spherical harmonics require $l\in\mathbb{N}$, whereas for half-integer j we have spinors, which under a 2π rotation acquire a minus sign, $\psi(r,\theta,2\pi) = -\psi(r,\theta,0)$, and return to themselves under a 4π rotation.
- For a free particle, since $\Delta_{\theta,\phi} = \vec{L}^2$, we have a centrifugal potential $\hbar^2 l(l+1)/2mr^2$, so the solution to the Schrödinger equation in spherical coordinates is $Y_{lm}(\theta,\phi)j_l(kr)$ (j_l is the spherical Bessel function), and indeed $e^{i\vec{k}\cdot\vec{r}}$ decomposes in these solutions as basis functions with coefficients $a_{lm}(\vec{k})$.

Further Reading

See [2], [3], [1] for more details.

Exercises

(1) Consider the symmetric traceless operator

$$\hat{X}_i\hat{X}_j - \frac{1}{3}\hat{\vec{X}}^2\delta_{ij}, \tag{16.73}$$

where $\hat{X}_i$ corresponds to the position of a particle. Is it a tensor operator? Justify your answer with a calculation.

(2) Consider a free particle. Rewrite the matrix elements

$$\langle E', l'm'| \left(\hat{P}_i\hat{P}_j - \frac{1}{3}\hat{\vec{P}}^2\delta_{ij}\right) |E, lm\rangle \tag{16.74}$$

in terms of Clebsch–Gordan coefficients and other quantities.

(3) Write down the differential equation satisfied by the reduced spherical harmonic $y_{lm}(\theta)$.

(4) Express the Schrödinger equation of the free particle in polar coordinates (r, θ, z), making use of L_z.

(5) Write down the relevant separation of variables for exercise 4.

(6) Consider two free particles, and write down the wave function (in coordinate space) of this system (of distinguishable particles) for the case where the two particles coincide, in terms of the total angular momentum $\vec{L}_1 + \vec{L}_2$.

(7) Find the normalization condition for orthonormal free particle solutions $Aj_l(kr)Y_{lm}(\theta, \phi)$.

17 Spin and $\vec{L} + \vec{S}$

Until now, we have assumed that $\vec{J} = \vec{L}$ is an orbital angular momentum, which would mean that states belong to representations of an *arbitrary* j, depending on the wave function. This reasoning was based on the classical theory. However, in relativistic theory, $SO(3)$ rotational invariance is enhanced to Lorentz $SO(3,1)$ invariance and, further, to Poincaré invariance under the group $ISO(3,1)$. The latter, however, leads to the existence of an *intrinsic* representation for an $SO(3)$ group, called "spin", that is associated with the type of particle, i.e., a given particle has a given spin s, that cannot be changed. Moreover, the statistics of identical particles, Bose–Einstein or Fermi–Dirac (to be studied further later on in the book), are associated with integer spin and half-integer spin, respectively, a statement known as the "spin–statistics theorem", which is however beyond the scope of this book (it is a statement in quantum field theory).

Therefore, a theory of spin arises from a relativistic theory, quantum field theory. For the electron, the simplest theory arises from the relativistic Dirac equation, to be studied towards the end of the book. At this point, however, we use nonrelativistic theory, which is based on classical intuition. Therefore we can think of the spin as an intrinsic "angular momentum" related to a rotation around an axis, i.e., to some precessing "currents". However, unlike $\vec{L}$, $\vec{S}$ is intrinsic, i.e., it is always the same for a given particle, whereas $\vec{L}$ varies.

For the electron, the spin has $s = 1/2$, so $\vec{S}^2 = \hbar^2 s(s+1) = (3/4)\hbar^2$, which corresponds to a two-dimensional Hilbert space for the simple two-state system described towards the beginning of the book. The states, as there, are written as

$$|1/2, +1/2\rangle = |+\rangle, \quad |1/2, -1/2\rangle = |-\rangle. \tag{17.1}$$

17.1 Motivation for Spin and Interaction with Magnetic Field

To motivate the study of spin, we look at the interaction of angular momenta with a magnetic field and in particular the Stern–Gerlach experiment described in Chapter 0.

Consider the coupling of a charged particle, say an electron, to an electromagnetic field. This is done through "minimal coupling", which replaces derivatives acting on states of the particle as follows;

$$\partial_i \to \partial_i - i\frac{q}{\hbar}A_i \quad \Rightarrow \quad \vec{\nabla} \to \vec{\nabla} - i\frac{q}{\hbar}\vec{A}, \tag{17.2}$$

where $\vec{A}$ is the vector potential of electromagnetism and q the electric charge of the particle. Then, in quantum mechanics, the canonical momentum operator acting on a wave function is replaced as follows:

$$\hat{\vec{p}} = \frac{\hbar}{i}\vec{\nabla} \to \hat{\vec{p}} - q\vec{A}. \tag{17.3}$$

The Hamiltonian operator of a free particle therefore is also changed by the electromagnetic interaction:

$$\hat{H}=\frac{\hat{\vec{p}}^2}{2m}\rightarrow\frac{1}{2m}\left(\hat{\vec{p}}-q\hat{\vec{A}}\right)^2=\frac{\hat{\vec{p}}^2}{2m}+\frac{q^2}{2m}\hat{\vec{A}}^2-\frac{q}{m}\frac{\hat{\vec{p}}\cdot\hat{\vec{A}}+\hat{\vec{A}}\cdot\hat{\vec{p}}}{2}. \tag{17.4}$$

If the electromagnetic potential $\vec{A}$ corresponds to a constant magnetic field $\vec{B}$ then in the Coulomb gauge we have

$$\vec{\nabla}\times\vec{A}=\vec{B},\quad \vec{\nabla}\cdot\vec{A}=0, \tag{17.5}$$

and we can choose a vector potential

$$\vec{A}=\frac{1}{2}\vec{B}\times\vec{r}. \tag{17.6}$$

Indeed, using

$$\sum_k \epsilon_{ijk}\epsilon_{klm}=\delta_{il}\delta_{jm}-\delta_{im}\delta_{jl}, \tag{17.7}$$

we obtain

$$\frac{1}{2}(\vec{\nabla}\times(\vec{B}\times\vec{r}))_i=\frac{1}{2}\epsilon_{ijk}\epsilon_{klm}\partial_j(B_l x_m)=\frac{1}{2}(B_i\delta_{jj}-B_j\delta_{ij})=B_i. \tag{17.8}$$

Then the (quantum) interaction Hamiltonian is

$$\begin{aligned}\hat{H}_{\text{int}}&=-\frac{q}{4m}[\hat{\vec{p}}\cdot(\vec{B}\times\hat{\vec{r}})+(\vec{B}\times\hat{\vec{r}})\cdot\hat{\vec{p}}]=-\frac{q}{4m}\epsilon_{ijk}[\hat{P}_iB_j\hat{X}_k+B_i\hat{X}_j\hat{P}_k]\\&=-\frac{q}{2m}\vec{B}\cdot\hat{\vec{L}}+\text{constant}.\end{aligned} \tag{17.9}$$

On the other hand, in general the interacting Hamiltonian for a closed current interactng with a magnetic field is

$$H_{\text{int}}=-\vec{\mu}\cdot\vec{B}, \tag{17.10}$$

where the *magnetic moment* $\vec{\mu}$ of a loop of current I surrounding a disk with area $A=\pi r^2$ and with unit vector $\vec{e}$ is

$$\vec{\mu}=IA\vec{e}. \tag{17.11}$$

The current through the loop is

$$I=\frac{qv}{2\pi r}, \tag{17.12}$$

which means that the magnetic moment is

$$\mu=\frac{qv}{2\pi r}\pi r^2=\frac{q}{2m}mvr=\frac{q}{2m}L\ \Rightarrow\ \vec{\mu}=\frac{q}{2m}\vec{L}. \tag{17.13}$$

We see that then, substituting this $\vec{\mu}$ in the interacting Hamiltonian $-\vec{\mu}\cdot\vec{L}$, we obtain the form of the quantum Hamiltonian given in (17.9).

Moreover, the angular momentum $\vec{L}$ suffers a precession around $\vec{B}$, situated at an angle θ to $\vec{L}$, of

$$\frac{d\vec{L}}{dt}=\vec{\mu}\times\vec{B}=\frac{q}{2m}LB\sin\theta\,\vec{e}_\phi, \tag{17.14}$$

but on the other hand we have (since $d\vec{L}$ moves on a circle of radius $L\sin\theta$)

$$\frac{d\vec{L}}{L\sin\theta\, dt} = \frac{d\phi}{dt} \equiv \omega, \tag{17.15}$$

the angular precession speed is therefore

$$\vec{\omega} = -\frac{q}{2m}\vec{B}. \tag{17.16}$$

The proportionality between the magnetic moment and the angular momentum in quantum mechanics leads to a quantized projection of the magnetic moment onto any direction Oz,

$$\vec{\mu} = \frac{q}{2m}\vec{L} \Rightarrow \mu_z = \frac{q}{2m}L_z, \tag{17.17}$$

and, since $L_z = m_l\hbar$, we obtain

$$\mu_z = \frac{e\hbar}{2m}m_l \equiv \mu_B m_l, \tag{17.18}$$

where μ_B is called the Bohr magneton, and $m_l = -l, \dots, +l$.

Thus, when constructing an electron beam traversing a constant B field (in the Oz direction) area, the electrons in the beam will have no orbital angular momentum, and yet in the Stern–Gerlach experiment we see a split of the beam into two, corresponding to a two-state system, therefore indicating that the electrons are in an angular momentum $j = 1/2$ state. This means that there must be an electron spin of $s = 1/2$, with $m_s = \pm 1/2$ and $S_z = \hbar m_s$. Moreover, then we can measure H_{int} and therefore μ_z, the projection of μ along the direction of B. We find

$$\mu_z = g\frac{e\hbar}{4m}(2m_s), \tag{17.19}$$

where $2m_s = \pm 1$, so that

$$\mu_z = \pm g\frac{\mu_B}{2}, \tag{17.20}$$

and we find that $g \simeq 2$. In fact, from the Dirac equation we find that $g = 2$ exactly, whereas in QED there are small corrections that take us slightly away from $g = 2$. This underscores the fact that spin is really a different type of angular momentum, unlike the orbital angular momentum. This was to be expected, since it is an intrinsic property of the particle.

For particles having a different spin, such as $s = 1$ (these are called "vector particles", photons are an example), we have different values for g but again the Hilbert space (the representation) can be described in the same way as for the angular momentum states.

17.2 Spin Properties

For any particle that has both orbital angular momentum $\vec{L}$ and spin $\vec{S}$, we can consider states characterized by both, as well as by the total angular momentum $\vec{J} = \vec{L} + \vec{S}$. If we have several particles composing the system, for instance an atom, with the nucleus, composed of nucleons, and the electrons around it, we can add up the angular momenta, as follows:

$$\vec{S}_{\text{total}} = \sum_i \vec{S}_i, \quad \vec{L}_{\text{total}} = \sum_i \vec{L}_i, \quad \vec{J}_{\text{total}} = \vec{L}_{\text{total}} + \vec{S}_{\text{total}}. \tag{17.21}$$

In the case of an atom, both the nucleons and the electrons are particles of spin $s = 1/2$, but we have several possibilities for S_{total}, by the general vector addition theory for angular momenta.

For one particle, we can consider eigenstates of $\vec{L}^2, L_z$, together with $\vec{S}^2, S_z$, as well as other observables, with eigenvalues generically called α:

$$|\alpha, lm\rangle \otimes |sm_s\rangle. \tag{17.22}$$

A generic state $|\psi\rangle$ can be described in a coordinate and spin basis $\langle \vec{r}, m_s|$ by a wave function

$$\langle \vec{r} m_s|\psi\rangle = \psi_{m_s}(\vec{r}), \quad \text{or} \quad \psi(\vec{r}, m_s). \tag{17.23}$$

In the case of $s = 1/2$, we describe the states by wave functions $\psi_{\pm}(\vec{r})$, corresponding to $m_s = \pm 1/2$. We can represent these wave functions only in the coordinate basis $\langle \vec{r}|$, so states in the spin Hilbert space are given by

$$\langle \vec{r}'|\psi\rangle = \psi_+(\vec{r}')|+\rangle + \psi_-(\vec{r}')|-\rangle \tag{17.24}$$

or

$$\psi = \begin{pmatrix} \psi_+(\vec{r}) \\ \psi_-(\vec{r}) \end{pmatrix}, \tag{17.25}$$

which is a (two-component) spinor field.

We can also represent *just the spin state* (in the spin Hilbert space) as

$$|\alpha\rangle = |+\rangle\langle+|\alpha\rangle + |-\rangle\langle-|\alpha\rangle \equiv C_+|+\rangle + C_-|-\rangle = \begin{pmatrix} C_+ \\ C_- \end{pmatrix}, \tag{17.26}$$

where we have used a completeness relation in the spin Hilbert space, $|+\rangle\langle+| + |-\rangle\langle-| = \mathbb{1}$, and defined $C_{\pm} \equiv \langle\pm|\alpha\rangle$. Further, we can factorize the state of the particle with spin, $\langle \vec{r}|\psi\rangle$ (see 17.24), and find

$$\psi_+(\vec{r}) = C_+\psi(\vec{r}), \quad \psi_-(\vec{r}) = C_-\psi(\vec{r}). \tag{17.27}$$

17.3 Particle with Spin 1/2

In view of the fact that the spin and orbital angular momentum components commute,

$$[\hat{L}_i, \hat{S}_j] = 0, \tag{17.28}$$

just as for $[J_{1i}, J_{2j}]$ in the general theory of addition of angular momenta, it means that we can have simultaneous eigenstates for spin and orbital angular momentum, as we have seen already.

Thus, considering a particle in a rotationally invariant system, we can use as observables $\vec{L}^2, L_z, \vec{S}^2, S_z$. In terms of eigenstates for these operators, we have found that we can separate variables in the wave function for the system without spin,

$$\psi(\vec{r}) \to \psi_l(r) Y_{lm}(\theta, \phi), \tag{17.29}$$

so in the presence of spin we can write states also characterized by s, m_s:

$$\psi_{lm,m_s}(r, \theta, \phi) = \psi_l(r) Y_{lm}(\theta, \phi) \begin{pmatrix} C_+ \\ C_- \end{pmatrix}. \tag{17.30}$$

To continue, we give more details on the spherical harmonics introduced in the previous chapter. As we saw, the spherical harmonics also factorize in terms of the dependences on θ and ϕ:

$$Y_{lm}(\theta,\phi) = P_{l,m}(\cos\theta)\Phi_m(\phi), \tag{17.31}$$

where the normalized eigenfunctions for the ϕ dependence are

$$\Phi_m(\phi) = \frac{1}{\sqrt{2\pi}}e^{im\phi}, \tag{17.32}$$

the $P_{l,m}$ are the associated Legendre functions of the second degree, defined for $m \geq 0$ by the formula

$$P_{l,m}(w) = (-1)^m\sqrt{\frac{(l-m)!}{(l+m)!}\frac{2l+1}{2}}\frac{1}{l!\,2^l}(1-w^2)^{m/2}\left(\frac{d}{dw}\right)^{l+m}(w^2-1)^l, \tag{17.33}$$

which take real values; for negative values of m they are defined by imposing

$$Y_{l,-m}(\theta,\phi) = (-1)^m Y^*_{lm}(\theta,\phi). \tag{17.34}$$

For $m = 0$, the spherical harmonics reduce to the Legendre polynomials,

$$Y_{l,0}(\theta,\phi) = \sqrt{\frac{2l+1}{4\pi}}P_l(\cos\theta). \tag{17.35}$$

The Legendre polynomials are defined by the relation

$$P_l(w) = \frac{1}{l!\,2^l}\left(\frac{d}{dw}\right)^l(w^2-1)^l, \tag{17.36}$$

and satisfy the differential equation

$$(1-w^2)P_l''(w) - 2wP_l'(w) + l(l+1)P_l(w) = 0; \tag{17.37}$$

their orthogonality condition is

$$\int_{-1}^{1} dw P_l(w)P_{l'}(w) = \frac{2}{2l+1}\delta_{ll'}. \tag{17.38}$$

For Y_{lm}, when $w = \cos\theta$ the differential equation (17.37) becomes

$$\left[\frac{d^2}{d\theta^2} + \cot\theta\frac{d}{d\theta} + l(l+1)\right]P_l(\cos\theta) = 0. \tag{17.39}$$

This equation is obtained from the eigenvalue equation

$$\frac{\vec{L}^2}{\hbar^2}Y_{lm}(\theta,\phi) = l(l+1)Y_{lm}(\theta,\phi), \tag{17.40}$$

when we take into account that (using the formulas for L_z, L_+, L_- from the previous chapter)

$$\begin{aligned}
\frac{\hat{\vec{L}}^2}{\hbar^2} &= \frac{1}{\hbar^2}\left(\hat{L}_3^2 + \frac{\hat{L}_+\hat{L}_- + \hat{L}_-\hat{L}_+}{2}\right)\\
&= -\frac{\partial^2}{\partial\phi^2} + \frac{1}{2}\left\{e^{i\theta}\left(\frac{\partial}{\partial\theta} + i\cot\theta\frac{\partial}{\partial\phi}\right), e^{-i\theta}\left(-\frac{\partial}{\partial\theta} + i\cot\theta\frac{\partial}{\partial\phi}\right)\right\}\\
&= -\left(\frac{\partial^2}{\partial\theta^2} + \cot\theta\frac{\partial}{\partial\theta} + \frac{1}{\sin^2\theta}\frac{\partial^2}{\partial\phi^2}\right).
\end{aligned} \tag{17.41}$$

Other properties of the spherical harmonics are as follows. From the completeness of the $|lm\rangle$ states in the Hilbert space for integer l, we find

$$\sum_{l=0}^{\infty}\sum_{m=-l}^{l}|lm\rangle\langle lm| = \mathbb{1}. \tag{17.42}$$

Multiplying from the left with $\langle\vec{n}|$ and from the right with $|\vec{n}'\rangle$, where $\vec{n}$ is a unit vector characterized by angles θ, ϕ so that $|\vec{n}\rangle$ is a coordinate basis for angles, we obtain the completeness relation for spherical harmonics,

$$\sum_{l=0}^{\infty}\sum_{m=-l}^{l}Y_{lm}^{*}(\vec{n}')Y_{lm}(\vec{n}) = \delta(\vec{n}-\vec{n}') = \delta(\cos\theta-\cos\theta')\delta(\phi-\phi'). \tag{17.43}$$

Also, the orthonormality of $|lm\rangle$,

$$\langle lm|l'm'\rangle = \delta_{ll'}\delta_{mm'}, \tag{17.44}$$

leads, by the insertion of the completeness relation of $|\vec{n}\rangle$,

$$\int d\Omega_{\vec{n}}|\vec{n}\rangle\langle\vec{n}| = \mathbb{1}, \tag{17.45}$$

into the orthonormality relation for spherical harmonics,

$$\int d\Omega_{\vec{n}}\, Y_{l'm'}^{*}(\vec{n})Y_{lm}(\vec{n}) = \delta_{ll'}\delta_{mm'}. \tag{17.46}$$

Besides these relations, we also have the *addition formula* for spherical harmonics,

$$\sum_{m=-l}^{l}\langle\vec{n}|lm\rangle\langle lm|\vec{n}'\rangle = \sum_{m=-l}^{l}Y_{lm}(\vec{n})Y_{lm}^{*}(\vec{n}') = \frac{2l+1}{4\pi}P_l(\vec{n}\cdot\vec{n}'). \tag{17.47}$$

Finally, the recursion relation for finding $Y_{l,m-1}$ from $Y_{l,m}$ is obtained from the recursion relation for $|lm\rangle$, where one uses $\hat{L}_-$, in the $\langle\vec{n}|$ basis, so that

$$\langle\vec{n}|l,m-1\rangle = \frac{\langle\vec{n}|\hat{L}_-/\hbar|lm\rangle}{\sqrt{(l+m)(l+1)}}, \tag{17.48}$$

from which we obtain the formula

$$Y_{l,m-1}(\vec{n}) = \frac{1}{(l+m)(l-m+1)}e^{-i\phi}\left(-\frac{\partial}{\partial\theta}+i\cot\theta\frac{\partial}{\partial\phi}\right)Y_{lm}(\vec{n}). \tag{17.49}$$

Using it, we find the formulas already quoted, (17.31) and (17.33).

17.4 Rotation of Spinors with $s = 1/2$

We can also find the effect of rotating spinors in three-dimensional space. A rotation by an angle ϕ around an arbitrary direction defined by $\vec{n}$ is generated by $\vec{\sigma}\cdot\vec{n}/2$, since the quantities $\vec{\sigma}/2$ (the σ_i are the Pauli matrices) are the generators of $SO(3)$ in the spinor ($s = 1/2$) representation. Then the group element (acting on Hilbert spaces) is

$$g = \exp\left(-i\frac{\vec{\sigma}\cdot\vec{n}\phi}{2}\right). \tag{17.50}$$

However, since $\sigma_i^2 = \mathbb{1}$ and $\vec{n}^2 = 1$, it follows that also $(\vec{\sigma} \cdot \vec{n})^2 = \mathbb{1}$, which means that

$$(\vec{\sigma} \cdot \vec{n})^m = \begin{cases} \mathbb{1}, & m = 2k \\ \vec{\sigma} \cdot \vec{n}, & m = 2k + 1. \end{cases} \tag{17.51}$$

Then the group element for rotations in the spin 1/2 representation becomes

$$\begin{aligned} g = e^{-i\vec{\sigma}\cdot\vec{n}\phi/2} &= \sum_{m=0}^{\infty} \frac{(-i\phi/2)^m}{m!} (\vec{\sigma} \cdot \vec{n})^m = \sum_{k=0}^{\infty} \frac{(-i\phi/2)^{2k}}{(2k)!} + \vec{\sigma} \cdot \vec{n} \sum_{k=0}^{\infty} \frac{(-i\phi/2)^{2k+1}}{(2k+1)!} \\ &= \cos \frac{\phi}{2} \mathbb{1} - i \sin \frac{\phi}{2} (\vec{\sigma} \cdot \vec{n}). \end{aligned} \tag{17.52}$$

Since $\vec{\sigma} \cdot \vec{n}$ is a vector operator, we have

$$\sigma_i' = g^{-1} \sigma_i g = \sum_j R_{ij} \sigma_j, \tag{17.53}$$

where R_{ij} is the rotation matrix (in the vector representation).

We can check that, for instance for a rotation of σ_1 around the third direction, we have

$$\begin{aligned} \sigma_1' = e^{i\sigma_3\phi/2} \sigma_1 e^{-i\sigma_3\phi/2} &= (\cos \phi/2 + i \sin \phi/2\, \sigma_3) \sigma_1 (\cos \phi/2 - i \sin \phi/2\, \sigma_3) \\ &= \sigma_1 \cos^2 \phi/2 + \sin^2 \phi/2\, \sigma_3 \sigma_1 \sigma_3 + i \sin \phi/2 \cos \phi/2\, [\sigma_3, \sigma_1] \\ &= \sigma_1 \cos \phi - \sigma_2 \sin \phi, \end{aligned} \tag{17.54}$$

where we have used

$$\sigma_1 \sigma_3 = -\sigma_3 \sigma_1 = -i\sigma_2. \tag{17.55}$$

The resulting transformation is indeed a rotation of σ_i, according to the vector operator formula.

We can use the same rotation formula to construct the eigenstate of a general vector operator $\vec{\sigma} \cdot \vec{n}$ with spin projection $+1/2$ onto the direction of $\vec{n}$.

We start with the eigenstate of σ_3 with spin projection $+1/2$ along the third direction, $|+\rangle = \begin{pmatrix} 1 \\ 0 \end{pmatrix}$. Then we rotate the system by an angle θ around direction 2 and after that by ϕ around direction 3. The resulting state is

$$\begin{aligned} |\psi\rangle = e^{-i\sigma_3\phi/2} e^{-i\sigma_2\theta/2} |+\rangle &= (\cos \phi/2 - i \sin \phi/2\, \sigma_3)(\cos \theta/2 - i \sin \theta/2\, \sigma_2) \begin{pmatrix} 1 \\ 0 \end{pmatrix} \\ &= \begin{pmatrix} \cos \theta/2\, e^{-i\phi/2} \\ \sin \theta/2\, e^{i\phi/2} \end{pmatrix}. \end{aligned} \tag{17.56}$$

17.5 Sum of Orbital Angular Momentum and Spin, $\vec{L} + \vec{S}$

In order to sum orbital angular momentum and spin, we follow the general theory of addition of angular momenta, with

$$\vec{J} = \vec{L} \otimes \mathbb{1} + \mathbb{1} \otimes \vec{S}. \tag{17.57}$$

We can use a basis of eigenstates of commuting operators (observables) $\vec{L}^2, L_z, \vec{S}^2, S_z$,

$$|lm\rangle \otimes |sm_s\rangle, \tag{17.58}$$

or a new basis of eigenstates of $\vec{L}^2, \vec{S}^2, \vec{J}^2, J_z$, namely

$$|lsjm_j\rangle. \tag{17.59}$$

In the case of $s = 1/2$, by multiplying the state with $\langle\vec{n}|\langle\alpha|$ we obtain

$$Y_{lm}(\vec{n})\begin{pmatrix} C_+ \\ C_- \end{pmatrix}, \tag{17.60}$$

leading to states $\psi_{l,1/2,j,m_j}$.

Whether $|lmsm_s\rangle$ or $|lsjm_j\rangle$ are the more appropriate basis states depends on whether the interaction Hamiltonian $\hat{H}_{\text{int}}$ has spin–orbit coupling terms $\vec{L}\cdot\vec{S}$ or not.

17.6 Time-Reversal Operator on States with Spin

Finally, we consider the effect of the time-reversal operator T on particles with spin (before, we only considered its effect on spinless particles).

We first note that the angular momentum, thought of as the generator of $SO(3)$, is odd under T,

$$TJ_iT^{-1} = -J_i. \tag{17.61}$$

We can see that this is true by for instance observing that group elements $g \in SO(3)$ should commute with T (rotations and time reversals are independent), even for infinitesimal $g \simeq 1 - i\vec{\alpha}\cdot\vec{J}$, so that

$$[e^{i\vec{\alpha}\cdot\vec{J}}, T] = 0 \Rightarrow T(iJ_i)T^{-1} = iJ_i. \tag{17.62}$$

But, since T is an anti-unitary operator (antilinear and unitary), $T(iJ_i)T^{-1} = -iT(J_i)T^{-1}$, giving the stated relation.

Moreover, as we saw, T acts on wave functions by complex conjugation. Applying to the spherical harmonics, we obtain

$$T : Y_{lm}(\theta,\phi) \to Y^*_{lm}(\theta,\phi) = (-1)^m Y_{l,-m}(\theta,\phi). \tag{17.63}$$

This means that for orbital angular momentum states,

$$T|lm\rangle = (-1)^m|l,-m\rangle, \tag{17.64}$$

at least for integer m.

But one can generalize to say (this involves a phase convention) that the same formula applies to eigenstates of $\vec{J} = \vec{L} + \vec{S}$,

$$T|jm\rangle = (-1)^m|j,-m\rangle, \quad m \in \mathbb{Z}. \tag{17.65}$$

Then we can generalize this further to any j, integer or half-integer, by writing it as

$$T|jm\rangle = i^{2m}|j,-m\rangle. \tag{17.66}$$

Applying T twice gives

$$
\begin{aligned}
T^2|jm\rangle &= +|jm\rangle, \quad j \in \mathbb{Z} \\
&= -|jm\rangle, \quad j \in \mathbb{Z}/2.
\end{aligned}
\tag{17.67}
$$

Note that this is quite unexpected, since reversing time twice gives the same system, and it is not clear why we should have a phase ± 1.

The above action of T on a state with general j can be proven more rigorously and generally, but we will not do it here.

Next, we consider the action of T on spin, the intrinsic angular momentum of a particle. From the above analysis, $\vec{S}$ should also be odd under time reversal. This is consistent with the semiclassical idea of an intrinsic "spinning" around the direction of the momentum. Indeed, reversing the direction of time, this spinning would reverse as well, thus changing the sign of the spin, $\vec{S} \to -\vec{S}$; thus $S_z \to -S_z$ for any direction z.

Taking into account also the action of T on $\vec{r}$, which is to leave it invariant (even under T, consistent with the idea that $\vec{r}$ is not related to time evolution), and on $\vec{p}$, $\vec{p} \to -\vec{p}$ (odd under T, consistent with the idea that changing the direction of time changes the direction of motion), we can conclude that the time operator corresponds to a rotation by π around the y axis (given that the z axis is the direction of momentum), plus the complex conjugation operator K_0 (the basic antilinear operator),

$$
T = e^{-i\pi S_y/\hbar} K_0. \tag{17.68}
$$

Indeed we could check that the correct transformation rules are obtained, though we leave it as an exercise. Here we just point out that a rotation by π around y, perpendicular to the momentum (which is in the z direction), reverses any vector in the z direction.

Important Concepts to Remember

- Besides the orbital angular momentum $\vec{L}$, there is an intrinsic spin angular momentum $\vec{S}$, which comes from quantum field theory (Poincaré invariant theory), and is associated with the type of particle (thus constant for each type of particle). We have the spin–statistics theorem relating integer spin with Bose–Einstein statistics and half-integer spin with Fermi–Dirac statistics.
- A charged particle in a magnetic field has a magnetic moment $\vec{\mu}$, such that $H_{\text{int}} = -\vec{\mu} \cdot \vec{B}$, and classically $\vec{\mu} = \frac{q}{2m}\vec{L}$, so quantum mechanically we expect $\mu_z = \frac{q\hbar}{2m} m_l$. For the electron, $\frac{e\hbar}{2m} = \mu_B$ is the Bohr magneton.
- For the electron, $\mu_z = g\mu_B m_s$ with $g \simeq 2$ (2 from the Dirac equation, and small corrections from QED), and for other particles we have different values of g that still differ from 1.
- For a particle with spin, the state is $|\alpha l m\rangle \otimes |s m_s\rangle$, with wave function in the $|\vec{r} m_s\rangle$ basis $\psi_{m_s}(\vec{r}) = \psi(\vec{r}, m_s)$, or $\langle \vec{r}|\psi\rangle = \psi_+(\vec{r})|+\rangle + \psi_-|-\rangle = \begin{pmatrix} \psi_+(\vec{r}) \\ \psi_-(\vec{r}) \end{pmatrix}$ with $\psi_+(\vec{r}) = C_+\psi(\vec{r})$, $\psi_-(\vec{r}) = C_-\psi(\vec{r})$, or otherwise $\psi_{lmm_s}(r, \theta, \phi) = \psi_l(r) Y_{lm}(\theta, \phi) \begin{pmatrix} C_+ \\ C_- \end{pmatrix}$.

- The spherical harmonics are $Y_{lm}(\theta, \phi) = P_{lm}(\cos\theta)e^{im\phi}/\sqrt{2\pi}$, with P_{lm} the associated Legendre functions of second degree, orthonormal and complete; the spherical harmonics satisfy the addition formula $\sum_{m=-l}^{l} Y_{lm}(\vec{n})Y^*_{lm}(\vec{n}') = \frac{2l+1}{4\pi}P_l(\vec{n}\cdot\vec{n}')$.
- A spinor (which describes a spin 1/2 particle), initially of spin +1/2 in the third direction, rotated with $\vec{n}(\theta, \phi)$, becomes $\begin{pmatrix} \cos\theta/2\, e^{-i\phi/2} \\ \sin\theta/2\, e^{+i\phi/2} \end{pmatrix}$.
- Under time reversal angular momentum is odd, $TJ_iT^{-1} = -J_i$, and states transform as $T|jm\rangle = i^{2m}|j,-m\rangle$.

Further Reading

See [2], [1], [3] for more details.

Exercises

(1) For an electron inside an atom, with angular momentum $l = 1$ (and spin 1/2), in a magnetic field, how many energy levels (from spectral lines) do we see? Assume there is a single energy level for l, s fixed.

(2) Write down the wave function for a free oscillating neutrino (using general parameters) in Cartesian coordinates.

(3) Write down the wave function for a free oscillating neutrino (using general parameters) in spherical coordinates.

(4) Prove the addition formula for spherical harmonics, (17.47).

(5) Consider an electron (with spin 1/2) in a magnetic field $\vec{B}$. Classically, the spin has a precession motion around $\vec{B}$ *(Larmor precession)*. Prove this. Then calculate the quantum mechanical wave function $|\psi(t)\rangle$ corresponding to this classical motion.

(6) Consider an electron with orbital angular momentum $l = 2$. Write down its possible states in the $\vec{J}$ basis.

(7) Prove that the time-reversal operator can be written as $T = e^{-i\pi S_y/\hbar}K_0$.

18 The Hydrogen Atom

In this chapter, we consider the simplest nontrivial case of a central potential, the negative $1/r$ potential, valid for the hydrogen atom and hydrogen-like atoms (those with a nucleus of charge $+Ze$ and an electron of charge $-e$ around it), one of the first and most important cases to be analyzed by quantum mechanics. It is also a very simple system, forming a central-potential substitute for the one-dimensional harmonic oscillator. This system teaches most of the important methods for solving a quantum system.

18.1 Two-Body Problem: Reducing to Central Potential

The first thing we will show is that for a two-body problem with a likewise two-body potential we can factorize out the center of mass behavior, and we are then left with reduced system in a central potential, as in classical mechanics. We will do this for a general two-body (central) potential $V(|\vec{r}_{ij}|)$.

The quantum two-body Hamiltonian acting on a wave function is

$$\hat{H} = -\frac{\hbar^2}{2m}\vec{\nabla}_1^2 - \frac{\hbar^2}{2m_2}\vec{\nabla}_2^2 + V(|\vec{r}_1 - \vec{r}_2|). \tag{18.1}$$

Defining

$$\begin{aligned} \vec{r} &\equiv \vec{r}_1 - \vec{r}_2, \qquad & \vec{R} &\equiv \frac{m_1\vec{r}_1 + m_2\vec{r}_2}{m_1 + m_2}, \\ M &\equiv m_1 + m_2, \qquad & \frac{1}{\mu} &\equiv \frac{1}{m_1} + \frac{1}{m_2}, \end{aligned} \tag{18.2}$$

where μ is known as the reduced mass, we obtain

$$\partial_{i1,2} = (\partial_{i1,2}r_j)\partial_{r_j} + (\partial_{i1,2}R_j)\partial_{R_j} = \pm\partial_{r_i} + \frac{m_{1,2}}{M}\partial_{R_j}, \tag{18.3}$$

leading to an alternative split of the two-body kinetic terms, into center of mass and relative motion terms:

$$\begin{aligned} \frac{1}{m_1}\vec{\nabla}_1^2 + \frac{1}{m_2}\vec{\nabla}_2^2 &= \frac{1}{m_1}\left(\vec{\nabla}_{\vec{r}} + \frac{m_1}{M}\vec{\nabla}_{\vec{R}}\right)^2 + \frac{1}{m_2}\left(\vec{\nabla}_{\vec{r}} - \frac{m_2}{M}\vec{\nabla}_{\vec{R}}\right)^2 \\ &= \frac{1}{\mu}\vec{\nabla}_{\vec{r}}^2 + \frac{1}{M}\vec{\nabla}_{\vec{R}}^2. \end{aligned} \tag{18.4}$$

Then the Schrödinger equation reduces to

$$\left[-\frac{\hbar^2}{2M}\vec{\nabla}_{\vec{R}}^2 - \frac{\hbar^2}{2\mu}\vec{\nabla}_{\vec{r}}^2 + V(r)\right]\psi(\vec{r},\vec{R}) = E\psi(\vec{r},\vec{R}). \tag{18.5}$$

We can separate the variables so that the wave function becomes a function of the center of mass position $\vec{R}$ times a function of the relative position $\vec{r}$,

$$\psi(\vec{r},\vec{R}) = \psi(\vec{r})\phi(\vec{R}). \tag{18.6}$$

Then the Schrödinger equation splits into a part depending only on $\vec{r}$, plus a part depending only on $\vec{R}$:

$$\left[-\frac{\hbar^2}{2M}\vec{\nabla}^2_{\vec{R}}\phi(\vec{R}) - E_{\rm CM}\phi(\vec{R})\right]\psi(\vec{r}) + \left[-\frac{\hbar^2}{2\mu}\vec{\nabla}^2_{\vec{r}}\psi(\vec{r}) + V(\vec{r})\psi(\vec{r})\right]\phi(\vec{R})$$
$$= (E - E_{\rm CM})\psi(\vec{r})\phi(\vec{R}). \tag{18.7}$$

This allows us to set the two parts to zero independently:

$$\begin{aligned}
-\frac{\hbar^2}{2M}\vec{\nabla}^2_{\vec{R}}\phi(\vec{R}) &= E_{\rm CM}\phi(\vec{R})\\
\left[-\frac{\hbar^2}{2\mu}\vec{\nabla}^2_{\vec{r}} + V(r)\right]\psi(\vec{r}\,) &= (E - E_{\rm CM})\psi(\vec{r}\,).
\end{aligned} \tag{18.8}$$

The first equation corresponds to the center of mass motion, and is of free particle type.

The second equation is the equation for the relative motion, with relative energy $E_{\rm rel} = E - E_{\rm CM}$, with a central potential $V(r)$.

18.2 Hydrogenoid Atom: Set-Up of Problem

Consider a "hydrogenoid" atom, with a negative Coulomb potential corresponding to the interaction of a nucleus of charge $Q = Ze$ and an electron of charge $q = -e$ around it. The reduced mass μ is approximately equal to the electron mass m_e, since $m_e \ll m_n$ (the nucleus mass m_n). Then

$$V(r) = -\frac{|Q||q|}{4\pi\epsilon_0}\frac{1}{r} \equiv -\frac{\tilde{Q}^2}{r}. \tag{18.9}$$

In the Schrödinger equation for this relative motion with central potential $V(r)$,

$$\left[-\frac{\hbar^2}{2\mu}\vec{\nabla}^2_{\vec{r}} + V(r)\right]\psi(\vec{r}) = E_{\rm rel}\psi(\vec{r}), \tag{18.10}$$

where from now on we will replace $E_{\rm rel}$ with E, we can separate variables further.

First, we consider the equation in spherical coordinates, $\psi(\vec{r}) \to \psi(r,\theta,\phi)$, and then, remembering that

$$\Delta_{\vec{r}} = \vec{\nabla}^2_{\vec{r}} = \frac{\partial^2}{\partial r^2} + \frac{2}{r}\frac{\partial}{\partial r} + \frac{1}{r^2}\Delta_{\theta,\phi}, \tag{18.11}$$

where the angular Laplacian equals the square of the angular momentum,

$$\Delta_{\theta,\phi} = -\frac{\vec{L}^2}{\hbar^2}, \tag{18.12}$$

we write

$$\psi(r,\theta,\phi) = R(r)Y_{lm}(\theta,\phi). \tag{18.13}$$

However, the spherical harmonics $Y_{lm}(\theta, \phi)$ are eigenfunctions of the angular Laplacian $\Delta_{\phi,\theta}$, or the angular momentum squared,

$$\Delta_{\theta,\phi} Y_{lm}(\theta, \phi) = -\frac{\vec{L}^2}{\hbar^2} Y_{lm}(\theta, \phi) = -l(l+1) Y_{lm}(\theta, \phi), \tag{18.14}$$

implying that the radial function $R(r)$ satisfies the following radial equation

$$\left\{ -\frac{\hbar^2}{2\mu} \left[\frac{d^2}{dr^2} + \frac{2}{r}\frac{d}{dr} - \frac{l(l+1)}{r^2} \right] + V(r) \right\} R(r) = ER(r). \tag{18.15}$$

The solution of this equation will depend on the parameters E and l only, so

$$R(r) = R_{E,l}(r) \quad \Rightarrow \quad \psi = \psi_{Elm}(r, \theta, \phi) = R_{El}(r) Y_{lm}(\theta, \phi). \tag{18.16}$$

We see that, with respect to the Schrödinger equation in one dimension, in the radial equation for $R(r)$ we have a term with a first derivative, $r^{-1} d/dr$. But we can reduce (18.16) to the same type of problem as in one dimension by redefining the radial wave function as

$$R(r) = \frac{\chi(r)}{r}, \tag{18.17}$$

which implies that

$$\frac{d^2\chi(r)}{dr^2} + \frac{2\mu}{\hbar^2} \left[-\frac{\hbar^2}{2\mu} \frac{l(l+1)}{r^2} + E - V(r) \right] \chi(r) = 0; \tag{18.18}$$

this is indeed of the form of the one-dimensional Schrödinger equation, with an "effective potential"

$$V_{\text{eff}}(r) = V(r) + \frac{\hbar^2}{2\mu} \frac{l(l+1)}{r^2}. \tag{18.19}$$

The only difference is that here $r \geq 0$, unlike in one dimension. In our particular case the effective potential is

$$V_{\text{eff}}(r) = -\frac{\tilde{Q}^2}{r} + \frac{\hbar^2}{2\mu} \frac{l(l+1)}{r^2}, \tag{18.20}$$

and at $r \to 0$ we have $V(r) \to +\infty$, so it is as though we have an infinite "wall" at $r = 0$, allowing us to have a complete analogy with the one-dimensional system. The potential is of the type depicted in Figs. 19.1b and 19.2a, with a negative minimum at a positive value for r,

$$r_{\min} = \frac{l(l+1)\hbar^2}{\tilde{Q}^2 \mu}; \tag{18.21}$$

the potential goes to zero at infinity, from below.

Boundary Conditions

Since at $r = 0$ we have an infinite "wall" for the motion, in $\chi(r)$, as in the one-dimensional case, we have the boundary condition

$$\chi(r = 0) = 0. \tag{18.22}$$

At $r \to \infty$, we must have at least $\chi(r \to \infty) \to 0$. Actually, though, due to the condition of (three-dimensional) normalizability, we have

$$\int_0^\infty |\psi|^2 r^2 dr d\Omega < \infty \Rightarrow \int_0^\infty \chi(r)^2 dr < \infty, \tag{18.23}$$

which means that the behavior at infinity of the "one-dimensional wave function" $\chi(r)$ must be reducing faster than

$$\chi(r) < \frac{A}{r^{1/2}}. \tag{18.24}$$

Case 1: $E > 0$

In this case, since for $r \geq r_0$ we have $V(r) < 0$, then, according to the general analysis of one-dimensional systems, the state behaves at infinity sinusoidally,

$$\chi(r) \sim e^{\pm ikr}, \quad r \to \infty, \tag{18.25}$$

since $V(r) \simeq 0$ there, and thus

$$\frac{\hbar^2 k^2}{2\mu} = E - V(r) \simeq E. \tag{18.26}$$

This sinusoidal behavior is not square integrable at infinity, as in the one-dimensional case, and corresponds to scattering states, to be studied in the second part of the book.

Case 2: $E < 0$

This is the more interesting case for us at present. By the general one-dimensional analysis we expect to find bound states, valid for a set of discrete values for the energy, and so for a quantization condition, obtained from the vanishing of a coefficient A of a possible solution:

$$A = 0 \Rightarrow E = E_n. \tag{18.27}$$

18.3 Solution: Sommerfeld Polynomial Method

To solve the radial equation, we use a variant of the same method as that used in the case of the one-dimensional harmonic oscillator to solve the equation.

First, we rescale the equation, both variables and parameters, in order to have a dimensionless equation. Define

$$\rho = 2\kappa r, \quad \kappa \equiv \sqrt{\frac{2\mu|E|}{\hbar^2}}, \quad \lambda \equiv \tilde{Q}^2 \sqrt{\frac{\mu}{2|E|\hbar^2}}, \tag{18.28}$$

so that the equation becomes (note that now we set $E = -|E|$)

$$\frac{d^2\chi(\rho)}{d\rho^2} + \left[-\frac{l(l+1)}{\rho^2} - \frac{1}{4} + \frac{\lambda}{\rho}\right]\chi = 0. \tag{18.29}$$

In order to solve this equation, we first find the behaviors at $r \to 0$ and $r \to \infty$ of the solution, and then factor them out, leaving us with an easier (and hopefully recognizable) equation.

At $r \to 0$, writing

$$\chi \sim A\rho^{\alpha}, \tag{18.30}$$

we find

$$\rho^{\alpha-2}[\alpha(\alpha-1) - l(l+1)] - \frac{1}{4}\rho^{\alpha} + \lambda\rho^{\alpha-1} = 0. \tag{18.31}$$

Setting to zero the coefficient of the leading power in order to solve the equation for the leading behavior, we find

$$\alpha(\alpha-1) = l(l+1), \tag{18.32}$$

which has two solutions,

$$\alpha = l+1 \quad \text{or} \quad -l \;\Rightarrow\; \chi \sim \rho^{l+1} \quad \text{or} \quad \rho^{-l}. \tag{18.33}$$

However, the behavior $\chi \sim \rho^{-l}$ is not normalizable at zero, so this solution is excluded.

At $r \to \infty$, from the general theory, since $V(r \to \infty) \to 0$ the solution is exponential with parameter κ,

$$\chi(\rho) \sim Be^{\pm\kappa r} = Be^{\pm\rho/2}. \tag{18.34}$$

However, as in the one-dimensional case, the positive exponential is not normalizable but the negative one, $\sim e^{-\rho/2}$, is normalizable.

According to the general procedure outlined earlier, first we factorize this behavior at infinity by redefining the radial wave function with it:

$$\chi(\rho) = e^{-\rho/2}H(\rho). \tag{18.35}$$

Then we have

$$\begin{aligned} \chi' &= -\frac{1}{2}e^{-\rho/2}H + e^{-\rho/2}H' \quad \Rightarrow \\ \chi'' &= e^{-\rho/2}\left(\frac{1}{4}H - H' + H''\right), \end{aligned} \tag{18.36}$$

so the radial equation becomes

$$H'' - H' - \left(\frac{l(l+1)}{\rho^2} - \frac{\lambda}{\rho}\right)H = 0. \tag{18.37}$$

We next factorize the behavior at $r \to 0$, again redefining the radial wave function:

$$H(\rho) = \rho^{l+1}F(\rho). \tag{18.38}$$

Since

$$\begin{aligned} H' &= (l+1)\rho^{l}F + \rho^{l+1}F' \\ H'' &= l(l+1)\rho^{l-1}F + 2(l+1)\rho^{l}F' + \rho^{l+1}F'', \end{aligned} \tag{18.39}$$

the radial equation becomes finally

$$F'' + \left(\frac{2l+2}{\rho} - 1\right)F' - (l+1-\lambda)\frac{F}{\rho} = 0. \tag{18.40}$$

But this is the equation for the *confluent hypergeometric function* ${}_1F_1(a, b; z)$, namely

$$F'' + \left(\frac{b}{z} - 1\right) F' - \frac{a}{z} F = 0, \tag{18.41}$$

with general solution

$$F = A\,{}_1F_1(a, b; z) + B\rho^{1-b}\,{}_1F_1(a - b + 1, 2 - b; z). \tag{18.42}$$

In our case, since $a = l + 1 - \lambda$ and $b = 2l + 2$, the general solution for the function F in the radial equation is

$$F = A\,{}_1F_1(l + 1 - \lambda, 2l + 2; \rho) + B\rho^{-2l-1}\,{}_1F_1(-\lambda - l, -2l; \rho). \tag{18.43}$$

However, since ${}_1F_1$ equals 1 at $z = \rho = 0$, the solution with coefficient B has the wrong behavior at $\rho = 0$, $F \sim \rho^{-2l-1}$, so that $\chi \sim \rho^{l+1}\rho^{-2l-1} - \rho^{-l}$, which has already been excluded on physical (boundary condition) grounds.

That means that only the solution with coefficient A is good at $\rho = 0$. On the other hand, at $\rho \to \infty$,

$${}_1F_1(a, b; \rho) \sim \alpha\rho^n + \beta e^{+\rho}, \tag{18.44}$$

so that the leading behavior of χ is $\chi \sim \alpha e^{-\rho/2} + \beta e^{\rho/2}$. Again, we know that the behavior with $e^{+\rho/2}$ is not normalizable, so must be excluded. That imposes a condition $\beta = 0$ which is a (quantization) condition *on the parameters* a, b (and so on l and on λ and therefore on the energy E) of the function ${}_1F_1(a, b; \rho)$.

The confluent hypergeometric function can be defined as an infinite series,

$${}_1F_1(a, b; \rho) = \sum_{k=0}^{\infty} \frac{1}{k!} \frac{a(a + 1) \cdots (a + k - 1)}{b(b + 1) \cdots (b + k - 1)} \rho^k. \tag{18.45}$$

We can check this by writing the solution to this equation as a series in z (or ρ) and then finding a recursion relation for the coefficients from the equation. Thus, write

$$F = \sum_{k=0}^{\infty} C_k \rho^k, \tag{18.46}$$

and substitute in the defining equation, to find

$$\sum_{k=0}^{\infty} \left[k(k - 1)C_k\rho^{k-2} + bkC_k\rho^{k-2} - kC_k\rho^{k-1} - aC_k\rho^{k-1}\right] = 0. \tag{18.47}$$

We can rewrite this by redefining the sums to have the same ρ^{k-1} factor in all terms, in other words, redefine k as $k + 1$ in the first two terms in the sum. This only affects the $k = 0$ term, which becomes $k = 1$, but then the new $k = 0$ term still vanishes owing to the k prefactor. Thus we get

$$\sum_{k=0}^{\infty} \rho^{k-1}[C_{k+1}(k + 1)(k + b) - C_k(a + k)] = 0. \tag{18.48}$$

This implies the recursion relation

$$\frac{C_{k+1}}{C_k} = \frac{a + k}{(k + 1)(k + b)}, \tag{18.49}$$

which is indeed satisfied by (18.45).

18.4 Confluent Hypergeometric Function and Quantization of Energy

Since (18.45) is written as an infinite sum of powers of ρ, the only way to avoid an exponential blow-up, and thus to put $\beta = 0$ in (18.44), is for the infinite series to terminate at a given $k = n_r \in \mathbb{N}$, since then $C_{k+1}/C_k = 0$. Then ${}_1F_1$ becomes a polynomial of order n_r for

$$a + n_r = 0. \tag{18.50}$$

This is a quantization condition, which implies that

$$\begin{aligned} a &= l + 1 - \lambda = -n_r \Rightarrow \\ \lambda &= n_r + l + 1 \equiv n \\ &= \tilde{Q}^2 \sqrt{\frac{\mu}{2|E|\hbar^2}} = \frac{Ze^2}{4\pi\epsilon_0}\sqrt{\frac{\mu}{2|E|\hbar^2}}, \end{aligned} \tag{18.51}$$

where we have defined the total energy quantum number $n = n_r + l + 1$, with n_r the radial quantum number, which takes values 0,1,2,...

That means that the energy levels are quantized, as

$$E_n = -\frac{(Ze^2/4\pi\epsilon_0)^2\mu}{2\hbar^2 n^2} = -\mu c^2 \frac{(\alpha Z)^2}{2n^2}, \tag{18.52}$$

where

$$\alpha \equiv \frac{e^2}{4\pi\epsilon_0 \hbar c} \tag{18.53}$$

is the fine structure constant, approximately equal to 1/137. Note that, since $l = 0, 1, 2, \ldots$ and $n_r = 0, 1, 2, \ldots$, it follows that $n = 1, 2, 3, \ldots$ Conversely, at fixed n, since $l = n - n_r - 1$, we have $l = 0, 1, 2, \ldots, n - 1$.

We can also rewrite (18.52) as

$$E_n = -\frac{Z^2 e_0^2}{2a_0}\frac{1}{n^2}, \tag{18.54}$$

where

$$e_0^2 \equiv \frac{e^2}{4\pi\epsilon_0} \tag{18.55}$$

and we have defined the *Bohr radius*

$$a_0 \equiv \frac{\hbar^2}{\mu e_0^2} = \frac{\hbar^2 4\pi\epsilon_0}{e^2\mu}. \tag{18.56}$$

The energy can be put into Rydberg's form, i.e., in terms of the Rydberg constant R,

$$E = -\frac{Z^2 R}{n^2} \Rightarrow R = \frac{\mu e_0^4}{2\hbar^2}. \tag{18.57}$$

Then we have also (see (18.28))

$$\kappa_n = \sqrt{\frac{2\mu|E_n|}{\hbar^2}} = \frac{Z}{na_0}, \tag{18.58}$$

and the minimum of the effective potential is

$$r_{\min} = \frac{l(l+1)\hbar^2}{Ze_0^2\mu} = \frac{l(l+1)}{Z}a_0 \propto a_0. \tag{18.59}$$

Therefore we can write the radial wave function as

$$R_{n,l}(r) = N_{n,l}e^{-\kappa_n r}(2\kappa_n r)^l {}_1F_1(-n+l+1, 2l+2; 2\kappa r), \tag{18.60}$$

where the normalization constant $N_{n,l}$ is found to be (by integrating over the square of the wave function)

$$N_{n,l} = (2\kappa_n)^{3/2}\frac{1}{(2l+1)!}\sqrt{\frac{(n+l)!}{(n-l-1)!\,2n}}. \tag{18.61}$$

18.5 Orthogonal Polynomials and Standard Averages over Wave Functions

When the confluent hypergeometric function terminates at a finite order n it becomes a polynomial, and since the constituent functions are orthonormal, it is an orthogonal polynomial. In fact, it is a classical orthogonal polynomial, the Laguerre polynomial $L_n^b(z)$ (defined in Chapter 8), up to a constant,

$${}_1F_1(-n, b+1; z) = \frac{n!\,\Gamma(b+1)}{\Gamma(b+n+1)}L_n^b(z). \tag{18.62}$$

The Laguerre polynomials for integer parameters can be defined by

$$\begin{aligned} L_m^0(z) &= e^z\frac{d^m}{dz^m}(e^{-z}z^m) \\ L_m^k &= (-)^k\frac{d^k}{dz^k}L_{m+k}^0(z), \end{aligned} \tag{18.63}$$

and obey the orthonormality condition

$$\int_0^\infty e^{-z}z^k L_m^k(z)L_n^k(z)dz = \frac{(m+k)!}{m!}\delta_{mn}. \tag{18.64}$$

Thus the radial wave function can be written in terms of the Laguerre polynomials:

$$R_{n,l} = \tilde{N}_{n,l}e^{-\kappa_n r}(2\kappa_n r)^l L_{n-l-1}^{2l+1}(2\kappa_n r), \tag{18.65}$$

where the new normalization factor is

$$\tilde{N}_{n,l} = \sqrt{\frac{(n-l-1)!}{2n(n+l)!}}(2\kappa_n)^{3/2}. \tag{18.66}$$

We can use the square of the radial wave function, giving the probability as a function of radius, to calculate average powers of the radius:

$$\langle r^k\rangle = \int_0^\infty r^2 dr r^k\left[R_{n,l}(r)\right]^2, \tag{18.67}$$

to obtain

$$\begin{aligned}\langle r\rangle_{nl} &= \frac{a_0}{2Z}[3n^2 - l(l+1)]\\ \langle r^2\rangle_{nl} &= \frac{a_0^2}{2Z^2}n^2[5n^2 + 1 - 3l(l+1)]\\ \left\langle \frac{1}{r}\right\rangle_{nl} &= \frac{1}{n^2}\frac{Z}{a_0}.\end{aligned} \tag{18.68}$$

We see that the average position of the electron goes as $a_0 n^2/Z$, increasing as n becomes larger, so in the higher energy levels (higher n), the electron is, on average, further away from the nucleus. Also, the average potential energy is

$$\langle V\rangle_{nl} = -\tilde{Q}^2\left\langle \frac{1}{r}\right\rangle_{nl} = -\frac{\tilde{Q}^2 Z}{n^2 a_0} = \frac{E_n}{2}, \tag{18.69}$$

as is somewhat expected from the equipartition of energy.

For the maximum value of l for fixed n, $l_{\max} = n-1$, the classical orbit would be the least eccentric. In this case, we find

$$\begin{aligned}\langle r^2\rangle_{n,n-1} &= \frac{a_0^2}{Z^2}n^2(n+1/2)(n+1),\\ \langle r\rangle_{n,n-1} &= \frac{a_0}{Z}(n^2 + n/2),\end{aligned} \tag{18.70}$$

implying that the relative error in the position vanishes at large n,

$$\begin{aligned}\Delta r \equiv \sqrt{\langle r^2\rangle - \langle r\rangle^2} = \frac{na_0}{2Z}\sqrt{2n+1} = \frac{\langle r\rangle_{n,n-1}}{\sqrt{2n+1}} \quad \Rightarrow\\ \frac{\Delta r}{\langle r\rangle} \to 0 \quad \text{as} \quad n\to\infty,\end{aligned} \tag{18.71}$$

so the electron becomes more classical in this limit.

Important Concepts to Remember

- The hydrogen, or hydrogenoid, atom is a very important and simple case that teaches most of the methods involved in solving a quantum system in three spatial dimensions.
- We can reduce the motion of such an atom, from the motion of the nucleus of charge Ze and the electron of charge $-e$ to a free center of mass motion and a relative motion in the central potential, with reduced mass μ.
- Separating variables so that $\psi_{nlm}(r,\theta,\phi) = R_{nl}(r)Y_{lm}(\theta,\phi)$, and with $R(r) = \chi(r)r$, we get a one-dimensional Schrödinger equation for $\chi(r)$ in an effective potential

$$V_{\text{eff}}(r) = V(r) + \frac{\hbar^2}{2\mu}\frac{l(l+1)}{r^2}.$$

- The effective potential has a wall at $r = 0$, goes to zero at infinity, and has a minimum at $r_{\min} > 0$, leading to bound states for $E < 0$ and scattering states for $E > 0$.

- The solution of the Schrödinger equation is found by the Sommerfeld polynomial method: we factor out the behavior at zero and at infinity, and thus obtain an equation for a confluent hypergeometric function ${}_1F_1(a,b;z)$, with good behavior at zero; however the behavior at infinity has two components, a polynomial and a rising exponential (giving normalizable and non-normalizable behaviors for the wave function).
- Then we find a quantization condition on the parameters a, b by imposing that the wave function at infinity is normalizable, or equivalently, that ${}_1F_1$ reduces to a polynomial (so in the recursion condition for coefficients we have $C_{k+1}/C_k = 0$ at some finite k); this turns out to be a Laguerre classical orthogonal polynomial.
- The resulting quantization condition for the energy is

$$E_n = -\mu c^2 \frac{(\alpha Z)^2}{2n^2} = -\frac{Ze^2}{2a_0 n^2} = -\frac{RZ^2}{n^2},$$

  with a_0 the Bohr radius, R the Rydberg constant, and $n = l + n_r + 1$, $n_r = 0, 1, 2, \ldots$ so $n = 1, 2, \ldots$
- The average position is

$$\langle r \rangle = \frac{3}{2}\frac{a_0}{Z}[n^2 - l(l+1)/3],$$

  and so increases for higher n, while $\Delta r/\langle r \rangle \to 0$ for $n \to \infty$.

Further Reading

See [2] for more details.

Exercises

(1) Consider the three-body problem, with the same two-body potential $V_{ij}(r_{ij})$ for all three pairs. Can we separate variables to factor out anything in this general case?

(2) Set up the problem for the same hydrogenoid atom, but living in two spatial dimensions instead of three (yet with the $1/r$ potential of three dimensions).

(3) Use the same Sommerfeld polynomial method to solve the problem in exercise 2.

(4) Prove that the normalization constant N_{nl} is given by equation (18.61).

(5) Prove the relations (18.68).

(6) Calculate the average radial momentum of the reduced orbiting particle in the state n, l of the hydrogenoid atom.

(7) If we introduce a small constant magnetic field $\vec{B}$ for the hydrogenoid atom, what happens to the energy levels E_n? Give an approximate quantitative description of how this happens, based on the quantization of energy at $\vec{B} = 0$.

19 General Central Potential and Three-Dimensional (Isotropic) Harmonic Oscillator

After having examined the most important example of a central potential, the Coulomb potential, and the corresponding hydrogen atom, we now consider the general case of a central potential, and apply it to a few other examples: a free particle, a radial square well, and finally three-dimensional isotropic harmonic oscillator.

19.1 General Set-Up

We have basically derived the general formulas for a central potential in the previous chapter, in the process of applying them to the Coulomb potential, but we review them here. The Schrödinger equation for the wave function in a general central potential,

$$\left[-\frac{\hbar^2}{2M}\vec{\nabla}^2 + V(r)\right]\psi(\vec{r}) = E\psi(\vec{r}), \tag{19.1}$$

reduces in spherical coordinates to

$$\left[-\frac{\hbar^2}{2M}\left(\frac{\partial^2}{\partial r^2} + \frac{2}{r}\frac{\partial}{\partial r} - \frac{\vec{L}^2/\hbar^2}{r^2}\right) + (V(r) - E)\right]\psi(\vec{r}) = 0. \tag{19.2}$$

This means that we can separate the variables in the equation,

$$\psi(\vec{r}) = R_{El}(r)Y_{lm}(\theta, \phi), \tag{19.3}$$

where Y_{lm} are eigenfunctions of the angular momentum,

$$\vec{L}^2 Y_{lm}(\theta, \phi) = -l(l+1)\hbar^2 Y_{lm}(\theta, \phi), \tag{19.4}$$

to obtain

$$\left[\frac{d^2}{dr^2} + \frac{2}{r}\frac{d}{dr} - \frac{l(l+1)}{r^2} + \frac{2m}{\hbar^2}(E - V(r))\right]R_{El}(r) = 0. \tag{19.5}$$

Further, to reduce this equation to one-dimensional Schrödinger-type equation, we need to get rid of the $(2/r)d/dr$ term, by defining

$$R_{El}(r) = \frac{\chi_{El}(r)}{r}, \tag{19.6}$$

obtaining

$$\left[\frac{d^2}{dr^2} + \frac{2m}{\hbar^2}\left(E - V(r) - \frac{\hbar^2 l(l+1)}{2mr^2}\right)\right]\chi_{El}(r) = 0. \tag{19.7}$$

We see that indeed the equation is the one-dimensional Schrödinger equation for an effective potential

$$V_{\text{eff}}(r) = V(r) + \frac{\hbar^2 l(l+1)}{2mr^2}. \tag{19.8}$$

Normalization Conditions

In order to have normalizable solutions, we need to have a finite integral over the square of the wave function:

$$\int |\psi|^2 d^3r = \int |\psi|^2 r^2 dr\, d\Omega = \int |\chi|^2 dr\, d\Omega. \tag{19.9}$$

Then the condition at infinity,

$$\int^{\infty} |\chi(r)|^2 dr < \infty, \tag{19.10}$$

implies that, as $r \to \infty$, the function $\chi(r)$ vanishes faster than $1/r^{1/2}$,

$$|\chi(r)| < \frac{1}{r^{1/2}}. \tag{19.11}$$

The condition at zero,

$$\int_0 |\chi(r)|^2 < \infty, \tag{19.12}$$

which means that at $r \to 0$, the function $\chi(r)$ either blows up or vanishes more slowly than $1/r^{1/2}$,

$$|\chi(r)| > \frac{1}{r^{1/2}}. \tag{19.13}$$

In the case where we have a Laurent expansion for $\chi(r)$, this would mean that $|\chi(r)| \sim r^k$, $k \geq 0$, at $r \to 0$.

19.2 Types of Potentials

Case I

We will consider first the case with

$$\lim_{r\to\infty} V(r) = \text{constant} \equiv 0, \tag{19.14}$$

where we have set the constant to zero, since we can rescale the constant in the energy. This condition is true in many cases of interest, but not in all.

Case II

In particular, the case of the three-dimensional harmonic oscillator is not of the above type, and then it will be automatically described as having case II behavior.

Case I can be split into subcases, depending on the behavior at $r \to 0$. These subcases will be listed and then discussed separately.

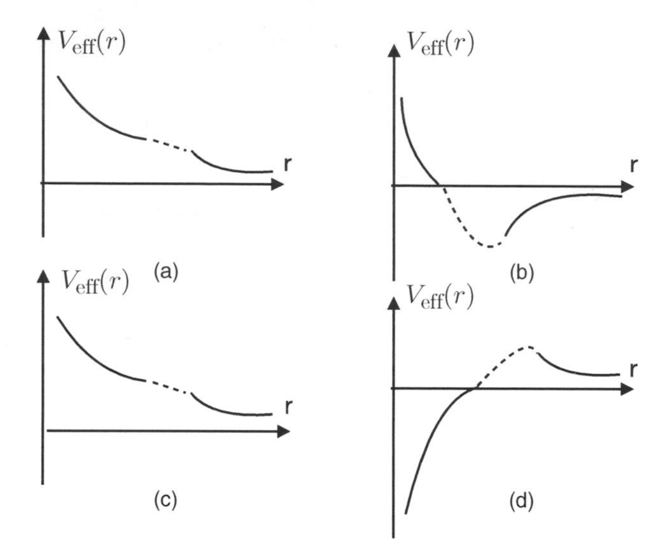

Figure 19.1 (a) Case Ia, with $\alpha > 0$. (b) Case Ia, with $\alpha < 0$, if the behavior at 0 and at ∞ is the same. (c) Cases Ib with $\alpha > 0$ and Ic with $\beta > 0$. (d) Cases Ib with $\alpha < 0$ and Ic with $\beta < 0$.

Case Ia. When

$$V(r \to 0) \sim \frac{\alpha}{r^s}, \quad s < 2, \tag{19.15}$$

the one-dimensional effective potential is dominated by the $l(l+1)/r^2$ term and thus blows up at zero,

$$V_{\rm eff}(r \to 0) \to +\infty. \tag{19.16}$$

In this case, we have an infinite "wall" at zero, as in Figs. 19.1a and b, which means that we have the boundary condition $\chi(0) = 0$.

Case Ib. When

$$V(r \to 0) \sim \frac{\alpha}{r^s}, \quad s > 2, \tag{19.17}$$

we obtain that the one-dimensional effective potential is dominated by the potential itself and thus blows up at zero, with a sign depending on the sign of α, as in Figs. 19.1c and d,

$$V_{\rm eff}(r \to 0) \to sgn(\alpha)\infty. \tag{19.18}$$

The more interesting case is, however, when $\alpha < 0$, so that $V_{\rm eff} \to -\infty$.

Case Ic. Finally, the potential could have the same behavior as the angular momentum (centrifugal) term,

$$V(r \to 0) \sim \frac{\alpha}{r^2}. \tag{19.19}$$

Case Ia

We consider first case Ia.

(1) $E > 0$. For positive energy, since we have an infinite wall at $r = 0$ and at $r = \infty$, then for $E > V(r)$, according to the general theory in Chapter 7, we have unbound states, with imaginary exponentials (or sinusoidal and cosinusoidal),

$$\chi(r) \sim Ae^{ikr} + Be^{-ikr}, \tag{19.20}$$

where

$$k = \sqrt{\frac{2mE}{\hbar^2}}. \tag{19.21}$$

The function (19.20) is not square integrable at infinity. On the other hand, at zero, as we said previously, we need $\chi(0) = 0$, which imposes a constraint, thus selecting a single solution and leading to a continuous nondegenerate spectrum.

Indeed, at $r \to 0$, the equation for χ is approximately

$$\frac{d^2\chi}{dr^2} = \frac{l(l+1)}{r^2}\chi, \tag{19.22}$$

which has solutions r^α with $\alpha = l + 1$ or $\alpha = -l$; thus

$$\chi = Cr^{l+1} + Dr^{-l}. \tag{19.23}$$

Because of the boundary condition at zero, we need to impose $D/C = 0$, which is the constraint we mentioned earlier.

(2) $E < 0$. For negative energy, we have bound states, *if there is a region where* $V(r) < E < 0$, by the same general analysis from Chapter 7. At $r \to 0$, we have the solution in (19.23) again. On the other hand, at $r \to \infty$ we have exponential solutions,

$$\chi \sim Ae^{-\kappa r} + Be^{+\kappa r}, \tag{19.24}$$

where

$$\kappa \equiv \sqrt{\frac{2m|E|}{\hbar^2}}. \tag{19.25}$$

Imposing $D/C = 0$ in (19.23) selects a single solution, but then imposing $B/A = 0$ gives a quantization condition, selecting only discrete energies E_n.

In this case, we can extend Sommerfeld's polynomial method. Defining $\rho \equiv 2\kappa r$, we find the equation for χ,

$$\frac{d^2\chi}{d\rho^2} + \left[-\frac{l(l+1)}{\rho^2} - \frac{1}{4} - \frac{2m}{\hbar^2}V(\rho)\right]\chi(\rho) = 0. \tag{19.26}$$

Then, at $\rho \to 0$, the equation becomes $d^2\chi/d\rho^2 \simeq +l(l+1)\chi/\rho^2$, which as we saw has the physical solution

$$\chi \sim \rho^{l+1}, \tag{19.27}$$

whereas at $\rho \to \infty$ the equation is $d^2\chi/d\rho^2 \simeq 1/4\chi(\rho)$, with physical solution

$$\chi \sim e^{-\rho/2}. \tag{19.28}$$

We then factor out the two behaviors, at zero and infinity, write

$$\chi = e^{-\rho/2}H(\rho), \quad H(\rho) = \rho^{l+1}F(\rho), \tag{19.29}$$

and find an equation for $F(\rho)$ that is a polynomial at zero; we impose that it is a polynomial also at infinity, which turns into a quantization condition.

Further, in this case, with $s < 2$, consider a wave function that is only nonzero for $r < r_0$, so that $\Delta r = r_0$. Then, by Heisenberg's uncertainty relation,

$$p \sim \Delta p \sim \frac{\hbar}{\Delta r} = \frac{\hbar}{r_0}. \tag{19.30}$$

The energy of the state with this wave function is

$$E \sim \frac{\hbar^2}{2mr_0^2} - \frac{\alpha}{r_0^s}, \tag{19.31}$$

and it is always positive for small enough r_0. This means that we cannot have $E < 0$ in this case; the bound states must start at some minimum distance from the center $r = 0$, and we cannot have arbitrarily small energies: the ground state has a finite E and is at a finite r. That means that the particle cannot fall into $r = 0$ quantum mechanically in this case.

On the other hand, if the potential behaves in the same way, $\sim -\alpha/r^s$, but with $s < 2$, then at $r \to \infty$, for large enough r_0, with a wave function centered on such an r_0 and with $\Delta r \ll r_0$, the energy

$$E \sim \frac{\hbar^2}{2m(\Delta r)^2} - \frac{\alpha}{r_0^s}, \tag{19.32}$$

becomes negative at large enough r_0. Thus there are bound states, stationary states of negative energy, situated *at large distances* from the origin, which implies they have arbitrarily small negative energy. In particular, this implies that there is an infinite number of (bound) energy levels, becoming infinitely dense at $E = 0$.

Case Ib

In this case, if $\alpha < 0$, $V_{\text{eff}}(r \to 0) \to -\infty$, which suggests that we would have a large probability at $r = 0$: $\chi(0)$ need not be zero and in fact χ can even be non-normalizable. The energy of a state bounded by $r_0 \to 0$,

$$E \sim \frac{\hbar^2}{2mr_0^2} - \frac{\alpha}{r_0^s}, \tag{19.33}$$

is negative for small enough r_0, meaning that there are states of arbitrarily large $|E|$ for $E < 0$. The interpretation is that the quantum particle can "fall" to $r = 0$, where it has $E = -\infty$ (which is an eigenvalue). We can see this directly too, since at $r \to 0$ the Schrödinger equation is approximately

$$\frac{d^2\chi}{dr^2} = \left(-\frac{\alpha}{r^s} + \frac{l(l+1)}{r^2}\right)\chi(r), \tag{19.34}$$

with solution

$$\chi \sim \frac{Ae^{if(r)} + Be^{-if(r)}}{r^b}. \tag{19.35}$$

The second derivative is

$$\chi'' \sim -(f'(r))^2\chi + 0 + \frac{b(b+1)}{r^2}\chi, \tag{19.36}$$

and the leading term in this equation of motion implies that $(f'(r))^2 = \alpha/r^s$, the zero in the middle term amounts to the vanishing of the imaginary part of the solution, and fixes B/A, and the matching of the subleading term gives $b(b+1) = l(l+1)$, selecting $b = l$ in this case i.e., a non-normalizable solution at zero. Therefore the probability of finding the particle outside $r = 0$ is negligible, and the particle falls into $r = 0$ in quantum mechanics, as expected.

If the potential behaves in the same way, $V \sim -\alpha/r^s$ but with $s > 2$, at $r \to \infty$ then the energy is

$$E \sim \frac{\hbar^2}{2m(\Delta r)^2} - \frac{\alpha}{r_0^s}, \tag{19.37}$$

where $s > 2$ and $r_0 \to \infty$, implying that $E > 0$ at large enough r_0. Thus, there is no state of arbitrarily small negative energy and large r_0. The spectrum terminates at a finite and nonzero negative energy.

Case Ic

When the potential $V \simeq -\alpha/r^2$, with $\alpha > 0$, the effective potential is

$$V_{\text{eff}}(r) = \frac{\hbar^2 l(l+1) - 2m\alpha}{2mr^2} \equiv \frac{\hbar^2\beta}{2mr^2}. \tag{19.38}$$

The Schrödinger equation at $r \to 0$ is

$$\frac{d^2\chi}{dr^2} = \frac{\beta}{r^2}\chi, \tag{19.39}$$

with solution $\chi \sim r^a$, which implies $a(a-1) = \beta$, i.e.,

$$a_{1,2} = \frac{1 \pm \sqrt{1+4\beta}}{2}. \tag{19.40}$$

(1) If $\beta > -1/4$, both solutions $a_{1,2}$ are real but, for reasons of normalization, we must choose the slowest behavior at $r \to 0$, namely

$$\chi(r) \sim r^{a_1} = r^{\left(1+\sqrt{1+4\beta}\right)/2}, \tag{19.41}$$

for which $\int_0 |\chi(r)|^2 dr < \infty$.

(2) If $\beta < -1/4$, both solutions have imaginary parts and we find

$$R(r) \sim \frac{1}{r^{1/2}}\left(Ae^{i\sqrt{-1-4\beta}\ln r} + Be^{-i\sqrt{-1-4\beta}\ln r}\right). \tag{19.42}$$

Imposing the condition that the solution must be real, we find one that has an infinite number of zeroes,

$$R(r) \sim \frac{1}{r^{1/2}}\cos\left(\sqrt{-1-4\beta}\ln r + \delta\right). \tag{19.43}$$

But the *ground* state cannot have zeroes, which means that the wave function (19.43) is wrong, and in the ground state the particle is actually at $r = 0$ (one can produce an argument that is a bit more rigorous for this statement, but the result is the same).

On the other hand, if the potential is the same, i.e., $V = -\alpha/r^2$, at $r \to \infty$, then:

for $\beta > -1/4$, we have (at most) a finite number of negative energy levels: the solutions $\chi \sim r^{a_{1,2}}$ are valid also for $r \to \infty$, but these solutions have no zeroes. Thus there can be zeroes for, at most, r less than some r_0, implying a finite number of energy levels, each associated with an interval between zeroes, according to the general one-dimensional analysis.
for $\beta < -1/4$, the solutions $R \sim \cos\left(\sqrt{-1-4\beta}\ln r + \delta\right)/r^{1/2}$ are valid also at $r \to \infty$, and have an infinite number of zeroes, thus an infinite number of energy levels.

19.3 Diatomic Molecule

Consider a molecule made up of two atoms. Reducing the Schrödinger equation to the relative motion of the atoms with reduced mass μ, we find a central potential that has a minimum at some r_0. If the molecule is stable then $V''(r_0) > 0$, so the minimum is stable, as in Fig. 19.2a. This means that we can expand the potential around r_0,

$$V(r) \simeq V(r_0) + V''(r_0)\frac{(r-r_0)^2}{2}, \tag{19.44}$$

and we define

$$V''(r_0) \equiv \mu\frac{\omega^2}{2}. \tag{19.45}$$

Also expanding the angular momentum term,

$$\frac{\hbar^2}{2m\mu}\frac{l(l+1)}{r^2} \simeq \frac{\hbar^2}{2\mu}\frac{l(l+1)}{r_0^2} + \cdots, \tag{19.46}$$

and defining the inertial momentum $I \equiv \mu r_0^2$, we find the following Schrödinger equation:

$$\left[\frac{d^2}{dr^2} + \frac{2\mu}{\hbar^2}\left(E - V(r_0) - \frac{\hbar^2 l(l+1)}{2I} - \frac{\mu\omega^2}{2}\rho^2\right)\right]\chi(\rho) = 0. \tag{19.47}$$

However, this is nothing other than the Schrödinger equation for a linear harmonic oscillator, of shifted energy $E - V(r_0) - \hbar^2 l(l+1)/2I$. Thus the eigenenergies are

$$E_n = V(r_0) + \frac{\hbar^2 l(l+1)}{2I} + \hbar\omega\left(n + \frac{1}{2}\right). \tag{19.48}$$

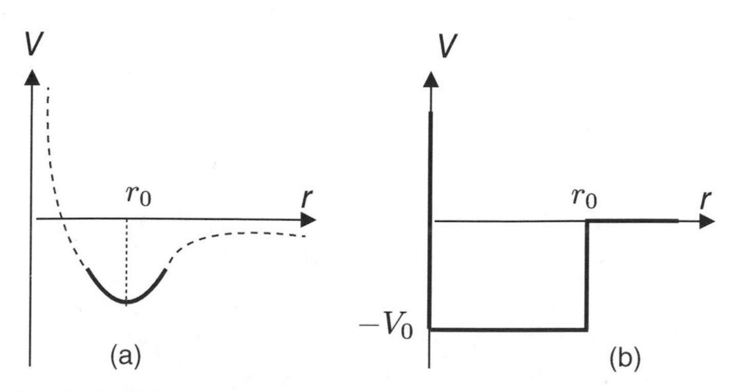

Figure 19.2 (a) Potential for diatomic molecule; (b) Spherical square well potential.

19.4 Free Particle

We have already looked at the free particle in spherical coordinates, but we review it here for completeness. For $E > 0$, the Schrödinger equation in spherical coordinates reduces to

$$\frac{d^2\chi}{d\rho^2} + \left[\frac{1}{4} - \frac{l(l+1)}{\rho^2}\right]\chi(\rho) = 0. \tag{19.49}$$

The equation above has a unique solution that is everywhere bounded, the Bessel function $j_l(\rho/2) = j_l(kr)$ for $R = \chi/r$, meaning that the full wave function is

$$\psi(\vec{r}) = j_l(kr)Y_{lm}(\theta, \phi). \tag{19.50}$$

Then the known solution in Cartesian coordinates, $e^{i\vec{k}\cdot\vec{r}}$, can be expanded in terms of the spherical solutions above,

$$e^{i\vec{k}\cdot\vec{r}} = \sum_{l=0}^{\infty}\sum_{m=-l}^{l} a_{lm}(\vec{k})j_l(kr)Y_{lm}(\theta, \phi). \tag{19.51}$$

Choosing the momentum $\vec{k}$ to be in the direction Oz, and defining $\cos\theta \equiv u$ (and since $\rho/2 = kr$),

$$e^{i\rho u/2} = e^{ikr\cos\theta} = \sum_{l=0}^{\infty} c_l j_l(\rho/2)P_l(u). \tag{19.52}$$

From the recurrence relations obtained for the spherical Bessel functions j_l and the Legendre polynomials P_l , we find that

$$c_l = (2l+1)i^l c_0, \tag{19.53}$$

and with the normalization choice $c_0 = 1$ we find that

$$e^{ikr\cos\theta} = \sum_{l=0}^{\infty}(2l+1)i^l j_l(kr)P_l(\cos\theta). \tag{19.54}$$

19.5 Spherical Square Well

Consider the important case where the potential is piecewise constant, specifically smaller for $r \le r_0$ than outside it as in Fig. 19.2b,

$$V(r) = \begin{cases} -V_0, & r \le r_0 \\ 0, & r > r_0. \end{cases} \tag{19.55}$$

For negative energy, $-V_0 < E < 0$, by the general theory we have bound states (since $E > V(r)$ over a finite region). For $r < r_0$ we have a "free wave" solution like that above, $R_l(r) = j_l(kr)$, but with

$$k = \sqrt{\frac{2m(E+V_0)}{\hbar^2}}. \tag{19.56}$$

For $r > r_0$, we have an exponentially decaying solution (the only one that vanishes at infinity),

$$\chi_l(\kappa r) \sim e^{-\kappa r}, \tag{19.57}$$

where

$$\kappa = \sqrt{\frac{2m|E|}{\hbar^2}}. \tag{19.58}$$

Then, the condition of continuity of the logarithmic derivative (and so of the function and of its derivative) at $r = r_0$,

$$\frac{d}{dr} \ln \frac{\chi_l(\kappa r)}{r}\bigg|_{r=r_0} = \frac{d}{dr} \ln j_l(kr)\bigg|_{r=r_0}, \tag{19.59}$$

gives a quantization condition that gives a finite number of energy levels E_n.

19.6 Three-Dimensional Isotropic Harmonic Oscillator: Set-Up

This is an example of Case II, where the potential grows without bound as $r \to \infty$.

Consider a harmonic oscillator in d dimensions, with a priori different frequencies in each dimension,

$$\begin{aligned} \hat{H} &= \sum_{i=1}^{d} \hat{H}_i \\ \hat{H}_i &= \frac{p_i^2}{2m} + m\frac{\omega_i^2 q_i^2}{2}. \end{aligned} \tag{19.60}$$

Then the Hilbert space is a tensor product of the harmonic oscillator Hilbert spaces for each dimension,

$$\mathcal{H} = \mathcal{H}_1 \otimes \cdots \otimes \mathcal{H}_d, \tag{19.61}$$

so the states (in the "Fock space") are tensor product of states of occupation numbers n_i,

$$|\{n_i\}\rangle = |n_1\rangle \otimes \cdots \otimes |n_d\rangle; \tag{19.62}$$

in particular, the vacuum is a tensor product of the vacua for each dimension,

$$|0\rangle \equiv |0\rangle_1 \otimes \cdots \otimes |0\rangle_d. \tag{19.63}$$

For these states the Hamiltonian for each dimension is diagonal,

$$\hat{H}_i|n_i\rangle = \left(n_i + \frac{1}{2}\right)\hbar\omega_i|n_i\rangle, \tag{19.64}$$

so that

$$\hat{H}|\{n_i\}\rangle = \sum_{i=1}^{d}\left[\hbar\omega_i\left(n_i + \frac{1}{2}\right)\right]|\{n_i\}\rangle. \tag{19.65}$$

In the case of an *isotropic* harmonic oscillator, with the same frequency in all directions, $\omega_i = \omega$, we have

$$\hat{H} = \frac{\hat{\vec{p}}^2}{2m} + \frac{m\omega^2}{2}\hat{\vec{r}}^2. \tag{19.66}$$

In terms of the number operators $\hat{N}_i = \hat{a}_i^\dagger \hat{a}_i$, we have

$$\hat{N} = \sum_{i=1}^{d} \hat{N}_i, \tag{19.67}$$

so the total Hamiltonian is

$$\hat{H} = \left(\hat{N} + \frac{d}{2}\right)\hbar\omega, \tag{19.68}$$

and the eigenenergies are

$$\hat{H}|\{n_i\}\rangle = \left(\sum_{i=1}^{d} n_i + \frac{d}{2}\right)\hbar\omega|\{n_i\}\rangle = \left(n + \frac{d}{2}\right)\hbar\omega|\{n_i\}\rangle. \tag{19.69}$$

The degeneracy of these energies equals the number of different n_i that give the same $n = \sum_i n_i$. We are mostly interested in $d = 3$, in which case the degeneracy is

$$g_n = \frac{(n+1)(n+2)}{2}. \tag{19.70}$$

19.7 Isotropic Three-Dimensional Harmonic Oscillator in Spherical Coordinates

Instead of solving each dimension independently, as above (each as a harmonic oscillator), we can go to spherical coordinates, noting that the central potential is

$$V(r) = \frac{m\omega^2}{2}r^2. \tag{19.71}$$

Then the Sommerfeld polynomial method works as before. The Schrödinger equation becomes

$$\frac{d^2\chi}{dr^2} + \left(\frac{2mE}{\hbar^2} - \frac{m^2\omega^2}{\hbar^2}r^2 - \frac{l(l+1)}{r^2}\right)\chi(r) = 0. \tag{19.72}$$

Define

$$\alpha = \sqrt{\frac{m\omega}{\hbar}}, \quad \frac{2E}{\hbar\omega} = \lambda, \quad \xi = \alpha r, \tag{19.73}$$

leading to the equation

$$\frac{d^2\chi}{d\xi^2} + \left(\lambda - \frac{l(l+1)}{\xi^2} - \xi^2\right)\chi = 0. \tag{19.74}$$

Then, the equation at $r \to 0$ or $\xi \to 0$ is

$$\frac{d^2\chi}{d\xi^2} = \frac{l(l+1)}{\xi^2}\chi, \tag{19.75}$$

which has the solution

$$\chi = A\xi^{l+1} + B\xi^{-l}; \tag{19.76}$$

this implies that the normalizable solution has $B = 0$.

The equation at $r \to \infty$ or $\xi \to \infty$ is

$$\frac{d^2\chi}{d\xi^2} = \xi^2\chi, \tag{19.77}$$

with the leading, normalizable, solution at infinity

$$\chi = e^{-\xi^2/2} \times (\text{subleading terms}). \tag{19.78}$$

We now factor out the behaviors at infinity and zero. First, we write

$$\chi = e^{-\xi^2/2} H(\xi), \tag{19.79}$$

which implies

$$\chi'' = e^{-\xi^2/2}\left(H'' - 2\xi H' + (\xi^2 - 1)H\right), \tag{19.80}$$

so that the equation for H is (dividing by $e^{-\xi^2/2}$)

$$H'' - 2\xi H' + \left(\lambda - 1 - \frac{l(l+1)}{\xi^2}\right) H = 0. \tag{19.81}$$

Next, we write

$$H = \xi^{l+1} F(\xi), \tag{19.82}$$

which implies

$$\begin{aligned} H' &= \xi^{l+1} F' + (l+1)\xi^l F \\ H'' &= \xi^{l+1} F'' + 2(l+1)\xi^l F' + l(l+1)\xi^{l-1} F, \end{aligned} \tag{19.83}$$

so that the equation for F is (dividing by ξ^{l+1})

$$F'' + 2\left(\frac{l+1}{\xi} - \xi\right) F' + (\lambda - 2l - 3)F = 0. \tag{19.84}$$

Finally, change variables by setting $z = \xi^2$, so that

$$\frac{dF}{d\xi} = 2\xi\frac{dF}{dz}, \quad \frac{d^2F}{d\xi^2} = 2\frac{dF}{dz} + 4z\frac{d^2F}{dz^2}, \tag{19.85}$$

and we obtain the equation

$$z\frac{d^2F}{dz^2} + \left(l + \frac{3}{2} - z\right)\frac{dF}{dz} + \frac{\lambda - 2l - 3}{4} F = 0. \tag{19.86}$$

But this is the same equation as that for the confluent hypergeometric function ${}_1F_1(a, b; z)$ from the previous chapter, with

$$-a = \frac{\lambda - 2l - 3}{4}, \quad b = l + \frac{3}{2}. \tag{19.87}$$

Therefore, as in the previous chapter, we find the solution

$$F(z) = {}_1F_1\left(\frac{2l + 3 - 2E/(\hbar\omega)}{4}, l + \frac{3}{2}; z\right), \tag{19.88}$$

with quantization condition

$$\frac{2l + 3 - 2E/(\hbar\omega)}{4} = -n_r \in \mathbb{N}. \tag{19.89}$$

Finally then, the eigenenergies are

$$\frac{E_n}{\hbar\omega} = l + \frac{3}{2} + 2n_r \equiv n + \frac{3}{2}, \tag{19.90}$$

where we have defined $n = 2n_r + l$. This is the same formula as that obtained from the Cartesian form. Moreover, the degeneracy is again

$$g_n = \frac{(n+1)(n+2)}{2}, \tag{19.91}$$

meaning that the possible states are indeed the same. In spherical coordinates, the radial functions are (putting together the formulas derived before)

$$R_{n,l}(r) = N_{n_r,l} e^{-\alpha^2 r^2/2} (\alpha r)^l {}_1F_1(-n_r, l + 3/2; \alpha^2 r^2), \tag{19.92}$$

where

$$N_{n,l} = \frac{\alpha\sqrt{2\alpha}}{\Gamma(l + 3/2)} \left(\frac{n_r!}{\Gamma(n_r + l + 3/2)}\right)^{1/2}. \tag{19.93}$$

19.8 Isotropic Three-Dimensional Harmonic Oscillator in Cylindrical Coordinates

For completeness, we now show how to solve the same oscillator in cylindrical coordinates. Then the Hamiltonian is

$$\begin{aligned} \hat{H} &= -\frac{\hbar^2}{2m_0}\left[\frac{1}{\rho}\frac{\partial}{\partial\rho}\left(\rho\frac{\partial}{\partial\rho}\right) + \frac{1}{\rho^2}\frac{\partial^2}{\partial\phi^2} + \frac{\partial^2}{\partial z^2}\right] + \frac{m_0\omega^2}{2}(\rho^2 + z^2) \\ &= \hat{H}_z + \hat{H}_\rho + \frac{\hat{L}_z^2}{2m_0\rho^2}, \end{aligned} \tag{19.94}$$

where we have denoted the mass by m_0 in order not to confuse it with m, the eigenvalue of $L_z/\hbar$. We can easily check that H, H_z, L_z form a complete set of mutually compatible observables,

$$0 = [\hat{H}, \hat{L}_z] = [\hat{H}, \hat{H}_z] = [\hat{L}_z, \hat{H}_z]. \tag{19.95}$$

Then we separate variables by writing the wave function as

$$\psi_{n_z,n_\rho,|m|}(z, \phi, \rho) = Z_{n_z}(z)\Phi_m(\phi)R_{n_\rho,|m|}(\rho). \tag{19.96}$$

The equation of motion for $Z_{n_z}(z)$ is that for a harmonic oscillator in one dimension,

$$H_z Z_{n_z} = \left(-\frac{\hbar^2}{2m_0}\frac{\partial^2}{\partial z^2} + \frac{m_0\omega^2 z^2}{2}\right) Z_{n_z} = E_z Z_{n_z}, \tag{19.97}$$

with the standard solution

$$Z_{n_z} = N(\alpha)e^{-\alpha^2 z^2/2} H_{n_z}(\alpha z)$$
$$E_z = \hbar\omega(n_z + 1/2). \tag{19.98}$$

The equation of motion for $\Phi_m(\phi)$ is even simpler,

$$L_z \Phi_m(\phi) = m\hbar\Phi_m(\phi), \tag{19.99}$$

so that the eigenfunction is

$$\Phi_m(\phi) = \frac{1}{\sqrt{2\pi}} e^{im\phi}. \tag{19.100}$$

Then, finally, the equation for the two-dimensional radial function (in the plane) is

$$\frac{d^2R}{d\rho^2} + \frac{1}{\rho}\frac{dR}{d\rho} + \left\{-\frac{m^2}{\rho^2} + \frac{2m_0}{\hbar^2}\left[E - \hbar\omega\left(n_z + \frac{1}{2}\right)\right] - \frac{m_0^2\omega^2}{\hbar^2}\rho^2\right\} R = 0. \tag{19.101}$$

Redefining the variable R, in order to get rid of the $dR/d\rho$ term, by setting

$$R = \frac{\chi(\rho)}{\sqrt{\rho}}, \tag{19.102}$$

so that

$$\frac{d^2R}{d\rho^2} + \frac{1}{\rho}\frac{dR}{d\rho} = \frac{1}{\sqrt{\rho}}\left(\frac{d^2\chi}{d\rho^2} + \frac{1}{4\rho^2}\chi\right), \tag{19.103}$$

we find the equation

$$\frac{d^2\chi}{d\rho^2} + \left\{-\frac{m^2 - 1/4}{\rho^2} + \frac{2m_0}{\hbar^2}\left[E - \hbar\omega\left(n_z + \frac{1}{2}\right)\right] - \frac{m_0\omega^2}{2}\rho^2\right\} \chi = 0. \tag{19.104}$$

This is the same equation as (19.72), with the following changes,

$$E \to E - \hbar\omega\left(n_z + \frac{1}{2}\right), \quad l(l+1) \to m^2 - \frac{1}{4}, \tag{19.105}$$

implying

$$l^2 + l - m^2 + \frac{1}{4} = 0 \;\Rightarrow\; l = -\frac{1}{2} \pm |m|, \tag{19.106}$$

and eigenenergies

$$\frac{E}{\hbar\omega} - \left(n_z + \frac{1}{2}\right) = l + \frac{3}{2} + 2n_\rho = 1 \pm |m| + 2n_\rho \;\Rightarrow$$
$$E = \hbar\omega\left(n_z + 2n_\rho \pm |m| + \frac{3}{2}\right) \equiv \hbar\omega\left(n + \frac{3}{2}\right). \tag{19.107}$$

We see that again we obtain the same spectrum, and in fact also the same degeneracy, as expected.

Important Concepts to Remember

- For motion in a central potential in three dimensions, we can reduce the Schrödinger equation to an equation in one dimension, for $\chi(r)$, in an effective potential

$$V_{\text{eff}}(r) = V(r) + \frac{\hbar^2}{2m}\frac{l(l+1)}{r^2},$$

thus including the centrifugal term.
- For $V \sim \alpha/r^s$ with $s < 2$, we can have bound states if there is a region where $V(r) < E < 0$. If $V \sim -|\alpha|/r^s$ with $s < 2$ at $r \to \infty$, we have an infinite number of bound states, of vanishingly small energy.
- For $V \sim -|\alpha|/r^s$ with $s > 2$, we can have bound states of arbitrarily large $|E|$, meaning there is an arbitrarily large probability for the particle to be at $r = 0$, implying that the particle "falls into $r = 0$ quantum mechanically".
- For $V \sim -|\alpha|/r^2$, $\beta = l(l+1) - 2m|\alpha|/\hbar^2 < -1/4$, the ground state is at $r = 0$.
- The diatomic molecule has a potential with a minimum at r_0, approximated by a shifted harmonic oscillator, so $E = V(r_0) + \hbar^2 l(l+1)/2I + \hbar\omega(n + 1/2)$.
- The spherical square well has a finite number of bound states for $E < 0$.
- The three-dimensional isotropic harmonic oscillator is equivalent to three one-dimensional harmonic oscillators, with tensor product states, giving an $(n+1)(n+2)/2$ degeneracy.
- Equivalently, in spherical coordinates, we have a central potential $m\omega^2 r^2/2$, which can be solved by the Sommerfeld polynomial method in terms of the same ${}_1F_1$, giving a quantized energy $E_n = \hbar\omega(l + 2n_r + 3/2) \equiv \hbar\omega(n + 3/2)$, with the same $(n+1)(n+2)/2$ degeneracy.
- In cylindrical coordinates, we obtain a one-dimensional harmonic oscillator in z times a one-dimensional radial harmonic oscillator for planar motion, with the same energy levels and degeneracy.

Further Reading

See [4] for more details on the various central potential cases and also [2].

Exercises

(1) Consider the central potential $V(r) = \alpha/r \cos\beta r$. How many bound states are there?

(2) Consider the central potential $V(r) = |\alpha|/r^2 \ln r/r_*$. In what region of space are the bound states confined?

(3) Consider the attractive Yukawa central potential, $V(r) = -|\alpha|/re^{-mr}$. Is the bound state spectrum finite or infinite? Can we use the same approximation as that used for the diatomic molecule, and if so, under what conditions?

(4) For the free particle in spherical coordinates, prove relation (19.53).

(5) Complete the proof of the fact that the number of bound states of the spherical square well is finite.

(6) Show that the normalization constant for the isotropic harmonic oscillator in spherical coordinates is given by (19.93).

(7) Solve the isotropic harmonic oscillator in two spatial dimensions, using polar coordinates.

20 Systems of Identical Particles

In this chapter we start considering a more physical case, with several particles. According to the general theory discussed earlier in the book, if the particles are independent, so that the observable operators for different particles commute (in particular the Hamiltonians, $\hat{H} = \sum_{i=1}^{N} \hat{H}_i$, with $[\hat{H}_i, \hat{H}_j] = 0$), we can write the Hilbert space as a tensor product,

$$\mathcal{H} = \mathcal{H}_1 \otimes \cdots \otimes \mathcal{H}_N. \tag{20.1}$$

Correspondingly, for a state with particles at positions $x_1, x_2, \ldots, x_N$, we have

$$|\psi\rangle = |x_1\rangle \otimes \cdots \otimes |x_N\rangle. \tag{20.2}$$

We have seen something similar before, in the case where we had several systems. For instance, in the previous chapter we considered a d-dimensional harmonic oscillator, with the oscillation in each dimension acting as an independent harmonic oscillator, so that $\hat{H} = \sum_{i=1}^{d} \hat{H}_i$ and the general state in the Cartesian occupation number picture was $|n_1\rangle \otimes \cdots \otimes |n_N\rangle$.

20.1 Identical Particles: Bosons and Fermions

However, we want to ask what happens when the particles are not just similar, but *identical*. In classical mechanics that is not a problem because we can always track the position and momentum of a particle, so we can always distinguish it from another particle. In quantum mechanics, however, the particles have probabilities of being anywhere, including at the same point, so we cannot distinguish them by position: the particles are *indistinguishable*.

Consider first a state with only two particles, and specifically the state with one particle at x_1 and one at x_2, denoted $|x_1 x_2\rangle$. Because of what we have just said, this state should be indistinguishable from the state $|x_2 x_1\rangle$, which means the two states should be proportional,

$$|x_2 x_1\rangle = C_{12} |x_1 x_2\rangle. \tag{20.3}$$

Define the *permutation operator*, which exchanges particles 1 and 2, by

$$\hat{P}_{12} |x_1 x_2\rangle = |x_2 x_1\rangle. \tag{20.4}$$

Then the two previous equations imply that $\hat{P}_{12}$ must be diagonal, and its eigenvalue is C_{12}.

Moreover, it seems obvious that we must have $\hat{P}_{12} = \hat{P}_{21}$ and $\hat{P}_{12}\hat{P}_{34} = \hat{P}_{34}\hat{P}_{12}$, i.e., the permutations are Abelian (commutative) and $\hat{P}_{12}^2 = \mathbb{1}$, meaning that two permutations will take us back to the original (unpermuted) case. This latest consideration implies that

$$C_{12}^2 = 1 \;\Rightarrow\; C_{12} = \pm 1. \tag{20.5}$$

Now, the conditions that permutations are Abelian (commutative), and that two permutations take us back to the original case sound harmless and obvious, but in fact they are not. These are assumptions that are broken in some possible cases, leading to what are known as "non-Abelions" and "anyons", which will be analyzed at the end of this book.

For the moment however, we note that we have only two possible symmetries, associated with the statistics describing particles: $C_{12} = +1$ leads to *Bose–Einstein statistics*, for *bosons*, while $C_{12} = -1$ leads to *Fermi–Dirac statistics*, for *fermions*.

Since

$$\hat{P}_{12}|x_1\rangle \otimes |x_2\rangle = |x_2\rangle \otimes |x_1\rangle \neq |x_1\rangle \otimes |x_2\rangle, \tag{20.6}$$

it follows that states such as $|x_1\rangle \otimes |x_2\rangle$ are actually not satisfactory states for identical particles. We must have either *symmetric* (for bosons) or *antisymmetric* (for fermions) combinations of the above states and their permutations,

$$|x_1 x_2; S/A\rangle = \frac{1}{\sqrt{2}}(|x_1\rangle \otimes |x_2\rangle \pm |x_2\rangle \otimes |x_1\rangle), \tag{20.7}$$

corresponding to $C_{12} = \pm 1$ (bosons/fermions). The factor $1/\sqrt{2}$ appears because of normalization, as the two states $|x_1\rangle \otimes |x_2\rangle$ and $|x_2\rangle \otimes |x_1\rangle$ are orthogonal so their sums or differences must have a coefficient $1/\sqrt{2}$.

More generally, for states a, b characterizing particles 1 and 2, instead of x_1, x_2, for bosons/fermions we have

$$|ab\rangle = \frac{1}{\sqrt{2}}(|a\rangle \otimes |b\rangle \pm |b\rangle \otimes |a\rangle). \tag{20.8}$$

In the case of fermions, when $a = b$ (the same state for particles 1 and 2), the two-particle state is

$$|aa\rangle = \frac{1}{\sqrt{2}}(|a\rangle \otimes |a\rangle - |a\rangle \otimes |a\rangle) = 0. \tag{20.9}$$

This is a mathematical expression of the *Pauli exclusion principle*, stating that we *cannot have two fermions occupying the same state*, unlike for bosons or classical particles.

Next, consider symmetric or antisymmetric states $|\psi_{S/A}\rangle$ in the coordinate representation, which means that we have to take the scalar product with the states $|x_1 x_2; S/A\rangle$. The wave function is thus

$$\psi_{S/A}(x_1, x_2) \equiv \langle x_1 x_2; S/A|\psi_{S/A}\rangle. \tag{20.10}$$

If the state $|\psi_{S/A}\rangle$ corresponds to eigenstates for observables $\hat{\Omega}$ (with eigenvalues ω), measured for particles 1 and 2, then we must have

$$|\psi_{S,A}\rangle = |\omega_1 \omega_2; S/A\rangle = \frac{1}{\sqrt{2}}(|\omega_1\rangle \otimes |\omega_2\rangle \pm |\omega_2\rangle \otimes |\omega_1\rangle). \tag{20.11}$$

Then, in the antisymmetric (A) or fermion case, we have the wave function

$$\psi_A(x_1, x_2) = \frac{1}{\sqrt{2}}\left|\begin{pmatrix} \psi_{\omega_1}(x_1) & \psi_{\omega_2}(x_1) \\ \psi_{\omega_1}(x_2) & \psi_{\omega_2}(x_2) \end{pmatrix}\right| = \frac{1}{\sqrt{2}}\left(\psi_{\omega_1}(x_1)\psi_{\omega_2}(x_2) - \psi_{\omega_2}(x_1)\psi_{\omega_1}(x_2)\right). \tag{20.12}$$

This is known as a Slater determinant.

20.2 Observables under Permutation

Consider (generalized) coordinates $\hat{q}_1, \hat{q}_2$ for particles 1 and 2, and let an observable depending on these coordinates be represented by the operator $\hat{A}(\hat{q}_1, \hat{q}_2)$. Then the fact that particles are indistinguishable means that the operators $\hat{A}(\hat{q}_1, \hat{q}_2)$ and $\hat{A}(\hat{q}_2, \hat{q}_1)$ must be equal (while the states can differ by a phase).

On the other hand, consider separate operator observables associated with the two particles, $\hat{A}_1$ and $\hat{A}_2$, which generalize $\hat{q}_1, \hat{q}_2$, and consider eigenstates for the two operators, $|a_1\rangle \otimes |a_2\rangle$. Then

$$\begin{aligned}\hat{A}_1|a_1\rangle \otimes |a_2\rangle &= a_1|a_1\rangle \otimes |a_2\rangle \\ \hat{A}_2|a_1\rangle \otimes |a_2\rangle &= a_2|a_1\rangle \otimes |a_2\rangle.\end{aligned} \tag{20.13}$$

Acting with $\hat{P}_{12}\hat{A}_1$ on the states, and introducing $\mathbb{1} = \hat{P}_{12}^{-1}\hat{P}_{12}$ before the states, we obtain (by calculating in two different ways)

$$\begin{aligned}\hat{P}_{12}\hat{A}_1\hat{P}_{12}^{-1}\hat{P}_{12}|a_1\rangle \otimes |a_2\rangle &= a_1\hat{P}_{12}|a_1\rangle \otimes |a_2\rangle = a_1|a_2\rangle \otimes |a_1\rangle \\ &= (\hat{P}_{12}\hat{A}_1\hat{P}_{12}^{-1})|a_2\rangle \otimes |a_1\rangle.\end{aligned} \tag{20.14}$$

Since this relation is valid for any states $|a_1\rangle \otimes |a_2\rangle$ (and the eigenstates of observables make a complete set in Hilbert space), we have also the operatorial equation

$$\hat{P}_{12}\hat{A}_1\hat{P}_{12}^{-1} = \hat{A}_2. \tag{20.15}$$

In particular, the same is true when $\hat{A}_i = \hat{q}_i$.

Applying relation (20.15) to the operator observable $\hat{A}(\hat{q}_1, \hat{q}_2)$, and calculating in two ways (the second using the fact that the operator for indistinguishable particles is the same when we switch them), we obtain

$$\begin{aligned}\hat{A}(\hat{q}_1, \hat{q}_2) &= \hat{A}(\hat{P}_{12}\hat{q}_1\hat{P}_{12}^{-1}, \hat{P}_{12}\hat{q}_2\hat{P}_{12}^{-1}) = \hat{P}_{12}\hat{A}(\hat{q}_1, \hat{q}_2)\hat{P}_{12}^{-1} \\ &= \hat{A}(\hat{q}_1, \hat{q}_2),\end{aligned} \tag{20.16}$$

thus leading to

$$\hat{P}_{12}\hat{A}(\hat{q}_1, \hat{q}_2)\hat{P}_{12}^{-1} = \hat{A}(\hat{q}_1, \hat{q}_2), \tag{20.17}$$

or $[\hat{P}_{12}, \hat{A}] = 0$ (multiparticle observables commute with permutations).

Another proof of the same relation is obtained by acting with $\hat{A}\hat{P}_{12}$ on states,

$$\begin{aligned}\hat{A}(\hat{q}_2, \hat{q}_1)\hat{P}_{12}|q_1\rangle \otimes |q_2\rangle &= \hat{A}(\hat{q}_2, \hat{q}_1)|q_2\rangle \otimes |q_1\rangle = A(q_1, q_2)|q_2\rangle \otimes |q_1\rangle \\ &= \hat{P}_{12}(A(q_2, q_1)|q_1\rangle \otimes |q_2\rangle) = \hat{P}_{12}\hat{A}(\hat{q}_2, \hat{q}_1)|q_1\rangle \otimes |q_2\rangle,\end{aligned} \tag{20.18}$$

and since it is true for states, it is also true as an operator relation.

In particular, the Hamiltonian operator must be of this type (it must commute with $\hat{P}_{12}$, i.e., be symmetric). However, in general we have

$$\hat{H} = \frac{\hat{P}_1^2}{2m_1} + \frac{\hat{P}_2^2}{2m_2} + V(\hat{X}_1, \hat{X}_2). \tag{20.19}$$

(a) One possibility is that the two particles don't interact at all, corresponding to a separable system, i.e.,

$$V(\hat{X}_1, \hat{X}_2) = V(\hat{X}_1) + V(\hat{X}_2) \;\Rightarrow\; \hat{H} = \hat{H}_1 + \hat{H}_2. \tag{20.20}$$

In this case, the solutions to the Schrödinger equation (the states) are separable,

$$|\psi\rangle = |\psi_1\rangle \otimes |\psi_2\rangle, \tag{20.21}$$

leading to separable wave functions (as we saw),

$$\psi(x_1, x_2) = \psi_1(x_1)\psi_2(x_2). \tag{20.22}$$

(b) Another possibility is that the potential depends only on the relative distance of the particles,

$$V(\hat{X}_1, \hat{X}_2) = V(|\hat{X}_1 - \hat{X}_2|), \tag{20.23}$$

which, as we saw, means that the Schrödinger equation reduces to separate equations for the center of mass motion and for the relative motion, with the relative motion corresponding to a central potential.

In this case, we do indeed have $\hat{P}_{12}\hat{H}\hat{P}_{12}^{-1} = \hat{H}$, as the relative motion is the same when we interchange the particles.

20.3 Generalization to *N* Particles

In the case of N particles, we must consider the behavior under a generic permutation,

$$P = \begin{pmatrix} 1 & 2 & \cdots & N \\ i_1 & i_2 & \cdots & i_N \end{pmatrix}. \tag{20.24}$$

But consider in particular just the transposition of particles i and j,

$$P_{ij} \equiv \begin{pmatrix} 1 & \cdots & i & \cdots & j & \cdots & N \\ 1 & \cdots & j & \cdots & i & \cdots & N \end{pmatrix}. \tag{20.25}$$

In particular, the action of P_{ij} on a state is

$$\hat{P}_{ij}|12\ldots i\ldots j\ldots N\rangle = |12\ldots j\ldots i\ldots N\rangle. \tag{20.26}$$

However, in this case, we are back to the previous case (since only two particles are interchanged, the others are left untouched), so by the same argument we must also have

$$\hat{P}_{ij}|12\ldots i\ldots j\ldots N\rangle = C_{ij}|12\ldots i\ldots j\ldots N\rangle, \tag{20.27}$$

and again we must have $C_{ij}^2 = 1$ (see (20.3)), so

$$C_{ij} = \pm 1, \tag{20.28}$$

corresponding as before to either Bose–Einstein or Fermi–Dirac statistics. Moreover, the same argument leads to the Pauli exclusion principle, saying that for fermions we can have at most one particle per state.

The above analysis must be true for any i, j, which means that in fact, the state $|12\ldots N\rangle$ must be *totally symmetric or antisymmetric*, corresponding respectively to bosons and fermions. Thus, for any generic permutation P, obtained as a product of transpositions P_{ij}, the same analysis as above must hold. For bosons, we have sums over permutations, whereas for fermions, we sum with a sign equal to the sign of the permutation P (which is given by the product of the minus signs corresponding to each transposition that composes P). We say that

$$|12\dots N\rangle = \frac{1}{\sqrt{N_{\text{perms}}}} \sum_{\text{perms. } P} (\text{sgn}(P))\hat{P}|1\rangle \otimes |2\rangle \otimes \cdots \otimes |N\rangle. \tag{20.29}$$

Wave functions in the coordinate representation are defined by

$$\psi_{S/A}(x_1, \dots, x_N) = \langle x_1 \dots x_N; S/A|\psi_{S/A}\rangle, \tag{20.30}$$

and thus have the same symmetry properties. They are totally symmetric for bosons and totally antisymmetric for fermions.

For a state that corresponds to individual states a_i for each particle i, we have the multiparticle state

$$|\psi_{S/A}\rangle = \frac{1}{\sqrt{N_{\text{perms}}}} \sum_{\text{perms. } P} (sgn(P))\hat{P}|a_1\rangle \otimes \cdots \otimes |a_n\rangle, \tag{20.31}$$

which means that the wave functions can be written as a determinant generalizing the two-particle case,

$$\psi_{S/A}(x_1, \dots, x_N) = \frac{1}{\sqrt{N_{\text{perms}}}} \left| \begin{pmatrix} \psi_{a_1}(x_1) & \cdots & \psi_{a_N}(x_1) \\ \vdots & & \vdots \\ \psi_{a_N}(x_1) & \cdots & \psi_{a_N}(x_N) \end{pmatrix} \right|. \tag{20.32}$$

20.4 Canonical Commutation Relations

For bosons, as we saw, we had commuting fundamental (canonical) operators, so in particular

$$[\hat{q}_i, \hat{q}_j] = [\hat{p}_i, \hat{p}_j] = 0, \quad [\hat{q}_i, \hat{p}_j] = i\hbar\delta_{ij}. \tag{20.33}$$

But if we have (extended) phase space variables ψ_i that are truly fermionic, i.e., anticommuting, their canonical anticommutation relations will be

$$\{\psi_i, \psi_j\} = \{p_{\psi_i}, p_{\psi_j}\} = 0, \quad \{\psi_i, p_{\psi_j}\} = i\hbar\delta_{ij}. \tag{20.34}$$

The simplest system will still be the harmonic oscillator, but now a fermionic version of it whose Hamiltonian can be written in terms of the above phase space variables,

$$\hat{H} = \frac{\hat{p}_\psi^2}{2m} + \frac{m\omega^2}{2}\hat{\psi}^2, \tag{20.35}$$

which again can be written in terms of creation and annihilation operators $b, b^\dagger$ but now obeying anticommutation relations,

$$\{b, b\} = \{b^\dagger, b^\dagger\} = 0, \quad \{b, b^\dagger\} = 1. \tag{20.36}$$

Then the quantum Hamiltonian can be written in terms of them as

$$\hat{H} = \hbar\omega\left(\frac{-bb^\dagger + b^\dagger b}{2}\right) = \hbar\omega\left(b^\dagger b - \frac{1}{2}\right). \tag{20.37}$$

Observations:

- We note that there exists a unique totally symmetric or antisymmetric wave function, a fact that should be obvious since we can start from one state and symmetrize it. Also, the totally symmetric or antisymmetric state is in a given representation of $SO(N)$, so is uniquely defined by group theory.
- The other important observation is that we have considered only N particles in order to (anti-) symmetrize, but we can ask: why not consider all the particles in the Universe? The point is that the particles considered must have some kind of interaction, even if a small one, in which case we are forced to use the symmetrization.

  But if we can ignore the coupling to an "exterior" made up of the rest of the Universe, we can ignore symmetrization with respect to it, and consider separation of variables instead:

$$\psi = \psi_{\text{system}} \psi_{\text{rest}}. \tag{20.38}$$

  Then the result is the same, as if we had not considered the exterior at all, since the probability is given by

$$P = |\psi|^2 = |\psi_{\text{system}}|^2 |\psi_{\text{rest}}|^2, \tag{20.39}$$

  and by summing (integrating) over the rest of the Universe, with $\int |\psi_{\text{rest}}|^2 = 1$, we obtain that

$$P_{\text{system}} = |\psi_{\text{system}}|^2, \tag{20.40}$$

  the same result as if we had just ignored the rest of the Universe.

20.5 Spin–Statistics Theorem

Now, a natural question can be asked: how do we know whether a particle is a boson or a fermion, other than by measuring its statistics (i.e., its behavior under measurement)?

The answer is that the statistics obeyed by a particle is related to its spin: for integer spin the particle is a boson, whereas for half-integer spin $j \in (\mathbb{N}/2)\backslash\mathbb{N}$, the particle is a fermion. This statement goes under the name of the *spin–statistics theorem*; however, it is not so much a single mathematical proof for this statement as it is a set of different proofs, of various degrees of rigor.

Some of the more rigorous proofs are based on quantum field theory, and will not be reproduced here, but only stated:

(1) *Lorentz invariance of the S-matrix* for the scattering of particles implies the need to correlate spin with statistics, in order not to break the symmetry at the quantum level.

(2) *Vacuum stability*: if we consider the wrong statistics for the fields associated with particles with spin, then we find that there are contributions of negative energy, implying an arbitrarily negative potential. Then the vacuum is unstable, which cannot hold for a true vacuum, implying that the assumption about the statistics was wrong.

A third proof based on quantum field theory will be sketched, being reasonably simple to understand (though some facts will have to be taken for granted by the reader):

(3) *Causality criterion*: For a spacelike separation between two points, $(x-y)^2 > 0$, such that there is no causal contact between the points and we can make a Lorentz transformation to a reference system where the two points are at the same time t, the observables measured at points x and y

should be independent. But in quantum mechanics, that amounts to the operators associated with them commuting,

$$[\mathcal{O}(x), \mathcal{O}(y)] = [\mathcal{O}(\vec{x}, t), \mathcal{O}(\vec{y}, t)] = 0. \tag{20.41}$$

In quantum field theory, however, a quantum field is (up to a normalization factor N_k) a sum over modes described by $e^{ik\cdot x}$ and associated creation operators,

$$\phi(\vec{x}, t) = \int d^3k \left(N_k a(\vec{k}) e^{-i\omega t + i\vec{k}\cdot\vec{x}} + \bar{N}_k a^\dagger(\vec{k}) e^{+i\omega t - i\vec{k}\cdot\vec{x}} \right), \tag{20.42}$$

where $\omega = \sqrt{\vec{k}^2 + m^2}$. The commutation relation for the creation and annihilation operators is

$$[a(\vec{k}), a^\dagger(\vec{k}')] = \delta^3(\vec{k} - \vec{k}'), \tag{20.43}$$

leading to commutation relation between the fields themselves at different points (that are spacelike separated),

$$[\phi(\vec{x}, t = 0), \phi(\vec{y}, t = 0)] = \int d^3k |N_k|^2 \left(e^{i\vec{k}\cdot(\vec{x}-\vec{y})} - e^{-i\vec{k}\cdot(\vec{x}-\vec{y})} \right) = 0, \tag{20.44}$$

where in the last equality we have used the fact that we can make the change $\vec{k} \to -\vec{k}$ in the integration. This is indeed what we expect, and since this relation is true for the basic operators (the fields), it will also be true for composite operators (arbitrary observables) made up from them. Note, however, that if we didn't have commutation relations for $a, a^\dagger$, but anticommutation relations instead (as for $b, b^\dagger$), we would get a nonzero result, thus proving the relation between spin (zero in this case) and statistics (bosonic, in this case). The relation can be proven case by case, though we do not have a general proof valid for all spins.

(4) However, the simplest (yet least rigorous) proof of the spin–statistics theorem relies on just the quantum mechanics we have defined so far (though it becomes slightly more rigorous when using a relativistic formulation, which will not be given here).

The generator of rotations around Oz is $J_3/\hbar$, as we saw in Chapter 16, so the matrix element for these rotations (with an angle θ) is

$$g = e^{i\theta J_3/\hbar}. \tag{20.45}$$

But the interchange of two particles can be obtained as follows: we can make a rotation by π in the plane perpendicular to Oz, and then a translation (depending on where the origin is, with respect to the midline between the particles). This amounts to a rotation by π of *each* of the two particles, which is the same as a rotation by 2π of a single particle.

However, a particle of spin j, with $J_3 = j\hbar$ (its maximum value), rotated by $\theta = 2\pi$, gives

$$g(2\pi) = e^{2\pi i j} = (-1)^{2j}, \tag{20.46}$$

which is indeed the formula we were looking for: symmetric ($g = +1$) for $j \in \mathbb{N}$, and antisymmetric ($g = -1$) for $j \in (\mathbb{N}/2)\backslash\mathbb{N}$ under this exchange of particles.

For $j = 1/2$, and considering a single particle, we must rotate by $\theta = \pi$, so that

$$g = e^{i\pi\sigma_3/2} = (\cos\pi/2)\mathbb{1} + i(\sin\pi/2)\,\sigma_3 = i\sigma_3 = \begin{pmatrix} i & 0 \\ 0 & -i \end{pmatrix}. \tag{20.47}$$

Thus, for a single particle in a state $|\psi_1\rangle$, we have

$$\hat{g}(\pi)|\psi_1\rangle = \pm i|\psi_1\rangle, \tag{20.48}$$

and for a two-particle state,

$$\hat{g}(\pi)|\psi_1\rangle \otimes |\psi_2\rangle = (\pm i)^2|\psi_1\rangle|\psi_2\rangle = -|\psi_1\rangle \otimes |\psi_2\rangle$$
$$= |\psi_2\rangle \otimes |\psi_1\rangle, \tag{20.49}$$

as promised.

Moreover, for $j = (2k+1)/2$ and $\theta = \pi$, we obtain

$$g^2(\pi) = g(2\pi) = (-1)^{2j}, \tag{20.50}$$

which is what we obtained from our first way of deriving the interchange factor.

20.6 Particles with Spin

Next, we consider what happens when the identical particles have a spin degree of freedom. Consider two electrons, with spin $s = 1/2$. Then the two-particle states are characterized by the individual positions and spin projections on Oz, with Hilbert space basis

$$|x_1, m_{s_1}\rangle \otimes |x_2, m_{s_2}\rangle \rightarrow |x_1, m_{s_1}; x_2, m_{s_2}\rangle. \tag{20.51}$$

If the total Hamiltonian commutes with the total spin squared, $[\vec{S}^2_{\rm tot}, \hat{H}] = 0$, i.e., if the Hamiltonian is independent of the total spin, then we can write the states as the product of separate states for position and spin,

$$|\psi\rangle = |\phi\rangle_{\rm position} \otimes |\chi\rangle_{\rm spin}. \tag{20.52}$$

The wave functions in the coordinate–spin basis are

$$\psi(x_1, x_2; m_{s_1}, m_{s_2}) = \langle x_1, m_{s_1}; x_2, m_{s_2}|\psi\rangle, \tag{20.53}$$

and they are separable into pure coordinate and pure spin wave functions,

$$\psi(x_1, x_2; m_{s_1}, m_{s_2}) = \phi(x_1, x_2)\chi(m_{s_1}, m_{s_2}). \tag{20.54}$$

Since the addition of the two angular momenta (of the electrons with spin $s = 1/2$) gives

$$\vec{1/2} \otimes \vec{1/2} = \vec{1} \oplus \vec{0}, \tag{20.55}$$

meaning the four states of the product space decompose into three states of total spin 1 (the triplet representation), $|jm\rangle = |1,-1\rangle, |1,0\rangle, |1,1\rangle$, plus a state of total spin zero (the singlet representation), $|jm\rangle = |0,0\rangle$, it follows that we can construct spin wave functions that correspond to these representations,

$$\begin{aligned}\chi(m_{s_1}, m_{s_2}) &= \chi_{++} \\ &= \frac{1}{\sqrt{2}}(\chi_{+-} + \chi_{-+}) \\ &= \chi_{--} \\ &= \frac{1}{\sqrt{2}}(\chi_{+-} - \chi_{-+}),\end{aligned} \tag{20.56}$$

where the first three states are in the triplet representation, and the last in the singlet representation. We note that indeed the triplet states are even under permutations (symmetric), as expected, since it implies that the electron angular momenta are parallel, whereas the singlet state is odd under permutations (antisymmetric), since it implies that the electron angular momenta are antiparallel.

Since in the case where both angular momenta are positive, we expect a permutation to give back the same (symmetric) state (since the state is symmetric),

$$\hat{P}_{12}\chi_{++} = +\chi_{++}, \tag{20.57}$$

the other two states in the triplet representation are obtained by acting with the total spin annihilation operator $S_- = S_{1,-} + S_{2,-}$. Since we also have $[\hat{P}_{12}, S_-] = 0$, the property of being symmetric is maintained by acting with S_-.

Because the wave function factorizes (into separate variables), we have that the permutation operator also factorizes,

$$\hat{P}_{12} = \hat{P}_{12}^{\text{space}} \cdot \hat{P}_{12}^{\text{spin}}. \tag{20.58}$$

Since we only need the total permutation to be antisymmetric (as is required for a fermion),

$$\hat{P}_{12}\psi(x_1, x_2) = -\psi(x_1, x_2) = \psi(x_2, x_1), \tag{20.59}$$

it means we have two possibilities:

- We can have a symmetric spatial wave function, $\phi(x_1, x_2) = \phi(x_2, x_1)$, and an antisymmetric spin wave function, $\chi(m_{s_1}, m_{s_2}) = -\chi(m_{s_2}, m_{s_1})$. This is the case when the spins are antiparallel, and so corresponds to the singlet representation.
- We can have an antisymmetric spatial wave function, $\phi(x_1, x_2) = -\phi(x_2, x_1)$, and a symmetric spin wave function, $\chi(m_{s_1}, m_{s_2}) = +\chi(m_{s_2}, m_{s_1})$. This is the case when the spins are parallel, and so corresponds to the triplet representation.

In the case of a separable Hamiltonian, $\hat{H} = \hat{H}_1 + \hat{H}_2$, we can further separate the spatial wave function $\phi(x_1, x_2)$, which thus has solutions of the type

$$\phi(x_1, x_2) = \phi_A(x_1)\phi_B(x_2), \tag{20.60}$$

where $\phi_A(x_1)$ corresponds to $\hat{H}_1$ and $\phi_B(x_2)$ corresponds to $\hat{H}_2$. But then we can write down a spatial symmetric or antisymmetric wave function,

$$\phi(x_1, x_2) = \frac{1}{\sqrt{2}}[\phi_A(x_1)\phi_B(x_2) \pm \phi_A(x_2)\phi_B(x_1)], \tag{20.61}$$

corresponding to the triplet and singlet representations, respectively.

In that case, the probability density of finding one electron at x_1 and the other at x_2 is

$$\begin{aligned} |\phi(x_1, x_2)|^2 = \frac{1}{2}\Big\{&|\psi_A(x_1)|^2|\phi_B(x_2)|^2 + |\psi_B(x_1)|^2|\psi_A(x_2)|^2 \\ &\pm 2\,\text{Re}[\psi_A(x_1)\psi_B(x_2)\psi_A^*(x_2)\psi_B^*(x_1)]\Big\}. \end{aligned} \tag{20.62}$$

The last term in the above is called the exchange probability density. We note that for the same state, $A = B$, when $x_1 = x_2$ we find that $|\phi(x_1, x_2)|^2_{\text{singlet}} = 0$, which is a statement of the Pauli exclusion principle. On the other hand, for the triplet, the probability when $A = B$ for $x_1 = x_2$ is maximal.

Important Concepts to Remember

- In quantum mechanics identical particles are indistinguishable, which means the result of acting with the permutation operator on a state with two particles must be related to the unpermuted state by a phase, $|x_2x_1\rangle = C_{12}|x_1x_2\rangle$.
- Excluding the nontrivial cases of anyons and nonabelions, that means we can either have Bose–Einstein statistics (bosons) for $C_{12} = +1$, or Fermi–Dirac statistics (fermions), $C_{12} = -1$.
- For fermions, this leads to the Pauli exclusion principle, which states that we cannot have two fermions in the same state: $|xx\rangle = 0$.
- Multiparticle operators commute with the permutation operator; for the Hamiltonian, it is either separable and symmetric, or has a potential depending only on the *relative* distances of the particles.
- A multiparticle state is either totally symmetric or totally antisymmetric. In the antisymmetric case, this leads to a Slater determinant for the wave function.
- Fermionic phase space variables have anticommutation relations rather than commutation relations, indicated by $\{,\}$. In particular, the fermionic harmonic oscillator has operators $b, b^\dagger$ with $\{b, b^\dagger\} = 1$.
- The spin–statistics theorem says that integer spin particles are bosons, and half-integer spin particles are fermions; this can be understood from the rotation matrix $g = e^{i\theta J_3/\hbar}$, implying that $g(2\pi) = (-1)^{2j}$.
- For particles with spin, the total wave function is a product of the coordinate wave function times the spinor wave function, both of which need to be totally symmetric or totally antisymmetric; only the product wave function is governed by the spin–statistics theorem.
- Thus a fermion has either a symmetric spatial wave function and an antisymmetric spin wave function, or an antisymmetric spatial wave function and a symmetric spin wave function.
- For a separable Hamiltonian, the probability density has a direct term and an exchange term.

Further Reading

See [2], [1], [3] for more details.

Exercises

(1) Write down an antisymmetric (fermionic) state for three particles, the energy eigenstates E_1, E_2, E_3, and the corresponding Slater determinant.

(2) Consider a three-body problem with two-body interactions, i.e., with the potential

$$\hat{V} = \hat{V}_{\text{int}}(|\hat{X}_1 - \hat{X}_2|) + \hat{V}_{\text{int}}(|\hat{X}_1 - \hat{X}_3|) + \hat{V}_{\text{int}}(|\hat{X}_2 - \hat{X}_3|). \tag{20.63}$$

Is this quantum potential suitable for indistinguishable particles?

(3) Consider a state with N indistinguishable particles. What is the phase that relates $|12\ldots N\rangle$ with $|N\ldots 21\rangle$?

(4) Write down a general state for fermionic harmonic oscillator. Also write down general state for N identical fermionic harmonic oscillators.

(5) We saw in the text that $[\phi(\vec{x},t),\phi(\vec{y},t)]=0$ for a bosonic field. Show also that $\{\psi(\vec{x},t),\psi(\vec{y},t)\}=0$ for a fermion (Hint: think about the fact that ψ and $\psi^\dagger$ are conjugate for fermions, unlike for bosons). What relation will we have for the product $\phi(\vec{x},t)\psi(\vec{y},t)$?

(6) Consider three particles of spin 1/2. Write the possible spin wave functions for the three-particle states, and construct the possible total wave functions, for the spatial wave function times the spin wave function.

(7) For the case in exercise 6, write down the probability and identify the exchange terms.

21 Application of Identical Particles: He Atom (Two-Electron System) and H_2 Molecule

In this chapter, we will apply the formalism of the previous chapter to two cases with two electrons: the He atom (and helium-like atoms, with an arbitrary Z but two electrons only) and the H_2 molecule.

In order to do these calculations, we will introduce approximation methods that will be formalized and generalized later in the book.

21.1 Helium-Like Atoms

Helium-like atoms have a nucleus of charge $+Ze$, around which two electrons move. Of course, in the case of helium $Z = 2$, but we can consider general Z. The nucleus will be considered as approximately fixed. As in the case of the hydrogen atom, the center of mass motion factorizes, and the relative motions are the only ones of importance: since the mass of the nucleus is much larger than that of the electrons, for the relative motions of the electrons we can consider a fixed nucleus situated at the origin, around which the electrons move (with reduced mass approximately equal to the electronic mass).

The Hamiltonian for the system contains the potentials for the interactions of the two electrons with the nucleus and with each other, thus being given by

$$\begin{aligned} \hat{H} &= \frac{\hat{\vec{p}}_1^2}{2m} + \frac{\hat{\vec{p}}_2^2}{2m} - \frac{Ze_0^2}{r_1} - \frac{Ze_0^2}{r_2} + \frac{e_0^2}{r_{12}} \\ &= \hat{H}_1 + \hat{H}_2 + \hat{H}_{12}, \end{aligned} \tag{21.1}$$

where $e_0^2 = e^2/(4\pi\epsilon_0)$, as before.

As in the previous chapter, the state of the system factorizes into a spin state for spin and a coordinate state, and correspondingly the wave function is the product of a wave function for positions and one for spins,

$$\psi(x_1, x_2; m_{s_1}, m_{s_2}) = \phi(x_1, x_2)\chi(m_{s_1}, m_{s_2}), \tag{21.2}$$

where the spin wave function $\chi(m_{s_1}, m_{s_2})$ is either in the triplet representation,

$$\left(\chi_{++} = \chi_{1+}\chi_{2+};\ \frac{1}{\sqrt{2}}(\chi_{+-} + \chi_{-+}) = \frac{1}{\sqrt{2}}(\chi_{1+}\chi_{2-} + \chi_{1-}\chi_{2+});\ \chi_{--} = \chi_{1-}\chi_{2-}\right), \tag{21.3}$$

or in the singlet representation

$$\frac{1}{\sqrt{2}}(\chi_{+-} - \chi_{-+}) = \frac{1}{\sqrt{2}}(\chi_{1+}\chi_{2-} - \chi_{1-}\chi_{2+}). \tag{21.4}$$

The position wave function $\phi(x_1, x_2)$ is either antisymmetric, corresponding to the triplet spin wave function, or symmetric, corresponding to the singlet spin wave function. The case with triplet spin (when the spins of the individual electrons are parallel) and antisymmetric $\phi(x_1, x_2)$ is a state that is known as "ortho" (so in this case, we have ortho-helium), while the case with singlet spin (when the electron spins are antiparallel) and symmetric $\phi(x_1, x_2)$ is a state known as "para" (so in this case, we have para-helium). To proceed, we will make some approximations.

Approximation 1

We neglect the interaction between the electrons, $\hat{H}_{12} = e_0^2/r_{12}$, for the wave functions, and so consider eigenfunctions of the noninteracting Hamiltonian,

$$\hat{H} \simeq \hat{H}_1 + \hat{H}_2, \tag{21.5}$$

and then, as in the noninteracting case, we can separate variables, obtaining eigenfunctions

$$\psi(\vec{r}_1, \vec{r}_2) = \psi(\vec{r}_1)\psi(\vec{r}_2). \tag{21.6}$$

For these eigenfunctions, the eigenenergies are

$$E \simeq E^{(1)} + E^{(2)}, \tag{21.7}$$

and $(\psi(\vec{r}_1), E^{(1)})$, $(\psi(\vec{r}_2), E^{(2)})$ are solutions of a hydrogen-like (noninteracting) Schrödinger equation,

$$\left(-\frac{\hbar^2}{2m}\Delta_{\vec{r}_s} - \frac{Ze_0^2}{r_s}\right)\psi^{(s)}(\vec{r}_s) = E^{(s)}\psi^{(s)}(\vec{r}_s). \tag{21.8}$$

Then the individual electron wave functions are

$$\psi^{(s)}(\vec{r}_s) = \psi_{n_s l_s m_s}(\vec{r}_s) \equiv \psi_{q_s}(\vec{r}_s), \tag{21.9}$$

and their energies are

$$E^{(s)} = E_{n_s} = -\frac{mZ^2e_0^4}{2\hbar^2}\frac{1}{n_s^2} = -\frac{Z^2e_0^2}{2a_0}\frac{1}{n_s^2}. \tag{21.10}$$

The corresponding symmetric (spin singlet) and antisymmetric (spin triplet) coordinate wave functions are

$$\begin{aligned}
\phi_s(\vec{r}_1, \vec{r}_2) &= \frac{1}{\sqrt{2}}[\psi_{q_1}(\vec{r}_1)\psi_{q_2}(\vec{r}_2) + \psi_{q_2}(\vec{r}_1)\psi_{q_1}(\vec{r}_2)] \\
\phi_a(\vec{r}_1, \vec{r}_2) &= \frac{1}{\sqrt{2}}[\psi_{q_1}(\vec{r}_1)\psi_{q_2}(\vec{r}_2) - \psi_{q_2}(\vec{r}_1)\psi_{q_1}(\vec{r}_2)].
\end{aligned} \tag{21.11}$$

Approximation 2

We now suppose that the energy of the interaction-less case above is corrected by the interaction term $\hat{H}_{12}$ evaluated in the above (noninteracting, S or A) state,

$$\begin{aligned}
E &= E^{(1)} + E^{(2)} + \Delta E \\
\Delta E &= \left\langle \phi_{S/A} \left| \frac{e_0^2}{r_{12}} \right| \phi_{S/A} \right\rangle \\
&= \int d^3r_1 \int d^3r_2 \frac{e_0^2}{r_{12}} |\phi_{S/A}(\vec{r}_1, \vec{r}_2)|^2.
\end{aligned} \tag{21.12}$$

The probability density in the integral can be written as

$$\begin{aligned}
|\phi_{s.a}(\vec{r}_1,\vec{r}_2)| &= \left|\frac{\psi_{q_1}(\vec{r}_1)\psi_{q_2}(\vec{r}_2) \pm \psi_{q_2}(\vec{r}_1)\psi_{q_1}(\vec{r}_2)}{\sqrt{2}}\right|^2 \\
&= \frac{1}{2}\left\{|\psi_{q_1}(\vec{r}_1)|^2|\psi_{q_2}(\vec{r}_2)|^2 + |\psi_{q_2}(\vec{r}_1)|^2|\psi_{q_1}(\vec{r}_2)|^2\right. \\
&\quad \left.\pm 2\,\mathrm{Re}[\psi_{q_1}(\vec{r}_1)\psi^*_{q_1}(\vec{r}_2)\psi_{q_2}(\vec{r}_2)\psi^*_{q_2}(\vec{r}_1)]\right\}.
\end{aligned} \tag{21.13}$$

The last term is called the exchange probability density, as we said in the previous chapter.

The integrals involving the terms above can be rewritten in a useful way. Consider the integrals of the first two terms:

$$\begin{aligned}
C &\equiv \frac{1}{2}\int d^3r_1 \int d^3r_2 \frac{e_0^2}{r_{12}}\left[|\psi_{q_1}(\vec{r}_1)|^2|\psi_{q_2}(\vec{r}_2)|^2 + |\psi_{q_2}(\vec{r}_1)|^2|\psi_{q_1}(\vec{r}_2)|^2\right] \\
&= \int d^3r_1 \int d^3r_2 \frac{e_0^2}{r_{12}}|\psi_{q_1}(\vec{r}_1)|^2|\psi_{q_2}(\vec{r}_2)|^2 \\
&= \int d^3r_1 \int d^3r_2 \frac{\rho_{q_1}(\vec{r}_1)\rho_{q_2}(\vec{r}_2)}{4\pi\epsilon_0 r_{12}},
\end{aligned} \tag{21.14}$$

where in the second line we have made the redefinition $\vec{r}_1 \leftrightarrow \vec{r}_2$ in the integral of the second term, in order to show that it equals the first term, and in the third line we have defined the charge density as the charge times the probability density, $\rho_q(\vec{r}) = e|\psi_q(\vec{r})|^2 = e_0\sqrt{4\pi\epsilon_0}|\psi_q(\vec{r})|^2$. As we can see, this integral is a Coulomb-type integral; it is just the integral of the Coulomb potential for charge densities, hence the name C for the contribution.

Next consider the last term in (21.13) (with $\pm 2\,\mathrm{Re}$ in front):

$$\begin{aligned}
A &\equiv \int d^3r_1 \int d^3r_2 \frac{e_0^2}{r_{12}}\frac{1}{2}[\psi_{q_1}(\vec{r}_1)\psi^*_{q_1}(\vec{r}_2)\psi_{q_2}(\vec{r}_2)\psi^*_{q_2}(\vec{r}_1) + \psi_{q_1}(\vec{r}_2)\psi^*_{q_1}(\vec{r}_1)\psi_{q_2}(\vec{r}_1)\psi^*_{q_2}(\vec{r}_2)] \\
&= \int d^3r_1 \int d^3r_2 \frac{e_0^2}{r_{12}}\psi_{q_1}(\vec{r}_1)\psi^*_{q_1}(\vec{r}_2)\psi_{q_2}(\vec{r}_2)\psi^*_{q_2}(\vec{r}_1),
\end{aligned} \tag{21.15}$$

where we have used again that the two terms (which are otherwise complex conjugates of each other) are equal, as can be seen by making the redefinition $\vec{r}_1 \leftrightarrow \vec{r}_2$. This means that the term is actually real, as needed since it is a term in the energy. It also means that we can define $f(\vec{r}) \equiv \psi_{q_1}(\vec{r})\psi^*_{q_2}(\vec{r})$, and write

$$A = \int d^3r_1 \int d^3r_2 \frac{e_0^2}{r_{12}} f(\vec{r}_1) f(\vec{r}_2). \tag{21.16}$$

Here A is the *exchange integral* or *exchange energy*, and the initial comes from the German "Austausch", meaning exchange.

Finally then, the energy is

$$E = E_{q_1} + E_{q_2} + C \pm A. \tag{21.17}$$

Moreover, we can show that A is not only real, but positive. Indeed, we can see that the largest contribution to the integral comes from $r_{12} \to 0$, when the two particles are close to each other. But A is an integral of $f(\vec{r}_1)f\vec{r}_2)/r_{12}$, and if $\vec{r}_1, \vec{r}_2$ are close together then, by the continuity of the function f, $f(\vec{r}_1) \simeq f(\vec{r}_2)$, or at the very least they have the same sign, meaning that their product is positive, and so this leading contribution to A is positive.

But that means that the energy of the ortho-helium state (antisymmetric position wave function) is smaller than the energy of the para-helium state (symmetric position wave function), meaning that it is more stable. Thus we have a tendency to have parallel spins, i.e., triplet states, even though we don't have any interaction that depends on spin (this tendency follows simply from the statistics of the fermions).

Also note that when $q_1 = q_2$ (both electrons have the same quantum numbers n, l, m), the analysis is special (the previous analysis does not apply). Then there is no antisymmetric (ortho-helium, or spin triplet) state, only a symmetric one (para-helium, or spin singlet), corresponding to antiparallel spins for the electrons.

21.2 Ground State of the Helium (or Helium-Like) Atom

We will calculate only the ground state of the helium (or helium-like) atom, so we need only consider the ground states of hydrogen-like atoms as individual wave functions. Then we have

$$\psi_q = \psi_{100}(\vec{r}) = R_{10}(r)Y_{00}(\theta, \phi). \tag{21.18}$$

But the ground state spherical harmonic is just a normalization constant, since the state has no angular momentum, and this means that it is spherically symmetric,

$$Y_{00}(\theta, \phi) = \frac{1}{\sqrt{4\pi}}, \tag{21.19}$$

so that $\int d\Omega |Y_{00}| = 1$. In orbital notation, the ground state of a hydrogen-like atom is $(1s)^2$, meaning that $n = 1, l = 0$ (denoted as an "s" orbital) and the index 2 refers to there being two electrons in this state (necessarily of opposite spins by the Pauli exclusion principle, since they are in the same state for coordinates). The radial wave function is $R_{10}(r) \sim e^{-kr}$, where $k = Z/a_0$, and the normalization constant for it comes from

$$\int_0^\infty e^{-2kr} r^2 dr = \frac{1}{4k^3}, \tag{21.20}$$

leading to

$$\psi_q = \left(\frac{Z}{a_0}\right)^{3/2} \frac{2}{\sqrt{4\pi}} e^{-Zr/a_0}. \tag{21.21}$$

Then the para (spin singlet) state of a helium-like atom is

$$\psi_s = \psi_q(\vec{r}_1)\psi_q(\vec{r}_2)\chi_{\text{singlet}} = \frac{Z^3}{\pi a_0^3} e^{-Z(r_1+r_2)/a_0}, \tag{21.22}$$

and the electron interaction energy is

$$\Delta E = e_0^2 \left\langle \frac{1}{r_{12}} \right\rangle_{\psi_s} = \int d^3 r_1 \int d^3 r_2 \frac{Z^6}{\pi^2 a_0^6} e^{-2k(r_1+r_2)} \frac{e_0^2}{r_{12}}. \tag{21.23}$$

To do this integral, we first define

$$\Phi(\vec{r}_1) \equiv \int d^3 r_2 \frac{e^{-2kr_2}}{r_{12}}, \tag{21.24}$$

and note that it has the form of a Coulomb potential at the point $\vec{r}_1$ for a distribution of charge with density e^{-2kr_2}. It then satisfies the Poisson equation

$$\Delta_{\vec{r}_1}\Phi(\vec{r}_1) = -4\pi e^{-2kr_1}. \tag{21.25}$$

Because of the spherical symmetry of the equation, $\vec{L}^2\Phi = 0$, we obtain just the radial equation,

$$\left(\frac{d^2}{dr_1^2} + \frac{2}{r_1}\frac{d}{dr_1}\right)\Phi(\vec{r}_1) = -4\pi e^{-2kr_1} \tag{21.26}$$

or

$$\frac{d^2}{dr_1^2}(r_1\Phi(r_1)) = -4\pi r_1 e^{-2kr_1}. \tag{21.27}$$

We see that it has a *particular* solution of the form $r_1\Phi(r_1) = (ar_1 + b)e^{-2kr_1}$ and, by substituting it into (21.27) we obtain

$$[(2k)^2(ar_1 + b) - 4ka]e^{-2kr_1} = -4\pi r_1 e^{-2kr_1}, \tag{21.28}$$

which fixes the coefficients as follows:

$$a = -\frac{\pi}{k^2}, \quad b = \frac{a}{k} = -\frac{\pi}{k^3}. \tag{21.29}$$

However, the general solution of a second-order differential equation with a source (the Poisson equation) is equal to the particular solution plus the general solution of the homogenous (without source) equation, which is $r_1\Phi(r_1) = C_1r_1 + C_2$. To fix the coefficients C_1, C_2, we impose the physical conditions that the potential $\Phi(r_1)$ should vanish at infinity, $\Phi(r_1 \to \infty) \to 0$, which gives $C_1 = 0$, and that it is nonsingular at $r_1 \to 0$ since in the original definition (21.24) there is nothing special about $r_1 = 0$. This last condition implies that the coefficient of $1/r_1$ must vanish at $r_1 = 0$, leading to $C_2 = -b = \pi/k^3$.

Finally, then, we have

$$\Phi(r_1) = -\pi\left(\frac{1}{k^2}e^{-2kr_1} + \frac{1}{k^3}\frac{e^{-2kr_1}}{r_1} - \frac{1}{k^3}\frac{1}{r_1}\right). \tag{21.30}$$

Integrating over $\vec{r}_1$ as well, we get

$$\begin{aligned}
\int \Phi(r_1)e^{-2kr_1}d^3r_1 &= \int_0^\infty \Phi(r_1)e^{-2kr_1}4\pi r_1^2 dr_1 \\
&= -4\pi^2\left[\frac{1}{k^2}\int_0^\infty e^{-4kr_1}r_1^2dr_1 + \frac{1}{k^3}\int_0^\infty e^{-4kr_1}r_1dr_1 - \frac{1}{k^3}\int_0^\infty e^{-2kr_1}r_1dr_1\right] \\
&= -4\pi^2\left[\frac{1}{k^2}\frac{1}{32k^3} + \frac{1}{k^3}\frac{1}{16k^2} - \frac{1}{k^3}\frac{1}{4k^2}\right] \\
&= +\frac{5}{8}\frac{\pi^2}{k^5}.
\end{aligned} \tag{21.31}$$

Thus the correction ΔE to the energy is

$$\Delta E = \frac{e_0^2Z^6}{\pi^2a_0^6}\frac{5\pi^2}{8k^5} = \frac{5}{8}\frac{Ze_0^2}{a_0}, \tag{21.32}$$

and the total energy is the sum of two ground state energies for a hydrogen-like atom, plus the above correction,

$$E = 2E_0 + \Delta E = -\frac{Z^2 e_0^2}{a_0} + \frac{5}{8}\frac{Ze_0^2}{a_0} = -\frac{Ze_0^2}{a_0}\left(Z - \frac{5}{8}\right). \tag{21.33}$$

Using $Z = 2$ for helium, we obtain $E \simeq -74.8\,\text{eV}$, which may be compared with the experimental value of $E_{\text{exp}} = -78.8\,\text{eV}$. That is quite good, but we can in fact do better.

21.3 Approximation 3: Variational Method, "Light Version"

We can use a variational method to account for the fact that, when considering the interaction of the individual electrons with the nucleus, the electrons do not see the full charge $+Ze$ of the nucleus but, rather, a smaller effective charge, owing to the partial screening of the nuclear charge by the charge density cloud created by the second electron. To find this effective charge Z_{eff}, we use a *variational method* (studied in detail later, in Chapter 41).

Namely, we consider a modified wave function, with $Z \to Z_{\text{eff}}$, so that

$$\tilde{\psi}_s = \frac{Z_{\text{eff}}^3}{\pi a_0^3} e^{-Z_{\text{eff}}(r_1+r_2)/a_0}, \tag{21.34}$$

and evaluate the Hamiltonian in this effective state,

$$E = \left\langle \tilde{\psi}_s \middle| \left(\frac{\hat{\vec{p}}_1^2}{2m} + \frac{\hat{\vec{p}}_2^2}{2m} \right) \middle| \tilde{\psi}_s \right\rangle + \left\langle \tilde{\psi}_s \middle| \left(-\frac{Ze_0^2}{r_1} - \frac{Ze_0^2}{r_2} \right) \middle| \tilde{\psi}_s \right\rangle + \left\langle \tilde{\psi}_s \middle| \frac{e^2}{r_{12}} \middle| \tilde{\psi}_s \right\rangle. \tag{21.35}$$

But we saw in Chapter 18 that

$$\begin{aligned} \left\langle \frac{1}{r} \right\rangle_n &= \frac{Z}{a_0}\frac{1}{n^2} \Rightarrow \left\langle \tilde{\psi}_s \middle| \frac{1}{r_1} \middle| \tilde{\psi}_s \right\rangle = \frac{Z_{\text{eff}}}{a_0} \\ \left\langle \frac{\hat{\vec{p}}^2}{2m} \right\rangle_n &= \frac{Z^2 e_0^2}{2a_0}\frac{1}{n^2} \Rightarrow \left\langle \tilde{\psi}_s \middle| \frac{\hat{\vec{p}}^2}{2m} \middle| \tilde{\psi}_s \right\rangle = \frac{Z_{\text{eff}}^2 e_0^2}{2a_0}, \end{aligned} \tag{21.36}$$

so that the energy functional to be minimized is

$$E = 2\frac{Z_{\text{eff}}^2 e_0^2}{2a_0} - 2\frac{ZZ_{\text{eff}} e_0^2}{a_0} + \frac{5}{8}\frac{Z_{\text{eff}} e_0^2}{a_0^2}. \tag{21.37}$$

Minimizing this over Z_{eff} at fixed Z by $\delta E/\delta Z_{\text{eff}} = 0$, we obtain

$$Z_{\text{eff}} = Z - \frac{5}{16}. \tag{21.38}$$

Replacing it back in (21.37), we find the correct minimized energy as

$$E = -\frac{e_0^2}{a_0}\left(Z - \frac{5}{16}\right)^2 = -\frac{Z_{\text{eff}}^2 e_0^2}{a_0}. \tag{21.39}$$

An equivalent way to obtain this result is to note that (as seen in Chapter 18)

$$\left\langle \frac{1}{r} \right\rangle_n = \frac{Z}{a_0}\frac{1}{n^2} \Rightarrow \left\langle \frac{1}{r_{12}} - \frac{5}{16}\frac{1}{r_1} - \frac{5}{16}\frac{1}{r_2} \right\rangle = 0, \tag{21.40}$$

and use this to rewrite the Hamiltonian in the form

$$\hat{H} = -\frac{\hat{\vec{p}}_1^2}{2m} - \frac{\hat{\vec{p}}_2^2}{2m} - \left(Z - \frac{5}{16}\right)\frac{e_0^2}{r_1} - \left(Z - \frac{5}{16}\right)\frac{e_0^2}{r_2} + e_0^2\left(\frac{1}{r_{12}} - \frac{5}{16}\frac{1}{r_1} - \frac{5}{16}\frac{1}{r_2}\right), \tag{21.41}$$

where the term in the last line has zero quantum average, so can be dropped. But then the remaining terms are the same as for two decoupled electrons in the potential of a nucleus with $Z_{\text{eff}} = Z - 5/16$, so the total energy is

$$E = -2\frac{e_0^2}{2a_0}\left(Z - \frac{5}{16}\right)^2. \tag{21.42}$$

21.4 H_2 Molecule and Its Ground State

Another two-electron system is the hydrogen molecule H_2, with two protons (nuclei), called A and B, and two electrons, called 1 and 2. The Hamiltonian corresponds to the Coulomb interactions between the electrons and each of the nuclei, between themselves, and between the two nuclei, so is given by

$$\hat{H} = -\frac{\hbar^2}{2m}\Delta_1 - \frac{\hbar^2}{2m}\Delta_2 - \frac{e_0^2}{r_{1A}} - \frac{e_0^2}{r_{1B}} - \frac{e_0^2}{r_{2A}} - \frac{e_0^2}{r_{2B}} + \frac{e_0^2}{r_{12}} + \frac{e_0^2}{R}, \tag{21.43}$$

where $R = |\vec{r}_A - \vec{r}_B|$ is the distance between the nuclei.

It is clear that in the ground state, at least for $R \gg a_0$, one electron is near nucleus A, and the other is near nucleus B.

Indeed, consider the opposite situation, with both electrons near one nucleus; thus we have two ionized hydrogen atoms, H^+ and H^-. Here H^- is just a bare nucleus, so doesn't contribute in the first approximation, whereas H^+ can be thought of as a helium-like atom (with two electrons), but with $Z = 1$. Then we can apply the helium-like energy formulas, obtaining that the electron energy of the system is

$$E_e = -\frac{e_0^2}{a_0}\left(1 - \frac{5}{16}\right)^2 > -\frac{e_0^2}{a_0} = 2E_{H,0}, \tag{21.44}$$

which is larger than the energy of the electrons in the configuration where we have two neutral hydrogen atoms.

This means that for the ground state in the $R \gg a_0$ case we have one electron near nucleus A and the other near nucleus B, with two neutral hydrogen atoms, H + H. For this configuration, we have two eigenstates, depending on which electron is near which nucleus,

$$\begin{aligned} \psi_1 &= \psi(r_{1A})\psi(r_{2B}) \\ \psi_2 &= \psi(r_{1B})\psi(r_{2A}). \end{aligned} \tag{21.45}$$

But while these two states are normalized, they are not orthogonal because they admit a nonzero product,

$$\langle\psi_1|\psi_2\rangle = \int d^3r_1\psi^*(r_{1A})\psi(r_{1B})\int d^3r_2\psi(r_{2A})\psi^*(r_{2B}) = |S|^2, \tag{21.46}$$

where we have defined the "overlap integral"

$$S \equiv \int d^3 r \psi^*(r_A)\psi(r_B). \tag{21.47}$$

Since we have the products

$$\langle \psi_1 \pm \psi_2 | \psi_1 \pm \psi_2 \rangle = \langle \psi_1 | \psi_1 \rangle + \langle \psi_2 | \psi_2 \rangle \pm 2\langle \psi_1 | \psi_2 \rangle = 2(1 \pm |S|^2), \tag{21.48}$$

the normalized symmetric and antisymmetric coordinate wave functions for the two-electron system are

$$\psi_{s/a} = \frac{\psi_1 \pm \psi_2}{\sqrt{2}\sqrt{1 \pm |S|^2}} = \frac{\psi(r_{1A})\psi(r_{2B}) \pm \psi(r_{1B})\psi(r_{2A})}{\sqrt{2}\sqrt{1 \pm |S|^2}}. \tag{21.49}$$

Then the matrix elements of the Hamiltonian are

$$\begin{aligned}
\langle \psi_1 | \hat{H} | \psi_1 \rangle &= \left\langle \psi_1 \middle| \left(-\frac{\hbar^2}{2m}\Delta_1 - \frac{e_0^2}{r_{1A}} \right) \middle| \psi_1 \right\rangle + \left\langle \psi_1 \middle| \left(-\frac{\hbar^2}{2m}\Delta_2 - \frac{e_0^2}{r_{2B}} \right) \middle| \psi_1 \right\rangle \\
&\quad + \int d^3 r_1 \int d^3 r_2 \left(-\frac{e_0^2}{r_{2A}} - \frac{e_0^2}{r_{1B}} + \frac{e_0^2}{r_{12}} + \frac{e_0^2}{R} \right) |\psi(r_{1A})|^2 |\psi(r_{2B})|^2 \\
&= 2E_0 + \int d^3 r_1 \int d^3 r_2 \left(-\frac{e_0^2}{r_{2A}} - \frac{e_0^2}{r_{1B}} + \frac{e_0^2}{r_{12}} + \frac{e_0^2}{R} \right) |\psi(r_{1A})|^2 |\psi(r_{2B})|^2 \\
\langle \psi_2 | \hat{H} | \psi_2 \rangle &= \left\langle \psi_2 \middle| \left(-\frac{\hbar^2}{2m}\Delta_1 - \frac{e_0^2}{r_{1A}} \right) \middle| \psi_2 \right\rangle + \left\langle \psi_2 \middle| \left(-\frac{\hbar^2}{2m}\Delta_2 - \frac{e_0^2}{r_{2B}} \right) \middle| \psi_2 \right\rangle \\
&\quad + \int d^3 r_1 \int d^3 r_2 \left(-\frac{e_0^2}{r_{2A}} - \frac{e_0^2}{r_{1B}} + \frac{e_0^2}{r_{12}} + \frac{e_0^2}{R} \right) |\psi(r_{2A})|^2 |\psi(r_{1B})|^2 \\
&= 2E_0 + \int d^3 r_1 \int d^3 r_2 \left(-\frac{e_0^2}{r_{2A}} - \frac{e_0^2}{r_{1B}} + \frac{e_0^2}{r_{12}} + \frac{e_0^2}{R} \right) |\psi(r_{2A})|^2 |\psi(r_{1B})|^2 \\
\langle \psi_1 | \hat{H} | \psi_2 \rangle &= \left\langle \psi_1 \middle| \left(-\frac{\hbar^2}{2m}\Delta_1 - \frac{e_0^2}{r_{1A}} \right) \middle| \psi_2 \right\rangle + \left\langle \psi_1 \middle| \left(-\frac{\hbar^2}{2m}\Delta_2 - \frac{e_0^2}{r_{2B}} \right) \middle| \psi_2 \right\rangle \\
&\quad + \int d^3 r_1 \int d^3 r_2 \left(-\frac{e_0^2}{r_{2A}} - \frac{e_0^2}{r_{1B}} + \frac{e_0^2}{r_{12}} + \frac{e_0^2}{R} \right) \psi^*(r_{1A})\psi(r_{1B})\psi(r_{2A})\psi^*(r_{2B}) \\
&= 2E_0 S^2 + \int d^3 r_1 \int d^3 r_2 \left(-\frac{e_0^2}{r_{2A}} - \frac{e_0^2}{r_{1B}} + \frac{e_0^2}{r_{12}} + \frac{e_0^2}{R} \right) \psi^*(r_{1A})\psi(r_{1B})\psi(r_{2A})\psi^*(r_{2B}) \\
\langle \psi_2 | \hat{H} | \psi_1 \rangle &= \left\langle \psi_2 \middle| \left(-\frac{\hbar^2}{2m}\Delta_1 - \frac{e_0^2}{r_{1A}} \right) \middle| \psi_1 \right\rangle + \left\langle \psi_2 \middle| \left(-\frac{\hbar^2}{2m}\Delta_2 - \frac{e_0^2}{r_{2B}} \right) \middle| \psi_1 \right\rangle \\
&\quad + \int d^3 r_1 \int d^3 r_2 \left(-\frac{e_0^2}{r_{2A}} - \frac{e_0^2}{r_{1B}} + \frac{e_0^2}{r_{12}} + \frac{e_0^2}{R} \right) \psi(r_{1A})\psi^*(r_{1B})\psi^*(r_{2A})\psi(r_{2B}) \\
&= 2E_0 S^2 + \int d^3 r_1 \int d^3 r_2 \left(-\frac{e_0^2}{r_{2A}} - \frac{e_0^2}{r_{1B}} + \frac{e_0^2}{r_{12}} + \frac{e_0^2}{R} \right) \psi(r_{1A})\psi^*(r_{1B})\psi^*(r_{2A})\psi(r_{2B}).
\end{aligned} \tag{21.50}$$

Then, renaming the integration variables by exchanging $\vec{r}_1$ and $\vec{r}_2$, we find that

$$\begin{aligned}\langle\psi_1|\hat{H}|\psi_1\rangle &= \langle\psi_2|\hat{H}|\psi_2\rangle\\ \langle\psi_1|\hat{H}|\psi_2\rangle &= \langle\psi_2|\hat{H}|\psi_1\rangle.\end{aligned} \tag{21.51}$$

Finally then, the energies of the symmetric (spin singlet) and antisymmetric (spin triplet) H_2 states are

$$\begin{aligned}E_\pm(R) &= \left\langle\frac{\psi_1\pm\psi_2}{\sqrt{2(1\pm|S|^2)}}\right|\hat{H}\left|\frac{\psi_1\pm\psi_2}{\sqrt{2(1\pm|S|^2)}}\right\rangle\\ &= \frac{1}{2(1+|S|^2)}\left[\langle\psi_1|\hat{H}|\psi_1\rangle+\langle\psi_2|\hat{H}|\psi_2\rangle\pm\langle\psi_1|\hat{H}|\psi_2\rangle\pm\langle\psi_2|\hat{H}|\psi_1\rangle\right]\\ &= \frac{1}{1+|S|^2}\left[\langle\psi_1|\hat{H}|\psi_1\rangle\pm\langle\psi_2|\hat{H}|\psi_1\rangle\right]\\ &= 2E_0+\frac{C\pm A}{1\pm|S|^2},\end{aligned} \tag{21.52}$$

where we have defined, as in the helium case, the Coulomb and exchange integrals,

$$\begin{aligned}C &= \int d^3r_1\int d^3r_2\left(-\frac{e_0^2}{r_{2A}}-\frac{e_0^2}{r_{1B}}+\frac{e_0^2}{r_{12}}+\frac{e_0^2}{R}\right)|\psi(r_{1A})|^2|\psi(r_{2B})|^2\\ A &= \int d^3r_1\int d^3r_2\left(-\frac{e_0^2}{r_{2A}}-\frac{e_0^2}{r_{1B}}+\frac{e_0^2}{r_{12}}+\frac{e_0^2}{R}\right)\psi^*(r_{1A})\psi(r_{1B})\psi(r_{2A})\psi^*(r_{2B}).\end{aligned} \tag{21.53}$$

We can calculate numerically and plot the two energy functions $E_-(R)$ and $E_+(R)$, and we find that $E_-(R)-2E_0$ is positive and uniformly decreases to zero at infinity, $E_+(R)-2E_0$ decreases to a negative minimum, reached at $R=R_0$, stays negative and at infinity tends to zero from below.

We also note that the van der Waals interaction between two hydrogen atoms at a large distance, $\sim e_0^4/r^6$, is obtained when calculating $E_+(R)$ at large R, but this is in second-order perturbation theory (which will be introduced later on in the book).

Important Concepts to Remember

- For helium (two electrons and a nucleus), we can have an ortho- state, for triplet spin and antisymmetric spatial wave function, and a para- state, for singlet spin and symmetric spatial wave function.
- As an approximation, we calculate the states and wave functions of the electrons as if there were no interaction between the electrons, and then evaluate the interaction Hamiltonian for these states.
- We find that the extra energy depends on the probability of finding the electrons near one or the other atom, so has a direct part, and an exchange part.
- The direct energy becomes

$$A=\int d^3r_1\int d^3r_2\frac{\rho_{q_1}(\vec{r}_1)\rho_{q_2}(\vec{r}_2)}{4\pi\epsilon_0 r_{12}},$$

and the exchange part of the energy is

$$C = \int d^3r_1 \int d^3r_2 \frac{e_0^2}{r_{12}} f(\vec{r}_1) f(\vec{r}_2),$$

where $f(\vec{r}) = \psi_{q_1}(\vec{r})\psi^*_{q_2}(\vec{r})$.

- Using the above approximation, the ground state of helium is found from $\Delta E = 5/8\, Ze_0^2/a_0$, which gives a good approximation.
- A better approximation is found using a variational method, which in this (simplified) case means writing a modified wave function, in which we make the change $Z \to Z_{\text{eff}}$, while keeping Z in the Hamiltonian, and then minimizing over Z_{eff} at fixed Z. We find $Z_{\text{eff}} = Z - 5/16$, and $E_0 = -Z_{\text{eff}}e_0^2/a_0$.
- For the H_2 molecule (two protons and two electrons), if $R \gg a_0$, where R is the distance between the protons, we can prove that in the minimum energy configuration each electron is around a proton.
- One finds

$$E_\pm(R) = 2E_0 + \frac{C \pm A}{1 \pm S},$$

where C is the direct contribution and A is the exchange contribution, for r_{1A} and r_{2B}, while $S = \int d^3r\ \psi^*(r_A)\psi(r_B)$.

Further Reading

See [1] for more details.

Exercises

(1) Consider a helium atom with one electron in the $n = 1, l = 0$ state, and another in the $n = 2, l = 0$ state. Calculate the energy of the ortho-helium state (the lowest energy state for these quantum numbers), using the two approximations in the text (Approximation 1 and Approximation 2).

(2) Consider a helium atom with two electrons in the $n = 2, l = 0$ state. Calculate their ground state, using the two approximations in the text.

(3) Apply the variational method (Approximation 3) to the case in exercise 1.

(4) Apply the variational method (Approximation 3) to the case in exercise 2.

(5) Argue that, for the H_2 molecule, $E_+(R) - 2E_0$ tends to infinity from below, and $E_-(R)$ tends to infinity from above.

(6) Consider a *potential* H_3 molecule, in an equilateral triangle of side R. Give an argument why this molecule would be unstable.

(7) If we have four H atoms, write down the Hamiltonian, and describe two *potentially* stable configurations.

22 Quantum Mechanics Interacting with Classical Electromagnetism

In this chapter we will learn how to deal with quantum mechanics in the presence of a classical electromagnetic field. We started this analysis in Chapter 17, but here we treat everything methodically.

22.1 Classical Electromagnetism plus Particle

To start with, we consider classical mechanics for a particle interacting with a classical electromagnetic field. We already mentioned in Chapter 17 that in quantum mechanics the electromagnetic interaction is obtained by a "minimal coupling" that amounts to a replacement of derivatives with derivatives plus fields,

$$\begin{aligned} \partial_i &\to \partial_i - i\frac{q}{\hbar}A_i \\ \partial_0 &\to \partial_0 - i\frac{q}{\hbar}A_0. \end{aligned} \tag{22.1}$$

In the Lagrangian of classical mechanics, correspondingly, the replacements are

$$\begin{aligned} \vec{p} &= m\vec{v} \to m\vec{v} + q\vec{A} \\ p^0 &= mc^2 \to mc^2 - qA_0, \end{aligned} \tag{22.2}$$

to be substituted into

$$L = -mc^2 + m\frac{\vec{v}^2}{2}, \tag{22.3}$$

thus leading to (the electric potential is $-A_0$)

$$L = -mc^2 + qA_0 + \frac{\left(m\vec{v} + q\vec{A}\right)^2}{2m}. \tag{22.4}$$

Then the Lagrange equations,

$$\frac{d}{dt}\frac{\partial L}{\partial v_r} = \frac{\partial L}{\partial x_r}, \tag{22.5}$$

become (neglecting the term quadratic in the fields, $q^2\vec{A}^2/2m$)

$$\frac{d}{dt}(mv_r + qA_r) = q\partial_r A_0 + qv_i\partial_r A_i, \tag{22.6}$$

and, since we have (using the definition of $\vec{E}, \vec{B}$ in terms of the vector potential $\vec{A}$ and the scalar potential A_0)

$$\begin{aligned}\frac{d}{dt}A_r &= \frac{\partial A_r}{\partial t} + v_s \partial_s A_r\\ \partial_r A_s - \partial_s A_r &= \epsilon_{rsk} B_k\\ \partial_r A_0 - \partial_0 A_r &= E_r,\end{aligned} \tag{22.7}$$

we obtain

$$\begin{aligned}m\dot{v}_r &= q[\partial_r A_0 - \partial_0 A_r + (\partial_r A_s - \partial_s A_r)v_s]\\ &= q(E_r + \epsilon_{rsk} v_s B_k) \Rightarrow\\ m\dot{\vec{v}} &= q(\vec{E} + \vec{v} \times \vec{B}).\end{aligned} \tag{22.8}$$

This is the Lorentz force law, so the Lagrangian we wrote down was indeed correct.

From this Lagrangian, we can find the *canonically conjugate momentum*

$$p_r = \frac{\partial L}{\partial v_r} = mv_r + qA_r. \tag{22.9}$$

Then the Hamiltonian is (again neglecting the term $q^2\vec{A}^2/2m$)

$$H = \sum_r p_r v_r - L = \frac{mv^2}{2} + mc^2 - qA_0 = \frac{1}{2m}\left(\vec{p} - q\vec{A}\right)^2 - qA_0 + mc^2. \tag{22.10}$$

22.2 Quantum Particle plus Classical Electromagnetism

Next, we consider how to use quantum mechanics for the particle, while leaving the electromagnetism classical. This means that we replace $\vec{p}$ with the quantum operator $\hat{\vec{p}}$, while the vector and scalar potentials remain scalar functions times the identity operator, $\vec{A}\mathbb{1}$ and $A_0\mathbb{1}$. The Hamiltonian operator is then

$$\hat{H} = \frac{1}{2m}\left(\hat{\vec{p}} - q\vec{A}\mathbb{1}\right)^2 - qA_0\mathbb{1}. \tag{22.11}$$

Further, it is still the canonically conjugate momentum $\vec{p}$ that is replaced by $\frac{\hbar}{i}\vec{\nabla}$, since we still have $[x_i, p_j] = i\hbar\delta_{ij}$, where p_j are the canonically conjugate momenta. That means that we still have the replacement

$$\hat{\vec{p}} = \frac{\hbar}{i}\vec{\nabla} \to \hat{\vec{p}} - q\vec{A} = \frac{\hbar}{i}\vec{\nabla} - q\vec{A}, \tag{22.12}$$

which implies the "minimal coupling" expressed in (22.1).

Then the Schrödinger equation $i\hbar\partial_t|\psi\rangle = \hat{H}|\psi\rangle$ becomes, acting on the wave function,

$$i\hbar\partial_t\psi(t,\vec{r}) = \left[\frac{1}{2m}\left(\hat{\vec{p}} - q\vec{A}\mathbb{1}\right)^2 - qA_0\mathbb{1}\right]\psi(t,\vec{r}). \tag{22.13}$$

This Hamiltonian contains an interaction part, for the interaction between the particle (via its operator $\hat{\vec{p}}$) and the classical electromagnetic field,

$$\hat{H}_{\text{int}} = -\frac{q}{m}\hat{\vec{p}} \cdot \vec{A}. \tag{22.14}$$

As we saw in Chapter 17, *in the gauge* $\vec{\nabla} \cdot \vec{A} = 0$, *and for a constant magnetic field* $\vec{B}$, given that $\vec{B} = \vec{\nabla} \times \vec{A}$, we can choose (considering also $\partial_t \vec{A} = 0$, a stationary field)

$$\vec{A} = \frac{1}{2}\vec{B} \times \vec{r}, \tag{22.15}$$

leading to an interaction Hamiltonian

$$\hat{H}_{\text{int}} = -\frac{q}{2m}\vec{B} \cdot \hat{\vec{L}} = -\vec{B} \cdot \hat{\vec{\mu}}, \tag{22.16}$$

where the magnetic moment can be written as an integral involving the current density $\vec{j}$, or the current differential $d\vec{I} = \vec{j}d^3r$, as

$$\vec{\mu} = \int d^3r \frac{1}{2}\vec{r} \times \vec{j} = \int \frac{\vec{r} \times d\vec{I}}{2}. \tag{22.17}$$

Note that in this relation, both $\vec{\mu}$ and $\vec{j}$ are understood as expectation values, quantities that can be measured in experiments, and not as quantum operators. Indeed, $\vec{j}$ is constructed out of a wave function, and so is $\vec{\mu}$.

We next show that a change in the wave function by a phase amounts to a change of gauge for the gauge field (A_0, A_i). The redefinition of the wave function by a phase,

$$\psi = \psi' e^{-iq\Lambda/\hbar}, \tag{22.18}$$

implying

$$\begin{aligned} \frac{\hbar}{i}\partial_t \psi &= \left(\frac{\hbar}{i}\partial_t \psi' - q(\partial_t \Lambda)\psi'\right) e^{-iq\Lambda/\hbar} \\ \hat{p}_i = \frac{\hbar}{i}\partial_i \psi &= \left(\frac{\hbar}{i}\partial_i \psi' - q(\partial_i \Lambda)\psi'\right) e^{-iq\Lambda/\hbar}, \end{aligned} \tag{22.19}$$

means that the Schrödinger equation for ψ, written as

$$\left\{\frac{\hbar}{i}\partial_t \psi + \frac{1}{2m}\left[\frac{\hbar}{i}\partial_i - qA_i\right]^2 \psi - qA_0\right\}\psi = 0, \tag{22.20}$$

can be rewritten in terms of ψ' as

$$\left\{\frac{\hbar}{i}\partial_t \psi' + \frac{1}{2m}\left[\frac{\hbar}{i}\partial_i - q(A_i + \partial_i \Lambda)\right]^2 \psi - q(A_0 + \partial_t \Lambda)\right\}\psi = 0. \tag{22.21}$$

This is the Schrödinger equation for the gauge transformed fields

$$A_i' = A_i + \partial_i \Lambda, \quad A_0' = A_0 + \partial_t \Lambda, \tag{22.22}$$

proving the statement above (22.18).

However, given the fact that

$$[p_r, A_s] = \frac{\hbar}{i}[\partial_r, A_s] \neq 0, \tag{22.23}$$

we have to be careful about the order of operators. Expanding the Schrödinger equation, we obtain

$$\frac{\hbar}{i}\partial_t\psi - \frac{\hbar^2}{2m}\Delta\psi - \frac{q}{m}\frac{\hbar}{i}\vec{A}\cdot\vec{\nabla}\psi - \frac{q}{2m}\frac{\hbar}{i}(\vec{\nabla}\cdot\vec{A})\psi + \frac{q^2}{2m}\vec{A}^2\psi - qA_0\psi = 0, \tag{22.24}$$

with complex conjugate

$$-\frac{\hbar}{i}\partial_t\psi^* - \frac{\hbar^2}{2m}\Delta\psi^* + \frac{q}{m}\frac{\hbar}{i}\vec{A}\cdot\vec{\nabla}\psi^* + \frac{q}{2m}\frac{\hbar}{i}(\vec{\nabla}\cdot\vec{A})\psi^* + \frac{q^2}{2m}\vec{A}^2\psi^* - qA_0\psi^* = 0. \tag{22.25}$$

Multiplying the Schrödinger equation (22.24) by ψ^* and its complex conjugate by ψ, and subtracting the two, we obtain

$$\frac{\hbar}{i}\partial_t(|\psi|^2) - \frac{\hbar}{2m}(\psi^*\Delta\psi - \psi\Delta\psi^*) - \frac{q}{m}\frac{\hbar}{i}\vec{A}\cdot(\psi^*\vec{\nabla}\psi + \psi\vec{\nabla}\psi^*) - \frac{q}{m}\frac{\hbar}{i}(\vec{\nabla}\cdot\vec{A})|\psi|^2 = 0, \tag{22.26}$$

which can be rewritten as

$$\partial_t(|\psi|^2) + \vec{\nabla}\cdot\left[\frac{\hbar}{2mi}(\psi^*\vec{\nabla}\psi - \psi\vec{\nabla}\psi^*) - \frac{q}{m}\vec{A}|\psi|^2\right] = 0. \tag{22.27}$$

This has the form of a continuity equation for the probability density, adding the interaction with the electromagnetic field to the previously derived equation from Chapter 7. Indeed, defining the probability density

$$\rho(\vec{r},t) = |\psi(\vec{r},t)|^2, \tag{22.28}$$

we see that we can define the probability current density in the presence of the vector potential $\vec{A}$ as

$$\vec{j} = \frac{\hbar}{2mi}(\psi^*\vec{\nabla}\psi - \psi\vec{\nabla}\psi^*) - \frac{q}{m}\vec{A}|\psi|^2 \equiv \vec{j}_0 + \vec{j}_1, \tag{22.29}$$

so that we have the standard continuity equation

$$\partial_t\rho + \vec{\nabla}\cdot\vec{j} = 0. \tag{22.30}$$

We also define the current $\vec{j}_0$ as the current in the absence of $\vec{A}$,

$$\vec{j}_0 = \frac{\hbar}{2mi}(\psi^*\vec{\nabla}\psi - \psi\vec{\nabla}\psi^*), \tag{22.31}$$

as well as the new contribution $\vec{j}_1$, depending on $\vec{A}$,

$$\vec{j}_1 = -\frac{q}{m}\vec{A}|\psi|^2 = -\frac{q\rho}{m}\vec{A}. \tag{22.32}$$

Note that the probability density ρ and probability current density $\vec{j}$ are related to the charge density and current density by multiplying with the individual charge of the particle,

$$\rho_{\text{charge}} = q\rho, \quad \vec{j}_{\text{charge}} = q\vec{j}. \tag{22.33}$$

We can then split the (expectation value for the) magnetic moment associated with each current density individually, first the current density in the absence of $\vec{A}$, denoted $\vec{\mu}_0$,

$$\begin{aligned}\vec{\mu}_0 &= \int d^3r\frac{1}{2}\vec{r}\times\vec{j}_{0,\text{charge}}\\ &= \frac{q}{4m}\int d^3r\left[\psi^*\left(\vec{r}\times\frac{\hbar\vec{\nabla}}{i}\psi\right) + \left(\vec{r}\times\frac{\hbar}{i}\vec{\nabla}\psi\right)^*\psi\right]\\ &= \frac{q}{4m}\left[\langle\psi|(\hat{\vec{L}}|\psi\rangle) + (\langle\psi|\hat{\vec{L}})|\psi\rangle\right] = \frac{q}{2m}\langle\psi|\hat{\vec{L}}|\psi\rangle,\end{aligned} \tag{22.34}$$

where in the second equality we have used the fact that $\hat{\vec{L}} = \hat{\vec{r}} \times \hat{\vec{p}} = \hat{\vec{r}} \times \frac{\hbar}{i}\vec{\nabla}$, and in the last equality we have used the fact that $\hat{\vec{L}}$ is a self-adjoint operator, like all observables. Finally, then, since we have a relation valid for any state $|\psi\rangle$, we obtain the same relation between an abstract operator (an operator that is not in a particular representation) describing the magnetic moment $\hat{\vec{\mu}}$ and the angular momentum operator $\hat{\vec{L}}$ as at the classical level,

$$\hat{\vec{\mu}} = \frac{q}{2m}\hat{\vec{L}}. \tag{22.35}$$

However, the same is not true for the new term in $\vec{\mu}$, called $\vec{\mu}_1$, that comes from the new term in the current density, $\vec{j}_1$. For it, we obtain

$$\begin{aligned}\vec{\mu}_1 &= \int d^3r \frac{1}{2}\vec{r} \times \vec{j}_{1,\text{charge}} = -\frac{q^2}{2m}\int d^3r \psi^*(\vec{r} \times \vec{A})\psi \\ &= -\frac{q^2}{2m}\langle\psi|(\hat{\vec{r}} \times \vec{A})|\psi\rangle.\end{aligned} \tag{22.36}$$

Again, though the relation is valid for any state $|\psi\rangle$, so we can write a relation valid for abstract operators,

$$\hat{\vec{\mu}}_1 = -\frac{q^2}{2m}(\hat{\vec{r}} \times \vec{A}). \tag{22.37}$$

22.3 Application to Superconductors

As an application of the previous formalism, we will consider the quantum theory of superconductors. In a superconductor there are superconducting electrons (which means electrons that don't experience collisions with the nuclei that can reduce their speed) described by a wave function $\psi(\vec{r})$ and with a local density equal to the probability density $n_s = |\psi(\vec{r})|^2$.

The only difference with respect to the previous analysis is that instead of the mass m of the electrons, we have an "effective mass" m_*, an effective description of the mutual interaction of the electrons. Indeed, in a "Fermi liquid", describing electrons in most of these materials, the effect of the interactions between the electrons is just to replace the mass of the free electrons (described as a "Fermi gas") m by the renormalized mass m_*, which can be 10%–50% larger. Actually, in heavy fermions it can even be the case that $m_* \gg m$. Therefore, the electric current density of the superconducting electrons, $\vec{j}_s = q\vec{j}$, is then

$$\vec{j}_s(\vec{r}) = -i\frac{q\hbar}{2m_*}(\psi^*\vec{\nabla}\psi - \psi\vec{\nabla}\psi^*) - \frac{q^2}{m_*}\psi^*\psi\vec{A}. \tag{22.38}$$

Deep inside the superconductor, the wave function of the superconducting electrons approaches a constant absolute value (so that the probability density is constant) that minimizes the thermodynamic free energy. Calling it ψ_0 ($\in \mathbb{R}$), we have that

$$\psi \simeq \psi_0 e^{i\alpha}. \tag{22.39}$$

Then the electric current density of the superconducting electrons is

$$\vec{j}_s \simeq -\frac{q^2}{m_*}\psi_0^2\vec{A}. \tag{22.40}$$

Further, the superconductor is described in the London–London model (named after H. London and F. London) as a sum of two components, a normal electron component $\vec{j}_n$, that experiences collisions with nuclei, so has a finite conductivity σ, and current density $\vec{j}_n = \sigma\vec{E}$, and a superconducting component $\vec{j}_s$ as above. The normal electrons have density n_n and velocity $\vec{v}$ and the superconducting ones have density $n_s = |\psi|^2 \simeq \psi_0^2$ and velocity $\vec{v}_s$, so we have a total density $n = n_n + n_s$, and

$$\begin{aligned}\vec{j}_n &= -en_n\vec{v}_n\\ \vec{j}_s &= -en_s\vec{v}_s\\ \vec{j} &= \vec{j}_n + \vec{j}_s.\end{aligned} \tag{22.41}$$

The superconducting electrons have mass m_*, and on them only the electric force $-e\vec{E}$ acts, without any collisions to slow them down, so it increases their velocity by

$$-e\vec{E} = m_*\frac{d\vec{v}_s}{dt}. \tag{22.42}$$

But that in turn means that the superconducting electric current density changes as

$$\frac{d\vec{j}_s}{dt} = \frac{n_s e^2}{m_*}\vec{E} = -\frac{n_s e^2}{m_*}\frac{d\vec{A}}{dt}, \tag{22.43}$$

which is the first London equation. It really should be *postulated* not derived, as are the usual Maxwell's equations, since it is based on observations but doesn't quite follow from what we already know about (normal) electrons.

The second London equation is also postulated, though it is based on the observation we made that, in the superconductor, $\vec{j}_s \simeq -(q^2/m^*)\psi_0^2\vec{A}$, where $n_s = |\psi|^2 \simeq \psi_0^2$, so by taking the cross product of $\vec{\nabla}$ with it we get

$$\vec{\nabla}\times\vec{j}_s = -\frac{n_s e^2}{m_*}\vec{B} = -\frac{n_s e^2}{m_*}\vec{\nabla}\times\vec{A}. \tag{22.44}$$

However, once it is postulated, we have to further postulate that $\vec{j}_s \propto \vec{A}$ *in a simply connected superconductor* (we will see in the next chapter why this is needed), in order to obtain

$$\vec{j}_s = -\frac{n_s e^2}{m_*}\vec{A} \equiv -\frac{1}{\mu_0\lambda_L^2}\vec{A}, \tag{22.45}$$

where the last equality defines λ_L. On the other hand, from the Maxwell equation $\vec{\nabla}\times\vec{B} = \mu_0\vec{j}$ and using $\vec{\nabla}^2\vec{C} = -\vec{\nabla}\times(\vec{\nabla}\times\vec{C}) - \vec{\nabla}(\vec{\nabla}\cdot C)$ (true for any vector $\vec{C}$) and that $\vec{\nabla}\cdot\vec{B} = 0$, we obtain

$$\vec{\nabla}^2\vec{B} = -\vec{\nabla}\times(\vec{\nabla}\times\vec{B}) = -\mu_0\vec{\nabla}\times\vec{j}_s = \frac{\mu_0 n_s e^2}{m_*}\vec{B} = \frac{1}{\lambda_L^2}\vec{B}. \tag{22.46}$$

The solution of this equation is an exponentially decaying magnetic field inside the superconductor, when going inside from the boundary,

$$\vec{B}(r) = \vec{B}(0)e^{-r/\lambda_L}. \tag{22.47}$$

This is the mathematical description of the observed phenomenon that superconductors expel magnetic field, known as the *Meissner effect*: here we see that the magnetic field only penetrates a distance of λ_L inside the material.

22.4 Interaction with a Plane Wave

As a first step towards quantizing the electromagnetic field, we consider a simple space dependence for $\vec{A}$, and promote $\vec{x}$ to be a quantum operator. This is a simple generalization of the previously considered case of $\vec{B}$ constant, with $\vec{A} = (\vec{B} \times \vec{r})/2$, where we also promoted $\vec{r}$ to be an operator. It is also an obvious thing to do since, in the coordinate representation for wave functions, x acts trivially (since it is replaced by its eigenvalue).

Consider a vector potential that is a plane wave, with sinusoidal form,

$$\vec{A} = 2A_{(0)}\vec{\epsilon}\cos\left(\frac{\omega}{c}\vec{n}\cdot\vec{x} - \omega t\right), \tag{22.48}$$

where $\vec{\epsilon}$ is a polarization vector, transverse to the propagation direction $\vec{n}$, so that $\vec{\epsilon}\cdot\vec{n} = 0$.

We can rewrite (22.48) as the sum of two exponentials,

$$\vec{A} = A_{(0)}\vec{\epsilon}\left[\exp\left(i\frac{\omega}{c}\vec{n}\cdot\vec{x} - i\omega t\right) + \exp\left(-i\frac{\omega}{c}\vec{n}\cdot\vec{x} + i\omega t\right)\right], \tag{22.49}$$

where, as we will see in the second part of the book, the first term is associated with absorption of light and the second with stimulated emission. This is replaced inside the interaction Hamiltonian,

$$\hat{H}_{\rm int} = -\frac{q}{m}\hat{\vec{P}}\cdot\vec{A}, \tag{22.50}$$

in order to calculate its matrix element $\langle n|\hat{H}_{\rm int}|m\rangle$ in between eigenstates $|n\rangle$ of the free Hamiltonian, for which $\hat{H}_0|n\rangle = E_n|n\rangle$. Defining the frequency of transition between two states,

$$\omega_{nm} = \frac{E_n - E_m}{\hbar}, \tag{22.51}$$

and using the fact that for the x component of the kinetic momentum we have

$$\hat{P}_x = \frac{m}{i\hbar}[\hat{X}, \hat{H}_0], \tag{22.52}$$

where X is the kinetic position operature, we find that

$$\langle n|\hat{P}_x|m\rangle = \frac{m}{i\hbar}\langle n|(\hat{X}\hat{H}_0 - \hat{H}_0\hat{X})|m\rangle = im\omega_{nm}\langle n|\hat{X}|m\rangle. \tag{22.53}$$

Moreover, as a first approximation, we can replace the exponentials in the expansion of $\vec{A}$ (the plane wave) with their leading term, 1. That amounts, of course, to having just a constant $\vec{A}$; however it must be thought of as the leading approximation to a plane wave, and gives

$$\langle n|\hat{H}_{\rm int}|m\rangle = -qA_{(0)}i\omega_{nm}\langle n|\hat{X}|m\rangle. \tag{22.54}$$

22.5 Spin–Magnetic-Field and Spin–Orbit Interaction

Finally, we consider an interaction term in the Hamiltonian that really comes from the relativistic theory (quantum field theory), but, like the electron spin itself, has a simple nonrelativistic description, as an interaction between spin and orbital angular momentum.

First, we consider the interaction of the electron spin with the magnetic field. We remember that, for an electron, the spin generates a magnetic moment with $g = 2$. Considering also that $q = -e$, we have

$$\vec{\mu}_S = -\frac{e}{m_e}\vec{S}. \tag{22.55}$$

But the spin (intrinsic angular momentum) is given by $\vec{S} = \hbar\vec{\sigma}/2$, where σ^i are the Pauli matrices, acting on the two-dimensional spin vector space. As we showed earlier, the Pauli matrices satisfy

$$\sigma_i\sigma_j = \delta_{ij} + i\epsilon_{ijk}\sigma_k. \tag{22.56}$$

Multiplying this with $\hat{P}_i\hat{P}_j$, where $\hat{P}_i$ is the kinetic momentum, we obtain

$$(\vec{\sigma}\cdot\hat{\vec{P}})(\vec{\sigma}\cdot\hat{\vec{P}}) = \hat{\vec{P}}\cdot\hat{\vec{P}} + i\vec{\sigma}\cdot(\hat{\vec{P}}\times\hat{\vec{P}}). \tag{22.57}$$

Now using the fact that $\hat{P}_i$ is related to the conjugate momentum $\hat{p}_i$ by

$$\hat{\vec{P}} = \hat{\vec{p}} - q\vec{A} = \hat{\vec{p}} + e\vec{A}, \tag{22.58}$$

we obtain (since $\hat{\vec{p}} = \frac{\hbar}{i}\vec{\nabla}$)

$$(\vec{\sigma}\cdot\hat{\vec{P}})^2 = \left(\hat{\vec{p}} + e\vec{A}\right)^2 + \hbar e\vec{\sigma}\cdot\vec{B}. \tag{22.59}$$

On the other hand, the Hamiltonian *in the absence of* $\vec{A}$ can be rewritten using (22.56), multiplied by $\hat{p}_i\hat{p}_j$, as

$$\hat{H} = \frac{\hat{\vec{p}}\cdot\hat{\vec{p}}}{2m} - eA_0 = \frac{1}{2m}\left(\hat{\vec{p}}\cdot\hat{\vec{p}} + i\vec{\sigma}\cdot(\vec{p}\times\vec{p})\right) - eA_0 = \frac{1}{2m}(\vec{\sigma}\cdot\hat{\vec{p}})^2 - eA_0, \tag{22.60}$$

and in the last form we can again replace $\hat{\vec{p}}$ by $\hat{\vec{P}}$, obtaining that the Hamiltonian is that previously derived for interaction with a classical electromagnetic field, plus a term giving the interaction of the spin with the magnetic field,

$$\hat{H} = \frac{1}{2m}(\vec{\sigma}\cdot\hat{\vec{P}})^2 - eA_0 = \left(\hat{\vec{p}} + e\vec{A}\right)^2 + \hbar e\vec{\sigma}\cdot\vec{B} - eA_0. \tag{22.61}$$

To obtain the interaction between spin and orbital angular momentum (the spin–orbit interaction), we have to consider the relativistic Dirac equation and expand it in $1/c^2$, obtaining the interaction term

$$H_{\text{int,LS}} = -\frac{e\hbar}{4m_0^2c^2}\vec{\sigma}\cdot(\vec{\nabla}A_0\times\vec{p}). \tag{22.62}$$

We will not derive this here, though we will do so towards the end of the book, when considering the Dirac equation. For spherically symmetric systems, we find

$$H_{\text{int,LS}} = -\frac{e\hbar}{4m_0^2c^2}\frac{1}{r}\frac{dA_0}{dr}\vec{\sigma}\cdot(\vec{r}\times\hat{\vec{p}}) = \frac{e}{2m_0^2c^2}\frac{1}{r}\frac{dA_0}{dr}\vec{S}\cdot\vec{L}. \tag{22.63}$$

Important Concepts to Remember

- Minimal coupling of a particle to an electromagnetic field amounts to the replacement of the momentum $\vec{p}$ (usually both the kinetic momentum, and that conjugate to x) by $\vec{p} - q\vec{A} \equiv \vec{P}$, where $\vec{p}$ is now the canonically conjugate momentum, and $\vec{P}$ is the kinetic momentum.
- At the quantum level for the particle (while $\vec{A}$ may be kept classical), we replace the canonically conjugate momentum $\vec{p}$ with $\frac{\hbar}{i}\vec{\nabla}$, so $\vec{P} = \vec{p} - q\vec{A}$ with $\frac{\hbar}{i}\vec{\nabla} - q\vec{A}\mathbb{1}$.
- The resulting interacting Hamiltonian, in the gauge $\vec{\nabla} \cdot \vec{A} = 0$, for constant $\vec{B}$ and stationary $\vec{A}$ is $-\vec{\mu} \cdot \vec{B}$, where the magnetic moment is $\vec{\mu} = \int \vec{r} \times d\vec{I}/2$.
- A gauge transformation of $A_\mu = (A_0, A_i)$ by Λ corresponds to multiplying the wave function by the phase $e^{iq\Lambda/\hbar}$.
- The gauge transformation of ψ does not change the probability density ρ, but does change the current of probability $\vec{j}$ by the addition of $\vec{j}_1 = -\frac{q\rho}{m}\vec{A}$, and the magnetic moment operator by the addition of $\hat{\vec{\mu}}_1 = -\frac{q^2}{2m}(\hat{\vec{r}} \times \vec{A})$, while $\vec{\mu}_0 = \frac{q}{2m}\langle\psi|\hat{\vec{L}}|\psi\rangle$.
- In a superconductor, we have electrons of effective mass m_* and of two types (the London–London model): normal, with n_n and $\vec{j}_n = -en_n\vec{v}_n = \sigma\vec{E}$, and superconducting, with n_s and $\vec{j}_s = -en_s\vec{v}_s$, satisfying

$$\frac{d\vec{j}_s}{dt} = -\left(\frac{n_s e^2}{m_*}\right)\frac{d\vec{A}}{dt} \quad \text{and} \quad \vec{\nabla} \times \vec{j}_s = -\left(\frac{n_s e^2}{m_*}\right)\vec{\nabla} \times \vec{A}.$$

- A superconductor expels magnetic field, meaning it penetrates the superconductor only a distance $\lambda_L = \sqrt{\mu_0 n_s e^2/m_*}$, $B(x) = B_0 e^{-x/\lambda_L}$. This is the Meissner effect.
- For a sinusoidal electromagnetic wave, with the position variable replaced by the $\hat{X}$ operator, we find that the matrix elements of the interaction Hamiltonian are related to the transition frequency $\omega_{mn} = (E_m - E_n)/\hbar$ and the matrix elements of $\hat{X}$ as $H_{\text{int},mn} = -qA_{(0)}i\omega_{mn}X_{mn}$.
- Minimal coupling in the presence of spin leads to a spin–magnetic field interaction, and from the Dirac equation we also find a spin–orbit interaction $\frac{e}{2m_0^2c^2}\frac{1}{r}\frac{dA_0}{dr}\vec{S} \cdot \vec{L}$.

Further Reading

See [2] and [3] for more details.

Exercises

(1) The relativistic Lagrangian for particle is $L = -mc^2\sqrt{1 - \vec{v}^2/c^2}$. Does minimal coupling at the classical level based on it work? Justify.

(2) Write down formally the quantum level version for exercise 1, without bothering about what ψ means now.

(3) Consider a conductor forming a closed loop C in a magnetic field, with nonzero flux $\Phi = \int_S \vec{B} \cdot d\vec{S}$ through it. Show that there is a nonzero probability current loop $\oint_C d\vec{l} \cdot \vec{I}$.

(4) In the London–London theory, show that by writing down the energy of the magnetic field and the kinetic term for the superconducting electrons, we obtain the free energy

$$f = \frac{1}{2\mu_0}\int_0^\infty (\vec{B}^2 + \lambda_L^2\mu_0^2\vec{j}^2)2\pi r dr, \tag{22.64}$$

and from it we obtain the London equation $\lambda_L^2\vec{\nabla}^2\vec{B} = \vec{B}$.

(5) Explain why a small permanent magnet brought down along the (vertical) axis of a superconducting ring levitates (i.e., it doesn't fall under gravity).

(6) Consider a plane electromagnetic wave incident in a perpendicular direction onto a planar material containing electrons bound in it. Does the wave induce (leading-order) transitions in the electronic states?

(7) Calculate the spin–orbit interaction for a hydrogen atom in the ground state.

23 Aharonov–Bohm Effect and Berry Phase in Quantum Mechanics

In this chapter, we consider relevant *geometric phases* that appear in the wave function: first, in the coupling to electromagnetic fields, the Aharonov–Bohm phase, and then, in general, the Berry phase.

23.1 Gauge Transformation in Electromagnetism

In the previous chapter, we saw that the canonically conjugate momentum $\hat{\vec{p}}$ is represented by $\frac{\hbar}{i}\vec{\nabla}$ as usual, but when the particle couples to electromagnetic fields, the kinetic momentum is represented by

$$\hat{\vec{p}}_{\text{kin}} = \hat{\vec{p}} - q\vec{A}\mathbb{1}, \tag{23.1}$$

and is the operator appearing in the Hamiltonian,

$$\hat{H} = \frac{\hat{\vec{p}}_{\text{kin}}^{\,2}}{2m} - qA_0\mathbb{1} + \cdots = \frac{\left(\hat{\vec{p}} - q\vec{A}\mathbb{1}\right)}{2m} - qA_0\mathbb{1} + \cdots \tag{23.2}$$

But this means that the canonically conjugate momentum acts on wave functions as (a constant times) a translation operator,

$$\hat{\vec{p}}\psi = \frac{\hbar}{i}\frac{\partial}{\partial\vec{x}}\psi, \tag{23.3}$$

which in turn means that in the wave functions there is a phase factor

$$\exp\left(\frac{i}{\hbar}\int_P q\vec{A}\cdot d\vec{x}\right), \tag{23.4}$$

where P is a path in coordinate space. Indeed, in terms of the kinetic momentum, the translational phase (going along P from x to x' in order to find the wave function at x' from that defined at x) is

$$\exp\left(\frac{i}{\hbar}\int_{P:x}^{x'}\vec{p}\cdot d\vec{x}\right) = \exp\left[\frac{i}{\hbar}\int_{P:x}^{x'}\left(\vec{p}_{\text{kin}} + q\vec{A}\right)\cdot d\vec{x}\right] \equiv \exp\left(\frac{i}{\hbar}\int_{P:x}^{x'}\vec{p}_{\text{kin}}\cdot d\vec{x}\right)e^{i\delta}, \tag{23.5}$$

where the kinetic momentum is taken to be gauge invariant (since, classically, it corresponds to $m\vec{v}$), which means that we have a gauge-dependent phase

$$\delta = \frac{q}{\hbar}\int_{P:x}^{x'}\vec{A}\cdot d\vec{x}.$$

More precisely, we saw in the previous chapter that the gauge transformation acts on the solution of the Schrödinger equation as follows:

$$\begin{aligned} A_i &\to A'_i = A_i + \partial_i \Lambda \\ A_0 &\to A'_0 = A_0 + \partial_0 \Lambda \quad \Rightarrow \\ \psi &= \psi' e^{-iq\Lambda/\hbar}, \end{aligned} \tag{23.6}$$

so that

$$\psi' = e^{+iq\Lambda/\hbar}\psi, \tag{23.7}$$

which is consistent with the phase factor $e^{i\delta}$ in the wave function ψ.

Another way to obtain the same result is to require on physical grounds that, under a gauge transformation on the wave function given by $|\psi\rangle \to |\psi'\rangle = \hat{U}|\psi\rangle$, we have:

- The norm of the state is invariant,

$$\langle\psi'|\psi'\rangle = \langle\psi|\psi\rangle \quad \Rightarrow \quad \hat{U}^\dagger = \hat{U}^{-1}, \tag{23.8}$$

  so $\hat{U}$ is unitary.
- The expectation value of $\hat{X}$ is invariant,

$$\langle\psi'|\hat{X}|\psi'\rangle = \langle\psi|\hat{X}|\psi\rangle \quad \Rightarrow \quad \hat{U}^\dagger \hat{X} \hat{U} \to X. \tag{23.9}$$

- The kinetic momentum is gauge invariant,

$$\begin{aligned} \left\langle \psi' \middle| \left(\hat{\vec{p}} - q\vec{A}'\mathbb{1}\right) \middle| \psi' \right\rangle &= \left\langle \psi \middle| \left(\hat{\vec{p}} - q\hat{A}\right) \middle| \psi \right\rangle \quad \Rightarrow \\ \hat{U}^\dagger \left[\hat{\vec{p}} - q\left(\vec{A} + \vec{\nabla}\Lambda\right)\right] \hat{U} &= \hat{\vec{p}} - q\vec{A}. \end{aligned} \tag{23.10}$$

We then see that the only solution for the unitary operator $\hat{U}$ is

$$\hat{U} = e^{iq\Lambda/\hbar}, \tag{23.11}$$

which is the same as that found before.

Yet another way to obtain the same result is to consider the path integral. In the propagator, we have the path integral expression

$$U(t', t) = \int \mathcal{D}q(t) \exp\left(i\frac{S}{\hbar}\right) = \int \mathcal{D}q(t) \exp\left(\frac{i}{\hbar}\int_t^{t'} L\, dt\right). \tag{23.12}$$

But in the previous chapter, we saw that the classical Lagrangian, appearing in the path integral exponent, is

$$L = \frac{1}{2}mv^2 + q\vec{A}\cdot\vec{v} + qA_0 + \cdots \tag{23.13}$$

Thus, there is a phase factor

$$e^{i\delta} = \exp\left(\frac{iq}{\hbar}\int_t^{t'} \vec{A}\cdot\frac{d\vec{r}}{dt}dt\right) = \exp\left(\frac{iq}{\hbar}\int_P \vec{A}\cdot d\vec{r}\right), \tag{23.14}$$

the same as before.

23.2 The Aharonov–Bohm Phase δ

Now, in general a phase factor in the state $|\psi\rangle$ or wave function is irrelevant, since we can always redefine it by a phase factor, while keeping the probability, $||\,|\psi\rangle||^2 = \langle\psi|\psi\rangle$ invariant. However, we cannot redefine the phase *difference* between the phases of two different particles. This is the same as saying that the phase on a *closed path* (formed by two different paths in between the same initial and final points, as in Fig. 23.1a) cannot be redefined away.

Consider a double slit experiment, but with an electromagnetic field in between the two paths of the particles, as in Fig. 23.1b. That is, there is a unique source for particles (they have to be charged particles, not photons, in order to couple with electromagnetism), and two slits in a screen, and then we measure the interference in a unique point P behind the screen. The two generic (quantum) particle paths, going through the two slits, are called P_1 and P_2. We can measure the phase difference through the interference between the waves traveling along paths P_1 and P_2. The corresponding time dependent states are

$$\begin{aligned}|\psi_1(t)\rangle &= \exp\left(\frac{iq}{\hbar}\int_{P_1}\vec{A}\cdot d\vec{r}\right)|\psi_1\rangle\\ |\psi_2(t)\rangle &= \exp\left(\frac{iq}{\hbar}\int_{P_2}\vec{A}\cdot d\vec{r}\right)|\psi_2\rangle \equiv e^{i\delta}\left[\exp\left(\frac{iq}{\hbar}\int_{P_1}\vec{A}\cdot d\vec{r}\right)|\psi_2\rangle\right],\end{aligned} \tag{23.15}$$

where $|\psi_2\rangle \sim |\psi_1\rangle$ times perhaps an extra phase factor (independent of the electromagnetic field).

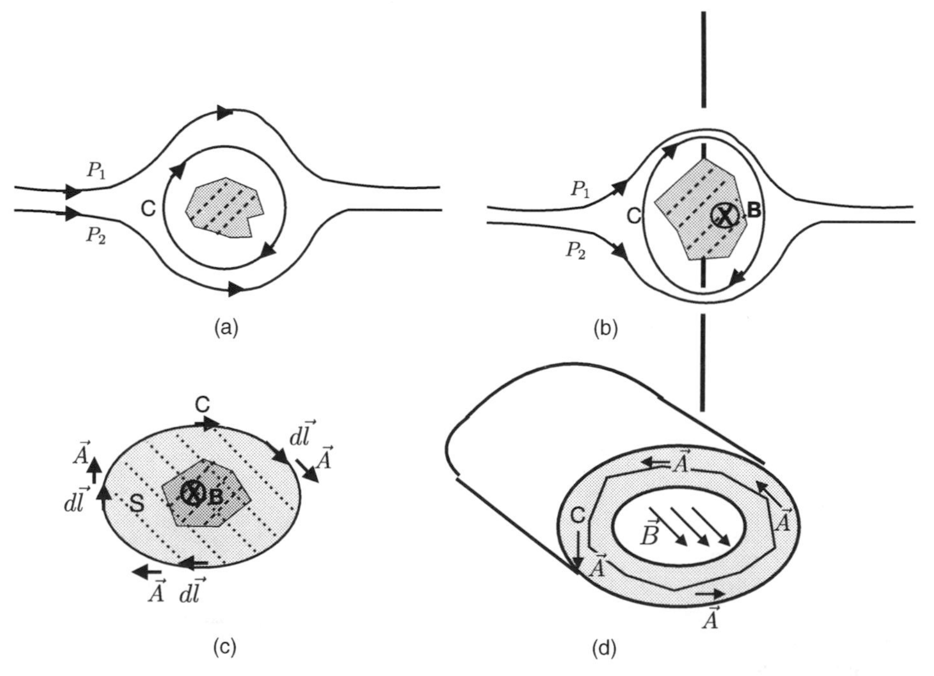

Figure 23.1 (a) For two different paths in between the same initial and final points, there is a closed path $C = P_1 - P_2$, with a magnetic field $\vec{B}$ inside it. (b) The same for a double slit experiment; the region with $\vec{B}$ (into the page) is in between the slits. (c) The surface S bounded by the closed path C, with the region with $\vec{B}$ inside it. There is a nonzero field $\vec{A}$ around C, even though $\vec{B}$ is 0. (d) A physical construction: a conducting cylinder, with a field $\vec{B}$ only in the void inside the cylinder; the path C is inside the conductor.

Then we can measure the phase difference

$$e^{i\delta} = \exp\left\{\frac{iq}{\hbar}\int_{P_1-P_2}\vec{A}\cdot d\vec{r}\right\} = \exp\left\{\frac{iq}{\hbar}\oint_C \vec{A}\cdot d\vec{r}\right\}, \tag{23.16}$$

where $C = P_1 - P_2$ is a closed loop. Then, if $C = \partial S$ is the boundary of a surface S, by Gauss's law (or Stokes' law), we find

$$e^{i\delta} = \exp\left[\frac{iq}{\hbar}\int_S \vec{B}\cdot d\vec{S}\right] = \exp\left[\frac{iq}{\hbar}\Phi_m\right], \tag{23.17}$$

where Φ_m is the magnetic flux going through the surface S.

The implication seems to be that we are observing $\vec{A}$, and not $\vec{B}$ itself. In favor of this interpretation, consider the case where there is no field $\vec{B}$ on the paths P_1 and P_2, therefore none on $C = P_1 - P_2$, see Fig. 23.1c, yet we observe nontrivial interference between the two paths, due to the phase δ. However, there is certainly a field $\vec{A}$ on it, so $\oint_C \vec{A}\cdot d\vec{r} \neq 0$. As a more precise set-up for this case, consider a cylindrical sheet of metal (a larger cylinder with a smaller cylinder in the middle), and electrons moving in the material, around the center. Consider also that there is a magnetic field parallel to the axis of the cylinder, but only in the void in the middle; see Fig. 23.1d. So there is a magnetic flux going through the surface bounded by any closed contour C in the section of the material (around the axis), and then there is also a field $\vec{A}$ on the path C itself. But there is no $\vec{B}$ on C itself, so it seems that we can measure $\vec{A}$, not $\vec{B}$.

However, this doesn't mean that we are breaking gauge invariance, since, as we saw above, δ depends only on the gauge invariant $\vec{B}$, through Φ_m, so it is a gauge invariant phase. This leads to a sort of philosophical debate: is this nonlocality, whereby δ depends on the full closed contour C, or the fact that δ depends on the value of $\vec{A}$ on C, more important for the interpretation?

In the path integral formalism, it appears that the nonlocality is more important. Indeed, consider the path integral for the wave function in the double slit experiment (or, equivalently, in the cylinder set-up, with electron paths going either clockwise or anticlockwise around the circle, between two points on it), and write it as a sum over paths P_1 going through one slit, and paths P_2 going through the other,

$$\begin{aligned}
\psi(r,t') &= \int \mathcal{D}q(t)\exp\left[\frac{i}{\hbar}\int_0^{t'} L\,dt\right]\psi(t=0) = \sum_{\text{paths P}}\exp\left[\frac{iq}{\hbar}\int_P \vec{A}\cdot d\vec{r} + \frac{i}{\hbar}S(\vec{A}=0)\right]\\
&= \left(\int_{\text{above}=\{P_1\}}\mathcal{D}q(t)e^{iS_0(P_1)/\hbar}\right)\exp\left[\frac{iq}{\hbar}\int_{P_1}\vec{A}\cdot d\vec{r}\right]\psi_1(r)\\
&\quad+\left(\int_{\text{below}=\{P_2\}}\mathcal{D}q(t)e^{iS_0(P_2)/\hbar}\right)\exp\left[\frac{iq}{\hbar}\int_{P_2}\vec{A}\cdot d\vec{r}\right]\psi_2(r)\\
&= \exp\left[\frac{iq}{\hbar}\int_{P_1}\vec{A}\cdot d\vec{r}\right]\left(\psi_1(r) + e^{i\delta}\psi_2(r)\right).
\end{aligned} \tag{23.18}$$

Again, we obtain interference due to the phase factor $e^{i\delta}$ but now we see explicitly that the nonlocality associated with the contour C is the key issue.

Finally, note that if the phase δ is a multiple of 2π then the phase factor $e^{i\delta} = 1$ is unobservable. This happens for a magnetic flux through the surface S bounded by C that is quantized,

$$\Phi[C] = n\Phi_0,\quad \Phi_0 = \frac{2\pi\hbar}{q} = \frac{h}{q}, \tag{23.19}$$

where Φ_0 is called a flux quantum.

23.3 Berry Phase

The Aharonov–Bohm phase factor $e^{i\delta}$ can be generalized to a much more important case, the observable *geometric phase*, valid for any quantum mechanical system undergoing adiabatic changes. This was derived in a seminal paper by Michael Berry in 1983.

Indeed, suppose that the Hamiltonian depends on a (set of) parameter(s) K that change slowly in time, $K = K(t)$. However, suppose that the change is slow enough (adiabatic) that the system, initially in an eigenstate of the initial Hamiltonian, continues to be in an eigenstate $|n(K(t))\rangle$ of the Hamiltonian for any later time t, $\hat{H}(t)$, so that

$$\hat{H}(K(t))|n(K(t))\rangle = E_n(K(t))|n(K(t))\rangle. \tag{23.20}$$

But the Schrödinger equation is (considering the state $|n(K_0), t_0\rangle$ at time t_0)

$$\hat{H}(K(t))|n(K_0), t_0; t\rangle = i\hbar\frac{d}{dt}|n(K_0), t_0; t\rangle, \tag{23.21}$$

which means that its time-dependent solution is of the form

$$|\psi(t)\rangle = \exp\left[-\frac{i}{\hbar}\int_0^t E_n(t')dt'\right]e^{i\gamma_n(t)}|n(K(t))\rangle. \tag{23.22}$$

This is obvious except for the phase factor $e^{i\gamma_n(t)}$, but this phase factor must exist since it compensates the fact that d/dt can also act on $K(t)$. Indeed, taking this derivative into account in the Schrödinger equation, we obtain

$$0 = i\hbar\frac{d}{dt}\left[e^{i\gamma_n(t)}|n(K(t))\rangle\right] = \hbar e^{i\gamma_n(t)}\left[-\frac{d\gamma_n}{dt}|n(K(t))\rangle + i\frac{d}{dt}|n(K(t))\rangle\right], \tag{23.23}$$

which implies that (multiplying with $\langle n(K(t))|$ from the left and using the normalization of states)

$$\frac{d\gamma_n(t)}{dt} = i\langle n(K(t))|\nabla_k|n(K(t))\rangle\frac{dK}{dt}, \tag{23.24}$$

which can be integrated to give the phase γ_n as

$$\gamma_n = i\int\langle n(K(t))|\nabla_k|n(K(t))\rangle dK. \tag{23.25}$$

This phase γ_n is the *Berry phase*, or *geometric phase*. We can express it in a more familiar way by defining the *Berry "connection"* (or, generalized gauge field)

$$A_n(K) \equiv i\langle n(K(t))|\nabla_k|n(K(t)\rangle. \tag{23.26}$$

Then, if K is a vector, such as a position $\vec{R}$ (more on that later), we have, more precisely (considering the factors of $\hbar$ as well in order to have a more precise analog of a gauge field),

$$\begin{aligned}\vec{A}_n(\vec{R}) &\equiv i\hbar\langle n(\vec{R}(t))|\vec{\nabla}_{\vec{R}}|n(\vec{R}(t))\rangle \Rightarrow\\ \gamma_n &= \frac{1}{\hbar}\int\vec{A}_n(\vec{R})\cdot d\vec{R}.\end{aligned} \tag{23.27}$$

Moreover, in general, as for a gauge field, we have a gauge transformation that leaves the system invariant. Indeed, as in the case of the Aharonov–Bohm phase, we realize that we can change the state $|\psi(t)\rangle$ by an arbitrary phase $e^{\chi(K)}$ without affecting the observables and probabilities, so

$$|\tilde{n}(K(t))\rangle = e^{i\chi(K)}|n(K(t))\rangle \tag{23.28}$$

is as good as $|n(K(t))\rangle$ in describing the system. Therefore

$$i\left\langle \tilde{n}(K(t))\left|\frac{d}{dt}\right|\tilde{n}(K(t))\right\rangle = i\left\langle n(K(t))\left|\frac{d}{dt}\right|n(K(t))\right\rangle - \frac{d\chi(t)}{dt} = A_n(K)\frac{dK}{dt} - \nabla_K\chi\frac{dK}{dt} \tag{23.29}$$

also serves as a Berry connection term; thus the Berry connection transforms as

$$A_n(K) \to A_n(K) - \nabla_K\chi, \tag{23.30}$$

exactly like a gauge field. In the case $K = \vec{R}$, we have the standard vector potential transformation law,

$$\vec{A}_n(\vec{R}) \to \vec{A}_n(\vec{R}) - \vec{\nabla}_{\vec{R}}\chi(\vec{R}(t)). \tag{23.31}$$

We note that γ_n must be a phase, since it is real. Indeed, differentiating the normalization relation of the time-dependent state $|n(K(t))\rangle$, we have

$$\begin{aligned}
&\frac{d}{dt}\left[\langle n(K(t))|n(K(t))\rangle = 1\right] \Rightarrow \\
0 &= \left\langle n(K(t))\left|\frac{d}{dt}\right. n(K(t))\right\rangle + \left\langle \frac{d}{dt}n(K(t))|n(K(t))\right\rangle \\
&= 2\mathrm{Re}\left(\left\langle n(K(t))\left|\frac{d}{dt}\right|n(K(t))\right\rangle\right),
\end{aligned} \tag{23.32}$$

where we have used that

$$\left\langle \frac{d}{dt}n(K(t))|n(K(t))\right\rangle = \left(\left\langle n(K(t))\left|\frac{d}{dt}\right. n(K(t))\right\rangle\right)^*,$$

so we obtain that the phase is real (note the extra i in the phase),

$$\gamma_n \equiv i\left\langle n(K(t))\left|\frac{d}{dt}\right|n(K(t))\right\rangle \in \mathbb{R}. \tag{23.33}$$

But, if the Hamiltonian is periodic, i.e., it returns to the initial point after a time T,

$$\hat{H}(K(T)) = \hat{H}(K(0)), \tag{23.34}$$

then this means that the gauge transformation must also be single valued, $e^{i\chi(T)} = e^{i\chi(0)}$, in order to have a well-defined physical system, so

$$\chi(K(T)) = \chi(K(0)) + 2\pi m, \tag{23.35}$$

where $m \in \mathbb{Z}$. Therefore the Berry phase *on a closed path* C, modulo $2\pi m$, is invariant, and cannot be removed by a gauge transformation:

$$\gamma_n(C) + 2\pi m = \oint_C A_n(K)dK + \chi(T) - \chi(0) = \oint_C A'_n(K)dK = \gamma'_n(C). \tag{23.36}$$

23.4 Example: Atoms, Nuclei plus Electrons

A standard example, and one that was implicit when we said that the parameters K can be a vector $\vec{R}$ for the position of some quantity, is the case of several atoms interacting (close by), composed of nuclei and electrons. The nuclei are slow variables, since the time variations of their positions are small, whereas the electrons are fast variables since the time variations of their positions are large. Consider for simplicity a diatomic molecule, with distance $\vec{R}$ between the nuclei (determining relative motion of the nuclei). Factoring out the center of mass motion, the total Hamiltonian for the system then splits into a Hamiltonian for the nuclei depending on $\vec{R}$, $H_N(\vec{R})$, a Hamiltonian for the electrons $H_{\rm el}(\vec{r}_1, \ldots, \vec{r}_N)$, and a potential depending on the positions of all of the electrons, $V(\vec{R}, \vec{r}_1, \ldots, \vec{r}_N)$,

$$\hat{H}_{\rm tot} = \hat{H}_N(\vec{R}) + \hat{H}_{\rm el}(\vec{r}_1, \ldots, \vec{r}_N) + V(\vec{R}, \vec{r}_1, \ldots, \vec{r}_N). \tag{23.37}$$

This is a slight generalization of the hydrogen molecule H_2, considered in Chapter 21. There we implicitly used the adiabatic approximation $\vec{R} \simeq$ constant: the nuclei have only a small motion (relative oscillation) with respect to the distance R between them. This is called the *Born–Oppenheimer approximation*, after the people who considered it first.

We can then assume, adiabatically, that the positions of the nuclei are approximately constant as the distance $\vec{R}$ changes in time, as far as the electrons are considered. Then, for the electrons, we can write down the eigenenergy equation with electron eigenstate $|n(\vec{R})\rangle$, where $\vec{R}$ appears as just a parameter,

$$\left(H_{\rm el}(\vec{r}_1, \ldots, \vec{r}_N) + V(\vec{R}, \vec{r}_1, \ldots, \vec{r}_N)\right) |n(\vec{R})\rangle = U_n(\vec{R})|n(\vec{R})\rangle. \tag{23.38}$$

We see then that this is of the type considered in the Berry phase above.

For the total system, the state is approximately (assuming the separation of variables) the tensor product of a state for the nuclei and the electron state,

$$|\psi\rangle = |\psi_N\rangle \otimes |n(\vec{R})\rangle. \tag{23.39}$$

Neglecting the back reaction of the electrons on the nuclei equation, we obtain the simple separated-variable equation

$$\left(H_N(\vec{R}) + U_n(\vec{R})\right) |\psi_N\rangle = E_n|\psi_N\rangle, \tag{23.40}$$

and corresponding time-dependent Schrödinger equation, whose solution classically corresponds to the motion of the relative position of the nuclei, $\vec{R} = \vec{R}(t)$; this could be plugged back into the adiabatic eigenenergy equation for the electrons and a Berry phase obtained.

We can ask: why is it then that Born and Oppenheimer missed this Berry phase in their analysis? One answer is that the total Hamiltonian in this case is real, $\hat{H} \in \mathbb{R}$, so by the general theory we can choose wave functions that are real as well (sines and cosines), so we can neglect phases; this is what Born and Oppenheimer did and therefore missed the Berry phase.

23.5 Spin–Magnetic Field Interaction, Berry Curvature, and Berry Phase as Geometric Phase

We next aim to understand why the Berry phase is also called a geometric phase. We will examine the property of this phase of being defined by geometry for another classic case, that of a spin interacting with a magnetic field that depends slowly on time. Then, we can again consider an adiabatic approximation, and write the interaction of the spin with $\vec{B}(t)$ as giving a time-dependent Hamiltonian, in which the magnetic field appears as a time-dependent parameter. The interaction term is

$$H_{\rm int}(\vec{B}) = -g\mu_B \frac{\vec{S}}{\hbar}\cdot \vec{B}(t). \tag{23.41}$$

Then the Berry connection, defined with respect to the parameters $\vec{B}(t)$, is

$$\vec{A}_B = i\langle n(\vec{B}(t))|\vec{\nabla}_{\vec{B}}|n(\vec{B}(t))\rangle. \tag{23.42}$$

However, associated with it, we can define a "Berry curvature", meaning the field strength of the Berry connection (the gauge field), in the usual way. Writing it for the general case (not necessarily for $K = \vec{B}$, but for general parameters) we have

$$\vec{F}_n(K) = \vec{\nabla}_K \times \vec{A}_n(K), \tag{23.43}$$

which means that in components we have

$$F^n_{ij} = \partial_i A^n_j - \partial_j A^n_i \equiv \frac{\partial}{\partial K_i}A_j - \frac{\partial}{\partial K_j}A_i, \tag{23.44}$$

as for a regular gauge field, and then we can define the "magnetic field" $\vec{F}_n(K)$ in the usual way (as for electromagnetism), by

$$F^n_{ij} = \epsilon_{ijk}F^n_k. \tag{23.45}$$

Then, moreover, using Gauss's (Stokes') law for this gauge field, we have that (as for the Aharonov–Bohm phase)

$$\oint_{C=\partial S} \vec{A}_n(k)\cdot d\vec{K} = \int_S \vec{B}_n(K)\cdot d^2\vec{K}. \tag{23.46}$$

The field strength of the gauge field can be rewritten as

$$\begin{aligned} F^n_{ij} &= i\left[\partial_{K_i}\langle n(K(t))|\partial_{K_j}|n(K(t))\rangle - \partial_{K_j}\langle n(K(t))|\partial_{K_i}|n(K(t))\rangle\right] \\ &= i\left[\langle \partial_{K_i} n(K(t))|\partial_{K_j} n(K(t))\rangle - \langle \partial_{K_j} n(K(t))|\partial_{K_i} n(K(t))\rangle\right]. \end{aligned} \tag{23.47}$$

Now we can insert the identity, rewritten using the completeness relation for the adiabatic states,

$$\mathbb{1} = \sum_m |m(K(t))\rangle\langle m(K(t))|, \tag{23.48}$$

inside the scalar products, obtaining

$$\begin{aligned} F_{ij} = i\sum_{m\neq n} \Big[&\langle \partial_{K_i} n(K(t))|m(K(t))\rangle\langle m(K(t))|\partial_{K_j}|n\rangle \\ &- \langle \partial_{K_j} n(K(t))|m(K(t))\rangle\langle m(K(t))|\partial_{K_i}|n\rangle\Big]. \end{aligned} \tag{23.49}$$

For $m \neq 0$, due to the fact that $\hat{H}|m\rangle = E_m|m\rangle$ and $\langle n|m\rangle = 0$, which is still valid when $K = K(t)$, we take ∇_K and obtain

$$\nabla_K\langle n(K)|\hat{H}|m(K)\rangle = 0. \tag{23.50}$$

Moreover, also using the fact that $\nabla_K\langle n(K(t))|m(K(t))\rangle = 0$ (in the adiabatic case), which implies that

$$\langle\nabla_K n(K(t))|m(K(t))\rangle = -\langle n(K(t))|\nabla_K m(K(t))\rangle, \tag{23.51}$$

and acting with ∇_K on both states $\langle n|$ and $|m\rangle$, and on $\hat{H}$ in (23.50), we obtain

$$0 = \langle n(K(t))|(\nabla_K\hat{H})|m(K(t))|\rangle - (E_m - E_n)\langle n(K(t))|\nabla_K m(K(t))\rangle. \tag{23.52}$$

This implies that

$$\langle n(K(t))|\nabla_K|m(K(t))\rangle = \frac{\langle n(K(t))|(\nabla_K\hat{H})|m(K(t))\rangle}{E_m - E_n}. \tag{23.53}$$

Substituting this in (23.49), we obtain that the Berry curvature is expressed as

$$\begin{aligned} F^n_{ij} = i\sum_{m\neq n}\Bigg[&\frac{\langle n(K(t))|(\partial_i\hat{H})|m(K(t))\rangle\langle m(K(t))|(\partial_j\hat{H})|n(K(t))\rangle}{(E_m - E_n)^2} \\ &-\frac{\langle n(K(t))|(\partial_j\hat{H})|m(K(t))\rangle\langle m(K(t))|(\partial_i\hat{H})|n(K(t))\rangle}{(E_m - E_n)^2}\Bigg]. \end{aligned} \tag{23.54}$$

Finally, we can use this formula in the case of the interaction between a spin and a time-dependent magnetic field $\vec{B}(t)$, acting as parameters $K(t)$. Then, we have

$$\partial_i\hat{H} = \frac{\partial}{\partial B_i}\hat{H} = -g\mu_B\frac{\hat{S}_i}{\hbar}(= -\mu_B\sigma_i), \tag{23.55}$$

where in parenthesis we have written an expression for the right-hand side in terms of the Pauli matrices acting on a two-dimensional Hilbert space for spin 1/2 ith projection $m\hbar/2$ ($m = \pm 1$) for the spin in the z direction, using $S_i = \sigma_i\hbar/2$ and $g = 2$. Then, after a calculation (left as an exercise), we find

$$B^n_i = m\hbar\frac{\hat{B}_i}{B^2}, \tag{23.56}$$

where $\hat{B}_i$ stands for the i component of the magnetic field. Then the Berry phase is (using the fact that $K_i = B_i$ and Gauss's, or Stokes', law)

$$\gamma_n = \frac{1}{\hbar}\oint_{C=\partial S} A^n_i dB_i = \frac{1}{\hbar}\int_{S(C)} B^n_i d^2B_i = -m\int_{S(C)}\frac{\hat{B}_i}{B^2}d^2B_i = -m\int_C d\Omega. \tag{23.57}$$

Here in the second equality we have used Stokes' law to convert the integral over the closed path $C = \partial S$ in B_i space to an integral over the surface $S(C)$ bounded by C in B_i space, and in the last equality we have noted that $(\hat{B}_i/B^2)d^2B_i$ is a solid angle $d\Omega$ on the surface S bounded by the contour C, with respect to the origin O in $\vec{B}$ space; see Fig. 23.2. Then we see that

$$\gamma_n = -m\Delta\Omega \tag{23.58}$$

is a solid angle in parameter ($\vec{B}$) space, thus being entirely geometric in nature, as we set out to show.

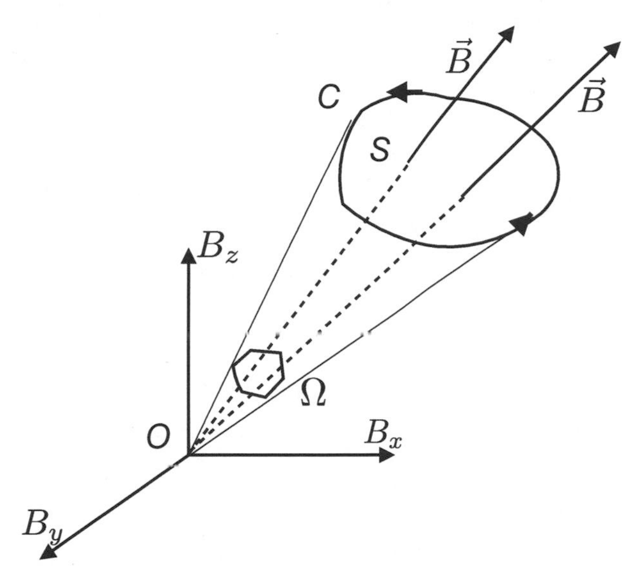

Figure 23.2 A surface S in $\vec{B}$ space, bounded by a contour C and making a solid angle Ω from the point of view of the origin O.

23.6 Nonabelian Generalization

We can also define a nonabelian generalization of the Berry connection. Indeed, now suppose that we have N *degenerate* eigenstates $|n\rangle$, with $n = 1, \dots, N$, of the Hamiltonian, which depends on time-dependent parameters $K(t)$, so we denote the eigenstates by $|n, K\rangle$.

Then we can still define the Abelian Berry connection as the diagonal matrix element

$$A(K) = i \sum_n \langle n, K(t)|\nabla_K|n, K(t)\rangle, \tag{23.59}$$

but now we can define also the *nonabelian Berry connection* as the off-diagonal matrix element in the space of the same energy,

$$A^{(n,m)}(K) = i\langle n, K(t)|\nabla_K|m, K(t)\rangle, \tag{23.60}$$

which is an element in the unitary group $U(N)$ (it is a unitary $N \times N$ matrix), so that $A(K)$ is its trace,

$$A(K) = \sum_{n=1}^{N} A^{(n,n)}(K), \tag{23.61}$$

which is known to be the Abelian component in the decomposition $U(N) \simeq (U(1) \times SU(N))/\mathbb{Z}_N$.

23.7 Aharonov–Bohm Phase in Berry Form

We can rewrite the Aharonov–Bohm phase δ in the Berry phase form as follows. Consider (electron) states at a point $\vec{R}$, translated along the curve C (for example, a curve inside a conductor) until $\vec{r}$. Then the position-space wave function at point $\vec{r}$ is

$$\langle \vec{r}|n(\vec{R})\rangle = \exp\left[\frac{iq}{\hbar}\int_{\vec{R}}^{\vec{r}} \vec{A}(\vec{r}\,')\cdot d\vec{r}\,'\right]\psi_n(\vec{R}), \tag{23.62}$$

which leads to

$$\langle n(\vec{R})|\vec{\nabla}_{\vec{R}} n(\vec{R})\rangle = -iq\vec{A}(\vec{R}), \tag{23.63}$$

meaning we have rewritten the vector potential $\vec{A}(\vec{R})$ as a Berry connection, and thus the Aharonov–Bohm phase as a Berry phase.

Important Concepts to Remember

- Since the canonically conjugate momentum $\vec{p} = \vec{p}_{\rm kin} - q\vec{A}$ translates in x $\left(\text{so is equal to } \frac{\hbar}{i}\partial_x\right)$, we find that when we go along a path P from x to x' we acquire a phase $e^{i\delta}$, with $\delta = \int_{P:x}^{x'} q\vec{A}\cdot d\vec{x}$.
- In turn, this means that on a *closed path* C, we have a gauge-invariant phase proportional to the magnetic flux, since $\delta = \frac{q}{\hbar}\oint_{C=\partial S}\vec{A}\cdot d\vec{x} = \frac{q}{\hbar}\int_S \vec{B}\cdot d\vec{S} = \frac{q}{\hbar}\Phi_m$. This is the Aharonov–Bohm phase.
- The Aharonov–Bohm phase would appear to mean that $\vec{A}$ is observable (we can have $\vec{B} = 0$ on C, but $\vec{A} \neq 0$); the phase itself is unobservable ($e^{i\delta} = 1$) for integer flux, $\Phi_n = n\Phi_0$, $\Phi_0 = h/q$.
- If the Hamiltonian depends on a set of parameters that change slowly in time, $K = K(t)$, with adiabatic energy eigenstates $|n(K(t))\rangle$, we can define a Berry connection $\vec{A}_n(K) = i\langle n(K(t))|\vec{\nabla}_K|n(K(t))\rangle$, and the states have an additional Berry phase $e^{i\gamma_n}$, with $\gamma_n = \int \vec{A}_n(K)\cdot d\vec{K}$.
- The Berry phase is real and gauge invariant (under generalized gauge transformations for $\vec{A}_n(K)$) modulo $2\pi m$, so is physically observable.
- For atoms (nuclei plus electrons) in the Born–Oppenheimer approximation (nuclei with small motion), we have a Berry phase in terms of $\vec{R}$ (the nuclear separation).
- The Berry curvature, the field strength of the Berry connection, can be written in terms of the matrix elements of the derivatives of the Hamiltonian.
- In the case of an interaction between a spin 1/2 and a time-dependent magnetic field $\vec{B}(t)$, with projection $m\hbar/2$ of the spin on $\vec{B}$, the Berry phase is a geometric phase in parameter space ($\vec{B}$ space), specifically $-m$ times the solid angle bounded by a moving direction $\vec{B}(t)/|\vec{B}(t)|$ along a closed path.
- In the case of N degenerate eigenstates of the Hamiltonian, $|n\rangle$, we can define a nonabelian Berry connection, $A^{(nm)}(K) = i\langle n, K(t)|\nabla_K|m, K(t)\rangle$, an element in $U(N)$, while the Abelian Berry connection is its trace, $\sum_n A^{(n,n)}$.
- The Aharonov–Bohm phase is a particular example of the Berry phase, since it can be put in the form of the latter.

Further Reading

See [1] and the original paper of Michael Berry [9] for more details.

Exercises

(1) Consider a figure-eight loop made of conductor, mostly in a single plane (with just enough nonplanarity for the loop to not cross itself, but to slightly avoid it), and a constant magnetic field perpendicular to this plane. Do we have an Aharonov–Bohm phase for the electrons in the conducting loop?

(2) What happens to the wave function of the electrons moving around the loop in exercise 1 (or to the interference pattern if we shoot electrons at a point on the loop in opposite directions on the loop and measure their interference when they cross again) as we undo the figure-eight into a circle in the same plane?

(3) Consider a system depending on N time-dependent vectors $\vec{R}_i$, $i = 1, 2, \ldots, N$, for instance a molecule with N nuclei with such positions. How do we construct its generic Berry phase?

(4) Consider an H_2 molecule moving slowly, in the Born–Oppenheimer approximation, in a constant electric field $\vec{E}$. What Berry connection does one define, and how can we have a nontrivial Berry phase?

(5) Consider a molecule with fixed dipole $\vec{d}$ in a time-dependent electric field $\vec{E}(t)$. Calculate the Berry connection and Berry magnetic field in this case. Do we have a geometrical phase in this case?

(6) Prove that for a spin 1/2 particle in a magnetic field the Berry curvature is given by (23.56).

(7) How would you define the Berry curvature in the nonabelian case, in such a way that it transforms *covariantly* under $U(N)$, i.e., with U acting from the left and U^{-1} from the right?

24 Motion in a Magnetic Field, Hall Effect and Landau Levels

In this chapter we consider particles with and without spin, and later atoms, in a magnetic field and calculate the resulting quantum effects. For a particle, we find that there are "Landau levels" for the states, and for particles in a conductor these lead to a quantum version of the classical Hall effect. Finally, for the atom, we find a first-order form for the Zeeman effect.

24.1 Spin in a Magnetic Field

As we saw in previous chapters, for the interaction of a spin with magnetic field, we have the Hamiltonian

$$\hat{H}(B) = \hat{H}_0 - g\mu_B \vec{B} \cdot \frac{\vec{S}}{\hbar}. \tag{24.1}$$

In the case of a spin 1/2 particle, we can rewrite this as

$$\hat{H}(B) = \hat{H}_0 - \mu_B \vec{B} \cdot \vec{\sigma}. \tag{24.2}$$

For a magnetic field in the z direction,

$$\vec{B} \cdot \vec{\sigma} = B_z \sigma_3 = \begin{pmatrix} +B_z & 0 \\ 0 & -B_z \end{pmatrix}, \tag{24.3}$$

which means that the time-independent Schrödinger equation $\hat{H}(B)|\psi\rangle = E(B)|\psi\rangle$ gives

$$E(B) = E_0 \mp \mu_B B_z, \tag{24.4}$$

which is the result of the Stern–Gerlach experiment for $S_z = \pm\hbar/2$ (it is the spin-1/2-particle splitting energy as a function of S_z in a transverse magnetic field).

24.2 Particle with Spin 1/2 in a Time-Dependent Magnetic Field

We consider a magnetic field $\vec{B} = \vec{B}(t)$, i.e., constant in space, but time dependent. In this case, we can separate variables and write a wave function that is a tensor product of a position state and a spin state (as in Chapters 17 and 20),

$$|\psi\rangle = \psi(\vec{r}, t)|s(t)\rangle. \tag{24.5}$$

The Schrödinger equation splits into independent Schrödinger equations. The solution for $|\psi\rangle$ is unique, though the separation into parts is not. We can add a time dependence with a function $F(t)$, split between the two equations,

$$\left(\hat{H}_0 - F(t)\right)\psi'(\vec{r},t) + \frac{\hbar}{i}\frac{\partial}{\partial t}\psi'(\vec{r},t) = 0$$
$$\frac{\hbar}{i}\frac{d}{dt}|s'\rangle + (-\mu_B \vec{B}(t)\cdot\vec{\sigma} + F(t))|s'\rangle = 0. \tag{24.6}$$

The solution of the two equations factorizes the dependence on $F(t)$,

$$\psi'(\vec{r},t) = \psi(\vec{r},t)\exp\left[\frac{i}{\hbar}\int_0^t F(t')dt'\right]$$
$$|s'(t)\rangle = |s(t)\rangle\exp\left[-\frac{i}{\hbar}\int_0^t F(t')dt'\right]. \tag{24.7}$$

But, as we said, the total state is independent of $F(t)$,

$$|\psi'\rangle = \psi'(\vec{r},t)|s'(t)\rangle = \psi(\vec{r},t)|s(t)\rangle = |\psi\rangle. \tag{24.8}$$

So we can use the solution without $F(t)$ (i.e., put $F(t) = 0$), which is unique.

However, as explained in the previous chapter, in this situation there is also a Berry phase that is actually geometrical; we will not repeat the argument here.

24.3 Particle with or without Spin in a Magnetic Field: Landau Levels

We saw in the previous chapter that the coupling of a particle with a vector potential describing an electromagnetic field is given by

$$\hat{H} = \frac{1}{2m}\left(\hat{\vec{p}} - q\vec{A}\right)^2 - qA_0. \tag{24.9}$$

Consider a magnetic field in the z direction, $\vec{B} = B\vec{e}_z$, with no electric field, so we can put the gauge $A_0 = 0$. Then we have the vector potential

$$\vec{A} = Bx\vec{e}_y, \tag{24.10}$$

which indeed gives $F_{xy} = \epsilon_{xyz}B_z = B$. Then the Hamiltonian becomes

$$\hat{H} = \frac{1}{2m}\left[\hat{p}_x^2 + \left(\hat{p}_y - qBx\right)^2 + \hat{p}_z^2\right]. \tag{24.11}$$

To solve the time-independent Schrödinger equation, we write a solution with separated variables, where in the y, z directions we have just the free particle ansatz

$$\psi(x,y,z) = u_{n,p_y,p_z}(x,y,z) = X_n(x)e^{i(yp_y+zp_z)/\hbar}. \tag{24.12}$$

The time-independent Schrödinger equation becomes

$$\begin{aligned}\hat{H}\psi &= E\psi\\ &= -\frac{\hbar^2}{2m}\left[\partial_x^2 + \left(\partial_y - \frac{iq}{\hbar}Bx\right)^2 + \partial_z^2\right]\psi\\ &= -\frac{\hbar^2}{2m}\left[\partial_x^2 + \left(\frac{ip_y}{\hbar} - \frac{iq}{\hbar}Bx\right)^2 - \frac{p_z^2}{\hbar^2}\right]X_n(x)e^{i(yp_y+zp_z)/\hbar}.\end{aligned} \tag{24.13}$$

If the particle also has a spin, say spin 1/2, then we saw that there is an extra term $-\mu_B \vec{B} \cdot \vec{\sigma}$ in $\hat{H}$, corresponding to a term $\mp\mu_B B$ in the energy. With this extra term, the equation for a particle with spin 1/2 in a magnetic field is given by

$$X_n''(x) + \frac{2m}{\hbar^2}\left[E \pm \mu_B B - \frac{p_z^2}{2m} - \frac{q^2B^2}{2m}\left(x - \frac{p_y}{qB}\right)^2\right] X_n(x) = 0. \tag{24.14}$$

Then, defining

$$\begin{aligned} x_0 &= \frac{p_y}{qB} \\ \frac{q^2B^2}{m^2} &\equiv \omega^2(B) \Rightarrow \omega(B) = \omega_c = \frac{|q|B}{m}, \end{aligned} \tag{24.15}$$

where ω_c is called the cyclotron frequency and is the angular frequency for the motion of a particle of mass m in a field B, we find

$$X_n''(x) + \frac{2m}{\hbar^2}\left[E \pm \mu_B B - \frac{p_z^2}{2m} - \frac{m\omega_c^2}{2}(x - x_0)^2\right] X_n(x) = 0. \tag{24.16}$$

This is the equation for a harmonic oscillator, with coordinate $x - x_0$ and frequency ω_c, so the x dependent wave function is

$$X_n(x) = N_n e^{-\alpha^2(x-x_0)^2/2} H_n(\alpha(x - x_0)), \tag{24.17}$$

and the eigenenergies are

$$E_{n;|p_z|,m_s} = \frac{p_z^2}{2m} \mp \mu_B B + \hbar\omega_c\left(n + \frac{1}{2}\right). \tag{24.18}$$

We note that

$$\mu_B B = \frac{|e|\hbar B}{2m} = \frac{\hbar\omega_c}{2} = |m_s|\hbar\omega_c, \tag{24.19}$$

where $m_s = \pm 1/2$ gives the values of the spin projection onto the z direction. Thus the eigenenergies can be rewritten as

$$E_{n;|p_z|,m_s} = \frac{p_z^2}{2m} + \hbar\omega_c\left(n + m_s + \frac{1}{2}\right). \tag{24.20}$$

The parameter α in the harmonic oscillator is rewritten as

$$\alpha = \sqrt{\frac{m\omega_c}{\hbar}} = \sqrt{\frac{|q|B}{\hbar}} \equiv \frac{1}{l}, \tag{24.21}$$

where the distance l becomes $\frac{250\ \mathring{A}}{\sqrt{B(\text{tesla})}}$, and the position x_0 is

$$x_0(p_y) = \pm\frac{l^2}{\hbar}p_y = \pm\frac{p_y}{m\omega_c}. \tag{24.22}$$

The harmonic oscillator index n (listing the energy levels) is called now a *Landau level*. We note that the harmonic oscillation:

- is shifted by $x_0(p_y)$ and rescaled by l;
- is a motion in a plane perpendicular to $\vec{B}$, in this case the plane defined by (x, y).

To compare the behavior obtained in quantum mechanics with that in classical mechanics, note that in classical mechanics we have a circular motion around the direction of $\vec{B}$. The electron is under the influence of the Lorentz force $\vec{F} = q(\vec{E} + \vec{v} \times \vec{B})$, with $\vec{E} = 0$ and a velocity perpendicular to $\vec{B}$. Since the Lorentz force acts as a centripetal force, or balances the centrifugal force, we obtain

$$F = |q|vB = F_{\text{centrifugal}} = mv\omega, \tag{24.23}$$

leading to a motion with the cyclotron angular frequency

$$\omega = \frac{|q|B}{m} = \omega_c. \tag{24.24}$$

The particle thus describes circles of radius

$$r = \frac{v}{\omega_c} = \frac{p_y}{m\omega(B)} = |x_0(p_y)|. \tag{24.25}$$

Thus in the quantum case we have a shifted oscillator instead of circular motion, though the parameters are the same as in the classical case.

We also note that the energy is independent of p_y, so there is a degeneracy of the levels, corresponding to all possible p_y. In a physical situation, the area of the system in the (x, y) plane is finite. Consider a rectangle in this plane, with length L_x in x and L_y in y, so with area $S = L_x L_y$. Then, the momenta p_y are quantized as $p_{y,m}$, since we need to have periodicity of the wave function (which should vanish at the boundaries of the sample) $\exp\left(\frac{i}{\hbar} p_{y,m} L_y\right) = e^{2\pi i m} = 1$, implying

$$p_{y,m} = \hbar \frac{2\pi m}{L_y}. \tag{24.26}$$

But, since for each p_y the level is shifted by $x_0(p_y)$, there is a maximum value for p_y, $p_{y,\text{max}}$, such that we shift all the way to the end of the system, $x_0(p_{y,\text{max}}) = L_x$,

$$\frac{p_{y,\text{max}}}{qB} = \frac{2\pi\hbar m_{\text{max}}}{qBL_y} = L_x, \tag{24.27}$$

which implies that there is a maximum number of states (degeneracy) in a Landau level,

$$N_{\text{max}} = m_{\text{max}} = \frac{qBL_y L_x}{h} = \frac{L_x L_y}{2\pi l^2}. \tag{24.28}$$

Since $S = L_x L_y$, the magnetic flux is $\Phi = BL_x L_y$, and thus we have

$$N_{\text{max}} = \frac{\Phi}{\Phi_0}, \tag{24.29}$$

where, as before, $\Phi_0 = h/e$ is the *flux quantum*. Therefore the maximum number of electron states in a Landau level is independent of the level n. And the maximum number of electron states per unit area $L_x L_y$ is

$$n_B = \frac{N_{\text{max}}}{L_x L_y} = \frac{1}{2\pi l^2} = \frac{eB}{h}. \tag{24.30}$$

24.4 The Integer Quantum Hall Effect (IQHE)

The Landau levels have a very important application to the physics of conducting materials, the Hall effect.

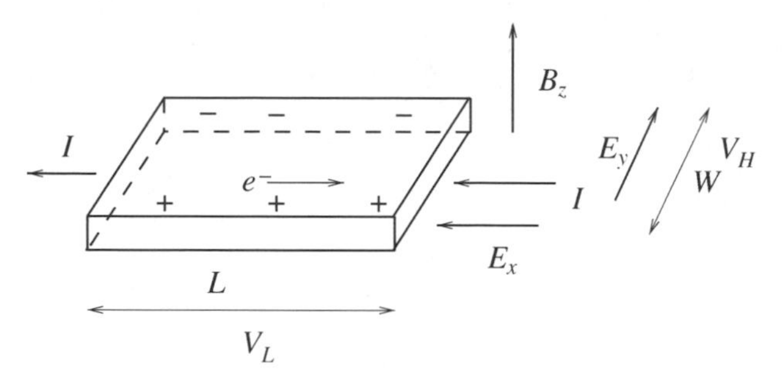

Figure 24.1 The classical Hall effect: a current in the x direction and magnetic field in the z direction result in a Hall potential in the y direction.

The *classical Hall effect* is as follows. Consider a quasi-two-dimensional sample of conducting material of size $L_x = L$ and $L_y = W$ and magnetic field B in the z direction. Add a voltage V_L in the x direction, implying an electric field E_x, so that $V_L = LE_x$; see Fig. 24.1.

Conduction electrons move sideways in the material under the Lorentz force until there is a compensating electric field E_y, generating a "Hall voltage" $V_H = L_y E_y$. The equilibrium of the Hall voltage generating E_y with the Lorentz force gives, from $\vec{F} = -e(\vec{E} + \vec{v} \times \vec{B}) = 0$,

$$E_y = v_x B_z. \tag{24.31}$$

This is the classical Hall effect, which we will quantify next.

Conduction electrons move in the material by accelerating between collisions, which stop them and bring the velocity down to zero. The time between collisions is the "relaxation time" τ. This means that the average velocity in the conduction direction x is

$$v_x = a\tau = -\frac{eE_x\tau}{m}. \tag{24.32}$$

Then, considering the current density per unit length j_x (note that we have a quasi-two-dimensional sample, with a transverse area that is quasi-one-dimensional, so the appropriate density is given per unit length in the y direction instead of per unit area) generated by a number of electrons per unit area n_e, we have

$$j_x = n_e e v_x = \frac{e^2 \tau n_e E_x}{m}, \tag{24.33}$$

which means that the compensating electric field in the y direction is

$$E_y = v_x B_z = -\frac{eB}{m}\tau E_x = -\omega_c \tau E_x. \tag{24.34}$$

We define the transverse, or Hall resistance R_H as the ratio of the Hall voltage V_H (in the y direction) and the applied current I (in the x direction),

$$R_H = \frac{V_H}{I} = \frac{L_y E_y}{L_y j_x} = -\frac{m\omega_c \tau}{e^2 \tau n_e} = -\frac{m\omega_c}{n_e e^2} = -\frac{B}{n_e e}. \tag{24.35}$$

Here the minus sign is conventional, to emphasize that the Hall voltage is a compensating effect that opposes the applied current.

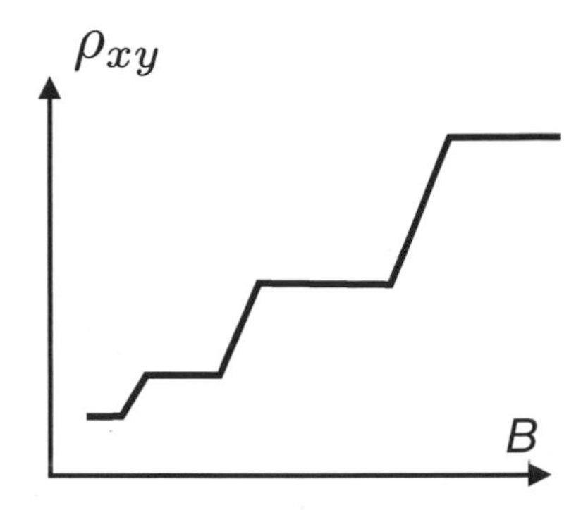

Figure 24.2 R_H, or the Hall resistivity ρ_{xy}, versus magnetic field B has plateaux, while $\rho_{xx} \simeq 0$ on the plateaux.

The Hall conductance is defined as the inverse of the resistance,

$$\sigma_H = \frac{1}{R_H} = -\frac{n_e e}{B}. \tag{24.36}$$

However, in 1980, Klaus von Klitzing measured experimentally the Hall effect in a two-dimensional electron gas confined at the interface between an oxide, SiO_2, and the Si semiconductor MOSFET (metal-oxide-semiconductor field-effect transistor, a standard electronic device). He found that in this case, the B dependence of the Hall voltage is not linear, as suggested by the above relation, but rather has *plateaux*, whose position is independent of the sample (i.e., is universal), as in Fig. 24.2. He also found that actually the Hall conductance σ_H is *quantized* as

$$\sigma_H = -n\frac{e^2}{h} = n\sigma_0. \tag{24.37}$$

Moreover, von Klitzing found that at the plateaux the longitudinal (normal) resistance vanishes, $R_L = V_L/I = 0$. This behavior appears in various $(2+1)$-dimensional strongly coupled systems (systems where the effective interactions between the degrees of freedom are strong).

The above phenomenon is called the integer quantum Hall effect, or IQHE, and von Klitzing got a Nobel Prize for it in 1985 (which is a record time between research and Nobel Prize). Actually, in 1982 Daniel Tsui and Horst Störmer had found that, at much higher B and much lower temperature T, for similar systems we have a fractional quantum Hall effect (FQHE), with

$$\sigma_H = \nu\frac{e^2}{h}, \quad \nu = \frac{k}{2p+1}. \tag{24.38}$$

The FQHE was (partially) explained by Robert Laughlin in terms of a "Laughlin wave function", but we will not address it here. For this explanation, he (together with Tsui and Störmer) also got a Nobel Prize, in 1998.

Coming back to the IQHE, and applying the Landau level theory, if we have a maximum number n_B of states per unit area in a Landau level, we can have n fully occupied levels, so the number of (conducting) electron states per unit area is given by

$$\nu = \frac{n_e}{n_B} = n \in \mathbb{N}. \tag{24.39}$$

Then the Hall conductance is, as given earlier,

$$\sigma_H = -\frac{n_e e}{B} = -\frac{n n_B e}{B} = -\frac{ne^2}{h}. \tag{24.40}$$

24.5 Alternative Derivation of the IQHE

We will now consider an alternative derivation of the IQHE, which also explains an important property of the quantum Hall effect, the fact that at the plateaux we have $R_L = 0$ (zero normal, or longitudinal, resistance).

As we said, at equilibrium the Lorentz force vanishes, $\vec{v} \times \vec{B} = -\vec{E}$, so

$$v_i B_k = E_j \epsilon_{ijk}. \tag{24.41}$$

If we take B to be in the z direction, and $k = z$, we find

$$v_i = \frac{\epsilon_{ij} E_j}{B_z}. \tag{24.42}$$

If we have n fully occupied Landau levels, therefore each with $N_{\max}$ states in them, so that $Q = nN_{\max}$, for electrons ($q = -e$) we find the current density

$$j_i = \frac{eQ}{L_x L_y} v_i = \frac{eQ}{BL_x L_y} \epsilon_{ij} E_j \equiv \sigma_{ij} E_j, \tag{24.43}$$

where we have defined a *conductance tensor* by the standard relation $j_i = \sigma_{ij} E_j$. Then we have

$$\sigma_{ij} = \frac{eQ}{\Phi} \epsilon_{ij} = \frac{enN_{\max}}{N_{\max} h/e} = n\frac{e^2}{h} \epsilon_{ij} = \sigma_H \epsilon_{ij}, \tag{24.44}$$

so the conductance matrix is

$$\sigma = \begin{pmatrix} 0 & ne^2/h \\ -ne^2/h & 0 \end{pmatrix}. \tag{24.45}$$

We see that σ_{xx}, the diagonal or normal (longitudinal) conductance, vanishes. If there were no off-diagonal (Hall) conductance, but only a longitudinal conductance, $\sigma_{xx} = 0$ would imply infinite resistance, $R_{xx} = \infty$. However, now σ is a matrix, so has to be inverted as a matrix, giving the resistance matrix

$$R = \sigma^{-1} = \begin{pmatrix} 0 & -1/\sigma_{xy} \\ 1/\sigma_{xy} & 0 \end{pmatrix}, \tag{24.46}$$

and we see that now in fact $R_{xx} = 0$! This means that we have zero longitudinal resistance on the plateaux, the other important experimental observation about the integer quantum Hall effect.

24.6 An Atom in a Magnetic Field and the Landé g-Factor

Consider now the next case in terms of complexity of the system, a multi-electron atom in a magnetic field. The electrons have spins that add up to a total electronic spin $\vec{S} = \sum_{i=1}^{N} s_i$ (for N electrons), and these electrons interact with $\vec{B}$ through both the vector potential $\vec{A}$ and the spin–magnetic field interaction $\vec{B} \cdot \vec{S}$. Specifically, the Hamiltonian will be

$$\hat{H} = \frac{1}{2m} \sum_{i=1}^{N} \left(\hat{\vec{p}}_i + |e| \vec{A}(\vec{r}_i) \right)^2 + U(\vec{r}_1, \ldots, \vec{r}_N) - \mu_B \frac{\vec{B} \cdot \vec{S}}{\hbar}, \tag{24.47}$$

where $U(\vec{r}_1, \dots, \vec{r}_N)$ is the interaction potential between the electrons and nucleus and between the electrons themselves.

Expanding, we get

$$\hat{H} = \hat{H}(A=0) + \frac{|e|}{m}\sum_{i=1}^{N} \vec{A}(\vec{r}_i)\cdot\hat{\vec{p}} + \frac{e^2}{2m}\sum_{i=1}^{N} \vec{A}^2(\vec{r}_i) + \frac{|e|}{m}\vec{B}\cdot\vec{S}. \tag{24.48}$$

If the magnetic field is constant, we can choose a gauge where the vector potential is $\vec{A} = (\vec{r}\times\vec{B})/2$, so the Hamiltonian is then rewritten as

$$\begin{aligned}\hat{H} &= \hat{H}(A=0) + \frac{e}{2m\hbar}\vec{B}\cdot\sum_{i=1}^{N}(\vec{r}_i\times\hat{\vec{p}}_i) + \frac{e^2}{8m}\sum_{i=1}^{N}(\vec{B}\times\vec{r}_i)^2 + \frac{|e|}{m}\vec{B}\cdot\vec{S}\\ &= \hat{H}(A=0) + \mu_B(\vec{L}+2\vec{S})\cdot\vec{B} + \frac{e^2}{8m}\sum_{i=1}^{N}(\vec{r}_i\times\vec{B})^2,\end{aligned} \tag{24.49}$$

where we have used the individual orbital angular moment $\vec{L}_i = \vec{r}_i\times\vec{p}_i$ and the total orbital angular momentum $\vec{L} = \sum_{i=1}^{N}\vec{L}_i$. The last term is independent of L, S, and the middle term is (since $\vec{L}$ is the total angular momentum),

$$\vec{\mu}_{\rm tot} = \mu_B(\vec{L}+2\vec{S}). \tag{24.50}$$

For the electrons, we can define the total angular momentum

$$\vec{J} = \vec{L} + \vec{S}, \tag{24.51}$$

and similarly for the projections onto the z direction, $J_z = L_z + S_z$. The projection J_z is defined by an eigenvalue $M_J\hbar$, so the average for a state is $\langle J_z\rangle = M_J\hbar$. Moreover, we also have $\langle L_z\rangle = M_L\hbar$ and $\langle S_z\rangle = M_S\hbar$.

Then the energy split due to the angular momenta is

$$\Delta E = \frac{\mu_B B}{\hbar}(\langle L_z\rangle + 2\langle S_z\rangle) = \frac{\mu_B B}{\hbar}(\langle J_z\rangle + \langle S_z\rangle), \tag{24.52}$$

and we can write it in terms of a total Landé, or gyromagnetic, factor g for the whole atom, as

$$\Delta E = g\mu_B B M_J. \tag{24.53}$$

This is the *Zeeman effect*, of the splitting of the atomic levels under a magnetic field. Note that there are further terms, coming from the nonrelativistic expansion for the Dirac equation, including the spin–orbit coupling, but we will not consider them here.

In order to calculate g, we first square the relation $\vec{J} = \vec{L} + \vec{S}$, to obtain

$$2\vec{L}\cdot\vec{S} = \vec{J}^2 - \vec{L}^2 - \vec{S}^2. \tag{24.54}$$

Then, we take the quantum average over a state of given J, L, S, for which $\vec{J}^2 = J(J+1)\hbar^2$, $\vec{L}^2 = L(L+1)\hbar^2$ and $\vec{S}^2 = S(S+1)\hbar^2$, to obtain

$$\langle\vec{S}\cdot\vec{J}\rangle = \frac{\hbar^2}{2}\left[J(J+1) - L(L+1) - S(S+1)\right]. \tag{24.55}$$

Next, we note that, for a spin–orbit coupling that aligns the spin and total angular momentum, we obtain

$$\langle S_z\rangle = \langle J_z\rangle\frac{\langle\vec{J}\cdot\vec{S}\rangle}{\langle\vec{J}^2\rangle} = \frac{M_J\hbar}{2J(J+1)}\left[J(J+1) - L(L+1) - S(S+1)\right]. \tag{24.56}$$

We finally write

$$gM_J\hbar = \langle J_z\rangle + \langle S_z\rangle = M_J\hbar\left[1 + \frac{J(J+1) - L(L+1) - S(S+1)}{2J(J+1)}\right], \tag{24.57}$$

allowing us to write the Landé factor as

$$g = \left[1 + \frac{J(J+1) - L(L+1) - S(S+1)}{2J(J+1)}\right]. \tag{24.58}$$

Important Concepts to Remember

- For a quantum particle (perhaps with spin 1/2) in a constant magnetic field, we find that it behaves as a shifted and rescaled harmonic oscillator with cyclotron frequency $\omega_c = |q|B/m$ in a direction perpendicular to $\vec{B}$, leading to Landau levels n, $E_n = p_z^2/(2m) + \hbar\omega_c(n + m_s + 1/2)$.
- Classically, we have a motion at ω_c on a circle of radius $r = p_y/(m\omega_c)$ (with momentum perpendicular to the radial direction), while quantum mechanically we have motion in a direction x in the same plane as the circle, shifted by $x_0(p_y) = r$ (the x direction is perpendicular to the p_y direction), as that of a one-dimensional harmonic oscillator with the same ω_c, and rescaled by $l = \sqrt{\hbar/(|q|B)}$.
- In a physical situation, for each Landau level we have a degeneracy determined by the quantized momentum p_y, which is bounded because it leads to a shift in the perpendicular direction. This leads to a maximum number of states (degeneracy) of each Landau level equal to Φ/Φ_0, where $\Phi_0 = h/e$ is the flux quantum, and to $n_B = eB/h$ states per unit area in a Landau level.
- The classical Hall effect is as follows: if we have a B field in the z direction, then an E field in the y direction is correlated with the current in the x direction (either the current induces E_y, i.e., the Hall voltage V_H, or E_y induces I_x), leading to a Hall conductivity $\sigma_H = -n_e e/B$.
- The integer quantum Hall effect (IQHE) is the phenomenon whereby, for certain (2 + 1)-dimensional systems, the relation between V_H and B (which classically is $V_H = I/\sigma_H \propto B$) is not linear, but has *plateaux*, whose position is universal, and we also have an integer conductivity, $\sigma_H = -ne^2/h$, and $R_H = 0$ (zero longitudinal resistance).
- The fractional quantum Hall effect (FQHE) is the phenomen whereby, at much higher B fields and lower temperature T, the conductivity is fractional, $\sigma_H = \nu e^2/h$, with $\nu = k/(2p+1)$, (partially) explained by Laughlin in terms of a Laughlin quantum wave function.
- The Hall conductance σ_H of IQHE comes from n fully occupied Landau levels, with $n_e/n_B = n$, which also means $\sigma_{ij} = \sigma_H\epsilon_{ij}$, such that $\sigma_{xx} = 0$, yet because $R = \sigma^{-1}$ is a matrix inverse we also have $R_{xx} = 0$ (zero longitudinal resistance).
- For a multi-electron atom, the Landé g factor is found from writing $\Delta E = \frac{\mu_B B}{\hbar}(\langle J_z\rangle + \langle S_z\rangle)$ as $\frac{\mu_B B}{\hbar} g\langle J_z\rangle$.

Further Reading

See for instance [6] for more details on the Hall effects (IQHE and FQHE), and [2] for more on the Landé g factor.

Exercises

(1) Consider a stationary gauge transformation $\Lambda(x, y)$ in the plane transverse to a magnetic field B_z. Calculate the effect it has on the Schrödinger equation and its solutions.

(2) For the case in exercise 1, specialize to an infinitesimal gauge transformation that rotates $\vec{A}$ in its plane (transverse to $\vec{B}$), and interpret this in terms of the classical motion.

(3) For a $(2+1)$-dimensional system of sides L_x and L_y under the IQHE, calculate the length of the plateau in R_H as B is increased.

(4) In the FQHE, how many occupied states can there be in a Landau level? What do you deduce about the description of the states of the electron?

(5) In some more complicated materials, we can have both a Hall conductivity and a longitudinal, usually isotropic, conductivity. Calculate the resistivity matrix R_{ij}. Consider a transformation that acts by exchanging the values of the electric field components E_i and those of the current densities j_i, and write it as a transformation on the complex value $\sigma \equiv \sigma_{xy} + i\sigma_{xx}$.

(6) For a multi-electron atom, with total angular momentum $L > 1$ and total spin 1, what are the possible values of the Landé g factor? Specialize to the classical limit of large L.

(7) Consider a multi-electron atom, with spin S, orbital angular momentum L, and total angular momentum J in a slowly varying magnetic field $B(t)$ moving on a closed curve C. Calculate the Berry phase of the quantum state of the atom.

25 The WKB; a Semiclassical Approximation

We now go back to the WKB method, which first appeared in Chapter 11, where it was related to the classical limit as an expansion around the Hamilton–Jacobi formulation of classical theory. The WKB method is a method of approximation, and as such we will also go back to it in Part II of the book, but here we make a second pass at it since it will lead, in the next chapter, to the Bohr–Sommerfeld quantization, which was one of the first formulations of quantum mechanics.

25.1 Review and Generalization

The wave function solution of the Schrödinger equation was written as

$$\begin{aligned}\psi(\vec{r},t) &= e^{iS(\vec{r},t)/\hbar}\\ S(\vec{r},t) &= W(\vec{r}) - Et,\end{aligned} \tag{25.1}$$

which means that the time-independent wave function is

$$\psi(\vec{r}) = e^{iW(\vec{r})/\hbar}. \tag{25.2}$$

Moreover, we wrote $W(\vec{r})$ as

$$W(\vec{r}) = s(\vec{r}) + \frac{\hbar}{i}T(\vec{r}), \tag{25.3}$$

where $s(\vec{r})$ and $T(\vec{r})$ are uniquely defined as *even* functions in $\hbar$. Then

$$\psi(\vec{r}) = e^{T(\vec{r})}e^{is(\vec{r})/\hbar} \equiv Ae^{is(\vec{r})/\hbar}. \tag{25.4}$$

We can expand the even functions s, T in $\hbar$,

$$\begin{aligned}s &= s_0 + \left(\frac{\hbar}{i}\right)^2 s_2 + \left(\frac{\hbar}{i}\right)^4 s_4 + \cdots\\ T &= t_0 + \left(\frac{\hbar}{i}\right)^2 t_2 + \left(\frac{\hbar}{i}\right)^4 t_4 + \cdots .\end{aligned} \tag{25.5}$$

If we define $t_n = s_{n+1}$, it means that $W(\vec{r})$ has an expansion in all powers of $\hbar/i$, with coefficients s_n.

Then, in one spatial dimension, the Schrödinger equation becomes

$$\left(\frac{dW}{dx}\right)^2 - 2m(E - V(x)) + \frac{\hbar}{i}\frac{d^2W}{dx^2} = 0, \tag{25.6}$$

and expanding it in powers of $\hbar/i$ we obtain

$$\left[\sum_{n\geq 0}\frac{ds_n}{dx}\left(\frac{\hbar}{i}\right)^n\right]^2 - 2m(E - V(x)) + \sum_{n\geq 0}\frac{d^2}{dx^2}s_n\left(\frac{\hbar}{i}\right)^{n+1} = 0. \tag{25.7}$$

We can set to zero the coefficients of each power of $\hbar/i$, obtaining an infinite set of coupled equations for s_n.

Then, as we have seen, keeping only s_0 and $s_1 = t_0$ the WKB solution in one dimension is

$$\psi(x) = \frac{\text{const.}}{\sqrt{p(x)}} \exp\left[\pm\frac{i}{\hbar}\int_{x_0}^{x} dx' p(x') - \frac{iEt}{\hbar}\right], \tag{25.8}$$

where, if $E > V(x)$,

$$p(x) = \sqrt{2m(E - V(x))}. \tag{25.9}$$

However, for the solution in the classically forbidden region, $E < V(x)$, if we write

$$p(x) = i\chi(x) = i\sqrt{2m(V(x) - E)} \tag{25.10}$$

then the solution is

$$\psi(x,t) = \frac{\text{const}'}{\sqrt{\chi(x)}} \exp\left[\pm\frac{1}{\hbar}\int_{x_0}^{x} dx' \chi(x') - \frac{iEt}{\hbar}\right]. \tag{25.11}$$

The condition for the validity of this WKB approximation is $|\delta_\lambda \lambda/\lambda| \ll 1$, or

$$|\delta_\lambda \lambda| \ll \frac{\hbar}{\sqrt{2m(V(x) - E)}} \Leftrightarrow \frac{\hbar}{\sqrt{2m|V(x) - E|}} \ll \frac{2(E - V(x))}{|dV/dx|}, \tag{25.12}$$

where the last expression is the characteristic distance for the variation of the potential.

However, we note that, while being valid mostly everywhere, the above condition is certainly not valid near the turning points for the potential, where $E = V(x)$. Indeed, there we have $\psi \to \infty$, and

$$\frac{\hbar}{\sqrt{2m(V(x) - E)}} \to \infty \gg \frac{2|E - V(x)|}{|dV/dx|} \to 0. \tag{25.13}$$

Instead, we must use a different approximation.

25.2 Approximation and Connection Formulas at Turning Points

We consider first the case when $V(x)$ decreases through E at $x = x_1$, so the barrier (the classically forbidden region) is on the left-hand side of the potential diagram.

Then we can approximate the function

$$f(x) = \frac{2m}{\hbar^2}(E - V(x)) \tag{25.14}$$

by a Taylor expansion to the first order, i.e., a linear approximation,

$$f(x) = f'(x_1)(x - x_1) = -\frac{2m}{\hbar^2}V'(x_1)(x - x_1). \tag{25.15}$$

This can then be analytically continued through the complex-x plane, in order to avoid the point x_1, where $f(x_1) = 0$. One can make a rigorous analysis in this way, but it is rather complicated.

Instead, there is a much simpler shortcut, where we replace the function $f(x)$ with a step function:

$$f(x) = \begin{cases} -\alpha^2, & x < x_1 \\ +\alpha^2, & x > x_1. \end{cases} \tag{25.16}$$

This function is discontinuous, but that is not a problem as we have already dealt with step functions in the general chapter on one-dimensional problems.

With this $f(x)$, the time-independent Schrödinger equation becomes

$$\begin{aligned} \psi_-'' - \alpha^2 \psi_- &= 0, \quad x < x_1 \\ \psi_+'' + \alpha^2 \psi_+ &= 0, \quad x > x_1. \end{aligned} \tag{25.17}$$

The solutions $\psi_\pm$ on the different segments can be glued using the usual joining conditions, of continuity at x_1 for $\psi(x)$ and $\psi'(x)$.

The solution in the region $x < x_1$ that goes to zero at $x \to -\infty$ is $e^{-\alpha(x-x_1)}$; introducing a factor $1/\sqrt{\alpha}$ for consistency with the WKB approximation, we obtain

$$\psi_-(x) = \frac{1}{\sqrt{\alpha}} e^{-\alpha(x-x_1)} = \frac{1}{\sqrt{\alpha}} e^{-\int_x^{x_1} dx\, \alpha}, \tag{25.18}$$

so for this solution, we have

$$\psi_-(x_1) = \frac{1}{\sqrt{\alpha}}, \quad \psi_-'(x_1) = \sqrt{\alpha}. \tag{25.19}$$

On the other hand, the general solution in the region $x > x_1$ is

$$\psi_+(x) = \frac{1}{\sqrt{\alpha}} A \sin[\alpha(x - x_1) + \delta], \tag{25.20}$$

so for this solution, we have

$$\psi(x_1) = \frac{A \sin\delta}{\sqrt{\alpha}}, \quad \psi_+'(x_1) = A\sqrt{\alpha}\cos\delta. \tag{25.21}$$

The joining conditions are

$$\begin{aligned} \psi_+(x_1) = \psi_-(x_1) &\Rightarrow A \sin\delta = 1 \\ \psi_+'(x_1) = \psi_-'(x_1) &\Rightarrow A \cos\delta = 1. \end{aligned} \tag{25.22}$$

The ratio of the conditions gives $\tan\delta = 1$, so $\delta = \pi/4$, and thus we obtain $A = \sqrt{2}$. Then, since

$$\alpha = \sqrt{\frac{2m}{\hbar^2}|E - V(x)|}, \tag{25.23}$$

we obtain the condition for replacement (when going from the left to the right of the point) at the turning point:

$$\begin{aligned} &\frac{1}{[V(x) - E]^{1/4}} \exp\left[-\frac{1}{\hbar}\int_x^{x_1} dx' \sqrt{2m[V(x') - E]}\right] \\ &\to \frac{\sqrt{2}}{[E - V(x)]^{1/4}} \sin\left[\frac{1}{\hbar}\int_{x_1}^{x} dx' \sqrt{2m[E - V(x')]} + \frac{\pi}{4}\right] \\ &= \frac{\sqrt{2}}{[E - V(x)]^{1/4}} \cos\left[\frac{1}{\hbar}\int_{x_1}^{x} dx' \sqrt{2m[E - V(x')]} - \frac{\pi}{4}\right]. \end{aligned} \tag{25.24}$$

Similarly, for the opposite turning point, for an increasing $V(x)$ (thus for the barrier, or classically forbidden region, on the right-hand side), we obtain the condition for replacement (from the right to the left of x_2),

$$\frac{1}{[V(x)-E]^{1/4}}\exp\left[-\frac{1}{\hbar}\int_{x_2}^{x}dx'\sqrt{2m[V(x')-E]}\right]$$
$$\to\frac{\sqrt{2}}{[E-V(x)]^{1/4}}\sin\left[\frac{1}{\hbar}\int_{x}^{x_2}dx'\sqrt{2m[E-V(x')]}-\frac{\pi}{4}\right]$$
$$=\frac{\sqrt{2}}{[E-V(x)]^{1/4}}\cos\left[-\frac{1}{\hbar}\int_{x}^{x_2}dx'\sqrt{2m[E-V(x')]}+\frac{\pi}{4}\right]. \tag{25.25}$$

25.3 Application: Potential Barrier

Consider a potential well with depth $-V_0$ until $x = x_1$, after which $V(x)$ jumps up to $+V_{\max}$, and then decreases monotonically to zero at infinity, as in Fig. 25.1. The relevant case is that of the potential in nuclear α-decay, meaning nuclear fragmentation. Then the energy E of the decaying nucleus is higher than zero, representing the energy of the fragmented pieces. The potential well at $x < x_1$ corresponds to the nuclear force, while the monotonically decreasing barrier for $x > x_1$ is a Coulomb potential.

We define region I for $x < x_1$, setting $E = V(x)$ at $x = x_2 > x_1$; we have region II for $x_1 \leq x \leq x_2$, and region III for $x > x_2$. We further write, as usual,

$$\sqrt{\frac{2m(E-V(x))}{\hbar^2}} \equiv k(x), \quad x \in \text{III}$$
$$\sqrt{\frac{2m(V(x)-E)}{\hbar^2}} \equiv \kappa(x), \quad x \in \text{II} \tag{25.26}$$
$$\sqrt{\frac{2m(E+V_0)}{\hbar^2}} \equiv k, \qquad x \in \text{I}.$$

Then the wave function in region III is

$$\psi_{\text{III}}(x) = \frac{\sqrt{2}}{\sqrt{k(x)}}\cos\left[\int_{x_2}^{x}k(x)\,dx-\frac{\pi}{4}\right]$$
$$=\frac{1}{\sqrt{k(x)}}\left\{\exp\left[i\int_{x_2}^{x}k(x)\,dx-i\frac{\pi}{4}\right]+\exp\left[-i\int_{x_2}^{x}k(x)\,dx+i\frac{\pi}{4}\right]\right\}. \tag{25.27}$$

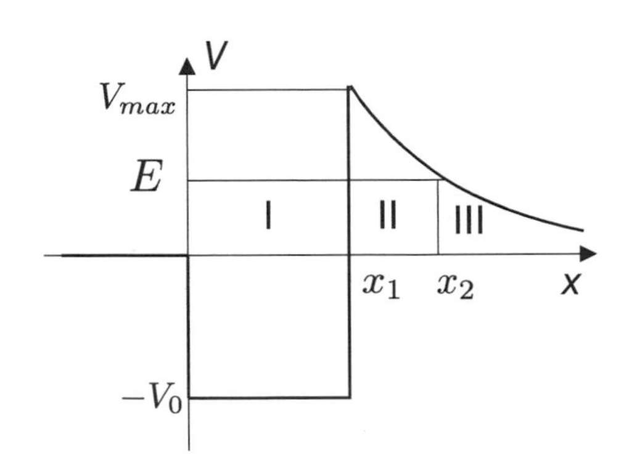

Figure 25.1 Potential barrier for α decay.

According to the previous WKB rules for continuity at the turning point, we have for the wave function in region II (except that the wave function should be now *increasing* away from the transition point, not decreasing)

$$\begin{aligned}\psi_{II}(x) &= \frac{1}{\sqrt{\kappa(x)}} \exp\left(+\int_x^{x_2} dx' \kappa(x')\right) \\ &= \frac{1}{\sqrt{\kappa(x)}} e^{\tau} \exp\left(-\int_{x_1}^{x} dx' \kappa(x')\right),\end{aligned} \tag{25.28}$$

where we have defined

$$\tau = \int_{x_1}^{x_2} dx' \kappa(x'). \tag{25.29}$$

Finally, in region I, where the potential is constant, $V(x) = -V_0$, we can solve the Schrödinger equation exactly (rigorously, without a WKB approximation):

$$\begin{aligned}\psi_I(x) &= A \sin[k(x - x_1) + \beta] \\ &= A \frac{e^{i[k(x-x_1)+\beta]} + e^{-i[k(x-x_1)+\beta]}}{2i}.\end{aligned} \tag{25.30}$$

The conditions for the continuity of $\psi(x)$ and $\psi'(x)$ at $x = x_1$ are

$$\begin{aligned}\psi_I(x_1) = \psi_{II}(x_1) &\Rightarrow \frac{1}{\sqrt{\kappa(x_1)}} e^{\tau} = A \sin \beta \\ \psi_I'(x_1) = \psi_{II}'(x_1) &\Rightarrow k \cot \beta \quad = -\kappa(x_1).\end{aligned} \tag{25.31}$$

But, given that we have written the wave functions ψ_I and ψ_{III} in terms of exponentials $e^{i\theta(x)}$ in (25.30) and (25.27), we can use the general formulas from Chapter 7 for the incident and transmitted currents,

$$\begin{aligned}\vec{j}_{\text{inc}} &= \frac{\hbar k}{m} \left(\frac{|A|}{2}\right)^2 \vec{e}_x \\ \vec{j}_{\text{trans}} &= \frac{\hbar k(x)}{m} \frac{1}{(\sqrt{k(x)})^2} \vec{e}_x = \frac{\hbar}{m} \vec{e}_x.\end{aligned} \tag{25.32}$$

Then the transmission coefficient is

$$T = \frac{|j_{\text{trans}}|}{|j_{\text{inc}}|} = \frac{4}{k|A|^2}. \tag{25.33}$$

But from (25.31), we obtain

$$\begin{aligned}\sin^2 \beta &= \frac{1}{1 + \cot^2 \beta} = \frac{1}{1 + \kappa^2(x_1)/k^2} \Rightarrow \\ |A|^2 &= \frac{e^{2\tau}}{\kappa(x_1)} \frac{1}{\sin^2 \beta} = \frac{e^{2\tau}}{\kappa(x_1)} \left(1 + \frac{\kappa^2(x_1)}{k^2}\right),\end{aligned} \tag{25.34}$$

so the transmission coefficient becomes

$$T = 4e^{-2\tau} \frac{k\kappa(x_1)}{k^2 + \kappa^2(x_1)} = 4e^{-2\tau} \frac{\sqrt{(V_{\max} - E)(E + V_0)}}{V_{\max} + V_0}. \tag{25.35}$$

25.4 The WKB Approximation in the Path Integral

The wave function is written in terms of the propagator (evolution operator),

$$\psi(x,t) = U(t,t_0)\psi(x,t_0), \tag{25.36}$$

and the propagator can be written as a path integral,

$$U(t,t_0) = \int \mathcal{D}x(t) e^{iS[x(t)]/\hbar}. \tag{25.37}$$

In terms of the path integral, the expansion in $\hbar$ of the exponent in $U(t,t_0) = e^{iW/\hbar}$ is related to the expansion of the action around its classical value. To a first approximation, the action is equal to the classical value plus a quadratic fluctuation around it,

$$S \simeq S_{\rm cl}[x_{\rm cl}(t)] + \int_0^t dt' \frac{(x(t') - x_{\rm cl}(t'))^2}{2} \frac{\delta^2 S}{\delta x^2}[x_{\rm cl}], \tag{25.38}$$

which corresponds to the expansion of $W(x) \simeq s_0(x) + \frac{\hbar}{i} s_1(x)$, which leads to $\psi(x) = e^{s_1} e^{is_0/\hbar}$. The classical part does not need to be integrated (since it is constant), while the path integration acts on the quadratic fluctuation part.

Since (putting $t_0 = 0$, and using the fact that $L = p\dot{x} - H$)

$$S_{\rm cl} = \int_0^t dt' \left(p(x)\frac{dx}{dt'} - E \right) = \int_{x_0}^{x_f} dx\, p(x) - Et, \tag{25.39}$$

and since

$$S = \int_0^t dt \left(\frac{m\dot{x}^2}{2} - V(x) \right) \Rightarrow \frac{p^2(x)}{2m} + V(x) = E, \tag{25.40}$$

we find that the zeroth-order term in the propagator becomes

$$\exp\left(\frac{i}{\hbar} S_{\rm cl}\right) = \exp\left[\frac{i}{\hbar}\int dx \sqrt{2m(E - V(x))} - \frac{i}{\hbar} Et\right], \tag{25.41}$$

which is the zeroth-order term ($e^{is_0/\hbar}$) in the WKB approximation.

On the other hand, the quadratic fluctuation part of S can be path integrated, as it is the form of a Gaussian integral type,

$$\int_{-\infty}^{+\infty} dx e^{-\alpha x^2} = \sqrt{\frac{\pi}{\alpha}}. \tag{25.42}$$

Since we can write

$$S \simeq S_{\rm cl}[x_{\rm cl}] + \frac{1}{2}\int_0^t dt' \,(x(t) - x_{\rm cl}(t)) \left(-m\frac{d^2}{dt^2} + (E - V(x))'' \right) (x(t) - x_{\rm cl}(t)), \tag{25.43}$$

the Gaussian integration over the quadratic fluctuation part gives the prefactor

$$e^{s_1} = \left\{ \det\left[-m\frac{d^2}{dt^2} - (V(x) - E)'' \right] \right\}^{-1/2} \equiv [\det A]^{-1/2}. \tag{25.44}$$

Now define $V''(x) \equiv m\omega^2$, and factor out the m prefactor from the operator A, since it gives an overall constant (independent of ω).

To calculate the determinant of the operator A, we must choose a basis of eigenfunctions for it, as follows.

In general, it is hard to obtain the WKB approximation term s_1 from the path integral calculation, and so from $[\det A]^{-1/2}$. Instead, we can obtain the eigenenergies, e.g., for periodic paths, as in the harmonic oscillator case. In this case we choose $x(t_f) = x(t_i) = x_0$ and then we integrate over it, $\int dx_0$. On these periodic paths, with period $T = t_f - t_i$, a good (complete) basis of eigenfunctions is $\sin \frac{n\pi t}{T}$. For this basis we have

$$\left(-\frac{d^2}{dt^2} - \omega^2\right) \sin \frac{n\pi t}{T} = \left(\frac{n^2\pi^2}{T^2} - \omega^2\right) \sin \frac{n\pi t}{T}. \tag{25.45}$$

Then, the determinant of the operator equals the product of the eigenvalues,

$$\det \mathcal{O} \equiv \det \frac{A}{m} = \prod_{n=1}^{\infty} \left(\frac{n^2\pi^2}{T^2} - \omega^2\right) \equiv K(T) \prod_{n=1}^{\infty} \left(1 - \frac{\omega^2 T^2}{n^2\pi^2}\right) = K(T) \frac{\sin \omega T}{\omega T}, \tag{25.46}$$

where we have factored out another kinematic, i.e., ω-independent, factor, $K(T)$.

Now one can do the integral over x_0 for the combined classical and first quantum terms. After a calculation that will not be repeated here (it can be found for instance in [10], Chapter 6.1), we obtain

$$\int dx_0 e^{iS_{\rm cl}/\hbar} \frac{1}{\sqrt{\det \mathcal{O}}} = \tilde{K}(T) \sum_{n=0}^{\infty} e^{-i(n+1/2)\omega T} = \tilde{K}(T) \sum_{n=0}^{\infty} e^{-iE_n T/\hbar}, \tag{25.47}$$

where $\tilde{K}(T)$ is another constant that is independent of the dynamics (i.e., of ω). The above result is the correct propagator for the harmonic oscillator.

Important Concepts to Remember

- In the expansion $\psi(\vec{r}, t) = \exp\left[\frac{i}{\hbar}(W(\vec{r}) - Et)\right]$, we write $W(\vec{r}) = s(\vec{r}) + \frac{\hbar}{i} T(\vec{r})$ and expand $s(\vec{r})$ and $T(\vec{r})$ in $(\hbar/i)^2$, obtaining a series expansion of the Schrödinger equation that corresponds to quantum corrections to the Hamilton–Jacobi equation.
- The WKB solution (the solution in the WKB approximation) in one dimension is still formally a solution in the classically forbidden region $E - V(x) < 0$, by writing $p(x) = i\chi(x)$ (where $p(x) = \sqrt{2m(E - V(x))}$), but the solution (the WKB approximation) is not valid near the turning points in the potential $E = V(x)$.
- At the turning points in the potential, we can write connection formulas, which connect a solution in terms of $p(x)$ into a solution in terms of $\chi(x)$.
- In the path integral, the WKB approximation amounts to the quadratic approximation for the action S in the exponent e^{iS} around its classical (minimum) value $S_{\rm cl}$. We are left with an integral over x_0, in the case of periodic paths that begin and start at the same x, $x_f = x_i = x_0$, as for the harmonic oscillator, obtaining $\sum_{n=0}^{\infty} e^{-iE_n T/\hbar}$, where E_n are the eigenenergies of the harmonic oscillator.

Further Reading

See [3], [2] for more details on WKB in one dimension, and [10] for more details on WKB in the path integral.

Exercises

(1) Write down the equation for s_2 in terms of s_0 and $s_1 = t_0$, and substitute into it the values of s_0 and s_1 from the WKB solution in one dimension.

(2) Write down the connection formulas at the turning points for a harmonic oscillator in one dimension.

(3) Write down the connection formulas at the turning points for a potential $V = V_0 \cos(ax)$ $(V_0 > 0)$, for $E < V_0$.

(4) Consider the potential barrier given by $V_{\text{eff}}(r)$ for the radial motion in a potential $V = -|\alpha|/r^3$ with a positive energy smaller than the barrier. Write down the connection formulas or the turning points in this case.

(5) In the case in exercise 4, calculate the transmission coefficient.

(6) Consider the potential $V = -V_0 \cos(ax)$ $(V_0 > 0)$ with a small initial condition $(x_0 < \pi/(2a))$. Calculate the WKB approximation in the path integral.

(7) Can we truncate the sum over n in the propagator (25.47) to a finite order, as an approximation?

26 Bohr–Sommerfeld Quantization

In this chapter we use the WKB method on *bound states* to derive a modified version of the quantization condition put forward initially by Bohr and Sommerfeld, which was one of the first calculational methods of quantum mechanics. We then apply this modified Bohr–Sommerfeld quantization condition to a set of examples, including the harmonic oscillator and the hydrogen atom, in order to derive the eigenenergies, and wave functions for the states.

26.1 Bohr–Sommerfeld Quantization Condition

We consider a potential well, where for $x < x_1$ we have $V(x) > E$, called region I (classically forbidden), for $x_1 \leq x \leq x_2$ we have $V(x) \leq E$, called region II (the classically allowed region, the potential well), and for $x > x_2$ we also have $V(x) > E$, called region III, as in Fig. 26.1.

We can now use the connection formulas at the turning points x_1, x_2 derived in the previous chapter. In region I, we have the WKB solution

$$\psi_{\text{I}}(x) = \frac{1}{[2m(V(x) - E)]^{1/4}} \exp\left[-\frac{1}{\hbar}\int_x^{x_1} dx'\sqrt{2m(V(x') - E)}\right], \tag{26.1}$$

that transitions into the solution in region II,

$$\psi_{\text{II}}(x) = \frac{\sqrt{2}}{[2m(E - V(x))]^{1/4}} \cos\left[\frac{1}{\hbar}\int_{x_1}^{x} dx'\sqrt{2m(E - V(x'))} - \frac{\pi}{4}\right]. \tag{26.2}$$

On the other hand, the correct WKB solution in region III is

$$\psi_{\text{III}}(x) = \frac{1}{[2m(V(x) - E)]^{1/4}} \exp\left[-\frac{1}{\hbar}\int_{x_2}^{x} dx'\sqrt{2m(V(x') - E)}\right], \tag{26.3}$$

which transitions into the solution in region II.

$$\tilde{\psi}_{\text{II}}(x) = \frac{\sqrt{2}}{[2m(E - V(x))]^{1/4}} \cos\left[-\frac{1}{\hbar}\int_{x}^{x_2} dx'\sqrt{2m(E - V(x'))} + \frac{\pi}{4}\right]. \tag{26.4}$$

In order for the two solutions in region II to be the same, $\psi_{\text{II}}(x) = \tilde{\psi}_{\text{II}}(x)$, we must have the quantization condition

$$\frac{1}{\hbar}\int_{x_1}^{x_2} dx'\sqrt{2m(E - V(x'))} - \frac{\pi}{2} = n\pi. \tag{26.5}$$

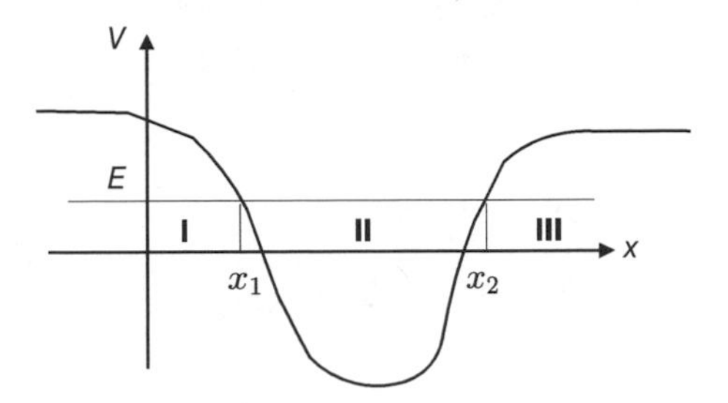

Figure 26.1 Potential regions.

Note that for odd n, we have that the argument of cos in $\psi_{\rm II}$ is minus the argument in $\tilde{\psi}_{\rm II}$ (which still gives the same function), whereas for even n, it is the same argument. Then the quantization condition becomes

$$\int_{x_1}^{x_2} dx' \sqrt{2m(E - V(x'))} = \left(n + \frac{1}{2}\right) \pi\hbar. \tag{26.6}$$

Considering the integral over a closed path, $\int_{x_1}^{x_2} + \int_{x_2}^{x_1} = \oint_C$, corresponding to a full oscillation between x_1 and x_2, and generalizing x to a general variable q, and $p(x)$ to its canonically conjugate momentum p, we obtain

$$\oint_C dx\, p(x) = \oint_C p\, dq = \left(n + \frac{1}{2}\right) 2\pi\hbar = \left(n + \frac{1}{2}\right) h, \tag{26.7}$$

where $n = 0, 1, 2, \ldots$

We note that, except for the constant (1/2 instead of 1), this is the same as the original Bohr–Sommerfeld quantization condition, which was postulated (not derived),

$$\oint_C p\, dq = (n + 1)h, \tag{26.8}$$

where again $n = 0, 1, 2, \ldots$ We have thus obtained an improved version of Bohr–Sommerfeld quantization that includes the classical and first semiclassical terms, and thus is expected to give the correct result either in very simple cases (such as the harmonic oscillator and the hydrogen atom), or for large quantum numbers n, when we are closer to classicality.

We will thus apply it to find the eigenenergies E_n and eigenfunctions ψ_n, and we expect in general to find very good agreement at large n.

Finally, another observation is that the WKB wave function has n nodes for the quantum number n, just as in the general (exact) analysis of one-dimensional problems. To see this, note the form of the function $\psi_{\rm II}(x)$ in the classically allowed region, written in terms of the cos therefore having n nodes for the quantum number n.

26.2 Example 1: Parity-Even Linear Potential

We consider the potential

$$V(x) = k|x|, \tag{26.9}$$

which applies for instance, to the $q\bar{q}$ potential in QCD: when pulling the quark and antiquark apart, there is a constant force pulling them back together. This example can also be solved exactly, so we can compare how well the result works.

The turning points are given by

$$E - V(x_{1,2}) = 0 \Rightarrow x_{1,2} = \mp\frac{E}{k}. \tag{26.10}$$

Then, integrating between them and noting that $x_1 = -x_2$ for an even integrand, we obtain

$$\begin{aligned}\int_{x_1}^{x_2} dx' \sqrt{2m(E - V(x'))} &= 2\int_0^{x_2} dx' \sqrt{2m(E - kx')} \\ &= 2\frac{\sqrt{2mE}E}{k}\int_0^1 dy\sqrt{1-y} \\ &= \frac{4}{3}\frac{E\sqrt{2mE}}{k},\end{aligned} \tag{26.11}$$

where we have used $\int_0^1 dy\sqrt{1-y} = 2/3$. Since this integral equals $(n + 1/2)\pi\hbar$ by the quantization condition, we obtain the eigenenergies

$$E_n = \left[\frac{3\pi k\hbar(n + 1/2)}{4\sqrt{2m}}\right]^{2/3}. \tag{26.12}$$

But in this case, the exact eigenenergies are known, and are given by

$$E_n = \left[\lambda_n \frac{k\hbar}{\sqrt{2m}}\right]^{2/3}, \tag{26.13}$$

where λ_n are the zeroes of the Airy function,

$$Ai(-\lambda_n) = 0. \tag{26.14}$$

Then we find excellent agreement, starting at $n = 1$, where the error is at about 1%, after which it gets better. The only poor agreement is for the ground state, E_0 (for $n = 0$). So in this case, the eigenenergies are getting better at large n, as expected. Also as expected, the wave functions of these states are also improving at large n except near the turning points, where we must use the approximation from the previous chapter.

26.3 Example 2: Harmonic Oscillator

Consider next the harmonic oscillator potential,

$$V(x) = \frac{m\omega^2}{2}x^2. \tag{26.15}$$

In this case, the integral appearing in the quantization condition is (note that now we also have $x_1 = -x_2$, and at the turning points we have $2mE - m^2\omega^2 x_{1,2}^2 = 0$)

$$
\begin{aligned}
I &= \int_{x_1}^{x_2} dx' \sqrt{2m(E - V(x'))} = \int_{x_1}^{x_2} dx' \sqrt{2mE - m^2\omega^2 x'^2} \\
&= x\sqrt{2mE - m^2\omega^2 x^2}\Big|_{x_1}^{x_2} + \int_{x_1}^{x_2} x' dx' \frac{m^2\omega^2 x'}{\sqrt{2mE - m^2\omega^2 x'^2}} \\
&= -\int_{x_1}^{x_2} dx' \sqrt{2mE - m\omega^2 x'^2} + 2mE \int_{x_1}^{x_2} \frac{dx'}{\sqrt{2mE - m^2\omega^2 x'^2}},
\end{aligned}
\tag{26.16}
$$

which leads to

$$
2I = 2mE \int_{x_1}^{x_2} \frac{dx}{\sqrt{2mE - m^2\omega^2 x^2}} = \frac{2mE}{m\omega}\pi = \frac{2\pi E}{\omega}. \tag{26.17}
$$

Therefore the quantization condition is

$$
\frac{E}{\omega} = \hbar\left(n + \frac{1}{2}\right) \Rightarrow E = \hbar\omega\left(n + \frac{1}{2}\right). \tag{26.18}
$$

This is in fact the exact quantization condition for the harmonic oscillator, so it works at all n.

This result is consistent with the path integral result from the end of the previous chapter, where we saw that the WKB propagator contained the exact eigenenergies for the harmonic oscillator,

$$
U(T) = \sum_{n=0}^{\infty} e^{-iE_n T/\hbar}. \tag{26.19}
$$

Wave Functions

The wave functions in this case are not satisfactory at intermediate values of x, but become better at large x ($x \to \pm\infty$), which means in the classically forbidden region. In this classically forbidden region, the WKB wave function is (for instance, for $x > x_2$, $x \to \infty$)

$$
\psi(x) = \frac{1}{[m^2\omega^2 x^2 - 2mE]^{1/4}} \exp\left[-\frac{1}{\hbar}\int_{x_2}^{x} dx' \sqrt{m^2\omega^2 x^2 - 2mE}\right]. \tag{26.20}
$$

However, at large x,

$$
\sqrt{m^2\omega^2 x^2 - 2mE} = m\omega x \sqrt{1 - \frac{2E}{m\omega^2 x^2}} \simeq + m\omega x - \frac{E}{\omega x}, \tag{26.21}
$$

so the exponent in the wave function is

$$
-\frac{1}{\hbar}\int_{x_2}^{x} dx' \sqrt{m^2\omega^2 x^2 - 2mE} \simeq \text{const} - \frac{m\omega^2}{\hbar}\frac{x^2}{2} + \frac{E}{\hbar\omega}\ln x + \mathcal{O}\left(\frac{1}{x^2}\right). \tag{26.22}
$$

Since $E/\hbar\omega = n + 1/2$, we obtain the approximate wave function

$$
\begin{aligned}
\psi(x) &\simeq \frac{\text{const.}}{[m^2\omega^2 x^2 - 2mE]^{1/4}} \exp\left[-\frac{m\omega^2}{\hbar}\frac{x^2}{2} + \left(n + \frac{1}{2}\right)\ln x\right] \\
&\simeq \frac{\text{const}}{\sqrt{m\omega x}} x^{n+1/2} \exp\left(-\frac{m\omega^2}{\hbar}\frac{x^2}{2}\right) = \text{const}' \times x^n \exp\left(-\frac{m\omega^2}{\hbar}\frac{x^2}{2}\right).
\end{aligned}
\tag{26.23}
$$

But these are indeed the leading (exponential) and first subleading (power law) terms in the exact wave functions, since x^n is the leading term in the Hermite polynomial $P_n(x)$.

26.4 Example 3: Motion in a Central Potential

Consider a central potential in three dimensions, $V = V(r)$. Then the time-independent Schrödinger equation is

$$-\hbar^2 \Delta\psi - 2m(E - V(r))\psi = 0. \tag{26.24}$$

Writing the time-independent wave function as the usual exponential (since $S(\vec{r}, t) = W(\vec{r}) - Et$),

$$\psi(\vec{r}) = e^{iW(\vec{r})/\hbar}, \tag{26.25}$$

we obtain an equation for W that is a quantum-corrected version of the Hamilton–Jacobi equation,

$$(\vec{\nabla}W)^2 - 2m(E - V(r)) + \frac{\hbar}{i}\Delta W = 0. \tag{26.26}$$

But, in spherical coordinates, we have

$$\begin{aligned}(\vec{\nabla}W)^2 &= \left(\frac{\partial W}{\partial r}\right)^2 + \frac{1}{r^2}\left(\frac{\partial W}{\partial \theta}\right)^2 + \frac{1}{r^2 \sin^2\theta}\left(\frac{\partial W}{\partial \phi}\right)^2 \\ \Delta W &= \frac{\partial^2 W}{\partial r^2} + \frac{2}{r}\frac{\partial W}{\partial r} + \frac{1}{r^2}\left[\frac{\partial^2 W}{\partial \theta^2} + \cot\theta \frac{\partial W}{\partial \theta} + \frac{1}{\sin^2\theta}\frac{\partial^2 W}{\partial \phi^2}\right].\end{aligned} \tag{26.27}$$

Then, as we saw in Chapter 11, expanding W in $\hbar$ in the usual way,

$$W = s_0 + \frac{\hbar}{i}s_1 + \left(\frac{\hbar}{i}\right)^2 s_2 + \cdots, \tag{26.28}$$

and substituting in the Schrödinger equation, we obtain an equation that is also an infinite series in $\hbar$, where setting to zero each coefficient leads to an independent equation. Keeping only the leading and first subleading terms, we obtain

$$\begin{aligned}&\left(\frac{\partial s_0}{\partial r}\right)^2 + \frac{1}{r^2}\left(\frac{\partial s_0}{\partial \theta}\right)^2 + \frac{1}{r^2\sin^2\theta}\left(\frac{\partial s_0}{\partial \phi}\right)^2 - 2m(E - V(r)) \\ &\quad + \frac{\hbar}{i}\left[2\frac{\partial s_0}{\partial r}\frac{\partial s_1}{\partial r} + \frac{2}{r^2}\frac{\partial s_0}{\partial \theta}\frac{\partial s_1}{\partial \theta} + \frac{2}{r^2\sin^2\theta}\frac{\partial s_0}{\partial \phi}\frac{\partial s_1}{\partial \phi} + \Delta s_0\right] = 0.\end{aligned} \tag{26.29}$$

The leading (order-one) term in the equation is just the Hamilton–Jacobi equation,

$$\left(\frac{\partial s_0}{\partial r}\right)^2 + \frac{1}{r^2}\left(\frac{\partial s_0}{\partial r}\right)^2 + \frac{1}{r^2\sin^2\theta}\left(\frac{\partial s_0}{\partial \phi}\right)^2 - 2m(E - V(r)) = 0. \tag{26.30}$$

We separate variables, as usual in the Hamilton–Jacobi equation, by writing in spherical coordinates

$$s_0(r, \theta, \phi) = R_0(r) + \Theta_0(\theta) + \Phi_0(\phi), \tag{26.31}$$

which leads to three independent equations, one for each separated function,

$$\begin{aligned} \left(\frac{d\Phi_0}{d\phi}\right)^2 &= L_3^2 \\ \left(\frac{d\Theta_0}{d\theta}\right)^2 + \frac{L_3^2}{\sin^2\theta} &= L^2 \\ \left(\frac{dR_0}{dr}\right)^2 + \frac{L^2}{r^2} - 2m[E - V(r)] &= 0. \end{aligned} \tag{26.32}$$

The first equation is solved directly as

$$\Phi_0(\phi) = L_3\phi, \tag{26.33}$$

the second is integrated as

$$\Theta_0(\theta) = \int d\theta \sqrt{L^2 - \frac{L_3^2}{\sin^2\theta}}, \tag{26.34}$$

and the third is also integrated as

$$R_0(r) = \int dr \sqrt{2m(E - V(r)) - \frac{L^2}{r^2}}. \tag{26.35}$$

The equation for s_1 is obtained by setting to zero the coefficient for $\hbar/i$ in (26.29), which, after substituting the separated $s_0(r, \theta, \phi)$, gives

$$2\left[\frac{dR_0}{dr}\frac{\partial s_1}{\partial r} + \frac{1}{r^2}\frac{d\Theta_0}{d\theta}\frac{\partial s_1}{\partial \theta} + \frac{1}{r^2\sin^2\theta}\frac{d\Phi_0}{d\phi}\frac{\partial s_1}{\partial \phi}\right] + \Delta s_0 = 0. \tag{26.36}$$

We can separate variables also in s_1,

$$s_1 = R_1(r) + \Theta_1(\theta) + \Phi_1(\phi). \tag{26.37}$$

Since Δs_0 contains also $d^2\Phi_0/d\phi^2 = 0$, we do not have any ϕ dependence in the equation for s_1, which means that we can *choose* to put $\Phi_1(\phi) = 0$, in which case the equation becomes

$$2\left[\frac{dR_0}{dr}\frac{dR_1}{dr} + \frac{1}{r^2}\frac{d\Theta_0}{d\theta}\frac{d\Theta_1}{d\theta}\right] + \frac{d^2R_0}{dr^2} + \frac{2}{r}\frac{dR_0}{dr} + \frac{1}{r^2}\left[\frac{d^2\Theta_0}{d\theta^2} + \cot\theta\frac{d\Theta_0}{d\theta}\right] = 0. \tag{26.38}$$

But the separation of variables gives the equations

$$\begin{aligned} 2\frac{dR_0}{dr}\frac{dR_1}{dr} + \frac{d^2R_0}{dr^2} + \frac{2}{r}\frac{dR_0}{dr} &= 0 \\ 2\frac{d\Theta_0}{d\theta}\frac{d\Theta_1}{d\theta} + \frac{d^2\Theta_0}{d\theta^2} + \cot\theta\frac{d\Theta_0}{d\theta} &= 0, \end{aligned} \tag{26.39}$$

or equivalently (denoting by a prime the derivative with respect to the coordinate, no matter what that is)

$$\begin{aligned} \Theta_1'(\theta) + \frac{1}{2}\left(\frac{\Theta_0''(\theta)}{\Theta_0'(\theta)} + \cot\theta\right) &= 0 \\ R_1'(r) + \frac{1}{2}\left(\frac{R_0''(r)}{R_0'(r)} + \frac{2}{r}\right) &= 0. \end{aligned} \tag{26.40}$$

The solutions of these equations are given by

$$\Theta_1(\theta) = \text{const} - \frac{1}{2}\ln\left[\Theta_0'(\theta)\sin\theta\right]$$
$$R_1(r) = \text{const} - \frac{1}{2}\ln\left[r^2 R_0'(r)\right], \tag{26.41}$$

which implies for the WKB wave function the solution

$$\psi(r,\theta,\phi) = \frac{e^{iR_0(r)/\hbar}}{r\sqrt{R_0'(r)}}\frac{e^{i\Theta_0(\theta)/\hbar}}{\sqrt{\Theta_0'(\theta)\sin\theta}}e^{iL_3\phi/\hbar}. \tag{26.42}$$

The requirement for 2π periodicity in the angle ϕ means that we need to have

$$L_3 = n_\phi \hbar. \tag{26.43}$$

Substituting the integral formula for $\Theta_0(\theta)$ into the WKB wave function, the θ dependence in it is

$$\frac{1}{[L^2\sin^2\theta - L_3^2]^{1/4}}\exp\left[\pm\frac{i}{\hbar}\int d\theta\sqrt{L^2 - \frac{L_3^2}{\sin^2\theta}}\right]. \tag{26.44}$$

Considering it as a WKB wave function in one dimension, for which we have the WKB quantization condition, or, equivalently, taking a Bohr–Sommerfeld quantization condition modified by the addition of 1/2, we obtain the condition

$$\int_{\theta_1}^{\theta_2} d\theta'\sqrt{L^2 - \frac{L_3^2}{\sin^2\theta'}} = \left(n_\theta + \frac{1}{2}\right)\pi\hbar, \tag{26.45}$$

where $\theta_{1,2}$ are the turning points,

$$|\sin\theta_{1,2}| = \frac{|L_3|}{L}. \tag{26.46}$$

To obtain the quantization condition, we will use an integral that we can calculate:

$$I(a,b,c) = \int_{x_1}^{x_2}\frac{dx}{\sqrt{a + 2bx - c^2x^2}}, \tag{26.47}$$

where $x_{1,2}$ are the zeroes of the square root in the denominator (the turning points). It can be rewritten by shifting the integral over x in such a way that the endpoints $x_{1,2}$ become symmetric with respect to the origin, namely $\pm|d|/|c|$,

$$I(a,b,c) = \int_{-|d|/|c|}^{+|d|/|c|}\frac{dx}{\sqrt{d^2 - c^2x^2}} = \int_{-\pi/2}^{+\pi/2} d\theta\frac{|d/c|\cos\theta}{\sqrt{d^2(1-\sin^2\theta)}} = \frac{1}{|c|}\int_{-\pi/2}^{+\pi/2} d\theta = \frac{\pi}{|c|}, \tag{26.48}$$

where we have used the change of variables $x = |d/c|\sin\theta$, meaning that the integral from $-|d|/|c|$ to $+|d|/|c|$ becomes an integral from $-\pi/2$ to $+\pi/2$.

Then we can calculate the integral in the quantization condition, as

$$\begin{aligned}\int_{\theta_1}^{\theta_2} d\theta\sqrt{L^2-\frac{L_3^2}{\sin^2\theta}} &= \int_{\theta_1}^{\theta_2} d\theta\frac{L^2}{\sqrt{L^2-L_3^2/\sin^2\theta}} - \int_{\theta_1}^{\theta_2}\frac{d\theta}{\sin^2\theta}\frac{L_3^2}{\sqrt{L^2-L_3^2/\sin^2\theta}}\\ &= -\int_{\theta_1}^{\theta_2} d(\cos\theta)\frac{L^2}{\sqrt{L^2-L_3^2-L^2\cos^2\theta}} + \int_{\theta_1}^{\theta_2}\frac{L_3^2 d(\cot\theta)}{\sqrt{L^2-L_3^2-L_3^2\cot^2\theta}}\\ &= \pi L - \pi|L_3|,\end{aligned} \tag{26.49}$$

where we have used $d(\cot\theta) = -d\theta/\sin^2\theta$, and then the formula (26.48) in both terms (i.e., for the terms in $\cos\theta$, and in $\cot\theta$).

Finally then, the quantization condition is

$$L - |L_3| = \hbar\left(n_\theta + \frac{1}{2}\right), \tag{26.50}$$

and, since $L_3 = n_\phi\hbar$, we find

$$L = \hbar\left(n_\theta + n_\phi + \frac{1}{2}\right) \equiv \hbar\left(l + \frac{1}{2}\right), \tag{26.51}$$

where $l \geq |n_\phi|$. But then we have, in the WKB approximation,

$$L^2 = \hbar^2\left(l + \frac{1}{2}\right)^2, \tag{26.52}$$

which can be compared with the exact solution,

$$L^2 = \hbar^2 l(l+1) = \hbar^2\left[\left(l + \frac{1}{2}\right)^2 - \frac{1}{4}\right]. \tag{26.53}$$

Note that, strictly speaking, since $\theta \in [0,\pi]$ we cannot use the previous WKB analysis on $\mathbb{R}$. But the condition we used was that the function does not blow up at $\theta = 0$ or π, and that is indeed true. Using the connection formulas at a turning point, for $0 \leq \theta \leq \theta_1$, we have the angular wave function

$$\frac{1}{[L_3^2 - L^2\sin^2\theta]^{1/4}}\exp\left[-\frac{1}{\hbar}\int_0^{\theta_1} d\theta\sqrt{\frac{L_3^2}{\sin^2\theta} - L^2}\right]. \tag{26.54}$$

At $\theta \to 0$, this becomes

$$\frac{1}{\sqrt{|L_3|}}\exp\left[-\frac{|L_3|}{\hbar}\int\frac{d\theta}{\sin\theta}\right] = \frac{1}{\sqrt{|L_3|}}\exp\left(|n_\phi|\ln\tan\frac{\theta}{2}\right) = \frac{1}{\sqrt{|L_3|}}\left(\tan\frac{\theta}{2}\right)^{|n_\phi|} \to 0, \tag{26.55}$$

so it means we actually have the correct boundary conditions, despite this not being obvious a priori.

Radial Wave Function

Finally we move on to the radial wave function in $\psi(r,\theta,\phi)$, which is

$$\frac{1}{r\left[2m(E-V(r)) - L^2/r^2\right]^{1/4}}\exp\left[\frac{i}{\hbar}\int_{r_1}^{r} dr'\sqrt{2m(E-V(r)) - \frac{L^2}{r^2}}\right]. \tag{26.56}$$

However, the radial quantization condition,

$$\int_{r_1}^{r_2} dr\sqrt{2m(E-V(r))-\frac{L^2}{r^2}}=\pi\hbar\left(n_r+\frac{1}{2}\right), \tag{26.57}$$

depends on the potential $V(r)$, so we must choose a potential in order to obtain a WKB result and compare with the exact result.

26.5 Example: Coulomb Potential (Hydrogenoid Atom)

The potential for the hydrogenoid atom is the usual

$$V(r)=-\frac{Ze_0^2}{r}, \tag{26.58}$$

where $e_0^2 = e^2/(4\pi\epsilon_0)$. Also, the energy of bound states is negative, so $E = -|E|$. Then the quantization condition is written as

$$J=\int_{r_1}^{r_2} dr\sqrt{-2m|E|+\frac{2mZe_0^2}{r}-\frac{L^2}{r^2}}=\pi\hbar\left(n_r+\frac{1}{2}\right), \tag{26.59}$$

and the integral J is calculated by integrating by parts,

$$\begin{aligned} J &= r\sqrt{-2m|E|+\frac{2mZe_0^2}{r}-\frac{L^2}{r^2}}\Bigg|_{r_1}^{r_2} \\ &\quad -\int_{r_1}^{r_2} r\,dr\frac{-2mZe_0^2/r^2+2L^2/r^3}{2\sqrt{-2m|E|+2mZe_0^2/r-L^2/r^2}} \\ &= -\int_{r_1}^{r_2} dr\frac{-mZe_0^2/r+L^2/r^2}{\sqrt{-2m|E|+2mZe_0^2/r-L^2/r^2}} \\ &= \int_{r_1}^{r_2} dr\frac{mZe_0^2}{\sqrt{-L^2+2mZe_0^2r-2m|E|r^2}} \\ &\quad -\int_{r_1}^{r_2}\frac{dr}{r^2}\frac{L^2}{\sqrt{-2m|E|+2mZe_0^2/r-L^2/r^2}} \\ &= \frac{mZe_0^2\pi}{\sqrt{2m|E|}}+L^2\int_{z_2}^{z_1}\frac{dz}{\sqrt{-2m|E|+2mZe_0^2z-L^2z^2}} \\ &= \frac{mZe_0^2}{\sqrt{2m|E|}}\pi-\pi L, \end{aligned} \tag{26.60}$$

where in the fourth equality we have used the integral (26.48) in the first term and defined $1/r = z$ in the second, and in the last line, we have used the integral (26.48) in the second term. Then, the WKB quantization means that

$$J = \pi\hbar\left(n_r + \frac{1}{2}\right), \tag{26.61}$$

and, using the fact that $L = \hbar(l + 1/2)$, as we have seen before, we obtain the eigenenergies

$$E = -|E| = -\frac{mZ^2 e_0^4}{2\hbar^2(n_r + l + 1)^2}, \tag{26.62}$$

which are in fact the exact eigenenergies (at all $n = n_r + l + 1$, independently of whether n is large or small).

For the wave functions, we have the same observation that we made for the θ dependence: the radial direction is only on half the real line, $0 \leq r < \infty$, so technically speaking the WKB analysis would not apply. Nevertheless, in the classically forbidden region $0 \leq r \leq r_1$, we have the wave function

$$\begin{aligned}\psi(r) &\simeq \frac{1}{r\left(L^2/r^2\right)^{1/4}} \exp\left[-\frac{L}{\hbar}\int_r^{r_1}\frac{dr}{r}\right] = \frac{1}{\sqrt{Lr}}\exp\left[-(l+1/2)\ln\frac{r_1}{r}\right] \\ &= \frac{(r/r_1)^{l+1/2}}{\sqrt{Lr}} \sim r^l \to 0,\end{aligned} \tag{26.63}$$

where we have used the fact that $L/\hbar = l + 1/2$. So, as before, this is the correct boundary condition at $r = 0$ (finiteness of the wave function), even though the wave function doesn't extend to $r = -\infty$, so the WKB analysis actually applies.

Important Concepts to Remember

- The Bohr–Sommerfeld quantization condition (the version modified by 1/2 from its original form) is obtained by applying the WKB approximation, and the resulting turning-point connection formulas, to a closed path, $\int_{x_1}^{x_2} + \int_{x_2}^{x_1} = \oint_C$. This results in $\oint_C dx\, p(x) = \oint_C dq\, p = (n+1/2)2\pi\hbar = (n + 1/2)h$, with $n = 0, 1, 2, \ldots$ (compared with the original, for which $(n + 1)h$).
- Applying Bohr–Sommerfeld quantization to some simple cases, we can either get an exact result or obtain the exact result at large n (in the classical limit).
- For $V(x) = k|x|$, we get a wrong result for the energies only for $n = 0$; from $n = 1$ onward we get less than 1% error.
- For the harmonic oscillator, we get the correct eigenenergies, but only the leading result at (small x and) large x for the wave functions is correct.
- For the motion in a central potential, writing as usual $\psi = e^{iW}$ with $W = s(\vec{r}) - Et$ with expanding $s(\vec{r})$ into $\hbar/i$ we get the quantum-corrected Hamilton–Jacobi equation, which can be solved by the separation of variables, $s_0(r, \theta, \phi) = R_0(r) + \Theta_0(\theta) + \Phi_0(\phi)$, $s_1 = R_1(r) + \Theta_1(\theta) + \Phi_1(\phi)$, etc.
- One thus obtains a WKB solution (by directly solving for R_0, Θ_0, Φ_0 and then R_1, Θ_1, Φ_1 in terms of them); the Bohr–Sommerfeld quantization conditions for $\theta \in [0, \pi]$ and $r \in [0, \infty)$ (given that $L_3 = n_\phi\hbar$) lead to $L = (l + 1/2)\hbar$ with $l = n_\theta + n_\phi$ and another quantization condition, depending on $V(r)$.
- For the hydrogenoid atom, n_r quantization (for $R(r)$) gives the correct eigenenergies and the correct boundary conditions for the wave function at $\theta = 0$ and $r = 0$, though not the exact wave function.

Further Reading

See [2] for more details.

Exercises

(1) Consider the Lagrangian

$$L = \frac{\dot{q}^4}{4} + \frac{\alpha \dot{q}^2}{2} - \lambda q^4. \tag{26.64}$$

Write down the explicit (improved) Bohr–Sommerfeld quantization condition for periodic motion in this Lagrangian.

(2) For the potential $V = k|x|$ in Bohr–Sommerfeld quantization, calculate the leading and first subleading terms for the eigenfunctions at large x and small x.

(3) Apply the Bohr–Sommerfeld quantization method to particle in a box (an infinitely deep square well) to find the eigenenergies and eigenfunctions, and compare with the exact results.

(4) For the wave function of the harmonic oscillator in the Bohr–Sommerfeld quantization method, what happens in the classical, large-n, limit? Is there any sense in which we can consider that the result matches the exact result?

(5) For the case of motion in a central potential, write down the equations for R_2 and Θ_2.

(6) Calculate the Bohr–Sommerfeld quantization condition for $R(r)$ for motion in the central potential $V = -\alpha/r^2$.

(7) For the hydrogenoid atom in Bohr–Sommerfeld quantization, compare the wave function as $r \to \infty$ (the leading and first subleading terms) with the exact wave function. Do they match? Is there any sense in which the wave function matches the exact one in the classical limit $n \to \infty$?

27 Dirac Quantization Condition and Magnetic Monopoles

In this chapter we consider a quantization condition found by Dirac that gives an argument for the existence of magnetic monopoles. The quantization condition is found in several different ways, including a semiclassical way. We also give an argument for the existence of magnetic monopoles based on a duality symmetry of the Maxwell equations.

27.1 Dirac Monopoles from Maxwell Duality

The Maxwell equations in vacuum are

$$\begin{aligned}\vec{\nabla}\times\vec{E} &= -\frac{\partial}{\partial t}\vec{B}, \quad \vec{\nabla}\cdot\vec{E}=0\\ \vec{\nabla}\times\vec{B} &= +\frac{1}{c^2}\frac{\partial}{\partial t}\vec{E}, \quad \vec{\nabla}\cdot\vec{B}=0,\end{aligned} \tag{27.1}$$

or, in relativistically covariant form, writing

$$\begin{aligned}E_i &= E^i = F^{0i} = -F_{0i}\\ B_i &= B^i = \frac{1}{2}\epsilon^{ijk}F_{jk},\end{aligned} \tag{27.2}$$

the second and third Maxwell equations become the equation of motion

$$\partial_\mu F^{\mu\nu} = 0, \tag{27.3}$$

while the first and the fourth Maxwell equations become the Bianchi identities,

$$\partial_{[\mu}F_{\nu\rho]} = 0, \tag{27.4}$$

where, because of the total antisymmetrization, this is equivalent to

$$F_{\mu\nu} = \partial_\mu A_\nu - \partial_\nu A_\mu, \tag{27.5}$$

with $A_\mu = (-\phi, \vec{A})$ or $A^\mu = (\phi, \vec{A})$.

It can be seen that the Maxwell equations have a symmetry, called *Maxwell duality*, on making the exchanges

$$\vec{E}\to\vec{B}, \quad \vec{B}\to-\vec{E}, \tag{27.6}$$

or, in relativistic form,

$$F_{\mu\nu}\to *F_{\mu\nu} \equiv \frac{1}{2}\epsilon_{\mu\nu\rho\sigma}F^{\rho\sigma}, \tag{27.7}$$

which takes $\partial_\mu F^{\mu\nu} = 0$ into $\partial_{[\mu} F_{\nu\rho]} = 0$. Note that $*^2 = -\mathbb{1}$, so (as we can also see from the vector components), applying the duality symmetry twice gives $-\mathbb{1}$, so we obtain an overall sign for the fields.

Adding sources means adding terms on the right-hand side of the second and third Maxwell equations,

$$\begin{aligned} \vec{\nabla} \cdot \vec{E} &= \frac{\rho}{\epsilon_0} \\ \vec{\nabla} \times \vec{B} - \frac{1}{c^2} \frac{\partial}{\partial t} \vec{E} &= \mu_0 \vec{j}, \end{aligned} \tag{27.8}$$

or, in relativistic notation, the equation of motion acquires a source term,

$$\partial_\mu F^{\mu\nu} + j^\nu = 0. \tag{27.9}$$

Here the 4-current is

$$j^\mu = \left(\frac{\rho}{\epsilon_0}, \mu_0 \vec{j} \right). \tag{27.10}$$

But now we have a problem, since there is no magnetic source, so Maxwell duality is broken. We note, however, that it would be fixed if we also introduce a magnetic 4-current, made up of a magnetic charge density and current,

$$k^\mu = \left(\mu_0 \rho_m, \frac{\vec{j}_m}{\epsilon_0} \right). \tag{27.11}$$

Then the Bianchi identity has a source term,

$$\frac{1}{2} \epsilon^{\mu\nu\rho\sigma} \partial_\mu F_{\rho\sigma} + k^\nu = \partial_\mu * F_{\mu\nu} + k^\nu = 0. \tag{27.12}$$

In terms of 3-vectors, we have the modified Maxwell equations

$$\begin{aligned} \vec{\nabla} \cdot \vec{B} &= \mu_0 \rho_m \\ \vec{\nabla} \times \vec{E} + \frac{\partial}{\partial t} \vec{B} &= \frac{\vec{j}_m}{\epsilon_0}. \end{aligned} \tag{27.13}$$

Thus we again have Maxwell duality, if we extend it to act on sources as well, as

$$j^\mu \leftrightarrow k^\mu, \quad \rho_e \to \rho_m, \quad q = \int d^3x \rho_e \leftrightarrow g = \int d^3x \rho_m. \tag{27.14}$$

Considering a pointlike electric source, an electron with $\rho_e = q\delta^3(x)$, we have the Maxwell equation

$$\vec{\nabla} \cdot \vec{E} = \frac{q}{\epsilon_0} \delta^3(x). \tag{27.15}$$

But applying Maxwell duality to it, we find a pointlike magnetic source with $\rho_m = g\delta^3(x)$, and a Maxwell equation

$$\vec{\nabla} \cdot \vec{B} = \mu_0 g \delta^3(x). \tag{27.16}$$

Then, using a magnetic Gauss's law, namely integrating over a sphere, or rather a ball, centered on the charge, and converting the integral over the volume to an integral over the 2-sphere of radius R, S^2_R, we obtain

$$\int_{B_R} d^3x \vec{\nabla} \cdot \vec{B} = \oint_{S^2_R} \vec{B} \cdot d\vec{S} = \mu_0 g, \tag{27.17}$$

and, since the $\vec{B}$ field must be radial, we finally obtain the magnetic field of the pointlike magnetic source,

$$\vec{B} = \frac{\mu_0 g}{4\pi} \frac{\hat{r}}{r^2}. \tag{27.18}$$

This magnetic pointlike source is called a "Dirac magnetic monopole". Usually, a magnetic field has only a dipole mode, or higher (quadrupole, etc.). The reason is that we have not yet found a magnetic monopole in nature; if there were one, we would say that the magnetic field also starts at the monopole, as for an electric field.

In the presence of magnetic sources, we can also act with Maxwell duality on the Lorentz force law, in order to find the full law, at $q \neq 0, g \neq 0$. Since $\vec{B} \to -\vec{E}$, $\vec{E} \to \vec{B}$, $q \to g$, we find the general law

$$\vec{F} = q(\vec{E} + \vec{v} \times \vec{B}) + g(\vec{B} - \vec{v} \times \vec{E}). \tag{27.19}$$

In relativistic notation, since $F^{\mu\nu} \to *F^{\mu\nu}$, we have

$$\frac{dp^\mu}{d\tau} = (qF^{\mu\nu} + g * F^{\mu\nu})\, u_\nu. \tag{27.20}$$

We also note that, for a Dirac magnetic monopole, the energy density diverges at $r \to 0$ since the magnetic field has density

$$\mathcal{E} = \frac{\vec{B}^2}{2\mu_0} = \frac{g^2 \mu_0}{32\pi^2 r^4} \propto \frac{1}{r^4} \to \infty. \tag{27.21}$$

Moreover, also the *total* energy of the magnetic field of the Dirac magnetic monopole diverges,

$$E = \int d^3x \mathcal{E} = \int_{r \to 0} 4\pi r^2 dr\, \mathcal{E} \sim \frac{g^2 \mu_0}{8\pi} \frac{1}{r} \to \infty. \tag{27.22}$$

But this is just the same as in the case of the electron. For the electron, the particle is fundamental, but there are quantum corrections to the field at $r \to 0$ which cut off the divergence in E. For the monopole, the particle is usually a large-r approximation of a *nonabelian* monopole (i.e., a 't Hooft monopole), which also cuts off the divergence in energy, though in a different way.

27.2 Dirac Quantization Condition from Semiclassical Nonrelativistic Considerations

The existence of the Dirac monopole implies, at the quantum level, an important quantization condition known as the Dirac quantization condition. It can be derived in different ways, but (given the analysis in the previous chapters) we will start with a semiclassical nonrelativistic method.

Consider a nonrelativistic system of an electric charge q in the magnetic field of a magnetic charge g. Indeed, the magnetic charge has the fields

$$\vec{E}_{\text{mon}} = 0, \quad \vec{B}_{\text{mon}} = \frac{\mu_0 g}{4\pi} \frac{\hat{r}}{r^2}. \tag{27.23}$$

Then the Lorentz force on the electric charge is

$$\dot{\vec{p}} = m\ddot{\vec{r}} = q\dot{\vec{r}} \times \vec{B} = \frac{\mu_0 q g}{4\pi} \frac{\dot{\vec{r}} \times \vec{r}}{r^3}. \tag{27.24}$$

The time derivative of the orbital angular momentum is

$$\frac{d\vec{L}}{dt} = \frac{d}{dt}(m\vec{r} \times \dot{\vec{r}}) = m\vec{r} \times \ddot{\vec{r}} = \frac{\mu_0 q g}{4\pi r^3}\vec{r} \times (\dot{\vec{r}} \times \vec{r}) = \frac{d}{dt}\left(\frac{\mu_0 q g}{4\pi}\frac{\vec{r}}{r}\right), \tag{27.25}$$

where we have used

$$\begin{aligned} \frac{1}{r^3}[\vec{r} \times (\dot{\vec{r}} \times \vec{r})]_i &= \frac{1}{r^3}\epsilon_{ijk}x^j(\epsilon^{klm}v_l x_m)) = \frac{1}{r^3}(\delta_i^l\delta_j^m - \delta_i^m\delta_j^l)x^j v_l x_m \\ &= \frac{v_i}{r} - \frac{x_i}{r^3}(\vec{r} \cdot \dot{\vec{r}}) = \left[\frac{\dot{\vec{r}}}{r} - \frac{\vec{r}(\vec{r} \cdot \dot{\vec{r}})}{r^3}\right]_i = \frac{d}{dt}\left(\frac{\vec{r}}{r}\right)_i. \end{aligned} \tag{27.26}$$

We finally obtain the conservation law

$$\frac{d}{dt}\left(\vec{L} - \frac{\mu_0 q g}{4\pi}\frac{\vec{r}}{r}\right) \equiv \frac{d}{dt}\vec{J} = 0. \tag{27.27}$$

Since the first term is the orbital angular momentum, the full expression (the one that is conserved) must be a *total angular momentum*, which is why we called it $\vec{J}$. But the total angular momentum is quantized in units of $\hbar/2$ so, now considering the opposite regime, when $\vec{L} = 0$, we obtain the quantization condition

$$\frac{\mu_0 q g}{4\pi} = \frac{N\hbar}{2} \Rightarrow \mu_0 q g = 2\pi\hbar N = Nh. \tag{27.28}$$

This is the *Dirac quantization condition*.

Note then that, if there is a single magnetic charge g anywhere in the Universe, the Dirac quantization condition means that electric charge is *quantized*, which we know experimentally to be true. The above argument is the only available theoretical explanation for the quantization of the electric charge, which is a strong indirect argument for the existence of magnetic charges somewhere in our Universe even if we have not observed them experimentally yet.

27.3 Contradiction with the Gauge Field

When we wrote down the duality-invariant Maxwell equations, with both electric and magnetic sources, by introducing the magnetic sources we obtained a contradiction with the existence of the gauge field $A_\mu = (-\phi, \vec{A})$. Indeed, since we define $\vec{B} = \vec{\nabla} \times \vec{A}$, we can use Gauss's law, or rather its generalization, Stokes' law, *twice*, and find

$$\mu_0 Q_m = \int_{M^3} d^3x \vec{\nabla} \cdot \vec{B} = \oint_{\partial M^3 = S^2} \vec{B} \cdot d\vec{S} = \oint_{S^2(\text{closed})} (\vec{\nabla} \times \vec{A}) \cdot d\vec{S} = \oint_{\partial S^2(\text{closed})} \vec{A} \cdot d\vec{l} = 0. \tag{27.29}$$

Indeed, note that the 2-sphere S^2 is a closed surface, so it has no boundary, which means that the "line integral over the boundary" is zero. But this relation is a contradiction, since it implies that the magnetic charge must be zero.

In relativistically covariant notation the contradiction is easier to understand, since $F_{\mu\nu} = \partial_\mu A_\nu - \partial_\nu A_\mu$ means that $\partial_{[\mu} F_{\nu\rho]} = 0$, but this is in contradiction with the Bianchi identity with sources,

$$\frac{1}{2}\epsilon^{\mu\nu\rho\sigma}\partial_\mu F_{\rho\sigma} + k^\nu = 0. \tag{27.30}$$

The solution of this contradiction is that $F_{\mu\nu} = \partial_\mu A_\nu - \partial_\nu A_\mu$ or, in the $\phi = 0$ gauge, $\vec{B} = \vec{\nabla} \times \vec{A}$, is valid *only on patches*. That is, it is valid only on open surfaces that do not intersect the magnetic charges, so it is not valid on the closed surfaces S that surround the magnetic charge.

27.4 Patches and Magnetic Charge from Transition Functions

In order to use patches for the closed surface S^2, we divide it and similar closed surfaces into two overlapping patches $\mathcal{O}_\alpha$ and $\mathcal{O}_\beta$, such that $\mathcal{O}_{\alpha\beta} = \mathcal{O}_\alpha \cap \mathcal{O}_\beta$ and $\mathcal{O}_\alpha \cup \mathcal{O}_\beta = S^2$, as in Fig. 27.1a. Then, we define A_μ on each patch, i.e., $A_\mu^{(\alpha)}$ and $A_\mu^{(\beta)}$, such that

$$F_{\mu\nu} = \partial_\mu A_\nu^{(\alpha)} - \partial_\nu A_\mu^{(\beta)} = \partial_\mu A_\nu^{(\beta)} - \partial_\nu A_\mu^{(\beta)}. \tag{27.31}$$

But this means that on the common patch $\mathcal{O}_{\alpha\beta}$, the two gauge fields must differ by a gauge transformation,

$$A_\mu^{(\alpha)} = A_\mu^{(\beta)} + \partial_\mu \lambda^{(\alpha\beta)}, \tag{27.32}$$

where the gauge parameter $\lambda^{(\alpha\beta)}$ is called a transition function.

In the $A_0 = 0$ gauge, we have

$$\vec{B} = \vec{\nabla} \times \vec{A}^{(\alpha)} = \vec{\nabla} \times \vec{A}^{(\beta)}, \tag{27.33}$$

so that the vector potentials differ by a gauge transformation,

$$\vec{A}^{(\alpha)} = \vec{A}^{(\beta)} + \vec{\nabla}\lambda^{(\alpha\beta)}. \tag{27.34}$$

We can consider explicit patches as in Fig. 27.1b, i.e.,

$$\begin{aligned} &(\alpha): S^2 - \text{north pole}, \ \theta = \pi \\ &(\beta): S^2 - \text{south pole}, \ \theta = 0. \end{aligned} \tag{27.35}$$

In the gauge $A_0 = 0$, on the patch (α) the vector potential is

$$\vec{A}^{(\alpha)} = \frac{\mu_0 g}{4\pi r} \frac{(-1 + \cos\theta)}{\sin\theta} \vec{e}_\phi, \tag{27.36}$$

where $\vec{e}_\phi$ is the unit vector in the direction of increasing ϕ, in Cartesian coordinates:

$$\vec{e}_\phi = (-\sin\phi, \cos\phi, 0). \tag{27.37}$$

The formula for $\vec{A}^{(\alpha)}$ is singular at $\sin\theta = 0$ but $\cos\theta \neq 1$, which means at $\theta = \pi$. Thus, indeed, $\vec{A}^{(\alpha)}$ is defined only on (α).

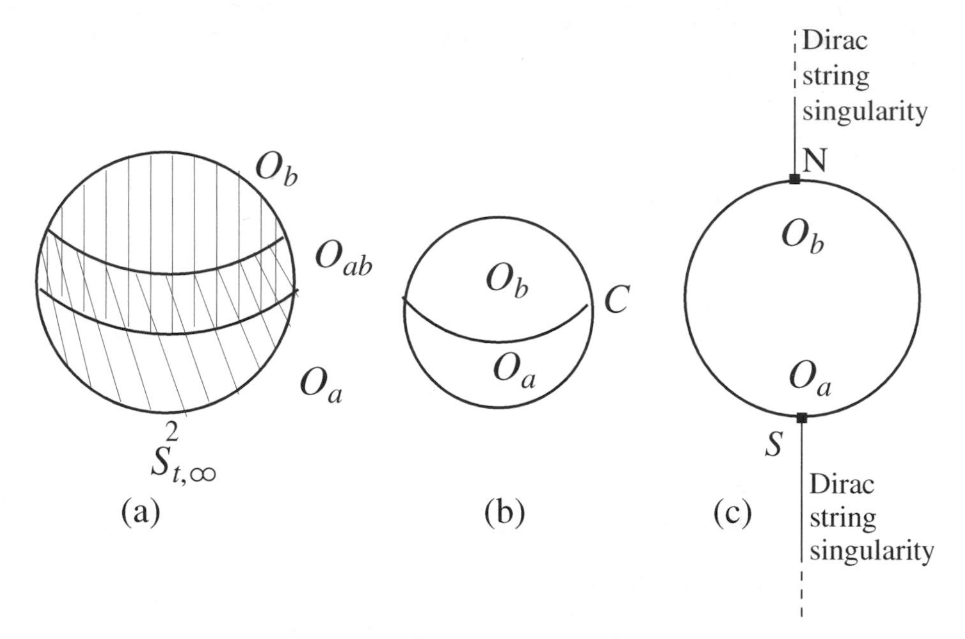

Figure 27.1 (a) Two overlapping patches O_a and O_b for the 2-sphere S^2, with common patch O_{ab}. (b) The case when the overlap O_{ab} is an equator, C. (c) The case when $O_a = S^2$ less the north pole, and $O_b = S^2$ less the south pole. The Dirac string singularity is a line going to infinity either at the north or the south pole.

We can check that the formula for $\vec{A}^{(\alpha)}$ is correct, since we have, in spherical coordinates (note that we have replaced θ with $\pi - \theta$ with respect to the usual definition),

$$\begin{aligned}\vec{\nabla} \times \vec{A} &= \frac{1}{r \sin\theta}\left(-\frac{\partial}{\partial\theta}(A_\phi \sin\theta) + \frac{\partial A_\theta}{\partial\phi}\right)\hat{r} - \frac{1}{r}\left(\frac{1}{\sin\theta}\frac{\partial A_r}{\partial\phi} - \frac{\partial}{\partial r}(rA_\phi)\right)\vec{e}_\theta \\ &+ \frac{1}{r}\left(-\frac{\partial}{\partial r}(rA_\theta) + \frac{\partial A_r}{\partial\theta}\right)\vec{e}_\phi.\end{aligned} \tag{27.38}$$

On the patch (β), the vector potential (again in the gauge $A_0 = 0$) is

$$\vec{A}^{(\beta)} = \frac{\mu_0 g}{4\pi r}\frac{(+1 + \cos\theta)}{\sin\theta}\vec{e}_\phi. \tag{27.39}$$

This potential is singular at $\sin\theta = 0$ but $\cos\theta \neq -1$, so at $\theta = 0$. Therefore it is indeed defined only on (β).

Then, moreover, the difference between the two potentials is a gauge transformation,

$$\vec{A}^{(\alpha)} - \vec{A}^{(\beta)} = \frac{\mu_0}{2\pi}\frac{-g}{r\sin\theta}\vec{e}_\phi = \vec{\nabla}\lambda^{(\alpha\beta)}, \tag{27.40}$$

where the transition function is

$$\lambda^{(\alpha\beta)} = -\frac{\mu_0 g}{2\pi}\phi. \tag{27.41}$$

This is correct since in spherical coordinates

$$\vec{\nabla} = \frac{\partial}{\partial r}\hat{r} + \frac{1}{r}\frac{\partial}{\partial\theta}\vec{e}_\theta + \frac{1}{r\sin\theta}\frac{\partial}{\partial\phi}\vec{e}_\phi. \tag{27.42}$$

Then the gauge transformation $\vec{\nabla}\lambda^{(\alpha\beta)}$ is single valued around the circle parametrized by $\phi \in [0, 2\pi]$, but the gauge parameter $\lambda^{(\alpha\beta)}$ is not. Indeed, around it we have

$$\lambda^{(\alpha\beta)}(\phi = 2\pi) - \lambda^{(\alpha\beta)}(\phi = 0) = -\mu_0 g. \tag{27.43}$$

In general, consider a common one-dimensional patch that is comprised by an equator: $\mathcal{O}_{\alpha\beta} = C$, $\mathcal{O}^\alpha_C \cup \mathcal{O}^\beta_C = S^2$, and $\mathcal{O}^\alpha_C \cap \mathcal{O}^\beta_C = C$, as in Fig. 27.1b. Then we can again use Stokes' law (or the generalized Gauss's law) twice, but now on the sum of patches, so

$$\begin{aligned} \mu_0 Q_m &= \oint_{S^2} \vec{B} \cdot d\vec{S} - \int_{\mathcal{O}^\alpha_C} \vec{\nabla} \times \vec{A}^{(\alpha)} + \int_{\mathcal{O}^\beta_C} \vec{\nabla} \times \vec{A}^{(\beta)} \\ &= \oint_{\partial\mathcal{O}^\alpha_C = +C} \vec{A}^{(\alpha)} \cdot d\vec{l} + \oint_{\partial\mathcal{O}^\beta_C = -C} \vec{A}^{(\beta)} \cdot d\vec{l} = \oint_C (\vec{A}^{(\alpha)} - \vec{A}^{(\beta)}) \cdot d\vec{l}. \end{aligned} \tag{27.44}$$

Here we have used the fact that $\mathcal{O}^\alpha_C$ has an outward normal, associated with the right-hand screw rule, whose direction is on C, and so $\mathcal{O}^\beta_C$ has an outward normal, associated with the right-hand screw rule, with the opposite direction on C; hence when using the same outward normal, i.e., the same integral on C, we have difference of integrands.

Finally, we obtain that the magnetic charge is the integral of the gauge transformation, or the non-single-valuedness of the gauge parameter,

$$\mu_0 Q_m = \oint_C (\vec{\nabla}\lambda^{(\alpha\beta)}) \cdot d\vec{l} = \lambda^{(\alpha\beta)}(\phi = 2\pi) - \lambda^{(\alpha\beta)}(\phi = 0). \tag{27.45}$$

27.5 Dirac Quantization from Topology and Wave Functions

There is an alternative derivation of the Dirac quantization condition that builds upon the formalism of patches described above.

If there are electrically charged particles in the system then we have complex fields, in particular complex wave functions, that are *minimally coupled* to the electromagnetic potential $\vec{A}$ through the electric charge q, as we saw in Chapter 23,

$$\vec{D}\psi = \left(\vec{\nabla} - \frac{iq}{\hbar}\vec{A}\right)\psi. \tag{27.46}$$

But we saw that gauge transformations with parameters λ act on the wave functions as

$$\psi = \psi' \exp\left(-i\frac{q\lambda}{\hbar}\right). \tag{27.47}$$

We must then impose that the gauge transformation on the wave function is single-valued around the circle C in Fig. 27.1, i.e., that

$$\exp\left[-\frac{iq}{\hbar}\lambda^{(\alpha\beta)}(\phi = 2\pi)\right] = \exp\left[-\frac{iq}{\hbar}\lambda^{(\alpha\beta)}(\phi = 0)\right], \tag{27.48}$$

which means that we must have

$$\frac{q}{\hbar}\left(\lambda^{(\alpha\beta)}(\phi = 2\pi) - \lambda^{(\alpha\beta)}(\phi = 0)\right) = 2\pi n. \tag{27.49}$$

Substituting the non-single-valuedness of λ (for a magnetic charge $Q_m = g$), we obtain

$$\frac{2\pi n\hbar}{q} = \mu_0 g \Rightarrow \mu_0 g q = 2\pi\hbar n = hn, \tag{27.50}$$

which is again the Dirac quantization condition.

27.6 Dirac String Singularity and Obtaining the Dirac Quantization Condition from It

There is yet another way to derive the Dirac quantization condition, which will expose another important feature in the presence of magnetic monopoles: the Dirac string.

For the Dirac monopole, we found that (on both patches), the vector potential $\vec{A}$ is defined everywhere except on a line. In the case of $\vec{A}^{(\alpha)}$, there was a singularity on the north pole $\theta = \pi$ on S^2, which is a "string" at $\theta = \pi$ extending from the monopole ($r = 0$) to infinity ($r = \infty$). Similarly, $\vec{A}^{(\beta)}$ was defined everywhere except on the "string" on the south pole $\theta = 0$, extending from the monopole to infinity; see Fig. 27.1c.

This singularity is known as the Dirac string singularity. We see that by gauge transformations (like that taking us between $\vec{A}^{(\alpha)}$ and $\vec{A}^{(\beta)}$) we can move the Dirac string singularity around, which means that it is not physical.

Dirac's interpretation of this singularity was that we can find a regular, i.e., nonsingular, magnetic field $\vec{B}_{\text{reg}}$, that can be written in terms of an everywhere-defined gauge field (vector potential) $\vec{A}$. This magnetic field is formed by the monopole field $\vec{B}_{\text{mon}}$ plus the field of the Dirac string singularity,

$$\vec{B}_{\text{string}} = \mu_0 g\theta(-z)\delta(x)\delta(y)\hat{z}, \tag{27.51}$$

so that

$$\vec{B}_{\text{reg}} = \vec{B}_{\text{mon}} + \vec{B}_{\text{string}}. \tag{27.52}$$

The above field $\vec{B}_{\text{string}}$ is for the (β) patch, when the singularity is on the south pole, from $r = 0$ to $r = \infty$, i.e, at $z < 0$, hence we have the Heaviside function $\theta(-z)$. The regular field $\vec{B}_{\text{reg}}$ obeys the usual rules, so that

$$\vec{\nabla} \cdot \vec{B}_{\text{reg}} = 0 \Rightarrow \vec{B}_{\text{reg}} = \vec{\nabla} \times \vec{A}. \tag{27.53}$$

The Dirac quantization condition now arises by imposing that the Aharonov–Bohm effect from a contour encircling the Dirac string is trivial, as it should be for an unphysical singularity: this means that the Dirac string must be unobservable.

Since the Dirac string is associated with a vector potential $\vec{A}$ generating the delta function magnetic field $\vec{B}_{\text{string}}$, we have

$$\oint_{C=\partial S} \vec{A} \cdot d\vec{l} = \int_S \vec{B} \cdot d\vec{S} \neq 0, \tag{27.54}$$

yet it must be unobservable in an Aharonov–Bohm effect. Therefore, when going around a circle encircling the Dirac string, the change in the wave function, its "monodromy",

$$\psi \to \exp\left[\frac{iq}{\hbar} \oint_{C=\partial S} \vec{A} \cdot d\vec{l}\right] \psi, \tag{27.55}$$

must be trivial, i.e.,

$$\exp\left[\frac{iq}{\hbar}\oint_{C=\partial S}\vec{A}_{\text{string}}\cdot d\vec{l}\right]=\exp\left[\frac{iq}{\hbar}\int_S\vec{B}_{\text{string}}\cdot d\vec{S}\right]=1. \tag{27.56}$$

This means that $e^{i\mu_0 qg/\hbar}=1$, so we obtain again the Dirac quantization condition,

$$\mu_0 qg=2\pi\hbar N=Nh. \tag{27.57}$$

Important Concepts to Remember

- Maxwell duality refers to the invariance of the Maxwell equations without source under $\vec{E}\to\vec{B}$, $\vec{B}\to-\vec{E}$. If we add a magnetic current source $k^\mu=(\mu_0\rho_m,\vec{j}_m/\epsilon_0)$ similar to the electric current source $j^\mu=(\rho_e/\epsilon_0,\mu_0\vec{j}_e)$, we can have Maxwell duality even with sources.
- A magnetic current source of delta function type (like the electron in relation to the electric current) gives a Dirac magnetic monopole,

$$\vec{B}=\frac{\mu_0 g}{4\pi}\frac{\hat{r}}{r^2}.$$

- We can obtain the Dirac quantization condition $\mu_0 qg=2\pi\hbar N=hN$ from the conservation of the total angular momentum, such that N is an angular momentum quantum number.
- If there is a single magnetic monopole with charge g in the Universe, we obtain the quantization of electric charge, which is an experimental fact but otherwise theoretically unexplained. This is good indirect evidence for the existence of magnetic monopoles.
- In the presence of magnetic charges, $F_{\mu\nu}=\partial_\mu A_\nu-\partial_\nu A_\mu$ cannot be valid everywhere (in particular, not on a closed surface enclosing the charge), but must be true *only on patches*.
- The magnetic charge is equal to the monodromy on C (the difference in the field when going around a closed loop) of the transition function (the gauge parameter for a transformation) between two patches with C as a common loop.
- Dirac quantization can also be obtained from the single-valuedness of the gauge transformation defined by the transition function on the wave function minimally coupled to the electromagnetic potential.
- The Dirac string singularity is a spurious singularity in the vector potential $\vec{A}$ of a monopole, starting at the monopole and going to infinity. It can be moved around by gauge transformations. The Dirac quantization condition also arises from the condition that the Aharonov–Bohm effect from a contour encircling the Dirac string is unobservable (trivial).

Further Reading

See [7] for more details.

Exercises

(1) Can Maxwell duality be derived from the classical action for electromagnetism (plus electric and magnetic sources)? Is there a simple way to generalize this duality to the quantum theory, if there are particles minimally coupled to the electromagnetic fields? Explain.

(2) Consider an electron and a monopole at the same point at a distance R from a perfectly conducting infinite plane. What are $\vec{E}$ and $\vec{B}$ on the plane at the minimum-distance point from the charges?

(3) In the presence of so-called *dyons*, particles that carry both electric and magnetic charges, the Dirac quantization condition for a particle with charges (q_1, g_1) and another particle with charges (q_2, g_2) is generalized to the *Dirac–Schwinger–Zwanziger quantization condition*

$$q_1 g_2 - q_2 g_1 = 2\pi\hbar N, \quad N \in \mathbb{N}. \tag{27.58}$$

Prove this relation using a generalization of the argument for the quantization of the total angular momentum of system of two particles.

(4) Consider two dyons satisfying the Dirac–Schwinger–Zwanziger quantization condition given in exercise 3, the minimum value for the integer, $N = 1$, occurring when one of the dyons has $q_1 = e, g_1 = h/e$. Calculate the total relative force between the dyons.

(5) Suppose that the magnetic field for a particle at the point 0 is a delta function,

$$\vec{B}(x) = \frac{2\theta}{e}\delta(\vec{x}). \tag{27.59}$$

Take two such particles and rotate one around the other. Calculate the Aharonov–Bohm phase $\exp\left[i\oint_C \vec{A}\cdot d\vec{l}\right]$ of the moving particle. Since the particles are identical, how would you interpret this result?

(6) Consider two monopoles of opposite charge (monopole and antimonopole) situated at a distance $2R$ from each other, and a sphere of radius $2R$ centered on the midpoint between the monopole and antimonopole. Is there a unique vector potential $\vec{A}$ that is valid over the whole sphere?

(7) Suppose in our Universe there is only a monopole and antimonopole pair, situated at a distance d of each other that is much smaller than the distance to any other atom in the Universe. Can we still infer that the electron charge is quantized? Explain.

28 Path Integrals II: Imaginary Time and Fermionic Path Integral

In this chapter we will extend the quantum mechanical formalism of path integrals, in order to describe finite temperature, and fermions. We will first study path integrals in Euclidean space, which, as we will see, are better defined than those in Minkowski space, and also give their relation to statistical mechanics at finite temperature. Moreover, the usual case is defined as a limit of the Euclidean space path integral. Finally, we will also define path integrals for fermionic variables.

28.1 The Forced Harmonic Oscillator

In Chapter 10, we obtained the path integral for the transition amplitude (for the propagator)

$$\begin{aligned} M(q',t';q,t) &\equiv {}_H\langle q',t';q,t\rangle_H = \langle q'|\exp\left[-\frac{i}{\hbar}\hat{H}(t-t')\right]|q\rangle \\ &= \int \mathcal{D}p(t)\mathcal{D}q(t)\exp\left\{\frac{i}{\hbar}\int_{t_i}^{t_f} dt\,[p(t)\dot{q}(t) - H(q(t),p(t))]\right\} \end{aligned} \tag{28.1}$$

and, for a Hamiltonian quadratic in momenta, $H = p^2/2 + V(q)$, we obtained the path integral in terms of the Lagrangian,

$$M(q',t';q,t) = \mathcal{N}\int \mathcal{D}q\exp\left(\frac{i}{\hbar}S[q]\right). \tag{28.2}$$

We also saw that we can define N-point correlation functions, and calculate them in terms of path integrals, as

$$\begin{aligned} G_N(\bar{t}_1,\ldots,\bar{t}_N) &= {}_H\langle q',t'|T\{\hat{q}(\bar{t}_1)\cdots\hat{q}(\bar{t}_N)\}|q,t\rangle \\ &= \int_{q(t)=q,q(t')=q'} \mathcal{D}q(t)\exp\left(\frac{i}{\hbar}S[q]\right)q(\bar{t}_1)\cdots q(\bar{t}_N), \end{aligned} \tag{28.3}$$

where we have the boundary conditions $q(t) = q, q(t') = q'$. These correlation functions are obtained from their generating functional, which is

$$Z[J] = \int_{q(t)=q,q(t')=q'} \mathcal{D}q\exp\left(\frac{i}{\hbar}S[q;J]\right) = \int_{q(t)=q,q(t')=q'} \exp\left[\frac{i}{\hbar}S[q] + \frac{i}{\hbar}\int dt J(t)q(t)\right], \tag{28.4}$$

by means of multiple derivatives at zero,

$$G_N(\bar{t}_1,\ldots,\bar{t}_N) = \frac{\delta}{\frac{i}{\hbar}\delta J(\bar{t}_1)}\cdots\frac{\delta}{\frac{i}{\hbar}\delta J(\bar{t}_N)}Z[J]\bigg|_{J=0}. \tag{28.5}$$

But, while $Z[J]$, called the partition function, is just a generating functional (a mathematical construct), we note that $J(t)$ can be interpreted as a *source* term for the classical variable $q(t)$. In the case of a harmonic oscillator, this would be an external driving force, giving rise to a forced (driven) harmonic oscillator, as seen in the equation of motion,

$$0 = \frac{\delta S[q;J]}{\delta q(t)} = \frac{\delta S[q]}{\delta q(t)} + J(t), \tag{28.6}$$

so $Z[J]$ and its derivatives make sense also at nonzero $J(t)$.

The action of the driven harmonic oscillator is then

$$S[q;J] = \int dt \left[\frac{1}{2}\left(\dot{q}^2 - \omega^2 q^2\right) + J(t)q(t) \right]. \tag{28.7}$$

In this case, the path integral is still Gaussian, since we are just adding a linear term to the quadratic one. But, in order to calculate it, we still have one issue: we need to understand the boundary conditions for $q(t)$. In the absence of boundary terms, we can partially integrate and obtain

$$S[q;J] = \int dt \left\{ -\frac{1}{2} q(t) \left[\frac{d^2}{dt^2} + \omega^2 \right] q(t) + J(t)q(t) \right\}. \tag{28.8}$$

This is schematically of the type

$$\frac{i}{\hbar} S = -\frac{1}{2\hbar} q \cdot \Delta^{-1} \cdot q + \frac{i}{\hbar} J \cdot q, \tag{28.9}$$

where

$$\Delta^{-1} q(t) \equiv i \left[\frac{d^2}{dt^2} + \omega^2 \right] q(t). \tag{28.10}$$

In this case, the Gaussian path integral can be calculated as

$$\begin{aligned} Z[J] &= \mathcal{N} \int \mathcal{D}q \exp\left(-\frac{1}{2\hbar} q \cdot \Delta^{-1} \cdot q + \frac{i}{\hbar} J \cdot q \right) \\ &\equiv \int d^n x \exp\left[-\left(\frac{1}{2} x^T \cdot A \cdot x + b^T \cdot x \right) \right] \\ &= (2\pi)^{n/2} (\det A)^{-1/2} \exp\left(\frac{1}{2} b^T \cdot A^{-1} \cdot b \right), \end{aligned} \tag{28.11}$$

where $b = -\frac{i}{\hbar} J(t)$ and $A = \Delta^{-1}/\hbar$, so we finally obtain

$$Z[J] = \mathcal{N}' \exp\left(-\frac{1}{2\hbar} J \cdot \Delta \cdot J \right), \tag{28.12}$$

where $\mathcal{N}'$ contains constants (expressions independent of J), including $(\det \Delta)^{-1/2}$.

In the above, Δ is called the *propagator*, which seems unrelated to the previous notion of a propagator, as $U(t,t')$ relating $|\psi(t')\rangle$ to $|\psi(t)\rangle$. In the case of the driven harmonic oscillator, the propagator is actually related to the two-point correlation function since then, from (28.12), we have

$$\begin{aligned} G_2(t_1,t_2) &= -\hbar^2 \frac{\delta^2}{\delta J(t_1)\delta J(t_2)} Z[J] \bigg|_{J=0} \\ &= \hbar \frac{\delta}{\delta J(t_1)} (J \cdot \Delta)(t_2) \exp\left(-\frac{1}{2\hbar} J \cdot \Delta \cdot J \right) \bigg|_{J=0} = \hbar \Delta(t_1,t_2). \end{aligned} \tag{28.13}$$

Also from (28.12), we find

$$\Delta(t,t') = i\int \frac{dp}{2\pi}\frac{e^{-ip(t-t')}}{p^2-\omega^2}, \tag{28.14}$$

since then

$$\begin{aligned}\Delta^{-1}\Delta(t,t') &= i\left[\frac{d^2}{dt^2}+\omega^2\right] i\int \frac{dp}{2\pi}\frac{e^{-ip(t-t')}}{p^2-\omega^2}\\ &= -\int \frac{dp}{2\pi}\frac{-p^2+\omega^2}{p^2-\omega^2}e^{-ip(t-t')}\\ &= \int \frac{dp}{2\pi}e^{-ip(t-t')} = \delta(t-t').\end{aligned} \tag{28.15}$$

However, the expression we have found for Δ is ill defined, since we have a singularity at $p^2 = \omega^2$, where the denominator vanishes, and we must somehow avoid this singularity. Its avoidance is related to the question of how to invert Δ^{-1}: this depends, as for any operator, on the space of functions on which the operator acts. In quantum mechanics, this is the Hilbert space of states and, in this continuous case, must be better defined. The relevant issue is that the Hilbert space for Δ^{-1} has zero modes, where

$$\Delta^{-1}q_0(t) = \left[\frac{d^2}{dt^2}+\omega^2\right]q_0(t) = 0. \tag{28.16}$$

Having zero modes (eigenstates of zero eigenvalue), the operator on the full Hilbert space is not invertible (think of a matrix with zero eigenvalues, therefore with zero determinant). Moreover, these are not some pathological states but rather are the solutions of the classical equations of motion of the harmonic oscillator. To obtain an invertible operator, we must therefore find a way to exclude these classical states from the Hilbert space, by imposing some boundary conditions that contradict them.

We will find that the correct result is described in terms of an integral over *slightly complex* momenta p, by avoiding the pole just slightly in the complex plane, according to the formula

$$\Delta_F(t,t') = \int \frac{dp}{2\pi}\frac{e^{-ip(t-t')}}{p^2-\omega^2+i\epsilon}. \tag{28.17}$$

Here F stands for Feynman; this is the *Feynman propagator*.

If an operator A has eigenstates $|q\rangle$ with eigenvalues a_q,

$$A|q\rangle = a_q|q\rangle, \tag{28.18}$$

and the states are orthonormal,

$$\langle q|q'\rangle = \delta_{qq'}, \tag{28.19}$$

then the operator can be written as

$$A = \sum_q a_q|q\rangle\langle q|. \tag{28.20}$$

Thus the inverse operator is

$$A^{-1} = \sum_q \frac{1}{a_q}|q\rangle\langle q|. \tag{28.21}$$

In our case, if Δ_F^{-1} has orthonormal eigenfunctions $\{q_i(t)\}$,

$$\int dt q_i^*(t) q_j(t) = \delta_{ij}, \tag{28.22}$$

with eigenvalues λ_i, then

$$\Delta_F^{-1}(t,t') = \sum_i \lambda_i q_i(t) q_i^*(t'), \tag{28.23}$$

so the inverse is

$$\Delta_F(t,t') = \sum_i \frac{1}{\lambda_i} q_i(t) q_i^*(t'). \tag{28.24}$$

In order to get (28.17), we can make the identifications $q_i(t) \sim e^{-ipt}$, $\sum_i \sim \int dp/2\pi$, and $\lambda_i \sim (p^2 - \omega^2 + i\epsilon)$.

We can be more precise than the above, however. Since in $\Delta_F(t,t')$ there are poles in the complex plane at $p = \pm(\omega - i\epsilon)$, we can calculate the integral with the residue theorem on the complex plane. In order to do that, we must form a closed contour from the integral over the real line, by adding a piece of contour whose integral vanishes. Such a contour is a semicircle at infinity, provided the factor $e^{-ip(t't')}$ is exponentially small.

Thus if $t > t'$, by considering $\text{Im}(p) < 0$, we obtain a factor that vanishes exponentially, $e^{-|\text{Im}(p)|(t-t')} \to 0$, on the semicircle at infinity in the *lower* half of the complex plane. We can thus add, without affecting the result, the integral over this contour, obtaining a closed total contour C such that the pole $p_1 = +(\omega - i\epsilon)$ is inside C, as in Fig. 28.1. By the residue theorem, we obtain

$$\Delta_F(t,t') = \frac{1}{2\omega} e^{-i\omega(t-t')}. \tag{28.25}$$

Similarly, if $t < t'$, by considering $\text{Im}(p) > 0$, we obtain an exponentially vanishing factor $e^{\text{Im}(p)(t'-t)} \to 0$ on the semicircle at infinity in the *upper* half of the complex plane. Adding

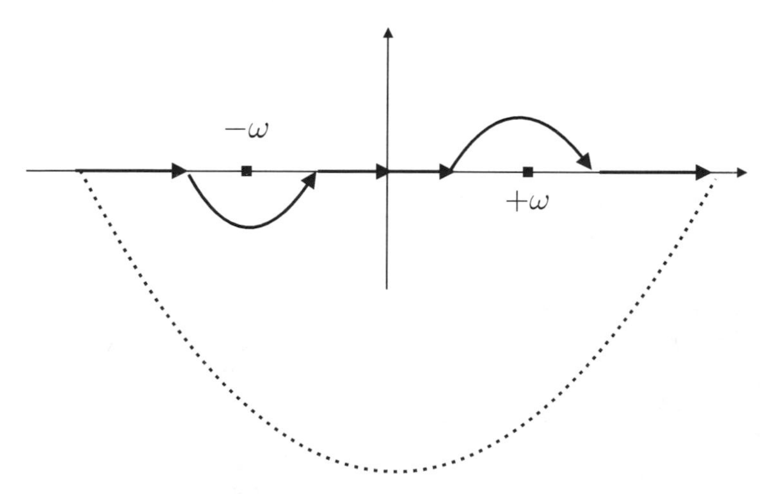

Figure 28.1 The contour for the Feynman propagator avoids $-\omega$ from below and $+\omega$ from above. It is closed in the lower half-plane (see the dotted contour "at infinity"), for $t > t'$.

again without affecting the result the integral over this contour, we obtain a closed total contour C that encircles the pole $p_2 = -(\omega - i\epsilon)$, so by the residue theorem we get

$$\Delta_F(t,t') = \frac{1}{2\omega} e^{-i\omega(t'-t)}. \tag{28.26}$$

Putting together the two cases, we have at general t, t',

$$\Delta_F(t,t') = \frac{1}{2\omega} e^{-i\omega|t-t'|}. \tag{28.27}$$

Now we can calculate the boundary conditions, since from (28.24) we see that Δ_F has the same boundary conditions for t as $q(t)$. Since at $t \to +\infty$, $\Delta_F \sim e^{-i\omega t}$ and at $t \to -\infty$, $\Delta_F \sim e^{+i\omega t}$, it follows that, in order to obtain (28.17) we need to impose the boundary conditions

$$\begin{aligned} q(t) &\sim e^{-i\omega t}, \quad t \to +\infty \\ q(t) &\sim e^{+i\omega t}, \quad t \to -\infty. \end{aligned} \tag{28.28}$$

We can use these boundary conditions to calculate more precisely the Gaussian path integral, and find the boundary term correction to the partition function,

$$Z[J] = \mathcal{N}'' \exp\left[-\frac{1}{2\hbar} J \cdot \Delta \cdot J + \frac{i}{\hbar} q_{\text{cl}}(t) \cdot J\right], \tag{28.29}$$

though we will not show the details. The same result can also be obtained, a bit more rigorously, from the harmonic phase-space path integral of Chapter 10. Moreover, the harmonic phase-space path integral gives the same boundary conditions for the functions $q_i(t)$.

28.2 Wick Rotation to Euclidean Time and Connection with Statistical Mechanics Partition Function

We have shown above how to calculate the partition function in the case of a quadratic action, by doing a kind of Gaussian integral. However, the calculation is not very rigorous, since the Gaussian is of an imaginary kind while the result is strictly speaking only correct in the real case, $\int_{-\infty}^{+\infty} e^{-\alpha x^2}$. Indeed, in the case of an imaginary integral, $\int dx e^{-i\alpha x}$, the integrand oscillates rapidly, which makes it impossible for it to converge at infinity. If we take a cut-off for the integral at a large Λ, the difference between $\int_{-\Lambda}^{+\Lambda} dx\, e^{-i\alpha x}$ and $\int_{-\Lambda-C}^{+\Lambda+C} dx\, e^{-i\alpha x}$ is $2\int_0^C dx\, e^{-i\alpha x}$, which is finite and oscillatory for finite C, so we cannot even define the limit properly when $\Lambda \to \infty$.

A simple solution would be to replace $e^{iS/\hbar}$ with $e^{-S/\hbar}$, in which case the exponential at infinity would be decaying instead of oscillatory, and we could define the integral $\int_{-\infty}^{+\infty} dx$ without any problem. This is what happens when we go to *Euclidean time* (i.e., work in Euclidean spacetime).

For a Hamiltonian $\hat{H}$ that is independent of time, and has a complete set $\{|n\rangle\}$ of eigenstates (so that $\mathbb{1} = \sum_n |n\rangle\langle n|$), with positive eigenenergies $E_n > 0$, we can write

$$\begin{aligned} {}_H\langle q', t'|q, t\rangle_H &= \langle q'|e^{-i\hat{H}(t-t')/\hbar}|q\rangle = \sum_n \sum_m \langle q'|n\rangle\langle n|e^{-i\hat{H}(t-t')/\hbar}|m\rangle\langle m|q\rangle \\ &= \sum_n \langle q'|n\rangle\langle n|q\rangle e^{-iE_n(t-t')/\hbar} \\ &= \sum_n \psi_n(q')\psi_n^*(q) e^{-iE_n(t-t')/\hbar}. \end{aligned} \tag{28.30}$$

In the second line, we have used the fact that the matrix element on energy eigenstates is

$$\langle n|e^{-i\hat{H}(t-t')/\hbar}|m\rangle = \delta_{nm}e^{-iE_n(t-t')/\hbar}, \tag{28.31}$$

as well as the definition $\psi_n(q) \equiv \langle q|n\rangle$.

At this point, we can make the required *analytical continuation to Euclidean time*, also known as a "Wick rotation", where we just make the replacement $\Delta t \to -i\hbar\beta$. We then obtain

$$\langle q', \beta|q, 0\rangle = \sum_n \psi_n(q')\psi_n^*(q)e^{-\beta E_n}. \tag{28.32}$$

If moreover $q = q'$ and we integrate over it, we obtain an expression for the statistical mechanics partition function of a quantum system at temperature T, with $k_BT = 1/\beta$. Indeed, then

$$\int dq\langle q, \beta|q, 0\rangle = \int dq \sum_n |\psi_n(q)|^2 e^{-\beta E_n} = \text{Tr}[e^{-\beta\hat{H}}] \equiv Z[\beta], \tag{28.33}$$

where we sum over the Boltzmann factors $e^{-\beta E_n}$ times the probability $|\psi_n(q)|^2$ of the state, and then sum over q and n. The procedure above implies that we are considering closed (periodic) paths in Euclidean time, of period $\beta = 1/k_BT$, with

$$q' = q(t_E = \hbar\beta) = q(t_E = 0) = q. \tag{28.34}$$

For a Lagrangian in the usual (Minkowski) spacetime, with the canonical kinetic term,

$$L(q, \dot{q}) = \frac{1}{2}\left(\frac{dq}{dt}\right)^2 - V(q), \tag{28.35}$$

the action can be rewritten in terms of the Euclidean time t_E ($t = -it_E$) as

$$iS[q] = i\int_0^{t_E=\hbar\beta}(-idt_E)\left[\frac{1}{2}\left(\frac{dq}{d(-it_E)}\right)^2 - V(q)\right] \equiv -S_E[q], \tag{28.36}$$

where we have defined the *Euclidean action* $S_E[q]$, obtaining

$$S_E[q] = \int_0^{\hbar\beta} dt_E\left[\frac{1}{2}\left(\frac{dq}{dt_E}\right)^2 + V(q)\right] = \int_0^{\hbar\beta} dt_E\mathcal{L}_E(q, \dot{q}). \tag{28.37}$$

In this way, we obtain the *Feynman–Kac formula*, relating the statistical mechanics partition function to the partition function from the path integral in Euclidean space,

$$Z(\beta) = \text{Tr}[e^{-\beta\hat{H}}] = \int_{q(t_E+\hbar\beta)=q(t_E)} \mathcal{D}q \exp\left(-\frac{1}{\hbar}S_E[q]\right). \tag{28.38}$$

We can also introduce nonzero sources $J(t)$ into the Euclidean-time partition function,

$$Z[\beta, J] = \int \mathcal{D}q \exp\left[-\frac{1}{\hbar}S_E[\hbar\beta] + \frac{1}{\hbar}\int_0^{\hbar\beta} d\tau\ J_E(\tau)q_E(\tau)\right], \tag{28.39}$$

where the source term is the analytical continuation of the Minkowski-time source term,

$$i\int dtJ(t)q(t) = i\int d(-it_E)J(-it_E)q(-it_E) = \int dt_E J_E(t_E)q_E(t_E). \tag{28.40}$$

From the partition function (28.39) we can calculate correlation functions as before, for instance the two-point function,

$$\frac{\hbar^2}{Z(\beta)}\frac{\delta^2 Z[\beta;J]}{\delta J(\tau_1)\delta J(\tau_2)} = \frac{1}{Z(\beta)}\int \mathcal{D}q(\tau)q(\tau_1)q(\tau_2)e^{-1S_E(\beta)/\hbar}, \tag{28.41}$$

which now is also a statistical mechanics trace,

$$\frac{1}{Z(\beta)}\text{Tr}[e^{-\beta\hat{H}}T(\hat{q}(-i\tau_1)\hat{q}(-i\tau_2))], \tag{28.42}$$

where the Heisenberg operators in Minkowski time become the analytical continuation in Euclidean time,

$$\hat{q}(t) = e^{i\hat{H}t/\hbar}\hat{q}e^{-i\hat{H}t/\hbar} \Rightarrow \hat{q}(-i\tau) = e^{1\hat{H}\tau/\hbar}\hat{q}e^{-1\hat{H}\tau/\hbar}. \tag{28.43}$$

However, the path integrals are still taken over periodic Euclidean paths.

To go back to the regular case in Minkowski space, from the Euclidean formulation, we must both undo the Wick rotation, and take the limit $\beta \to \infty$ (infinite periodicity), corresponding to $T \to 0$ in statistical mechanics (zero temperature). In this limit, inside the trace remains only the vacuum contribution, for E_0, as

$$\sum_n \psi_n(q')\psi_n^*(q)e^{-\beta E_n} \to \psi_0(q')\psi_0(q)e^{-\beta E_0}. \tag{28.44}$$

Harmonic Oscillator Example

We will apply the Euclidean time formalism to the simplest example: the forced harmonic oscillator. Its Euclidean partition function is

$$\begin{aligned}
Z_E[J] &= \int \mathcal{D}q \exp\left\{-\frac{1}{2\hbar}\int dt\left[\left(\frac{dq}{dt}\right)^2 + \omega^2 q^2\right] + \frac{1}{\hbar}\int dt\, J(t)q(t)\right\} \\
&= \int \mathcal{D}q \exp\left\{-\frac{1}{2\hbar}\int dt q(t)\left[-\frac{d^2}{dt^2} + \omega^2\right]q(t) + \frac{1}{\hbar}\int dt\, J(t)q(t)\right\} \\
&= \mathcal{N}\exp\left\{\frac{1}{2\hbar}\int dt\int dt' J(t)\Delta_E(t,t')J(t')\right\},
\end{aligned} \tag{28.45}$$

where in the second line we have used partial integration, now *without boundary terms*, since the Euclidean-time path integral is over periodic paths (without boundary), and in the third line the *Gaussian integration is now well defined*, as it is a real integral, of the type $\int dte^{-S(x)}$, and not an oscillatory imaginary one, and the Euclidean-time propagator is defined as before through a Fourier transform in (Euclidean) energy E_E,

$$\Delta_E(t,t') = \left(-\frac{d^2}{dt^2} + \omega^2\right)^{-1}(t,t') = \int \frac{dE_E}{2\pi}\frac{e^{-iE_E(t-t')}}{E_E^2 + \omega^2}. \tag{28.46}$$

We finally note that there are *no poles* in this expression, since, for real E_E, $E_E^2 + \omega^2 > 0$. We have thus solved all three problems that we found with the result in Minkowski space: there are no boundary terms, the Gaussian integration is well defined, and there are no poles in the expression for the propagator.

When Wick-rotating back to Minkowski time, via $Et = E_E t_E$, so that (since $t = -it_E$) $E_E = -iE$, we can consider the procedure as a *rotation* in the complex E plane by $\pi/2$. However, for a Wick rotation of a full $\pi/2$ ($E_E = -iE$), we would be back to having poles in the integrand for $\Delta(t, t')$, since $E_E^2 + \omega^2 \to -E^2 + \omega^2$. To avoid obtaining a pole, we must instead do a rotation by $\pi/2 - \epsilon$, so

$$E_E \to e^{-i(\pi/2-\epsilon)} E = -i(E + i\epsilon'). \tag{28.47}$$

Then the propagator becomes

$$\Delta_E(t_E = it) = -i \int_{-\infty}^{+\infty} \frac{dE}{2\pi} \frac{e^{-iEt}}{-E^2 + \omega^2 - i\epsilon} = \Delta_F(t), \tag{28.48}$$

so it turns exactly into the Feynman propagator from before.

28.3 Fermionic Path Integral

Until now, we have considered path integrals appropriate for bosons, as in the case of the particle with position $x(t)$, but that is not the only possibility. Indeed, we have seen that the path integral for a harmonic oscillator is best defined as a harmonic phase-space path integral, in terms of $a, a^\dagger$ instead of q, p.

However, for fermions, we saw in Chapter 20 that we have anticommuting variables, with canonical quantization conditions

$$\{\psi_i, \psi_j\} = \{p_{\psi_i}, p_{\psi_j}\} = 0, \quad \{\psi_i, p_{\psi_j}\} = i\hbar\delta_{ij}, \tag{28.49}$$

and we can define fermionic raising and lowering operators $b, b^\dagger$, satisfying anticommutation relations

$$\{b, b\} = \{b^\dagger, b^\dagger\} = 0, \quad \{b, b^\dagger\} = 1. \tag{28.50}$$

In this case, we can define a sort of "classical limit" for the fermions, by taking $\hbar \to 0$. This is not quite a classical limit, since strictly speaking there is no classical fermion since for fermions there is a single particle per state, and we need a (macroscopically) large number of particles in the same state to obtain a classical limit. But one defines the limit in the abstract, and leaves the interpretation for later. In this case, remembering that (like $a, a^\dagger$), we have defined $b, b^\dagger$ in terms of q, p with coefficients that have $1/\sqrt{\hbar}$ in front, so in fact we have $b = \tilde{b}/\sqrt{\hbar}$, $b^\dagger = \tilde{b}^\dagger/\sqrt{\hbar}$, and then

$$\{\tilde{b}, \tilde{b}^\dagger\} = \hbar \to 0. \tag{28.51}$$

Together with the fact that $\{b, b\} = \{b^\dagger, b^\dagger\} = 0$, we see that the classical limit does not give the usual commuting functions but rather a *Grassmann algebra*, for anticommuting objects, with complex coefficients. This is defined in terms of regular, commuting objects (complex numbers), called "even" (or bosonic) elements, and anticommuting objects such as $b, b^\dagger$, the "odd" (or fermionic) part of the algebra. We then have the usual product relations, *bose* $\times$ *bose* $=$ *bose*, *fermi* × *fermi* = *bose* and *bose* × *fermi* = *fermi* × *bose* = *fermi*. For example, the product of two odd objects is even, for example $[bb^\dagger, bb^\dagger] = 0$.

Having defined the Grassmann algebra for fermionic objects, we must next define the path integral over it.

Definitions

We consider a Grassmann algebra with N objects x_i, $i = 1, \dots, N$, $\{x_i, x_j\} = 0$, together with the identity $\mathbb{1}$, which commutes with the rest, $[x_i, \mathbb{1}] = 0$. As we said, it is an algebra over the complex numbers, which means that the coefficients $c \in \mathbb{C}$. Since $(x_i)^2 = 0$, the Taylor expansion in these x_i will end after N terms,

$$f(\{x_i\}) = f(x_0) + \sum_i f_i^{(1)} x_i + \sum_{i<j} f_{ij}^{(2)} x_i x_j + \sum_{i<j<k} f_{ijk}^{(3)} x_i x_j x_k + \dots + f_{12\dots N}^{(N)} x_1 \cdots x_N. \quad (28.52)$$

Since we cannot add an even element to an odd element, it means that the functions are either even or odd. But, in order for the functions to be nontrivial, we must consider as x_i only a subset of the anticommuting elements, and we can use the rest for the coefficients $f^{(k)}_{i_1 \dots i_k}$. For instance, for $N = 1$, $f(x) = a + bx$, which means that (if the function $f(x)$ is even) a is even and b is odd, but then we can have $b = cy$, where $c \in \mathbb{C}$, and y is another odd element, and $f(x) = a + cyx$.

In more generality, consider an even number of odd elements, and use half of them for the Taylor expansion (as x_i) and the other half as coefficients.

We define differentiation using the same basic relation as in the commuting case,

$$\frac{d}{dx_i} x_j = \delta_{ij}. \quad (28.53)$$

The only difference is in the differentiation of a product of elements, since the derivative is also anticommuting (is fermionic), so we get extra minus signs compared with the commuting (bosonic) case,

$$\frac{\partial}{\partial x_i}(x_j \dots) = \delta_{ij}(\dots) - x_j \frac{\partial}{\partial x_i}(\dots). \quad (28.54)$$

Then for even functions $f(x), g(x)$,

$$\frac{\partial}{\partial x_i}(f(x)g(x)) = \left(\frac{\partial}{\partial x_i} f(x)\right) g(x) + f(x) \left(\frac{\partial}{\partial x_j} g(x)\right), \quad (28.55)$$

whereas for an odd function $f(x)$ and arbitrary $g(x)$, we have

$$\frac{\partial}{\partial x_i}(f(x)g(x)) = \left(\frac{\partial}{\partial x_i} f(x)\right) g(x) - f(x) \left(\frac{\partial}{\partial x_j} g(x)\right). \quad (28.56)$$

Next we need to define integration. However, on a Grassmann algebra, we cannot define an integration in the usual Riemann or Lebesgue sense, since there is no "Riemann sum" or nontrivial measure. This means that we cannot define a definite integral (with variable limits of integration), but only an indefinite integral.

For a single odd element x, as the basis elements of the Grassmann algebra are $\mathbb{1}$ and x, we must define the indefinite integral over them. The result of the integral for both elements must be a c-number, whereas $\int dx$ must also be odd (fermionic), which defines them rather uniquely as

$$\int dx \mathbb{1} = 0, \quad \int dx\, x = 1. \tag{28.57}$$

For several odd elements x_i, owing to the anticommuting nature of the elements we have (for $i \neq j$)

$$\{dx_i, dx_j\} = 0, \quad \{x_i, dx_j\} = 0. \tag{28.58}$$

But then we note that integration is the same as differentiation, as the rules for both are the same. For instance, we also obtain translational invariance for the integral,

$$\int dx f(x+a) = \int dx [f^0 + f^1(x+a)] = \int dx f^1 x = \int dx f(x). \tag{28.59}$$

The delta function in the Grassmann algebra must be defined as well. We will prove that $\delta(x) = x$. Indeed, for a single x, it is enough to prove the relation by integrating with a general function, $f(x) = f^0 + f^1 x$. We obtain

$$\begin{aligned} \int dx \delta(x-y) f(x) &= \int dx (x-y)(f^0 + f^1 x) = \int dx (x f^0 - y f^1 x) \\ &= f^0 + y \int dx f^1 x = f^0 - y f^1 = f^0 + f^1 y = f(y), \end{aligned} \tag{28.60}$$

which is the correct result, proving that indeed $\delta(x) = x$.

Changing the integration variable by rescaling with a complex number a, setting $y = ax$, we obtain

$$1 = \int dx\, x = \int dy\, y = a \int dy\, x \Rightarrow dy = \frac{1}{a} dx. \tag{28.61}$$

28.4 Gaussian Integration over the Grassmann Algebra

We will consider a real $n \times n$ antisymmetric matrix A, such that A^2 has negative, nonzero eigenvalues. We must have $n = 2m$, and then we obtain for the real Gaussian integration,

$$\int d^n x \exp\left(x^T A x\right) = 2^m \sqrt{\det A}, \tag{28.62}$$

though we will not prove it here. Instead, it is simpler to consider the case of a complex Gaussian integration. Consider independent x_i, y_i (such that $y_i \neq x_i^*$), in which case we have

$$\int d^n x \int d^n y \exp\left(y^T A x\right) = \det A. \tag{28.63}$$

To prove this relation, we need to realize that in the exponential (written as an infinite sum of power laws) only the term of order n contributes, since $(x_i)^2 = 0$, so terms with more than n factors $y^T A x$ have at least one x_i repeated, whereas terms with fewer than n factors $y^T A x$ will give zero since at least one x_i will be missing, giving $\int dx_i \mathbb{1} = 0$.

Then the only nonzero terms are of order n, and the only difference between them is in the permutation Q of the x_i and P of the y_i, and the integral gives

$$\begin{aligned}
&\frac{1}{n!}\sum_P\sum_Q (y_{P(1)}A_{P(1)Q(1)}x_{Q(1)})\cdots(y_{P(n)}A_{P(n)Q(n)}x_{Q(n)})\\
&=\frac{1}{n!}\sum_P\sum_Q \left(y_1A_{1QP^{-1}(1)}x_{QP^{-1}(1)}\right)\cdots\left(y_nA_{nQP^{-1}(n)}x_{QP^{-1}(n)}\right)\\
&=\sum_{Q'}(y_1A_{1Q'(1)}x_{Q'(1)})\cdots(y_nA_{nQ'(n)}x_{Q'(n)})\\
&=\epsilon(y_1\cdots y_n)(x_1\cdots x_n)\sum_Q\epsilon_Q A_{1Q(1)}\cdots A_{nQ(n)}\\
&=\epsilon(y_1\cdots y_n)(x_1\cdots x_n)\det A,
\end{aligned} \tag{28.64}$$

where in the second line we have relabeled the x's, as $x_{P(i)}\to x_i$, in the third we define $Q'=QP^{-1}$, in the fourth we have ordered the y's in front, in the correct order $(y_1\cdots y_n)$, followed by the x's in the correct order $(x_1\cdots x_n)$, generating a constant sign ϵ that depends on n, times a sign ϵ_Q that depends on the permutation Q.

28.5 Path Integral for the Fermionic Harmonic Oscillator

Consider the fermionic harmonic oscillator Hamiltonian,

$$\hat{H}_F=\hbar\omega\left(\hat{b}^\dagger\hat{b}-\frac{1}{2}\right). \tag{28.65}$$

Analogously to the bosonic case, we define coherent fermionic ket states,

$$|\psi\rangle\equiv e^{\hat{b}^\dagger\psi}|0\rangle=(1+\hat{b}^\dagger\psi)|0\rangle=(1-\psi\hat{b}^\dagger)|0\rangle, \tag{28.66}$$

that have eigenvalue ψ under b,

$$b|\psi\rangle=\psi|0\rangle=\psi(1-\psi\hat{b}^\dagger)|0\rangle=\psi|\psi\rangle, \tag{28.67}$$

as well as corresponding bra states, $\langle\psi^*|$. Together, they satisfy the completeness relation

$$\mathbb{1}=\int d\bar{\psi}d\psi|\psi\rangle\langle\psi^*|e^{-\bar{\psi}\psi}. \tag{28.68}$$

Following the same steps as in the bosonic case, we find the fermionic transition amplitude as a path integral over the coherent (harmonic) phase space:

$$\langle\bar{\psi},t'|\psi,t\rangle=\int\mathcal{D}\bar{\psi}\mathcal{D}\psi\exp\left\{\frac{i}{\hbar}\int_t^{t'}d\tau\left[-i\hbar\partial_\tau\bar{\psi}(\tau)\psi(\tau)-H\right]+\bar{\psi}(t)\psi(t)\right\}. \tag{28.69}$$

Doing a partial integration in τ (without boundary terms), and introducing sources $\bar{\eta},\eta$ for $\psi,\bar{\psi}$, in order to obtain the partition function $Z[\eta,\bar{\eta}]$, the generating functional of correlation functions, we obtain (this time writing the source term without an $\hbar$, since the kinetic term does not have it explicitly either)

$$Z[\eta, \bar{\eta}] = \int \mathcal{D}\bar{\psi} \int \mathcal{D}\psi \exp\left\{i \int dt \left[\bar{\psi}(i\partial_t - \omega)\psi + \bar{\psi}\eta + \bar{\eta}\psi\right]\right\}. \quad (28.70)$$

Doing the Gaussian integration on the transition amplitude, we find

$$\begin{aligned}\langle \bar{\psi}, t' | \psi, t\rangle &= \mathcal{N} \exp\left\{-\int_{-\infty}^{+\infty} d\tau \int_{\tau}^{+\infty} d\tau' e^{i\omega(\tau-\tau')} \bar{\eta}(\tau')\eta(\tau)\right\} \\ &= \mathcal{N} \exp\left[-\int_{-\infty}^{+\infty} d\tau \int_{-\infty}^{+\infty} d\tau' \bar{\eta}(\tau') D_F(\tau, \tau')\eta(\tau)\right],\end{aligned} \quad (28.71)$$

where we have used the formula for the Feynman propagator,

$$D_F(\tau, \tau') \equiv (-i(i\partial_t - \omega))^{-1}(\tau, \tau') = i \int \frac{dE}{2\pi} \frac{e^{-iE(\tau'-\tau)}}{E - \omega + i\epsilon} = \theta(\tau - \tau')e^{-i\omega(\tau'-\tau)}. \quad (28.72)$$

The fermionic path integral can be defined better in Euclidean time, similarly to the bosonic case, but we will not do it here.

Important Concepts to Remember

- The partition function $Z[J]$ is a generating functional, but $J(t)$ can be understood as more than a mathematical artifact, in fact, rather as a source for the classical variable $q(t)$. In the case of a harmonic oscillator, we obtain the driven harmonic oscillator.
- The kinetic operator $i\left(d^2/dt^2 + \omega^2\right)$ is inverted to the propagator $\Delta(t, t')$, via a Fourier transform, as

$$i \int \frac{dp}{2\pi} \frac{e^{-ip(t-t')}}{p^2 - \omega^2},$$

  but this is not invertible since it has zero modes, the classical solutions. The correct formula is then the Feynman propagator, avoiding the singularity in the complex plane, by making the replacement $\omega^2 \to \omega^2 - i\epsilon$.
- The Feynman propagator ends up being $\Delta_F(t, t') = \frac{1}{2\omega} e^{-i\omega|t-t'|}$, which is consistent with a space of eigenfunctions (on which we invert the kinetic propagator) with boundary conditions $q(t) \sim e^{-i\omega t}$ for $t \to -\infty$ and $q(t) \sim e^{+i\omega t}$ for $t \to +\infty$. Then $Z[J] = \mathcal{N}'' \exp\left[-\frac{1}{2\hbar} J \cdot \Delta \cdot J + \frac{i}{\hbar} q_{\text{cl}}(t) \cdot J\right]$.
- In order to do the Gaussian integration correctly (since the integrand is complex, $e^{iS/\hbar}$, and is thus highly oscillatory), one does a Wick rotation to imaginary time (analytical continuation to Euclidean time), such that $e^{iS} \to e^{-S_E/\hbar}$, where

$$S_E = \int dt \left[\frac{1}{2}\left(\frac{dq}{dt_E}\right)^2 + V(q)\right]$$

  is the Euclidean action.
- From the Wick rotation of the quantum amplitude, written as a path integral, for a periodic path $q = q'$ and integration over q, we obtain the Feynman–Kac formula, relating it to the statistical mechanics partition function, $Z[\beta] = \text{Tr}[e^{-\beta\hat{H}}] = \int_{q(t_E + \hbar\beta) = q(t_E)} e^{-S_E[q]/\hbar}$.

- We can introduce sources also in the Euclidean formulation, and calculate n-point functions of $\hat{q}(-i\tau) = e^{\hat{H}\tau/\hbar}\hat{q}e^{-\hat{H}\tau/\hbar}$ via derivatives.
- For a harmonic oscillator, besides the correct Gaussian integration with $e^{-S_E/\hbar}$, we have no boundary terms (owing to the periodic paths in the path integral), and moreover the propagator is well defined, as its Fourier transform is $1/(E_E^2 + \omega^2)$.
- Since the formal classical limit, $\hbar \to 0$, of the fermionic harmonic oscillator leads to $\{b, b^\dagger\} = 0$, that is, anticommuting variables, we have a Grassmann algebra, over which we need to define a path integral.
- The functions over the Grassmann algebra are at most linear in each independent element x_i, but the coefficients are also Grassmann objects (so that half the Grassmann objects are coordinates, half are coefficients).
- We can define an indefinite integral $\int dx$ that is the same as the derivative. Gaussian integration gives $\int d^n x \int d^n x \exp\left(y^T A x\right) = \det A$.
- The path integral for the transition amplitude of a harmonic oscillator gives $\exp\left[-\int \bar{\eta} \cdot D_F \cdot \eta\right]$, where in Fourier space we have $D_F = i/(E - \omega + i\epsilon)$.

Further Reading

See [5] for more details.

Exercises

(1) Consider the propagator obtained by the replacement $\omega^2 \to \omega^2 + i\epsilon$ ("anti-Feynman") in the Fourier transform. Calculate the integral, and find the corresponding boundary conditions for the (Hilbert space of) functions over which we invert the kinetic operator.

(2) Repeat the above exercise for the case where both the $p = \pm\omega$ poles in the integral are avoided in the complex space from above (via a complex integration contour that is slightly above the real line at the level of both poles).

(3) Prove that

$$\left[-\frac{d^2}{d\tau^2} + \omega^2\right] K(\tau, \tau') = \delta(\tau - \tau'), \tag{28.73}$$

where $K(\tau, \tau') = \Delta_{\text{free}}(\tau - \tau')$, and $\Delta(\tau - \beta) = \Delta(\tau)$ has a *unique* solution: if $\tau \in [0, \beta]$, the solution is

$$\Delta_{\text{free}}(\tau) = \frac{1}{2\omega}[(1 + n(\omega))e^{-\omega\tau} + n(\omega)e^{\omega\tau}], \tag{28.74}$$

where

$$n(\omega) = \frac{1}{e^{\beta|\omega|} - 1}. \tag{28.75}$$

(4) Wick-rotate the formula

$$I(E_E) = \int \frac{dE_{1,E}}{2\pi} \int \frac{dE_{2,E}}{2\pi} \frac{1}{(E_{1,E}^2 + \omega_1^2)(E_{2,E}^2 + \omega_2^2)[(E_{1,E} + E_{2,E} - E_E)^2 + \omega_3^2]}. \tag{28.76}$$

(5) For the driven harmonic oscillator, Wick-rotated to Euclidean space, calculate the four-point function $G_{4,E}(t_1, t_2, t_3, t_4)$.

(6) Consider Grassmann variables θ^α, $\alpha = 1, 2$, and the even function

$$\Phi(x, \theta) = \phi(x) + \sqrt{2}\theta^\alpha \psi_\alpha(x) + \theta^2 F(x), \tag{28.77}$$

where $\theta^2 = \epsilon_{\alpha\beta}\theta^\alpha\theta^\beta$. Calculate

$$\int d^2\theta(a_1\Phi + a_2\Phi^2 + a_3\Phi^3), \tag{28.78}$$

where

$$\int d^2\theta \equiv -\frac{1}{4}\int d\theta^\alpha d\theta^\beta \epsilon_{\alpha\beta}. \tag{28.79}$$

(7) Fill in the details omitted in the text for the calculation of

$$Z[\bar{\eta}, \eta] = Z[0, 0] \exp\left(-i \int d\tau \int d\tau' \, \bar{\eta}(\tau) D_F(\tau, \tau') \eta(\tau')\right). \tag{28.80}$$

29 General Theory of Quantization of Classical Mechanics and (Dirac) Quantization of Constrained Systems

In this chapter, we will analyze the answer to the question of how to quantize general classical mechanics theories, in the presence of constraints. The answer was given by Dirac in his famous book *Lectures on Quantum Mechanics* [11], published in 1964, and based on four lectures given at Yeshiva University. It is probably the most influential book in theoretical physics, for the number of pages. His presentation was so well conceived that there is little left to change even more than 50 years afterwards, so my presentation is largely based on his original one.

There were several potential issues that needed solving by Dirac's picture.

- One is the presence of physical constraints, such as for instance in the case of a motion in a circle. One solution is to solve the classical motion in terms of independent variables, but that may not always be available or practical. Instead, we can still quantize in terms of the original variables, without solving the constraints.
- Sometimes there are no independent variables: indeed, we will see that constraints can involve spatial and/or momentum variables, q_i and p_i, so constraints that involve both at the same time are harder to understand.
- Finally, there is the existence of gauge invariance symmetries, i.e., local redundancies in the description. For instance, in the case of electromagnetism, there is the invariance $\delta A_\mu = \partial_\mu \lambda$. In this case we can fix it by a gauge choice that acts as a constraint on the A_μ (phase space) variables. The constraints are defined on phase space, i.e., $\phi(p, q) = 0$, which means that we must use the Hamiltonian formalism and find a specific form for the Hamiltonian. In the case of the Lagrangian formalism, we can introduce constraints by Lagrange multipliers, $L \to L + \lambda\phi$. Then in the Hamiltonian formalism, we should also add constraints to H, but it will happen in a different way.

29.1 Hamiltonian Formalism

For completeness, we start with a review of the Hamiltonian formalism. From a Lagrangian $L(q_n, \dot{q}_n)$, we define the canonically conjugate momenta

$$p_n = \frac{\partial L}{\partial \dot{q}_n}, \tag{29.1}$$

after which we define the (naive, see later) Hamiltonian

$$H = \sum_{n=1}^{N} (p_n \dot{q}_n) - L = H(\{q_n\}, \{p_n\}), \tag{29.2}$$

where we replace the $\dot{q}_n$ with their expressions in terms of p_n in order to find $H(\{q_n\}, \{p_n\})$. The Hamiltonian equations of motion are

$$\dot{q}_n = \frac{\partial H}{\partial p_n}, \quad \dot{p}_n = -\frac{\partial H}{\partial q_n}. \tag{29.3}$$

We can rewrite the Hamiltonian formalism in a way appropriate for quantization by introducing Poisson brackets for two functions of phase space $f(q,p)$ and $g(q,p)$ as

$$\{f, g\}_{P.B.} = \sum_n \left(\frac{\partial f}{\partial q_n} \frac{\partial g}{\partial p_n} - \frac{\partial f}{\partial p_n} \frac{\partial g}{\partial q_n} \right). \tag{29.4}$$

In terms of them, the Hamiltonian equations of motion become

$$\dot{q}_n = \{q_n, H\}_{P.B.}, \quad \dot{p}_n = \{p_n, H\}_{P.B.}, \tag{29.5}$$

or, more generally, for an arbitrary function of phase space $g(q,p)$,

$$\dot{g} = \{g, H\}_{P.B.}. \tag{29.6}$$

29.2 Constraints in the Hamiltonian Formalism: Primary and Secondary Constraints, and First and Second Class Constraints

We start by assuming a set of M constraints on phase space,

$$\phi_m(q,p) = 0, \quad m = 1, \ldots, M. \tag{29.7}$$

We will call them *primary constraints*; we will see shortly why.

In order to understand the constraints, we give two examples.

Example 1 Consider the Lagrangian

$$L = q\dot{q} - \alpha q^2. \tag{29.8}$$

Then the momentum p canonically conjugate to q is

$$p = \frac{\partial L}{\partial \dot{q}} = q, \tag{29.9}$$

which means there is a constraint relating momenta and coordinates,

$$p - q = 0. \tag{29.10}$$

Example 2 Consider the Lagrangian

$$L = \frac{1}{2}\dot{q}_2^2 - \alpha q_2^2 - \beta q_1 q_2'^2. \tag{29.11}$$

In this case, the momentum canonically conjugate to p_2 is $\dot{q}_2$, which is standard, but the momentum canonically conjugate to q_1 is

$$p_1 = 0, \tag{29.12}$$

which is a constraint on phase space.

We define $\approx$ as an equality that holds only after using the constraints $\phi_m = 0$; it is called a *weak equality*. It means that we use the constraints only at the end of the calculation, after for instance evaluating all the Poisson brackets. Indeed, the Poisson brackets are defined by assuming that $\{q_n, p_n\}$ are independent variables that have nonzero brackets. By definition, we have

$$\phi_m \approx 0, \tag{29.13}$$

so these brackets are not thought of as being identically zero, only weakly zero. Since $\phi_m \approx 0$, from the point of view of time evolution there is no difference between the Hamiltonian H and $H + u_m\phi_m$. The equations of motion associated with the latter are

$$\begin{aligned} \dot{q}_n &= \frac{\partial H}{\partial p_n} + u_m \frac{\partial \phi_m}{\partial p_n} \approx \{q_n, H + u_m\phi_m\}_{P.B.} \\ \dot{p}_n &= -\frac{\partial H}{\partial q_n} - u_m \frac{\partial \phi_m}{\partial q_n} \approx \{p_n, H + u_m\phi_m\}_{P.B.}, \end{aligned} \tag{29.14}$$

where in the last $\approx$ in both equations we have taken into account that terms with $\partial u_m/\partial p_n$ and $\partial u_m/\partial q_n$ are both multiplied by ϕ_m, so are ≈ 0.

For a general function $g(q, p)$, then, we obtain the new equation of motion

$$\dot{g} = \{g, H_T\}_{P.B.}, \tag{29.15}$$

where

$$H_T = H + u_m\phi_m \tag{29.16}$$

is called the *total Hamiltonian*.

We can apply the above time evolution to the primary constraints ϕ_m themselves. But if the ϕ_m are good constraints, it follows that they must be respected by time evolution as well, and must stay on the constraint surface $\phi_m \simeq 0$, so that $\dot{\phi}_m \approx 0$. This gives the equation

$$\{\phi_n, H_T\}_{P.B.} \simeq \{\phi_n, H\}_{P.B.} + u_m\{\phi_n, \phi_m\}_{P.B.} \approx 0. \tag{29.17}$$

In general, this equation produces other, potentially independent, constraints, called *secondary constraints*. Then we repeat the process, and find the equation for time evolution of the above equations, etc., until we find no more new constraints. This is then the full set of secondary constraints,

$$\phi_k, \quad k = M+1, \ldots, M+K. \tag{29.18}$$

Together, the primary and secondary constraints are defined as

$$\phi_j \approx 0, \quad j = 1, \ldots, M+K. \tag{29.19}$$

We will consider them together, since the difference between the two is not really important.

There is however a different separation of constraints that is useful. We call a general function of phase space $R(q, p)$ *first class*, if

$$\{R, \phi_j\}_{P.B.} \approx 0, \quad \forall j = 1, \ldots, M+K, \tag{29.20}$$

and *second class* if there exists at least one ϕ_j such that $\{R, \phi_j\}_{P.B.}$ is not ≈ 0. Considering the case where R is a constraint $\phi_{j'}$ itself, we can define first-class and second-class constraints. The separation of constraints into first class and second class is of relevance, unlike that between primary and secondary.

For the full set of constraints, we can write the equations for time evolution in the form

$$\dot{\phi}_j = \{\phi_j, H_T\}_{P.B.} \approx \{\phi_j, H\}_{P.B.} + u_m\{\phi_j, \phi_m\}_{P.B.} \approx 0, \quad j = 1, \dots, M + K. \tag{29.21}$$

In view of the previous definition, the equation also says that H_T is first class. The equations (29.21) comprise $M + K$ equations for the M coefficients u_j, which is an overconstrained system. However, on physical grounds, since there must exist a true time evolution we must have at least one solution, call it U_m. However, if U_m is a particular solution of (29.21), then

$$u_m = U_m + v_a V_{am} \tag{29.22}$$

is the general solution, where V_{am} solves the homogenous equation,

$$V_{am}\{\phi_j, \phi_m\}_{P.B.} \approx 0. \tag{29.23}$$

Then we can split the first-class Hamiltonian H_T as

$$H_T = H + U_m\phi_m + v_a V_{am}\phi_m = H' + v_a\phi_a, \tag{29.24}$$

where $H' = H + U_m\phi_m$ is also first class, since U_m is a particular solution of (29.21) for u_m, and

$$\phi_a = V_{am}\phi_m \tag{29.25}$$

are first-class primary constraints, since the ϕ_m are primary, and $V_{am}\{\phi_j, \phi_m\}_{P.B.} \approx 0$ implies that $V_{am}\phi_m$ are first class.

Theorem If R and S are first class, i.e., $\{R.\phi_j\}_{P.B.} = r_{jj'}\phi_{j'}$ and $\{S, \phi_j\}_{P.B.} = s_{jj'}\phi_{j'}$, then $\{R, S\}_{P.B.}$ is first class.

Proof Consider the Jacobi identity, which is an identity (so equivalent to $0 = 0$ as can be seen by writing all the terms explicitly), derived from the antisymmetry of the Poisson brackets, namely

$$\{\{R, S\}_{P.B.}, P\}_{P.B.} + \{\{P, R\}_{P.B.}, S\}_{P.B.} + \{\{S, P\}_{P.B.}, R\}_{P.B.} = 0. \tag{29.26}$$

The power of the Jacobi identity comes from its use in calculating explicitly all the Poisson brackets. The case of interest is $P = \phi_j$, in which case we obtain

$$\begin{aligned}\{\{R, S\}_{P.B.}, \phi_j\}_{P.B.} &= \{\{R, \phi_j\}_{P.B.}, S\}_{P.B.} - \{\{S, \phi_j\}_{P.B.}, R\}_{P.B.}\\ &= \{r_{jj'}\phi_{j'}, S\}_{P.B.} - \{s_{jj'}\phi_{j'}, R\}_{P.B.}\\ &= r_{jj'}s_{j'j''}\phi_{j''} - \{r_{jj'}, S\}_{P.B.}\phi_{j'} - s_{jj'}(-r_{j'j''}\phi_{j''}) - \{s_{jj'}, R\}_{P.B.}\phi_{j'} \approx 0.\end{aligned} \tag{29.27}$$

But this is then the definition of $\{R, S\}_{P.B.}$ being first class. *q.e.d.*

Adding the first-class secondary constraints, denoted $\phi_{a'}$, to the Hamiltonian (since, as we said, there is no difference between primary and secondary, only between first-class and second-class), we obtain the *extended Hamiltonian*

$$H_E = H_T + v_{a'}\phi_{a'}. \tag{29.28}$$

To summarize, the first-class constraints generate motion tangent to the constraint hypersurface (so can be added to the Hamiltonian), whereas the second-class constraints generate motion away from it, so cannot be added.

29.3 Quantization and Dirac Brackets

The standard quantization procedure is to replace the Poisson brackets $\{,\}_{P.B.}$ with $\frac{1}{i\hbar}[,]$. But the presence of constraints introduces a subtlety.

The simplest way to deal with the constraints is to introduce them as an operator constraint (since $f(q,p)$ becomes $\hat{f}(\hat{q},\hat{p})$) acting on physical states $|\psi\rangle$,

$$\hat{\phi}_j|\psi\rangle = 0. \tag{29.29}$$

However, that leads to a potential problem, since acting twice with $\hat{\phi}_j$ and taking the commutator leads to

$$[\hat{\phi}_j, \hat{\phi}_{j'}]|\psi\rangle = 0. \tag{29.30}$$

In quantum mechanics, for this to be zero, considering that ϕ_j is a full set of constraints, the commutator must be proportional to the constraints themselves, i.e., the constraint operators must satisfy an algebra,

$$[\hat{\phi}_j, \hat{\phi}_{j'}] = c_{jj'j''}\hat{\phi}_{j''}. \tag{29.31}$$

Note that $c_{jj'j''}$ could be an operator itself, and since operators do not in general commute, we must write it to the left of $\phi_{j''}$, so that this vanishes when acting on $|\psi\rangle$.

Since time evolution should be a constraint too (it should keep us within the constraint hypersurface), we must also have

$$[\hat{\phi}_j, \hat{H}] = b_{jj'}\hat{\phi}_{j'}. \tag{29.32}$$

If there are only first-class constraints, there is no problem since then the first-class condition is

$$\{\phi_j, \phi_{j'}\}_{P.B.} \approx 0 \;\Rightarrow\; \{\phi_j, \phi_{j'}\}_{P.B.} = c_{jj'j''}\phi_{j''}, \tag{29.33}$$

which in quantum theory turns into (29.31).

However, if there is a second-class constraint, we do have a problem, since then $\{\phi_j, \phi_{j'}\}_{P.B.}$ is not ≈ 0.

Example

Consider an example, to help us understand the issue and a possible solution. Consider a system with N degrees of freedom, and constraints $q_1 \approx 0$ and $p_1 \approx 0$. But then $\{q_1, p_1\}_{P.B.} = 1 \neq 0$, so the constraints are not first class. We quantize by imposing on physical states $|\psi\rangle$ that

$$\hat{q}_1|\psi\rangle = 0, \quad \hat{p}_1|\psi\rangle = 0. \tag{29.34}$$

However taking the commutator results in $[\hat{q}_1, \hat{p}_1]|\psi\rangle = 0$, which is in contradiction with the quantization of the Poisson bracket $\{q_1, p_1\}_{P.B.}$, which leads to $[\hat{q}_1, \hat{p}_1] = i\hbar$.

In conclusion, either $\hat{\phi}_j|\psi\rangle = 0$ is not a good way to impose the constraint or the quantization prescription of replacing the Poisson bracket $\{,\}_{P.B.}$ with the commutator $\frac{1}{i\hbar}[,]$ does not work. In fact, it turns out to be the latter. We need to modify the Poisson bracket to a new bracket.

In the case above, the solution is simple: we just remove (q_1, p_1) from the phase space, so the modified Poisson bracket is now

$$\{f, g\}_{P.B.'} = \sum_{n=2}^{N} \left(\frac{\partial f}{\partial q_n} \frac{\partial g}{\partial p_n} - \frac{\partial f}{\partial p_n} \frac{\partial g}{\partial q_n} \right). \tag{29.35}$$

But we must find a way to extend this procedure to the general case, by introducing what is known as a *Dirac bracket*. Consider *independent* second-class constraints by taking out linear combinations that are first class and leaving only constraints all of whose linear combinations are second class, and call them χ_s.

Define the inverse matrix $c_{ss'}$ of Poisson brackets of the second-class constraints χ_s,

$$c_{ss'} \{\chi_{s'}, \chi_{s''}\}_{P.B.} = \delta_{ss''}. \tag{29.36}$$

Then we can define the Dirac brackets of two functions of phase space coordinates as

$$[f, g]_{D.B.} = \{f, g\}_{P.B.} - \{f, \chi_s\}_{P.B.} c_{ss'} \{\chi_{s'}, g\}_{P.B.}. \tag{29.37}$$

If we use Dirac brackets instead of Poisson brackets, the time evolution is not modified, since

$$\begin{aligned} [g, H_T]_{D.B.} &= \{g, H_T\}_{P.B.} - \{g, \chi_s\}_{P.B.} c_{ss'} \{\chi_{s'}, H_T\}_{P.B.} \\ &\approx \{g, H_T\}_{P.B.}, \end{aligned} \tag{29.38}$$

where we have used that H_T is first class, hence $\{\chi_{s'}, H_T\}_{P.B.} \approx 0$.

However, using Dirac brackets instead of Poisson brackets leads to the bracket of any function of phase space coordinates with a second-class constraint being equal to zero *strongly*,

$$[f, \chi_{s''}]_{D.B.} = \{f, \chi_{s''}\}_{P.B.} - \{f, \chi_s\}_{P.B.} c_{ss'} \{\chi_{s'}, \chi_{s''}\}_{P.B.} = 0, \tag{29.39}$$

where we have used the definition of $c_{ss'}$, $c_{ss'}\{\chi_{s'}, \chi_{s''}\}_{P.B.} = \delta_{ss''}$. Since the equality of the bracket to zero is exact (strongly), we can set χ_s to zero strongly if we use the Dirac bracket.

In quantum mechanics, we can still impose the constraints on states $\hat{\chi}_s|\psi\rangle = 0$, and not obtain any more contradictions, by replacing the Dirac bracket with $\frac{1}{i\hbar}[,]$. Indeed, then, for the commutator of second-class constraints with any function f of phase space coordinates, we obtain zero when acting on states, so without any more contradictions,

$$[\hat{f}, \hat{\chi}_s]|\psi\rangle = 0. \tag{29.40}$$

To understand this better, consider the previous example, of constraints $q_1 \approx 0$, $p_1 \approx 0$, and use this general formalism to see that indeed the Dirac bracket amounts to the same solution as the one we found. Since $\{q_1, p_1\}_{P.B.} = 1$, it means that both constraints are second class, so we need to use Dirac brackets. Define $\chi_1 = q_1$ and $\chi_2 = p_1$. Then $\{\chi_1, \chi_2\}_{P.B.} = 1$, which implies

$$\begin{aligned} c_{s1}\{\chi_1, \chi_2\}_{P.B.} = \delta_{s2} &\Rightarrow \chi_{21} = 1 \\ c_{s2}\{\chi_2, \chi_1\}_{P.B.} = \delta_{s1} &\Rightarrow \chi_{12} = -1. \end{aligned} \tag{29.41}$$

Then the Dirac bracket is

$$\begin{aligned} [f, g]_{D.B.} &= \{f, g\}_{P.B.} - \{f, \chi_2\}_{P.B.}\{\chi_1, g\}_{P.B.} + \{f, \chi_1\}_{P.B.}\{\chi_2, g\}_{P.B.} \\ &= \{f, g\}_{P.B.} - \frac{\partial f}{\partial q_1} \frac{\partial g}{\partial p_1} + \frac{\partial f}{\partial p_1} \frac{\partial g}{\partial q_1} \\ &= \sum_{n=2}^{N} \left(\frac{\partial f}{\partial q_n} \frac{\partial g}{\partial p_n} - \frac{\partial f}{\partial p_n} \frac{\partial g}{\partial q_n} \right), \end{aligned} \tag{29.42}$$

where in the second equality we have substituted $\chi_1 = q_1$ and $\chi_2 = p_1$, and then calculated the Poisson brackets.

Since it is this Dirac bracket that quantizes to $\frac{1}{i\hbar}[,]$, in quantum mechanics $[\hat{q}_n, \hat{p}_m] = i\hbar\delta_{nm}$ only for $n = 2, \dots, N$, but not for $n = 1$, so $[\hat{q}_1, \hat{p}_1] = 0$. Then imposing

$$\hat{q}_1|\psi\rangle = 0 = \hat{p}_1|\psi\rangle \tag{29.43}$$

leads to no contradiction.

Next, we consider in detail two examples that have complementary features and that were described in the general formalism.

Example 1

The first example is the first example from before, with $L = q\dot{q} - \alpha q^2$. The example is a bit pathological, but it has some of the features of the case of (Majorana, i.e., real) massive fermion(s) ψ_i, with Lagrangian

$$L = \frac{1}{2}\psi_i\dot{\psi}_i - \frac{m}{2}\psi_i\sigma_{ij}\psi_j,$$

and where the ψ_i are anticommuting and there is an antisymmetric matrix σ_{ij}, so that the mass term is nonzero. However, we will not treat Majorana fermions here, since they have some subtleties.

The equations of motion of the Lagrangian are

$$\frac{d}{dt}\frac{\partial L}{\partial \dot{q}} - \frac{\partial L}{\partial q} = 0 \Rightarrow \frac{d}{dt}q + 2\alpha q - \dot{q} = 0 \Rightarrow q = 0. \tag{29.44}$$

However, in the Hamiltonian formalism, the situation is more interesting. First, the Hamiltonian is

$$H = p\dot{q} - L = \alpha q^2, \tag{29.45}$$

since as we saw, $p = q$. This is actually a (primary) constraint, so we have

$$\phi_1 = p - q \approx 0. \tag{29.46}$$

But we have to consider the time evolution of this primary constraint, and set it to zero weakly, generating a secondary constraint ϕ_2,

$$\{\phi_1, H\}_{P.B.} = \{p - q, H\}_{P.B.} = -2\alpha q \approx 0 \Rightarrow \phi_2 = q \approx 0. \tag{29.47}$$

The bracket of $\phi_2 = q$ with H is zero, so there are no more secondary constraints.

The equations of motion of the Hamiltonian are

$$\dot{q} = 0, \quad \dot{p} = -2\alpha q. \tag{29.48}$$

Note that, since $p = q$, we have that the second equation is $\dot{q} = -2\alpha q$, which because of the first equation reduces to $q = 0$, as in the Lagrangian formalism.

Calculating the Poisson brackets of the constraints, we get

$$\begin{aligned} \{p - q, p - q\}_{P.B.} &= \{q, q\}_{P.B.} = 0 \\ \{\phi_1, \phi_2\}_{P.B.} &= \{p - q, q\}_{P.B.} = 1 \neq 0, \end{aligned} \tag{29.49}$$

so both ϕ_1 and ϕ_2 are second class, and there are no first-class constraints. On the other hand, ϕ_1 is a primary constraint and ϕ_2 is a secondary constraint. The total Hamiltonian is obtained by adding the primary constraints, so

$$H_T = H + u_m\phi_m = \alpha q^2 + u(p-q). \tag{29.50}$$

For this, we have two conditions: $\{\phi_j, H_T\}_{P.B.} \approx 0$, for $j = 1, 2$. The first is

$$\{p-q, \alpha q^2\} + u\{p-q, p-q\} \approx 0, \tag{29.51}$$

but since the first bracket is $-2\alpha q \approx 0$ (it is proportional to ϕ_2), and the second is identically zero, the equation is satisfied. The second condition is

$$\{q, \alpha q^2\}_{P.B.} + u\{q, p-q\}_{P.B.} \approx 0, \tag{29.52}$$

which becomes (since the first bracket is zero, and the second is one)

$$u \approx 0. \tag{29.53}$$

This is not surprising, since H_T is first class, and there are no first-class constraints. Further, $u = U + v_a\phi_a$, where ϕ_a are the first-class constraints, of which there are none, so $u = U$ and U is a particular solution; but since there is a particular solution $u = 0$, as we saw, we can put $U = 0$ as well. Then finally

$$H_T = H' = H = H_E = \alpha q^2. \tag{29.54}$$

To obtain the Dirac brackets, note first that all the constraints are second class, so $\chi_1 = \phi_1 = p - q$, and $\chi_2 = \phi_2 = q$. Since

$$\{\chi_1, \chi_2\}_{P.B.} = -1 = -\{\chi_2, \chi_1\}_{P.B.}, \tag{29.55}$$

we have $c_{12} = 1, c_{21} = -1$. Then the Dirac brackets are

$$\begin{aligned}[f, g]_{D.B.} &= \{f, g\}_{P.B.} + \{f, \chi_2\}_{P.B.}\{\chi_1, g\}_{P.B.} - \{f, \chi_1\}_{P.B.}\{\chi_2, g\}_{P.B.} \\ &= \frac{\partial f}{\partial q}\frac{\partial g}{\partial p} - \frac{\partial f}{\partial p}\frac{\partial g}{\partial q} + \frac{\partial f}{\partial p}\left(\frac{\partial g}{\partial q} + \frac{\partial g}{\partial p}\right) - \left(\frac{\partial f}{\partial q} + \frac{\partial f}{\partial p}\right)\frac{\partial g}{\partial p} = 0,\end{aligned} \tag{29.56}$$

where in the second equality we substituted $\chi_1 = p - q$ and $\chi_2 = q$, and then evaluated the Poisson brackets.

Since the Dirac bracket is zero, there is nothing to quantize.

29.4 Example: Electromagnetic Field

As a second example, consider the electromagnetic field A_μ, but without a gauge choice, so constraints will appear only from the form of the action itself. It is a field, so $A_\mu = A_\mu(\vec{x}, t)$, and in terms of quantum mechanics, we can think of $\vec{x}$ as an extra index i; otherwise we have the same classical story. However, we will not go into the quantization details, since they are in the domain of quantum field theory, beyond the scope of this book. Instead, we will just use it as an example of the constraint formalism in classical physics.

The action for electromagnetism is

$$S = -\int d^3x\, dt \frac{F_{\mu\nu}F^{\mu\nu}}{4} = \int d^3x\, dt \left[-\frac{1}{4}F_{ij}F^{ij} - \frac{1}{2}F_{0i}F^{0i}\right], \tag{29.57}$$

where $F_{\mu\nu} = \partial_\mu A_\nu - \partial_\nu A_\mu$, and A_μ is the field variable. For an arbitrary function f, we denote derivatives with a comma, $\partial_\mu f = f_{,\mu}$. Then the momentum canonically conjugate to A_μ is

$$P^\mu(\vec{x}) \equiv \frac{\delta L}{\delta A_{\mu,0}(\vec{x})} = F^{\mu 0}(\vec{x}), \tag{29.58}$$

where the functional derivative δ amounts to a regular partial derivative after removing the integral over $\vec{x}$ (since $\vec{x}$ is like an index i, and this index sits on the variable $A_\mu(\vec{x})$, so the integral stands for $\sum_i$). But since $F^{\mu\nu}$ is antisymmetric we have $F^{00} = 0$, so $P^0 = 0$, which is a (primary) constraint,

$$P^0 \approx 0. \tag{29.59}$$

Since $\vec{x}$ is an index i like μ, the fundamental Poisson brackets for A_μ and P^ν are

$$\{A_\mu(\vec{x},t), P^\nu(\vec{x}\,',t)\}_{P.B.} = \delta_\mu^\nu \delta^3(\vec{x} - \vec{x}\,'). \tag{29.60}$$

The Hamiltonian is

$$\begin{aligned} H &= \int d^3x P^\mu A_{\mu,0} - L = \int d^3x \left[F^{i0}A_{i,0} + \frac{1}{4}F_{ij}F^{ij} + \frac{1}{2}F_{i0}F^{i0}\right] \\ &= \int d^3x \left[F^{i0}A_{0,i} + \frac{1}{4}F_{ij}F^{ij} - \frac{1}{2}F_{i0}F^{i0}\right] \\ &= \int d^3x \left[\frac{1}{4}F_{ij}F^{ij} + \frac{1}{2}P^iP^i - A_0 P^i{}_{,i}\right], \end{aligned} \tag{29.61}$$

where in the second equality we have used $F_{i0} = A_{0,i} - A_{i,0}$, and in the third one, we have used $P^i = F^{0i}$ and partial integration.

The secondary constraints are obtained from the time evolution,

$$\{P^0, H\}_{P.B.} = P^i{}_{,i} \approx 0, \tag{29.62}$$

where we have partially integrated under the integral sign in the Hamiltonian. We see that we obtain a secondary constraint. Taking one more time evolution with H,

$$\{P^i{}_{,i}, H\}_{P.B.} = 0, \tag{29.63}$$

we obtain no new secondary constraints. We find that both constraints (both the primary one, and the secondary one) are first class, since the Poisson brackets of the constraints vanish,

$$\{P^0(\vec{x},t), P^0(\vec{x}\,',t)\}_{P.B.} = 0 = \{P^0(\vec{x},t), P^i{}_{,i}(\vec{x}\,',t)\}_{P.B.} = \{P^i{}_{,i}(\vec{x},t), P^j{}_{,j}(\vec{x}\,',t)\}_{P.B.}. \tag{29.64}$$

Thus there are no Dirac brackets, and we can use Poisson brackets.
The constraints are all first class, but they divide into primary and secondary,

$$\phi_m : \phi_1 = P^0, \quad \phi_k : \phi_2 = P^i{}_{,i}. \tag{29.65}$$

Then the first-class Hamiltonian H' is

$$H' = H + U_m\phi_m = H + \int d^3x\, U P^0, \tag{29.66}$$

but we can choose $U = 0$ as a solution, and thus $H' = H$.

To obtain the total Hamiltonian, we add the primary first-class constraints with coefficients, $v_a\phi_a$, namely P^0, so

$$H_T = H' + \int d^3x\, v(\vec{x})P^0(\vec{x}) = \int d^3x \left(\frac{1}{2}F_{ij}F^{ij} + \frac{1}{2}P_iP^i\right) + \int d^3x \left[-A_0 P^i{}_{,i} + vP^0\right]. \quad (29.67)$$

Note that since $P^i = F^{0i} = E^i$, $\phi_2 = P^i{}_{,i} = \vec{\nabla}\cdot\vec{E}$ is a Gaussian constraint, and that, in H_T, $-A_0$ is its Lagrange multiplier.

To obtain the extended Hamiltonian H_E, we add the secondary first-class constraint $P^i{}_{,i}$ with an arbitrary coefficient,

$$H_E = H_T + \int d^3x\, u(\vec{x})P^i{}_{,i}(\vec{x}). \quad (29.68)$$

However, we can put $P^0 = 0$ in H_T, since its only purpose is to set the time evolution of A_0 to zero, as $\dot{A}_0 = 0$. But that means we can get rid of A_0 by redefining u, as $u' = u - A_0$. Then the extended Hamiltonian is

$$H_E = \int d^3x \left[\frac{1}{4}F_{ij}F^{ij} + \frac{1}{2}P_iP^i\right] + \int d^3x\, u'(\vec{x})P^i{}_{,i}(\vec{x}). \quad (29.69)$$

In this case, in quantum mechanics there is nothing new since the Dirac brackets equal the Poisson brackets, but there are still subtleties due to gauge invariance, which will not be addressed here as they relate to quantum field theory.

Important Concepts to Remember

- On a system in Hamiltonian formalism we can impose some primary constraints $\phi_m(q,p)$ and consider weak equality $\approx$ as an equality that holds only after using the constraints.
- Time evolution is weakly unchanged by adding the constraints to the Hamiltonian, $H \to H_T = H + u_m\phi_m$, and from the time evolution of the primary constraints (and further secondary constraints) we get secondary constraints.
- The set of all (primary and secondary) constraints is divided into first-class $\phi_{j'}$, those that have Poisson brackets weakly zero with all constraints, $\{\phi_{j'}, \phi_j\}_{P.B.} \approx 0$, so $= c_{j'ck}\phi_k$, and second-class, those for which there is at least a ϕ_j such that $\{\phi_{j'}, \phi_j\}_{P.B.} \neq c_{j'jk}\phi_k$ (more generally, any function of phase space coordinates can be first or second class).
- The time evolution equations of the constraints are $\{\phi_j, H\}_{P.B.} + u_m\{\phi_j, \phi_m\}_{P.B.} \approx 0$ and have a particular solution U_m and solutions V_{am} of the homogenous equation, so $u_m = U_m + V_{am}v_a$ is the general solution.
- The first-class total Hamiltonian is $H_T = H' + v_a\phi_a$, where $H' = H + U_m\phi_m$ is first class and $\phi_a = V_{am}\phi_m$ are first-class constraints, and the extended Hamiltonian adds also first-class secondary constraints, $H_E = H_T + v_{a'}\phi_{a'}$.
- In order to quantize, we can only introduce operator constraints acting on states, $\hat{\phi}_j|\psi\rangle = 0$, if we use Dirac brackets instead of Poisson brackets when quantizing the commutator, $[,]_{D.B.} \to \frac{1}{i\hbar}[,]$.
- Dirac brackets are obtained by removing the independent second-class constraints, which would introduce contradictions (such as $0 = (\{\phi_m, \phi_j\}_{P.B.})_{\text{quantized}}|\psi\rangle \sim \text{const.}|\psi\rangle$), via $[f,g]_{D.B.} = \{f,g\}_{P.B.} - \{f,\chi_s\}_{P.B.}(\{\chi_s,\chi_{s'}\}_{P.B.})^{-1}_{ss'}\{g,\chi_{s'}\}_{P.B.}$.

- The Dirac brackets of functions of phase space with second-class constraints equal to zero strongly, meaning there are no contradictions coming from imposing second-class constraints as operators acting on states during quantization, since now their commutator with anything vanishes.
- For electromagnetism we have only first-class constraints, $P^0 \approx 0$ being the primary one and $P^i{}_{,i} \approx 0$ the unique secondary one. In the extended Hamiltonian $H_E = H + \int (u - A_0)P^i{}_{,i} + \int vP^0$, we can put $v = 0$ and redefine $u' = u - A_0$. Since $P^i = E^i$, $P^i{}_{,i} = \vec{\nabla} \cdot \vec{E}$ is the Gaussian constraint and, in H_T, $-A_0$ is its Lagrange multiplier.

Further Reading

See Dirac's book [11] for more details.

Exercises

(1) Consider the Lagrangian

$$L = \frac{1}{2}(\dot{q}_1^2 + \dot{q}_2^2) - \alpha(q_1^2 + q_2^2)^2, \tag{29.70}$$

and the (primary) constraint $q_1 + \beta q_2 = 0$. Calculate the secondary constraints, and find out which of the constraints is first class and which is second class.

(2) Consider a particle of mass m in an (approximately) constant gravitational field (such as that of the Earth), but constrained to move on a circle within a vertical plane. Using Cartesian coordinates and the constraint formalism, write down the Lagrangian and the primary constraint and calculate the secondary constraints, and see which is first class, and which is second class.

(3) Consider the Lagrangian

$$L = \frac{1}{2}\dot{q}_1^2 + \alpha\dot{q}_2 q_1 - \beta q_1 q_2^2. \tag{29.71}$$

Find the constraints of the system, write down the total Hamiltonian, and solve for U_m, v_a.

(4) Use Dirac quantization to quantize the system in exercise 3.

(5) Consider the Born–Infeld action for nonlinear eletromagnetism,

$$S = -L^{-4}\int d^3x\, dt \left[\sqrt{1 + L^4\frac{F_{\mu\nu}F^{\mu\nu}}{2} - L^8\left(\frac{1}{8}\epsilon^{\mu\nu\rho\sigma}F_{\mu\nu}F_{\rho\sigma}\right)^2} - 1\right], \tag{29.72}$$

where L is a constant length.

Calculate the Hamiltonians H, H', H_T, H_E in a similar way to the Maxwell electromagnetism case in the text, and find the Dirac brackets.

(6) Consider the action for a (Dirac) spinor,

$$S = \int dt(i\psi^*\dot{\psi}), \tag{29.73}$$

written in terms of independent variables ψ and ψ^*. Calculate the primary constraints and the total Hamiltonian H_T. Check that there are no secondary constraints, and then from

$$\{\phi_m, H_T\}_{P.B.} \approx 0 \tag{29.74}$$

solve for U_m, v_A. Note that classical fermions are anticommuting, so we defined p by taking the derivatives from the left, for example $\frac{\partial}{\partial\psi}\left(\psi\,\chi\right) = \chi$, so that $\{p^A, q_B\} = -\delta^A_B$. In general, we must define

$$\{f,g\}_{P.B.} = -(\partial f/\partial p^\alpha)\left(\frac{\partial}{\partial q_\alpha}g\right) + (-)^{fg}(\partial g/\partial p^\alpha)\left(\frac{\partial}{\partial q_\alpha}f\right), \tag{29.75}$$

where $\partial f/\partial p^\alpha$ is the right derivative, for example $\frac{\partial}{\partial\psi}(\chi\psi) = \chi$, and $(-)^{fg} = -1$ if f and g are both fermions and $+1$ otherwise (if f and/or g is bosonic). This bracket is antisymmetric if f and g are bose–bose or bose–fermi, and symmetric if f and g are both fermionic.

(7) (Continuation of exercise 6) Show that all constraints are second class, thus writing also H' and H_E. Write down the Dirac brackets, and find a (potentially!) new expression for H_T and the resulting Dirac quantization relations.

PART II_a

ADVANCED FOUNDATIONS

30 Quantum Entanglement and the EPR Paradox

In Part II$_a$, we will study advanced topics related to the foundations of quantum mechanics. In this chapter, we will start by understanding entanglement, and associated notions and explaining the EPR paradox for entangled states.

Entanglement is a notion related to a total system composed of two subsystems, A and B, not necessarily with any physical division though as we shall see, the case of a macroscopic distance between the states is quite important. We consider not only two subsystems, but also independent observers for each, one called "Alice" for A and the other "Bob" for B. The total Hilbert space is the product of Hilbert spaces, $\mathcal{H} = \mathcal{H}_A \otimes \mathcal{H}_B$.

But if system A is correlated, or "entangled", with system B, in the quantum state of the total system, we say that we have an "entangled state". In particular, this means that we cannot write the state of the total system as a product state of the states in each subsystem. To understand these concepts better, we consider the simplest system, a spin 1/2 or two-state system.

30.1 Entanglement: Spin 1/2 System

The spin 1/2 system is the standard example used, since it contains all necessary ingredients, besides being the simplest system.

Each spin 1/2 system has two states, denoted by $|\uparrow\rangle$ and $|\downarrow\rangle$. Consider two such systems, A and B. Then the general entangled states in total Hilbert space $\mathcal{H} = \mathcal{H}_A \otimes \mathcal{H}_B$ are

$$
\begin{aligned}
|\psi\rangle &= a|\uparrow_A\uparrow_B\rangle + b|\downarrow_A\downarrow_B\rangle \\
|\phi\rangle &= a'|\uparrow_A\downarrow_B\rangle + b'|\downarrow_A\uparrow_B\rangle,
\end{aligned}
\tag{30.1}
$$

with $a, b \neq 0$, where the spins in $|\psi\rangle$ are correlated, and in $|\phi\rangle$ are anticorrelated. The standard examples of these entangled states are

$$
\begin{aligned}
|\psi_{1,2}\rangle &= \frac{1}{\sqrt{2}}\left(|\uparrow_A\uparrow_B\rangle \pm |\downarrow_A\downarrow_B\rangle\right) \\
|\phi_{1,2}\rangle &= \frac{1}{\sqrt{2}}\left(|\uparrow_A\downarrow_B\rangle \pm |\downarrow_A\uparrow_B\rangle\right).
\end{aligned}
\tag{30.2}
$$

In both the $|\psi\rangle$ and $|\phi\rangle$ states, if $a = \pm b$ and $a' = \pm b'$ we have the same probabilities for the two cases, since

$$
\begin{aligned}
\mathcal{P}_{\uparrow\uparrow}(\psi) &= |a|^2, \quad \mathcal{P}_{\downarrow\downarrow}(\psi) = |b|^2 \\
\mathcal{P}_{\uparrow\downarrow}(\phi) &= |a'|^2, \quad \mathcal{P}_{\downarrow\uparrow}(\phi) = |b'|^2.
\end{aligned}
\tag{30.3}
$$

On the other hand, the nonentangled states in $\mathcal{H}$ are the states with $a = 0$ or $b = 0$, or with $a' = 0$ or $b' = 0$, i.e.,

$$|\uparrow\rangle_A \otimes |\uparrow\rangle_B, \quad |\uparrow\rangle_A \otimes |\downarrow\rangle_B, \quad |\downarrow\rangle_A \otimes |\uparrow\rangle_B, \quad |\downarrow\rangle_A \otimes |\downarrow\rangle_B, \tag{30.4}$$

where we have written the tensor product explicitly in order to show that the states are generic (tensor product) states,

$$|\psi\rangle = |\psi_A\rangle \otimes |\psi_B\rangle. \tag{30.5}$$

In fact, for any $|\psi_A\rangle$ state for A and $|\psi_B\rangle$ state for B (including states that are linear combinations of the basis A or B states), the states are nonentangled.

The relevant point about an entangled state is best explained in $|\psi_{1,2}\rangle$ and $|\phi_{1,2}\rangle$: if say, Alice measures the spin in A (which is random, that is, is obtained with probability 1/2 to be either up or down), then she also knows what B will measure afterwards, *if he measures the same spin in the same direction*. That is so, since measuring the spin of A, say to be in $|\uparrow_A\rangle$, collapses the state $|\psi_{1,2}\rangle$ to $|\uparrow_A\uparrow_B\rangle$, which has the spin up for B.

In order to understand further the effects of entanglement on a subsystem, consider a Hermitian operator acting only on A, $M_A \otimes \mathbb{1}_B$. Then the expectation value of the operator (the average measured value) in the state $|\psi\rangle$ is

$$\langle M_A\rangle = \langle\psi|M_A \otimes \mathbb{1}_B|\psi\rangle = |a|^2\langle\uparrow|M_A|\uparrow\rangle_A + |b|^2\langle\downarrow|M_A|\downarrow\rangle_A. \tag{30.6}$$

In terms of the total Hilbert space $\mathcal{H}$, this entangled state is pure, with density matrix

$$\rho_{\text{tot}} = |\psi\rangle\langle\psi| = (a|\uparrow_A\uparrow_B\rangle + b|\downarrow_A\downarrow_B\rangle)\left(a^*\langle\uparrow_A\uparrow_B| + b^*\langle\downarrow_A\downarrow_B|\right). \tag{30.7}$$

Yet in terms of the Hilbert space of A only, the entangled state $|\psi\rangle$ is mixed, with density matrix

$$\rho_A = |a|^2|\uparrow\rangle\langle\uparrow| + |b|^2|\downarrow\rangle\langle\downarrow|, \tag{30.8}$$

since the expectation value can be rewritten in the density matrix formalism as

$$\langle M_A\rangle = \text{Tr}(M_A\rho_A) = \langle\uparrow|M_A\rho_A|\uparrow\rangle + \langle\downarrow|M_A\rho_A|\downarrow\rangle, \tag{30.9}$$

which is equal to the result in (30.6).

But then the density matrix of A, ρ_A, is obtained as the trace over only the degrees of freedom of B of the total density matrix ρ_{tot},

$$\begin{aligned}
\rho_A &= \text{Tr}_B\, \rho_{\text{tot}} \equiv \langle\uparrow_B|\rho_{\text{tot}}|\uparrow_B\rangle + \langle\downarrow_B|\rho_{\text{tot}}|\downarrow_B\rangle \\
&= \langle\uparrow_B|\left(a|\uparrow_A\uparrow_B + b|\downarrow_A\downarrow_B\rangle\right)\left(a^*\langle\uparrow_A\uparrow_B| + b^*\langle\downarrow_A\downarrow_B|\right)|\uparrow_B\rangle \\
&\quad + \langle\downarrow_B|\left(a|\uparrow_A\uparrow_B + b|\downarrow_A\downarrow_B\rangle\right)\left(a^*\langle\uparrow_A\uparrow_B| + b^*\langle\downarrow_A\downarrow_B|\right)|\downarrow_B\rangle \\
&= a|\uparrow_A\rangle a^*\langle\uparrow_A| + b|\downarrow_A\rangle b^*\langle\downarrow_A|,
\end{aligned} \tag{30.10}$$

which is exactly the mixed state ρ_A from before. Therefore finding the trace over a subsystem of an entangled state leads to a mixed state, at least in this example.

On the other hand, if the state is nonentangled, i.e., separable,

$$|\psi\rangle = |\psi_A\rangle|\psi_B\rangle, \tag{30.11}$$

where $|\psi_A\rangle = a|\uparrow_A\rangle + b|\downarrow_A\rangle$ and $|\psi_B\rangle = c|\uparrow_B\rangle + d|\downarrow_B\rangle$, so that

$$|\psi\rangle = (a|\uparrow_A\rangle + b|\downarrow_A\rangle) \otimes (c|\uparrow_B\rangle + d|\downarrow_B\rangle), \tag{30.12}$$

then finding the trace over the subsystem B leads to a pure state instead, since

$$\begin{aligned}
\mathrm{Tr}_B\, \rho_{\rm tot} &= \mathrm{Tr}_B\, |\psi\rangle\langle\psi| \\
&= \langle\uparrow_B|\,(a|\uparrow_A\rangle + b|\downarrow_A\rangle)\otimes(c|\uparrow_B\rangle + d|\downarrow_B\rangle)\,(a^*\langle\uparrow_A| + b^*\langle\downarrow_A|)\otimes(c^*\langle\uparrow_B| + d^*\langle\downarrow_B|)\,|\uparrow_B\rangle \\
&\quad + \langle\downarrow_B|\,(a|\uparrow_A\rangle + b|\downarrow_A\rangle)\otimes(c|\uparrow_B\rangle + d|\downarrow_B\rangle)\,(a^*\langle\uparrow_A| + b^*\langle\downarrow_A|)\otimes(c^*\langle\uparrow_B| + d^*\langle\downarrow_B|)\,|\downarrow_B\rangle \\
&= c\,(a|\uparrow_A\rangle + b|\downarrow_A\rangle)\,c^*\,(a^*\langle\uparrow_A| + b^*\langle\downarrow_B|) + d\,(a|\uparrow_A\rangle + b|\downarrow_A\rangle)\,d^*\,(a^*\langle\uparrow_A| + b^*\langle\downarrow_B|) \\
&= (|c|^2 + |d|^2)|\psi_A\rangle\langle\psi_A| \equiv \rho_A.
\end{aligned} \tag{30.13}$$

Here we have used the (probability) normalization condition $|c|^2 + |d|^2 = 1$.

30.2 Entanglement: The General Case

Now we consider a general system for A and B, with orthonormal basis $\{|i\rangle\}$ for A and orthonormal basis $\{|m\rangle\}$ for B. Then a general entangled state is written as

$$|\psi\rangle_{AB} = \sum_{i,m} C_{im}|i\rangle_A \otimes |m\rangle_B = \sum_i |i\rangle_A \otimes \left(\sum_m C_{im}|m\rangle_B\right) \equiv \sum_i |i\rangle_A \otimes |\tilde{i}\rangle_B, \tag{30.14}$$

where we define a new basis for B, by linear combinations

$$|\tilde{i}\rangle_B = \sum_m C_{im}|m\rangle_B, \tag{30.15}$$

but which is not necessarily orthonormal at this point. Choose however the basis $|i\rangle$ such that it also diagonalizes the density matrix of A obtained by finding the trace over the subsytem B,

$$\rho_A = \sum_i p_i |i\rangle\langle i|, \tag{30.16}$$

where p_i is the probability for state $|i\rangle$ in the ensemble.

Then we can calculate ρ_A from (30.14) as

$$\rho_A = \mathrm{Tr}_B\, |\psi\rangle_{AB\,AB}\langle\psi| = \mathrm{Tr}_B\left[\sum_{i,j} (|i\rangle_{AA}\langle j|) \otimes \left(|\tilde{i}\rangle_{BB}\langle\tilde{j}|\right)\right]. \tag{30.17}$$

To compute it further, note that

$$\mathrm{Tr}_B\, |\tilde{i}\rangle_{BB}\langle\tilde{j}| = \sum_k {}_B\langle k|\tilde{i}\rangle_{BB}\langle\tilde{j}|k\rangle_B = \langle\tilde{j}|\tilde{i}\rangle, \tag{30.18}$$

where we have used $\sum_k |k\rangle\langle k| = \mathbb{1}$. Using this relation, we finally obtain

$$\rho_A = \sum_{i,j} {}_B\langle\tilde{j}|\tilde{i}\rangle_B |i\rangle_{AA}\langle j|. \tag{30.19}$$

We can now compare this formula with (30.16), and equating the terms implies

$${}_B\langle\tilde{j}|\tilde{i}\rangle_B = p_i\delta_{ij}, \tag{30.20}$$

which means that the states

$$|i'\rangle_B = \frac{1}{\sqrt{p_i}}|\tilde{i}\rangle_B \tag{30.21}$$

form an orthonormal set as well, just as the $|m\rangle_B$ do. Then, finally, we can rewrite the pure state $|\psi\rangle_{AB}$ in the form

$$|\psi\rangle_{AB} = \sum_i \sqrt{p_i}\,|i\rangle_A \otimes |i'\rangle_B, \tag{30.22}$$

known as the *Schmidt decomposition* of a bipartite pure state $|\psi\rangle_{AB}$.

This decomposition can be performed for any bipartite pure state $|\psi\rangle_{AB}$; however, the bases $|i\rangle_A$ and $|i'\rangle_B$ do depend on the explicit form of $|\psi\rangle$.

The density matrix for the B subsystem is obtained in the same way as that for A,

$$\begin{aligned}\rho_B &\equiv \mathrm{Tr}_A\,\rho_{\text{tot}} = \mathrm{Tr}_A\,|\psi\rangle_{AB\,AB}\langle\psi| \\ &= \sum_{i,j,k} \sqrt{p_i}\langle k|i\rangle_A \otimes |i'\rangle_B \sqrt{p_i}\,{}_B\langle j'| \otimes {}_A\langle j|k\rangle \\ &= \sum_i p_i |i'\rangle_{B\,B}\langle i'|.\end{aligned} \tag{30.23}$$

We see then that ρ_A and ρ_B have the same set of nonzero eigenvalues p_i.

30.3 Entanglement: Careful Definition

We must now define more precisely the entanglement of a state $|\psi\rangle_{AB}$. Define the number of nonzero eigenvalues p_i in ρ_A (or, equivalently, in ρ_B) as the *Schmidt number n*. If $n > 1$ then the state is called entangled, or nonseparable, whereas if $n = 1$ then the state is called unentangled, or separable.

For the separable state

$$|\psi\rangle = |\psi_A\rangle \otimes |\psi_B\rangle, \tag{30.24}$$

the density matrices of the subsystems are still of the pure state type,

$$\rho_A = |\psi_A\rangle\langle\psi_A|, \quad \rho_B = |\psi_B\rangle\langle\psi_B|. \tag{30.25}$$

Returning to the spin 1/2 example, the entangled state has quantum correlations, since for instance in $|\psi\rangle$ states, $|\uparrow\rangle_A$ implies $|\uparrow\rangle_B$ and $|\downarrow\rangle_A$ implies $|\downarrow\rangle_B$. But these quantum correlations *cannot be created locally*, in A or B independently, but rather have to involve interactions between A and B. On the other hand, $|\uparrow\rangle_A \otimes |\uparrow\rangle_B$ can be prepared by communicating classically (and separately) between A and B.

In terms of a quantum transformation, however, we can entangle states by a unitary transformation U_{AB} in $\mathcal{H}$ that is not of the product type $U_A \otimes U_B$, with U_A and U_B independent unitary transformations in $\mathcal{H}_A$ and $\mathcal{H}_B$, respectively. Indeed, we can act on the separable states $|1\rangle = |\uparrow_A\uparrow_B\rangle$ and $|2\rangle = |\downarrow_A\downarrow_B\rangle$ by rotating them,

$$U_{AB}\begin{pmatrix}|1\rangle \\ |2\rangle\end{pmatrix} = \frac{1}{\sqrt{2}}\begin{pmatrix}|1\rangle + |2\rangle \\ |1\rangle - |2\rangle\end{pmatrix} = \begin{pmatrix}|\psi_1\rangle \\ |\psi_2\rangle\end{pmatrix}, \tag{30.26}$$

in the same way that we act on individual up and down states in each subsystem to obtain

$$U_A \begin{pmatrix} |\uparrow\rangle_A \\ |\downarrow\rangle_A \end{pmatrix} = \frac{1}{\sqrt{2}} \begin{pmatrix} |\uparrow\rangle_A + |\downarrow\rangle_A \\ |\uparrow\rangle_A - |\downarrow\rangle_A \end{pmatrix}$$
$$U_B \begin{pmatrix} |\uparrow\rangle_B \\ |\downarrow\rangle_B \end{pmatrix} = \frac{1}{\sqrt{2}} \begin{pmatrix} |\uparrow\rangle_B + |\downarrow\rangle_B \\ |\uparrow\rangle_B - |\downarrow\rangle_B \end{pmatrix}. \tag{30.27}$$

The *maximally entangled states* are $|\psi_{1,2}\rangle$ or $|\phi_{1,2}\rangle$, in the following sense. We can calculate the density matrix ρ_A, and find

$$\rho_A = \mathrm{Tr}_B\, |\psi_{1,2}\rangle\langle\psi_{1,2}| = |a|^2 |\uparrow_A\rangle\langle\uparrow_A| + |b|^2 |\downarrow_A\rangle\langle\downarrow_A| = \frac{1}{2}\left(|\uparrow_A\rangle\langle\uparrow_A| + |\downarrow_A\rangle\langle\downarrow_A|\right), \tag{30.28}$$

and similarly for the A$|\phi\rangle$ states,

$$\rho_A = \mathrm{Tr}_B\, |\phi_{1,2}\rangle\langle\phi_{1,2}| = |a|^2 |\uparrow_A\rangle\langle\uparrow_A| + |b|^2 |\downarrow_A\rangle\langle\downarrow_A| = \frac{1}{2}\left(|\uparrow_A\rangle\langle\uparrow_A| + |\downarrow_A\rangle\langle\downarrow_A|\right), \tag{30.29}$$

and moreover $\rho_A = \rho_B$ for all these pure states, which means that

$$\rho_A = \rho_B = \frac{1}{d}\mathbb{1}, \tag{30.30}$$

where d is the number of basis states in the Hilbert space of A or B (the dimension of the Hilbert space); here $d = 2$.

Obtaining the above density matrices for the subsystems A or B, with equal probabilities in all states, is the definition of a *maximally entangled state*: namely, we find a maximally mixed state by taking the trace over a subsystem B.

The Schmidt basis decomposition is not unique, and, in order to define it uniquely, we would need more information. In the case of a maximally entangled state, with $\rho_A = \rho_B = (1/d)\mathbb{1}$, with probabilities $p_i = 1/d$ in each state, rotating the basis states $|i\rangle$ and $|j'\rangle$ with a unitary matrix U_{ij} in the Schmidt basis decomposition,

$$|\psi_{AB}\rangle = \frac{1}{\sqrt{d}} \sum_{i,j} |i\rangle_A U_{ij} |j'\rangle_B, \tag{30.31}$$

gives the same $\rho_A = \rho_B$ density matrices. Moreover, rotating $|i\rangle$ and $|j'\rangle$ with complex conjugate (so opposite) unitary matrices,

$$|i\rangle_A = \sum_a |a\rangle_A V_{ai}$$
$$|i'\rangle_B = \sum_b |b'\rangle_B V^*_{bi}, \tag{30.32}$$

leads to the same state in the total Hilbert space,

$$|\psi\rangle_{AB} = \frac{1}{\sqrt{d}} \sum_i |i\rangle_A \otimes |i'\rangle_B = \frac{1}{\sqrt{d}} \sum_{i,a,b} |a\rangle_A V_{ai} \otimes |b'\rangle_B V^\dagger_{ib} = \frac{1}{\sqrt{d}} \sum_a |a\rangle_A \otimes |a'\rangle_B. \tag{30.33}$$

30.4 Entanglement Entropy

Entanglement between the two subsystems of a system is associated with a kind of entropic measure, corresponding to the classical notion of entropy as relating to many possible outcomes, with various probabilities. The entropy in the classical case of various states i, arising with classical probabilities p_i, was defined by Gibbs as

$$S = -k_B \sum_i p_i \ln p_i. \tag{30.34}$$

In this formula for the Gibbs entropy, the probabilities are distributed according to the Boltzmann distribution,

$$p_i = \frac{1}{Z} \exp\left(-\frac{\epsilon_i}{k_B T}\right), \tag{30.35}$$

for energies ϵ_i for the states, so

$$S = -k_B \sum_i p_i \left(-\frac{\epsilon_i}{k_B T} - \ln Z\right) = \sum_i p_i \left(\frac{\epsilon_i}{T} + k_B \ln Z\right), \tag{30.36}$$

which varies as

$$\delta S = \sum_i \left(\frac{\delta p_i \epsilon_i}{T} + k_B \delta p_i \ln Z\right) = \sum_i \frac{\delta p_i \epsilon_i}{T}, \tag{30.37}$$

since $\sum_i p_i = 1$ is preserved by the variation, and this implies that $\sum_i \delta p_i = 0$.

In the case of a quantum system, we can define an entropy that extends the classical Gibbs entropy, the *von Neumann entropy* (for $k_B = 1$)

$$S = -\operatorname{Tr} \rho \ln \rho, \tag{30.38}$$

where ρ is the density matrix, and $\ln \rho$ is understood as the infinite Taylor series applied for a matrix (the same as for the exponential function). For a density matrix that is diagonal,

$$\rho = \sum_i p_i |i\rangle\langle i|, \tag{30.39}$$

the von Neumann entropy reduces to the Gibbs formula,

$$S = -\sum_i p_i \ln p_i. \tag{30.40}$$

In particular, for a pure state $|\psi\rangle$ (a single quantum state), thus for

$$\rho = |\psi\rangle\langle\psi|, \tag{30.41}$$

the probabilities are $p_i = \delta_{i1}$, so the entropy vanishes, $S = 0$, as expected. This is the minimum value of S. On the other hand, for the density matrix corresponding to the most entropic situation, with all states having equal probabilities,

$$\rho = \frac{1}{d}\mathbb{1}, \tag{30.42}$$

we obtain the maximum value of the entropy,

$$S_{\max} = -\operatorname{Tr} p_i \ln p_i = -\frac{1}{d} d \ln \frac{1}{d} = \ln d. \tag{30.43}$$

In particular, for two-state system, with $d = 2$, we have $S_{\max} = \ln 2 \simeq 0.69$.

We can now specialize to the case when we have a total system that can be divided into two systems A and B, $\mathcal{H} = \mathcal{H}_A \otimes \mathcal{H}_B$, whether the division is physical (some sort of separation, either a wall, or a distance) or not. Then we define as before

$$\rho_A = \text{Tr}_B\, \rho_{\text{tot}}, \quad \rho_B = \text{Tr}_A\, \rho_{\text{tot}}, \tag{30.44}$$

and we define the entanglement entropy of A in the system as the von Neumann entropy of ρ_A,

$$S_A = -\,\text{Tr}\, \rho_A \ln \rho_A. \tag{30.45}$$

This, however, is not the usual extensive entropy, since for instance for a pure state $|\psi\rangle_{AB}$, with total density matrix $\rho_{\text{tot}} = |\psi\rangle_{AB\,AB}\langle\psi|$, the entanglement entropy is the same for A and B,

$$\rho_A = \rho_B \;\Rightarrow\; S_A = S_B, \tag{30.46}$$

contradicting the extensive property.

Going back to the generic von Neumann entropy, it has the following properties:

- It is invariant under unitary transformations, $|i\rangle \to U|i\rangle$, implying $\rho \to U\rho U^\dagger = U\rho U^{-1}$. Indeed, then

$$S(\rho) \to -\,\text{Tr}[U\rho U^{-1} \ln(U\rho U^{-1})] = S(\rho), \tag{30.47}$$

  where we have used the fact that the natural log is written as an infinite Taylor series, leading to $\ln(U\rho U^{-1}) = U \ln \rho U^{-1}$.
- It is additive for *independent* systems (it is not additive for interacting subsystems of a larger system, as we have seen) ρ_A, ρ_B:

$$S(\rho_A \otimes \rho_B) = S(\rho_A) + S(\rho_B). \tag{30.48}$$

- It has strong subadditivity. Considering three systems A, B, C, we have

$$S_{ABC} + S_B \le S_{AB} + S_{BC}. \tag{30.49}$$

  This is less obvious, and the proof of this property is quite involved. We also obtain that

$$S_A + S_C \le S_{AB} + S_{BC}. \tag{30.50}$$

  Putting $B = 0$ in the first relation, we obtain the standard subadditivity relation,

$$S_{AC} \le S_A + S_C. \tag{30.51}$$

Since the entanglement entropy is the von Neumann entropy of a subsystem, it also satisfies the strong subadditivity inequality, as the inequality does not depend on S_A being calculated from a subsystem of a system.

30.5 The EPR Paradox and Hidden Variables

The EPR paradox was put forward in the Einstein–Podolsky–Rosen 1935 paper, which described a “paradoxical situation”, with correlations over potentialy infinite distances. The original paper dealt with measurements of p and x, but the “paradox” is easier to understand in the situation of measuring spins 1/2, in line with what we have already done in this chapter. This description was put forward by David Bohm.

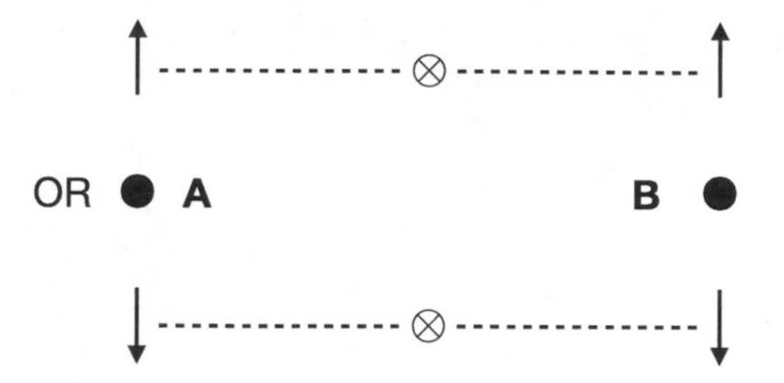

Figure 30.1 The EPR paradox from entanglement. What is different in quantum mechanics is that the measurement of the spin is in *any* direction. "OR" means that the upper and lower parts of the figure are alternatives. The encircled cross indicates that the spins are connected via a tensor product.

We consider a physical situation in which a particle of total spin 0 decays (or disintegrates) into two particles of spin 1/2, in which case the total final spin, the vector sum of the two spins 1/2, must vanish. If the particles continue to move apart, even after they are at a large (macroscopic) distance, we still need to have the total spin zero, since the total state is entangled. An example of the above decay is the decay of the spin zero η particle into two oppositely charged muons,

$$\eta \to \mu^+ + \mu^- . \tag{30.52}$$

The result of the decay is then a state of the type $|\phi\rangle$,

$$|\phi_{1,2}\rangle = \frac{1}{\sqrt{2}}\left(|\uparrow_A\downarrow_B\rangle \pm |\downarrow_A\uparrow_B\rangle\right), \tag{30.53}$$

which means that if Alice measures the spin projection, she knows for sure what Bob would measure if later he measured the spin projection in the same direction; see Fig. 30.1.

However, if this were all that there was to say the situation would not be quantum mechanical in nature, since in a classical measurement we could get the same thing. Consider a measurement that corresponds to Alice picking out at random a ball from a hat containing a black ball (spin up) and a white ball (spin down). Then seeing (measuring) what the spin (ball) is by Alice implies that you would know what Bob would measure when he subsequently pulls the remaining ball (spin) from the hat.

What is different about quantum mechanics, though is that we have total spin zero *for the spin projection in any direction*, for instance in the x and z directions. But, on the other hand, the states of given spin in x can be decomposed as linear combinations of the states of given spin in z,

$$|S_x;\pm\rangle = \frac{1}{\sqrt{2}}\left(|S_z+\rangle \pm |S_z-\rangle\right), \tag{30.54}$$

with inverse

$$|S_z;\pm\rangle = \frac{1}{\sqrt{2}}\left(|S_x+\rangle \pm |S_x-\rangle\right). \tag{30.55}$$

Then the entangled spin-singlet total state for spin measured along the z direction,

$$|\phi\rangle_{AB} = \frac{1}{\sqrt{2}}\left(|S_z+\rangle_A \otimes |S_z-\rangle_B - |S_z-\rangle_A \otimes |S_z+\rangle_B\right), \tag{30.56}$$

can be rewritten in terms of the spin-singlet total state for spin measured along the x direction,

$$|\phi\rangle_{AB} = -\frac{1}{\sqrt{2}}\left(|S_x+\rangle_A \otimes |S_x-\rangle_B - |S_x-\rangle_A \otimes |S_x+\rangle_B\right). \tag{30.57}$$

The minus sign in front is irrelevant, since it represents the same state. Then the expression above shows the ambiguity of the Schmidt basis for a maximally entangled state such as $|\phi\rangle_{AB}$. And the physical consequence is that, *unlike in the classical case*, Alice and Bob can measure the spin in different directions, and still obtain probabilistic results. For instance, Alice can measure the spin on z, and afterwards Bob can measure the spin on x, in which case Bob will obtain spin up or down with the same probability 1/2. This is then definitely not a classical result!

The measurement by Alice of a spin in A, which dictates what Bob would find if he then measures the same spin in the B system, seems to suggest an instantaneous (with $v > c$) travel of information, so is the information transmitted at superluminal speeds? The answer is NO. The point is that it is not a message (information) being transmitted, since we have no way of determining whether Bob actually *made* that measurement (which would amount to information about Bob's actions). It would seem as though we would need something like a simultaneous measurement on x and z, which is not quite possible.

To generalize the previous case, consider an arbitrary direction $\vec{n}$; changing $|\phi\rangle$ states to $|\psi\rangle$ states for simplicity, we find, by changing the states on A with the unitary matrix V, and the states on B with $V^\dagger$,

$$\begin{aligned}|\psi\rangle_{AB} &= \frac{1}{\sqrt{2}}\left(|S_z+\rangle_A \otimes |S_z+\rangle_B + |S_z-\rangle_A \otimes |S_z-\rangle_B\right)\\ &= \frac{1}{\sqrt{2}}\left(|S_{\vec{n}}+\rangle_A \otimes S_{\vec{n}}+\rangle_B + |S_{\vec{n}}-\rangle_A \otimes |S_{\vec{n}}-\rangle_B\right).\end{aligned} \tag{30.58}$$

Given this transformation of the entangled state between the directions of spin measured, we might hope that we could use the relation to send signals, with the message encoded in whichever direction has been used for measurement by Bob (for instance, x or z). However, in fact there is no way to do this, as we can check. The only way to do it is to use extra information, which needs to be passed between Alice and Bob classically, so it must travel at subluminal speeds.

In conclusion, entanglement does not imply a violation of causality, as *information* cannot be exchanged at superluminal speeds. In this sense, the EPR experiment is not paradoxical. Einstein knew this, as shown in the EPR paper. However, he felt that there was a paradox nevertheless, since he defined a theory as something with a *complete* description of physical reality, which would need to satisfy the stronger condition later called *Einstein locality*, or *local realism*, namely that:

An action on A must not modify the description of B.

Clearly the EPR experiment involves a paradox, as it breaks the Einstein locality criterion: while information does not travel superluminally, measurement of A changes the description of B by collapsing the state of the system to a state of given spin for B.

Einstein then devised possible descriptions of the EPR paradox experiment that mimic quantum mechanics but satisfy Einstein locality, by having some classical description, with some *hidden variables* (that cannot be known by experiment). The description is classical, but the appearance of quantum probabilities is due to our not knowing some degrees of freedom, the hidden variables, which can for instance be some $\lambda \in (0, 1)$. It seemed as though this was a solution, and quantum mechanics would not be needed.

But Bell showed that we can distinguish *experimentally* between a hidden variable theory and quantum mechanics, by calculating the satisfaction or not of some inequalities, called Bell's inequalities, which will be studied in the next chapter.

Important Concepts to Remember

- In the case of two (sub)systems A and B, a separable state is a state of tensor product form, $|\psi_A\rangle \otimes |\chi_B\rangle$, while an entangled state is a state in $\mathcal{H}_A \otimes \mathcal{H}_B$ that cannot be written as such product (is not separable).
- The formal definition of entanglement comes from the Schmidt decomposition of the bipartite state $|\psi\rangle_{AB}$, as $\sum_i \sqrt{p_i}|i\rangle_A \otimes |i'\rangle_B$. If the Schmidt number n (the number of nonzero eigenvalues p_i) is > 1, we have an entangled state, and if $n = 1$, a separable state.
- We can entangle states by a unitary transformation U_{AB} that is not a product of unitaries $U_A \otimes U_B$.
- A maximally entangled state is a state that, when traced over either A or B, leads to a constant (and equal) density matrix, $\rho_A = \rho_B = \frac{1}{d}\mathbb{1}$.
- The quantum mechanical von Neumann entropy, $S_{vN} = -\operatorname{Tr} \rho \ln \rho$, extends the classical Gibbs entropy $S = -k_B \sum_i p_i \ln p_i$.
- For a system formed by two subsystems, we define the entanglement entropy as the von Neumann entropy of A, $S_E = -\operatorname{Tr} \rho_A \ln \rho_A$, where $\rho_A = \operatorname{Tr}_B \rho_{\text{tot}}$ and $\rho_B = \operatorname{Tr}_A \rho_{\text{tot}}$.
- The EPR "paradox" arises when we make a measurement in A of an entangled state, thus modifying what B can see. In the quantum spin 1/2 case, unlike the classical case, knowing the spin in one direction means a given state for the spin in another direction, thus still implies something nontrivial for a different measurement.
- In the EPR experiment *information* does not travel at superluminal speeds, but a measurement of A changes *the description of* B, thus we violate Einstein locality, or local realism.

Further Reading

See Preskill's lecture notes on quantum information [12].

Exercises

(1) Consider the state $|\chi^a_{1,2}\rangle = |\psi_{1,2}\rangle + a|\phi_{1,2}\rangle$, for an arbitrary $a \in \mathbb{R}$. When is it separable? Calculate ρ_A and ρ_B to check that you find the correct results.

(2) Apply Schmidt decomposition to the general case in exercise 1.

(3) Consider the same general state as that in exercise 1, in the generic entangled case. What is the unitary matrix that disentangles them?

(4) Consider the density matrix

$$\rho = C(|1\rangle\langle 2| + |2\rangle\langle 1| + a|2\rangle\langle 2| + |2\rangle\langle 3| + |3\rangle\langle 2|), \tag{30.59}$$

where $|1\rangle, |2\rangle, |3\rangle$ are orthonormal states. Calculate C, and thus the von Neumann entropy of this mixed state.

(5) For the generic state in exercise 1, calculate the entanglement entropy.

(6) Consider three spin 1/2 systems, A, B, and C, and a state

$$C(|\uparrow\uparrow\uparrow\rangle + |\downarrow\uparrow\downarrow\rangle + |\uparrow\downarrow\uparrow\rangle). \tag{30.60}$$

Check that the strong subadditivity condition is satisfied.

(7) Consider an EPR state $|\phi\rangle_{AB}$; Alice measures the spin on z, then Bob measures it on x, and then Alice measures it again on z. Classify the possible answers for the second measurement of Alice, with their probabilities, depending on the initial measurement of Alice (and without knowledge of the measurement of Bob). What do you deduce?

31 The Interpretation of Quantum Mechanics and Bell's Inequalities

In this chapter we describe the Bell's inequalities that apply for hidden variable theories in EPR paradox-type situations; these theories are violated by the quantum mechanics results, thus providing an experimental test, which has been verified to give quantum mechanics as the correct description in all cases. Finally, we describe some of the leading interpretations of quantum mechanics.

31.1 Bell's Original Inequality

As we described at the end of the previous chapter, Einstein considered a deterministic theory, a "hidden variable" theory, in which there are some variables that cannot be observed that generate the observed statistical results in an a priori deterministic model. The assumption is that the hidden variable theory will reproduce the quantum mechanical results perfectly. But, in his seminal paper, John Bell showed that such a hidden variable model would lead to an inequality that could be violated by quantum mechanics, and thus could be experimentally distinguished from it.

Consider measuring the reduced spin, $\vec{\sigma} \equiv \frac{2}{\hbar}\vec{S}$ in two different directions, defined by unit vectors $\vec{a}$ and $\vec{b}$, so we have $(\vec{\sigma} \cdot \vec{a})$ for system A, i.e., measured by Alice, and $(\vec{\sigma} \cdot \vec{b})$ for system B, i.e., measured by Bob.

According to the hidden variable model, with hidden variable λ, Alice measures $A(\vec{a}; \lambda)$ and Bob measures $B(\vec{b}; \lambda)$, obtaining possible values $A = \pm 1$ and $B = \pm 1$.

In the case of Bell's original inequality, applied to David Bohm's (now standard) version of the EPR paradox, with a system of total spin 0, meaning that if $\vec{a} = \vec{b}$ (Alice and Bob measure spin in the same direction) then $A = -B$ (and if $\vec{a} = -\vec{b}$, then $A = B$), we thus get

$$A(\vec{a}; \lambda) = -B(\vec{a}; \lambda). \tag{31.1}$$

The hidden variable λ appears (since we cannot measure it) as a statistical ensemble, with probability distribution $\rho(\lambda)$, normalized to one,

$$\int_\Lambda d\lambda \; \rho(\lambda) = 1. \tag{31.2}$$

Then define, *in this hidden variable theory*, the correlation function of the measurements of Alice and Bob,

$$C_{\text{hid}}(\vec{a}, \vec{b}) = \int_\Lambda d\lambda \; \rho(\lambda) A(\vec{a}; \lambda) B(\vec{b}; \lambda). \tag{31.3}$$

Given (31.1), we rewrite this as

$$C_{\rm hid}(\vec{a},\vec{b}) = -\int_\Lambda d\lambda\ \rho(\lambda)A(\vec{a};\lambda)A(\vec{b};\lambda). \tag{31.4}$$

In order to obtain a relevant inequality, we consider a situation involving spin measurements in three directions $\vec{a}, \vec{b}, \vec{c}$. We then calculate

$$\begin{aligned} C_{\rm hid}(\vec{a},\vec{b}) - C_{\rm hid}(\vec{a},\vec{c}) &= -\int_\Lambda d\lambda\ \rho(\lambda)\left[A(\vec{a};\lambda)A(\vec{b};\lambda) - A(\vec{a};\lambda)A(\vec{c};\lambda)\right] \\ &= \int_\Lambda d\lambda\ \rho(\lambda)A(\vec{a};\lambda)A(\vec{b};\lambda)\left[A(\vec{b};\lambda)A(\vec{c};\lambda) - 1\right], \end{aligned} \tag{31.5}$$

where in the second equality we have used the fact that $A = \pm 1$, so $[A(\vec{b};\lambda)]^2 = 1$, which we introduced in the second term.

Since $A = \pm 1$, this means that also $A(\vec{a};\lambda)A(\vec{b};\lambda) = \pm 1$. Then we obtain an inequality,

$$\begin{aligned} |C_{\rm hid}(\vec{a},\vec{b}) - C_{\rm hid}(\vec{a},\vec{c})| &= \left|\int_\Lambda d\lambda\ \rho(\lambda)A(\vec{a};\lambda)A(\vec{b};\lambda)\left[A(\vec{b};\lambda)A(\vec{c};\lambda) - 1\right]\right| \\ &\le \left|\int_\Lambda d\lambda\ \rho(\lambda)\left[A(\vec{b};\lambda)A(\vec{c};\lambda) - 1\right]\right|\left|A(\vec{a};\lambda)A(\vec{b};\lambda)\right| \\ &= \left|\int_\Lambda d\lambda\ \rho(\lambda)\left[A(\vec{b};\lambda)A(\vec{c};\lambda) - 1\right]\right|. \end{aligned} \tag{31.6}$$

But $\rho(\lambda) \ge 0$ and (since $A = \pm 1$)

$$1 - A(\vec{b};\lambda)A(\vec{c};\lambda) \ge 0, \tag{31.7}$$

which means we obtain

$$\begin{aligned} |C_{\rm hid}(\vec{a},\vec{b}) - C_{\rm hid}(\vec{a},\vec{c})| &\le \int_\Lambda d\lambda\ \rho(\lambda)\left[1 - A(\vec{b};\lambda)A(\vec{c};\lambda)\right] \\ &= 1 + C_{\rm hid}(\vec{b},\vec{c}), \end{aligned} \tag{31.8}$$

where in the second equality we used the normalization $\int_\Lambda d\lambda\ \rho(\lambda) = 1$ and the definition of $C_{\rm hid}$. The resulting inequality is called *Bell's (original) inequality*.

On the other hand, in quantum mechanics, we define the corresponding correlation function of the measurements of Alice and Bob. Since in this case $A = (\vec{\sigma}\cdot\vec{a})_{(A)}$ and $B = (\vec{\sigma}\cdot\vec{b})_{(B)}$, we obtain the expectation value of the product AB, in a state $|\psi\rangle$ for the total Hilbert space $\mathcal{H} = \mathcal{H}_A \otimes \mathcal{H}_B$,

$$C_{\rm QM}(\vec{a},\vec{b}) = \langle\psi|(\vec{\sigma}\cdot\vec{b})_{(B)}(\vec{\sigma}\cdot\vec{a})_{(A)}|\psi\rangle. \tag{31.9}$$

This correlation function will violate the Bell inequality, showing that quantum mechanics is different from the hidden variable theory.

We choose an entangled state of total spin 0, meaning a $|\phi\rangle$ state, in the nomenclature of the previous chapter, and one that is antisymmetric in the opposite spins for A and B, uniquely identifying the state as $|\phi_2\rangle$,

$$|\psi\rangle \equiv |\phi_-\rangle = |\phi_2\rangle = \frac{1}{\sqrt{2}}(|\uparrow_A\downarrow_B\rangle - |\downarrow_A\uparrow_B\rangle). \tag{31.10}$$

But, as we have seen, for such a state the density matrix obtained for the subsystem A by finding the trace over B is $\rho_A = \frac{1}{2}\mathbb{1}$. Then we obtain for the quantum mechanical correlation function

$$\begin{aligned}
C_{\text{QM}}(\vec{a}, \vec{b}) &= \langle\psi|(\vec{\sigma}\cdot\vec{b})_{(B)}(\vec{\sigma}\cdot\vec{a})_{(A)}|\psi\rangle \\
&= -\langle\psi|(\vec{\sigma}\cdot\vec{b})_{(A)}(\vec{\sigma}\cdot\vec{a})_{(A)}|\psi\rangle \\
&= -a_i b_j \langle\psi|\sigma_{j(A)}\sigma_{i(A)}|\psi\rangle \\
&= -a_i b_j \,\text{Tr}_A\left(\rho_A \sigma_j^{(A)}\sigma_i^{(A)}\right) \\
&= -\vec{a}\cdot\vec{b} = -\cos\theta(\vec{a}, \vec{b}).
\end{aligned} \tag{31.11}$$

In the fourth equality, we have used the fact that, since there is no operator acting on the B subsystem, the correlation function equals the result obtained from expectation value in A for the density matrix of the subsystem A. In the fifth equality we have used the fact that

$$\text{Tr}_A\left(\rho_A \sigma_j^{(A)}\sigma_i^{(A)}\right) = \frac{1}{2}\,\text{Tr}_A\left(\sigma_j^{(A)}\sigma_i^{(A)}\right) = \delta_{ij}, \tag{31.12}$$

and in the last equality we have defined $\theta(\vec{a}, \vec{b})$ as the angle between the two unit vectors.

In order to easily see that this correlation function violates the Bell inequalities, we consider the simpler case of small angles. First, if

$$|\theta(\vec{b}, \vec{c})| \ll 1, \tag{31.13}$$

then the right-hand side of the Bell inequality in the quantum mechanics case is

$$1 + C_{\text{QM}}(\vec{b}, \vec{c}) = 1 - \cos\theta(\vec{b}, \vec{c}) \simeq \frac{1}{2}\theta^2(\vec{b}, \vec{c}). \tag{31.14}$$

Then, if

$$|\theta(\vec{a}, \vec{b})| \ll 1, \quad |\theta(\vec{a}, \vec{c}| \ll 1 \tag{31.15}$$

as well, we obtain that the quantum mechanical version of the Bell inequality becomes

$$\frac{1}{2}\left|\theta^2(\vec{a}, \vec{b}) - \theta^2(\vec{a}, \vec{c})\right| \leq \frac{1}{2}\theta^2(\vec{b}, \vec{c}). \tag{31.16}$$

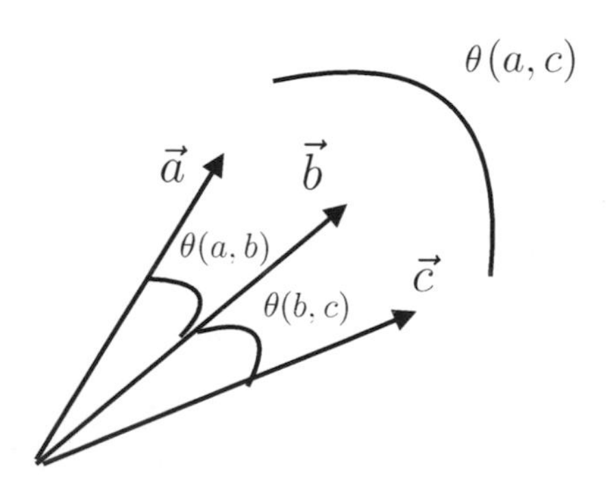

Figure 31.1 Set-up for violation of the Bell inequalities.

However, this relation can be easily violated. Consider coplanar unit vectors $\vec{a}, \vec{b}, \vec{c}$; see Fig. 31.1. Then $\theta(\vec{a}, \vec{c}) = \theta(\vec{a}, \vec{b}) + \theta(\vec{b}, \vec{c})$, so the inequality becomes

$$|(\theta(\vec{a}, \vec{b}) + \theta(\vec{b}, \vec{c}))^2 - \theta^2(\vec{a}, \vec{b})|^2 = \theta^2(\vec{b}, \vec{c}) + 2\theta(\vec{b}, \vec{c})\theta(\vec{a}, \vec{b}) \leq \theta^2(\vec{b}, \vec{c}), \tag{31.17}$$

which is clearly violated.

In principle, this Bell inequality can be tested experimentally. Alice measures A and gets $\pm$, and Bob measures B and gets $\pm$. Since the correlation function is a sum of products AB weighted by probabilities (so sum over events, and divide by the total number of events), or statistical expectation value, then $A = +, B = +$ and $A = -, B = -$ both contribute with a plus in the sum, whereas $A = +, B = -$ and $A = -, B = +$ both contribute with a minus in the sum, so the experimental correlation function is (here the Ns are numbers of events)

$$C_{\exp}(\vec{a}, \vec{b}) = \frac{N_{++} + N_{--} - N_{+-} - N_{-+}}{N_{++} + N_{--} + N_{+-} + N_{-+}}. \tag{31.18}$$

We could then see if the Bell inequality is experimentally violated or not. Unfortunately, the case of perfectly anti-aligned spins for A and B is hard to measure, so we also need other Bell inequalities that are easier to measure.

Before that, however, we will derive other, even simpler to understand, Bell inequalities, though they are less relevant for the possible hidden variable theories (they eliminate fewer hidden variable theories).

31.2 Bell–Wigner Inequalities

We now consider the Bell inequalities in a simple model by Wigner from 1970 [17]. This is also described in *Modern Quantum Mechanics* by Sakurai and Tuan [1], and a variant of the same was presented by Preskill in his lecture notes [12].

Consider the case of a hidden variable class of models, considered by Wigner, in which we have classically random emitters of spins, which emit spins of a given value in several directions at the same time (though it is impossible to measure the spin in several directions simultaneously, in consistency with experimental results that agree with quantum mechanics). Suppose that in the direction of unit vector $\vec{a}$ we get the possible values $\sigma_a = \pm$, and in the direction of unit vector $\vec{b}$ we get the possible values $\sigma_b = \pm$. All are pre-ordained, but not knowable.

For the EPR experiment, Alice and Bob must *always* measure opposite spins in a given direction. Then consider states that have given spins in two directions, $(\vec{a}\sigma_a, \vec{b}\sigma_b)$, or in three directions $(\vec{a}\sigma_a, \vec{b}\sigma_b, \vec{c}\sigma_c)$, etc. The EPR assumption means that Alice and Bob measure totally opposite states.

For spins in three directions, we obtain Table 31.1 for the various possibilities, each with an associated number N_i of events, thus with probability

$$P_i = \frac{N_i}{\sum_k N_k}. \tag{31.19}$$

We denote by $P(\vec{a}\sigma_a, \vec{b}\sigma_b)$ the probability for Alice to measure σ_a on $\vec{a}$ and for Bob to measure σ_b on $\vec{b}$.

Table 31.1 Possible measurements by Alice and Bob, and the corresponding numbers of events.

	Alice (system A)	Bob (system B)
Numbers of events	$\sigma_a\sigma_b\sigma_c$	$\sigma_a\sigma_b\sigma_c$
N_1	$+++$	$---$
N_2	$++-$	$--+$
N_3	$+-+$	$-+-$
N_4	$+--$	$-++$
N_5	$-++$	$+--$
N_6	$-+-$	$+-+$
N_7	$--+$	$++-$
N_8	$---$	$+++$

Then, by inspection of Table 31.1, we see that $P(\vec{a}+,\vec{b}+)$ corresponds to N_3 and N_4, $P(\vec{a}+,\vec{c}+)$ corresponds to N_2 and N_4, and $P(\vec{c}+,\vec{b}+)$ corresponds to N_3 and N_7, so we obtain

$$\begin{aligned}
P(\vec{a}+,\vec{b}+) &= P_3 + P_4 = \frac{N_3 + N_4}{\sum_k N_k} \\
P(\vec{a}+,\vec{c}+) &= P_2 + P_4 = \frac{N_2 + N_4}{\sum_k N_k} \\
P(\vec{c}+,\vec{b}+) &= P_3 + P_7 = \frac{N_3 + N_7}{\sum_k N_k}.
\end{aligned} \tag{31.20}$$

But then, since

$$N_3 + N_4 \leq (N_2 + N_4) + (N_3 + N_7), \tag{31.21}$$

it means that we obtain the inequality

$$P(\vec{a}+,\vec{b}+) \leq P(\vec{a}+,\vec{c}+) + P(\vec{c}+,\vec{b}+), \tag{31.22}$$

which is one form of the Bell–Wigner inequality.

Other inequalities are possible. One example is obtained by defining the probability to obtain the *same result for the same observer* (say, Alice) in a measurement of two directions $\vec{i},\vec{j}$, chosen among $\vec{a},\vec{b},\vec{c}$, the probability being denoted by $P_{\text{same}}(\vec{i},\vec{j})$. Note that this probability corresponds to having *opposite* spins (as in the EPR paradox) measured by Alice and Bob on the same pair of directions.

Then, again by a simple inspection of Table 31.1, we obtain that

$$\begin{aligned}
P_{\text{same}}(\vec{a},\vec{b}) &= \frac{N_1 + N_2 + N_7 + N_8}{\sum_k N_k} \\
P_{\text{same}}(\vec{a},\vec{c}) &= \frac{N_1 + N_5 + N_6 + N_8}{\sum_k N_k} \\
P_{\text{same}}(\vec{b},\vec{c}) &= \frac{N_1 + N_4 + N_5 + N_8}{\sum_k N_k}.
\end{aligned} \tag{31.23}$$

Summing the three probabilities, we obtain

$$P_{\text{same}}(\vec{a},\vec{b}) + P_{\text{same}}(\vec{a},\vec{c}) + P_{\text{same}}(\vec{b},\vec{c}) = \frac{\sum_k N_k + 2N_1 + 2N_8}{\sum_k N_k} \geq 1. \tag{31.24}$$

This is the new type of Bell–Wigner inequality.

We next consider the results corresponding to the Bell–Wigner identities in quantum mechanics. In quantum mechanics, in order to calculate the probability $P(\vec{a}+,\vec{b}+)$, we need to insert the projection operator along the spin up or spin down direction on $\vec{n} = \vec{a},\vec{b}$, i.e.,

$$E(\vec{n},\pm) = \frac{\mathbb{1} \pm \vec{n}\cdot\vec{\sigma}}{2}. \tag{31.25}$$

Specifically, we obtain

$$P(\vec{a}+,\vec{b}+) = \langle\phi_-|E^{(A)}(\vec{a},+)E^{(B)}(\vec{b},+)|\phi_-\rangle. \tag{31.26}$$

But, substituting the expression for $E(\vec{n},\pm)$, and using the fact that

$$\langle\phi_-|\mathbb{1}|\phi_-\rangle = 1,\quad \langle\phi_-|\vec{n}\cdot\vec{\sigma}|\phi_-\rangle = 0,\quad \langle\phi_-|(\vec{\sigma}\cdot\vec{a})(\vec{\sigma}\cdot\vec{b})|\phi_-\rangle = -\cos\theta(\vec{a},\vec{b}), \tag{31.27}$$

we get

$$P(\vec{a}+,\vec{b}+) = \frac{1}{4}(1-\cos\theta(\vec{a},\vec{b})) = \frac{1}{2}\sin^2\frac{\theta(\vec{a},\vec{b})}{2}. \tag{31.28}$$

Similarly, we obtain for the other three possibilities for measurement,

$$\begin{aligned}
P(\vec{a}-,\vec{b}-) &= \langle\phi_-|E^{(A)}(\vec{a},-)E^{(B)}(\vec{b},-)|\phi_-\rangle = \frac{1}{4}(1-\cos\theta(\vec{a},\vec{b}))\\
P(\vec{a}+,\vec{b}-) &= \langle\phi_-|E^{(A)}(\vec{a},+)E^{(B)}(\vec{b},-)|\phi_-\rangle = \frac{1}{4}(1+\cos\theta(\vec{a},\vec{b}))\\
P(\vec{a}-,\vec{b}+) &= \langle\phi_-|E^{(A)}(\vec{a},-)E^{(B)}(\vec{b},+)|\phi_-\rangle = \frac{1}{4}(1+\cos\theta(\vec{a},\vec{b})).
\end{aligned} \tag{31.29}$$

Then, we see that indeed we have probabilities normalized to one, since

$$P(\vec{a}+,\vec{b}+) + P(\vec{a}-,\vec{b}-) + P(\vec{a}+,\vec{b}-) + P(\vec{a}-,\vec{b}+) = 1. \tag{31.30}$$

The (first) Bell–Wigner inequality in the quantum mechanical version becomes

$$\begin{aligned}
&\frac{1}{4}(1-\cos\theta(\vec{a},\vec{c}) + 1 - \cos\theta(\vec{c},\vec{b})) \geq \frac{1}{4}(1-\cos\theta(\vec{a},\vec{b})) \Rightarrow\\
&\frac{1}{2}\left(\sin^2\frac{\theta(\vec{a},\vec{c})}{2} + \sin^2\frac{\theta(\vec{c},\vec{b})}{2}\right) \geq \frac{1}{2}\sin^2\frac{\theta(\vec{a},\vec{b})}{2}.
\end{aligned} \tag{31.31}$$

But this inequality is easily violated if $\vec{a},\vec{c},\vec{b}$ are coplanar and ordered in this way, so

$$\theta(\vec{a},\vec{b}) = \theta(\vec{a},\vec{c}) + \theta(\vec{c},\vec{b}) \tag{31.32}$$

is small, as before, so that

$$\begin{aligned}
\theta^2(\vec{a},\vec{b}) = (\theta(\vec{a},\vec{c}) + \theta(\vec{c},\vec{b}))^2 &= \theta^2(\vec{a},\vec{c}) + \theta^2(\vec{c},\vec{b}) + 2\theta(\vec{a},\vec{c})\theta(\vec{c},\vec{b})\\
&> \theta^2(\vec{a},\vec{c}) + \theta^2(\vec{c},\vec{b}).
\end{aligned} \tag{31.33}$$

As in the case of the other Bell–Wigner inequality, we first note that in quantum mechanics we have

$$P_{\text{same}}(\vec{a},\vec{b}) = P(\vec{a}+,\vec{b}+) + P(\vec{a}-,\vec{b}-) = \frac{1}{2}(1-\cos\theta(\vec{a},\vec{b})) = \sin^2\frac{\theta(\vec{a},\vec{b})}{2}. \tag{31.34}$$

Then the left-hand side of the Bell–Wigner inequality is

$$P_{\text{same}}(\vec{a},\vec{b}) + P_{\text{same}}(\vec{b},\vec{c}) + P_{\text{same}}(\vec{a},\vec{c}) = \sin^2\frac{\theta(\vec{a},\vec{b}}{2} + \sin^2\frac{\theta(\vec{b},\vec{c})}{2} + \sin^2\frac{\theta(\vec{a},\vec{c})}{2}, \tag{31.35}$$

which can be easily arranged to be less than 1.

However, again, as in the case of the original Bell's inequality, it is experimentally difficult to arrange since we again have opposite spins for Alice and Bob. This means that we need a generalization of the Bell inequality that can be tested experimentally.

31.3 CHSH Inequality (or Bell–CHSH Inequality)

A generalization of the Bell inequality that can be tested experimentally is provided by the inequality derived by John Clauser, Michael Horne, Abner Shimony and R. A. Holt: the CHSH inequality.

We consider the situation where Alice can measure two variables, a and a', that take values in $\{\pm 1\}$, and Bob measures other two variables, b and b', that also take values in $\{\pm 1\}$. In a hidden variable theory, depending on the hidden variables λ, the measurements of Alice are $A(a,\lambda) \equiv A$ and $A'(a',\lambda) \equiv A'$, and the measurements of Bob are $B(b,\lambda) \equiv B$ and $B'(b',\lambda) \equiv B'$. Note that a, a', b, b' can be *any variables*, not necessarily spins in some direction.

We define the correlation function in the hidden variable theory as before, by

$$C_{\text{hid}}(a,b) = \int_\Lambda d\lambda \rho(\lambda) A(a,\lambda) B(b,\lambda) \equiv \langle AB \rangle, \tag{31.36}$$

where the final notation is understood as the statistical (classical) average. Note that this definition has no assumption that $B(\vec{a},\lambda) = -A(\vec{a},\lambda)$, as for spins in the EPR experiment.

Then, since $A, A' = \pm 1$, it follows that

$$A + A' = 0,\quad A - A' = \pm 2,\quad \text{or}\quad A - A' = 0,\quad A + A' = \pm 2. \tag{31.37}$$

This implies that we have

$$\begin{aligned}(A+A')B + (A-A')B' &= \pm 2\\ &= AB + A'B + AB' - A'B' \equiv M.\end{aligned} \tag{31.38}$$

But then we obtain an inequality by using the statistical average and modulus,

$$\begin{aligned}\left|\int_\Lambda d\lambda\ \rho(\lambda) M(\lambda)\right| = |\langle M\rangle| \leq \langle |M|\rangle &= \int_\Lambda d\lambda\ \rho(\lambda)|M(\lambda)|\\ &= 2\int_\Lambda \rho(\lambda) = 2,\end{aligned} \tag{31.39}$$

where in the last equality we have used the normalization of the probability distribution $\rho(\lambda)$.

Now replacing M by its definition in (31.38), we obtain

$$
\begin{aligned}
&|\langle AB\rangle + \langle A'B\rangle + \langle AB'\rangle - \langle A'B'\rangle| \leq 2 \Rightarrow \\
&|C_{\text{hid}}(a,b) + C_{\text{hid}}(a',b) + C_{\text{hid}}(a,b') - C_{\text{hid}}(a',b')| \leq 2.
\end{aligned}
\tag{31.40}
$$

This is the CHSH, or Bell–CHSH inequality.

To calculate its equivalent in quantum mechanics, we choose the variables to be spins in various directions,

$$
\begin{aligned}
a &= \vec{\sigma}^{(A)} \cdot \vec{a}, \qquad a' = \vec{\sigma}^{(A)} \cdot \vec{a}' \\
b &= \vec{\sigma}^{(B)} \cdot \vec{b}, \qquad b' = \vec{\sigma}^{(B)} \cdot \vec{b}'.
\end{aligned}
\tag{31.41}
$$

But since, as we saw, we have

$$
\langle \phi_- | (\vec{\sigma}^{(A)} \cdot \vec{a})(\vec{\sigma}^{(B)} \cdot \vec{b}) | \phi_- \rangle = -\vec{a} \cdot \vec{b} = -\cos\theta(\vec{a},\vec{b}),
\tag{31.42}
$$

we can choose coplanar unit vectors, arranged as before in rotational order (when the origin is the same), $\vec{a}', \vec{b}, \vec{a}, \vec{b}'$, at successive $\pi/4$ intervals of angle. Then

$$
\langle AB\rangle = \langle A'B\rangle = \langle AB'\rangle = -\cos\frac{\pi}{4} = -\frac{1}{\sqrt{2}}, \quad \langle A'B'\rangle = -\cos\frac{3\pi}{4} = \frac{1}{\sqrt{2}}.
\tag{31.43}
$$

Thus the equivalent of the CHSH inequality in quantum mechanics is

$$
4 \times \frac{1}{\sqrt{2}} = 2\sqrt{2} \leq (?)2,
\tag{31.44}
$$

which is clearly violated.

In fact, one can show that this is the maximal violation of the classical CHSH inequality.

31.4 Interpretations of Quantum Mechanics

We end this chapter with a short analysis of the main possible interpretations of quantum mechanics. There are very many such interpretations, proof that, more than a hundred years after its start, quantum mechanics is still a mystery to us, but most of these possibilities are shaky and less standard, so we will stay with the leading ones.

The Standard, "Copenhagen", Interpretation

This is the most common interpretation of quantum mechanics, the one that has been implicit in most of the book so far. It was developed by Niels Bohr and Werner Heisenberg around 1927 in Copenhagen, extending the previous work of Max Born. In this view of quantum mechanics, we have a probabilistic interpretation for the outcome of measurements based on wave functions evolving in time, i.e., we simply have probabilities of obtaining various results. On the other hand, questions about non-measured things, such as the paths previous to one measurement experiment, are meaningless. We have only a wave function and measurements that have meaning. In particular, the interaction of a system with the observer in obtaining a measurement collapses the wave function.

The "Many Worlds" Interpretation

In this interpretation, there is no wave function collapse associated with measurements. Measurements occur via "decoherence", where the system "interacts with itself" without the need of an observer, and moreover at each measurement, the Universe splits into multiple, *mutually unobservable* Universes, or "alternative histories". These splittings occur at each moment in time corresponding to some measurement, leading to a "Multiverse", a distribution of Universes with slightly different histories at each point, splitting further as time goes by.

Note that this interpretation of quantum mechanics is the basis for a story device in science fiction movies and books that is quite common yet gets things mostly wrong. In the Sci-Fi version one can travel between Universes in the Multiverse (that can never happen, an essential part of this interpretation being that the Universes are mutually unobservable); the people are the same, but the events are slightly different (there is no difference between what happens to sentient people and objects or events with respect to branching: everything is slightly different, so the people should be too).

The "Consistent Histories" Interpretation

This is an interpretation that is somewhat related to the formulation of quantum mechanics in terms of path integrals, which are sums over quantum (generically nonsmooth) histories for the propagator. In this interpretation, we view the wave function as a sum over consistent histories for the particles, with some probabilities.

This interpretation is useful for dealing with the problem of quantum cosmology. Quantum cosmology refers to the quantum mechanical evolution of the whole Universe near the Big Bang (the time origin of the Universe), when all things are quantum mechanical, and the whole Universe is one big system. In that case, we have a problem with the standard Copenhagen interpretation, since there is a single Universe, there is no "ensemble of Universes" to measure, and there is only one experiment, the evolution of our Universe. Thus we have to find a way to deal with the quantum mechanics of a single experiment.

Of course, the many worlds interpretation could also be useful for quantum cosmology, since in it we imagine a Multiverse (a collection of alternative Universes) instead of ensembles, even in regular experiments on Earth so even more so for quantum cosmology. The unobservability of the Multiverse is part of the point: the ensemble is always an ensemble of Universes, not of successive experiments.

Ensemble Interpretation

The ensemble interpretation is the most minimalist, even more so than the Copenhagen interpretation. The quantum mechanics probabilistic interpretation is valid only for ensembles, not for single particles, or single Universes (so it definitely tells us we cannot apply quantum mechanics to cosmology, i.e., to the Universe itself). This is very agnostic, and restrictive, so it is perhaps going too far.

There are many other interpretations of quantum mechanics, but these are less standard, and more controversial, so we will not mention them here.

Important Concepts to Remember

- All or most of the variations of Bell's inequalities refer to three consecutive measurements by Alice and Bob, in the EPR paradox, relative to three directions $\vec{a}, \vec{b}, \vec{c}$, and in a classical, hidden variable, theory we obtain an inequality violated by the quantum mechanical analog of the same.
- The original Bell inequality is hard to measure, since it requires the measurement of exactly anti-aligned spins for Alice and Bob (assuming that the total spin is zero). The Wigner model, in which we have preordained numbers for each classical possibility, leading to Bell-Wigner inequalities, is also hard to measure, for the same reason.
- The Bell–CHSH inequalities are better experimentally, since we do not assume total spin zero, and we consider *four* arbitrary directions of measurement.
- The Copenhagen interpretation of quantum mechanics is the most common: it is a probabilistic interpretation based on wave functions and measurements (involving interaction with an observer and so leading to collapse of the wave function).
- In the many worlds interpretation there is no collapse of the wave function, and at each (self-) interaction, the Universe splits into multiple, mutually unobservable Universes, corresponding to each possibility. In the consistent histories interpretation, the wave function arises as the sum over all consistent histories (as in path integral), and allows us to deal with the quantum cosmology of a single Universe (i.e., a single experiment). In the ensemble interpretation, the most restrictive, we can deal only with ensembles, not with a single particle or Universe (it is the most agnostic description, i.e., it is the description that is maximal concerning what we cannot know).

Further Reading

For Bell's inequalities, see Preskill's lecture notes on quantum information [12], as well as the analysis of Sakurai and Tuan in [1]. The original articles cited here are [16] and [17].

Exercises

(1) Explore whether the original Bell inequality can be violated at large angles as well.

(2) Analyzing Table 1, find another example of a Bell–Wigner inequality that is violated in quantum mechanics.

(3) Find an example of (angles corresponding to) another violation of the Bell–CHSH inequality by quantum mechanics, which is not maximal but is more generic.

(4) Instead of the Bell–CHSH inequality in the text, consider an inequality obtained in the same way from $\tilde{M} = (A + A')B' + (A - A')B$ instead of M. Is it still violated by quantum mechanics?

(5) If we consider five measurements instead of the four in the Bell–CHSH inequality, namely a, a', a'' for Alice and b, b' for Bob, can we extend the Bell–CHSH inequality to include this case, such that the inequality is still violated by quantum mechanics?

(6) In quantum cosmology, one can define a "wave function of the Universe" $\Psi[a(t)]$, whose variable is the expanding scale factor of the Universe, $a(t)$, and which satisfies a general relativity (Einstein) version of the Schrödinger equation, called the Wheeler–DeWitt equation. Yet in the Copenhagen interpretation of quantum mechanics, there is only one "experiment" (our expanding Universe, since our Big Bang) that we can see, and no "outside observer". How would you slightly extend this interpretation to make sense of the results of the Wheeler–DeWitt equation (there is much debate, so there is no unique good answer to this question at present)?

(7) In some TV shows, instigated by the "many worlds interpretation" of quantum mechanics, one sees an "evil parallel Universe", where things have gone very bad in some sense, and the same characters behave in an evil way. Leaving aside philosophical speculations on good and evil, and the possibility of accessing (even just to see) this Universe, explain why this is inconsistent with the "many worlds interpretation" of quantum mechanics.

32 Quantum Statistical Mechanics and "Tracing" over a Subspace

In this chapter, we will describe the basics of quantum statistical mechanics as a natural extension of the most general formalism, for a density matrix (i.e., a mixed state) instead of a pure state. A density matrix is a classical distribution of quantum states, and, as such, it allows for a satisfactory statistical mechanics interpretation, since classical statistical mechanics deals with a classical distribution of classical states. On the other hand, we have also seen that a nontrivial $\hat{\rho}$ appears when we take the trace over a subspace in a total space.

Thus we can interpret mixed states in two possible ways:

- Perhaps we can ignore (meaning, we cannot measure) some part of a Hilbert space for a total *isolated* system: we have a "macroscopic" description only, not a microscopic description. This is in fact at the core of the statistical mechanics interpretation, even in the classical case: statistics comes from ignorance of (or "averaging over") the unknown parts of the state of the total system.
- Alternatively, perhaps we have a system in contact with a reservoir, schematically denoted as $\mathcal{S} \cup \mathcal{R}$. Then, in a similar manner, there is only a total state for the system plus reservoir, $\mathcal{S} \cup \mathcal{R}$, but not for $\mathcal{S}$ or $\mathcal{R}$ independently, since there is no such thing as a state, or wave function for only a part of a system, except for very special states (product, or separable, states).

32.1 Density Matrix and Statistical Operator

Either way (in both of the above interpretations), the system $\mathcal{S}$ is in a state chosen randomly from the set $\{|\psi_j\rangle\}$, with classical probability w_j, which is the definition of a mixed state.

Consider a complete and orthonormal set $\{|\psi_\alpha\rangle\}$ of eigenstates of a complete set of compatible observables. That means that for any state of the system, we can decompose it in the $|\psi_\alpha\rangle$ states,

$$|\psi\rangle = \sum_\alpha c_\alpha |\psi_\alpha\rangle. \tag{32.1}$$

Multiplying by $\langle\psi_\beta|$ from the left, we obtain as usual $c_\alpha = \langle\psi_\alpha|\psi\rangle$. Applying the decomposition to the states in the mixed state set, we have

$$|\psi_j\rangle = \sum_\alpha c_\alpha^{(j)} |\psi_\alpha\rangle. \tag{32.2}$$

The expectation value of an observable A in the state $|\psi\rangle$ is found to be

$$\begin{aligned}\langle A\rangle_\psi &= \langle\psi|\hat{A}|\psi\rangle\\ &= \sum_{\alpha,\beta}\langle\psi|\psi_\alpha\rangle\langle\psi_\alpha|\hat{A}|\psi_\beta\rangle\langle\psi_\beta|\psi\rangle\\ &= \sum_{\alpha,\beta} c_\alpha^* c_\beta A_{\alpha\beta},\end{aligned} \tag{32.3}$$

where we have inserted two completeness relations $\mathbb{1} = \sum_\alpha |\psi_\alpha\rangle\langle\psi_\alpha|$.

But then, for the classical statistics of the quantum states $|\psi_\alpha\rangle$ with probabilities w_j, we can calculate the average value of an observable A by taking first a quantum average, and then the classical statistical average over states:

$$\begin{aligned}\overline{A} \equiv \left\langle\langle A\rangle_{\rm qu}\right\rangle_{\rm stat} &= \sum_j w_j\langle A\rangle_{\psi_j} = \sum_{\alpha,\beta}\left(\sum_j w_j c_\alpha^{(j)*}c_\beta^{(j)}\right)A_{\alpha\beta}\\ &= \sum_{\alpha,\beta}\rho_{\beta\alpha}A_{\alpha\beta} = \mathrm{Tr}(\hat{\rho}\hat{A}).\end{aligned} \tag{32.4}$$

Here we have defined the *density matrix*

$$\begin{aligned}\rho_{\beta\alpha} &= \sum_j w_j c_\beta^{(j)}c_\alpha^{(j)*} = \sum_j w_j\langle\psi_\beta|\psi_j\rangle\langle\psi_j|\psi_\alpha\rangle\\ &= \langle\psi_\beta|\left(\sum_j w_j|\psi_j\rangle\langle\psi_j|\right)|\psi_\alpha\rangle\\ &\equiv \langle\psi_\beta|\hat{\rho}|\psi_\alpha\rangle,\end{aligned} \tag{32.5}$$

where

$$\hat{\rho} = \sum_j w_j|\psi_j\rangle\langle\psi_j| \tag{32.6}$$

is an operator that, at least in the context of quantum statistical mechanics, will be called the statistical operator.

The normalization of the statistical operator and its associated density matrix is found as follows:

$$\begin{aligned}\mathrm{Tr}\,\hat{\rho} &= \sum_\alpha\sum_j w_j\langle\psi_\alpha|\,|\psi_j\rangle\langle\psi_j|\,|\psi_\alpha\rangle\\ &= \sum_j w_j\langle\psi_j|\sum_\alpha|\psi_\alpha\rangle\langle\psi_\alpha|\,|\psi_j\rangle = \sum_j w_j = 1,\end{aligned} \tag{32.7}$$

where in the third equality we have used the completeness relation of $|\psi_\alpha\rangle$ and the orthonormality of $|\psi_j\rangle$.

Considering the statistical operator for a pure state $|\psi\rangle$ as a particular example,

$$\hat{\rho} = |\psi\rangle\langle\psi|, \tag{32.8}$$

the density matrix is

$$\rho_{\beta\alpha} = \langle\psi_\beta|\hat{\rho}|\psi_\alpha\rangle = c_\alpha^* c_\beta. \tag{32.9}$$

The formalism of mixed states applies to states that are not completely known, so that there are classical probabilities w_j, in which case the mixed states, defined by John von Neumann, appear in a macroscopic description.

Indeed, in Chapter 30 we saw that, for a pure total state, either a system plus reservoir, $\mathcal{S} \cup \mathcal{R}$, or an observed state plus an unobservable state, we sum over diagonal components in the unobservable Hilbert space B, i.e., *we take the trace*, with the result

$$\begin{aligned} \rho_B &= \text{Tr}_A\, \rho_{\text{tot}} = \text{Tr}_A\, |\psi\rangle_{AB}\ {}_{AB}\langle\psi| = \sum_i p_i |i'\rangle_{B\,B}\langle i'| \\ \rho_A &= \text{Tr}_B\, \rho_{\text{tot}} = \sum_i p_i |i\rangle_A\ {}_A\langle i|. \end{aligned} \tag{32.10}$$

Note that in this case, the resulting state for a reduced system (i.e., either system A or system B but not both) is not of the type $\sum_j w_j|\psi_j\rangle$, but simply is not a (pure) state $|\psi'\rangle$ at all!

32.2 Review of Classical Statistics

Before we turn to the description of quantum statistical mechanics, we review the classical version, in order to see how it can be generalized.

We start with the definition of a *statistical ensemble*, due to Boltzmann and Gibbs. It is a collection of identical systems to the real system $\mathcal{S}$, in the same conditions as the real one, independent of each other and each system being in any of the possible states (that are compatible with the external conditions on the system).

We also define the distribution function of the statistical ensemble,

$$\rho = \frac{d\mathcal{N}}{\mathcal{N}d\Gamma}, \tag{32.11}$$

where $\mathcal{N}$ refers to the number of systems in the ensemble and $d\Gamma$ is the differential of the total number of states with energy $\leq E$,

$$\Gamma(E) = \int_{\mathcal{H}\leq E} d\Gamma. \tag{32.12}$$

We also define the number of states in a region between E and $E + \Delta E$,

$$\Omega(E, \Delta E) = \int_{E\leq\mathcal{H}\leq E+\Delta E} d\Gamma, \tag{32.13}$$

and the energy distribution,

$$\omega(E) = \frac{\partial\Gamma(E)}{\partial E}, \tag{32.14}$$

so that

$$\Omega(E, \Delta E) \simeq \omega(E)\Delta E. \tag{32.15}$$

All quantities, including Γ and ρ, are functions of the phase space $(\mathbf{p}, \mathbf{q})$ of the total number of particles (so there is a very large number of variables, giving a very large dimension of the phase space), so we have $\rho(\mathbf{p}, \mathbf{q})$ and $d\Gamma(\mathbf{p}, \mathbf{q})$. Then the ensemble average of an observable A is

$$\langle A \rangle = \int_{\Gamma} A \, \rho \, d\Gamma. \tag{32.16}$$

The Ergodic Hypothesis

The ergodic hypothesis is due to Boltzmann. It states that for an isolated system, a point in the phase space of the system goes through all the points in the phase space on an isoenergetic (E = constant) hypersurface. But it is not valid.

A better variant is the *quasi-ergodic hypothesis*: that a point in the phase space of the system goes arbitrarily close to every point on the isoenergetic hypersurface. However, it is not valid and/or provable in general.

That means that we need a set of postulates about observables in the system, to replace any need for explanations or proofs.

Postulate 1

The first such postulate states that the observed expectation value of a quantity A equals the temporal average,

$$A_{\rm obs} = \overline{A} = \lim_{\tau \to \infty} \frac{1}{\tau} \int_0^{\tau} A(t) \, dt. \tag{32.17}$$

Postulate 2

The second postulate, called the ergodic postulate, or Gibbs–Tollman postulate, states that the statistical average (the average over the ensemble) equals the time average,

$$\langle A \rangle = \overline{A}. \tag{32.18}$$

Postulate 3

The third postulate, of a priori equal probabilities, known as the Tollman postulate, says that for an isolated system, the probability density in phase space is constant for all the *accessible* states (i.e., those consistent with the external boundary conditions). Mathematically, we write

$$\rho = \begin{cases} \text{constant} & \text{in } \mathcal{D} \\ 0, & \text{outside it.} \end{cases} \tag{32.19}$$

32.3 Defining Quantum Statistics

In the quantum mechanical case, we use classical statistics over quantum states, plus quantum statistics over a state. Moreover, states of the system are now (generically) discrete, labeled by an index n for the energy plus a degeneracy g_n for different states of the same energy.

Now we define the total number of states with energy less than E, similarly to the classical case,

$$\Gamma(E) = \sum_{n,g_n(E_n<E)} 1 = \sum_{n(E_n<E)} g_n, \tag{32.20}$$

and then define $\Omega(E, \Delta E)$ and $\omega(E)$ as in the classical case.

We now define postulates for quantum statistical mechanics that are equivalent to the classical ones above in the appropriate limit.

Postulate 1

This is the same as before, equating the observed value and the temporal average,

$$\overline{A}_{\text{obs}} = \overline{A(t)}. \tag{32.21}$$

Postulate 2

This is also the same as before, equating the statistical average (over the ensemble) with the time average (or the observed value), but now the statistical average is different,

$$A_{\text{obs}} = \langle A \rangle_{\text{stat}} = \text{Tr}(\hat{\rho}\hat{A}). \tag{32.22}$$

Postulate 3

This is a postulate that the amplitudes are a priori equal and the phases are a priori random. Then the amplitudes of $c_\alpha^{(j)}$ (for an expansion in the basis $|\psi_\alpha\rangle$) are a priori equal, and the phases of the same are a priori random,

$$c_\alpha^{(j)} = r^{(j)} e^{i\phi_\alpha^{(j)}}, \tag{32.23}$$

where $r^{(j)}$ is independent of α and the ϕ_α are random as a function of α. Defining the classical ensemble average (denoted by an overbar) as an average over the states j, the conditions are

$$\overline{r_\alpha^2} = \rho_0, \quad \overline{\cos(\phi_\alpha - \phi_\beta)} = \overline{\sin(\phi_\alpha - \phi_\beta)} = 0. \tag{32.24}$$

Then the density matrix in the basis $|\psi_\alpha\rangle$ is constant and diagonal,

$$\rho_{\alpha\beta} = \begin{cases} \rho_0 \delta_{\alpha\beta} & \text{in } \mathcal{D} \\ 0, & \text{for } (\alpha, \beta) \text{ outside } \mathcal{D}. \end{cases} \tag{32.25}$$

But on the other hand, the diagonal elements are

$$\begin{aligned}\rho_{\alpha\alpha} &= \rho_0 \\ &= \sum_j |c_\alpha^{(j)}|^2 w_j = w_\alpha,\end{aligned} \tag{32.26}$$

where w_α is the probability in the basis state α, and we have used that $\sum^\alpha$ of the above quantity should be equal to 1, hence we can identify it with w_α.

The properties of the statistical operator $\hat{\rho}$ are as follows.

(1) It is a Hermitian operator (which we have seen implicitly).
(2) It is normalized by $\text{Tr}\,\hat{\rho} = 1$.
(3) The eigenvalues of $\hat{\rho}$ are semi-positive-definite, i.e., ≥ 0. This must be so, since they represent probabilities.
(4) $\hat{\rho}$ is bounded, meaning that $|\rho_{\alpha\beta}| \leq 1$.

Further, to determine $\hat{\rho}$ for an isolated system (for which case we can prove the following statement) and in general (for which it needs to be postulated), we have the *Liouville–von Neumann equation*,

$$i\hbar \frac{\partial \hat{\rho}}{\partial t} = [\hat{H}, \hat{\rho}]. \tag{32.27}$$

For an isolated system, this is proven as follows. Substituting the diagonal form for $\hat{\rho}$, we find

$$\begin{aligned}i\hbar\partial_t\left(\sum_j w_j|\psi_j\rangle\langle\psi_j|\right) &= \sum_j w_j\left[(i\hbar\partial_t|\psi_j\rangle)\langle\psi_j| + |\psi_j\rangle(i\hbar\partial_t\langle\psi_j|)\right] \\ &= i\hbar\sum_j w_j\left(\hat{H}|\psi_j\rangle\langle\psi_j| - |\psi_j\rangle\langle\psi_j|\hat{H}\right) \\ &= i\hbar\left[\hat{H}\hat{\rho} - \hat{\rho}\hat{H}\right] = i\hbar[\hat{H}, \hat{\rho}],\end{aligned} \tag{32.28}$$

where in the second equality we have used the Schrödinger equation $i\hbar\partial_t|\psi\rangle = \hat{H}|\psi_j\rangle$ and its complex conjugate.

Statistical Ensemble at Equilibrium

In the quantum case, a statistical ensemble at equilibrium gives observables that are time independent, so

$$A_{\text{obs}} = \langle A\rangle = \text{Tr}[\hat{\rho}\hat{A}] \tag{32.29}$$

is time independent, implying that ρ is time independent. But, by the Liouville–von Neumann equation, we have

$$\frac{\partial\hat{\rho}}{\partial t} = 0 \Rightarrow [\hat{\rho}, \hat{H}] = 0. \tag{32.30}$$

Since these two Hermitian operators commute, it means that we can always define them (and measure them) at the same time. But then, classically, ergodicity would mean that the distribution depends on the energy only,

$$\rho(\mathbf{p}, \mathbf{q}) = \rho(\mathcal{H}(\mathbf{p}, \mathbf{q})). \tag{32.31}$$

The same statement in quantum mechanics is that now the statistical operator is a function of the Hamiltonian,

$$\hat{\rho} = \rho(\hat{H}). \tag{32.32}$$

We can now define the (quantum version of) ensembles and their associated distribution.

Quasi-Microcanonical Ensemble (Distribution)

In this case, we define the ensemble by saying that it is composed of isolated systems each with energy in a very small region $E' \in (E, E + \Delta E)$.

Since the statistical operator is a function of the Hamiltonian, $\hat{\rho} = \rho(\hat{H})$, for the eigenvalues (corresponding to eigenstates of energy) we have

$$\rho_{nm} = \rho(E_n)\delta_{nm}. \tag{32.33}$$

Defining a domain

$$\mathcal{D} = \{n | E < E_n < E + \Delta E\}, \tag{32.34}$$

and a function

$$\Delta_a(x) = \begin{cases} 1, & x \in \mathcal{D} \\ 0, & x \notin \mathcal{D}, \end{cases} \tag{32.35}$$

the quasi-microcanonical distribution is

$$\rho(E_n) = \begin{cases} \text{constant} & \text{in } \mathcal{D} \\ 0 & \text{outside it.} \end{cases} \tag{32.36}$$

This can be rewritten as

$$\rho_{nm} = \frac{\Delta_{\Delta E}(E_n - E)}{\Omega(E, \Delta E)} \delta_{nm}, \tag{32.37}$$

now with the correct normalization for the matrix (so that $\sum_n \rho_{nn} = 1$). This can be extended in the quantum case to the formula

$$\hat{\rho} = \frac{\Delta_{\Delta E}(\hat{H} - E\mathbb{1})}{\Omega(E, \Delta E)}, \tag{32.38}$$

where $\Delta_{\Delta E}(\hat{H} - E\mathbb{1})$ is a projector onto the domain $\mathcal{D}$.

Then the expectation value for an observable A is given by

$$\langle A \rangle = \frac{1}{\Omega(E, \Delta A)} \sum_{n \in \mathcal{D}} \langle n | \hat{A} | n \rangle. \tag{32.39}$$

We can now add another postulate to the axiomatic system for statistical mechanics:

Postulate 4: the Boltzmann formula

We define the entropy of an isolated system in terms of the number of states in the domain $\mathcal{D}$ (of energy in $(E, E+\Delta E)$), Ω, as

$$S = k_B \ln \Omega. \tag{32.40}$$

This defines the *statistical entropy*, and can be "heuristically proven", meaning it is not quite a proof, but a strong argument. The formula for the entropy S should satisfy the condition that, if the system is made of two subsystems, $\mathcal{S} = \mathcal{S}^a \cup \mathcal{S}^b$, then

$$\omega(E) \simeq \omega^{(a)}(E^{(a)})\omega^{(b)}(E^{(b)})\Delta E. \tag{32.41}$$

In the thermodynamical limit, of an infinite number of particles, volume, and energy, with fixed ratios,

$$N \to \infty, V \to \infty, E \to \infty, \quad \frac{E}{N} = \text{fixed}, \quad \frac{V}{N} = \text{fixed}, \tag{32.42}$$

we have other formulas for the entropy that are equivalent to the above,

$$S = k_B \ln \omega, \quad S = k_B \ln \Gamma, \quad S = -k_B \langle \ln \rho \rangle. \tag{32.43}$$

These formulas are based on the fact that, for a very large dimension of the phase space, $N \to \infty$, $\omega \Delta E \simeq \Gamma \simeq \Omega$. Moreover, the last relation, based on the distribution ρ, becomes in quantum mechanics (using the standard averaging with $\hat{\rho}$)

$$S = -k_B \, \text{Tr}[\hat{\rho} \ln \hat{\rho}], \tag{32.44}$$

which is the *von Neumann entropy*, generalizing the Gibbs entropy from Chapter 30. We also note that the entropy is only additive in the thermodynamic limit.

Canonical Ensemble (Distribution)

The (quasi-)microcanonical distribution, discussed above, is rather hard to use; and the simplest and most standard one is the canonical distribution, in which case the system is connected to a reservoir of temperature, $\mathcal{S} \cup \mathcal{R}_T$.

Then, since the variation of the entropy is related to heat (energy variation) divided by temperature, we obtain

$$\delta S = \frac{\delta Q}{T} = -k_B \langle \log \rho \rangle. \tag{32.45}$$

This can be satisfied by a Maxwell-type distribution in the thermodynamic limit,

$$\rho \propto e^{-\mathcal{H}/k_B T}. \tag{32.46}$$

Including normalization, the formula is

$$\rho = \frac{1}{Z} e^{-\mathcal{H}/k_B T}, \tag{32.47}$$

where we define $\beta = 1/(k_B T)$ and the *partition function* Z is the sum of the exponential factors,

$$Z = \sum_E e^{-\beta E}. \tag{32.48}$$

In quantum mechanics, the formula becomes more useful, since the statistical operator,

$$\hat{\rho} = \rho(\hat{H}) = \frac{1}{Z(\beta, V, N)} e^{-\beta \hat{H}}, \tag{32.49}$$

becomes discrete for states,

$$\rho_n(E_n) = \frac{e^{-\beta E_n}}{Z(\beta, V, N)}. \tag{32.50}$$

The partition function also becomes discrete,

$$Z(\beta, V, N) = \sum_n{}' g_n e^{-\beta E_n}, \tag{32.51}$$

such that ρ_n is normalized to one,

$$\sum_n g_n \rho_n = 1. \tag{32.52}$$

In this ensemble, we define the internal energy as the average of the energies

$$\begin{aligned} U \equiv \langle E \rangle &= \sum_n \rho_n E_n \\ &= \sum_{n, g_n} \frac{E_n e^{-\beta E_n}}{Z} = -\frac{1}{Z} \frac{\partial}{\partial \beta} \sum_n g_n e^{-\beta E_n} = -\frac{\partial}{\partial \beta} \ln Z. \end{aligned} \tag{32.53}$$

But since in thermodynamics

$$U = \frac{\partial}{\partial \beta} [\beta F(T, V, N)] = F + \beta \frac{\partial}{\partial \beta} F, \tag{32.54}$$

we can identify the free energy F (the thermodynamic potential in this case) as

$$F(T, V, N) = -k_B T \ln Z. \tag{32.55}$$

Grand Canonical Ensemble (Distribution)

Another relevant ensemble is the grand canonical one, in which case there is a reservoir of heat and particles in contact with our system, keeping fixed the system's temperature T and chemical potential μ, $\mathcal{S} \cup \mathcal{R}_{T,\mu}$. Then the classical distribution is

$$\rho = \frac{1}{Z(\beta, \beta\mu)} e^{-\beta(\mathcal{H} - \mu N)}, \tag{32.56}$$

which at the quantum level leads to the operator

$$\hat{\rho} = \rho(\hat{H}, \hat{N}) = \frac{1}{Z(\beta, \beta\mu)} e^{-\beta(\hat{H} - \mu \hat{N})}, \tag{32.57}$$

with eigenvalues

$$\rho_n = \frac{1}{Z(\beta, \beta\mu)} e^{-\beta(E_n - \mu N)}. \tag{32.58}$$

Then the partition function is

$$Z(\beta, \beta\mu) = \sum_n g_n e^{-\beta(E_n - \mu N)}, \tag{32.59}$$

and the thermodynamic potential is now

$$\Omega(T, V, \mu) = -k_B T \ln Z(\beta, \beta\mu). \tag{32.60}$$

The total average energy and number of particles can be calculated from the thermodynamic potential, as

$$U = \langle \mathcal{H} \rangle = -\left(\frac{\partial \ln Z}{\partial \beta}\right)_{\beta\mu}, \quad \langle N \rangle = \left(\frac{\partial \ln Z}{\partial \beta\mu}\right)_{\beta}. \tag{32.61}$$

We could define other ensembles, but the general procedure should be clear by now.

32.4 Bose–Einstein and Fermi–Dirac Distributions

Consider next systems consisting of identical quantum particles, bosons or fermions, which are identical and indistinguishable and must therefore satisfy either Bose–Einstein or Fermi–Dirac statistics.

In the case of Bose–Einstein statistics, in each state we can have an arbitrary number of particles $n_\alpha = 0, 1, 2, \ldots, \infty$.

In the case of Fermi–Dirac statistics, because of the Pauli exclusion principle we can only have $n_\alpha = 0$ or 1 (the state is either occupied or not).

Then in both cases (considered together), the energy splits over the energy states, as does the total number of particles,

$$E = \sum_\alpha \epsilon_\alpha n_\alpha, \quad N = \sum_\alpha n_\alpha. \tag{32.62}$$

Moreover, then the partition function also factorizes,

$$Z = \prod_\alpha Z_\alpha, \tag{32.63}$$

where

$$Z_\alpha = \sum_{n_\alpha} e^{-\beta n_\alpha(\epsilon_\alpha - \mu)}. \tag{32.64}$$

(1) In the case of Bose–Einstein statistics, summing over $n_\alpha = 0, 1, \ldots, \infty$, we get

$$Z_\alpha = \frac{1}{1 - e^{-\beta(\epsilon_\alpha - \mu)}}, \tag{32.65}$$

so the grand canonical potential is

$$\Omega = -k_B T \ln Z = -\sum_\alpha k_B T \ln Z_\alpha. \tag{32.66}$$

Then we obtain the average number of particles

$$\langle N \rangle = -\frac{\partial \Omega}{\partial \mu} = \sum_\alpha \langle n_\alpha \rangle, \tag{32.67}$$

where the Bose–Einstein distribution function is

$$\langle n_\alpha \rangle = \frac{1}{e^{\beta(\epsilon_\alpha - \mu)} - 1} \equiv f_{BE}(\epsilon_\alpha). \tag{32.68}$$

(2) In the case of Fermi–Dirac statistics, summing over $n_\alpha = 0, 1$, we have

$$Z_\alpha = 1 + e^{-\beta(\epsilon_\alpha - \mu)}, \tag{32.69}$$

from which we similarly obtain the Fermi–Dirac distribution function,

$$\langle n_\alpha \rangle = \frac{1}{e^{\beta(\epsilon_\alpha - \mu)} + 1} \equiv f_{FD}(\epsilon_\alpha). \tag{32.70}$$

32.5 Entanglement Entropy

In Chapter 30, we also described a different kind of entropy, associated with two subsystems of a system, $\mathcal{S} = A \cup B$ (so that $\mathcal{H}_{\rm tot} = \mathcal{H}_A \otimes \mathcal{H}_B$). Then, we defined the reduced density matrix

$$\hat{\rho}_A = \mathrm{Tr}_B\, \hat{\rho}_{\rm tot}, \tag{32.71}$$

both for a pure state $\hat{\rho}_{\rm tot}$ $(= |\psi\rangle\langle\psi|)$, and for a mixed state. Then the von Neumann entropy of $\hat{\rho}_A$ is defined as the entanglement entropy,

$$S_A = -k_B\, \mathrm{Tr}_A(\hat{\rho}_A \ln \rho_A). \tag{32.72}$$

But we note that, if we have a finite temperature T, we can have a thermal total density matrix,

$$\hat{\rho}_{\rm tot} = \hat{\rho}_{\rm tot,thermal} = e^{-\beta \hat{H}}. \tag{32.73}$$

If moreover the second subsystem vanishes, so that $B = \{0\}$ and $\mathcal{S} = A$, then the thermal entanglement entropy equals the usual (von Neumann) thermal entropy.

But we can also choose a more general situation, with $\mathcal{S} = A \cup B$ at finite temperature, in which case we say we have a *thermal entanglement entropy*.

We can also generalize the entanglement entropy by means of an integer n, obtaining the *Renyi entropy*,

$$S_n(\rho) = \frac{1}{1-n} \ln\left(\sum_{i=1}^N p_i^n\right) = \frac{1}{1-n} \ln\left[\mathrm{Tr}(\hat{\rho}^n)\right]. \tag{32.74}$$

This quantity is simpler to define, since it does not contain the log of a matrix (which is tricky in general), yet we can obtain the entanglement entropy as a limit: one can prove that

$$S(\rho) = \lim_{n\to 1} S_n(\rho). \tag{32.75}$$

In both cases, we can consider the thermal density matrix,

$$\hat{\rho} = \frac{e^{-\beta \hat{H}}}{Z}, \tag{32.76}$$

which is the highest in entropy.

We can describe entanglement as the result of an *entanglement Hamiltonian*, which is different from the actual Hamiltonian of the system (as far as the dynamics is concerned). Consider a variation in the entanglement entropy due to a variation in the density matrix,

$$\begin{aligned} \delta S(\rho) &= -\delta\, \mathrm{Tr}[\hat{\rho} \log \hat{\rho}] = -\mathrm{Tr}[\delta\hat{\rho} \log \hat{\rho}] - \mathrm{Tr}[\hat{\rho}\delta \log \hat{\rho}] \\ &= -\mathrm{Tr}[\delta\hat{\rho} \log \hat{\rho}] - \mathrm{Tr}\, \delta\hat{\rho}. \end{aligned} \tag{32.77}$$

Defining the entanglement Hamiltonian as

$$\hat{H}_E = -\log \hat{\rho} \Rightarrow \hat{\rho} = e^{-\hat{H}_E}, \tag{32.78}$$

it follows that we have

$$\delta S(\hat{\rho}) = \delta\,\mathrm{Tr}[\hat{\rho}\hat{H}_E] = \delta\langle \hat{H}_E\rangle. \tag{32.79}$$

However, if $\hat{\rho}$ describes a thermal (mixed) state,

$$\hat{\rho} = \frac{e^{-\beta\hat{H}}}{Z} \Rightarrow \delta S = \beta\delta\langle\hat{H}\rangle \Rightarrow dE = TdS. \tag{32.80}$$

Thus, indeed, for the entanglement Hamiltonian, we have the expected thermodynamic relation for $\beta = 1$ (unit temperature).

For this same thermal state, we can "purify it", meaning we can define a pure state, in a product Hilbert space $\mathcal{H} = \mathcal{H}_A \otimes \mathcal{H}_B$, such that the reduced density matrix ρ is thermal, and its entanglement entropy defines the von Neumann entropy of the state. The pure state in $\mathcal{H}$ is called the *thermofield double state*, and is defined as

$$|\psi\rangle = \frac{1}{Z}\sum_i e^{-\beta E_i/2}|i\rangle_A \otimes |i\rangle_B, \tag{32.81}$$

where $|i\rangle$ are the eigenstates of the Hamiltonian. We can easily check that taking the trace over system B gives the thermal density matrix.

We note that the entanglement entropy is hard to measure, or calculate, yet is interesting and has been the subject of much research and many developments.

Important Concepts to Remember

- The density matrix, the matrix element of the statistical operator $\hat{\rho}$, describes a mixed state. It arises from taking the trace over another system, in contact with this one, or from contact with a reservoir (which amounts to the same, except we don't know the description of the reservoir states).
- Classical statistics is based on four postulates, replacing the ergodic hypothesis, which is incorrect: (1) $A_{\text{obs}} = \overline{A(t)}$; (2) $\langle A\rangle = \overline{A}$; (3) for an isolated system, ρ is contant inside the domain $\mathcal{D}$ and 0 outside it; (4) the Boltzmann formula, $S = k_B \ln \Omega$, which is heuristically proven only.
- Quantum statistics is based on four postulates, replacing the above, with ρ replaced by $\hat{\rho}$: (1) $A_{\text{obs}} = \overline{A(t)}$; (2) $A_{\text{obs}} = \langle A\rangle_{\text{stat}} = \mathrm{Tr}[\hat{\rho}\hat{A}]$; (3) the phases and amplitudes of $c_\alpha^{(j)}$ are a priori random; (4) the von Neumann formula, $S = -k_B\,\mathrm{Tr}[\hat{\rho}\ln\hat{\rho}]$.
- The statistical operator obeys the Liouville–von Neumann equation, $i\hbar\,\partial\hat{\rho}/\partial t = [\hat{H}, \hat{\rho}]$, proven for an isolated system and postulated in general.
- At equilibrium, we have time independence, so $[\hat{H}, \hat{\rho}] = 0$; so classically $\rho = \rho(\mathcal{H}(\mathbf{p}, \mathbf{q}))$ and quantum mechanically $\hat{\rho} = \hat{\rho}(\hat{H})$.
- The Boltzmann formula is equivalent, but only in the thermodynamic limit, to $S = k_B \ln \omega$, $S = k_B \ln \Gamma$, $S = -k_B\langle \ln \rho\rangle$, with the latter motivating the von Neumann formula, $S = -k_B\,\mathrm{Tr}[\hat{\rho}\ln\hat{\rho}]$, in the quantum case.

- In the canonical distribution (ensemble), with $\mathcal{S} \cup \mathcal{R}_T$, $\rho = Z^{-1}e^{-\beta\mathcal{H}}$, with the partition function $Z = \sum_E e^{-\beta E}$ classically, and $\hat{\rho} = Z^{-1}(\beta, V, N)e^{-\beta\hat{H}}$, and $Z(\beta, V, N) = \sum_n g_n e^{-\beta E_n}$ quantum mechanically, and free energy $F(T, V, N) = -k_B T \ln Z$.
- In the grand canonical distribution (ensemble), with $\mathcal{S} \cup \mathcal{R}_{T,\mu}$, classically $\rho = Z^{-1}e^{-\beta(\mathcal{H}-\mu N)}$ and quantum mechanically $\hat{\rho} = Z^{-1}(\beta, \beta\mu)e^{-\beta(\hat{H}-\mu\hat{N})}$, with $Z(\beta, \beta\mu) = \sum_n g_n e^{-\beta(E_n - \mu N)}$ and thermodynamic potential $\Omega(T, V, \mu) = -k_B T \ln Z(\beta, \beta\mu)$.
- From the Bose–Einstein and Fermi–Dirac statistics, we obtain the corresponding distributions, $\langle n_\alpha \rangle = f_{BE}(\epsilon_\alpha) = 1/(e^{\beta(\epsilon_\alpha - \mu)} - 1)$ and $\langle n_a \rangle = f_{FD}(\epsilon_\alpha) = 1/(e^{\beta(\epsilon_\alpha - \mu)} + 1)$.
- One can define the entanglement entropy as before, $S_A = \text{Tr}_A \rho_{\text{tot}}$, but we can define it also in the case where the total system is at finite temperature, $\mathcal{S}_{AB} \cup \mathcal{R}_T$, giving the thermal entanglement entropy. We can also define the Renyi entropy, $S = (1-n)^{-1} \ln[\text{Tr}(\hat{\rho}^n)]$, such that the entanglement entropy is obtained in the $n \to 1$ limit.
- For entanglement, we can define the entanglement Hamiltonian via $\hat{\rho} = e^{-\hat{H}_E}$, so that $\delta S(\rho) = \delta\langle H_E \rangle$, similar to the thermodynamic relation $\delta S = \beta\delta\langle \hat{H} \rangle$, or $dE = TdS$.
- We can "purify" any thermal state by writing it as the trace, over a different system, of a pure state, called the thermofield double state, $|\psi_{TFD}\rangle = \frac{1}{Z}\sum_i e^{-\beta E_i/2}|i\rangle_A \otimes |i\rangle_B$.

Further Reading

See [2, 1, 3] and any advanced (quantum) statistical mechanics book.

Exercises

(1) Consider the bipartite state in $\mathcal{H}_A \otimes \mathcal{H}_B$

$$|\psi\rangle = \frac{1}{2}\left[|1\rangle \otimes |1\rangle + |2\rangle \otimes |2\rangle + |3\rangle \otimes |4\rangle + |2\rangle \otimes |3\rangle\right], \tag{32.82}$$

where $|1\rangle, |2\rangle, |3\rangle, |4\rangle$ are orthonormal states.

Calculate the mixed state obtained by taking the trace over system A, or system B.

(2) If the states $|1\rangle, |2\rangle, |3\rangle, |4\rangle, |5\rangle$ are (orthonormal) eigenstates of the Hamiltonian of energies E_1, E_2, E_3, E_4, and E_5, respectively, and at time $t = 0$ the statistical operator is

$$\hat{\rho} = |1\rangle\langle 1| + |2\rangle\langle 3| + |3\rangle\langle 4| + |4\rangle\langle 5|, \tag{32.83}$$

then find the time evolution of $\hat{\rho}$ at small times.

(3) Consider a thermodynamic system of N (of the order of the Avogadro number N_A) spins 1/2. Calculate their classical entropy, for a (quasi-)microcanonical ensemble. Describe a quantum statistical operator $\hat{\rho}$ for the same system now using quantum mechanics, that gives the same von Neumann entropy.

(4) Consider N harmonic oscillators of frequencies ω_i, $i = 1, \ldots, N$, connected to a reservoir of temperature. Calculate (in the canonical ensemble) the free energy and the heat capacity C_V.

(5) Consider the harmonic oscillators from exercise 4, connected to a reservoir of temperature T and chemical potential μ. Calculate (in the grand canonical ensemble) the thermodynamic potential Ω and the heat capacity C_V.

(6) Consider a distribution of free, nondegenerate, relativistic fermionic particles of mass m, of arbitrarily large three-dimensional momentum. Calculate the number density and energy density as a function of temperature. Write down an explicit analytical form at large temperatures.

(7) Consider a system $A \cup B$, with total Hamiltonian diagonalized by states $|1\rangle_A \otimes |1\rangle_B$, $|2\rangle_A \otimes |2\rangle_B$, $|3\rangle_A \otimes |4\rangle_B$, $|4\rangle_A \otimes |3\rangle_B$, of energies E_1, E_2, E_3, E_4, respectively. Calculate the thermal entanglement entropy at temperature T.

33 Elements of Quantum Information and Quantum Computing

In this chapter, we introduce the ideas of quantum information theory and ways to do computing in quantum mechanics.

33.1 Classical Computation and Shannon Theory

Before we turn to quantum theory, we start with a review of classical computation theory, in order to generalize it to the quantum case.

To define classical information, we need to quantify the information and redundancy in the transmission of some message. Generically, a message amounts to a string of n letters, chosen from an alphabet of k letters,

$$\{a_1, \ldots, a_k\}. \tag{33.1}$$

Each letter a_x occurs with an a priori probability $p(a_x)$, such that $\sum_x p(a_x) = 1$. For instance, in the English language there are 26 letters, the most frequent being e, with approximately 12.7% frequency so $p(e) = 0.127$, and then t, with 9.3% frequency, so $p(t) = 0.093$, etc.

However, as we are mostly interested in computers and their working, it is worth considering binary messages, where the alphabet is composed of 0 and 1 only. In that case, we denote

$$p(1) = p, \quad p(0) = 1 - p. \tag{33.2}$$

We can assume that at large n, a message will have approximately pn 1's and $(1 - p)n$ 0's. Then the number of distinct binary strings of length n that can be sent (i.e., messages of n bits) corresponds to the number of ways in which we can pick np letters out of the n in order to put 1's in them. Thus, the number of distinct possible messages is of the order of

$$\sim \binom{n}{np} = \frac{n!}{(np)!\,(n(1-p))!} \simeq 2^{nH(p)}, \tag{33.3}$$

where we have used the Stirling approximation at large N, $N! \simeq \sqrt{2\pi} N^{N-1/2} e^{-N}$, which results in a quantity

$$H(p) \simeq -p \log_2 p - (1 - p) \log_2(1 - p). \tag{33.4}$$

We can extend this analysis to the case of an alphabet of an arbitrary length, where the letter a_x is represented by the label x, with probability $p(x)$. Then the number of distinct strings of length n is of the order of the ways in which we can pick groups of $p(x)n$ out of n, namely

$$\sim \frac{n!}{\prod_x (np(x))!} \sim 2^{nH(X)}, \tag{33.5}$$

where again we have used the Stirling approximation formula to find an $H(X)$ (depending on the set of x's and their probabilities, defining an ensemble called X), as

$$H(X) \simeq -\sum_{x=1}^{k} p(x) \log_2 p(x). \tag{33.6}$$

This is called the *Shannon entropy*. Note that we have used logarithms to the base 2, in order to write the number as an exponent of 2. But we could have used any other number (including e, leading to ln) instead. Note that $\log_2 q = \log_2 n \log_n q$.

This Shannon entropy is then a measure of the total redundancy in the message, and is the way to encode the message in bits: we need $nH(X)$ bits to be able to put any message into them (p bits lead to 2^p positions).

Another way to compute the Shannon entropy is as follows. A message is a string of letters of size n, $x_1 \ldots x_n$. Then the a priori probability for this string is

$$P(x_1, \ldots, x_n) = p(x_1) \cdots p(x_n), \tag{33.7}$$

with

$$\log_2 P = \sum_{i=1}^{n} \log_2 p(x_i). \tag{33.8}$$

But this means that we can obtain, in the large-n limit, if there are $p(x)n$ instances for each letter a_x,

$$-\frac{1}{n} \log_2 P(x_1, \ldots, x_n) \sim \langle -\log_2 p(x) \rangle = -\sum_{i=1}^{k} p(x_i) \log_2 p(x_i) = H(X). \tag{33.9}$$

So the Shannon entropy is defined by the probability of the string being $2^{-nH(X)}$. Thus the optimal code compresses a letter onto $H(X)$ bits, so that a message of length n is compressed into $nH(X)$ bits, with $2^{nH(X)}$ states. This $H(X)$ depends on the ensemble X, defined by the probabilities $p(x)$.

Mutual Information

We can ask, how correlated are two different messages, $(x_1, \ldots, x_n) \equiv x$ and $(y_1, \ldots, y_n) \equiv y$, drawn from different ensembles, X and Y? The measure of this correlation is the *mutual information* $I(X, Y)$.

If $p(x, y)$ denotes the probability that both messages will occur, then the probability of message x, given that message y has occurred, is

$$p(x|y) = \frac{p(x, y)}{p(y)}. \tag{33.10}$$

Then we can define the *conditional Shannon entropy* as the entropy defined from $p(x|y)$, namely

$$\begin{aligned} H(X|Y) &= \langle -\log_2 p(x|y) \rangle = \langle -\log_2 p(x, y) \rangle + \langle \log_2 p(y) \rangle \\ &= H(X, Y) - H(Y), \end{aligned} \tag{33.11}$$

where $p(x, y)$ is the probability that both messages will occur.

Then we define

$$I(X; Y) \equiv H(X) - H(X|Y) = H(X) + H(Y) - H(X, Y) \tag{33.12}$$

called the *mutual information*. We can see that it is positive definite, $I(X;Y) \geq 0$, since $H(X|Y)$ contains less information than $H(X)$.

33.2 Quantum Information and Computation, and von Neumann Entropy

We are now ready to generalize to the quantum case.

In quantum mechanics, generically we have a mixed state, defined by a density matrix

$$\rho = \sum_x p_x \rho_x, \tag{33.13}$$

where p_x is a classical probability. As we saw, we can always diagonalize the matrix in an orthonormal basis $|i\rangle$, where

$$\rho = \sum_i p_i |i\rangle\langle i|. \tag{33.14}$$

Then $\log \rho$ is represented by its diagonal eigenvalues, $\log p_i$.

Thus we have an analog of the Shannon entropy at the quantum level,

$$\langle -\log \rho\rangle = -\operatorname{Tr}(\rho \log \rho) \equiv S(A) = H(i, p_i), \tag{33.15}$$

where A is an ensemble defining system A. This is the *von Neumann entropy*, and we see that it equals the Shannon entropy of the classical distribution of the states (i, p_i).

We can also define the mutual information in the same way as in the classical case, by

$$I(A;B) = S(A) + S(B) - S(A+B). \tag{33.16}$$

The von Neumann entropy satisfies several properties. The most relevant ones are:

(1) The entropy vanishes for a pure state,

$$S(\rho = |\psi\rangle\langle\psi|) = 0. \tag{33.17}$$

This is consistent, since the entropy of a single state should be zero.

(2) The entropy is invariant under unitary transformations, since

$$S(U\rho U^{-1}) = -\operatorname{Tr}[U\rho U^{-1}\log(U\rho U^{-1})] = -\operatorname{Tr}[\rho\log\rho] = S(\rho). \tag{33.18}$$

(3) Concavity of the entropy: if $p_1, p_2, \ldots, p_n \geq 0$ and $\sum_i p_i = 1$, then

$$S\left(\sum_i p_i \rho_i\right) \geq \sum_i p_i S(\rho_i), \tag{33.19}$$

which follows from the same property of the negative log function.

(4) Subadditivity of the entropy: for a bipartite system AB,

$$S(A+B) \leq S(A) + S(B). \tag{33.20}$$

As before,

$$\rho_A = \operatorname{Tr}_B \rho_{AB}, \quad \rho_B = \operatorname{Tr}_A \rho_{AB}, \tag{33.21}$$

so we have

$$S(\rho_{AB}) \leq S(\rho_A) + S(\rho_B). \tag{33.22}$$

This describes the fact that there is nontrivial information in AB encoded in the correlations of A and B, which is a reasonable assumption.

(5) Strong subadditivity of the entropy: for a tripartite system ABC, we have the inequalities

$$\begin{aligned} S(A+B+C) + S(B) &\leq S(A+B) + S(B+C) \\ S(A+C) &\leq S(A+B) + S(B+C). \end{aligned} \tag{33.23}$$

This is rather difficult to prove in the quantum case of the von Neumann entropy. The name refers to the fact that subadditivity is obtained from it in the special case of $B = 0$.

33.3 Quantum Computation

Having seen that we can define quantum information, it follows that we can store information in quantum systems, and do computations with it.

Instead of information encoded in classical bits of 0 and 1, we can now encode information in the *quantum bits*, or *qubits*, of two possible states, for instance the states $|+\rangle$ and $|-\rangle$ of a two-state system (such as a spin 1/2 system).

Then for n qubits, we have states in the product Hilbert space $\mathcal{H} = \mathcal{H}_{A_1} \otimes \cdots \otimes \mathcal{H}_{A_n}$, with basis

$$|\psi_{\pm,\pm,\ldots,\pm}\rangle \equiv |\pm\rangle \otimes |\pm\rangle \otimes \cdots \otimes |\pm\rangle. \tag{33.24}$$

A more useful set of states is the set of entangled states in $\mathcal{H}$, since we can use entanglement to our advantage in order to do things that are not possible with a classical computer.

In order to do computations on a state, we must pass it through a *circuit* that is constructed out of basic *gates*, which are (linear) transformations that act on two or more bits simultaneously, among the n bits of the original message (state). A minimal gate is then an action on two bits x and y, as

$$\begin{pmatrix} x \\ y \end{pmatrix} \to \begin{pmatrix} x' \\ y' \end{pmatrix} = M \begin{pmatrix} x \\ y \end{pmatrix}. \tag{33.25}$$

Classically, the action of a gate is on bits $x, y = 0$ or 1, and a circuit is a product of gates $M_1 \cdots M_p$.

In quantum mechanics, quantum gates are objects that act linearly on the quantum states. That means that quantum gates are unitary transformations U_i, acting on a fixed number of qubits at a time (a minimum of 2) out of the general n qubits of the message. A quantum circuit is, as before, a product of the individual gates (see Fig. 33.1),

$$U = U_1 \cdots U_n. \tag{33.26}$$

A quantum computation is the action of a circuit on a quantum state, according to the relation

$$|\psi\rangle \to |\psi'\rangle = U|\psi\rangle. \tag{33.27}$$

At the end of the computation, we make a *measurement*, in order to have a classical result. This means that we select a state in each of the basis sets (perhaps in a subset of qubits).

Note that it is of great use to consider *entangled qubits*, of the type $|\phi\pm\rangle, |\psi\pm\rangle$, in a product system $A \cup B$, instead of the nonentangled qubit states $|\pm\rangle, |\pm\rangle$.

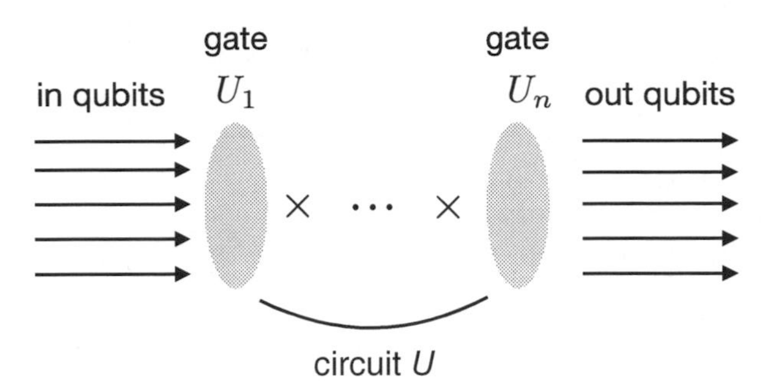

Figure 33.1 A quantum circuit doing a computation.

We can ask, why is it that we bother with quantum computation instead of the standard classical computation? One answer is that, with the increase in density of processors, and the decrease of the size onto which we can store a bit, eventually we will reach a size of the basic circuitry of the order of the quantum scale, meaning that quantum computational must become relevant then. This is a practical constructional reason.

But a better, theoretical, reason is that we can do more with a quantum computer than with a classical one. Can we do calculations that are not possible on a classical computer? No, that would be a contradiction, since we can simulate a quantum computer on a classical one.

However, we can do the same calculation *faster* on a quantum computer, meaning that the way in which they scale with the size of the problem is different. Generally, it is thought that if a specific computation can be done in a time that is a polynomial (approximately a power law) of the size of the data, i.e., that "it can be solved in polynomial time", then the problem is solvable by classical computers (since we will solve it in a reasonable time). These problems are in a set called P.

Nevertheless there are very important problems that are not in P, meaning that they involve a time that grows faster than a power law with the size of the problem. Then the problem is thought effectively unsolvable by a classical computer, since we can easily choose a problem size that leads to an unreasonable time to solve it.

There is a standard example, namely the problem of the factorization of large numbers. For a large number N, with k digits, we can write it as a product of n factors $N = n_1 \cdots n_n$. Then finding the factors n_i is a problem that is not in P, the time to find $\{n_i\}$ being related to the size k by $\sim \exp(f(k))$. However, if we have a solution, we can check that it is correct (by multiplying the factors n_i) in a polynomial time, so the check of the solution is in P.

This problem is related to cryptography: cryptography (the encryption of messages everywhere, including banking) is based on the existence of a key that can be related to the factors in a very large number. Then solving the factorization problem in polynomial time will result in a way to break encryptions in polynomial time.

Coming back to quantum computation, we note that a quantum gate is a unitary transformation, and as such is by definition reversible. But a classical computation is in general not reversible. That means that a classical reversible computer is mapped to a special case of a quantum computer.

To understand further the differences between a classical and a quantum computer, we note that quantum computers can simulate probabilistic classical computers. Conversely, a classical computer can simulate a quantum computer. Then, why is it that a quantum computer is better? The reason is that the simulation takes increasingly large times, becoming quickly prohibitive. Indeed, we saw that there is no difference between the *problems* that can be solved by a classical and a quantum computer, only in the *times* it takes to solve the problems on the two computers.

The gain in time efficiency when going to a quantum computer is offset somewhat by the fact that, since the computation is a unitary action on a state, followed by projection onto other states, the quantum computer gives only probabilities for the measurement. That means that the result is given as a probability distribution, so we have errors and error bars.

But there is no problem with that, since as we said, we are mostly interested in solving a given problem that is not in P in polynomial time. And once we have a solution, we can *check it* in polynomial time, so the probability distribution of results is enough.

What is the root of the effectiveness of quantum computation as against classical computation? The important point is "quantum parallelism": for instance, in the case of a single qubit, we can choose the input to be a superposition of the basis states $|\uparrow\rangle$ and $|\downarrow\rangle$ (or of other $|\psi\rangle$ states). Then the computer does the computation on the $|\uparrow\rangle$ and $|\downarrow\rangle$ states in parallel, in effect doing two computations in the time it takes a computer to do one. Of course, then we must go back and repeat the computation several times, in order to get probabilities, and then we also must check the result (in polynomial time). But the end result is still an improvement over polynomial time, owing to the large number of parallel computations.

One potential problem is the existence of errors that appear due to interaction with the environment, which can lead to a change, or even a collapse, of the wave function via a classical measurement. Indeed, it is to be understood that, at least in some cases, measurements and classicalization (the transition from a quantum state to a classical one) are the result of interaction with the environment, through *decoherence* (a generalization of the decoherence of an electromagnetic wave, light, to the decoherence of the probability wave, the wave function). This effect will be studied in more detail later on, but here we just state it without details.

However, errors that appear can be partially corrected, via quantum error correction (quantum) algorithms, which will not be described here. So, one can delay for a long time the decoherence and loss of quantum information, giving enough time to do the quantum calculation.

The end result is *quantum supremacy*, the ability of a quantum computer to solve quickly problems that a classical computer can only do slowly. Experimentally, at the time of writing it has not been obtained yet but it is believed theoretically to exist.

33.4 Quantum Cryptography, No-Cloning, and Teleportation

We have seen that we can use quantum computation to decrypt standard classical encryption, which is based on factorization of a large number. But is the reverse also possible, namely can we use quantum phenomena to generate a new kind of encryption? The answer is yes, and this entails a way to send a quantum key securely from one person to another, as we shall see.

But then another question arises: can we copy a quantum state that we possess in order to send a copy of it to some other person? The answer to that is actually NO, expressed in the form of a principle, the no-cloning principle:

> **It is not possible to copy the state $|\psi\rangle$ of a (sub-)system by any quantum mechanical (or classical) process.**

Such quantum processes would be unitary transformations acting on the state of the total system (comprising Alice, from whom we want to copy the state, and Bob, to whom we wish to send

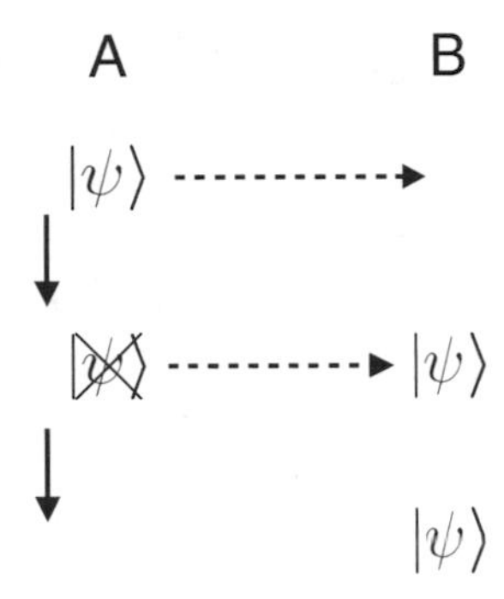

Figure 33.2 Principle of quantum teleportation.

the copy), as well as measurements collapsing the state of the (sub-)system. This is a principle, not a theorem, so it cannot be proven in any generality; one can only test that it cannot be done in any specific case that we consider.

Given this fact, it seems strange that we said we can send a state (a "key") from Alice to Bob. The process by which we send it is called "teleportation", after the name invented in science fiction (made popular mostly by the original "Star Trek" TV show) for a machine that sends something from point A to point B by erasing it from point A and making it appear at point B.

This "**quantum teleportation**" is similar to this general idea, since we need to erase the state at the original place (Alice), before (or rather, at the same time) recreating it at point B (Bob). We can "teleport" the state from Alice to Bob by using entanglement (in the EPR style) and measurement.

The procedure is as follows (see Fig. 33.2). We define a general state in the two-dimensional qubit Hilbert space,

$$|\psi\rangle = a|\uparrow\rangle + b|\downarrow\rangle, \tag{33.28}$$

and we want to send it from another location, C, to B, by use of entanglement with A. Defining as before

$$\begin{aligned}|\phi\pm\rangle &= \frac{1}{\sqrt{2}}(|\uparrow\uparrow\rangle \pm |\downarrow\downarrow\rangle)\\ |\psi\pm\rangle &= \frac{1}{\sqrt{2}}(|\uparrow\downarrow\rangle \pm |\downarrow\uparrow\rangle),\end{aligned} \tag{33.29}$$

we can show that we have the relation

$$\begin{aligned}|\psi\rangle_C|\phi+\rangle_{AB} = \frac{1}{2}\Big[&|\phi+\rangle_{AC}|\psi\rangle_B + |\psi+\rangle_{AC}\sigma_1|\psi\rangle_B\\ &+ |\psi-\rangle_{AC}(-i\sigma_2)|\psi\rangle_B + |\phi-\rangle_{AC}\sigma_3|\psi\rangle_B\Big].\end{aligned} \tag{33.30}$$

We leave the proof of this relation as an exercise (by simple substitution of the definitions of states and matrices in terms of the basis).

We can perform then a "Bell measurement" on the system AC, meaning that we measure onto an EPR style state; i.e., we project onto one of the states $|\phi\pm\rangle, |\psi\pm\rangle$. Then Alice can tell Bob *classically* (over a classical channel, like a telephone) what kind of measurement she did (what state she projected onto). Depending on the state, then all Bob has to do is to apply one of the operators $\mathbb{1}, \sigma_1, (-i\sigma_2)$, or σ_3 (for $|\phi+\rangle, |\psi+\rangle, |\psi-\rangle, |\phi-\rangle$, respectively), in order to obtain the state $|\psi\rangle$. Thus indeed, the state $|\psi\rangle$ has been teleported from C to B, having been erased at C (by the measurement), but created at B (by the measurement plus operation).

Based on these ideas of teleportation via entanglement, one can create a **quantum cryptographic** system that is invulnerable to attack. Indeed, for cryptography, we need to send a private key that is used to decode a message. In quantum mechanics this means that we need to send a quantum state (the key), after which the message is encoded in the *correlations* of a transmitted string and the private key. Then we can construct a *quantum key distribution* (a way to send the key) that is invulnerable to attack, since on the one hand the key is a state and, on the other, even knowing the key state is useless since the information is in the *correlation* of the string with the key, so having the string is irrelevant if it is not correlated to the key that you have.

Important Concepts to Remember

- In classical information theory, the Shannon entropy $H(p) = -\sum_i p_i \log_2 p_i$ is defined by the number of n-bit messages being $2^{nH(p)}$, or, put in another way, the probability for the occurrence of a single string being $2^{-nH(p)}$.
- In quantum computation, the analog of the Shannon entropy is the von Neumann entropy.
- In classical computation, a circuit is a product of gates acting on at least two bits. A quantum circuit is a product of gates (unitary transformations), acting on at least two qubits, and a quantum computation is the action of the circuit on a quantum state. At the end, we have made a measurement.
- A quantum computer can calculate faster: problems in NP (such problems cannot be solved in polynomial time by a classical computer) can be solved in polynomial time by a quantum computer. This is quantum supremacy, and would lead to the breaking of all classical encryption.
- This gain in efficiency is due partly to using entangled states and, as a result, to quantum parallelism: calculations are done in parallel, and then measurements are made and the result checked (in polynomial time).
- In quantum mechanics we have the no-cloning principle: we cannot copy a quantum state.
- We can have quantum teleportation, though: we erase the state from system A, and give it to system B. This leads to *quantum* cryptography.

Further Reading

See Preskill's lecture notes on quantum information [12].

Exercises

(1) Show the details of proving the formula (33.5) for the Shannon entropy in (33.6).

(2) Show that the mutual information is zero if and only if, in the probability of occurrence of the message x given the message y, $p(x, y)$, the presence of the message y is irrelevant (so we have independence of y).

(3) Show the details of the proof of the concavity of the von Neumann entropy.

(4) Give simple examples of when the von Neumann entropy satisfies $S(A + B) < S(A) + S(B)$, and when it satisfies $S(A + B) = S(A) + S(B)$.

(5) Give an example of a finite quantum circuit made up of a very large number of infinitesimal unitary gates acting on the same two qubits (and calculate the circuit).
(6) Prove the quantum teleportation relation (33.30).
(7) If a quantum computer can factorize a large number in polynomial time, thus breaking banking and internet security encryption based on the same, does a quantum computer mean the end of banking safety? Explain. If not, what problems do you see that need solving?

34 Quantum Complexity and Quantum Chaos

In this chapter we will define the notion of quantum complexity, related to quantum computations, and how to calculate it and its properties. Then, we will describe a quantum version of classical chaos, and the quantities that define it, and we will also show how to calculate them. The properties described here are general for quantum theories, and they help us to understand how to connect to general classical concepts.

34.1 Classical Computation Complexity

Before we consider quantum complexity, we define the easier classical complexity. In computer science, data is encoded in a set of bits, and a computation is an action on (a function of) initial data as bits. This function (the computation) is built out of building blocks called *gates*. A gate is a basic function that acts on several bits, and spits out a result that is one or more other bits. An example is the *NOT* gate, which acts on a single bit, and produces another, namely the opposite bit. Thus we can define the gate as:

$$NOT : 0 \to 1, \quad NOT : 1 \to 0. \tag{34.1}$$

An example of a gate that acts on two bits, and gives one bit is the gate *AND*, which acts as follows:

$$\begin{aligned} AND : \quad & (0,0) \to 0 \\ & (1,0) \to 0 \\ & (0,1) \to 0 \\ & (1,1) \to 1. \end{aligned} \tag{34.2}$$

Another one is the gate *OR*, which acts as follows:

$$\begin{aligned} OR : \quad & (0,0) \to 0 \\ & (1,0) \to 1 \\ & (0,1) \to 1 \\ & (1,1) \to 1. \end{aligned} \tag{34.3}$$

We say that a set of gates is a *universal set* if any general computation, i.e., any function of *all* possible input data, can be built out of a combination of, including repetitions, the set of gates (of course, any particular computation, from one input to one output, can be formed out of a smaller set of gates or even a single gate).

For example, a universal set of gates is $\{NOT, AND, OR,$ and $INPUT\}$, where $INPUT(x_i)$ inputs the ith bit (this allows us to recover the bits lost when acting with other gates which turn two bits into one).

Another example is the set $\{NOT, AND, OR,$ and $XOR\}$, where now we keep the first bit in the result, and the second bit is the result of the computation, thus having only two bits to two-bit gates, where XOR is defined as

$$\begin{aligned} XOR: \quad &(0,0) \to 0 \\ &(1,0) \to 1 \\ &(0,1) \to 1 \\ &(1,1) \to 0. \end{aligned} \tag{34.4}$$

Then we can define the notion of classical complexity, as the minimal number of gates that are needed in order to define a computation.

34.2 Quantum Computation and Complexity

In the quantum case, data is encoded in qubits, meaning as an action on a general qubit state

$$c_0|0\rangle + c_1|1\rangle. \tag{34.5}$$

Similarly to the classical case, we can define a quantum computation as a function of initial data on qubits. But quantum computations are unitary transformations on the state, $|\psi\rangle \to U|\psi\rangle$. A quantum computation can also be built from basic building blocks, i.e., quantum gates, which however now are defined as unitary matrices acting on the qubit.

For example, the quantum analog of the *NOT* gate is the matrix

$$X = \begin{pmatrix} 0 & 1 \\ 1 & 0 \end{pmatrix}. \tag{34.6}$$

Here the matrix acts on a column vector $\begin{pmatrix} |0\rangle \\ |1\rangle \end{pmatrix}$.

Then we can define the *Hadamard gate*,

$$H = \frac{1}{\sqrt{2}} \begin{pmatrix} 1 & 1 \\ 1 & -1 \end{pmatrix}, \tag{34.7}$$

which therefore acts on a general state as follows:

$$H(c_0|0\rangle + c_1|1\rangle) = \frac{1}{\sqrt{2}}(c_0 + c_1)|0\rangle + \frac{1}{\sqrt{2}}(c_0 - c_1)|1\rangle, \tag{34.8}$$

and is thought of as the "square root of the quantum *NOT* gate".

Another gate is the *phase gate*,

$$P = \begin{pmatrix} 1 & 0 \\ 0 & i \end{pmatrix}; \tag{34.9}$$

more generally, $e^{i\phi}$ replaces i.

We can define as in the classical case a universal set of gates as a set of gates into which we can decompose any quantum computation on any initial data. A universal set is composed of the

Hadamard gate, the phase gate, and the "Toffoli gate", defined as the matrix (acting on three qubits, with basis $|000\rangle, |001\rangle, |010\rangle, |011\rangle, |100\rangle, |101\rangle, |110\rangle, |111\rangle$)

$$T = \begin{pmatrix} 1 & 0 & 0 & 0 & 0 & 0 & 0 & 0 \\ 0 & 1 & 0 & 0 & 0 & 0 & 0 & 0 \\ 0 & 0 & 1 & 0 & 0 & 0 & 0 & 0 \\ 0 & 0 & 0 & 1 & 0 & 0 & 0 & 0 \\ 0 & 0 & 0 & 0 & 1 & 0 & 0 & 0 \\ 0 & 0 & 0 & 0 & 0 & 1 & 0 & 0 \\ 0 & 0 & 0 & 0 & 0 & 0 & 0 & 1 \\ 0 & 0 & 0 & 0 & 0 & 0 & 1 & 0 \end{pmatrix}. \tag{34.10}$$

Since quantum computations are merely unitary transformations on states, if we define a given *reference state* $|\psi_R\rangle$, we can define a notion of the quantum complexity of any *state* $|\psi_T\rangle$ as the quantum complexity of the computation from $|\psi_R\rangle$ to $|\psi_T\rangle$, namely the unitary matrix U defining

$$|\psi_T\rangle = U|\psi_R\rangle. \tag{34.11}$$

The minimum number of gates in the universal set that is required in order to build the computation, i.e., the matrix U, is called the quantum complexity.

But there is another catch now, with respect to the classical case. Since unitary transformations are continuous, whereas the product of gates is a discrete operation, we will rarely be able to obtain *exactly* the state $|\psi_T\rangle$ by a product of a finite number of gates. Instead, we must define the equality of the state made from gates with the actual state as being only up to a tolerance ϵ, i.e.,

$$\|\, |\psi_T\rangle - U|\psi_R\rangle\|^2 \leq \epsilon. \tag{34.12}$$

As before, there is no unique circuit that can give the wanted result. Instead, the complexity $C(U)$ is the *minimum* number of gates required to build U up to a desired tolerance, over the set of quantum circuits giving U. Since U is an action on n qubits, it is a $2^n \times 2^n$ matrix. One can show that $C(U) \geq 4^n$ for most cases, putting a lower bound on the complexity.

One can also have an upper bound, given the Solovay–Kitaev theorem, which states that, given a universal set of gates $\mathcal{G}$ and target unitary matrix U, the number N of gates needed to obtain equality up to the tolerance $\epsilon > 0$ is

$$N = \mathcal{O}(\log^c(1/\epsilon)), \quad \text{where} \quad 1 \leq c \leq 2. \tag{34.13}$$

34.3 The Nielsen Approach to Quantum Complexity

There is a geometrical approach to quantum complexity pioneered by M.A. Nielsen in [20], which was generalized to the quantum field theory case in [19].

In it, we can define the quantum unitary matrix U as a path ordered exponential of a "Hamiltonian" $H(s)$, as

$$U = \overleftarrow{P} \exp\left[-i \int_0^1 H(s)ds\right], \tag{34.14}$$

where the "Hamiltonian" is decomposed in the basic building blocks acting on qubits, the generalized Pauli matrices σ_I, understood as tensor products of the Pauli matrices σ_i acting on each qubit:

$$H(s) = \sum_I Y^I(s)\sigma_I. \tag{34.15}$$

Here the function $Y^I(s)$ is called the control function.

We can define things in a bit more generality, by not requiring an action on qubits, but rather on general states (since, in any case, we define complexity in terms of the relation of a general state to a reference state, neither of which *needs* to be in a multiple qubit state). Then we can replace the Pauli matrices σ_I with a general basis $\mathcal{O}_I$ of Hermitian generators. Furthermore we can also define an s-dependent matrix by integrating only up to s,

$$U(s) = \mathcal{P}\exp\left[-i\int_0^s Y^I(s')\mathcal{O}_I ds'\right]. \tag{34.16}$$

It satisfies a Schrödinger equation for evolution in s,

$$i\frac{dU(s)}{ds} = Y^I(s)\mathcal{O}_I U(s), \tag{34.17}$$

and can be solved, with boundary conditions $U(s=0) = \mathbb{1}$ and $U(s=1) = U$.

In order to define a quantum complexity, we need to define some path between these two extremes (i.e., boundary conditions) that can be minimized according to some rules. For this, we define a *cost function*, or *Finsler function*, $\mathcal{F}(Y^I(s), \partial_s Y^I(s))$, that can be used for the minimization. It must satisfy the following conditions:

(1) Positivity: $\mathcal{F} \geq 0$, and $\mathcal{F} = 0 \Leftrightarrow Y^I = 0$.
(2) Continuity and smoothness: $\mathcal{F} \in C^\infty$.
(3) Homogeneity: $F(\lambda Y^I) = \lambda F(Y^I)$.
(4) Triangle inequality: $\mathcal{F}(Y^I + Y^{I'}) \leq \mathcal{F}(Y^I) + \mathcal{F}(Y^{I'})$.

The simplest example is the function F_2,

$$F_2(Y^I) \equiv \sqrt{\sum_{IJ} \delta_{IJ} Y^I (Y^J)^*}. \tag{34.18}$$

It gives rise to a Riemannian geometry on the space of control functions.

Given a cost function, we can define the *distance functional*, or "circuit depth",

$$\mathcal{D}[U] = \int_0^1 ds\, \mathcal{F}(Y^I(s), \partial_s Y^I(s)). \tag{34.19}$$

Then its minimum, $\mathcal{D}_{\min}(U)$ gives the quantum complexity.

Moreover, using a theorem by Nielsen, we find that the complexity is also found from a geodesic in the Riemannian space, with metric

$$g_{\mu\nu}\frac{dx^\mu}{ds}\frac{dx^\nu}{ds} = \sum_I |Y^I(s)|^2. \tag{34.20}$$

In [21], it was found that the quantum complexity of k qubits is related to the entropy of a classical system with 2^k degrees of freedom, and one can find from it a (thermodynamic) second law of quantum complexity.

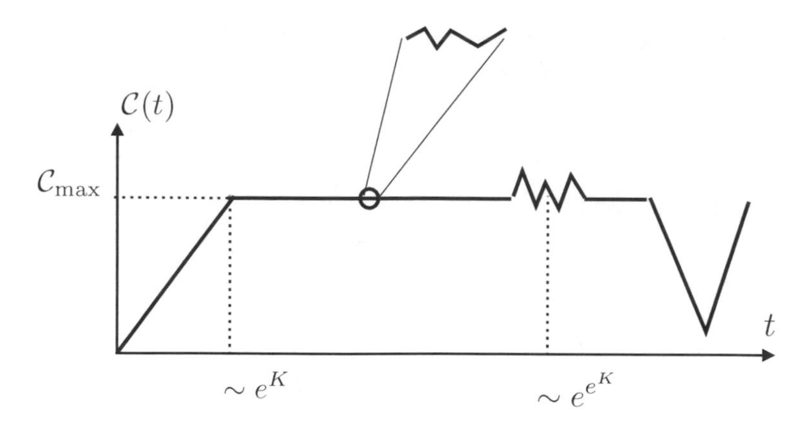

Figure 34.1 General behavior of the complexity $\mathcal{C}(t)$ with time. At e^K there is flattening and at e^{e^K}, quantum recurrences.

For the unitary operator $U(t) = e^{-iHt}$, we can define the time evolution of the complexity $\mathcal{C}(t)$ and find that it increases linearly,

$$\mathcal{C}(t) = Kt, \tag{34.21}$$

until a time of the order of $\sim e^K$ (with K the slope of the linear relation). After that, it remains flat at $\mathcal{C}_{\max}$, with only fluctuations around it. Then, at much larger times, around e^{e^K}, we obtain "quantum recurrences", where the state returns to previously passed values and $\mathcal{C}(t)$ has large downwards fluctuations, until eventually it dips to zero, and then comes back up, as in Fig. 34.1. That this picture is quite general is a conjecture, but there is a lot of evidence for it.

This behavior, for k qubits, is similar to that for the entropy of 2^k degrees of freedom.

The complexity in quantum field theories also has another, more concrete (both more precise and easily calculable) geometrical interpretation, in terms of the relation of quantum field theory to gravity, the AdS/CFT correspondence (or gauge/gravity duality). In it, the complexity can be calculated as an on-shell action for a classical gravity theory, or the volume of a geometrical hypersurface. We will, however, not explain these advanced concepts here.

34.4 Quantum Chaos

Classically, chaotical behavior is defined as the divergence of two nearby paths in phase space according to an exponential law in time, defined by the Lyapunov exponent λ_I,

$$\delta q(t) \sim e^{\lambda_I t}. \tag{34.22}$$

An exact analog for this in quantum mechanics is not well defined.

However, recently a very concrete measure of quantum chaos was proposed, in [18]. It relates to the "out of time ordered correlator (OTOC)".

To define this quantity, consider a state $|\psi\rangle$ and a Hamiltonian $\hat{H}$, for some general quantum system with Hilbert space $\mathcal{H}$. Consider also two operators, V and W, that can be defined as some local perturbations at different positions. For this, suppose that $\mathcal{H}$ defines something like a "spin chain", a discrete set of sites with two-state Hilbert spaces ("spins") at each site, with V and W

acting at different sites, separated by a distance x. Note that this description is equivalent to the description of quantum data in terms of (spatially ordered) qubits.

Now apply V to the state $|\psi\rangle$, evolve in time with $\hat{H}$ for a time t, then apply W and evolve backwards in time for $-t$, generating the state

$$|\psi_1\rangle = e^{iHt/\hbar} W e^{-iHt/\hbar} V |\psi\rangle. \tag{34.23}$$

Alternatively, evolve in time with $\hat{H}$ for a time t, then apply W, then evolve for a time $-t$, then apply V, obtaining

$$|\psi_2\rangle = V e^{iHt/\hbar} W e^{-iHt/\hbar} |\psi\rangle. \tag{34.24}$$

Then the overlap of the two states is the OTOC,

$$F(t) = \langle\psi_2|\psi_1\rangle = \langle\psi| e^{iHt/\hbar} W^\dagger e^{-iHt/\hbar} V^\dagger e^{iHt/\hbar} W e^{-iHt/\hbar} V |\psi\rangle. \tag{34.25}$$

The Heisenberg operator

$$\begin{aligned} W(t) &= e^{iHt/\hbar} W e^{-iHt/\hbar} \\ &= \sum_{n=0}^{\infty} \frac{(it/\hbar)^n}{n!} [H, \ldots [H, W] \ldots] \end{aligned} \tag{34.26}$$

is defined as an analog of the classical butterfly effect, when we go forwards in time, then act, and then go backwards in time, or, equivalently, as a sum of arbitrary nested commutators. Then if the operator W is local (defined at a single site) initially, for $W(t)$ it amounts to a "spreading out": the operator acts over a distance that increases in time around the initial site.

In terms of $W(t)$, the OTOC is

$$F(t) = \langle W^\dagger(t) V^\dagger W(t) V \rangle. \tag{34.27}$$

Note that here the spin chain gives another way to think of a qubit model for the data in a quantum computation. In that case, V and W are expressed as sums of products of basis operators, which are Pauli matrices σ_I.

Another relevant quantity for quantum chaos is the expectation value of the commutator of $W(t)$ and V, a more usual kind of correlator,

$$\begin{aligned} C(t) &= \langle [W(t), V]^\dagger [W(t), V] \rangle \\ &= \langle\psi| \left(V^\dagger W(t)^\dagger W(t) V + W(t)^\dagger V^\dagger V W(t) - V^\dagger W(t)^\dagger V W(t) - W(t)^\dagger V^\dagger W(t) V \right) |\psi\rangle. \end{aligned} \tag{34.28}$$

Then note that, *for unitary* V, W ($V^\dagger = V^{-1}, W^\dagger = W^{-1}$), we obtain

$$C(t) = 2(1 - \text{Re}\, F(t)). \tag{34.29}$$

We can also, more relevantly, generalize the above to finite temperature T, with $\beta = 1/(k_B T)$. Then the expectation value is defined using the thermal density matrix,

$$\langle \ldots \rangle = \frac{1}{Z} \text{Tr}(e^{-\beta H} \ldots). \tag{34.30}$$

For a chaotic system and two local operators V and W, the spread in time of the action in $W(t)$ is ballistic (unimpeded), with a "butterfly velocity" v_B which can be thought of as a kind of "emergent speed of information propagation". Then, if x denotes the distance between the points at which V and W act, we have, at fixed t,

$$\begin{aligned} F(t) &\simeq 1, \quad \text{for} \quad x \gg v_B t \\ F(t) &\simeq 0, \quad \text{for} \quad x \ll v_B t. \end{aligned} \tag{34.31}$$

On the other hand, as a function of time, $C(t) \simeq 0$ at small times and $C(t) \simeq 2$ at large times. This means that the correlation of the operators at large times is perfect, and at small times is negligible. Moreover, for the butterfly effect, we have

$$C(t) \sim 2\langle VV\rangle\langle WW\rangle \tag{34.32}$$

at large t.

For a local chaotic system, $C(t)$ becomes of order 1 at the so-called "scrambling (delocalization) time" t_*,

$$C(t) \sim 1 \quad \text{for} \quad t \sim t_*. \tag{34.33}$$

In terms of $F(t)$, this corresponds to the transition between 1 and 0, so we obtain

$$t_* \sim \frac{x}{v_B}. \tag{34.34}$$

There is another relevant time scale, the much smaller "dissipation time" t_d, (or "collision time" if there is a quasiparticle picture for dissipation through collisions), where $t_d \ll t_*$. At small times, but larger than t_d, the two-point function of operators decays exponentially, so

$$\langle V(0)V(t)\rangle \sim e^{-t/t_d}. \tag{34.35}$$

We also have

$$\langle V(0)V(0)W(t)W(t)\rangle \sim \langle VV\rangle\langle WW\rangle + \mathcal{O}(e^{-t/t_d}). \tag{34.36}$$

For strongly coupled systems, the dissipation time is $t_d \sim \hbar\beta = \hbar/k_B T$.

In systems at nonzero temperature T and with *large* number of degrees of freedom, we obtain

$$F(t) = 1 - \epsilon e^{\lambda_L t}, \tag{34.37}$$

where λ_L is called the *(quantum) Lyapunov exponent, defining the quantum counterpart to the classical chaotic behavior*. Both it and ϵ depend on the system. From the relation between $C(t)$ and the OTOC $F(t)$, we see that

$$t_* \sim \frac{1}{\lambda_L}. \tag{34.38}$$

The Lyapunov exponent is conjectured [18] to satisfy (for most quantum systems, at least) a bound,

$$\lambda_L \leq \frac{2\pi}{\hbar\beta} = \frac{2\pi k_B T}{\hbar}. \tag{34.39}$$

The bound is saturated for black holes.

Moreover, there is a conjectured form of the spatial spread in $C(t)$ [22],

$$C_x(t) \equiv C(x,t) \sim \exp\left[-\lambda_L \frac{(x - v_B t)^{1+p}}{c^{1+p} t^p}\right], \tag{34.40}$$

where, in $C_x(t)$, W acts at $x = 0$ and $V = V_x$ acts at x, and p is a universality class parameter that varies between 0 and 1, at least in known cases.

Important Concepts to Remember

- The classical computation complexity is the minimal number of gates necessary to do a computation.
- The quantum complexity is the minimal number of quantum gates in a universal set necessary to move from a given reference state $|\psi_R\rangle$ to the state we want, $|\psi_T\rangle$, as $|\psi_T\rangle = U|\psi_R\rangle$, more precisely up to a tolerance ϵ in norm.
- In the case of n qubits, U is $2^n \times 2^n$, so $\mathcal{C}(U) \geq 4^n$; for a tolerance ϵ, the number of gates is $\mathcal{O}(\log^c(1/\epsilon))$, with $1 \leq c \leq 2$.
- In Nielsen's approach to quantum computation, we build path-dependent unitary matrices as $U(s) = P\exp\left[-i\int_0^s Y^I(s')\mathcal{O}_I ds'\right]$, with $\mathcal{O}_I$ a basis of Hermitian generators and $Y^I(s)$ the control function making up the "Hamiltonian" $H(s) = \sum_I Y^I(s)\mathcal{O}_I$ for a Schrödinger equation, with boundary conditions $U(0) = \mathbb{1}$ and $U(1) = U$.
- One minimizes the integral over s of a cost (Finsler) function $\mathcal{F}(Y^I, \partial_s Y^I)$, called a distance functional or circuit depth, to give the quantum complexity at the minimum.
- The quantum complexity increases linearly with time, $\mathcal{C} = Kt$, until a time $t \sim e^K$, then it becomes flat, and then at e^{e^K} we have quantum recurrences.
- The measure of classical chaos is the Lyapunov exponent, from $\delta q(t) \sim e^{\lambda_I t}$, whereas the measure of quantum chaos comes from the out of time ordered correlator (OTOC), $F(t) \equiv \langle\psi|e^{iHt/\hbar}W^\dagger e^{-iHt/\hbar}V^\dagger e^{iHt/\hbar}We^{-iHt/\hbar}V|\psi\rangle$ and from $C(t) \equiv \langle\psi|[W(t), V]^\dagger[W(t), V]|\psi\rangle$, with $W(t) = e^{iHt/\hbar}We^{-iHt/\hbar}$. For unitary V, W, $C(t) = 2(1 - \mathrm{Re}\, F(t))$.
- For a chaotic system, the spread in time of the action of $W(t)$ is very fast, with "butterfly velocity" v_B (the emergent speed of information propagation), such that the OTOC is $F(t) \simeq 1$ for $x \ll v_B t$, and $F(t) \simeq 0$ for $x \gg v_B t$.
- Also for a chaotic system, $C(t)$ factorizes at large times, $C(t) \simeq 2\langle VV\rangle\langle WW\rangle$, and becomes (from 0 at $t = 0$) of order 1 at the scrambling time $t_* \sim x/v_B$.
- For a chaotic system at nonzero temperature T and with a large number of degrees of freedom, the OTOC $F(t) = 1 - \epsilon e^{\lambda_L t}$, with $\lambda_L \sim 1/t_*$ the quantum Lyapunov exponent. It is conjectured to satisfy the bound $\lambda_L \leq 2\pi k_B T/\hbar$, which is saturated by black holes.

Further Reading

For classical and quantum complexity, see Preskill's lecture notes on quantum information [12], as well as [19, 20, 21]. For quantum chaos, see [18, 22].

Exercises

(1) Consider the following classical computation on six-qubit space, $(1,0,1,0,1,0) \to (0,0,1,1,1,1)$. Choose your favorite universal set of gates, and find two examples of circuits (products of gates) giving this computation. Therefore deduce a bound on the complexity of the computation.

(2) Consider the following quantum computation on five qubit space, from reference state $|\psi_R\rangle$ to final state $|\psi_T\rangle$,

$$|\psi_R\rangle = \left(\frac{|0\rangle + |1\rangle}{\sqrt{2}}, \frac{|0\rangle + |1\rangle}{\sqrt{2}}, \frac{|0\rangle + |1\rangle}{\sqrt{2}} \frac{|0\rangle + |1\rangle}{\sqrt{2}}, \frac{|0\rangle + |1\rangle}{\sqrt{2}}\right) \to$$
$$|\psi_T\rangle = \left(\frac{|0\rangle - |1\rangle}{\sqrt{2}}, \frac{|0\rangle + |1\rangle}{\sqrt{2}}, \frac{|0\rangle - |1\rangle}{\sqrt{2}}, \frac{|0\rangle + |1\rangle}{\sqrt{2}}, \frac{|0\rangle - |1\rangle}{\sqrt{2}}\right), \tag{34.41}$$

with tolerance $\epsilon = 1/10$. Using the universal set of gates in the text, find one example of a circuit (products of gates) giving this computation, and therefore deduce a bound on the complexity of the computation.

(3) Consider a quantum computation on four-qubit space in the Nielsen approach. Find a basis $\mathcal{O}_I$ for the computation, and write down explicitly the Schrödinger equation for $U(s)$ in this basis.

(4) Consider the function $F(Y^I) = \sum_I |Y^I|$. Is it a good cost function (Finsler function)? Why?

(5) Consider a Heisenberg spin chain Hamiltonian, $H = -J\sum_i \vec{\sigma}_i \cdot \vec{\sigma}_{i+1}$, where $\vec{\sigma}_i$ are the Pauli matrices at site i on a spin chain or, equivalently, on the data for a quantum computation, and the operator $W = \sigma_k^1$, where k is a fixed site. Show that, indeed, $W(t)$ spreads along the spin chain away from k as time evolves.

(6) Is there a classical limit of the measure of quantum chaos described here? How would you compare the classical and quantum chaos descriptions?

(7) The conjectured bound (34.39) implies that at zero temperature $\lambda_L \to 0$. Does that imply that the scrambling time $t_* \to \infty$? Explain.

35 Quantum Decoherence and Quantum Thermalization

In this chapter, we will study the transition from quantum to classical, i.e., classicalization. Usually, when talking about a system interacting with an environment, we talk about decoherence, which will be studied first. In the case of an isolated system, we can also talk about decoherence since the mechanism is the same, but more commonly it is called thermalization, and as such one considers a separate analysis, which we will study next. Finally, we will describe a Bogoliubov transformation on a (system of) harmonic oscillator(s) that describes particle production. We can obtain a thermal particle production, and thus a finite temperature, in certain physically relevant cases.

35.1 Decoherence

In quantum mechanics, an obvious question that should have been asked from the beginning is: how do we transition to a classical state? Even if it is in terms of just increasing the system size it is a relevant question, since we mostly observe classical systems but, at the microscopic level, everything is quantum. However, the transition from a quantum state to a classical state is, for an isolated system, not a unitary transformation, so we need some other ingredient to describe it. A related issue is the question of measurement: in it, there is an interaction of a quantum system with a classical observer or apparatus, which changes the state of the quantum system in a nonunitary way. For the measurement issue, however, as we saw, this is connected with the interpretation of quantum mechanics. Since this is a controversial issue, we will not delve into it here even though decoherence theory has something to say about it.

The minimalist point of view, then, is to restrict ourselves to a description of the interaction of a system $\mathcal{S}$ with an environment $\mathcal{E}$ which will lead to *decoherence* of the state of the system $\mathcal{S}$, effectively leading to classical states. Even so, the field was regarded as somewhat disreputable, in as much as it was perceived as dealing with mostly philosophical issues (a viewpoint perhaps influenced by the connection with the interpretation of quantum mechanics alluded to above). This was the case until decoherence was experimentally verified, in particular by the work of French physicist Serge Haroche, and others, in 1996 and afterwards. As a result, Haroche gained a Nobel prize in 2012, for the "experimental manipulation of individual quantum systems".

35.2 Schrödinger's Cat

Erwin Schrödinger famously proposed a Gedanken experiment (thought experiment), the Schrödinger's cat paradox, to sharpen the problem with the interaction between classical and quantum, and (as he thought) with the standard Copenhagen interpretation of quantum mechanics. Consider an isolated system, a "black box", inside which we have a quantum trigger, for instance a Geiger counter, triggered by an individual radioactive decay, giving a classical effect, that of producing a killing procedure (spilling some poison, or shooting a pistol, etc.) at some time t on a cat inside the same box. If the quantum state of the radioactive atom is described as (since the atom decay has some probability per unit time, which can be integrated over some time to give 1/2)

$$|atom\rangle = \frac{1}{\sqrt{2}}(|decay\rangle + |no\ decay\rangle), \tag{35.1}$$

then it seems to imply that the quantum state of the Schrödinger cat is

$$|S.cat\rangle = \frac{1}{\sqrt{2}}(|dead\rangle + |alive\rangle). \tag{35.2}$$

This is almost certainly paradoxical, since a cat is either dead, or alive, but not in between. A question arises: why did Schrödinger consider a cat in the thought experiment? One answer is that the cat is macroscopic (so naturally classical) and alive. But is it an observer? If an observer is something related to consciousness, the experiment sharpens the issue, as the cat is not human but most people would assume it has consciousness. Schrödinger's point was that this question, in his view, makes the standard Copenhagen interpretation ludicrous.

But there is an obvious problem with the set-up of the Gedanken experiment: there are nonlocal correlations over macroscopic distances, with classical interactions in between the two points.

The solution to this issue is a solution to the paradox: we must eliminate these nonlocal "cat states" from the theory by decoherence, the loss of quantum coherence, which arises due to the interaction of the quantum state with the "environment" inside the box, meaning the classical cat and killing apparatus.

The issue of elimination of cat states through decoherence is also relevant for quantum computations. In a quantum computation, there are delicate superpositions of states of a large quantum system, the system of many qubits of information. As such, the quantum state will decay very rapidly (like a cat state of the macroscopic cat) through decoherence. Then the quantum information would be lost quickly. However, as we mentioned earlier, there are quantum error-correcting codes that delay decoherence for a long time.

35.3 Qualitative Decoherence

We first present the qualitative picture of decoherence, involving several points, as below.

1. What happens, very fast, is that a correlation between the $|S.cat\rangle$ state and the environment forms, in such a way that the $|S.cat\rangle$ state becomes inaccessible. We can say that, in effect, "the environment measures the cat", collapsing its state to a $|dead\rangle$ or $|alive\rangle$ one.

2. Then decoherence represents a (unitary, from the point of view of a larger system, of which our system is a subsystem) transition to entangled system + environment states.

3. A general state $|\psi\rangle$ of the quantum system can be expanded in different bases, such that the various basis elements interact with the environment in element-specific ways. Then the (unitary) evolution leads to a situation with no interaction between basis elements. That means that we lose phase coherence between these elements, i.e., we "decohere".

4. The environment selects only a particular basis of elements of the quantum system, one that interacts with the environment in a canonical way. We have then an "environment-induced superselection", or *Einselection*.

The theory of decoherence was initiated by H.D. Zeh in 1970 and developed by many people, though notably among them W.H. Zurek.

35.4 Quantitative Decoherence

Toy Model

To start making decoherence more quantitative, we will start with a three-qubit model, one qubit for each of the three elements, the system $\mathcal{S}$, the environment $\mathcal{E}$ and the apparatus $\mathcal{A}$. The two states of the system are denoted by $|\uparrow\rangle$ and $|\downarrow\rangle$, of the apparatus by $|A_0\rangle$ and $|A_1\rangle$, and of the environment by $|\epsilon_0\rangle$ and $|\epsilon_1\rangle$.

Consider a measurement, which is thought of as an interaction between the system and the apparatus, that leads to a (unitary) change of the total quantum system, with Hilbert space $\mathcal{H}_\mathcal{S} \otimes \mathcal{H}_\mathcal{A}$. Specifically, we start with the apparatus in the $|A_0\rangle$ state, so we have

$$\begin{aligned} |\uparrow\rangle \otimes |A_0\rangle &\rightarrow |\uparrow\rangle \otimes |A_1\rangle \\ |\downarrow\rangle \otimes |A_0\rangle &\rightarrow |\downarrow\rangle \otimes |A_0\rangle. \end{aligned} \tag{35.3}$$

Here the apparatus states are orthogonal, $\langle A_0|A_1\rangle = 0$.

On a general quantum state, the measurement means the evolution

$$(\alpha|\uparrow\rangle + \beta|\downarrow\rangle) \otimes |A_0\rangle \rightarrow \alpha|\uparrow\rangle \otimes |A_1\rangle + \beta|\downarrow\rangle \otimes |A_0\rangle \equiv |\Phi\rangle. \tag{35.4}$$

Then the final state (after the measurement) $|\Phi\rangle$ can be expanded in different bases, but the environment has not entered the discussion yet, so neither basis is preferred.

Next we must do a premeasurement, which is an evolution of the system, apparatus, and environment that aligns all of them,

$$\left(\alpha|\uparrow\rangle \otimes |A_1\rangle + \beta|\downarrow\rangle \otimes |A_0\rangle\right) \otimes |\epsilon_0\rangle \rightarrow \alpha|\uparrow\rangle \otimes |A_1\rangle \otimes |\epsilon_1\rangle + \beta|\downarrow\rangle \otimes |A_0\rangle \otimes |\epsilon_0\rangle = |\psi\rangle. \tag{35.5}$$

An important assumption of the decoherence model is

$$\langle \epsilon_0|\epsilon_1\rangle = 0, \tag{35.6}$$

which means that the orthogonal states of the environment are correlated with the apparatus in the basis in which the premeasurement was carried out.

Then the interaction with the environment, when we take the trace over the environment (since we don't know it), leads to a transition from a pure state to a mixed state that looks classical, *if the above condition holds*, eliminating interference terms.

At the beginning, the state in $\mathcal{S} \otimes \mathcal{A}$ is pure, as is the state in $\mathcal{E}$, so taking the trace over $\mathcal{E}$ is trivial, and does not change the pure state of the system. But after the premeasurement, we obtain a (reduced) density matrix

$$\rho_{\mathcal{AS}} = \mathrm{Tr}_\mathcal{E}\, |\psi\rangle\langle\psi| = |\alpha|^2 |\uparrow\rangle\langle\uparrow| \otimes |A_1\rangle\langle A_1| + |\beta|^2 |\downarrow\rangle\langle\downarrow| \otimes |A_0\rangle\langle A_0|, \tag{35.7}$$

which is a classical state, since it means that with probability $|\alpha|^2$ the apparatus shows 1 and the system is in the state $|\uparrow\rangle$ and with probability $|\beta|^2$ the apparatus shows 0 and the system is in the state $|\downarrow\rangle$, and the probabilities add up without quantum interference terms.

General Model

We can consider next a more general model. We define an "einselected" basis, namely an environment-induced selected basis $|i\rangle$. Expand a general state of the system $\mathcal{S}$ in it,

$$|\psi\rangle = \sum_i |i\rangle\langle i|\psi\rangle. \tag{35.8}$$

Start with the environment $\mathcal{E}$ in the state $|\epsilon\rangle$, so with a state for the total $\mathcal{S}$–$\mathcal{E}$ system of

$$|\psi\rangle \otimes |\epsilon\rangle = \sum_i |i\rangle \otimes |\epsilon\rangle\langle i|\psi\rangle. \tag{35.9}$$

Then the decoherence is a continuous phenomenon, on a scale between: total absorption of the quantum state by the system (as when a photon is absorbed by a cavity in which it resonates); and the case when the system is not disturbed at all by the environment (though the environment is changed by the system).

(1) Considering first the case of total absorption by the environment, this corresponds to an evolution of the special (einselected) basis elements as

$$|i\rangle \otimes |\epsilon\rangle \rightarrow |\epsilon_i\rangle, \tag{35.10}$$

leading to an evolution of a general state as

$$|\psi\rangle \otimes |\epsilon\rangle \rightarrow \sum_i |\epsilon_i\rangle\langle i|\psi\rangle. \tag{35.11}$$

In order to have unitary evolution for the total $\mathcal{S} - \mathcal{E}$ system, thus to preserve probability and thus state products, we must have

$$\langle\epsilon_i|\epsilon_j\rangle = \delta_{ij}, \tag{35.12}$$

since the einselected basis is orthonormal, $\langle i|j\rangle = \delta_{ij}$ and $\langle\epsilon|\epsilon\rangle = 1$.

(2) Considering next the case where the system is not disturbed by the environment, the evolution changes the environment by correlating its states with the einselected basis. Thus the action on the einselected basis is

$$|i\rangle \otimes |\epsilon\rangle \rightarrow |i\rangle \otimes |\epsilon_i\rangle, \tag{35.13}$$

leading to an action on a general state as

$$|\psi\rangle \otimes |\epsilon\rangle \rightarrow \sum_i |i\rangle \otimes |\epsilon_i\rangle\langle i|\psi\rangle. \tag{35.14}$$

In this case, unitarity requires again the conservation of probabilities, thus of state products, so

$$\begin{aligned}\langle i|j\rangle \otimes \langle\epsilon_i|\epsilon_j\rangle &= \delta_{ij}\langle\epsilon_i|\epsilon_j\rangle \\ &= \delta_{ij}.\end{aligned} \tag{35.15}$$

But if we also require *decoherence*, meaning einselection in this case, we need

$$\langle\epsilon_i|\epsilon_j\rangle = \delta_{ij}. \tag{35.16}$$

To consider the effect of decoherence, we use the density matrix formalism. Before the premeasurement, we start in the pure state $|\psi\rangle \otimes |\epsilon\rangle$, which means the density matrix

$$\rho_0 = |\psi\rangle\langle\psi| \otimes |\epsilon\rangle\langle\epsilon|. \tag{35.17}$$

Then the reduced matrix for the system $\mathcal{S}$ only is

$$\rho_{0,\mathcal{S}} = \text{Tr}_{\mathcal{E}}\, \rho_0 = |\psi\rangle\langle\psi|, \tag{35.18}$$

which indeed represents a pure state.

In this case, the expectation value in the $|\phi\rangle$ state, meaning the probability for being in the $|\phi\rangle$ state, is

$$\begin{aligned} \langle\phi|\rho_{0,\mathcal{S}}|\phi\rangle &= |\langle\psi|\phi\rangle|^2 \\ &= \sum_i |\langle\psi|i\rangle\langle i|\phi\rangle|^2 \\ &\quad + \sum_{i,j;i\neq j} \langle\psi|i\rangle\langle j|\phi\rangle\langle i|\phi\rangle\langle\psi|j\rangle. \end{aligned} \tag{35.19}$$

The last line contains the interference terms characteristic of a quantum state.

After the decoherence (evolution), we end up with a density matrix

$$\begin{aligned} \rho_{\mathcal{S}} = \text{Tr}_{\mathcal{E}}\, \rho &= \sum_{i,j,k} |i\rangle\langle j| \otimes \langle\epsilon_k|\epsilon_i\rangle\langle\epsilon_j|\epsilon_k\rangle\langle i|\psi\rangle\langle\psi|j\rangle \\ &= \sum_i |\langle\psi|i\rangle|^2 |i\rangle\langle i|, \end{aligned} \tag{35.20}$$

which instead represents a mixed state, of a classical nature. Indeed, the expectation value in a $|\phi\rangle$ state (and so the probability for the system to go to the $|\phi\rangle$ state) is

$$\langle\phi|\rho_{\mathcal{S}}|\phi\rangle = \sum_i |\langle\psi|i\rangle\langle i|\phi\rangle|^2, \tag{35.21}$$

without the interference terms, so this is a classical state.

35.5 Qualitative Thermalization

We next move to the issue of thermalization, the evolution of a large isolated quantum system to a thermal state, i.e., a classical statistical ensemble; we describe it qualitatively first. An essential factor is the large size of the system, in the thermodynamic limit, and the interaction between subsystems. In thermalization, the end result is the approach to the Gibbs ensemble, characterized by the classical density matrix

$$\hat{\rho}_A = \frac{e^{-\beta\hat{H}}}{Z}. \tag{35.22}$$

The effective temperature at the end of thermalization is calculated such that the average energy before and after thermalization is the same. Before we have a regular quantum average in the initial state $|\psi(0)\rangle$ and afterwards we have a quantum average with the Gibbs ensemble, so

$$\langle\psi(0)|\hat{H}|\psi(0)\rangle = \text{Tr}(\hat{H}e^{-\beta\hat{H}}). \tag{35.23}$$

In classical statistical mechanics, reaching thermodynamic equilibrium is guaranteed by the ergodic hypothesis, which says that we get arbitrarily close to all the points in the phase space (although as we noted before, it is strictly speaking – mathematically – not always correct). The latter is a consequence of the classical chaos of the system, since considering small variations in the initial conditions means that we map most of the phase space. Classical thermalization then means that an atypical initial state of the system goes over to the thermal ensemble at large times, $t \to \infty$.

In quantum mechanics, instead of the ergodic hypothesis we have the *eigenstate thermalization hypothesis (ETH)*, proposed independently by J.M. Deutch in 1991 and M. Srednicki in 1994, the latter being the standard version.

The hypothesis says that the expectation value of a few-body observable $\hat{A}$, $\langle m|\hat{A}|m\rangle$, in an eigenstate $|m\rangle$ of the Hamiltonian with energy E_m, in the case of a large, interacting, many-body system equals the thermal (quasi-)microcanonical average, $\langle A\rangle_{\text{microcanonical}}$ or (E_m), and so depends on the mean energy E_m.

This ETH is a hypothesis, meaning that it is not always satisfied. In particular, a system with "many body localization" (MBL) does not satisfy it [24].

Also, integrable systems do not satisfy it, since these are systems with an infinite number of conservation laws. These are analogs of non-chaotic systems in the classical version, so we expect non-ergodic behavior. However, these systems are described by a generalized Gibbs ensemble (GGE), in which there is a combination of conserved quantities in the exponent, $\exp\left[-\sum_i \beta_i P_i\right]$, not just the energy E (as in $e^{-\beta E} \equiv e^{-E/k_B T}$).

The bottom line of the ETH is that the knowledge of a *single* many-body eigenstate is sufficient to compute thermal averages: it can be any eigenstate in the (quasi-)microcanonical ensemble.

35.6 Quantitative Thermalization

We can make a more quantitative analysis as follows. Consider the initial (many-body) state expanded in the $|n\rangle$ eigenbasis for the Hamiltonian $\hat{H}$,

$$|\psi(0)\rangle = \sum_m C_m |m\rangle. \tag{35.24}$$

Its time evolution is

$$|\psi(t)\rangle = e^{-i\hat{H}t/\hbar}|\psi(0)\rangle = \sum_m C_m e^{-iE_m t/\hbar}|m\rangle. \tag{35.25}$$

The expectation value of an observable $\hat{A}$ in the time-dependent state is

$$\begin{aligned}\langle \hat{A}(t)\rangle = \langle\psi(t)|\hat{A}|\psi(t)\rangle &= \sum_{m,n} C_m^* C_n e^{i(E_m - E_n)t/\hbar} A_{mn} \\ &= \sum_m |C_m|^2 A_{mm} + \sum_{m\neq n} C_m^* C_n e^{i(E_n - E_m)t/\hbar} A_{mn},\end{aligned} \tag{35.26}$$

where $A_{mn} \equiv \langle m|\hat{A}|n\rangle$.

Then the eigenstate thermalization hypothesis (ETH) states that the time average equals the average in the diagonal ensemble,

$$\bar{A} = \lim_{T\to\infty} \frac{\int_0^T dt\, A(t)}{T} = \sum_m |C_m|^2 A_{mm} = \text{Tr}(\rho_d A), \tag{35.27}$$

where ρ_d is the diagonal ensemble. For an energy shell in the microcanonical ensemble we have

$$\langle E^2\rangle_{\psi(0)} - \langle E\rangle^2_{\psi(0)} \simeq 0. \tag{35.28}$$

More precisely, in order for (35.27) to hold, we have the proposed relation

$$A_{mn} = A(E)\delta_{mn} + e^{-S(\bar{E})/2} f_A(\bar{E}, \omega) R_{mn}, \tag{35.29}$$

where $\bar{E} = (E_m + E_n)/2$, $\omega = (E_n - E_m)/2\hbar$, $S(E)$ is the entropy, f_A is an arbitrary function, and R_{mn} is a random variable with a zero mean and unit variance. This relation was formalized by Srednicki.

Then the diagonal ensemble ρ_d is a generalized Gibbs ensemble (GGE) if we take *all* projection operators $\hat{P}_m = |m\rangle\langle m|$ as quantum operators corresponding to classical integrals of motion. Indeed, we find

$$\hat{\rho}_d = \exp\left[-\sum_{m=1}^{D} \lambda_m \hat{P}_m\right], \tag{35.30}$$

where D is the dimension of the Hilbert space and

$$\lambda_m = -\ln(|C_m|^2). \tag{35.31}$$

Indeed, to prove that this gives ρ_d, we expand the exponential as a sum of powers, and use $P_m P_m = |m\rangle\langle m|m\rangle\langle m| = P_m$. We find

$$\hat{\rho}_d = \prod_{m=1}^{D} \left[1 + (e^{-\lambda_m} - 1)\hat{P}_m\right] = \prod_{m=1}^{D} \left[\mathbb{1} + \left(|C_m|^2 - 1\right) P_m\right] = \sum_{m=1}^{D} \left(|C_m|^2 |m\rangle\langle m|\right), \tag{35.32}$$

where in the last equality we used $P_m P_n = 0$ for $m \neq n$, and $\sum_{m=1}^{D} P_m = \mathbb{1}$.

As an example of the ETH, Srednicki used Berry's conjecture to prove (something like) ETH in this case. Berry's conjecture says that the energy eigenfunctions are superpositions of plane waves in momentum space, with random phase and Gaussian random amplitudes.

Expanding the energy eigenfunction in position space in terms of the momentum space,

$$\psi_n(\vec{x}) \equiv \langle \vec{x}|n\rangle = \int d^{3N}p A_n(\vec{P}) e^{i\vec{P}\cdot\vec{X}/\hbar} \delta(\vec{p}^{\,2} - 2mE_n), \tag{35.33}$$

where $A_n(\vec{P}) \equiv \langle \vec{P}|n\rangle$ are Gaussian random variables, meaning that their two-point correlation in the eigenstate ensemble (EE) is

$$\langle A_m(\vec{P}) A_n(\vec{P}')\rangle_{EE} = \delta_{mn} \frac{\delta^{3N}(\vec{P} + \vec{P}')}{\delta(\vec{P}^2 - \vec{P}'^2)}. \tag{35.34}$$

Then, defining the integral over all momenta except one,

$$\phi_{mn}(\vec{p}_1) = \int d^2 p_2 \cdots d^3 p_n \psi^*_m(\vec{P}) \psi_n(\vec{P}), \tag{35.35}$$

we find that the diagonal elements, averaged in the eigenstate ensemble, give the Boltzmann thermalized distribution,

$$\langle \phi_{nn}(\vec{p}_1)\rangle_{EE} = (2\pi m k_B T_n)^{-3/2} \exp\left(-\frac{\vec{p}_1^{\,2}}{2mk_B T_n}\right), \tag{35.36}$$

and the variance for it, $\langle \Delta\phi_{mm}\rangle$ goes to zero as $N \to \infty$. Thus we have found that the ETH is obtained from the Berry conjecture.

35.7 Bogoliubov Transformation and Appearance of Temperature

There is another way to obtain a thermal quantum distribution from a $T = 0$ distribution. It is found in black holes, accelerated mirrors giving "Rindler spaces", and the expanding universe cosmology, all of which generate a temperature from the existence of "horizons" that cannot be penetrated by information.

The above-mentioned way is a *Bogoliubov transformation*, which is a transformation

$$\begin{aligned} b &= \alpha a + \beta a^\dagger \quad \Rightarrow \\ b^\dagger &= \beta^* a + \alpha^* a^\dagger, \end{aligned} \tag{35.37}$$

on the operators $a, a^\dagger$ of a harmonic oscillator, thus redefining the oscillator and, most importantly, its states. One must impose the normalization condition

$$|\alpha|^2 - |\beta|^2 = 1, \tag{35.38}$$

which means that we can find the inverse transformation

$$a = \alpha^* b - \beta b^\dagger. \tag{35.39}$$

Moreover, then the commutation relations for the $a, a^\dagger$ operators,

$$[a, a] = [a^\dagger, a^\dagger] = 0, \quad [a, a^\dagger] = 1, \tag{35.40}$$

are preserved,

$$\begin{aligned} [b, b] &= [\alpha a + \beta a^\dagger, \alpha a + \beta a^\dagger] = 0 \\ [b^\dagger, b^\dagger] &= [\beta^* a + \alpha^* a^\dagger, \beta^* a + \alpha^* a^\dagger] = 0 \\ [b, b^\dagger] &= [\alpha a + \beta a^\dagger, \beta^* a + \alpha^* a^\dagger] = |\alpha|^2 - |\beta|^2 = 1. \end{aligned} \tag{35.41}$$

The Hamiltonian

$$H = \frac{aa^\dagger + a^\dagger a}{2} \tag{35.42}$$

is not preserved, though, since

$$\frac{bb^\dagger + b^\dagger b}{2} = \alpha\beta^*(a)^2 + \alpha^*\beta(a^\dagger)^2 + (|\alpha|^2 + |\beta|^2)\frac{aa^\dagger + a^\dagger a}{2}, \tag{35.43}$$

which only equals H for $\beta = 0$.

Defining the vacuum state $|0\rangle_a$ of the original oscillator as being annihilated by a,

$$a|0\rangle_a = 0, \tag{35.44}$$

in terms of the redefined oscillator it is nontrivial,

$$b|0\rangle_a \neq 0. \tag{35.45}$$

Moreover, the new vacuum state, defined as before,

$$b|0\rangle_b = 0, \tag{35.46}$$

is a nontrivial function of the original variables, a coherent-type state,

$$|0\rangle_b = \exp\left[-\frac{\beta}{\alpha}(a^\dagger)^2\right]|0\rangle_a. \tag{35.47}$$

Indeed,

$$b|0\rangle_b = \left(\alpha a + \beta a^\dagger\right)\exp\left[-\frac{\beta}{\alpha}(a^\dagger)^2\right]|0\rangle_a, \tag{35.48}$$

but we have the commutator

$$\left\{a, \exp\left[-\frac{\beta}{\alpha}(a^\dagger)^2\right]\right\} = \sum_{n=0}^{\infty}\frac{n}{n!}\left(-\frac{\beta}{\alpha}\right)^n a^\dagger\left(a^{\dagger 2}\right)^n = -\frac{\beta}{\alpha}\exp\left[-\frac{\beta}{\alpha}(a^\dagger)^2\right], \tag{35.49}$$

where we have expanded the exponential as an infinite sum of powers, and then commuted. Then we obtain

$$b|0\rangle_b = \exp\left[-\frac{\beta}{\alpha}(a^\dagger)^2\right](\alpha a + \beta a^\dagger)0\rangle_a + \alpha\exp\left[-\frac{\beta}{\alpha}(a^\dagger)^2\right]\left(-\frac{\beta}{\alpha}\right)a^\dagger|0\rangle_a = 0, \tag{35.50}$$

where we have used $a|0\rangle_a = 0$.

Finally, then, considering the vacuum state of the original oscillator, $|0\rangle_a$, propagate it in time, and use it to evaluate the number operator of the redefined oscillator in this state. We find

$${}_a\langle 0|b^\dagger b|0\rangle_a = {}_a\langle 0|(\beta^* a + \alpha^* a^\dagger)(\alpha a + \beta a^\dagger)|0\rangle_a = |\beta|^2{}_a\langle 0|a\, a^\dagger|0\rangle_a = |\beta|^2. \tag{35.51}$$

The interpretation of this is that there is a number $|\beta|^2$ of b particles in the state, i.e., the vacuum state for a, a phenomenon known as *particle creation*.

The particles thus created can be distributed thermally in a specific case, relevant to black holes, accelerated mirrors, and cosmologies (all having horizons, "walls" that cannot be penetrated by information, and all implying a nonzero temperature).

We consider a "field" (the calculation is more sensible in full quantum field theory) that is a sum of harmonic oscillators i with some nontrivial functions for the coefficients f_i and f_i^*,

$$\phi = \sum_i (f_i a_i + f_i^* a_i^\dagger), \tag{35.52}$$

where the expansion is valid in some region of space A. Expanding the same "field" in terms of the basis $b_i, b_i^\dagger$ at a different region B,

$$\phi = \sum_i (p_i b_i + p_i^* b_i^\dagger), \tag{35.53}$$

with a relation between coefficients of the form

$$p_i = \sum_{i,j}(\alpha_{ij} f_j + \beta_{ij} f_j^*), \tag{35.54}$$

we also obtain a relation between oscillator operators, since then

$$\phi = \sum_{i,j} f_j(\alpha_{ij} b_i + \beta^*_{ij} b^\dagger_i) + f^*_j(\beta_{ij} b_i + \alpha^*_{ij} b^\dagger_i), \tag{35.55}$$

implying

$$a_j = \alpha_{ij} b_i + \beta^*_{ij} b^\dagger_i. \tag{35.56}$$

If the functions f_i in region A turn, in region B, into

$$p_j \sim \exp\left(-\frac{\epsilon_j}{2k_B T}\right), \tag{35.57}$$

where ϵ_j are the single-particle energies in region B, then we obtain a Boltzmann distributed number density,

$${}_a\langle 0|b^\dagger_i b_i|0\rangle_a = \sum_j |\beta_{ij}|^2 \sim \sum_j \exp\left(-\frac{\epsilon_i}{k_B T}\right), \tag{35.58}$$

which is thermally distributed.

Important Concepts to Remember

- Decoherence is understood as the interaction of a quantum system with an environment, leading to classical states as time evolves.
- The Schrödinger's cat Gedanken experiment consists of a cat in a black box, which may be killed by a quantum process such as radioactive decay trigger, leading to quantum states $|S.cat\rangle = \frac{1}{\sqrt{2}}(|dead\rangle + |alive\rangle)$. But these states (and analogous states in quantum computation) have nonlocal correlations over large distances, and must be eliminated from the theory, via decoherence.
- Decoherence represents a transition to an entangled system + environment state, with no interaction between the basis elements, where the environment selects the basis: this is environment-induced superselection, or einselection ("the environment measures the cat").
- Thermalization represents an (large) isolated quantum system evolving to a thermal state, with a classical ensemble.
- Thermalization is based on the eigenstate thermalization hypothesis (ETH), replacing the classical ergodic hypothesis and stating that for a few-body observable $\hat{A}$ in a large, interacting many-body system, $\langle m|\hat{A}|m\rangle = \langle A\rangle_{\text{microcanonical}}(E_m)$.
- The ETH implies a generalized Gibbs ensemble (GGE), $\rho_d = \exp[-\lambda_m |m\rangle\langle m|]$, with $\lambda_m = -\ln|C_m|^2$, and C_m the coefficients of the expansion of the initial many-body state in the eigenstates of the Hamiltonian, $|m\rangle$, only if *all* the $P_m = |m\rangle\langle m|$ are integrals of motion.
- A Bogoliubov transformation is a transformation on $a, a^\dagger$ for the harmonic oscillator, mixing them and giving $b = \alpha a + \beta a^\dagger$, which preserves the commutation relations $[a, a^\dagger] = [b, b^\dagger] = 1$ but takes a vacuum, $a|0\rangle_a$, into a nonvacuum, $b|0\rangle_a \neq 0$.
- In the case of horizons, coming from black holes, accelerated mirrors ("Rindler spaces") and cosmology, this transformation leads to particle creation, ${}_a\langle 0|b^\dagger_i b_i 0\rangle_a = \sum_j |\beta_{ij}|^2 \sim \sum_j \exp\left(-\frac{\epsilon_i}{k_B T}\right)$, a quantum thermal distribution.

Further Reading

For decoherence, see Zurek's review [25] and Preskill's lecture notes on quantum information [12]. For thermalization, see the original papers [23] and [26], and the review [24].

Exercises

(1) In the Schrödinger's cat Gedanken experiment, ignoring the issue of observers (which, admittedly, was what concerned Schrödinger), where would you simplify it, by eliminating pieces in its construction? Think then about how that relates to the quantum computation issue, as alluded to in the text.

(2) At the end of decoherence, from the point of view of the system $\mathcal{S}$, we no longer have a pure quantum state. But how is this result consistent with a classical picture (for instance, if we consider the Schrödinger cat Gedanken experiment)?

(3) In the quantitative decoherence general model, calculate what will happen to $\langle\phi|\rho_S|\phi\rangle$ if we did not have $\langle\epsilon_i|\epsilon_j\rangle = \delta_{ij}$, and explain why that would be unsatisfactory.

(4) Give an example of an integrable quantum mechanical model, and explain (without any calculation) why this is not expected to thermalize. One way to define an integrable quantum mechanical model is by saying that the three-point and higher-point scatterings can be reduced to just two-point scatterings and by satisfying the Yang–Baxter equations, which relate the possible orders in which we have two-point scatterings.

(5) If the dimension of the Hilbert space D in (35.30) is infinite, is the formula still true, and is this still a GGE? Can we deduce something about the system's behavior then?

(6) Diagonalize the Hamiltonian

$$H = \sum_{i=1}^{L} \left\{ \frac{a_i^\dagger a_i + a_i a_i^\dagger}{2} + \lambda \left[(a_i + a_i^\dagger) - (a_{i+1} + a_{i+1}^\dagger) \right]^2 \right\}, \tag{35.59}$$

where $a_i, a_i^\dagger$ are harmonic-oscillator creation/annihilation operators.

(7) The thermal particle creation in (35.58) arises from a state that was in a vacuum in region A. Considering energy conservation, what can we deduce about the space? Can this process happen in Minkowski space? Why?

PART II$_b$

APPROXIMATION METHODS

36 Time-Independent (Stationary) Perturbation Theory: Nondegenerate, Degenerate, and Formal Cases

In this Part II$_b$, we will describe approximation methods for the calculation of eigenenergies and eigenstates. In this chapter, we will start with time-independent perturbation theory. We will first treat the case of nondegenerate energy eigenstates, which is simpler. Then, we will consider the degenerate case, and finally a formal system for describing perturbation theory. We will end with an example that contains all the relevant features described here.

36.1 Set-Up of the Problem: Time-Independent Perturbation Theory

The case of interest is when the Hamiltonian contains a simple "free" part $\hat{H}_0$, which has large eigenvalues, and a (complicated) part $\hat{H}_1$ that has small eigenvalues and is proportional to a small parameter λ, so in effect we can write

$$\hat{H} = \hat{H}(\lambda) = \hat{H}_0 + \lambda \hat{H}_1, \tag{36.1}$$

where $\lambda \ll 1$. We want to consider a perturbation theory in λ, where both eigenenergies and eigenstates are expanded in it.

Consider in general an eigenvalue problem for $\hat{H}_0$ that has degeneracy parameter α for the energies E_n, so

$$\hat{H}_0|n^{(0)}, \alpha\rangle = E_n^{(0)}|n^{(0)}, \alpha\rangle. \tag{36.2}$$

Then the eigenvalue problem for $\hat{H}(\lambda)$ is

$$\hat{H}(\lambda)|n, \alpha; \lambda\rangle = E_n(\lambda)|n, \alpha; \lambda\rangle. \tag{36.3}$$

We want to calculate the eigenenergies $E_n(\lambda)$ and eigenstates $|n, \alpha; \lambda\rangle$ in perturbation theory in λ,

$$\begin{aligned} E_n(\lambda) &= E_n^{(0)} + \lambda E_n^{(1)} + \lambda^2 E_n^{(2)} + \cdots \\ |n, \alpha; \lambda\rangle &= |\tilde{n}^{(0)}, \alpha\rangle + \lambda|\tilde{n}^{(1)}, \alpha\rangle + \lambda^2|\tilde{n}^{(2)}, \alpha\rangle + \cdots, \end{aligned} \tag{36.4}$$

where we have denoted the eigenstate at $\lambda = 0$ by $|\tilde{n}^{(0)}, \alpha\rangle$ instead of $|n^{(0)}, \alpha\rangle$, since in principle it can be a different state. The basis can be defined for $\hat{H}$ in terms of eigenvalues for some operator(s) that commute with $\hat{H}$ and can be different from the operator(s) that commute with $\hat{H}_0$ (say, perhaps $\vec{L}^2, L_z$ for $\hat{H}_0$, but $\vec{S}^2, S_z$ for $\hat{H}$ so a basis with $E_n^{(0)}, l, m$ for H_0, but with E_n, j, m_j for $\hat{H}$).

We use two reasonable assumptions for the perturbation theory:

- that the zeroth-order energy is $E_n^{(0)}$, namely $E_n(\lambda = 0) = E_n^{(0)}$, which is an assumption on the smoothness of the $\lambda \to 0$ limit.
- that $|n^{(0)}, \alpha\rangle$ is a *complete* set for the full Hamiltonian $\hat{H}(\lambda)$, not just for $\hat{H}_0$, so that $\sum_{n,\alpha} |n^{(0)}, \alpha\rangle\langle n^{(0)}, \alpha| = \mathbb{1}$. This is less obvious, but it is also related to the smoothness of the $\lambda \to 0$ limit.

Then, to define perturbation theory, we substitute (36.4) in (36.3) and solve the equation order by order in λ.

At zeroth order in λ (the constant part of the equation), we obtain

$$(\hat{H}^{(0)} - E_n^{(0)})|\tilde{n}^{(0)}, \alpha\rangle = 0, \tag{36.5}$$

which is just the unperturbed Schrödinger equation, so is satisfied.

At first order in λ (the linear part of the equation), we obtain

$$(\hat{H}_0 - E_n^{(0)})|\tilde{n}^{(1)}, \alpha\rangle + (\hat{H}_1 - E_n^{(1)})|\tilde{n}^{(0)}, \alpha\rangle = 0. \tag{36.6}$$

At second order in λ (the quadratic part of the equation), we obtain

$$(\hat{H}_0 - E_n^{(0)})|\tilde{n}^{(2)}, \alpha\rangle + (\hat{H}_1 - E_n^{(1)})|\tilde{n}^{(1)}, \alpha\rangle - E_n^{(2)}|\tilde{n}^{(2)}, \alpha\rangle = 0. \tag{36.7}$$

Finally, at kth order in λ (for $k > 2$), (the term with λ^k in the equation), we obtain

$$(\hat{H}_0 - E_n^{(0)})|\tilde{n}^{(k)}, \alpha\rangle + (\hat{H}_1 - E_n^{(1)})|\tilde{n}^{(k-1)}, \alpha\rangle - E_n^{(2)}|\tilde{n}^{(k-2)}, \alpha\rangle - \cdots - E_n^{(k)}|\tilde{n}^{(0)}, \alpha\rangle = 0. \tag{36.8}$$

36.2 The Nondegenerate Case

We start with the simpler nondegenerate case, when there is no index α, so $|\tilde{n}, \alpha\rangle$ is replaced by just $|n\rangle$. The eigenstates of $\hat{H}_0$ are assumed to be orthonormal,

$$\langle n^{(0)}|m^{(0)}\rangle = \delta_{mn}. \tag{36.9}$$

The zeroth-order equation is satisfied, as we have seen.

The *first-order equation* becomes now simply

$$(\hat{H}_0 - E_n^{(0)})|n^{(1)}\rangle + (\hat{H}_1 - E_n^{(1)})|n^{(0)}\rangle = 0. \tag{36.10}$$

We multiply it with the ket of the free Hamiltonian, $\langle m^{(0)}|$, from the left.

In the case $m = n$, we obtain

$$\langle n^{(0)}|\hat{H}_0|n^{(1)}\rangle - E_n^{(0)}\langle n^{(0)}|n^{(1)}\rangle + \langle n^{(0)}|\hat{H}_1|n^{(0)}\rangle - E_n^{(1)} = 0. \tag{36.11}$$

Acting with $\hat{H}_0$ on the left (on the ket state) in the first term, we obtain that the first and second terms cancel, and we are left with a formula for the first correction to the energy,

$$E_n^{(1)} = \langle n^{(0)}|\hat{H}_1|n^{(0)}\rangle \equiv H_{1,nn}. \tag{36.12}$$

In the case $m \neq n$, we obtain

$$\langle m^{(0)}|\hat{H}_0|n^{(1)}\rangle - E_n^{(0)}\langle m^{(0)}|n^{(1)}\rangle + \langle m^{(0)}|\hat{H}_1|n^{(0)}\rangle = 0. \tag{36.13}$$

Again acting with $\hat{H}_0$ on the left in the first term, we now obtain

$$(E_n^{(0)} - E_m^{(0)})\langle m^{(0)}|n^{(1)}\rangle = \langle m^{(0)}|\hat{H}_1|n^{(0)}\rangle \equiv H_{1,mn}. \tag{36.14}$$

But we can now use the second assumption, of the completeness of the states $|n^{(0)}\rangle$ in the Hilbert space of the *full* Hamiltonian $\hat{H}(\lambda)$, in order to expand $|n^{(1)}\rangle$ in $|m^{(0)}\rangle$,

$$|n^{(1)}\rangle = \sum_m C_m^{n(1)}|m^{(0)}\rangle. \tag{36.15}$$

Multiplying from the left with $\langle p^{(0)}|$ gives $C_p^{n(1)} = \langle p^{(0)}|n^{(1)}\rangle$, so substituting this into the first order equation for $m \neq n$, we get

$$C_m^{n(1)} = \frac{H_{1,mn}}{E_n^{(0)} - E_m^{(0)}}. \quad (36.16)$$

Note that this equation is valid only for $m \neq n$. At nonzero λ, we will assume that there is no contribution to $|n^{(0)}\rangle$ itself for $|n\rangle$, $\langle n^{(0)}|n^{(1)}\rangle = 0$, which is a consistent choice, amounting to just a normalization condition. Then, substituting $C_m^{n(1)}$ into the expansion of $|n^{(1)}\rangle$, we obtain

$$|n^{(1)}\rangle = \sum_{m\neq n} \frac{H_{1,mn}}{E_n^{(0)} - E_m^{(0)}} |m^{(0)}\rangle. \quad (36.17)$$

The *second-order equation* is now simply

$$(\hat{H}_0 - E_n^{(0)})|n^{(2)}\rangle + (\hat{H}_1 - E_n^{(1)})|n^{(1)}\rangle - E_n^{(2)}|n^{(0)}\rangle = 0. \quad (36.18)$$

Again we multiply it with the free ket $\langle m^{(0)}|$ from the left.

In the $m = n$ case, we obtain

$$\langle n^{(0)}|\hat{H}_0|n^{(2)}\rangle - E_n^{(0)}\langle n^{(0)}|n^{(2)}\rangle + \langle n^{(0)}|\hat{H}_1|n^{(1)}\rangle - E_n^{(2)} = 0, \quad (36.19)$$

and, by acting with $\hat{H}_0$ on the left in the first term, we see that the first two terms cancel. But we can use the first-order result (the expansion of $|n^{(1)}\rangle$), to calculate

$$\langle n^{(0)}|\hat{H}_1|n^{(1)}\rangle = \sum_{m\neq n} \frac{\langle n^{(0)}|\hat{H}_1|m^{(0)}\rangle H_{1,mn}}{E_n^{(0)} - E_m^{(0)}}. \quad (36.20)$$

Substituting in the second-order equation, we obtain

$$E_n^{(2)} = \sum_{m\neq n} \frac{H_{1,nm}H_{1,mn}}{E_n^{(0)} - E_m^{(0)}} = \sum_{m\neq n} \frac{|H_{1,nm}|^2}{E_n^{(0)} - E_m^{(0)}}. \quad (36.21)$$

In the $m \neq n$ case, we obtain

$$\begin{aligned} &\langle m^{(0)}|(\hat{H}_0 - E_n^{(0)})|n^{(2)}\rangle + \langle m^{(0)}|(\hat{H}_1 - E_n^{(1)})|n^{(1)}\rangle \\ &= (E_m^{(0)} - E_n^{(0)})\langle m^{(0)}|n^{(2)}\rangle + \langle m^{(0)}|\hat{H}_1|n^{(1)}\rangle - E_n^{(1)}\langle m^{(0)}|n^{(1)}\rangle = 0. \end{aligned} \quad (36.22)$$

Now we can use the first-order result, with $E_n^{(1)} = H_{1,nn}$ and $|n^{(1)}\rangle = C_m^{n(1)}|m^{(0)}\rangle$, and obtain

$$(E_m^{(0)} - E_n^{(0)})\langle m^{(0)}|n^{(2)}\rangle + \sum_{p\neq n}\langle m^{(0)}|\hat{H}_1|p^{(0)}\rangle \frac{H_{1,pn}}{E_n^{(0)} - E_p^{(0)}} - H_{1,nn}\frac{H_{1,mn}}{E_n^{(0)} - E_m^{(0)}} = 0. \quad (36.23)$$

Again we expand in the complete set of zeroth-order states, as in the first order:

$$|n^{(2)}\rangle = \sum_m C_m^{n(2)}|m^{(0)}\rangle. \quad (36.24)$$

Then, we find $C_m^{n(2)} = \langle m^{(0)}|n^{(2)}\rangle$, so

$$C_m^{n(2)} = \langle m^{(0)}|n^{(2)}\rangle = \frac{1}{E_n^{(0)} - E_m^{(0)}}\left(\sum_{p\neq n} \frac{H_{1,mp}H_{1,pn}}{E_n^{(0)} - E_p^{(0)}} - \frac{H_{1,nn}H_{1,mn}}{E_n^{(0)} - E_m^{(0)}}\right). \quad (36.25)$$

36.3 The Degenerate Case

We next consider the degenerate case, where, as we said, in general we expect the basis at $\lambda \to 0$ for the degenerate Hilbert space $\mathcal{H}_n^{(0)}$ of given unperturbed energy to be different from the basis used for $\hat{H}_0$,

$$|\tilde{n}, \alpha; \lambda\rangle \stackrel{\lambda\to 0}{\to} |\tilde{n}^{(0)}, \alpha\rangle \equiv \sum_\beta C^{(0)}_{n,\alpha\beta}|n^{(0)}, \beta\rangle. \tag{36.26}$$

First Order

For the first perturbation order $m = n$ case, multiply (36.6) from the left with the bra $\langle \tilde{n}^{(0)}, \beta|$, obtaining

$$\langle \tilde{n}^{(0)}, \beta|\hat{H}_0|\tilde{n}^{(1)}, \alpha\rangle - E_n^{(0)}\langle \tilde{n}^{(0)}, \beta|\tilde{n}^{(1)}, \alpha\rangle + \langle \tilde{n}^{(0)}, \beta|\hat{H}_1|\tilde{n}^{(0)}, \alpha\rangle - E_n^{(1)}\delta_{\alpha\beta} = 0. \tag{36.27}$$

Acting with $\hat{H}_0$ from the left on the first term leads to the cancellation of the first two terms giving

$$H_{1,n\beta\alpha} \equiv \langle \tilde{n}^{(0)}, \beta|\hat{H}_1|\tilde{n}^{(0)}, \alpha\rangle = E_n^{(1)}\delta_{\alpha\beta}. \tag{36.28}$$

However, we see that this equation does not have a solution if $H_{1,n\beta\alpha} \neq 0$ for $\alpha \neq \beta$. In order for it to have a solution, we consider states that are linear combinations of the basis states,

$$|\psi_n\rangle = \sum_{\alpha=1}^{g_m} C_{n\alpha}|\tilde{n}, \alpha\rangle, \tag{36.29}$$

where both $|\psi_n\rangle$ and $|\tilde{n}, \alpha\rangle$ admit expansions in λ, or, more precisely, the above relation is true at each order in λ,

$$|\psi_n^{(s)}\rangle = \sum_{\alpha=1}^{g_m} C_{n\alpha}^{(s)}|\tilde{n}^{(s)}, \alpha\rangle. \tag{36.30}$$

Then we replace $|\tilde{n}^{(s)}, \alpha\rangle$ with $|\psi_n^{(s)}\rangle$ in the original perturbation theory equations, starting with (36.6). Then, again multiplying it with $\langle \tilde{n}^{(0)}, \beta|$, we obtain

$$\sum_{\alpha=1}^{g_m} H_{1,n\beta\alpha}C_{n\alpha}^{(0)} = E_n^{(1)}C_{n\beta}^{(0)}. \tag{36.31}$$

This is a matrix equation for diagonalization, $H_1 \cdot C = EC$, so the possible eigenvalues, for $E_n^{(1)}$, are solutions to the equation

$$\det(H_{1,n\beta\alpha} - E_n^{(1)}\delta_{\beta\alpha}) = 0. \tag{36.32}$$

Next we consider the $m \neq n$ case. Again replacing $|\tilde{n}^{(s)}, \alpha\rangle$ with $|\psi_n^{(s)}\rangle$ in (36.6), but multiplying from the left with $\langle \tilde{m}^{(0)}, \beta|$, we obtain

$$\sum_\alpha C_{n\alpha}^{(1)}\langle \tilde{m}^{(0)}, \beta|\hat{H}_0|\tilde{n}^{(1)}, \alpha\rangle - E_n^{(0)}\sum_\alpha C_{n\alpha}^{(1)}\langle \tilde{n}^{(0)}, \beta|\tilde{n}^{(1)}, \alpha\rangle + \sum_\alpha C_{n\alpha}^{(0)}\langle \tilde{n}^{(0)}, \beta|\hat{H}_1|\tilde{n}^{(0)}, \alpha\rangle = 0. \tag{36.33}$$

Acting with $\hat{H}_0$ from the left on the first term, we obtain

$$(E_m^{(0)} - E_n^{(0)}) \sum_\alpha C_{n\alpha}^{(1)} \langle \tilde{m}^{(0)}, \beta | \tilde{n}^{(1)}, \alpha \rangle + \sum_\alpha C_{n\alpha}^{(0)} H_{1,m\beta,n\alpha} = 0, \tag{36.34}$$

where we have defined

$$H_{1,m\beta,n\alpha} \equiv \langle \tilde{m}^{(0)}, \beta | \hat{H}_1 | \tilde{n}^{(0)}, \alpha \rangle \tag{36.35}$$

and, expanding $|\psi_n^{(1)}\rangle$ in the complete set of H_0 eigenstates $|\tilde{n}^{(0)}, \alpha\rangle$,

$$|\psi_n^{(1)}\rangle = \sum_\alpha C_{n\alpha}^{(1)} |n^{(1)}, \alpha\rangle = \sum_{m,\beta} \tilde{C}_{m\beta}^{(1)} |\tilde{m}^{(0)}, \beta\rangle, \tag{36.36}$$

we first find that (by multiplication with $\langle \tilde{m}^{(0)}, \beta|$)

$$\sum_\alpha C_{n\alpha}^{(1)} \langle \tilde{m}^{(0)}, \beta | \tilde{n}^{(1)}, \alpha \rangle = \tilde{C}_{m\beta}^{(1)}, \tag{36.37}$$

in which case the perturbation equation is solved by

$$\tilde{C}_{m\beta}^{(1)} = \frac{\sum_\alpha H_{1,m\beta,n\alpha} C_{n\alpha}^{(0)}}{E_n^{(0)} - E_m^{(0)}}. \tag{36.38}$$

Substituting back in the definition of $|\psi_n^{(1)}\rangle$, we find

$$|\psi_n^{(1)}\rangle = \sum_\alpha C_{n\alpha}^{(1)} |\tilde{n}^{(1)}, \alpha\rangle = \sum_{m,\beta} \sum_\alpha \frac{H_{1,m\beta,n\alpha} C_{n\alpha}^{(0)}}{E_n^{(0)} - E_m^{(0)}} |\tilde{m}^{(0)}, \beta\rangle. \tag{36.39}$$

Second Order

For the second perturbation order, replacing $|\tilde{n}^{(s)}, \alpha\rangle$ with $|\psi_n^{(s)}\rangle$ in (36.7), we find

$$(\hat{H}_0 - E_n^{(0)}) \sum_\alpha C_{n\alpha}^{(2)} |\tilde{n}^{(2)}, \alpha\rangle + (\hat{H}_1 - E_n^{(1)}) \sum_\alpha C_{n\alpha}^{(1)} |\tilde{n}^{(1)}, \alpha\rangle - E_n^{(2)} \sum_\alpha C_{n\alpha}^{(0)} |\tilde{n}^{(0)}, \alpha\rangle = 0. \tag{36.40}$$

The $m = n$ case. Multiplying from the left with the bra $\langle \tilde{n}^{(0)}, \beta|$, we find

$$\begin{aligned} &\sum_\alpha C_{n\alpha}^{(2)} \langle \tilde{n}^{(0)}, \beta | \hat{H}_0 | \tilde{n}^{(2)}, \alpha \rangle - E_n^{(0)} \sum_\alpha C_{n\alpha}^{(2)} \langle \tilde{n}^{(0)}, \beta | \tilde{n}^{(2)}, \alpha \rangle \\ &\quad + \sum_\alpha C_{n\alpha}^{(1)} \langle \tilde{n}^{(0)}, \beta | \hat{H}_1 | \tilde{n}^{(1)}, \alpha \rangle - E_n^{(1)} \sum_\alpha C_{n\alpha}^{(1)} \langle \tilde{n}^{(0)}, \beta | \tilde{n}^{(1)}, \alpha \rangle - E_n^{(2)} C_{n\beta}^{(0)} = 0. \end{aligned} \tag{36.41}$$

Acting with $\hat{H}_0$ from the left on the first term, the first two terms cancel, and we are left with

$$\sum_\alpha C_{n\alpha}^{(1)} \langle \tilde{n}^{(0)}, \beta | \hat{H}_1 | \tilde{n}^{(1)}, \alpha \rangle - E_n^{(1)} \sum_\alpha C_{n\alpha}^{(1)} \langle \tilde{n}^{(0)}, \beta | \tilde{n}^{(1)}, \alpha \rangle = E_n^{(2)} C_{n\beta}^{(2)}. \tag{36.42}$$

However, from the first-order analysis, replacing the form of $|\psi_n^{(1)}\rangle$ we have

$$\begin{aligned} &\sum_\alpha C_{n\alpha}^{(1)} \langle \tilde{m}^{(0)}, \beta | \hat{H}_1 | \tilde{n}^{(1)}, \alpha \rangle = \langle \tilde{m}^{(0)}, \beta | \hat{H}_1 | \psi_n^{(1)} \rangle = \sum_\alpha \sum_{p,\gamma} C_{n\alpha}^{(0)} \frac{H_{1,m\beta,p\gamma} H_{1,p\gamma,n\alpha}}{E_n^{(0)} - E_p^{(0)}} \\ &\sum_\alpha C_{n\alpha}^{(1)} \langle \tilde{m}^{(0)}, \beta | \tilde{n}^{(1)}, \alpha \rangle = \langle \tilde{m}^{(0)}, \beta | \psi_n^{(1)} \rangle = 0. \end{aligned} \tag{36.43}$$

Using these identities, we obtain that the second-order equation for $m = n$ becomes

$$E_n^{(2)} C_{n\beta}^{(0)} = \langle \tilde{m}^{(0)}, \beta | \hat{H}_1 | \psi_n^{(1)} \rangle = \sum_\alpha \sum_{p,\gamma} C_{n\alpha}^{(0)} \frac{H_{1,m\beta,p\gamma} H_{1,p\gamma,n\alpha}}{E_n^{(0)} - E_p^{(0)}}. \tag{36.44}$$

Then defining the matrix element of an abstract operator $\hat{K}^{(2)}$ by

$$\langle \tilde{n}^{(0)}, \beta | \hat{K}^{(2)} | \tilde{n}^{(0)}, \alpha \rangle \equiv \sum_{p,\gamma} \frac{\langle \tilde{m}^{(0)}, \beta | \hat{H}_1 | \tilde{p}^{(0)}, \gamma \rangle \langle \tilde{p}^{(0)}, \gamma | \hat{H}_1 | \tilde{n}^{(0)}, \alpha \rangle}{E_n^{(0)} - E_p^{(0)}}, \tag{36.45}$$

the equation for $E_n^{(2)}$ again becomes an eigenvalue problem for $\hat{K}^{(2)}$,

$$\sum_\alpha \langle \tilde{n}^{(0)}, \beta | \hat{K}^{(2)} | \tilde{n}^{(0)}, \alpha \rangle C_{n\alpha}^{(0)} = E_n^{(2)} C_{n\beta}^{(0)}. \tag{36.46}$$

For the case $m \neq n$ we could again find the second-order wave function, but the analysis is more complicated and will be skipped.

36.4 General Form of Solution (to All Orders)

Now we will show the main points of a different approach, which works to all orders in perturbation theory. Consider the eigenstate of the total Hamiltonian H, $|\tilde{n}, \alpha, \lambda\rangle$, and denote it by $|n, \alpha\rangle$.

Define the projector onto the Hilbert subspace of given energy E_n (the space of $|n, \alpha\rangle$ states for any α),

$$\hat{P}_n = \sum_\alpha |n, \alpha\rangle\langle n, \alpha|. \tag{36.47}$$

Since it is a projector, it satisfies

$$\hat{P}_n^2 = \hat{P}_n, \quad \text{and} \quad \hat{P}_n \hat{P}_m = 0 \Rightarrow \hat{P}_n \hat{P}_m = \delta_{mn} \hat{P}_n$$
$$\sum_n \hat{P}_n = \mathbb{1}, \tag{36.48}$$

as we can check.

Then the Schrödinger equation $\hat{H}|n, \alpha\rangle = E_n |n, \alpha\rangle$ becomes

$$\hat{H} \hat{P}_n = E_n \hat{P}_n. \tag{36.49}$$

Define the *resolvent* of this equation,

$$\hat{G}(z) \equiv (z - \hat{H})^{-1} \overset{\text{not.}}{=} \frac{1}{z - \hat{H}}, \tag{36.50}$$

where the label above the equals sign indicates that the notation with fractions will be used for convenience, and where z is a complex variable; the resolvent can be thought of as a sum of power laws,

$$\hat{G}(z) = \frac{1}{z} \sum_{n=0}^{\infty} \left(\frac{\hat{H}}{z} \right)^n. \tag{36.51}$$

Then, using the Schrödinger equation, we obtain

$$\hat{G}(z)\hat{P}_n = \frac{\hat{P}_n}{z - E_n}. \tag{36.52}$$

Summing over n, and using the fact that $\sum_n \hat{P}_n = \mathbb{1}$, we obtain

$$\hat{G}(z) = \sum_n \frac{\hat{P}_n}{z - E_n} = \sum_n \frac{\sum_\alpha |n\alpha\rangle\langle n\alpha|}{z - E_n}. \tag{36.53}$$

This means that we can use the residue theorem in the complex plane to identify a projector $\hat{P}_n$:

$$\hat{P}_n = \frac{1}{2\pi i}\oint_{\Gamma_n} G(z)dz, \tag{36.54}$$

where Γ_n is a contour in the complex z plane around the real pole at $z = E_n$.

Now we remember that $\hat{H} = \hat{H}_0 + \lambda\hat{H}_1$, which means that

$$\hat{G}(z) = \frac{1}{z - \hat{H}_0 - \lambda\hat{H}_1}, \tag{36.55}$$

and define the same resolvent for the free theory (just with $\hat{H}_0$),

$$\hat{G}_0(z) = \frac{1}{z - \hat{H}_0}. \tag{36.56}$$

Then we can write the identities

$$\begin{aligned}\hat{G}(z) = \frac{1}{z - \hat{H}_0 - \lambda\hat{H}_1} &= \frac{1}{z - \hat{H}_0}[(z - \hat{H}_0 - \lambda\hat{H}_1) + \lambda\hat{H}_1]\frac{1}{z - \hat{H}_0 - \lambda\hat{H}_1} \\ &= \frac{1}{z - \hat{H}_0} + \frac{1}{z - \hat{H}_0}\lambda\hat{H}_1\frac{1}{z - \hat{H}_0 - \lambda\hat{H}_1} \\ &= \hat{G}_0(z)\left(1 + \lambda\hat{H}_1\hat{G}(z)\right).\end{aligned} \tag{36.57}$$

We have thus obtained a self-consistent equation tailor-made for perturbation theory,

$$\hat{G}(z) = \hat{G}_0(z)\left(1 + \lambda\hat{H}_1\hat{G}(z)\right), \tag{36.58}$$

since the term with $\hat{G}(z)$ on the right-hand side is proportional to λ.

This means that we can iterate the equation, by putting $\hat{G} = \hat{G}_0$ on the right-hand side (so $\hat{G}^{(0)} = \hat{G}_0$), and then we have $\hat{G}$ up to the first order on the left-hand side. Next we put this on the right-hand side of the equation, obtaining $\hat{G}$ up to the second order on the left-hand side. We continue this procedure *ad infinitum*, thus obtaining the perturbative solution to all orders in perturbation theory,

$$\hat{G} = \sum_{n=0}^{\infty} \lambda^n \hat{G}_0(\hat{H}_1\hat{G}_0)^n. \tag{36.59}$$

Define a contour in the complex plane $\tilde{\Gamma}_n$ that encircles both $E_n^{(0)}$ (eigenvalue of $\hat{H}_0$) and the real values E_n. Then we have

$$\hat{P}_n = \frac{1}{2\pi i}\oint_{\tilde{\Gamma}_n} \hat{G}(z)dz \tag{36.60}$$

as before, but now we can also use the solution (36.59) inside the equation, and then use the same equation for $\hat{P}_{n,0}$ in terms of $\hat{G}_0$ to find

$$\hat{P}_n = \hat{P}_{n,0} + \sum_{n=1}^{\infty} \lambda^n A^{(n)}, \tag{36.61}$$

where

$$A^{(n)} = \frac{1}{2\pi i} \oint_{\tilde{\Gamma}_n} \hat{G}_0 (\hat{H}_1 \hat{G}_0)^n dz. \tag{36.62}$$

One can also find expressions for the expansions of $\hat{P}_n$ and $\hat{H}\hat{P}_n$ in λ^n from the above, but we will not do it here.

Then, acting with $\hat{H}\hat{P}_n$ on the eigenstates of $\hat{H}_0$, $|\tilde{n}^{(0)}, \alpha\rangle$, the operator $\hat{P}_n$ projects them onto the states of energy E_n for $\hat{H}$, so we can act with $\hat{H}$, and obtain

$$\hat{H}\hat{P}_n|\tilde{n}^{(0)}, \alpha\rangle = E_n \hat{P}_n |\tilde{n}^{(0)}, \alpha\rangle. \tag{36.63}$$

But then, without altering anything, we can put $\hat{P}_{n,0}$ in front of $|\tilde{n}^{(0)}, \alpha\rangle$ and multiply the equation from the left by $\hat{P}_{n,0}$.

Defining the following operators, which are Hermitian, as we can easily check,

$$\hat{H}_n = \hat{P}_{n,0}\hat{H}\hat{P}_n\hat{P}_{n,0}, \quad \hat{K}_n = \hat{P}_{n,0}\hat{P}_n\hat{P}_{n,0}, \tag{36.64}$$

the resulting equation becomes

$$\hat{H}_n|\tilde{n}^{(0)}, \alpha\rangle = E_n \hat{K}_n |\tilde{n}^{(0)}, \alpha\rangle. \tag{36.65}$$

In order for this equation to have a solution, the energy E_n must satisfy a secular equation,

$$\det(\hat{H}_n - E_n \hat{K}_n) = 0. \tag{36.66}$$

The expansion in λ of $\hat{H}_n$ and $\hat{K}_n$ can be derived from the previous expansion of $\hat{H}\hat{P}_n$ and $\hat{P}_n$, resulting in the eigenvalue equation for E_n in a λ expansion. We will not continue it here, however.

36.5 Example: Stark Effect in Hydrogenoid Atom

Consider the Stark effect, which is the interaction of a system having electric dipole moment d with a constant electric field E in the z direction, with interaction Hamiltonian

$$\hat{H}_1 = dEz = dEr\cos\theta. \tag{36.67}$$

Moreover, suppose that the unperturbed Hamiltonian $\hat{H}_0$ is that for the hydrogenoid atom. For perturbation theory to work, we need that the interaction energy $\ll E_{\text{nucleus}}$.

We first apply perturbation theory to the ground state, with energy quantum number $n = 1$, starting with the first order. For $n = 1$, $g_1 = 1$ the only state (nlm) is (100). But since $z = r\cos\theta$ is directional, whereas the (100) state is spherically symmetric, we have that

$$(H_1)_{100,100} = 0. \tag{36.68}$$

This means that for the first-order correction to the ground state energy E_1 we have

$$E_1^{(1)} = (H_1)_{100,100} = 0. \tag{36.69}$$

The first nonzero contribution is from the second order, which in this case is found from the general expression

$$E_1^{(2)} = \sum_{(nlm)\neq(100)} \frac{|H_{1,100,nlm}|^2}{E_1 - E_n}. \tag{36.70}$$

We cannot find a closed form for the solution; instead we must calculate term by term in (nlm) and then sum, so we will not go further in evaluation of this formula.

Next we apply perturbation theory to the first excited state, with $n = 2$. In this case $g_2 = 2^2 = 4$. The four states in this Hilbert subspace are $|2\alpha\rangle = |2lm\rangle$, specifically denoted by

$$|2;1\rangle \equiv |2,0,0\rangle,\quad |2;2\rangle \equiv |2,1,0\rangle,\quad |2;3\rangle \equiv |2,1,1\rangle,\quad |2;4\rangle \equiv |2,1,-1\rangle. \tag{36.71}$$

Then the matrix elements of the perturbation are

$$\langle nl_1m_1|\hat{H}_1|nl_2m_2\rangle = dE\langle l_1m_1|\cos\theta|l_2m_2\rangle\langle r\rangle_{nl_1,nl_2}. \tag{36.72}$$

In order for the result to be nonzero, we must have $m_1 = m_2$ and $l_1 = l_2 \pm 1$. Moreover, we find that

$$\langle l,m|\cos\theta|l-1,m\rangle = \langle l-1,m|\cos\theta|l,m\rangle = \sqrt{\frac{l^2-m^2}{4l^2-1}}, \tag{36.73}$$

which we leave as an exercise for the reader to prove.

Therefore in our case (with states 1, 2, 3, 4) the only nonzero matrix element is $H_{1,12} \neq 0$, with

$$\langle 2;1|\cos\theta|2;2\rangle = \sqrt{\frac{1}{3}}. \tag{36.74}$$

Then the matrix element is

$$H_{1,12} = -dE\frac{\langle r\rangle_{20,21}}{\sqrt{3}} = -3dE\frac{a_0}{Z}. \tag{36.75}$$

Next we need to diagonalize the matrix

$$\begin{pmatrix} 0 & H_{1,12} & 0 & 0 \\ H_{1,12} & 0 & 0 & 0 \\ 0 & 0 & 0 & 0 \\ 0 & 0 & 0 & 0 \end{pmatrix} = H_{1,12}\begin{pmatrix} 0 & 1 & 0 & 0 \\ 1 & 0 & 0 & 0 \\ 0 & 0 & 0 & 0 \\ 0 & 0 & 0 & 0 \end{pmatrix}. \tag{36.76}$$

The secular equation for the eigenvalues is (for $E_2^{(1)} = \lambda H_{1,12}$)

$$\lambda^2\left|\begin{pmatrix} -\lambda & H_{1,12} \\ H_{1,12} & -\lambda \end{pmatrix}\right| = 0 \Rightarrow \lambda^2(\lambda^2 - H_{1,12}) = 0, \tag{36.77}$$

with solutions

$$E_2^{(1)} = H_{1,12}(0, 0, +1, -1). \tag{36.78}$$

The eigenstates are also found from the diagonalization, namely

$$\begin{aligned}
|\psi_3\rangle &= |2;3\rangle = |2,1,1\rangle \\
|\psi_4\rangle &= |2;4\rangle = |2,1,-1\rangle \\
|\psi_1\rangle &= \frac{1}{\sqrt{2}}(|2;1\rangle + |2;2\rangle) = \frac{1}{\sqrt{2}}(|2,0.0\rangle + |2,1,0\rangle) \\
|\psi_2\rangle &= \frac{1}{\sqrt{2}}(|2;1\rangle - |2;2\rangle) = \frac{1}{\sqrt{2}}(|2,0,0\rangle - |2,1,0\rangle).
\end{aligned} \tag{36.79}$$

Important Concepts to Remember

- Time-independent perturbation theory arises for a time-independent Hamiltonian $\hat{H}(\lambda) = \hat{H}_0 + \lambda\hat{H}_1$, where $\hat{H}_0$ is the "free" part, with large eigenvalues, and $\lambda\hat{H}_1$ is a perturbation, with small eigenvalues.
- In the nondegenerate case, to first order we find $E_n^{(1)} = H_{1,nn}$ and $C_m^{n(1)} = H_{1,mn}/(E_n^{(0)} - E_m^{(0)})$, $E_n^{(2)} = \sum_{m\neq n} |H_{1,nm}|^2/(E_n^{(0)} - E_m^{(0)})$.
- In the degenerate case, the basis $|\tilde{n}^{(0)}, \alpha\rangle$ at $\lambda \to 0$ is generically different from the free basis, $|n^{(0)}, \alpha\rangle$. Moreover, the states are combinations of the basis states, $|\psi_n\rangle = \sum_n C_{n\alpha}|\tilde{n}, \alpha\rangle$.
- The equation for $E_n^{(1)}$ is a secular equation in the basis states, $\det(H_{1,n\beta\alpha} - E_n^{(1)}\delta_{\beta\alpha}) = 0$, which solves $\sum_\alpha [H_{1,n\beta\alpha} C_{n\alpha}^{(0)} - E_n^{(1)} C_{n\beta}^{(0)}] = 0$. Then $\tilde{C}_{m\beta}^{(1)} = \sum_\alpha H_{1,m\beta,n\alpha} C_{n\alpha}^{(0)}/[E_n^{(0)} - E_m^{(0)}]$.
- In the general case of perturbation theory, defining the resolvent

$$\hat{G}(z) = \frac{1}{z - \hat{H}} = \sum_n \left[\sum_\alpha |n\alpha\rangle\langle n\alpha|/(z - E_n)\right]$$

and the free resolvent,

$$\hat{G}_0(z) = \frac{1}{z - \hat{H}_0},$$

we have the self-consistent equation $\hat{G}(z) = \hat{G}_0(z)(1 + \lambda\hat{H}_1\hat{G}(z))$, solved by perturbation theory via iteration.
- The Stark effect for a hydrogenoid atom, $H_1 = dEz$, is trivial in first-order perturbation theory, and is nontrivial at second order.

Further Reading

See [2], [3], [1].

Exercises

(1) Consider a one-dimensional harmonic oscillator perturbed by a linear potential, $\lambda \hat{H}_1 = \lambda x$. Calculate the first-order perturbation theory corrections to the energy and wave functions, and compare with the exact solution for $\hat{H}_0 + \lambda \hat{H}_1$ (which is trivial to find in this case).

(2) Consider a hydrogenoid atom perturbed by a decaying exponential potential $\lambda \hat{H}_1 = \lambda e^{-\mu r}$. Calculate the first-order perturbation theory corrections to the energy and ground state wave function.

(3) Write down an explicit formula for the second-order perturbation contribution to the ground state energy in exercise 2, without calculating the terms and summing them.

(4) Calculate the first-order perturbation contribution to the energy of the first excited state of the hydrogenoid atom in exercise 2.

(5) Formally, to obtain the perturbative solution (36.59), we did not need the self-consistent equation (36.58), we just needed to expand the definition of $G(z)$ in λ: show this.

(6) Prove the relation (36.73), used in the analysis of the Stark effect.

(7) Use first-order perturbation theory to find the Stark effect splittings for the second excited state of the hydrogenoid atom, $n = 3$.

37 Time-Dependent Perturbation Theory: First Order

In this chapter, we will consider the time dependence of the Schrödinger equation, and solve it in perturbation theory. We will concentrate on the first order, leaving the second-order case and the general expansion for the next chapter.

Consider then the time-dependent Hamiltonian $\hat{H}$ split into a nonperturbed, time-independent part $\hat{H}_0$ and a time-dependent perturbation,

$$\hat{H}(t) = \hat{H}_0 + \lambda \hat{H}_1(t). \tag{37.1}$$

As before, for this to be a satisfactory perturbation theory, we need $\lambda \ll 1$, or rather that the matrix elements of $\lambda \hat{H}_1(t)$ are much smaller than the matrix elements of $\hat{H}_0$.

Thus, we will attempt to solve the time-dependent Schrödinger equation,

$$i\hbar\partial_t|\psi(t)\rangle = \hat{H}(t)|\psi(t)\rangle, \tag{37.2}$$

as an expansion in λ.

37.1 Evolution Operator

We defined the evolution operator $\hat{U}(t_2, t_1)$ by

$$|\psi(t_2)\rangle = \hat{U}(t_2, t_1)|\psi(t_1)\rangle, \tag{37.3}$$

which implies that the Schrödinger equation is now

$$i\hbar\partial_t\hat{U}(t, t_0) = \hat{H}(t)\hat{U}(t, t_0). \tag{37.4}$$

If the Hamiltonian is time independent, $\hat{H}(t) = \hat{H}$, we obtain $\hat{U}(t, t_0) = e^{-i\hat{H}(t-t_0)/\hbar}$. Defining the eigenstates of $\hat{H}$ as $|\psi_n\rangle$ with eigenvalues E_n, and if the state at $t = 0$ is

$$|\psi(t = 0)\rangle = \sum_n a_n|\psi_n(0)\rangle, \tag{37.5}$$

then the state at time t is

$$|\psi(t)\rangle = \sum_n a_n e^{-iE_n t/\hbar}|\psi_n(0)\rangle. \tag{37.6}$$

However, in general we have $\hat{H}(t)$ (a time-dependent Hamiltonian), so choosing at an initial time t_i a state among the eigenstates $|\psi_n\rangle$, namely

$$|\psi(t = 0)\rangle = |\psi_i\rangle, \tag{37.7}$$

we find that, at a final time, the state has become a superposition of all the $|\psi_n\rangle$ states,

$$|\psi(t = t_f)\rangle = \hat{U}(t_f, t_i)|\psi_i\rangle = \sum_n a_n(t_f)|\psi_n\rangle. \tag{37.8}$$

Multiplying from the left with the bra $\langle\psi_m|$, we find that

$$a_m(t_f) = \langle\psi_m|\hat{U}(t_f, t_i)|\psi_i\rangle \equiv U_{mi}(t_f, t_i). \tag{37.9}$$

Then the transition probability (during the period from t_i to t_f) between $|\psi_i\rangle$ and $|\psi_f\rangle$ is

$$P_{fi} = |U_{fi}(t_f, t_i)|^2 \tag{37.10}$$

37.2 Method of Variation of Constants

Consider the eigenvalue problem for the unperturbed Hamiltonian $\hat{H}_0$,

$$\hat{H}_0|\psi_n\rangle = E_n|\psi_n\rangle, \tag{37.11}$$

and an initial condition

$$|\psi\rangle = \sum_n c_n|\psi_n\rangle. \tag{37.12}$$

As in the previous analysis of the evolution operator, the time-dependent interaction $\hat{H}_1(t)$ can be used to calculate probabilities for transitions between the states $|\psi_n\rangle$. For the solution of the time-dependent Schrödinger equation for $\hat{H}(t)$, we write an *ansatz* by lifting to time dependence the constants c_n, so that

$$|\psi(t)\rangle = \sum_n c_n(t)|\psi_n\rangle. \tag{37.13}$$

Comparing with (37.8), we see that

$$a_n(t) = c_n(t) = U_{ni}(t, t_i). \tag{37.14}$$

Substituting this ansatz in the Schrödinger equation, we obtain

$$i\hbar\sum_n \dot{c}_n(t)|\psi_n\rangle = \hat{H}(t)\sum_n c_n(t)|\psi_n\rangle. \tag{37.15}$$

Multiplying from the left with the bra $\langle\psi_m|$, we obtain the system of equations

$$i\hbar\dot{c}_m(t) = \sum_n c_n(t)\langle\psi_m|\hat{H}(t)|\psi_n\rangle \equiv \sum_n c_n(t)H_{mn}(t). \tag{37.16}$$

This system of equations is equivalent to the original Schrödinger equation.

Again, we must use the assumption that $\{|\psi_n\rangle\}$ is a complete basis for the full $\hat{H}(t)$ as well. Then, we expand the $c_n(t)$ coefficients in λ,

$$c_n = c_n(t, \lambda) = \sum_{s=0}^{\infty} c_n^{(s)}\lambda^s. \tag{37.17}$$

Taking matrix elements of $\hat{H}(t) = \hat{H}_0 + \lambda\hat{H}_1(t)$ in the basis $|\psi_n\rangle$ of $\hat{H}_0$, we obtain

$$H_{mn}(t) = E_n\delta_{mn} + \lambda H_{1,mn}(t). \tag{37.18}$$

Moreover, we can factorize the time dependence of the unperturbed Hamiltonian from the coefficients $c_n(t)$,

$$c_n(t) = b_n(t)e^{-iE_nt/\hbar}, \tag{37.19}$$

and therefore find an expansion in λ of $b_n(t)$ as well,

$$b_n(t) = \sum_{s=0}^{\infty} b_n^{(s)} \lambda^s. \tag{37.20}$$

Then the Schrödinger equation for $c_n(t)$ becomes one for $b_n(t)$ that is simpler,

$$i\hbar\dot{b}_m(t) = \sum_n \lambda H_{1,mn}(t)e^{i(E_m-E_n)t/\hbar} = \sum_n \lambda H_{1,mn}(t)e^{i\omega_{mn}t}, \tag{37.21}$$

where we define the transition frequency

$$\omega_{mn} \equiv \frac{E_m - E_n}{\hbar}. \tag{37.22}$$

The boundary conditions for the equations for $b_n(t)$ are that $b_n^{(0)}$ is constant, and defines the initial (unperturbed) wave function. To solve the equations, we identify the coefficient of λ^{s+1} on both sides of the equation, resulting in

$$i\hbar\dot{b}_m^{(s+1)}(t) = \sum_m H_{1,mn}(t)b_n^{(s)}(t)e^{i\omega_{mn}t}. \tag{37.23}$$

This results in an iterative solution of the equation, given the boundary conditions. Put $b_n^{(0)}$ on the right-hand side, then find $b_n^{(1)}$ from the left-hand side. Then input $b_n^{(1)}$ on the right-hand side, and find $b_n^{(2)}$ from the left-hand side, etc.

Consider the boundary conditions given by an initial state $|\psi_i\rangle$, so that

$$\begin{aligned} b_n^{(0)}(t) &= \delta_{ni} \\ b_n^{(s)}(t=0) &= 0, \quad s > 0. \end{aligned} \tag{37.24}$$

The second condition means that there is no component for $n \neq i$ at all orders in λ, at the initial time.

Then, the equation for the *first order* in λ is

$$i\hbar\dot{b}_m^{(1)}(t) = H_{1,mi}(t)e^{i\omega_{mi}t}, \tag{37.25}$$

with solution

$$i\hbar b_m^{(1)}(\tau) = \int_0^\tau dt e^{i\omega_{mi}t} H_{1,mi}(t). \tag{37.26}$$

Finally, the time-dependent state to first order is

$$|\psi(t)\rangle = \sum_n (\delta_{ni} + b_n^{(1)}(t))e^{-iE_nt/\hbar}|\psi_n\rangle. \tag{37.27}$$

Then the transition probability for the system to go from a state with energy E_i to a state with energy E_n during the time t is

$$P_{ni}(t) = |b_n^{(1)}(t)|^2. \tag{37.28}$$

37.3 A Time-Independent Perturbation Being Turned On

Consider a time-independent perturbation turned on at time $t = 0$, so that

$$H_1(t) = \begin{cases} 0, & t < 0 \\ H_1, & t \geq 0. \end{cases} \tag{37.29}$$

Later we will see that this is an example of the "sudden approximation", and that it is a sensible thing to consider.

Then, during the period of integration, we have a time-independent H_1 or, more precisely, we have a time-independent matrix element $H_{1,mi}$. Thus we obtain

$$i\hbar b_m^{(1)}(\tau) = H_{1,mi} \int_0^\tau dt e^{i\omega_{mi}t} = -i\frac{H_{1,mi}}{\omega_{mi}} \left(e^{i\omega_{mi}\tau} - 1\right). \tag{37.30}$$

The probability of transition during the time τ is

$$P_{mi}(\tau) = |b_m^{(1)}(\tau)|^2 = \frac{|H_{1,mi}|^2}{\hbar^2\omega_{mi}^2}\left|e^{i\omega_{mi}t} - 1\right|^2 = \frac{4|H_{1,mi}(\tau)|^2}{(E_m - E_i)^2}\sin^2\frac{\omega_{mi}\tau}{2}, \tag{37.31}$$

where we have used that $\hbar\omega_{mi} = E_m - E_i$ and $\left|e^{i\omega_{mi}t} - 1\right|^2 = 2(1 - \cos\omega_{mi}\tau) = 4\sin^2\omega_{mi}\tau/2$.

Since however we have the limit formula

$$\lim_{a\to\infty} \frac{1}{\pi}\frac{\sin^2 ax}{ax^2} = \delta(x), \tag{37.32}$$

which we leave as an exercise for the reader to prove, then it means that at large times

$$\begin{aligned} |b_m^{(1)}(\tau)|^2 &= \frac{\tau}{2\hbar}4\pi|H_{1,mi}|^2\frac{1}{\pi}\frac{\sin^2(E_f - E_i)(\tau/2\hbar)}{(\tau/2\hbar)(E_f - E_i)^2} \\ &\overset{\tau\to\infty}{\to} \tau\frac{2\pi}{\hbar}|H_{1,mi}|^2\delta(E_f - E_i). \end{aligned} \tag{37.33}$$

Thus the probability per unit time at large times becomes

$$\frac{dP_{fi}}{dt} = \frac{2\pi}{\hbar}|H_{1,mi}|^2\delta(E_f - E_i). \tag{37.34}$$

This means that at large times, we can only have (i.e., there is a nonzero probability for) the same energy for the initial and final states of the transition.

37.4 Continuous Spectrum and Fermi's Golden Rule

One important particular case is when the final state $m = f$ belongs to the continuous spectrum of the Hamiltonian. For one such possibility, the initial state is in the discrete spectrum, which happens to overlap (in terms of energy, but not in position in space) with the continuous spectrum. For instance, we can have bound electrons in atoms, that can escape and transition to the continuous spectrum as free electrons, a process of "self-ionization".

We consider orthonormal final states $|\psi_f\rangle$, where $f = \{f_1, \ldots, f_r\}$ is a set of quantum numbers (parameters), out of which at least one, say f_1, is continuous, and perhaps there are also discrete parameters, so

$$\langle\psi_f|\psi_{f'}\rangle = \text{“}\delta(f - f')\text{”} \equiv \delta(f_1 - f_1')\delta(f_2 - f_2') \cdots \delta_{f_{r-1}f_{r-1}'}\delta_{f_r f_r'}. \tag{37.35}$$

Then in the general expansion

$$|\psi\rangle = \sum_n c_n(t)|\psi_n\rangle, \tag{37.36}$$

the index n runs over discrete values n', and continuous + discrete ones f, so that we have

$$\begin{aligned}\langle\psi|\psi\rangle &= \sum_{n'} |c_{n'}(t)|^2 + \int df_1 df_2 \cdots \sum_{f_{r-1}}\sum_{f_r} |c_f(t)|^2 \\ &= \sum_{n'} |b_{n'}(t)|^2 + \int df_1 df_2 \cdots \sum_{f_{r-1}}\sum_{f_r} |b_f(t)|^2.\end{aligned} \tag{37.37}$$

But we will have a delta function in energy, $\delta(E_f - E_i)$, that must be integrated over E, which means that we must replace one of the continuous final variables, say f_1, by the energy E. Note that all the f_k might depend on E, and E might depend on all the f_k, but still effectively only f_1 matters. Indeed, when doing this we will obtain the Jacobian of the transformation

$$J = \left\| \begin{pmatrix} \partial E/\partial f_1 & \partial E/\partial f_2 & \ldots & \ldots \\ 0 & 1 & 0 & \ldots \\ 0 & 0 & 1 & \ldots \\ \ldots & \ldots & \ldots & \ldots \\ 0 & 0 & \ldots & 1 \end{pmatrix} \right\| = \left|\frac{\partial E}{\partial f_1}\right|, \tag{37.38}$$

which in turn means that we have a “density of states” of energy in a window of dE around E_f,

$$\rho(E, f_2, \ldots) = \frac{1}{J} = \frac{1}{|\partial E/\partial f_1|}. \tag{37.39}$$

Then we can rewrite the sum of the probability over the final states as

$$\int df_1\, df_2 \cdots \sum_{f_{r-1}}\sum_{f_r} |b_f(t)|^2 = \int dE\, df_2 \cdots \sum_{f_{r-1}}\sum_{f_r} \rho(E, f_2, \ldots)|b_f(t)|^2. \tag{37.40}$$

Substituting the transition probability over a time t in (37.34), and summing over the final states, with the transformation to E, we get

$$\Delta P_{fi}(t) = t\frac{2\pi}{\hbar}|H_{1,fi}|^2 \int dE_f\, df_2 \cdots \sum_{f_r} \delta(E_f - E_i)\rho(E_f, f_2, \ldots). \tag{37.41}$$

Thus the transition probability density (in time) is given by

$$\frac{dP(i \to f)}{dt}(t) = \frac{2\pi}{\hbar}|H_{1,fi}|^2 \int df_2 \cdots \sum_{f_r} \rho(E_i, f_2, \ldots, f_r), \tag{37.42}$$

which is *Fermi's golden rule for transition processes*, and has applications in many domains.

There are two observations that we need to make:

(1) The perturbation expansion is valid only for small times t, meaning when $P(t)$ is small ($\ll 1$). Indeed, otherwise, eventually $P(t)$ would increase beyond 1 as time passes, which would be a contradiction.

(2) In order to have a well-defined probability, we need not only E_f to be near E_i but also that all the relevant f states are "near" each other. For instance, if there is a momentum direction characterizing the final state, say $f_2 = \vec{p}_f/p_f$, then we need the direction to be in a solid angle around a central value, so that df_2 corresponds to the solid angle.

37.5 Application to Scattering in a Collision

One important application of the Fermi golden rule, which also takes advantage of the second observation above, is the case of quantum scattering in collisions. In this case, we need to also generalize to the case where the *initial state* is in a continuous spectrum.

The standard quantum mechanical scattering process is the case of a monoenergetic (a given E_i) parallel beam of incoming particles of momentum $\vec{p}_0$, with direction $\vec{n}_0 = \vec{p}_0/p_0$ and with flux $\mathcal{J}$, incident on a transverse "foil" of target material and scattered at an arbitrary angle θ from the incoming axis, with arbitrary outgoing direction $\vec{n} = \vec{p}/p$. Then we have $f_2 = \vec{n}$, and $df_2 = d\Omega$ (where $d\Omega$ is an infinitesimal solid angle around $\vec{n}$); see Fig. 37.1.

We define the differential scattering cross section in terms of the number of particles incident on a surface element $dS_\perp$ perpendicular to the beam as

$$\begin{aligned} d\sigma &= \frac{\text{No. of parts. scatt. in } d\Omega \text{ and } dt}{\text{No. of incident parts. in } dS_\perp \text{ and } dt} \\ &= \frac{\text{prob. of scatt. in } d\Omega\, dt}{\text{Prob. of incidence through } dS_\perp \textit{ and } dt} \\ &= \frac{dP_{i\to f}/dt}{\mathcal{J}}. \end{aligned} \tag{37.43}$$

The current of the probability density is

$$\mathcal{J} = \vec{n}_0 \cdot \vec{j}_0 = |N_0|^2 v_0, \tag{37.44}$$

where N_0 is the normalization factor for the incident (free) wave function

$$\psi_0 = N_0 \exp\left[\frac{i}{\hbar}(\vec{p}_0 \cdot \vec{r} - Et)\right] \tag{37.45}$$

and $\vec{p}_0 = m\vec{v}_0$.

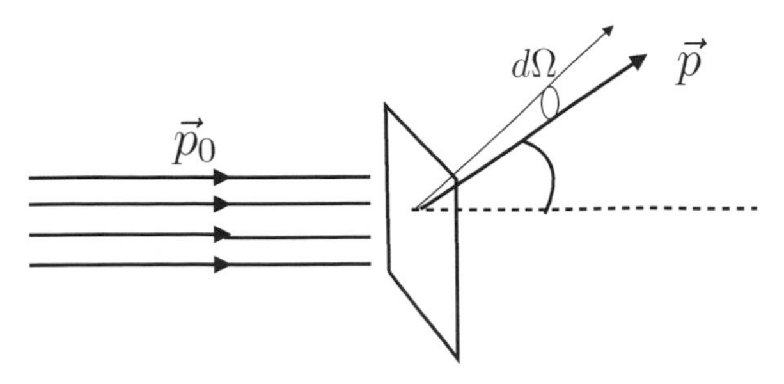

Figure 37.1 Scattering from a target foil.

The outgoing (free) wave function for particles of a given energy E and direction $\vec{n}$ is

$$\psi_{E,\vec{n}}(\vec{r},t) = \frac{\sqrt{m_0 p}}{(2\pi\hbar)^{3/2}} \exp\left[\frac{i}{\hbar}(\vec{p}\cdot\vec{r} - Et)\right], \tag{37.46}$$

which is normalized in such a way that it obeys a normalization condition over E and Ω,

$$\int_0^\infty dE \int d\Omega\, \psi^*_{E,\vec{n}}(\vec{r},t)\psi_{E,\vec{n}}(\vec{r}\,',t) = \delta(\vec{r} - \vec{r}\,'). \tag{37.47}$$

Indeed, in Chapter 5 we saw that the wave function for a free particle in one dimension, normalized over momentum p, is

$$\psi_{E,p}(x,t) = \frac{1}{\sqrt{2\pi\hbar}} \exp\left[\frac{i}{\hbar}(px - Et)\right], \tag{37.48}$$

and therefore the wave function for a three-dimensional free particle, normalized over momentum $\vec{p}$, is

$$\psi_{E,\vec{p}}(\vec{r},t) = \frac{1}{(2\pi\hbar)^{3/2}} \exp\left[\frac{i}{\hbar}(\vec{p}\cdot\vec{r} - Et)\right]. \tag{37.49}$$

But, since $\int d^3p = \int dp\, p^2 \int d\Omega = \int dE\, m_0 p \int d\Omega$, we have

$$|\psi_{E,\vec{n}}|^2 = m_0 p |\psi_{E,\vec{p}}|^2. \tag{37.50}$$

To apply perturbation theory for this scattering, we need the kinetic energy to be much larger than the interaction energy of the incoming particles with the target, $V(\vec{r})$,

$$H_0 = \frac{p^2}{2m} \gg H_1 = H_{\rm int} = V(\vec{r}). \tag{37.51}$$

If this condition is satisfied, we can use Fermi's golden rule, for the case when $f_1 = E_f$, so that $\rho = 1/J = 1$, and $f_2 = \vec{n}$, so that $df_2 = d\Omega$, obtaining

$$d\sigma = \frac{dP(i\to f)/dt}{N_0^2 v_0} = \frac{1}{N_0^2 v_0}\frac{2\pi}{\hbar}|V(\vec{r})_{fi}|^2 d\Omega, \tag{37.52}$$

where $v_0 = p_0/m_0$. The matrix element of the interaction Hamiltonian $V(\vec{r})$ is

$$\begin{aligned} V(\vec{r})_{fi} &= \int d^3r \psi^*_{E,\vec{n}}(\vec{r})V(\vec{r})\psi_{E,\vec{n}_0}(\vec{r}) \\ &= N_0 \frac{\sqrt{m_0 p}}{(2\pi\hbar)^{3/2}} \int d^3\vec{r} \exp\left[-\frac{i}{\hbar}(\vec{p} - \vec{p}_0)\cdot\vec{r}\right] V(\vec{r}) \\ &= N_0 \sqrt{m_0 p}\, \mathcal{V}(\vec{p} - \vec{p}_0), \end{aligned} \tag{37.53}$$

where we have defined the Fourier transform of the potential,

$$\mathcal{V}(\vec{p}) = \frac{1}{(2\pi\hbar)^{3/2}} \int d^3\vec{r} \exp\left(-\frac{i}{\hbar}\vec{p}\cdot\vec{r}\right) V(\vec{r}). \tag{37.54}$$

Then we obtain for the differential cross section per unit solid angle,

$$\frac{d\sigma}{d\Omega} = \frac{2\pi}{\hbar v_0} m_0 p |\mathcal{V}(\vec{p} - \vec{p}_0)|^2 = \frac{2\pi m_0^2}{\hbar}|\mathcal{V}(\vec{p} - \vec{p}_0)|^2, \tag{37.55}$$

which is the *Born approximation*, giving the *Born formula for elastic scattering*.

Coulomb Potential

The most important application of the Born formula is for the case of a Coulomb potential, obtained as a limit of the Yukawa potential (with an exponential, coming from the exchange of a massive particle, the pion, instead of a massless one, the photon),

$$V(r) = \frac{zZ|e_0|^2}{r} = \lim_{\lambda\to 0}\left(\frac{zZ|e_0|^2}{r}e^{-\lambda r}\right). \tag{37.56}$$

The Fourier transform of the Yukawa potential, in the limit, is

$$\mathcal{V}(\vec{p}) = \lim_{\lambda\to 0}\frac{2zZe_0^2}{\sqrt{2\pi\hbar}}\frac{1}{\hbar}\frac{1}{\vec{p}^{\,2}/\hbar^2+\lambda^2}. \tag{37.57}$$

The proof of this formula is left as an exercise.

Then, applying the Born formula to the case of the scattering of α particles in a Coulomb potential of a hydrogenoid atom with charge Z, with $z = 2$, gives

$$\frac{d\sigma}{d\Omega} = \left[\frac{4m_0 Ze_0^2}{(\vec{p}-\vec{p}_0)^2}\right]^2 = \left(\frac{2Ze_0^2}{4E_0}\right)^2\frac{1}{\sin^4\theta/2}, \tag{37.58}$$

where we have used the fact that $(\vec{p}-\vec{p}_0)^2 = p^2 2(1-\cos\theta) = 4p^2\sin^2\theta/2$, when $p = p_0$ (by energy conservation). This is the famous *Rutherford formula*, which can also be obtained classically (as Rutherford did) and was used to show that the target is made of atoms in which there are electrons around a nuclear core of charge $+Z|e|$ (i.e., the target is not made of atoms as solid neutral balls).

37.6 Sudden versus Adiabatic Approximation

There are two relevant approximations that are almost always considered when calculating the time-dependent perturbation theory.

The sudden approximation corresponds to having a sudden change in the Hamiltonian, by introducing the perturbation $\hat{H}_1$ over a time $\Delta t = 2\epsilon \to 0$. It is implicit in the time-independent perturbation analyzed before, where until $t_0-\epsilon$ we had $\hat{H}_1(t) = 0$ and after $t_0+\epsilon$ we had $\hat{H}_1(t) = \hat{H}_1$ (time independent).

Then, in this approximation we can integrate the Schrödinger equation over the time period of the change,

$$|\psi(+\epsilon)\rangle - |\psi(-\epsilon)\rangle = -\frac{i}{\hbar}\int_{-\epsilon}^{+\epsilon} dt\, H(t)|\psi(t)\rangle \to 0. \tag{37.59}$$

Indeed, since the integrand is finite and $\epsilon \to 0$, the integral vanishes. Then we obtain

$$|\psi(0+)\rangle = |\psi(0-)\rangle, \tag{37.60}$$

or that the state does not change over the sudden transition that introduces the perturbation Hamiltonian.

Adiabatic Perturbation

The opposite situation occurs when we introduce the perturbation very slowly, or *adiabatically*, meaning that we transition from the initial state $|n^{(0)}\rangle$, an eigenstate of $\hat{H}_0$, to the state $|n\rangle$, an eigenstate of the full Hamiltonian $\hat{H}$, which corresponds to $|n^{(0)}\rangle$ as $\lambda \to 0$.

The way in which we introduce the interaction $\hat{H}_1$ is to factorize the time dependence as follows:

$$\hat{H}_1(t) = e^{\gamma t}\hat{H}_1. \tag{37.61}$$

As we see, at $t = 0$ we introduce the perturbation ($\hat{H}_1(t = 0) = \hat{H}_1$), and we see that in fact we can start at $t_0 \to -\infty$, when $\hat{H}_1(t) \to 0$.

This formalism connects the time-dependent and time-independent perturbation theories, as we will now show.

From (37.26), substituting the above $\hat{H}_1(t)$, we obtain

$$\begin{aligned} i\hbar b_m^{(1)}(\tau) &= \int_{-\infty}^{\tau} dt\, e^{i\omega_{mi}t} e^{\gamma t} H_{1,mi} \\ &= H_{1,mi} \frac{e^{i\omega_{mi}\tau + \gamma\tau}}{i\omega_{mi} + \gamma}. \end{aligned} \tag{37.62}$$

Then the transition probability is

$$|b_m^{(1)}(\tau)|^2 = \frac{|H_{1,mi}|^2}{\hbar^2} \frac{e^{2\gamma\tau}}{\omega_{mi}^2 + \gamma^2}. \tag{37.63}$$

Moreover, the probability density in time is

$$\frac{d}{dt}|b_m^{(1)}(t)|^2 = \frac{2|H_{1,mi}|^2}{\hbar^2} \frac{\gamma e^{2\gamma t}}{\omega_{mi}^2 + \gamma^2}. \tag{37.64}$$

But since, by smoothly removing the perturbation, by taking $\gamma \to 0$, so that

$$\frac{1}{\pi} \lim_{\gamma \to 0} \frac{\gamma}{\gamma^2 + \omega_{mi}^2} = \delta(\omega_{mi}) = \hbar\delta(E_m - E_i), \tag{37.65}$$

we obtain

$$\frac{d}{dt}|b_m^{(1)}(t)|^2 = \frac{2\pi}{\hbar}|H_{1,mi}|^2 \delta(E_m - E_i), \tag{37.66}$$

which is Fermi's golden rule, as expected for a time-dependent perturbation (introduced suddenly, at a very early time, $t_0 \to -\infty$).

On the other hand, we can also obtain the time-independent perturbation formula. Indeed, first using the time-dependent formula, we have

$$c_m(t) = \langle m^{(0)}|\psi(t)\rangle = b_m(t) e^{-iE_m t/\hbar}. \tag{37.67}$$

Considering a negative initial time, $t_0 < 0$, and vanishing final time, $t = 0$, and restricting ourselves to first-order perturbation theory, we obtain

$$\begin{aligned} C_m(t = 0) &= -\frac{i}{\hbar}\int_{t_0}^{0} dt' e^{+i\omega_{mi}t'} H_{1,mi}(t') = -\frac{i}{\hbar}\int_{t_0}^{0} dt' e^{+i\omega_{mi}t'} H_{1,mi} e^{\gamma t'} \\ &= -\frac{i}{\hbar}\frac{1 - e^{\gamma t_0 + i\omega_{mi}t_0}}{i\omega_{mi} + \gamma} H_{1,mi}. \end{aligned} \tag{37.68}$$

Then, as we take $t_0 \to -\infty$ to introduce the perturbation adiabatically, we obtain

$$C_m(t=0) \to \frac{H_{1,mi}}{\hbar\omega_{mi}} = \frac{H_{1,mi}}{E_i - E_m}, \tag{37.69}$$

which is the time-independent perturbation formula.

Important Concepts to Remember

- Time-dependent perturbation theory corresponds to $\hat{H}(t) = \hat{H}_0 + \lambda \hat{H}_1(t)$, and can be solved with the method of variation of constants, $|\psi(t)\rangle = \sum_n c_n(t)|\psi_n\rangle$, where $|\psi_n\rangle$ are the eigenstates of $\hat{H}_0$.
- We expand the coefficients in λ, after factoring out the $\hat{H}_0$ time dependence: $c_n(t) = e^{-iE_n t/\hbar}\sum_s b_n^{(s)}\lambda^s$.
- At first order, we find $b_n^{(1)}(t) = (1/i\hbar)\int_0^t dt' e^{i\omega_{mi}t'} H_{1,mi}(t)$ and

$$|\psi(t)\rangle = \sum_n (\delta_{ni} + b_n^{(1)}(t)) e^{-iE_n t/\hbar}|\psi_n\rangle.$$

- In the sudden approximation ($\hat{H}_1$ turned on at $t = 0$), we find

$$\frac{dP_{if}}{d\tau} = \frac{d|b_m^{(1)}(\tau)|}{d\tau} = \frac{2\pi}{\hbar}|H_{1,mi}|^2\delta(E_f - E_i)$$

  at large times.
- For a transition to a continuous spectrum, we find Fermi's golden rule,

$$\frac{dP(i\to f)(t)}{dt} = \frac{2\pi}{\hbar}|H_{1,fi}|^2 \int df_2 \cdots \sum_{f_r} \rho(E_i, f_2, \cdots, f_r),$$

  where all relevant f states need to be "near each other".
- We can apply Fermi's golden rule to elastic scattering in a collision, and obtain the Born formula (Born approximation)

$$\frac{d\sigma}{d\Omega} = \frac{2\pi m_0^2}{\hbar}|\mathcal{V}(\vec{p} - \vec{p}_0)|^2,$$

  which for the hydrogenoid atom gives the Rutherford formula,

$$\frac{d\sigma}{d\Omega} = \left(\frac{2Ze_0^2}{4E_0}\right)^2 \frac{1}{\sin^4\theta/2}.$$

- The adiabatic approximation, with $\hat{H}_1$ introduced as $e^{\gamma t}\hat{H}_1$, for $t < 0$, allows us to connect the time-dependent and time-independent formalism.

Further Reading

See [2], [1].

Exercises

(1) Consider $\hat{H}_0$ corresponding to the hydrogen atom, with the initial state i being the ground state, and $\lambda H_{1,mi} = \lambda e^{-mA+\tilde{\lambda}t}$, for $t < 0$, where A is a constant. Calculate the probability of transition to the mth energy eigenstate as a function of time.

(2) Consider $\hat{H}_0$ corresponding to the hydrogen atom, the initial state i being the ground state, and $\hat{H}_1 = Kr$, with K constant, a perturbation introduced suddenly at $t = 0$. Calculate the transition probability to the state $n = 2, l = 0$ as a function of time. What happens at large times?

(3) Prove the formula (37.32) for the limit giving the delta function.

(4) Consider an electron in an atom, with a potential that can be approximated by a (radial) square well of radius r_0 above zero, in its ground state $i = 0$. Introduce a small perturbation $H_{mi} = H =$ constant. Calculate the rate for the electron to transition out of the square well and into the free space beyond.

(5) Calculate the differential cross section for scattering from a delta function potential, $V = V_0\delta(\vec{r})$.

(6) Prove the formula (37.57).

(7) Consider the adiabatic introduction of the perturbation H_1 with

$$\hat{H}_1(t) = \frac{\hat{H}_1}{1+\gamma^2 t^2},$$

for $t < 0$. Calculate the transition rate.

38 Time-Dependent Perturbation Theory: Second and All Orders

In this chapter, we continue with time-dependent perturbation theory, first with the second order, as applied to finding the Breit–Wigner distribution for a transition with an energy shift and decay width, and then with an all-orders formalism.

38.1 Second-Order Perturbation Theory and Breit–Wigner Distribution (Energy Shift and Decay Width)

As we saw in the previous chapter, the general recursion relation between the $(s+1)$th-order and the sth-order in perturbation theory is

$$i\hbar\dot{b}_m^{(s+1)}(t) = \sum_n H_{1,mn}(t)b_n^{(s)}(t)e^{i\omega_{mn}t}. \tag{38.1}$$

We are interested in the $s = 1$ time-independent (sudden approximation) case, which gives

$$i\hbar\dot{b}_m^{(2)}(t) = \sum_n H_{1,mn}b_n^{(1)}(t)e^{i\omega_{mn}t}. \tag{38.2}$$

Specializing for $m = i$, we get

$$i\hbar\dot{b}_i^{(2)}(t) = \sum_n H_{1,in}b_n^{(1)}(t)e^{i\omega_{in}t}, \tag{38.3}$$

but we saw in the previous chapter that

$$b_n^{(1)}(t) = \frac{1}{i\hbar}\int_0^t dt' e^{i\omega_{ni}t'}H_{1,ni}(t')b_i^{(0)}(t'), \tag{38.4}$$

for $n \neq i$, where, however, $b_i^{(0)}(t') = 1$ (more precisely, $b_m^{(0)}(t') = \delta_{mi}$). On the other hand, $b_i^{(1)}(t) = 0$, so the time dependence of the initial state only appears from the second order (i.e., in $b_i^{(2)}(t)$) onwards.

Considering the case of $H_{1,in} \neq 0$ and sudden approximation (constant $H_{1,in}$ in time), substituting $b_n^{(1)}$ in the formula for $b_i^{(2)}$, we obtain (note that $\omega_{in} = -\omega_{ni}$)

$$\begin{aligned} b_i^{(2)}(t) &= -\sum_n \frac{|H_{1,in}|^2}{\hbar^2}\int_0^t dt'' e^{i\omega_{in}t''}\int_0^{t''} dt' e^{i\omega_{ni}t'}b_i^{(0)} \\ &= \sum_n \frac{i|H_{1,in}|^2}{\hbar^2\omega_{ni}}\int_0^t dt''\left(1 - e^{-i\omega_{ni}t''}\right)b_i^{(0)}. \end{aligned} \tag{38.5}$$

Then we find a decay formula for the initial state, where $b_i(t) \simeq b_i^{(0)}$ $(=1)$ and $\dot{b}_i(t) \simeq \dot{b}_i^{(2)}$,

$$\frac{\dot{b}_i(t)}{b_i(t)} = \sum_n \frac{i|H_{1,in}|^2}{\hbar} \frac{1 - e^{-i(E_n - E_i)t/\hbar}}{E_n - E_i}. \tag{38.6}$$

To continue, we use the formula

$$\lim_{a\to\infty} \frac{1 - e^{-iax}}{x} = i\pi\delta(x) + \mathcal{P}\frac{1}{x}, \tag{38.7}$$

understood as a relation in distributions, which implies that (if $y < 0, z > 0$)

$$\begin{aligned} \lim_{a\to\infty} \int_y^z dx\, f(x) \frac{1 - e^{-iax}}{x} &= i\pi f(x) + \lim_{\epsilon\to 0}\left(\int_y^{-\epsilon} \frac{f(x)}{x}dx + \int_{+\epsilon}^z \frac{f(x)}{x}dx\right) \\ &= i\pi f(x) + \mathcal{P}\int_y^z \frac{f(x)}{x}dx. \end{aligned} \tag{38.8}$$

Here we have defined the *principal part* $\mathcal{P}$ of an integral as the integral minus its pole,

$$\mathcal{P}\int_y^z \frac{f(x)}{x}dx = \int_y^{-\epsilon} \frac{f(x)}{x}dx + \int_{+\epsilon}^z \frac{f(x)}{x}dx. \tag{38.9}$$

To prove the above relation, valid for distributions, we first note that

$$\frac{1 - e^{-iax}}{x} = \frac{i\sin ax}{x} + \frac{1 - \cos ax}{x}, \tag{38.10}$$

and, from the previous chapter, we know that if $a \to \infty$, then the first term goes to $i\pi\delta(x)$. Moreover,

$$\int_y^z dx\, f(x) \frac{1 - \cos ax}{x} = \left(\int_y^{-\epsilon} + \int_{-\epsilon}^{+\epsilon} + \int_\epsilon^z\right) \frac{1 - \cos ax}{x}, \tag{38.11}$$

and, since around $x = 0$ we have $(1 - \cos ax)/x \simeq a^2x/2$, we have that

$$\int_{-\epsilon}^{+\epsilon} \frac{1 - \cos ax}{x} \to 0, \tag{38.12}$$

which leaves only the principal part of the integral. Moreover, for the principal part (excluding the region near $x = 0$), we have

$$\lim_{a\to\infty} \int -\frac{\cos ax}{2} f(x)dx = 0, \tag{38.13}$$

since the integrand oscillates very fast, averaging to zero. Then finally we find

$$\lim_{a\to\infty} \int_y^z f(x) \frac{1 - e^{-iax}}{x} = i\pi f(x) + \mathcal{P}\int_y^z \frac{f(x)}{x}dx. \tag{38.14}$$

q.e.d.

Substituting into (38.6), with $a = t/\hbar$ and $x = E_n - E_i$, we get

$$\frac{\dot{b}_i(t)}{b_i(t)} \simeq \sum_n \frac{i}{\hbar}|H_{1,in}|^2 \left(i\pi\delta(E_n - E_i) + \mathcal{P}\frac{1}{E_n - E_i}\right). \tag{38.15}$$

More precisely, we have $\sum_n$ as a sum over final states, so

$$\frac{\dot{b}_i(t)}{b_i(t)} \simeq \int df_2 \cdots \sum_{f_n} \int dE_n \rho(E_n, f_2, \ldots, f_n) \frac{1}{\hbar} |H_{1,in}|^2 \left[-\pi\delta(E_n - E_i) + i\mathcal{P}\frac{1}{E_n - E_i} \right]$$

$$= \pi \int df_2 \cdots \sum_{f_n} \rho(E_i, f_2, \ldots, f_n) \frac{|H_{1,in}|^2}{\hbar} \tag{38.16}$$

$$+ \frac{i}{\hbar}\mathcal{P} \int dE_n \int df_2 \cdots \sum_{f_n} \rho(E_n, f_2, \ldots, f_n) \frac{|H_{1,ni}|^2}{E_n - E_i}.$$

The term on the second line is defined to be $\equiv -\Gamma/2$ and the term on the third line is defined to be $\equiv -(i/\hbar)\Delta E_i$, and in total the result is defined to be equal to

$$-\gamma \equiv -\left(\frac{\Gamma}{2} + \frac{i}{\hbar}\Delta E_i \right). \tag{38.17}$$

Thus

$$\Gamma = \frac{2\pi}{\hbar} \sum_n \int df_2 \cdots \sum_{f_n} |H_{1,ni}|^2 \rho(E, f_2, \ldots, f_n)$$
$$\Delta E_i = -\sum_n \mathcal{P} \int dE_n \int df_2 \cdots \sum_{f_n} \frac{|H_{1,ni}|^2}{E_n - E_i} \rho(E_n, f_2, \ldots, f_n), \tag{38.18}$$

and the decay of the initial state is given by

$$\frac{\dot{b}_i(t)}{b_i(t)} \simeq -\gamma \quad \Rightarrow \quad b_i(t) = e^{-\gamma t} = e^{-\Gamma t/2} e^{-i\Delta E t/\hbar}. \tag{38.19}$$

Note that the coefficient $b_i(t)$ calculated here, $\simeq 1 - \Gamma/2t - i/\hbar\Delta E t + \cdots$ amounts to a sum of the zeroth-order $b_i^{(0)} = 1$ and the second-order one in $-\Gamma/2 - i/\hbar\Delta E$, plus an infinite series of contributions from higher orders.

Thus the coefficients of the initial eigenstate of $\hat{H}_0$ are

$$c_i(t) = b_i(t) \exp\left(-\frac{i}{\hbar} E_i t \right) \simeq e^{-\Gamma t/2} e^{-i(E_i + \Delta E_i)t/\hbar}, \tag{38.20}$$

which means that Γ is a "decay width", whereas ΔE_i is an energy shift.

Indeed, the probability of finding the particle in the same state as the initial state is

$$P_i(t) = |c_i(t)|^2 \simeq e^{-\Gamma t}, \tag{38.21}$$

so Γ is indeed a decay width. Moreover, the lifetime of the particle is given by

$$\langle t \rangle = \int_0^\infty dt\, t \left(-\frac{dP}{dt} \right) = \int_0^\infty dt\, t \left(-\frac{d|c_i|^2}{dt} \right) = \int_0^\infty dt\, t\Gamma e^{-\Gamma t} = \frac{1}{\Gamma}. \tag{38.22}$$

Since we found that in (38.19) there are zeroth-order plus second-order contributions, substituting it into the right-hand side of (38.1), we obtain a sum of first-order and third-order terms for b_m. Moreover, the terms in the sum with $n \neq i$ are negligible compared with those with $n = i$, so only the term with $n = i$ remains, and we get

$$i\hbar \dot{b}_m^{(1)+(3)}(t) = H_{1,mi} b_i^{(0)+(2)}(t) e^{i\omega_{mi}t} = H_{1,mi} e^{-\gamma t} e^{i\omega_{mi}t}. \tag{38.23}$$

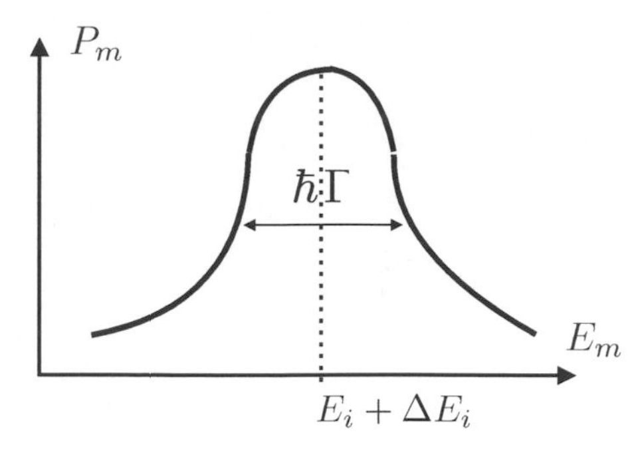

Figure 38.1 The Breit–Wigner distribution is a Lorentzian curve.

Integrating this equation, we obtain

$$\begin{aligned} b_m^{(1)+(3)}(\tau) &= \frac{1}{i\hbar} H_{1,mi} \int_0^\tau dt\, e^{(i\omega_{mi}-\gamma)t} \\ &= -\frac{H_{1,mi}\left(e^{i(E_m-E_i)\tau/\hbar} e^{-\gamma\tau} - 1\right)}{E_m - E_i + i\hbar\gamma}. \end{aligned} \tag{38.24}$$

At large times, $\tau \to \infty$, $e^{-\gamma\tau} \to 0$, so we obtain

$$b_m^{(1)+(3)}(\tau \to \infty) \to \frac{H_{1,mi}}{E_m - E_i + i\hbar\gamma} = \frac{H_{1,mi}}{E_m - E_i - \Delta E_i + i\hbar\Gamma/2}. \tag{38.25}$$

Therefore the probability of finding the state in the mth eigenstate at large times is

$$|c_m^{(1)+(3)}|^2 = |b_m^{(1)+(3)}|^2 = \frac{|H_{1,mi}|^2}{(E_m - E_i - \Delta E_i)^2 + (\hbar\Gamma/2)^2}, \tag{38.26}$$

which is the *Breit–Wigner distribution* in energies, a Lorentzian curve, as in Fig. 38.1. Indeed, here $\hbar\Gamma$ is the full width at half maximum value of the distribution.

Notice also that this distribution implies the energy–time uncertainty relation, since the uncertainty in time is $\Delta t \sim \langle t \rangle$, so

$$\Delta t \Delta E \sim \hbar. \tag{38.27}$$

38.2 General Perturbation

The general theory of perturbations (to all orders) is best described in the interaction picture, defined in Chapter 9. We review it here.

As before, we consider a nonperturbed Hamiltonian $\hat{H}_0$ and an interaction $\hat{H}_1$, so $\hat{H} = \hat{H}_0 + \hat{H}_1$. Then we go to the Heisenberg picture with the free Hamiltonian $\hat{H}_0$ only, so we change states and operators with respect to the Schrödinger picture,

$$\begin{aligned} |\psi_W\rangle &= \hat{W}|\psi\rangle \\ \hat{A}_W &= \hat{W}\hat{A}\hat{W}^{-1}, \end{aligned} \tag{38.28}$$

where

$$\hat{W}(t) = (U_S^{(0)}(t,t_0))^{-1} = \left\{\exp\left[-\frac{i}{\hbar}\hat{H}_0(t-t_0)\right]\right\}^{-1} = \exp\left[+\frac{i}{\hbar}\hat{H}_0(t-t_0)\right], \tag{38.29}$$

and $U_S^{(0)}(t,t_0)$ is the Schrödinger-picture evolution operator for propagation with $\hat{H}_0$. Then

$$\begin{aligned} |\psi_I(t)\rangle &= \exp\left[\frac{i}{\hbar}\hat{H}_0(t-t_0)\right]|\psi_S(t)\rangle \\ \hat{A}_I &= \exp\left[\frac{i}{\hbar}\hat{H}_0(t-t_0)\right]\hat{A}_S \exp\left[-\frac{i}{\hbar}\hat{H}_0(t-t_0)\right] \end{aligned} \tag{38.30}$$

Thus, now both the state $|\psi_I(t)\rangle$ and the operator $\hat{A}_I(t)$ are time dependent, with

$$\begin{aligned} i\hbar\frac{\partial}{\partial t}|\psi_I(t)\rangle &= \hat{H}_{1,I}|\psi_I(t)\rangle \\ i\hbar\frac{\partial}{\partial t}\hat{A}_I(t) &= [\hat{A}_I(t), \hat{H}_{0,I}], \end{aligned} \tag{38.31}$$

where the index I means that the operator is in the interaction picture.

The evolution in the interaction picture is given by

$$|\psi_I(t)\rangle = \hat{U}_I(t,t_0)|\psi_I(t_0)\rangle = \hat{U}_I(t,t_0)|\psi_S(t_0)\rangle, \tag{38.32}$$

since all the pictures must give the same result at time t_0. Then we find

$$\hat{U}_I(t,t_0) = \hat{W}(t)\hat{U}_S(t,t_0)\hat{W}^{-1}(t_0) = \exp\left[\frac{i}{\hbar}\hat{H}_0(t-t_0)\right]\hat{U}_S(t,t_0), \tag{38.33}$$

where we have used the fact that $\hat{W}(t_0) = 1$. This interaction-picture operator then satisfies the same equation as $|\psi_I(t)\rangle$, namely

$$i\hbar\frac{\partial}{\partial t}\hat{U}_I(t,t_0) = \hat{H}_{1,I}\hat{U}_I(t,t_0). \tag{38.34}$$

This equation is solved perturbatively, as

$$\begin{aligned} \hat{U}_I(t,t_0) &= 1 + \left(\frac{-i}{\hbar}\right)\int_{t_0}^t dt_1 H_{1,I}(t_1) + \left(\frac{-i}{\hbar}\right)^2\int_{t_0}^t dt_1\int_{t_0}^{t_1} dt_2 H_{1,I}(t_1)H_{1,I}(t_2) + \cdots \\ &= 1 + \left(\frac{-i}{\hbar}\right)\int_{t_0}^t dt_1 H_{1,I}(t_1) + \frac{1}{2!}\left(\frac{-i}{\hbar}\right)^2\int_{t_0}^t dt_1\int_{t_0}^{t} dt_2 T\left[H_{1,I}(t_1)H_{1,I}(t_2)\right] + \cdots \\ &= T\left\{\exp\left[-\frac{i}{\hbar}\int_{t_0}^t dt' H_{1,I}(t')\right]\right\}, \end{aligned} \tag{38.35}$$

where T is a time-ordering operator.

38.3 Finding the Probability Coefficients $b_n(t)$

Consider the initial interaction-picture state (which equals the Schrödinger-picture state) to be an eigenket of $\hat{H}_0$ of index i, as before, so that

$$|\psi_I(t)\rangle = \hat{H}_I(t,t_0)|\psi_I(t_0)\rangle = \hat{U}_I(t,t_0)|\psi_i\rangle. \tag{38.36}$$

Then the Schrödinger-picture state is

$$|\psi_S(t)\rangle = \exp\left[-\frac{i}{\hbar}\hat{H}_0(t-t_0)\right]|\psi_I(t)\rangle = \exp\left[-\frac{i}{\hbar}\hat{H}_0(t-t_0)\right]\hat{U}_I(t,t_0)|\psi_i\rangle. \tag{38.37}$$

In order to calculate transition probabilities, we insert on the left the identity expanded in a complete set of the eigenkets of $\hat{H}_0$, $\mathbb{1} = \sum_n |\psi_n\rangle\langle\psi_n|$. We obtain

$$\begin{aligned}|\psi_S(t)\rangle &= \sum_n |\psi_n\rangle\langle\psi_n| \exp\left[-\frac{i}{\hbar}|\hat{H}_0(t-t_0)\right]\hat{U}_I(t,t_0)|\psi_i\rangle \\ &= \sum_n \exp\left[-\frac{i}{\hbar}E_n(t-t_0)\right]|\psi_n\rangle\langle\psi_n|\hat{U}_I(t,t_0)|\psi_i\rangle,\end{aligned} \tag{38.38}$$

where we have acted with $\exp\left[-\frac{i}{\hbar}\hat{H}_0(t-t_0)\right]$ on the left, on $\langle\psi_n|$. Identifying the result with the general expansion in terms of $b_n(t)$ coefficients, we find that

$$\begin{aligned}b_n(t) &= \langle\psi_n|\hat{U}_I(t,t_0)|\psi_i\rangle \\ &= \left\langle\psi_n\right|\sum_{s=0}^{\infty}\left\{\frac{(-i/\hbar)^s}{s!}\int_{t_0}^{t}dt_1\cdots\int_{t_0}^{t}dt_s T\left[H_{1,I}(t_1)\dots H_{1,I}(t_s)\right]\right\}\left|\psi_i\right\rangle \\ &= \delta_{ni} + \sum_{s=1}^{\infty}\frac{(-i/\hbar)^s}{s!}\int_{t_0}^{t}dt_1\cdots\int_{t_0}^{t}dt_s\langle\psi_n|T\left\{H_{1,I}(t_1)\dots H_{1,I}(t_s)\right\}|\psi_i\rangle.\end{aligned} \tag{38.39}$$

Now insert the identity written as a complete set over m_k, $\mathbb{1} = \sum_{m_k}|\psi_{m_k}\rangle\langle\psi_{m_k}|$, after each Hamiltonian at t_k. Then, in

$$H_{1,I}(t_k) = \exp\left[\frac{i}{\hbar}\hat{H}_0(t_k-t_0)\right]H_{1,S}(t_k)\exp\left[-\frac{i}{\hbar}\hat{H}_0(t_k-t_0)\right], \tag{38.40}$$

we can act with the operator on the left on the left-hand $\langle\psi_{m_{k-1}}|$ and with the operator on the right on the right-hand $|\psi_{m_k}\rangle$, obtaining

$$\begin{aligned}b_n(t) = \delta_{ni} &+ \sum_{s=1}^{\infty}\frac{(-i/\hbar)^s}{s!}\int_{t_0}^{t}dt_1\cdots\int_{t_0}^{t}dt_s \\ &\times\sum_{\{m_k\}}T\left\{\left(\exp\left[\frac{i}{\hbar}E_n(t_1-t_0)\right]H_{1,S,nm_1}(t_1)\exp\left[-\frac{i}{\hbar}E_{m_1}(t_1-t_0)\right]\right)\right. \\ &\times\left(\exp\left[+\frac{i}{\hbar}E_{m_1}(t_2-t_0)\right]H_{1,S,m_1m_2}(t_2)\exp\left[-\frac{i}{\hbar}E_{m_2}(t_2-t_0)\right]\right)\cdots \\ &\times\left.\left(\exp\left[+\frac{i}{\hbar}E_{m_{s-1}}(t_s-t_0)\right]H_{1,S,m_{s-1}i}(t_s)\exp\left[-\frac{i}{\hbar}E_i(t_s-t_0)\right]\right)\right\}.\end{aligned} \tag{38.41}$$

This expansion can be identified with the expansion in λ of the coefficients,

$$b_n(t) = \sum_{s=0}^{\infty}b_n^{(s)}(t), \tag{38.42}$$

with $b_n^{(0)}(t) = \delta_{ni}$.

We have therefore re-obtained the zeroth-order, first-order, and second-order terms, and generalized to all orders, as promised.

Important Concepts to Remember

- In the second order of time-dependent perturbation theory, more precisely summing the zeroth-order, second-order, and higher-order contributions, we find, for the initial-state coefficient, $c_i(t) = b_i(t)\exp\left(-\frac{i}{\hbar}E_i t\right) = e^{-\frac{\Gamma}{2}t}\exp\left[-\frac{i}{\hbar}(E_i + \Delta E_i)t\right]$, where Γ is the decay width and ΔE_i is the energy shift.
- This leads to the decay of the original state, $P_i(t) \simeq e^{-\Gamma t}$, and to the Breit–Wigner distribution for the other states, $|c_m^{(1)+(3)}|^2 = |b_m^{(1)+(3)}|^2 = |H_{1,mi}|^2/[(E_m - E_i - \Delta E_i)^2 + (\hbar\Gamma/2)^2]$
- For a general perturbation, we can use the formalism of the evolution operator in the interaction picture, which is $\hat{U}_I(t,t_0) = T\left\{\exp\left[-\frac{i}{\hbar}\int_{t_0}^t dt' H_{1,I}(t')\right]\right\}$, and note that $b_n(t) = \langle\psi_n|\hat{U}_I(t,t_0)|\psi_i\rangle$.

Further Reading

See [1] and [2].

Exercises

(1) In general, can we have a zero-energy shift if we have a nonzero decay width (finite lifetime)? How about a zero decay width (infinite lifetime) for a nonzero energy shift?

(2) In a transition, if the density of final states $\rho(E_n)$ is an increasing exponential, $\rho(E_n) = Ae^{\alpha E_n}$, $\alpha > 0$, what does the (physical) condition of a finite energy shift imply for the transition Hamiltonian matrix $H_{1,in}$, if $E_i > 0$?

(3) If H_{ni} is independent of $E_n = E_i$, what physical condition can we impose on $\rho(E_n)$?

(4) Argue for the energy–time uncertainty relation, $\Delta E \Delta t \sim \hbar$, on the basis of the Breit–Wigner distribution.

(5) Calculate the interaction picture operator $\hat{U}_I(t,t_0)$ and $b_n(t)$ for the interaction in the sudden approximation.

(6) Show that the first-order formula for time-dependent perturbation theory can be recovered from (38.41).

(7) Show that the second-order formula for time-dependent perturbation theory can be recovered from (38.41).

39 Application: Interaction with (Classical) Electromagnetic Field, Absorption, Photoelectric and Zeeman Effects

In this chapter we will apply the formalisms of time-independent and time-dependent perturbation theory to the case of the interaction of electrons and atoms with an electromagnetic field. In particular, we will consider the Zeeman effect for bound electrons, and the absorption and photoelectric effects for atoms.

39.1 Particles and Atoms in an Electromagnetic Field

In Chapter 22, we described the interaction of a quantum mechanical particle with a classical electromagnetic field. We found that the interaction Hamiltonian is

$$\begin{aligned}\hat{H} &= \frac{1}{2m}\left(\vec{p} - q\vec{A}\right)^2 + V'(\vec{r}) - qA_0 \\ &= \frac{\vec{p}^{\,2}}{2m} - \frac{q}{m}\vec{A}\cdot\vec{p} - qA_0(\vec{r}) + V'(\vec{r}) + \frac{q^2\vec{A}^2}{2m}.\end{aligned} \tag{39.1}$$

Here $V'(\vec{r})$ is a non-electromagnetic part of the potential (more precisely, non-external electromagnetic-field), so

$$-qA_0(\vec{r}) + V'(\vec{r}) = V(\vec{r}), \tag{39.2}$$

and $\vec{p} = \frac{\hbar}{i}\vec{\nabla}$, as usual, so, in the gauge $\vec{\nabla}\cdot\vec{A} = 0$, we have

$$\hat{H} = -\frac{\hbar^2}{2m}\vec{\nabla}^2 + V'(\vec{r}) + \frac{iq\hbar}{m}\vec{A}\cdot\vec{\nabla} - qA_0 + \frac{q^2}{2m}\vec{A}^2. \tag{39.3}$$

Otherwise, there would be a $+\frac{iq\hbar}{2m}(\vec{\nabla}\cdot\vec{A})$ term.

Moreover, we can ignore the last term in $\hat{H}$. In a hydrogenoid atom, where $V'(\vec{r})$ is the interaction of the electron with the nucleus, we have

$$\hat{H}_0 = -\frac{\hbar^2}{2m}\vec{\nabla}^2 + V'(\vec{r}), \tag{39.4}$$

and the other terms make up $\hat{H}_1$.

If $\vec{A}, A_0$ constitute a radiation field, we can also put $A_0 = 0$ as a gauge choice, obtaining the radiation gauge. Neglecting the $\vec{A}^2$ term (as being small, of second order), we obtain

$$\hat{H}_1 = \frac{iq\hbar}{m}\vec{A}\cdot\vec{\nabla}. \tag{39.5}$$

39.2 Zeeman and Paschen–Back Effects

Consider an atom with strong spin–orbit coupling in a magnetic field. In Chapter 24, we found that in the presence of a magnetic field in the z direction, $B = B_z$, the interaction Hamiltonian between the electrons in the atom and the external magnetic field is

$$\hat{H}_{\text{B-int}} = -\vec{\mu}_{\text{tot}} \cdot \vec{B} = -\mu_B(\vec{L} + 2\vec{S}) \cdot \vec{B} = -\mu_B B(L_z + 2S_z) = -\mu_B B(J_z + S_z). \quad (39.6)$$

When continuing the analysis, the implicit assumption (not expressed) was that this $\hat{H}_{\text{B-int}}$ is a perturbation. Then, we can use time-independent perturbation theory in first order, which does not change the states, so that

$$|n^{(0)}\rangle = |E_n JM\rangle. \quad (39.7)$$

Therefore, the energy shift in first-order perturbation theory becomes

$$\begin{aligned} \Delta E_n^{(1)} &= H_{1,nn} = \langle n^{(0)}|\hat{H}_1|n^{(0)}\rangle \\ &= -\mu_B B\langle E_n JM|J_z + S_z|E_n JM\rangle. \end{aligned} \quad (39.8)$$

But, by the Wigner–Eckhart theorem (from Chapter 16), $\vec{S}$ is a vector operator like $\vec{J}$ so they are proportional, and so we have

$$\langle S_z\rangle = \frac{\langle \vec{J}\cdot\vec{S}\rangle}{\langle \vec{J}^2\rangle}\langle J_z\rangle. \quad (39.9)$$

Using this relation, we found that

$$\langle E_n JM|J_z + S_z|E_n JM'\rangle = gM\delta_{MM'}, \quad (39.10)$$

where g is the Landé g factor, calculated from the above relation. Then, finally, we found

$$\Delta E_n^{(1)} = -gM\mu_B B. \quad (39.11)$$

We must make one observation, however. This *linear* Zeeman effect in a hydrogenoid atom is valid only in the case where $j \neq 0$, so either $l \neq 0$ or $s \neq 0$. But, if $l = s = 0$ (so $j = 0$), then we need to consider the second-order term in $\hat{H}$,

$$\hat{H}_2 = \frac{e^2\vec{A}^2}{2m} = \frac{m\omega_L^2 r^2\sin^2\theta}{2}, \quad (39.12)$$

where

$$\omega_L = \frac{eB}{2m} \quad (39.13)$$

is the Larmor frequency. Then the energy shift in first-order perturbation theory, using this $\hat{H}_2$, is

$$\Delta E_n^{(1)} = \frac{m}{2}\left(\frac{eB}{2m}\right)^2\left(\frac{a_0}{Z}\right)^2\frac{1}{3}n^2(5n^2+1), \quad (39.14)$$

where we have used the fact that, for a hydrogenoid atom,

$$\langle r^2\sin^2\theta\rangle_n = \left(\frac{a_0}{Z}\right)^2\frac{1}{3}n^2(5n^2+1). \quad (39.15)$$

Paschen–Back Effect

This effect occurs in the opposite limit to the (linear) Zeeman effect, in which limit the LS (spin–orbit) coupling (described in Chapter 22) is $\ll \mu_B B$. This means that the spin–orbit interaction can be treated perturbatively, and does not couple $\vec{L}$ with $\vec{S}$ into $\vec{J}$ in this limit. Then the correct quantum states to consider are $|E_n LSM_L M_S\rangle$, instead of $|E_n LSJM_J\rangle$. Therefore the energy shift in first-order time-independent perturbation theory is

$$\begin{aligned}\Delta E_n^{(1)} &= \langle E_n LSM_L M_S|\hat{H}_1|E_n LSM_L M_S\rangle \\ &= -\mu_B B(M_L + 2M_S).\end{aligned} \tag{39.16}$$

However, we note that actually this $\hat{H}_1$ is *diagonal* in this basis, so it is actually a part of $\hat{H}_0$. Thus here the perturbation $\hat{H}_1$ is the spin–orbit coupling, which is a small effect.

39.3 Electromagnetic Radiation and Selection Rules

Before considering the absorption and emission of photons, we will consider the emission of classical electromagnetic radiation from an electric current, calculate the interaction Hamiltonian, and find the resulting selection rules for the transition between initial and final states interacting with the electromagnetic radiation.

We consider the same radiation gauge, $A_0 = 0$ and $\vec{\nabla} \cdot \vec{A} = 0$. Then the interaction Hamiltonian of the classical electromagnetic radiation with an electric current $\vec{j}$ is

$$H_i = \int d^3r \vec{j} \cdot \vec{A}. \tag{39.17}$$

In the radiation gauge, the equation of motion is the wave equation $\Box\vec{A} = 0$, with traveling wave solution (a retarded wave, with delayed time)

$$\vec{A} = \vec{A}\left(t - \frac{\vec{r}\cdot\vec{n}}{c}\right), \tag{39.18}$$

where $\vec{n}$ is the direction of propagation,

$$\vec{n} = \frac{\vec{k}}{k} = \frac{c\vec{k}}{\omega_k}. \tag{39.19}$$

The Fourier-transformed wave in the delayed time is $\vec{A}(\omega)$, and satisfies the condition

$$\vec{n} \cdot \vec{A}(\omega) = 0, \tag{39.20}$$

which comes from the gauge condition $\vec{\nabla} \cdot \vec{A} = 0$.

The Fourier-transformed vector potential is written in terms of a polarization vector $\vec{\epsilon}(\vec{k})$, as

$$\vec{A}(\omega) = \vec{\epsilon}(\vec{k})A(\omega) = A_1(\omega)\vec{\epsilon}_1 + A_2(\omega)\vec{\epsilon}_2, \tag{39.21}$$

where $\vec{\epsilon}_1$ and $\vec{\epsilon}_2$ are unit vectors perpendicular to $\vec{n}$ and each other, and

$$A(\omega) = \sqrt{1+a^2}A_1(\omega)$$
$$\vec{\epsilon}(\omega) = \frac{\vec{\epsilon}_1 + ae^{i\theta}\vec{\epsilon}_2}{\sqrt{1+a^2}} \tag{39.22}$$
$$\frac{A_1(\omega)}{A_2(\omega)} = ae^{i\theta}.$$

However, usually we consider a *monochromatic wave*, i.e., a wave with a single ω, so that instead of a Fourier-transformed wave, we have simply

$$\begin{aligned} \vec{A}(\vec{r},t) &= A_0\vec{\epsilon}(\vec{k})2\cos\left(\frac{\omega}{c}\vec{n}\cdot\vec{x} - \omega t\right) \\ &= A_0\vec{\epsilon}(\vec{k})\left[e^{i(\vec{k}\cdot\vec{x}-\omega t)} + e^{-i(\vec{k}\cdot\vec{x}-\omega t)}\right], \end{aligned} \tag{39.23}$$

where $\vec{k} = \omega\vec{n}/c$. Then, the interaction Hamiltonian between the radiation and the electric current presented by an atom, valid for both emission and absorption of electromagnetic radiation, is

$$H_i = A_0\int d^3r\vec{\epsilon}(\vec{k})\cdot\vec{j}\left[e^{i(\vec{k}\cdot\vec{x}-\omega t)} + e^{-i(\vec{k}\cdot\vec{x}-\omega t)}\right]. \tag{39.24}$$

We note that the current conservation law is

$$\frac{\partial\rho}{\partial t} + \vec{\nabla}\cdot\vec{j} = 0. \tag{39.25}$$

Next, we can proceed either classically or quantum mechanically, and obtain basically the same result.

Classically, we consider that the electric current generating or absorbing the radiation oscillates in syncronization with it, meaning the charge density $\rho = \rho(\omega)$ oscillates with the same monochromatic frequency ω as the wave, so that

$$\rho \propto e^{i\omega t} \Rightarrow \frac{\partial\rho}{\partial t} = i\omega\rho. \tag{39.26}$$

Quantum mechanically, the charge density is an operator, $\rho = \hat{\rho}$, and its oscillation in syncronization with the radiation induces a transition between atomic states, $|\psi_n\rangle \to |\psi_m\rangle$. Then, using the Heisenberg equation of motion for the operator ρ, we obtain

$$\begin{aligned} \left\langle\psi_m\left|\frac{\partial\hat{\rho}}{\partial t}\right|\psi_n\right\rangle &= \left\langle\psi_m\left|\frac{i}{\hbar}[\hat{H}_{\text{atom}},\hat{\rho}]\right|\psi_n\right\rangle \\ &= \frac{i}{\hbar}(E_m - E_n)\langle\psi_m|\hat{\rho}|\psi_n\rangle = i\omega_{mn}\langle\psi_m|\hat{\rho}|\psi_n\rangle. \end{aligned} \tag{39.27}$$

This is essentially the same result as in the classical case, except that it is an expectation value, between two states, and the frequency ω is generically quantized, $\omega = \omega_{mn}$.

Thus, either way we obtain

$$i\omega\rho + \vec{\nabla}\cdot\vec{j} = 0, \tag{39.28}$$

meaning we can replace $\vec{\nabla}\cdot\vec{j}$ with $-i\omega\rho$. But we can also use the relation

$$\begin{aligned} \vec{\nabla}\left[(\vec{\epsilon}\cdot\vec{r})\vec{j}\right] &= \partial_i[\epsilon_j x_j j_i] = \delta_{ij}\epsilon_j j_i + \epsilon_j x_j\partial_i j_i \\ &= \vec{\epsilon}\cdot\vec{j} + (\vec{\epsilon}\cdot\vec{r})\vec{\nabla}\cdot\vec{j} = \vec{\epsilon}\cdot\vec{j} - (\vec{\epsilon}\cdot\vec{r})i\omega\rho. \end{aligned} \tag{39.29}$$

Integrating over space, the left-hand side gives a boundary term, which is assumed to vanish, so we obtain

$$\int d^3r\, \vec{\epsilon}\cdot\vec{j} = i\omega \int d^3r \rho(\vec{\epsilon}\cdot\vec{r}). \tag{39.30}$$

We can now consider the term with $e^{i\vec{k}\cdot\vec{x}}$ in $\hat{H}_i$, namely the first term, and expand it as follows:

$$e^{i\vec{k}\cdot\vec{x}} \simeq 1 + i\vec{k}\cdot\vec{x} + \cdots \tag{39.31}$$

Considering only the zeroth-order term, i.e., 1, we have the interaction Hamiltonian at leading order,

$$H_i^{(1)} = A_0 \int d^3r\, \vec{\epsilon}(\vec{k})\cdot\vec{j} = i\omega A_0 \vec{\epsilon}\,(\vec{k})\cdot \int d^3r \rho\vec{r}, \tag{39.32}$$

where we have used the relation deduced above. But the integral

$$\vec{P} \equiv \int d^3r \rho\vec{r} \tag{39.33}$$

is the *electric dipole moment*, so

$$H_i^{(1)} = i\omega A_0 \vec{\epsilon}(\vec{k})\cdot\vec{P} \tag{39.34}$$

gives rise to electric dipole radiation.

Next, consider the first-order term in $e^{i\vec{k}\cdot\vec{x}}$, resulting in the next-to-leading order (NLO) Hamiltonian,

$$H_i^{(2)} = A_0 \int d^3r (\vec{\epsilon}(\vec{k})\cdot\vec{j})(\vec{k}\cdot\vec{r}). \tag{39.35}$$

There is now a similar relation to (39.29) that we can use to rewrite the NLO interacting Hamiltonian,

$$\begin{aligned}
\vec{\nabla}\cdot[(\vec{\epsilon}\cdot\vec{r})(\vec{k}\cdot\vec{r})\vec{j}] &= \partial_i[\epsilon_j x_j k_k x_k j_i] \\
&= \delta_{ij}\epsilon_j k_k x_k j_i + \epsilon_j x_j \delta_{ik} k_k j_i + \epsilon_j x_j k_k x_k \partial_i j_i \\
&= (\vec{k}\cdot\vec{r})(\vec{\epsilon}\cdot\vec{j}) + (\vec{\epsilon}\cdot\vec{r})(\vec{k}\cdot\vec{j}) + (\vec{\epsilon}\cdot\vec{r})(\vec{k}\cdot\vec{r})\vec{\nabla}\cdot\vec{j} \\
&= (\vec{k}\cdot\vec{r})(\vec{\epsilon}\cdot\vec{j}) + (\vec{\epsilon}\cdot\vec{r})(\vec{k}\cdot\vec{j}) - i\omega\rho(\vec{\epsilon}\cdot\vec{r})(\vec{k}\cdot\vec{r}),
\end{aligned} \tag{39.36}$$

where in the last line we used current conservation. Then, integrating over $\vec{r}$ as before, the left-hand side gives a boundary term, assumed to vanish, so that

$$\int d^3r \left[(\vec{k}\cdot\vec{r})(\vec{\epsilon}\cdot\vec{j}) + (\vec{\epsilon}\cdot\vec{r})(\vec{k}\cdot\vec{j}) - i\omega\rho(\vec{\epsilon}\cdot\vec{r})(\vec{k}\cdot\vec{r})\right] = 0. \tag{39.37}$$

But we still need to rewrite the middle term in (39.36). For that, we need to use yet another relation,

$$\begin{aligned}
(\vec{\epsilon}\times\vec{k})\cdot(\vec{r}\times\vec{j}) &= \epsilon_{ijk}\epsilon_i k_j \epsilon_{lmk} x_l j_m \\
&= (\delta_{il}\delta_{jm} - \delta_{im}\delta_{jl})\epsilon_i k_j x_l j_m \\
&= (\vec{\epsilon}\cdot\vec{r})(\vec{k}\cdot\vec{j}) - (\vec{\epsilon}\cdot\vec{j})(\vec{k}\cdot\vec{r}).
\end{aligned} \tag{39.38}$$

This allows us to rewrite the middle term resulting in a relation between integral quantities

$$\int d^3r(\vec{k}\cdot\vec{r})(\vec{\epsilon}\cdot\vec{j}) = \frac{i\omega}{2}\int d^3r\rho(\vec{\epsilon}\cdot\vec{r})(\vec{k}\cdot\vec{r}) - (\vec{\epsilon}\times\vec{k})\int d^3r\frac{(\vec{r}\times\vec{j})}{2}, \tag{39.39}$$

so that the interaction Hamiltonian is

$$H_i^{(2)} = A_0\left[\frac{i\omega}{2}\epsilon_i k_j Q_{ij} - (\vec{\epsilon}\times\vec{k})\cdot\vec{\mu}\right]. \tag{39.40}$$

Here we have defined the *electric quadrupole moment*

$$Q_{ij} \equiv \int d^3r\rho x_i x_j, \tag{39.41}$$

and the magnetic dipole moment

$$\vec{\mu} \equiv \int d^3r\frac{(\vec{r}\times\vec{j})}{2}. \tag{39.42}$$

In general, at lth subleading order, we have an electric 2^{l+1}th-polar perturbation and a magnetic 2^lth-polar perturbation. These are all *tensors* of $SO(3) \simeq SU(2)$, which means that they are quantities with a given angular momentum.

The electric dipole P_i is a vector, the magnetic dipole μ_i is also a vector (both with $j = 1$), while the electric quadrupole Q_{ij} is a two-index symmetric tensor (with $j = 2$). In general, for higher-order terms, we also obtain tensors, $T_q^{(k)}$. In quantum theory, these will be quantum operators, the same as those that were analyzed for the Wigner–Eckhart theorem in Chapter 16. The theorem states that

$$\langle\alpha' j' m|\hat{T}_q^{(k)}|\alpha, jm\rangle = \langle j, k; m, q|j, k; j', m'\rangle\frac{\langle\alpha', j'\|T^{(k)}\|\alpha, j\rangle}{\sqrt{2j+1}}, \tag{39.43}$$

and the matrix elements are nonzero only for

$$\begin{aligned} q &= m - m' \\ |j - j'| &\le k \le j + j'. \end{aligned} \tag{39.44}$$

These are then *selection rules* for the matrix element to be, resulting in a nonzero transition probability via first-order perturbation theory.

In the case of the leading, electric dipole, approximation, we have a vector operator P_i, with $q = 1$, so $m = 0, \pm 1$, but $m = 0$ is only possible when $\vec{P}$ is parallel to $\vec{\epsilon}$, which means a longitudinal mode (which is unphysical, in the radiation region). Otherwise, we have

$$\begin{aligned} m - m' &= \pm 1 \\ |j - j'| &\le 1 \le j + j', \end{aligned} \tag{39.45}$$

which means that

$$\begin{aligned} j - j' \le 1 &\Rightarrow j' \ge j - 1 \\ j' - j \le 1 &\Rightarrow j' \le j + 1, \end{aligned} \tag{39.46}$$

finally resulting in

$$j' = j - 1, j, j + 1, \quad \text{and} \quad m' = m \pm 1. \tag{39.47}$$

39.4 Absorption of Photon Energy by Atom

The direct application of the formalism is to the absorption of electromagnetic energy.

The first way to derive the transition amplitude is to use the interaction in the particle picture,

$$\hat{H}_1 = -\frac{q}{m}\vec{A}\cdot\vec{p} = -\frac{q}{m}A_0\vec{\epsilon}(\vec{k})e^{i\vec{k}\cdot\vec{x}}\cdot\vec{p}, \tag{39.48}$$

where the momentum operator is understood as $\frac{\hbar}{i}\vec{\nabla}$ acting on a wave function. In the electric dipole approximation, the exponential is replaced by 1, so

$$\hat{H}_1 = -\frac{q}{m}\vec{\epsilon}(\vec{k})\cdot\vec{p}. \tag{39.49}$$

The matrix element between an initial and a final state is

$$\begin{aligned} H_{1,fi} &= \langle\psi_f|\hat{H}_1|\psi_i\rangle = -\frac{q}{m}A_0\vec{\epsilon}(\vec{k})\cdot\langle\psi_f|\vec{p}|\psi_i\rangle \\ &= -\frac{q}{m}A_0\vec{\epsilon}(\vec{k})\cdot\int d^3r\psi_f^*(\vec{r})\frac{\hbar}{i}\vec{\nabla}\psi_i(\vec{r}). \end{aligned} \tag{39.50}$$

But, since for atoms we have $\hat{H}_0 = \vec{p}^2/2m + V(\vec{r})$,

$$\vec{p} = \frac{m}{i\hbar}[\vec{r},\hat{H}_0]. \tag{39.51}$$

Then we can replace $\vec{p}$ in the matrix element, and find

$$\begin{aligned} H_{1,fi} &= -\frac{q}{m}A_0\vec{\epsilon}(\vec{k})\cdot\langle\psi_f|\vec{p}|\psi_i\rangle = -\frac{q}{i\hbar}A_0\vec{\epsilon}(\vec{k})\cdot\langle\psi_f|[\vec{r},\hat{H}_0]|\psi_i\rangle \\ &= iqA_0\frac{E_i-E_f}{\hbar}\langle\psi_f|\vec{\epsilon}(\vec{k})\cdot\vec{r}|\psi_i\rangle \\ &= iqA_0\omega_{if}\int d^3r\psi_f^*(\vec{r})\vec{\epsilon}(\vec{k})\cdot\vec{r}\psi_i(\vec{r}). \end{aligned} \tag{39.52}$$

But, since ρ is the electric charge density, it equals q times the probability density $\psi^*\psi$, which means that the matrix element of ρ, ρ_{fi}, is given by

$$\rho_{fi}(\vec{r}) = q\psi_f^*(\vec{r})\psi_i(\vec{r}), \tag{39.53}$$

so that the interaction Hamiltonian can be rewritten as

$$H_{1,fi} = iA_0\omega_{if}\vec{\epsilon}\cdot\int d^3r\rho_{fi}\vec{r}, \tag{39.54}$$

which is the same formula as that obtained from (39.32) (which is true in the electric dipole approximation) in between the initial and final states.

We must also give a better definition of the electric dipole approximation for hydrogenoid atoms. In this case, the transition frequency ω_{ij} is of the order of the ground state energy (corresponding to the initial state being the ground state), so

$$\hbar\omega_{1f} \sim E_1 \sim \frac{Ze_0^2}{r_1} \sim \frac{Ze_0^2}{a_0/Z}, \tag{39.55}$$

where the ground state radius is $r_1 \sim a_0/Z$. Moreover, $e_0^2/(c\hbar) \sim 1/137$, so the wavelength λ associated with ω is given by

$$\frac{\lambda}{2\pi} = \frac{c}{\omega} \sim \frac{c\hbar a_0}{Z^2 e_0^2} \sim \frac{137 a_0/Z}{Z}. \tag{39.56}$$

But we require that the wavelength is much greater than the size of the atom, $\lambda \gg 2\pi r_1$, in order to use the dipole approximation, so

$$\frac{r_1}{\lambda/(2\pi)} = \frac{a_0/Z}{\lambda/(2\pi)} = \frac{Z}{137} \ll 1, \tag{39.57}$$

which is valid only if

$$Z \ll 137, \tag{39.58}$$

i.e., for light hydrogenoid atoms.

Next, we use Fermi's golden rule for absorption, when the initial state has the energy E_i for the atom and a photon of energy $\hbar\omega$ is to be absorbed by the atom, so the delta function for energy is $\delta(E_f - E_i - \hbar\omega)$, giving the probability density (in time)

$$\frac{dP}{dt} = \frac{2\pi}{\hbar}|H_{1,fi}|^2 \int df_2 \cdots \sum_{f_r} \rho(E_f, f_2, \ldots, f_r) dE_f \delta(E_f - E_i - \hbar\omega). \tag{39.59}$$

But we can use (39.50) to replace the matrix element in the above, obtaining

$$\frac{dP}{dt} = \frac{2\pi}{\hbar}\frac{q^2}{m^2}|A_0|^2 |\langle\psi_f|\vec{\epsilon}(\vec{k})\cdot\vec{p}|\psi_i\rangle|^2 \int df_2 \cdots \sum_{f_r} \rho(E_f, f_2, \ldots, f_r) dE_f \delta(E_f - E_i - \hbar\omega). \tag{39.60}$$

The absorption cross section is defined as the ratio of absorbed energy per unit time divided by the incident flux (the incident energy per unit time and unit transverse area),

$$\sigma_{\text{abs}} = \frac{dE/dt(\text{abs})}{dE_{\text{inc}}/dtdS} = \frac{\hbar\omega dP/dt(\text{abs})}{\mathcal{F}}, \tag{39.61}$$

where the absorbed energy per unit time equals the photon energy $\hbar\omega$ times the probability per unit time, and the incident flux $\mathcal{F}$ is written in terms of the incident energy density $\mathcal{E}$ as $\mathcal{F} = \mathcal{E}c$, so

$$\mathcal{F} = \frac{dE_{\text{inc}}}{dtdS} = \frac{c}{2}\left(\epsilon_0\langle\vec{E}^2\rangle + \frac{\langle\vec{B}^2\rangle}{\mu_0}\right), \tag{39.62}$$

and the electric and magnetic fields are written in terms of the traveling wave potential $\vec{A}$ as

$$\begin{aligned}\vec{B} &= \vec{\nabla}\times\vec{A}(\vec{r},t) = \vec{k}\times\vec{\epsilon}(\vec{k})A_0 e^{i(\vec{k}\cdot\vec{x}-\omega t)} \\ \vec{E} &= -\partial_t\vec{A}(\vec{r},t) = \omega\vec{\epsilon}(\vec{k})A_0 e^{i(\vec{k}\cdot\vec{x}-\omega t)},\end{aligned} \tag{39.63}$$

so

$$\mathcal{F} = \frac{ck^2|A_0|^2}{2} = \frac{\omega^2|A_0|^2}{2c}. \tag{39.64}$$

Thus the absorption cross section is

$$\sigma_{\rm abs} = \frac{4\pi\alpha\hbar}{m^2\omega_{fi}}|\langle\psi_f|\vec{\epsilon}(\vec{k})\cdot\vec{p}|\psi_i\rangle|^2 \int df_2\cdots\sum_{f_r}\rho(E_f, f_2, \ldots, f_r)dE_f\delta(E_f - E_i - \hbar\omega), \quad (39.65)$$

where we have used $\alpha = q^2/(\hbar c)$.

But we can also relate the $\vec{p}$ matrix elements to those for $-im\omega_{fi}\vec{r}$, as in (39.52), obtaining

$$\sigma_{\rm abs} = 4\pi\alpha\hbar\omega_{fi}|\langle\psi_f|\vec{\epsilon}(\vec{k})\cdot\vec{r}|\psi_i\rangle|^2 \int df_2\cdots\sum_{f_r}\rho(E_f, f_2, \ldots, f_r)dE_f\delta(E_f - E_i - \hbar\omega). \quad (39.66)$$

39.5 Photoelectric Effect

The photoelectric effect corresponds to an ionized final state of the atom, so the process is the absorption of a photon by the atom and the emission of a free electron, at a direction $\vec{n}$, with momentum $\vec{p}$. Then the final state is $|\psi_f\rangle = |\vec{k}_f\rangle$, a free electron with energy

$$E = \frac{\hbar^2 k_f^2}{2m_e}. \quad (39.67)$$

Thus the matrix element is

$$\langle\vec{k}_f|\vec{\epsilon}\cdot\vec{p}e^{i\vec{k}\cdot\vec{x}}|\psi_i\rangle = \vec{\epsilon}\cdot\int d^3r\,\psi^*_{\vec{k}_f}(\vec{r})e^{i\vec{k}\cdot\vec{r}}\frac{\hbar}{i}\vec{\nabla}\psi_i(\vec{r}), \quad (39.68)$$

where $\psi_i(\vec{r})$ corresponds to an electron inside the hydrogenoid atom, whereas $\psi_{\vec{k}_f}(\vec{r})$ is a free wave function, normalized over energy.

In the electric dipole approximation we replace $e^{i\vec{k}\cdot\vec{x}}$ with 1, so we then obtain a matrix element

$$H_{1,fi} = im\omega_{fi}\langle\psi_f|\vec{\epsilon}\cdot\vec{p}|\psi_i\rangle. \quad (39.69)$$

As in Chapter 37, where we analyzed the scattering wave collision, we have $f_1 = E$, $f_2 = \vec{n}$, so $df_2 = d\Omega$. The differential cross section is then

$$\frac{d\sigma}{d\Omega} = \frac{\hbar\omega dP/dt(\text{abs})}{\mathcal{F}}\int dE_f\rho(E_f)\delta(E_f - E_i - \hbar\omega), \quad (39.70)$$

but if the state ψ_f is normalized in energy, as in Chapter 37, then $\rho(E_f) = 1$. Therefore we finally obtain

$$\frac{d\sigma}{d\Omega} = \frac{4\pi\alpha\omega_{fi}\hbar}{(m\omega_{fi})^2}|\langle\psi_f|\vec{\epsilon}\cdot\vec{p}|\psi_i\rangle|^2. \quad (39.71)$$

In this formula, we would need to calculate the matrix element for a given state $|\psi_i\rangle$ of the hydrogenoid atom. Therefore, since the differential cross section depends on $|\psi_i\rangle$, we will not compute it further here.

Important Concepts to Remember

- The linear Zeeman effect corresponds to the linear part of the interaction Hamiltonian for interaction with electromagnetism, $\hat{H}_1 = \frac{iq\hbar}{m}\vec{A}\cdot\vec{\nabla}$, plus the interaction with the spin, leading to $\hat{H}_1 = -\mu_B(\vec{J}+\vec{S})\cdot\vec{B}$, and energy shift $\Delta E = -gM\mu_B B$. For $l = s = 0$, we must consider the quadratic part of $\hat{H}_1$ instead.
- The Paschen–Back effect corresponds to the opposite limit, in which the LS coupling is $\ll \mu_B B$, so the correct quantum states are $|E_n L S M_L M_S\rangle$, and $\Delta E_n = -\mu_B B(M_L + 2M_S)$.
- For electromagnetic radiation, $\vec{A} = \vec{A}(t - \vec{n}\cdot\vec{r}/c)$, or in particular $\vec{A} = A_0\vec{\epsilon}(k)[e^{i(\vec{k}\cdot\vec{x}-\omega t)} + e^{-i(\vec{k}\cdot\vec{x}-\omega t)}]$, in which case the charge density ρ also oscillates with ω, and $i\omega\rho + \vec{\nabla}\cdot\vec{j} = 0$.
- When expanding $e^{i\vec{k}\cdot\vec{x}}$ into $1 + i\vec{k}\cdot\vec{x} + \cdots$, the first-order interaction Hamiltonian depends on the electric dipole moment $\vec{P}$ as $H_i^{(1)} = i\omega A_0\vec{\epsilon}(k)\cdot\vec{P}$, and the second-order interaction Hamiltonian depends on the electric quadrupole moment Q_{ij} and the magnetic dipole moment $\vec{\mu}$ as $H_i^{(2)} = A_0[(i\omega/2)\epsilon_i k_j Q_{ij} - \epsilon^{ijk}\epsilon_i k_j \mu_k]$.
- The Wigner–Eckhart theorem leads to selection rules for the matrix elements for electromagnetic transition probabilities. In particular, the leading electric dipole approximation leads to $m' = m \pm 1$ and $j' = j-1, j, j+1$.
- For absorption of a photon by an atom, we obtain the matrix element $H_{1,fi} = iA_o\omega_{fi}\vec{\epsilon}\cdot\int d^3r\,\rho_{fi}\,\vec{r}$, and we can use it in Fermi's golden rule. For a hydrogenoid atom, this electric dipole approximation is valid for $Z \ll 137 = 1/\alpha$.
- For the photoelectric effect, we would apply the formalism for absorption of photon by an atom, with the final state an emitted free electron.

Further Reading

See [2], [1], [3], [4].

Exercises

(1) Consider the first relativistic correction to the energy of a particle, and introduce the coupling to the electromagnetic field. Write down the resulting extra terms that are linear and quadratic in $\vec{A}$.

(2) For the terms considered in exercise 1, find expressions for the first relativistic corrections to the linear and quadratic Zeeman effects (ignore any extra couplings to spin).

(3) Show that the splittings of energy levels according to the Zeeman or Paschen–Back effects lead to the same number of lines.

(4) Calculate the transition element $H_{1,fi}$ for the interaction of a hydrogenoid atom with a monochromatic electromagnetic wave, in the electric dipole approximation, from the $n = 1, l = 0$ state to the $n = 2, l = 1$ state.

(5) Consider a hydrogenoid atom in the state $n = 3, l = 2, m = 0$. What are the possible transitions induced by a monochromatic electromagnetic wave, to both first and second order?
(6) Calculate the absorption cross section for the transition in exercise 4.
(7) Calculate the differential cross section for the photoelectric effect for a monochromatic electromagnetic wave of energy of 14 eV incident on a hydrogen atom in the ground state.

40 WKB Methods and Extensions: State Transitions and Euclidean Path Integrals (Instantons)

In this chapter we pick up the previously described WKB (semiclassical) methods, and extend them for transitions between states in (time-dependent) perturbation theory, and we define "instanton" calculations in a Euclidean version of the path integral formalism.

40.1 Review of the WKB Method

The Schrödinger equation in one dimension, in the presence of a potential $V(x)$, with the wave function ansatz

$$\psi(x,t) = \exp\left[\frac{i}{\hbar}(W(x) - Et)\right], \tag{40.1}$$

is

$$\left(\frac{dW}{dx}\right)^2 - 2m(E - V(x)) + \frac{\hbar}{i}\frac{d^2W}{dx^2} = 0, \tag{40.2}$$

and has the quasiclassical solution

$$\psi(x) = \frac{\text{const}}{\sqrt{p(x)}} \exp\left[\pm\frac{i}{\hbar}\int_{x_0}^{x} dx' p(x') - \frac{i}{\hbar}Et\right] \tag{40.3}$$

for $E > V(x)$, where

$$p(x) = \sqrt{2m(E - V(x))}, \tag{40.4}$$

and the solution

$$\psi(x) = \frac{\text{const}'}{\sqrt{\chi(x)}} \exp\left[\pm\frac{1}{\hbar}\int_{x_0}^{x} dx' \chi(x') - \frac{i}{\hbar}Et\right] \tag{40.5}$$

for $E < V(x)$, where

$$\chi(x) = \sqrt{2m(V(x) - E)}. \tag{40.6}$$

These solutions are valid only asymptotically, meaning not very close to the "turning points" x_i, where $E - V(x_i) = 0$.

When going through these turning points, we match one type of solution to the other. If we have a decreasing potential $V(x)$, and we transition from left to right (from higher V to lower V), we are translating

$$\frac{1}{[V(x) - E]^{1/4}} \exp\left[-\frac{1}{\hbar}\int_{x}^{x_1} dx' \sqrt{2m(V(x') - E)}\right] \tag{40.7}$$

into

$$\begin{aligned}&\frac{\sqrt{2}}{[E-V(x)]^{1/4}}\sin\left[\frac{1}{\hbar}\int_{x_1}^{x}dx'\sqrt{2m(E-V(x'))}+\frac{\pi}{4}\right]\\&=\frac{\sqrt{2}}{[E-V(x)]^{1/4}}\cos\left[\frac{1}{\hbar}\int_{x_1}^{x}dx'\sqrt{2m(E-V(x'))}-\frac{\pi}{4}\right].\end{aligned} \tag{40.8}$$

On the other hand, for an increasing potential, and a transition from right to left (from higher V to lower V), we need to translate

$$\frac{1}{[V(x)-E]^{1/4}}\exp\left[-\frac{1}{\hbar}\int_{x}^{x_2}dx'\sqrt{2m(V(x')-E)}\right] \tag{40.9}$$

into

$$\begin{aligned}&\frac{\sqrt{2}}{[E-V(x)]^{1/4}}\sin\left[\frac{1}{\hbar}\int_{x_2}^{x}dx'\sqrt{2m(E-V(x'))}-\frac{\pi}{4}\right]\\&=\frac{\sqrt{2}}{[E-V(x)]^{1/4}}\cos\left[\frac{1}{\hbar}\int_{x_1}^{x}dx'\sqrt{2m(E-V(x'))}+\frac{\pi}{4}\right].\end{aligned} \tag{40.10}$$

In the path integral formalism, the propagator

$$U(t,t_0)=\int\mathcal{D}x(t)\exp\left\{\frac{i}{\hbar}S[x(t)]\right\}, \tag{40.11}$$

with boundary conditions $x(0)=x_0$ and $x(t)=x$, is approximated in the "saddle point approximation" by the action in the classical solution times a Gaussian integration for the quadratic fluctuation giving a quasi-classical determinant,

$$\begin{aligned}U(t,t_0)&\simeq\exp\left\{\frac{i}{\hbar}S[x_{\rm cl}(t)]\right\}[\det(\text{kinetic operator})]^{-1/2}\\&=\exp\left[\frac{i}{\hbar}\int_{x_0}^{x}dx'\sqrt{2m(E-V(x'))}-\frac{i}{\hbar}Et\right]\\&\quad\times\left\{\det\left[-m\frac{d^2}{dt^2}-(V(x)-E)''\right]\right\}^{-1/2}.\end{aligned} \tag{40.12}$$

After integrating over the initial position x_0, made to be identical with the final position x, we obtain the correct propagator $U(t,t_0)$ for the harmonic oscillator case.

40.2 WKB Matrix Elements

We want to calculate matrix elements

$$A_{12}\equiv\langle\psi_1|\hat{A}|\psi_2\rangle \tag{40.13}$$

where $|\psi_1\rangle,|\psi_2\rangle$ are eigenstates of the Hamiltonian $\hat{H}$ with energies E_1,E_2. Assume for concreteness that $E_2>E_1$. The analysis below follows the work of Landau (described in the book by Landau and Lifshitz).

We consider only operators $\hat{A}$ that are functions in the x representation, as for instance a potential $V(x)$. Then

$$A_{12} = \int dx \int dy \langle\psi_1|x\rangle\langle x|\hat{A}|y\rangle\langle y|\psi_2\rangle$$
$$= \int dx \int dy \psi_1^*(x)\psi_2(x)A_{xy} \quad (40.14)$$

and $A_{xy} = \delta(x-y)f(x)$. Assuming also that $\psi_1(x)$ and $\psi_2(x)$ are real, we have

$$A_{12} = f_{12} = \int dx \psi_1(x)f(x)\psi_2(x). \quad (40.15)$$

Consider a potential $V(x)$ that is decreasing, and consider a turning point a_1 for E_1 such that $V(a_1) = E_1$, and a turning point a_2 for E_2 such that $V(a_2) = E_2$, implying $a_1 > a_2$. Define further

$$p_1(x) = p_{E_1}(x), \quad \chi_1(x) = \chi_{E_1}(x)$$
$$p_2(x) = p_{E_2}(x), \quad \chi_2(x) = \chi_{E_2}(x). \quad (40.16)$$

Then, according to the general theory reviewed before, away from the turning point a_1 the wave function $\psi_1(x)$ corresponding to E_1 is

$$\psi_1 = \frac{C_1}{\sqrt{2}\sqrt{\chi_1(x)}} \exp\left[-\frac{1}{\hbar}\int_{a_1}^x \chi_1(x')dx'\right], \quad x < a_1$$
$$\psi_1 = \frac{C_1}{\sqrt{p_1(x)}} \cos\left[\frac{1}{\hbar}\int_{a_1}^x p_1(x')dx'\right], \quad x > a_1, \quad (40.17)$$

and similarly for $\psi_2(x)$.

However, we cannot use both these solutions to calculate f_{12}. First, the region near the turning points has a large contribution to the integral, since the wave function blows up there, and second, the turning point of one is not a turning point for the other, so it encroaches on the region that was good from the point of view of the other.

Instead, we write $\psi_2 = \psi_2^+ + \psi_2^-$, where ψ_2^+ is complex, and $\psi_2^- = (\psi_2^+)^*$. Then we write, for the solution for $x > a_2$, $2\cos[\ldots]$ as $e^{i[\ldots]} + e^{-i[\ldots]}$, which is a sum of complex conjugate terms, therefore corresponding to ψ_2^+ and ψ_2^-.

But now we can take advantage of an alternative correspondence rule through the turning points, for complex exponential solutions, namely (from $x > a_2$ to $x < a_2$)

$$\frac{C}{\sqrt{p(x)}} \exp\left[\frac{i}{\hbar}\int_{a_2}^x p(x')dx' + i\frac{\pi}{4}\right] \to \frac{C}{\sqrt{|p(x)|}} \exp\left[\frac{1}{\hbar}\left|\int_{a_2}^x p(x')dx'\right|\right], \quad (40.18)$$

thus defining ψ_2^+ also for $x < a_2$. We obtain (note that $-i = e^{-i\pi/2}$)

$$\psi_2^+ = \frac{-iC_2}{\sqrt{2}\sqrt{\chi_2(x)}} \exp\left[\frac{1}{\hbar}\left|\int_{a_2}^x p_2(x')dx'\right|\right], \quad x < a_2$$
$$\psi_2^+ = \frac{C_2}{2\sqrt{p_2(x)}} \exp\left[\frac{i}{\hbar}\int_{a_2}^x p_2(x')dx' - i\frac{\pi}{4}\right], \quad x > a_2. \quad (40.19)$$

Having defined ψ_2^+, and $\psi_2^- = (\psi_2^+)^*$, we can also split f_{12} similarly as $f_{12} = f_{12}^+ + f_{12}^-$, with

$$f_{12}^+ = \int_{-\infty}^{+\infty} \psi_1(x)f_{12}(x)\psi_2^+(x)dx. \quad (40.20)$$

Next, we consider moving x from the real line onto the complex plane, specifically the upper half-plane (the positive imaginary part). In that case, if the exponent is a function $g(x)$ that maps the upper half-plane to itself, $e^{ig(x)} \to e^{-\text{Im}g(x)}$ goes to zero at infinity, so is better defined near $|x| \to \infty$.

Then, we consider an integration contour C for x in the upper half-plane that is slightly above the real line, thus avoiding the turning points at real a_1, a_2 and so being well defined. For that, however, we must define, in the whole half-plane but away from a_1, a_2,

$$\psi_1(x) = \frac{C_1}{\sqrt{2}\sqrt{\chi_1(x)}} \exp\left[\frac{1}{\hbar}\int_{a_1}^{x} \chi_1(x')dx'\right]$$
$$\psi_2^+(x) = \frac{-iC_2}{\sqrt{2}\sqrt{\chi_2(x)}} \exp\left[-\frac{1}{\hbar}\int_{a_2}^{x} \chi_2(x')dx'\right]. \tag{40.21}$$

Then we can calculate the amplitude f_{12}^+ as

$$f_{12}^+ = \frac{-iC_1C_2}{2}\int_C dx \frac{f(x)}{\sqrt{\chi_1(x)\chi_2(x)}} \exp\left\{\frac{1}{\hbar}\left[\int_{a_1}^{x} dx'\chi_1(x') - \int_{a_2}^{x} dx'\chi_2(x')\right]\right\}. \tag{40.22}$$

The exponent in the above formula has an extremum x_0, i.e., a zero derivative with respect to x, *only* for $V(x_0) = \infty$ if $E_1 \neq E_2$. Indeed,

$$\frac{d}{dx}\left[\int_{a_1}^{x} dx'\chi_1(x') - \int_{a_2}^{x} dx'\chi_2(x')\right] = 0$$
$$\Rightarrow \sqrt{2m(V(x)-E_1)} - \sqrt{2m(V(x)-E_2)} \simeq \frac{E_1 - E_2}{\sqrt{2m(V(x)-E_1)}} = 0 \tag{40.23}$$

has only the solution $V(x) \to \infty$.

Assume that there is a single pole for $V(x)$ in the upper half-plane, i.e., $V(x_0) \to \infty$. Then, the integral over the contour C (slightly above the real line) is equal to the integral over a contour C' that goes vertically down from infinity to x_0, then encircles it from below (very close to it), then goes vertically back up again. Indeed, we can add for free two quarter-circles at infinity in the upper half-plane (since, as we just said, this contribution at infinity vanishes in the upper half-plane) that will connect C with $-C'$ at infinity, and there is no pole inside the resulting total contour, as in Fig. 40.1. We can thus use the residue theorem of complex analysis to say that the integral over the whole contour is zero (the integral equals the sum of residues at the poles inside the contour), or, since the added integral at infinity vanishes anyway, that the integral over C is equal to the integral over C'.

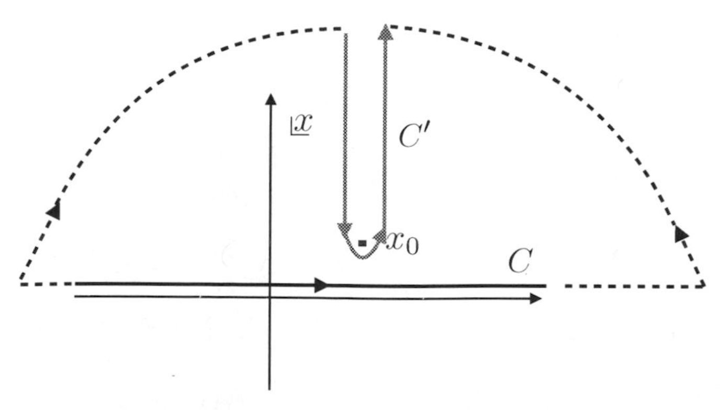

Figure 40.1 Integration over the contour C, slightly above the real axis in x, can be reduced to integration over the contour C', encircling x_0, down and up from infinity, indicated by the dashed curve.

Then f_{12}^+ is found to be the same as (40.22), but integrated over C', and can be thus approximated (since the further we go up in the imaginary plane, the more we add to the exponent in $e^{-\text{Im}[\dots]}$, making the contribution exponentially small) by the value of the exponential near the minimum x_0. Moreover, then

$$f_{12} = 2\,\text{Re} f_{12}^+ \propto 2\,\text{Im}\exp[\dots], \tag{40.24}$$

so that finally we have the approximation (note that $f(x_0)$ is a subleading part, which can be neglected compared with the exponential)

$$f_{12} \sim \exp\left\{-\frac{1}{\hbar}\text{Im}\left[\left[\int^{x_0} dx' p_2(x') - \int^{x_0} dx' p_1(x')\right]\right\}. \tag{40.25}$$

If the two energies E_1 and E_2 are close to each other, i.e.,

$$E_{1,2} \equiv E \pm \frac{1}{2}\hbar\omega_{21}, \tag{40.26}$$

with $\hbar\omega_{21} \ll E$, we can also make the approximations

$$\begin{aligned} p_2(x) &\simeq p_E(x) + \frac{\partial p_E(x)}{\partial E}(E)\frac{\hbar\omega_{21}}{2} \\ p_1(x) &\simeq p_E(x) - \frac{\partial p_E(x)}{\partial E}(E)\frac{\hbar\omega_{21}}{2}. \end{aligned} \tag{40.27}$$

Then, since

$$\frac{\partial p_E(x)}{\partial E}(E) = \sqrt{\frac{m}{2(E - V(x))}}, \tag{40.28}$$

we find that the transition amplitude matrix element is approximated by

$$f_{12} \sim \exp\left\{-\omega_{21}\text{Im}\int^{x_0} dx' \sqrt{\frac{m}{2(E - V(x))}}\right\} = e^{-\omega_{21}\text{Im}\,\tau} \tag{40.29}$$

where we have defined the *complex time*

$$\tau \equiv \int^{x_0} \frac{dx}{v(x)}, \tag{40.30}$$

and where also we have defined the *complex velocity*

$$v(x) \equiv \sqrt{\frac{2(E - V(x))}{m}}. \tag{40.31}$$

Thus we have a transition amplitude (a matrix element) that is approximated by a decay factor in complex time, found as an extremum over the complex plane of the action, i.e., the exponent of the WKB approximation.

40.3 Application to Transition Probabilities

The transition probability relations contain, as we saw in previous chapters, the modulus squared of transition amplitudes for the interaction Hamiltonian,

$$|\langle\psi_m|\hat{H}|\psi_i\rangle|^2, \tag{40.32}$$

and usually the interaction Hamiltonian $\hat{H}_1$ is a certain function of x, $f(x)$.

Then, time-dependent perturbation theory, in the sudden approximation (the interaction Hamiltonian jumps from 0 to a constant at t_0), gives

$$|b_m(t)|^2 = t\frac{2\pi}{\hbar}|\langle\psi_m|\hat{H}|\psi_i\rangle|^2\delta(E_f - E_i), \tag{40.33}$$

or Fermi's golden rule,

$$P(i \to m) = t\frac{2\pi}{\hbar}\int df_2 \cdots \sum_{f_r}\rho(E_i, f_2, \ldots, f_r)|\langle\psi_m|\hat{H}|\psi_i\rangle|^2 \sim |\langle\psi_m|\hat{H}|\psi_i\rangle|^2. \tag{40.34}$$

But then, we have for the amplitude

$$\langle\psi_m|\hat{H}_1|\psi_i\rangle \sim \exp\left\{-\frac{1}{\hbar}\mathrm{Im}\left[\left[\int_{x_1}^{x_0} dx' p_m(x') - \int_{x_2}^{x_0} dx' p_i(x')\right]\right]\right\}. \tag{40.35}$$

Since however, the integral in the exponent, according to the Bohr-Sommerfeld quantization, discussed in Chapter 26, is the action,

$$\int_{x_1}^{x_0} dx\, p(x) = \text{action} = S(x_1, x_0), \tag{40.36}$$

we obtain that the leading factor in the transition probability is

$$P(i \to m) \sim \exp\left\{-\frac{2}{\hbar}\mathrm{Im}\left[S(x_1, x_0) + S(x_0, x_2)\right]\right\}. \tag{40.37}$$

This means that we have an action in the exponent for a path in the complex plane in terms of variables (complex) time τ and (complex) position x, together giving the (complex) path $x(\tau)$, which starts at x_1 and goes to x_0, the extremum of the action S in complex time, then goes to x_2. Here x_1 and x_2 are positions corresponding to the two states ψ_i and ψ_m for which we are calculating the transition probability.

40.4 Instantons for Transitions between Two Minima

The previous explanation motivates a more precise analysis for the case of a transition between two local minima. For this, however, we need to define path integrals in Euclidean space first.

Path Integrals in Euclidean Space

In Chapter 28 we introduced path integrals in Euclidean (imaginary) time, though the motivation there was to connect with statistical mechanics. Here, the main motivation is to have a formalism that allows for a natural "classical path" in imaginary time.

Thus we consider the propagator written as a path integral,

$$U(t, t_0) = \int_{x_i=x(t_i)}^{x_f=x(t_f)} \mathcal{D}x(t)e^{iS[x(t)]/\hbar}, \tag{40.38}$$

and it can be Wick rotated to Euclidean (imaginary) time by setting $t_E = it$ (so we are considering $t = -it_E$, where now $t_E \in \mathbb{R}$), which changes the Minkowski metric of spacetime to the Euclidean one,

$$-c^2dt^2 + d\vec{x}^2 = +c^2dt_E^2 + d\vec{x}^2. \tag{40.39}$$

We define the exponent in the path integral, $iS[x]$, as $-S_E[x]$,

$$iS[x] = i\int(-idt_E)\left[\frac{1}{2}\left(\frac{dx}{d(-it_E)}\right)^2 - V(x)\right] \equiv -S_E[x], \tag{40.40}$$

which defines the Euclidean action as

$$S_E[x] = \int dt_E\left[\frac{1}{2}\left(\frac{dx}{dt_E}\right)^2 + V(x)\right] = \int dt_E L_E(x,\dot{x}). \tag{40.41}$$

This allows the propagator to be better defined, as

$$U(t,t_0) = \int \mathcal{D}x(t_E)\exp\left\{-\frac{1}{\hbar}S_E[x(t_E)]\right\}, \tag{40.42}$$

but more importantly, it still takes *real values* for real Euclidean time t_E instead of real normal time t.

Assume that $V(x)$ has two minima, x_1 and x_2, with the same value V_0, thus $V'(x_{1,2}) = 0$, with $V''(x_{1,2}) > 0$, which necessarily means that there is a local maximum x_0 in between them, $V'(x_0) = 0$ with $V''(x_0) < 0$.

These extrema for the potential are then also extrema of the Euclidean action $S_E[x]$, namely, $x(t) = x_1$ and $x(t) = x_2$ satisfy the classical equations of motion for the Euclidean action,

$$0 = \frac{\delta S_E}{\delta x(t_E)} = -\frac{d^2x}{dt_E^2} + \frac{dV}{dx} = 0. \tag{40.43}$$

With respect to the classical equations of motion for the normal action,

$$\frac{d^2x}{dt^2} + \frac{dV}{dx} = 0, \tag{40.44}$$

we see that we have effectively an inverted potential,

$$V_E(x) = -V(x). \tag{40.45}$$

Consider then the classical solution in this inverted potential $V_E(x)$, i.e., the classical solution of the Euclidean action, with the given boundary conditions; it is called an *instanton*. Note that $V_E(x)$ has two maxima at x_1 and x_2, and a local minimum x_0 in between them, as in Fig. 40.2.

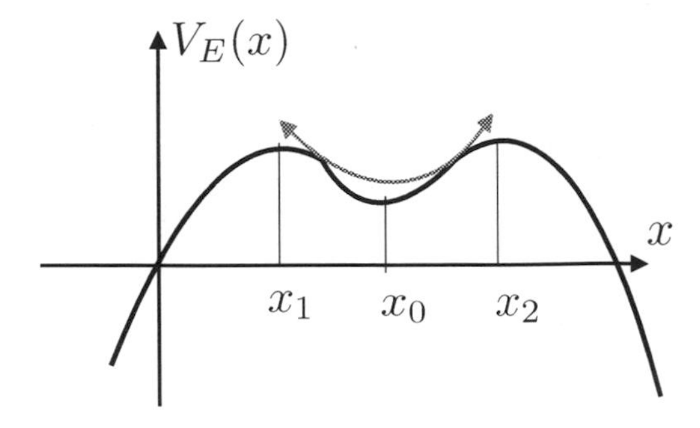

Figure 40.2 Instanton solution in between two maxima, x_1, x_2 of V_E, passing through the local minimum x_0 (equivalently, two minima of V, with the solution passing through the local maximum x_0).

Then we can consider an *instanton* solution that asymptotically goes to x_1 at $t_i \to -\infty$ and to x_2 at $t_f \to +\infty$, by passing through x_0 at a finite time. Physically, this motion in the inverted potential $V_E(x)$ corresponds to a particle initially (at $t_i \to -\infty$) sitting at the maximum (an unstable point) at x_1, and receiving an infinitesimal kick, after which it goes down the hill, through the local minimum at x_0, and on to the maximum at x_2, reached asymptotically when $t_f \to +\infty$.

If $t_i, t_f \in \mathbb{R}$ and are asymptotically at infinity, they would correspond to $t_{E,i}, t_{E,f} \in i\mathbb{R}$, which is not what we want. However, in the complex t plane this would correspond to integration over the contour C and, as we saw, this is equivalent to integration over the contour C', where x_0 is, not a pole of the potential, but, rather more generally now, the extremum of the action appearing in the exponent situated in between the initial and final points, meaning the same x_0 as was defined just above. Then $x(t)$ is the same complex path as that described in the WKB matrix element analysis.

In this case, now we can use the general formulas in Chapter 38 relating transition amplitudes with the propagator,

$$b_n(t) = \langle \psi_n | U_I(t, t_0) | \psi_i \rangle. \tag{40.46}$$

Here we take $|\psi_i\rangle$ to be the quantum state corresponding to the position x_1 at $t_i \to -\infty$ and $|\psi_m\rangle$ the quantum state corresponding to position x_2 at $t_f \to +\infty$. Then we obtain for the transition amplitude coefficient

$$b_m(t) \sim \int \mathcal{D}x(t) \exp\left\{\frac{i}{\hbar} S[x(t)]\right\} = \int \mathcal{D}x(t_E) \exp\left\{-\frac{1}{\hbar} S_E[x(t_E)]\right\}. \tag{40.47}$$

A saddle point approximation for the path integral gives just the exponent corresponding to the minimum of the Euclidean action S_E, meaning the Euclidean action evaluated on the classical solution (which extremizes S_E), i.e., the instanton, thus

$$b_m(t) \sim \exp\{-S_E[x_{\rm cl}(t_E)]\} = \exp\left(-S_E[x_1, x_0] - S_E[x_0, x_2]\right) = \exp\left(iS[x_1, x_0] + iS[x_0, x_2]\right), \tag{40.48}$$

where we have split the classical instanton solution into two halves, one going from x_1 to x_0 and the other from x_0 to x_2, and we have gone back to the normal action by setting $-S_E = iS$. Finally then, the probability of transition is

$$P(i \to m) = |b_m(t)|^2 \sim \exp\left(-\frac{2}{\hbar} {\rm Im}(S[x_1, x_0] + S[x_0, x_2])\right), \tag{40.49}$$

which is the same formula as that obtained above from the WKB method on matrix elements.

We can also consider the next correction to the above approximate formula, a further "quasi-classical approximation" around the classical Euclidean value. This corresponds to considering fluctuations around the classical path in complex space, i.e., around the instanton. We write the quadratic fluctuations around the instanton as follows:

$$S_E \simeq S_E[x_{\rm cl}(t_E)] + \frac{1}{2} \frac{\delta^2 S_E}{\delta x^2(t_E)} [x_{\rm cl}(t_E)](x - x_{\rm cl}(t_E))^2, \tag{40.50}$$

and then, doing Gaussian integration over x, we obtain the subleading fluctuations determinant contribution as well,

$$e^{-S_{E,0}} \left\{ \det\left[-\frac{d^2}{dt_E^2} + V''(x_{\rm cl}(t_E))\right]\right\}^{-1/2}. \tag{40.51}$$

We can also consider other contributions to the path integral, such as *multi-instanton solutions* and fluctuations around them.

In the simple case considered here, with only two minima of the potential, we only have an instanton solution, going from x_1 to x_2 in Euclidean time, and an anti-instanton solution, going from x_2 to x_1. We could have a quasi-solution that goes from x_1 to x_2 in a large, but not infinite time, and then back to x_1 in the same large time, which would be an instanton–anti-instanton solution.

However, if instead of the potential with two minima we had a periodic potential, with an infinite number of minima, we could consider a multi-instanton (quasi-)solution, where we go from one minimum to the next, then to the next, etc. In this case, in the path integral we must sum over all the possible multi-instantons, and the fluctuations around them, but we will not explore this possibility further here.

Important Concepts to Remember

- We can use the WKB method to calculate matrix elements $f_{12} \equiv \langle\psi_1|\hat{A}|\psi_2\rangle$ between two eigenstates of the Hamiltonian, by first going to the complex plane (more specifically, the imaginary line in the upper-half plane) for x; consequently, for the exponent in the WKB approximation for f_{12}, we have $e^{ig(x)} \to e^{-\text{Im }g(x)}$.
- Then, minimizing this exponent over the complex plane, we obtain

$$f_{12} \sim \exp\left[-\frac{E_2 - E_1}{\hbar}\text{Im }\tau\right],$$

  with $\tau = \int^{x_0} dx/v(x)$, where x_0 is the (unique) point in the upper-half plane where $V(x_0) \to \infty$, the integral only goes near it, and $v(x) = \sqrt{2(E - V(x))/m}$.
- Using Bohr–Sommerfeld quantization, the transition probability is found as $P(i \to m) \sim \exp\left[-\frac{2}{\hbar}\text{Im}[S(x_1, x_0) + S(x_0, x_2)]\right]$, for a path in complex plane for imaginary x and τ, from x_1 to x_2 via x_0, the extremum of the action in complex plane.
- In a more precise formalism, we define path integrals in Euclidean (imaginary) time, with propagator $U(t, t_0) = \int \mathcal{D}x(t_E) \exp\left[-\frac{1}{\hbar}S_E(x(t_E))\right]$.
- Motion in Euclidean time is motion in the inverted potential, $V_E(x) = -V(x)$, so motion between two minima for $V(x)$ via a maximum x_0 in between them becomes motion between two maxima of $V_E(x)$ via a minimum x_0 in between them, called an instanton.
- Since $b_m(t) = \langle\psi_m|U(t, t_0)|\psi_0\rangle$, the probability becomes the modulus squared of the Euclidean path integral for the instanton, which in first order gives the previous formula for $P(i \to m) \sim \exp\left\{-\frac{2}{\hbar}S_E[x_{\text{cl}}(t_E)]\right\}$, in terms of the instanton action. Fluctuations around the instanton, as well as multi-instanton contributions, give corrections to $P(i \to m)$.

Further Reading

For more on WKB methods for matrix elements, see [4], and for more on instantons, in both quantum mechanics and quantum field theory, see [10].

Exercises

(1) Use the WKB method, and Bohr–Sommerfeld quantization, to estimate the matrix element $\langle 2|\exp(-a\partial_x)|1\rangle$ for a harmonic oscillator.

(2) Consider the potential $V = -A/(x^2 + a^2)$, and states $|1\rangle$, $|2\rangle$, the first two eigenstates in this potential. Using the methods in the text, estimate the transition probability $f_{12} \equiv \langle 1|x|2\rangle$.

(3) Can the instanton be associated with the motion of a real particle, albeit nonclassical (perhaps a quantum path)?

(4) Consider a theory with only two minima of the potential, at x_1 and x_2. Sum the series of noninteracting multi-instantons that start at x_1 and end at x_2, having been through a continuous motion.

(5) Consider the Higgs-type potential $V = \alpha(x^2 - a^2)^2$, and a transition between the two vacua (minima of the potential) $x_1 = -a$ and $x_2 = +a$. Calculate the transition probability in the classical one-instanton approximation.

(6) Consider the periodic potential $V = A\cos^2(ax)$, and a transition between the first two vacua (minima) of $x > 0$. Calculate the transition probability in the classical one-instanton approximation.

(7) Calculate the (formal) fluctuation correction to the instanton calculation in exercise 6.

41 Variational Methods

In this chapter, we will study another type of approximation methods, variational methods. They give an approximation to eigenenergies and eigenstates by considering variations over a certain subset of functions, representing for states.

41.1 First Form of the Variational Method

The first form of the variational method applies to finding the energy of the ground state of a system.

Theorem This method starts from a simple theorem, expressed as follows. The energy of a state is always equal to or larger than that of the ground state:

$$\langle E\rangle_\psi \equiv \frac{\langle\psi|\hat{H}|\psi\rangle}{\langle\psi|\psi\rangle} \geq E_0, \quad \forall|\psi\rangle. \tag{41.1}$$

Proof Consider an orthonormal basis of exact eigenstates $|n\rangle$ of the Hamiltonian $\hat{H}$,

$$\hat{H}|n\rangle = E_n|n\rangle, \tag{41.2}$$

among which E_0 is the smallest, i.e., it is the ground state.

Since this is a basis, it is a complete set,

$$\sum_n |n\rangle\langle n| = \mathbb{1}. \tag{41.3}$$

Then expand the state $|\psi\rangle$ in this basis,

$$|\psi\rangle = \sum_n c_n|n\rangle = \sum_n |n\rangle\langle n|\psi\rangle. \tag{41.4}$$

Since the states $|n\rangle$ are an orthonormal set of eigenstates,

$$\langle m|\hat{H}|n\rangle = E_n\delta_{mn}. \tag{41.5}$$

Then it follows that

$$\begin{aligned}\langle E\rangle_\psi &= \frac{\sum_{m,n}\langle\psi|m\rangle\langle m|\hat{H}|n\rangle\langle n|\psi\rangle}{\sum_n\langle\psi|n\rangle\langle n|\psi\rangle}\\ &= \frac{\sum_n |\langle n|\psi\rangle|^2 E_n}{\sum_n |\langle n|\psi\rangle|^2}\\ &\geq \frac{\sum_n |\langle n|\psi\rangle|^2 E_0}{\sum_n |\langle n|\psi\rangle|^2} = E_0.\end{aligned} \tag{41.6}$$

q.e.d

From this theorem, it follows that we can choose *any* basis $|\alpha\rangle$ for the Hilbert space (not just an eigenbasis of $\hat{H}$), and expand in it,

$$|\psi\rangle = \sum_\alpha a_\alpha |\alpha\rangle = \sum_\alpha |\alpha\rangle\langle\alpha|\psi\rangle, \tag{41.7}$$

and then the ground state energy can be found as a minimum over the whole space of $\{a_\alpha\}$ coefficients,

$$\langle E\rangle_\psi = \frac{\sum_{\alpha,\beta} a_\alpha^* a_\beta H_{\alpha\beta}}{\sum_\alpha a_\alpha^* a_\alpha} \geq E_0 = \frac{\langle 0|\hat{H}|0\rangle}{\langle 0|0\rangle}. \tag{41.8}$$

The minimum conditions,

$$\left\{\frac{\partial E_\psi}{\partial a_\alpha} = 0, \ |\forall\alpha\right\} \tag{41.9}$$

will then select the ground state, $\langle E\rangle_{\psi,\min} = E_0 \Rightarrow |\psi\rangle = |0\rangle$.

41.2 Ritz Variational Method

However, the above is in general impractical, since in most cases the Hilbert space is either infinite dimensional, or of a very high dimension, and moreover one has to orthonormalize the basis states, making full minimization extremely hard if not impossible.

One particular variational method that *is* practical is the Ritz method, which amounts to considering a *subset* of states $\{|\psi_i\rangle\}_i$, and not even necessarily an orthonormal subset (a sub-basis), and *defining*

$$|\psi\rangle \equiv \sum_i c_i |\psi_i\rangle, \tag{41.10}$$

where c_i are arbitrary complex coefficients.

Then we define also

$$\langle E\rangle_\psi = \frac{\sum_{i,k} c_k^* c_i \langle\psi_k|\hat{H}|\psi_i\rangle}{\sum_{i,k} c_k^* c_i \langle\psi_k|\psi_i\rangle} = \frac{\sum_{i,k} c_k^* c_i H_{ki}}{\sum_{i,k} c_k^* c_i \Delta_{ki}}, \tag{41.11}$$

where we have denoted

$$\langle\psi_k|\hat{H}|\psi_i\rangle = H_{ki}, \ \ \langle\psi_k|\psi_i\rangle = \Delta_{ki}. \tag{41.12}$$

Treat the c_k and c_k^* as independent variables (this is always possible, being equivalent to writing $c_i = a_i + ib_i$ with a_i and b_i independent real variables). Then the stationarity (minimum) equations, with variables c_k^*, are

$$\frac{\partial\langle E\rangle_\psi}{\partial c_k^*} = 0, \tag{41.13}$$

which gives

$$\frac{1}{\sum_{i,k} c_k^* c_i \Delta_{ki}} \left[\sum_i c_i H_{ki} - \langle E \rangle_\psi \sum_i c_i \Delta_{ki} \right] = 0, \tag{41.14}$$

or equivalently

$$\sum_i c_i \left[H_{ki} - \langle E \rangle_\psi \Delta_{ki} \right] = 0, \tag{41.15}$$

where $\langle E \rangle_\psi$ is E at the minimum, leading to an eigenvalue equation for E,

$$\det(H_{ki} - E\Delta_{ki}) = 0. \tag{41.16}$$

Lemma The good thing about this method is that, even if we incur a significant error in finding the correct state (in this case $|0\rangle$), say

$$|\psi\rangle - |0\rangle \sim \mathcal{O}(\epsilon), \tag{41.17}$$

the error in finding the correct energy is much smaller,

$$E - E_0 \sim \mathcal{O}(\epsilon^2). \tag{41.18}$$

Proof We first trivially rewrite the form of the energy of the state as

$$\langle E \rangle_\psi = \frac{\sum_n |\langle n|\psi\rangle|^2 (E_n - E_0)}{\sum_n |\langle n|\psi\rangle|^2} + E_0. \tag{41.19}$$

We also use normalized states,

$$|\tilde{\psi}\rangle \equiv \frac{|\psi\rangle}{\sqrt{\langle\psi|\psi\rangle}} = \frac{|\psi\rangle}{\sqrt{\sum_n |\langle n|\psi\rangle|^2}}. \tag{41.20}$$

Then, if we have

$$\langle n|\tilde{\psi}\rangle \sim \mathcal{O}(\epsilon) \tag{41.21}$$

for $n \neq 0$, then

$$\langle E \rangle_\psi - E_0 = \sum_n |\langle n|\tilde{\psi}\rangle|^2 (E_n - E_0) \sim \mathcal{O}(\epsilon^2). \tag{41.22}$$

q.e.d.

41.3 Practical Variational Method

An even more practical version of the variational method is written explicitly in terms of wave functions, as opposed to states.

Consider a wave function depending on parameters $a_1, \ldots, a_n$,

$$\psi(\vec{x}) \equiv \langle \vec{x}|\psi\rangle = \psi(\vec{x}; a_1, \ldots, a_n). \tag{41.23}$$

Then minimize the energy of the state $\langle E\rangle_\psi$ over the parameters $a_1, \ldots, a_n$ (instead of over coefficients of states, or, given functions in an expansion, over parameters of a function):

$$\left\{\frac{\partial\langle E\rangle_\psi}{\partial a_i} = 0\right\}_i . \tag{41.24}$$

Thus, as in the case of the Ritz method, we obtain an approximation for the eigenenergy that is better, namely $\mathcal{O}(\epsilon^2)$, than the approximation for the eigenstates, which is $\mathcal{O}(\epsilon)$.

41.4 General Method

We can now obtain the most general form of the variational method, which is in fact valid for any Hermitian operator A, even though it is generally used for $\hat{H}$. As before, we define

$$\langle E\rangle_\psi = \langle\tilde{\psi}|\hat{H}|\tilde{\psi}\rangle = \frac{\langle\psi|\hat{H}|\psi\rangle}{\langle\psi|\psi\rangle}. \tag{41.25}$$

This general method is also based on a theorem.

Theorem

With the above definitions, and using normalized states $|\tilde{\psi}\rangle$, the theorem is written formally as

$$\frac{\delta\langle E\rangle_\psi}{\delta|\tilde{\psi}\rangle} = 0 \Leftrightarrow |\tilde{\psi}\rangle = |n\rangle, \tag{41.26}$$

i.e., the energy of the state is stationary, meaning there are no $\mathcal{O}(\epsilon)$ terms in it, only $\mathcal{O}(\epsilon^2)$ terms, if and only if the normalized state is an eigenstate of $\hat{H}$, with energy $E = E_n = \langle E\rangle_\psi$. Note that this is true *for all the eigenstates*, not just for the ground state, though note that the condition is of stationarity, not of a minimum (more on that later).

Proof in a basis. One way to prove it, is to consider a basis, i.e., a complete orthonormal set $|r\rangle$ of states, and expand the normalized state in it,

$$|\tilde{\psi}\rangle = \sum_r c_r|r\rangle. \tag{41.27}$$

We want to keep the normalization condition fixed, i.e.,

$$\langle\tilde{\psi}|\tilde{\psi}\rangle = 1 \Leftrightarrow \sum_{r,s} c_r^* c_r = 1. \tag{41.28}$$

This means that we need to minimize $\langle E\rangle_\psi$, understood as an action, while keeping the constraint, which is added to the "action" via Lagrange multipliers called E,

$$\delta\left[\langle E\rangle_\psi - E\left(\sum_r c_r^* c_r - 1\right)\right] = 0. \tag{41.29}$$

We rewrite it as

$$\delta\left[\sum_{r,s} c_r^* H_{rs} c_s - E\sum_{r,s} c_r^* \delta_{rs} c_s + E\right] = 0$$
$$\Rightarrow \quad \delta\left[\sum_{r,s} c_r^*(H_{rs} - E\delta_{rs})c_s\right] = 0. \tag{41.30}$$

Now we treat δc_r^* and δc_s as arbitrary and independent variations, just as in the Ritz variational method. Thus we rewrite the above condition as

$$\sum_{r,s} \delta c_r^*(H_{rs} - E\delta_{rs})c_s + \sum_{r,s} c_r(H_{rs} - E\delta_{rs})\delta c_s = 0, \tag{41.31}$$

but since the variations are arbitrary and independent, we obtain two equations,

$$(H_{rs} - E\delta_{rs})c_s = 0$$
$$(H_{rs} - E\delta_{rs})c_r^* = 0. \tag{41.32}$$

However, since $\hat{H}$ is Hermitian, $H_{rs} = H_{sr}^*$ and moreover the eigenergies are real, $E = E^*$, the second equation becomes

$$(H_{sr}^* - E^*\delta_{sr})c_r^* = 0, \tag{41.33}$$

which is just the complex conjugate of the first. But this is just the eigenstate condition,

$$H_{rs}c_s = Ec_r. \tag{41.34}$$

q.e.d.

We also note that here for the first time we used hermiticity, and this is the only property used about $\hat{H}$, which means that indeed, the result is valid for any Hermitian operator.

Proof in the general case In fact we don't need to expand in a basis in order to prove the theorem. We vary the energy of the state, with the normalization condition imposed with Lagrange multiplier E, as before (but without expanding in a basis),

$$\delta\left[\langle\tilde{\psi}|\hat{H}|\tilde{\psi}\rangle - E\langle\tilde{\psi}|\tilde{\psi}\rangle + E\right] = 0. \tag{41.35}$$

Then the variation becomes

$$\delta\left[\langle\tilde{\psi}|(\hat{H} - E\mathbb{1})|\tilde{\psi}\rangle\right] = 0$$
$$\Rightarrow \quad \langle\delta\tilde{\psi}|(\hat{H} - E\mathbb{1})|\tilde{\psi}\rangle + \langle\tilde{\psi}|(\hat{H} - E\mathbb{1})|\delta\tilde{\psi}\rangle = 0. \tag{41.36}$$

But then if we treat $\langle\delta\tilde{\psi}|$ and $|\delta\tilde{\psi}\rangle$ as arbitrary and independent variations, we obtain

$$(\hat{H} - E\mathbb{1})|\tilde{\psi}\rangle = 0$$
$$\langle\tilde{\psi}|(\hat{H} - E\mathbb{1}) = 0 \Rightarrow (\hat{H}^\dagger - E^*\mathbb{1})|\tilde{\psi}\rangle = 0. \tag{41.37}$$

Since $\hat{H}$ is Hermitian, $\hat{H}^\dagger = \hat{H}$ and $E^* = E$, so we obtain the same equation as the first, and moreover it is the eigenstate equation,

$$\hat{H}|\tilde{\psi}\rangle = E|\tilde{\psi}\rangle. \tag{41.38}$$

q.e.d

Moreover, again we can show that the energy error is $\mathcal{O}(\epsilon^2)$. Consider an error $|\delta\tilde{\psi}_n\rangle$, and divide it in a part parallel to $|n\rangle$, and a part $|\delta\tilde{\psi}'\rangle$ orthogonal to it,

$$|\tilde{\psi}_n\rangle = |n\rangle + |\delta\tilde{\psi}'_n\rangle = a|n\rangle + b|\delta\tilde{\psi}'\rangle, \tag{41.39}$$

where $\langle n|\delta\tilde{\psi}'\rangle = 0$. Then normalization means that

$$\langle\tilde{\psi}_n|\tilde{\psi}_n\rangle = 1 \Rightarrow a^2 + b^2 = 1. \tag{41.40}$$

But then if $b \sim \mathcal{O}(\epsilon)$, $a \sim 1 - \mathcal{O}(\epsilon^2)$. That leads to the following error in the energy,

$$\begin{aligned}\langle E\rangle_\psi &= \left(a^*\langle n| + b^*\langle\delta\tilde{\psi}'|\right)\hat{H}\left(a|n\rangle + b|\tilde{\psi}'\rangle\right)\\ &= |a|^2 E_n + |b|^2\langle\delta\tilde{\psi}'|\hat{H}|\delta\tilde{\psi}'\rangle\\ &\simeq E_n + \mathcal{O}(\epsilon^2),\end{aligned} \tag{41.41}$$

where in the second equality we have used $\hat{H}|n\rangle = E_n|n\rangle$ and $\langle\delta\tilde{\psi}'|n\rangle = 0$, and in the last equality we have used $|a|^2 = 1 - \mathcal{O}(\epsilon^2)$ and $|\delta\tilde{\psi}'\rangle \sim \mathcal{O}(\epsilon)$.

Note however that now the *sign* of the ϵ^2 correction is arbitrary, meaning that $\langle E\rangle_\psi$ reaches either a minimum *or* a maximum. Only for $E_n = E_0$ (the ground state energy) do we always have a minimum (since the sign of $\langle E\rangle - E_0$ is always plus).

41.5 Applications

We now apply the variational methods to some relevant cases. We first point out the pros and cons of the variational method.

Pro: The method gives a very good approximation to the energy, with $\mathcal{O}(\epsilon^2)$ error, even when the $\mathcal{O}(\epsilon)$ error in the state (or wave function) is relatively large.

Con: The method does not give us the error itself, since all we know is that we are minimizing over some functions, not how these functions relate to the exact ones (i.e., what the value of ϵ is).

But we can improve on the cons as follows:

(1) We can choose a wave function for $|\psi\rangle$ that has the same number of nodes as the state $|n\rangle$ which we want to reproduce (which is known by general theory, without knowing the state itself).
(2) We can impose the same asymptotic behavior at the extremes, $r = 0$ and/or $r = \infty$ in three dimensions.
(3) We can use eigenfunctions of operators that commute with $\hat{H}$, e.g., the angular momentum $\hat{\vec{L}}$.

The Hydrogenoid Atom

In this case, we actually know exactly the wave functions of the eigenstates, but we can pretend we don't, and try to use the variational method.

Using point (3) above, we can write the ansatz for the wave function in three dimensions,

$$\Phi = R(r)Y_{lm}(\theta,\phi). \tag{41.42}$$

Then we use point (2) above, and at $r = 0$, $R(r) \to$ const, whereas at $r \to \infty$, $R(r) \to 0$ really fast. For the ground state energy E_0, with spherical symmetry, we can then try (using the practical variational method) the wave function

$$R(r) = Ae^{-r/a}, \tag{41.43}$$

depending on the parameters A and a, and minimize the energy $\langle E \rangle_\psi$ over them. At the minimum, we find $a = a_0$ and $A = a_0^{-3/2}$, leading to the ground state wave function $|\psi_0\rangle$, and the corresponding ground state energy E_0. Thus in this way, we find the exact energy E_0.

But we could also try another function satisfying the same boundary conditions at $r = 0$ and $r = \infty$,

$$R(r) = \frac{A}{b^2 + r^2}, \tag{41.44}$$

depending on the parameters A and b, and minimize the energy over them. At the minimum, we find $E_{\text{min}} = -0.81|E_0|$, which is pretty close to the true value, even though the wave function is way off.

For the energy of an excited state E_n, $n > 0$, we must try wave functions that have the correct symmetry properties, the right number of nodes, etc. Otherwise, if we know the energy levels E_m and states for $m < n$, we then try functions that are orthogonal to them, leading to $E_n > E_m$.

Previously Studied Example: the Helium Atom (or Helium-Like Atom)

Helium-like atoms were studied in Chapter 21, and the variational method was used without much explanation, so here we return to it and streamline the presentation.

Consider a trial wave function for the electrons in a helium-like atom that is just a product of the states for a single electron, i.e., the zeroth-order approximation in the perturbation theory for the interaction between the electrons,

$$\Phi(\vec{r}_1, \vec{r}_2; a) = f(\vec{r}_1; a) f(\vec{r}_2; a) = \frac{1}{\pi a^3} \exp\left(-\frac{r_1 + r_2}{a}\right), \tag{41.45}$$

but where a is now a free parameter, and not the value $a = a_0/Z$ that would be appropriate for a helium-like atom of charge Z. Equivalently, write

$$a = \frac{a_0}{Z'}, \tag{41.46}$$

where Z' is an arbitrary *effective* charge felt by the electrons, and is not equal to Z.

Then, as shown in Chapter 21, the first-order (time independent) perturbation theory correction, the interaction potential (between the electrons) V_{12}, averaged in the zeroth-order state $|\Phi\rangle$, only now with charge Z' instead of Z, gives

$$\langle \Phi | \hat{V}_{12} | \Phi \rangle = \frac{5}{4} Z' |E_0|, \tag{41.47}$$

where $|E_0| = e_0^2/(2a_0)$ is the ground state energy of the hydrogen atom (the energy of a single electron in the field of a nucleus of charge one).

Then the energy functional of the helium-like atom is twice the energy of an electron in the field of a nucleus of charge Z', plus the above interaction energy, so

$$E(a) = 2|E_0|\left[Z'^2 - 2\left(Z - \frac{5}{16}\right)^2\right]. \tag{41.48}$$

Minimizing this over a, i.e., finding $\delta E(a)/\delta a = 0$, or equivalently minimizing over Z' by finding $\delta E/\delta Z' = 0$, we obtain

$$Z' = Z - \frac{5}{16}, \tag{41.49}$$

and the energy at this minimum is

$$E_{\Phi,\text{min}} = -2\left(Z - \frac{5}{16}\right)^2 |E_0|. \tag{41.50}$$

This is a better approximation than the first-order (time-independent) perturbation theory result, which is

$$\langle\Phi|\hat{V}_{12}|\Phi\rangle = \frac{5}{4}Z|E_0|, \tag{41.51}$$

leading to the total energy

$$E^{(0+1)} = -2|E_0|Z^2 + \frac{5}{4}Z|E_0| = E_{\Phi,\text{min}} + \frac{25}{128}|E_0|, \tag{41.52}$$

which is larger (and therefore a worse approximation) than $E_{\Phi,\text{min}}$.

The Ritz method will not be exemplified here, but rather will be revisited in the next chapter, where it will be used to work with atomic and molecular orbitals.

Important Concepts to Remember

- The energy of a state is always larger than its ground state, so we can find the ground state by minimizing the energy $\sum_{\alpha,\beta} a_\alpha^* a_\beta H_{\alpha\beta}/(\sum_\alpha |\alpha_\alpha|^2)$ over states expanded in an arbitrary basis, $|\psi\rangle = \sum_\alpha a_\alpha|\alpha\rangle$.
- In the Ritz variational method, we minimize over a subset of states in the Hilbert space, not necessarily orthonormal, $\langle E\rangle_\psi = \sum_{k,i} c_k^* c_i H_{ki}/(\sum_{k,i} c_k^* c_i \Delta_{ki})$, $\Delta_{ki} = \langle\psi_k|\psi_i\rangle$, with $|\psi\rangle = \sum_i c_i|\psi_i\rangle$. If the error in the state is $|\psi\rangle - |0\rangle = \mathcal{O}(\epsilon)$, then the error in energy is $E_\psi - E_0 = \mathcal{O}(\epsilon^2)$.
- In a practical variational method, one chooses wave functions depending on parameters, $\psi(x) = \langle x|\psi\rangle = \psi(\vec{x}; a_1, \ldots, a_n)$, and minimizes the energy $\langle E\rangle_\psi$ over the parameters, again finding a better error in the energy, $E_\psi - E_0 = \mathcal{O}(\epsilon^2)$, than in the state, $\psi - \psi_0 = \mathcal{O}(\epsilon)$.
- The most general variational method (valid for any operator A, not only for H), makes the energy stationary (no $\mathcal{O}(\epsilon)$ terms, only $\mathcal{O}(\epsilon^2)$ terms) over normalized states, i.e., $\delta\langle E\rangle_\psi/\delta|\psi\rangle = 0$ if and only if we have an eigenstate, $|\psi\rangle = |n\rangle$, for all states. For the ground state, stationarity implies a minimum.
- Variational methods improve the error, though we do not then know its value; but we can improve them by choosing wave functions with the correct number of nodes and behavior at infinity or zero, and that are eigenfunctions of operators that commute with $\hat{H}$.
- For a helium-like atom, the variational method is better than first-order time-independent perturbation theory.

Further Reading

See [2] and [1].

Exercises

(1) Consider a two-level system, with basis $|1\rangle, |2\rangle$, and in this basis, a Hamiltonian with elements $\begin{pmatrix} 1 & 1 \\ 1 & 1 \end{pmatrix}$. Use the first form of the variational method to find the ground state, and then check by finding the exact eigenstates.

(2) Use the Ritz variational method for the harmonic oscillator, with trial wave functions $\psi_1(x) = e^{-y^2/2}, \psi_2(x) = e^{-y^2}, \psi_3(x) = e^{-2y^2}$, where $y = x\sqrt{m\omega/\hbar}$, in order to find the best approximation to the ground state energy.

(3) Use the practical variational method for the same harmonic oscillator ground state energy, with trial wave function $\psi_a(x) = e^{-ay^2}$.

(4) Repeat exercise 3 for the third state (second excited state) $|3\rangle$ of the harmonic oscillator, namely, for $\psi_a(x) = y^2 e^{-ay^2}$.

(5) Consider the potential $V = kx^2 + \alpha|x|^3$, and a trial wave function $\psi(x; a, b) = |y|^a e^{-by^2}$. Find an estimate of the ground state energy (write down the equations for the minimum).

(6) Fill in the details in the text for minimization with the trial wave function $R(r) = Ae^{-r/a}$ for the hydrogenoid atom.

(7) Fill in the details in the text for minimization with the trial wave function $R(r) = A/(r^2 + b^2)$.

PART II$_c$

ATOMIC AND NUCLEAR QUANTUM MECHANICS

42 Atoms and Molecules, Orbitals and Chemical Bonds: Quantum Chemistry

In this chapter, we will build up from hydrogenoid atoms to multi-electron atoms, and onward to molecules, by defining atomic orbitals, molecular orbitals, the hybridization of orbitals, and the resulting chemical bonds, defining the field of quantum chemistry.

42.1 Hydrogenoid Atoms (Ions)

We start with a quick review of the hydrogenoid atoms, i.e., ions with a nucleus of charge Z and only one electron. The electronic quantum numbers are (n, l, m, m_s), where

$$l = 0, 1, \ldots, n-1, \quad m = -l, -l+1, \ldots, l-1, l, \quad m_s = \pm 1/2. \tag{42.1}$$

There is a spin–orbit $(\vec{l} \cdot \vec{s})$ coupling, meaning that there is such a term in the interaction Hamiltonian, where $\vec{l}$ is the angular momentum and $\vec{s}$ is the spin angular momentum. Then, at least if the interaction energy is sufficiently large, we must add the angular momenta to give the total angular momentum, $\vec{j} = \vec{l} + \vec{s}$, which means that in terms of quantum numbers we have $j = l \pm 1/2$. Moreover, the values of m_j are $-j, -j+1, \ldots, j-1, j$. We can then use the quantum numbers (m, l, j, m_j), which are more appropriate if the spin–orbit interaction energy is large enough.

The wave functions depend on the position $\vec{r}$ and the quantum numbers, and can be expanded as products:

$$\psi_{nlm}(r, \theta, \phi, m_s) = R_{nl}(r) Y_{lm}(\theta, \phi) \chi(m_s). \tag{42.2}$$

Since $Y_{lm}(\theta, \phi) = P_{lm}(\cos\theta) e^{im\phi}$, the probability is

$$P_{nlmm_s}(\vec{r}) = |\psi_{nlm}(r, \theta, \phi, m_s)|^2 = |R_{nl}(r) P_{nl}(\cos\theta)|^2 |\chi(m_s)|^2, \tag{42.3}$$

which means that it actually depends on r, θ and (n, l, m, m_s). This probability profile, more precisely the shape that contains most of it (say, 99% for instance), is defined to be an *orbital*, for a given (n, l, m, m_s) as above or a given (n, l, j, m_j).

For the orbitals, we have the *spectroscopic notation*, where we denote an orbital by nl^d, where d is the multiplicity of the state. More precisely, instead of the value of l, we use the notation where $l = 0$ is called s, $l = 1$ is called p, $l = 2$ is d and $l = 3$ is f. From $l = 4$ onwards, we have alphabetic notation, so $l = 4, 5, 6, 7$ are denoted by $g, h, i, j, \ldots$ The $l = 0$ or s orbital has spherical symmetry, so no θ dependence at all; the $l = 1$ or p orbital has axial (or dipole) symmetry, meaning one axis; the $l = 2$ or d orbital has quadrupole symmetry (or 2 axes); etc.

The above-defined hydrogenoid orbitals, i.e., atomic orbitals for hydrogenoid atoms, meaning the single-electron wave functions, are specifically (here $\rho = r/a_0$ as always)

$$
\begin{aligned}
1s: &\quad \phi_{100}(\rho,\theta,\phi) = Ce^{-\rho/2}\\
2s: &\quad \phi_{200}(\rho,\theta,\phi) = \frac{C}{\sqrt{32}}(2-\rho)e^{-\rho/2}\\
2p: &\quad \phi_{21i}(\rho,\theta,\phi) = \frac{C}{\sqrt{32}}e^{-\rho/2}\begin{pmatrix}\cos\theta\\ \sin\theta\cos\phi\\ \sin\theta\sin\phi\end{pmatrix} \propto \begin{pmatrix} Y_{1,0}\\ \frac{Y_{1,-1}-Y_{1,1}}{\sqrt{2}}\\ i\frac{Y_{1,-1}+Y_{1,1}}{\sqrt{2}}\end{pmatrix} \propto \frac{x_i}{r}\\
3s: &\quad \phi_{300}(\rho,\theta,\phi) = \frac{C}{\sqrt{972}}(6-6\rho+\rho^2)e^{-\rho/2}\\
3p: &\quad \phi_{31i}(\rho,\theta,\phi) = \frac{C}{\sqrt{972}}(4\rho-\rho^2)e^{-\rho/2}\begin{pmatrix}\cos\theta\\ \sin\theta\cos\phi\\ \sin\theta\sin\phi\end{pmatrix} \propto \begin{pmatrix} Y_{1,0}\\ \frac{Y_{1,-1}-Y_{1,1}}{\sqrt{2}}\\ i\frac{Y_{1,-1}+Y_{1,1}}{\sqrt{2}}\end{pmatrix} \propto \frac{x_i}{r}.
\end{aligned}
\tag{42.4}
$$

We make two observations:

- For the s orbitals, $\phi(\rho = 0) \neq 0$ so there is a direct interaction between the electrons and the nucleus, called a Fermi contact interaction.
- As we saw before, the average value of the radius is

$$\langle r\rangle = \frac{a_0}{2Z}[3n^2 - l(l+1)], \tag{42.5}$$

so it decreases with the charge Z of the nucleus, meaning that the hydrogenoid atom gets compressed for higher Z, rather than growing.

42.2 Multi-Electron Atoms and Shells

We next move on to multi-electron atoms. In this case,

- the other electrons screen the nucleus somewhat by an amount σ; they are described in terms of a hydrogenoid atom, but with an effective charge,

$$Z_{\text{eff}} = Z - \sigma. \tag{42.6}$$

- due to the other electrons, which are not spherically distributed (unless we only have s, i.e., spherically symmetric, orbitals for the other electrons), the potential seen by an electron is no longer central.

We have a *shell model*, where we fill up atomic orbitals, which are extensions of the hydrogenoid orbitals, with electrons, in order of increasing energy. We call a shell, the set of energy levels with a given n, and a sub-shell the set of energy levels of a given n and l; see Fig. 42.1.

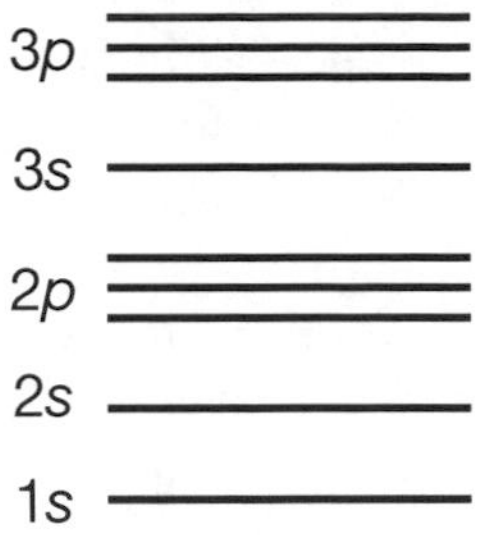

3p
3s
2p
2s
1s

Figure 42.1 The first sub-shells in the shell model, ignoring the spin (m_s) degree of freedom.

For a sub-shell, of given n, l and varying (m, m_s) or (j, m_j), if it is full, meaning all the energy levels in the sub-shell are occupied with electrons, then we have a spherical distribution of probability, and the total angular momenta vanish, $\vec{L}_{\rm tot} = 0 = \vec{S}_{\rm tot}$.

Hund Rules for the Atomic Orbitals:

(1) The electrons avoid being in the same orbitals if there is more than one orbital of equal l.
(2) Two electrons on equivalent (same value of l), but different, orbitals, have parallel ($\uparrow\uparrow$) spins in the ground state of the atom.

If we have a fully filled shell (all the energy levels of a given n are occupied by electrons), then for all calculations referring to further electrons, it is as if we have a hydrogenoid atom, with an effective charge $Z_{\rm eff} = Z - \sigma$.

42.3 Couplings of Angular Momenta

In an atom, in general we have many electrons, each with an orbital angular momentum $\vec{l}_i$ and a spin $\vec{s}_i$. We have various possibilities for coupling of these angular momenta, but we have two ideal situations:

(a) **LS coupling (normal, or Russell–Saunders)**. In this case, we first couple the orbital angular momenta to each other, and the spins to each other, and then we couple the resulting $\vec{L}$ and $\vec{S}$:

$$\vec{L} = \sum_i \vec{l}_i, \quad \vec{S} = \sum_i \vec{s}_i \quad \vec{J} = \vec{L} + \vec{S}. \tag{42.7}$$

This means that the possible values of J are

$$|L - S| \leq J \leq L + S. \tag{42.8}$$

The notation in this coupling is ${}^{2S+1}L_J$, or rather $n^{2S+1}L_J$, where usually $2S + 1$ is the multiplicity of states of given L, S. However, if $S > L$ then the multiplicity is $2L + 1$, replacing $2S + 1$ in the notation, and we say that the multiplicity is not fully developed. An example of the notation is the

$^2P_{3/2}$ state, which is read as "doublet P three halves". This LS coupling is predominant in the light atoms: electrostatic forces between the electrons couple the $\vec{l}_i$ into $\vec{L}$ and the $\vec{s}_i$ into $\vec{S}$, and then the smaller magnetic spin–orbit coupling couples $\vec{L}$ and $\vec{S}$ into $\vec{J}$.

(b) **jj coupling (spin–orbit)**. In this other idealized case, we first couple the orbital and spin angular momenta of each electron into their total angular momentum, and then the total angular momenta of the electrons to each other:

$$\vec{j}_i = \vec{l}_i + \vec{s}_i, \quad \vec{J} = \sum_i \vec{j}_i. \tag{42.9}$$

This case is predominantly used for heavy atoms, with large Z. In it, the magnetic spin–orbit coupling happens first, since it is due to the nuclear charge Z, which is large. Only after that do we couple the $\vec{j}_i$ to each other by the electrostatic forces between the electrons, which are Z-independent.

In general, a particular coupling lies somewhere in between the two idealized cases above. But the particular type of coupling does not modify the total number of levels, nor J, but rather changes the energy gaps between the various levels. This splitting of energy levels due to the coupling of the angular momenta of the electrons is called the "fine structure".

But there is also a "hyperfine structure" due to the nuclear angular momentum $\vec{I}$, more precisely its interaction with $\vec{J}$, leading to the total atomic angular momentum,

$$\vec{F} = \vec{J} + \vec{I}. \tag{42.10}$$

42.4 Methods of Quantitative Approximation for Energy Levels

There is an approximation method of introducing a self-consistent field, called the *Hartree–Fock method*, but we will leave it for later, when analyzing multiparticle states under some generality.

Method of Atomic Orbitals

The method we will describe here is an application of the Ritz variational method to atoms and their orbitals, called the atomic orbital method.

It is a way to introduce some interaction, usually between electrons. We split the Hamiltonian into free plus interaction parts,

$$\hat{H} = \hat{H}_0 + \hat{H}_{\text{int}}. \tag{42.11}$$

We use orthonormal eigenstates of $\hat{H}_0$, called $|\Phi_k\rangle$, which we can usually calculate (we choose $\hat{H}_0$ such that we can). If we consider all of them, for $k = 1$ to ∞, we obtain a basis of the Hilbert space of the system. But, rather than do that, we restrict to n terms only, for the expansion of a trial state,

$$|\psi_a\rangle = \sum_{k=1}^{n} c_k |\Phi_k\rangle \tag{42.12}$$

and use the full Hamiltonian $\hat{H}$ on it. Then, from the Ritz variational method, we want to obtain the eigenstates and eigenenergies for this system,

$$\sum_{k=1}^{n} c_k (\hat{H} - E)|\Phi_k\rangle = 0. \tag{42.13}$$

Multiplying with $\langle\Phi_i|$ from the left, we have

$$\sum_{k=1}^{n} c_k(H_{ik} - E\delta_{ik}) = 0. \quad (42.14)$$

Then we can find the secular equation for the eigenenergies,

$$\det(H_{ik} - E\delta_{ik}) = 0. \quad (42.15)$$

From it, we obtain n approximate values $E_1, \ldots, E_n$ for the eigenvalues of $\hat{H}$.

42.5 Molecules and Chemical Bonds

We next move on to molecules, which are several atoms bound together. The *chemical bonds* between the atoms are mostly between *two* atoms within a molecule, though there are exceptions such as the aromatic bonds, for instance in C_6H_6, explained briefly at the end of the chapter.

Chemical bonds usually fall into one of three categories:

- Electrovalent, or ionic, bonds, leading to a heteropolar molecule (meaning it has two different poles). In it, the electrons move between the two atoms, so one atom loses an electron (or more, or less; we are talking about probabilities in quantum mechanics, so fractional parts make sense) and the other gains it. For instance, in the ionic bond between an alkaline element (an element that has a single electron outside filled shells) and a halogen (an element for which a single electron is needed to have only filled shells), the alkaline element loses one electron, and the halogen gains it. The standard example of this is the ionic salt molecule, Na^+Cl^-; see Fig. 42.2a.
- Metallic bonds, in which the atoms have a crystalline structure, in which some electrons are delocalized within the crystalline structure. This means that the very many equal energy levels of the individual atoms split infinitesimally, creating almost continuous "bands" of electrons instead of discrete levels; see Fig. 42.2b.
- Covalent bonds, between neutral atoms, leading to a homopolar molecule (meaning with two like poles). The standard example is the H_2 molecule; see Fig. 42.2c.

A molecule is a *stable* structure, meaning it is a minimum of the energy (be it classical or quantum), associated with a given spatial structure. That in turn means that the nuclei are approximately fixed, and only the electrons move (or, rather, move fast compared with the nuclei).

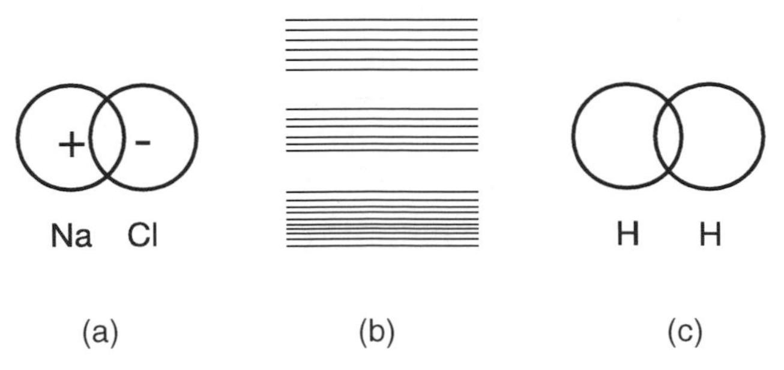

Figure 42.2 (a) Ionic bond: Na^+Cl^-. (b) Metallic bond: energy bands. (c) Covalent bond: H_2.

42.6 Adiabatic Approximation and Hierarchy of Scales

Since electrons are much lighter than nuclei,

$$\frac{m_e}{M_N} \ll 1, \tag{42.16}$$

as stated above we can consider that the nuclei are approximately fixed to their stable structure, and the electrons move fast within it, i.e., they "follow" the nuclei *adiabatically*; so, at all times we are able to treat the nuclei positions as mere parameters as far as the electrons are concerned.

Consider d as the characteristic distance between two atoms, or rather two nuclei, in the molecule. Then the electrons are delocalized over a distance $\Delta x \sim d$, meaning the electron momentum

$$p_{e^-} \sim \frac{\hbar}{d}. \tag{42.17}$$

Then the energy of an electron is about twice the kinetic energy, giving

$$E_e \sim 2T_e \sim \frac{p^2}{m} \sim \frac{\hbar^2}{md^2}. \tag{42.18}$$

On the other hand, the rotational energy associated with the molecule, or rather with the nuclei that compose it, is

$$E_{\rm rot} \sim \frac{\vec{L}^2}{I} \sim \frac{\hbar^2}{Md^2}, \tag{42.19}$$

where I is the moment of inertia of the nuclei. Thus

$$E_e \gg E_{\rm rot}. \tag{42.20}$$

Moreover, there is also a vibrational energy, associated with the vibrations of the nuclei around the stable structure. This means that the potential energy, approximately that of a harmonic oscillator,

$$E_{\rm pot} \sim \frac{M\omega^2 d^2}{2}, \tag{42.21}$$

should be balanced by the kinetic energy of an electron,

$$T_e \sim \frac{\hbar^2}{2md^2}, \tag{42.22}$$

so that

$$\omega \sim \frac{\hbar}{d^2\sqrt{Mm}} \Rightarrow E_{\rm vibr} \sim \hbar\omega \sim \frac{\hbar^2}{d^2\sqrt{Mm}}. \tag{42.23}$$

Then finally the hierarchy of energy scales for electronic, vibrational, and rotational modes is

$$E_{\rm rot} \ll E_{\rm vibr} \ll E_e. \tag{42.24}$$

While E_e is an atomic quantity, the rotational and vibrational modes are intrinsically molecular, have no atomic analogue, and are small. Concretely, the rotational energy $E_{\rm rot} \sim k_B T \sim 25$ meV, the vibrational energy $E_{\rm vibr} \sim 0.5$ eV, and the electronic energy is of a few eV.

The Schrödinger equation for a molecule is written as

$$\left(-\sum_{\alpha}\frac{\hbar^2}{2M_\alpha}\Delta_\alpha - \sum_i \frac{\hbar^2}{2m_i}\Delta_i + \sum_{\alpha,i} V_{\alpha i} + \sum_{\alpha,\beta} V_{\alpha\beta} + \sum_{i,j} V_{ij}\right)\psi_{n\{s\}}\left(\{\vec{r}_i\},\{\vec{R}_\alpha\}\right) = E_n\psi_{n\{s\}}\left(\{\vec{r}_i\},\{\vec{R}_\alpha\}\right). \tag{42.25}$$

Here m_i are the electron masses and $\vec{r}_i$ their positions, M_α are the masses of the nuclei and $\vec{R}_\alpha$ their positions, n is a generic index for energy states whereas $\{s\}$ is a degeneracy index. Moreover, T_N is the nuclear kinetic energy and T_e the electronic kinetic energy, where

$$T_N = -\sum_\alpha \frac{\hbar^2}{2M_\alpha}\Delta_\alpha, \quad T_e = -\sum_i \frac{\hbar^2}{2m_i}\Delta_i, \tag{42.26}$$

and V is the total potential energy, composed of nuclear–electronic part V_{Ne}, nucleus–nucleus part V_{NN}, and an electron–electron part V_{ee}, where

$$V_{Ne} = \sum_{\alpha,i} V_{\alpha i}, \quad V_{NN} = \sum_{\alpha,\beta} V_{\alpha\beta}, \quad V_{ee} = \sum_{i,j} V_{ij}, \quad V = V_{Ne} + V_{NN} + V_{ee}. \tag{42.27}$$

42.7 Details of the Adiabatic Approximation

We can consider a perturbation in which the unperturbed part $\hat{H}_0$ is the purely electronic part,

$$\hat{H}_0 = \hat{T}_e + \hat{V}, \tag{42.28}$$

neglecting T_N, so considering fixed nuclei. We can perhaps also neglect V_{NN}, which would mean neglecting the vibrational and rotational modes, thus assuming fixed nuclei at the positions in the stable structure. Then $\vec{R}_\alpha$ are just parameters, as in the practical variational method, and we can minimize over them.

Further, writing

$$\hat{H} = \hat{H}_0 + \hat{T}_N, \tag{42.29}$$

we can write down a solution that is not fully separated as an eigenfunction for the Schrödinger equation for $\hat{H}$,

$$\psi_{n\{s\}}\left(\{\vec{r}_i\},\{\vec{R}_\alpha\}\right) = \psi_{e,n\{s\}}\left(\{\vec{r}_i\},\{\vec{R}_\alpha\}\right)\psi_N\left(\{\vec{R}_\alpha\}\right), \tag{42.30}$$

where the ψ_e are eigenfunctions of $\hat{H}_0$,

$$\hat{H}_0\psi_{e,n\{s\}}\left(\{\vec{r}_i\},\{\vec{R}_\alpha\}\right) = E_n^{(0)}\psi_{e,n\{s\}}\left(\{\vec{r}_i\},\{\vec{R}_\alpha\}\right), \tag{42.31}$$

that depend on the positions $\vec{R}_\alpha$ of the nuclei only as parameters.

On the other hand, $\hat{T}_N$ acts on both ψ_e and ψ_N, so we can write

$$\left[\hat{T}_N(+\hat{V}_{NN})\right]\psi_{e,n\{s\}}\left(\{\vec{r}_i\},\{\vec{R}_\alpha\}\right)\psi_N\left(\{\vec{R}_\alpha\}\right) = E_{N,n}\psi_{e,n\{s\}}\left(\{\vec{r}_i\},\{\vec{R}_\alpha\}\right)\psi_N\left(\{\vec{R}_\alpha\}\right), \tag{42.32}$$

which is multiplied by the electron wave function ψ_e and integrated, leading to the kinetic energy plus an extra part,

$$\prod_i \int d^3\vec{r}_i\, \psi^*_{e,n\{s\}}\left(\{\vec{r}_i\},\{\vec{R}_\alpha\}\right)\left[\hat{T}_N(+\hat{V}_{NN})\right]\psi_{e,n\{s\}}\left(\{\vec{r}_i\},\{\vec{R}_\alpha\}\right)\psi_N\left(\{\vec{R}_\alpha\}\right)$$
$$= (T_N + \Delta E_N)\psi_N\left(\{\vec{R}_\alpha\}\right). \tag{42.33}$$

Then the free (electronic) Schrödinger equation is written, more explicitly, as

$$\left[\hat{T}_e + \hat{V}_{Ne} + \hat{V}_{ee}(+\hat{V}_{NN})\right]\psi_{e,n\{s\}}\left(\{\vec{r}_i\},\{\vec{R}_\alpha\}\right) = E_{e,n}\psi_{e,n\{s\}}\left(\{\vec{r}_i\},\{\vec{R}_\alpha\}\right). \tag{42.34}$$

The total energies split into electronic and nuclear parts,

$$E_n = E_{N,n} + E_{e,n} \equiv E_{N,n} + E_n^{(0)}. \tag{42.35}$$

The quantized vibrational energy is

$$E_{\text{vibr}} = \left(v + \frac{1}{2}\right)\hbar\omega_0, \tag{42.36}$$

and the quantized rotational energy is

$$E_{\text{rot}} = \frac{\vec{J}^2}{2I} = \frac{\hbar^2 J(J+1)}{2I}. \tag{42.37}$$

In order to find the stable structure, i.e., the molecular shape, we can apply the practical variational method for $\psi_{n\{s\}} = \psi_{e,n\{s\}}\psi_N$ and minimize the energy over $\{\vec{R}_\alpha\}$ as parameters.

42.8 Method of Molecular Orbitals

This method is an application of the Ritz variational method, similar to the method of atomic orbitals. We use the interaction between the electrons as a perturbation, $\hat{H}_1 = \hat{H}_{ee}$.

We first find an electronic wave function $\phi_{a_i}(\vec{r}_i)$ that is decoupled, in which one electron is in the field of all the nuclei, so the Schrödinger equation for it is

$$\left(-\frac{\hbar^2}{2m}\Delta_i - \sum_{\alpha=1}^{N}\frac{Z_\alpha e^2}{4\pi\epsilon_0 r_{\alpha i}}\right)\phi_{a_i}(\vec{r}_i) = E^{(i)}_{a_i}\phi_{a_i}(\vec{r}_i), \tag{42.38}$$

where $i = 1, \ldots, n_e$ is an index for the electrons.

Next, we write a *molecular orbital* as a product of individual wave functions for the electrons $\phi_{a_i}(\vec{r}_i)$,

$$\phi_r(\vec{r}_1, \ldots, \vec{r}_{n_e}) \equiv \phi_{a_1}(\vec{r}_1)\cdots\phi_{a_{n_e}}(\vec{r}_{n_e}), \tag{42.39}$$

where the total index r is made up of the $\{a_i\}$ indices, and we use the practical variational method and minimize over the parameters $\{\vec{R}_\alpha\}$.

Finally, we write the wave functions of the molecule as linear combinations of these molecular orbitals,

$$\psi_e\left(\{\vec{r}_i\};\{\vec{R}_\alpha\}\right) = \sum_r c_r\phi_r\left(\{\vec{r}_i\};\{\vec{R}_\alpha\}\right), \tag{42.40}$$

and use the Ritz variational method: we restrict to a finite subset of $r = 1, \ldots, k$. Then we obtain the eigenstate–eigenvalue equation,

$$\sum_{r=1}^{k} c_r \left(H_{sr} - E_e \delta_{rs}\right) = 0, \tag{42.41}$$

leading to the secular equation for the electronic energy,

$$\delta(H_{sr} - E_e \delta_{sr}) = 0. \tag{42.42}$$

42.9 The LCAO Method

A variant of the above method is the linear combination of atomic orbitals (LCAO) method.

In order to find the decoupled electronic wave function $\phi_{a_i}(\vec{r}_i)$, we use a type of Ritz method. We consider *atomic* orbitals for different nuclei inside the molecule, but the same electron, meaning the same position $\vec{r}_i$, $\chi_r(\vec{r}_i; \vec{R}_r)$, where $\vec{R}_r$ is a parameter, representing for the position of the rth nucleus. Then we make linear combinations of the orbitals for each nucleus, one combination per electron,

$$\phi_{a_i i}(\vec{r}_i) = \sum_{r=1}^{N} c_{ir} \chi_r \left(\vec{r}_i; \vec{R}_r\right), \tag{42.43}$$

where N is the number of nuclei in the molecule, i is an index for the electron, and a_i is the particular *molecular orbital*.

As an example, we can consider the NH molecular orbital, where the N nucleus has a position $\vec{R}_1$ and atomic orbital is P ($L = 1$), and the H nucleus has position $\vec{R}_2$ and atomic orbital is S ($L = 0$).

We minimize over c_{ir} (in the Ritz-like variational method) to find the molecular orbitals $\phi_{a_i i}$. The total energy is

$$E = \sum_{i=1}^{n_e} E_i, \tag{42.44}$$

where the individual electron energies are

$$E_i = \frac{\sum_{r,s} c_{ir}^* H_{rs} c_{is}}{\sum_{r,s} c_{ir}^* \Delta_{rs} c_{is}}, \tag{42.45}$$

and, as before, we use the notation

$$\begin{aligned} H_{rs} &= \langle \chi_r | \hat{H} | \chi_s \rangle \\ \Delta_{rs} &= \langle \chi_r | \chi_s \rangle. \end{aligned} \tag{42.46}$$

The eigenenergy–eigenstate equation is

$$\sum_{i=1}^{N} c_{ir} (H_{rs} - E_i \Delta_{rs}) = 0, \tag{42.47}$$

leading to a secular equation for the energies E_i,

$$\det(H_{rs} - E_i \Delta_{rs}) = 0. \tag{42.48}$$

42.10 Application: The LCAO Method for the Diatomic Molecule

Consider two nuclei only, with distance $\vec{R}$ between them, and electrons with positions $\vec{r}_1$ with respect to nucleus 1, and $\vec{r}_2$ with respect to nucleus 2, and generic position (with respect to some arbitrary origin) $\vec{r}$.

Then the molecular orbital is a linear combination of the atomic orbitals (MO = LCAO), meaning

$$\phi(\vec{r}) = A\phi_1(\vec{r}_1) + B\phi_2(\vec{r}_2). \tag{42.49}$$

The energy in this state is

$$\langle E\rangle_\phi = \frac{A^2 H_{11} + B^2 H_{22} + 2ABH_{12}}{A^2 + B^2 + 2ABS}, \tag{42.50}$$

where

$$S = \langle\phi_1|\phi_2\rangle = \int \phi_1^* \phi_2. \tag{42.51}$$

We minimize it with respect to A and B,

$$\frac{\partial\langle E\rangle_\phi}{\partial A} = \frac{\partial\langle E\rangle_\phi}{\partial B} = 0. \tag{42.52}$$

If the two atomic levels are equal, $H_{11} = H_{22}$, then at the minimum we find $A = \pm B$ and

$$E_{1,2} = \frac{H_{11} \pm H_{12}}{1 \pm S}, \tag{42.53}$$

meaning the level splits in two.

If $H_{11} \neq H_{22}$, but S is negligible ($S \ll 1$), then the split energy levels are

$$E_{1,2} = \frac{H_{11} + H_{22}}{2} \pm \frac{H_{11} - H_{22}}{2}\sqrt{1 + \left(\frac{2H_{12}}{H_{11} - H_{22}}\right)^2}. \tag{42.54}$$

We note that $E_1 < H_{11}, H_{22}$ and $E_2 > H_{11}, H_{22}$, and now $|B| \neq |A|$.

42.11 Chemical Bonds

Chemical bonds fall into two cases:

- Purely covalent bond, as in a homopolar molecule. It corresponds to $|A| = |B|$ ($A = \pm B$), and so to the first case above, with the MO being the symmetric LCAO,

$$\phi = a(\phi_1 \pm \phi_2). \tag{42.55}$$

  In it, we see that the electron is evenly distributed between the two nuclei (or, rather, atoms).
- An ionic bond, as in a heteropolar molecule. It corresponds to $|A| \neq |B|$, which is the second case above, with

$$\phi = a(\phi_1 + \lambda\phi_2), \tag{42.56}$$

  but where $\lambda \neq \pm 1$, so the electron is found mostly near one or the other nucleus.

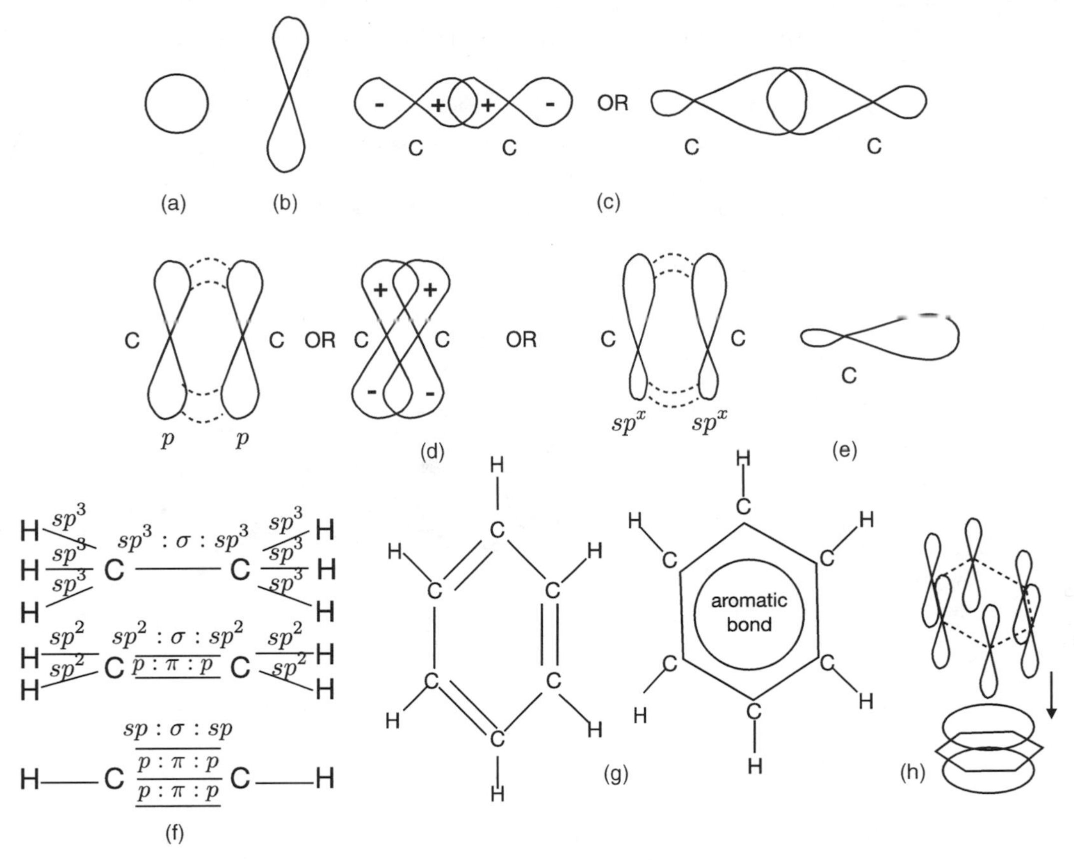

Figure 42.3 (a) s orbital. (b) p orbital. (c) σ bond in C–C between two $2p$ orbitals or sp^x orbitals. (d) π bond in C–C between two $2p$ orbitals or sp^x orbitals. (e) Hybridization: sp^x orbitals (sp, sp^2, sp^3). (f) C_2H_6, C_2H_4 and C_2H_2 structure, orbitals, and bonds. (g) Aromatic bond in C_6H_6, the naive structure and the correct structure. (h) Aromatic bond in C_6H_6, the p orbitals that make up the aromatic bond.

More precisely, describing the bonds between two atoms in a molecule (i.e, the chemical bonds), we first note that there is a preferred axis uniting the two nuclei. This means that $\vec{L}_{\rm tot}$ is not conserved, only its projection onto the axis, $L_{z,\rm tot}$, is conserved. The values are, as usual, $L_z = 0, 1, 2, \ldots$, denoted by $\sigma, \pi, \delta, \ldots$ instead of $s, p, d, f, \ldots$. These are molecular orbitals, which only have one axis of symmetry, not the full rotational invariance. See Fig. 42.3a, b, c, d for the spatial form of the s, p atomic orbitals and σ, π bonds, respectively (exemplified for carbon).

Hybridization

A very important example is the carbon atom, which has four free electrons (outside full shells), two electrons in S states, with opposite spins $\uparrow\downarrow$, and two electrons in P states. Specifically, the electron structure is $1s(\uparrow\downarrow), 2s(\uparrow\downarrow), 2p(\uparrow), 2p(\uparrow), 2p()$.

But in molecules that include a carbon atom, the split in energy between the S and the P states is much smaller than the binding energy of the molecule. That means that the relevant atomic orbitals of carbon are "hybridized", or effective, orbitals that are combinations of the S and P orbitals.

The particular hybridization depends on the (symmetry of the) molecule under study. In Fig. 42.3e we see the spatial form of the hybridized sp^x orbitals (with $x = 1, 2, 3$).

We now consider the CH_4 (methane) molecule. In it, each hydrogen has a bond with the carbon, all these bonds being equivalent. This means that we need to hybridize all four orbitals of carbon, obtaining sp^3 orbitals instead of the s and p orbitals.

The sp^3 hybridization defines a linear-combination wave function,

$$\begin{aligned}\phi_{sp^3} &= c_1\phi_s + c_2\phi_{p_x} + c_3\phi_{p_y} + c_4\phi_{p_z}\\ &= \frac{1}{\sqrt{1+c^2}}[\phi_s + c(u_1\phi_{p_x} + u_2\phi_{p_y} + u_3\phi_{p_z})].\end{aligned} \tag{42.57}$$

We have to define four sp^3 orbitals replacing the s and three p orbitals, which means that there are four different $\vec{u}$ vectors, $\vec{u}^{(1)}, \vec{u}^{(2)}, \vec{u}^{(3)}, \vec{u}^{(4)}$. The orbitals have to be equivalent, so we need to have the same c for all of them. Moreover, the states need to be orthonormal, so

$$\langle \phi^{(1)}_{sp^3} | \phi^{(2)}_{sp^3} \rangle = \frac{1}{1+c^2}[1 + c^2 \vec{u}^{(1)} \cdot \vec{u}^{(2)}] = 0, \quad \text{etc.} \tag{42.58}$$

But $\vec{u}^{(i)} \cdot \vec{u}^{(j)} = \cos\theta_{ij}$, thus the equation becomes

$$c^2 \cos\theta_{ij} = 1, \tag{42.59}$$

and since the angles between any two of the four vectors must be the same, the vectors end on the vertices of a regular tetrahedron, meaning

$$\cos\theta_{ij} = -\frac{1}{3} \Rightarrow c^2 = 3. \tag{42.60}$$

Each sp^3 orbital will join with a hydrogen atom and orbital, forming the CH_4 molecule.

Next we consider the C_2H_6 (ethane) molecule, H_3C–CH_3. In it, each of the two carbons has three bonds, each with a hydrogen atom, and there is one bond between the two carbons. In it, the same sp^3 hybridization works, and three such orbitals are used to bond with the hydrogen atoms. The last sp^3 orbital of each of the two carbons joins to form a σ ($l_z = 0$) bond. In it, the spins of the two electrons (from the two carbons) are antiparallel, $\uparrow\downarrow$, giving a bond (i.e., negative energy). See Fig. 42.3f for the structure, orbitals, and bonds of C_2H_6.

Next we consider the C_2H_4 (ethylene) molecule, $H_2C{=}CH_2$. In it, for each carbon atom, the s orbital and two of the p orbitals hybridize into three sp^2 orbitals, leaving a single p orbital unhybridized. The two bonds between the carbons are different. There is a σ bond ($L_z = 0$), with the two sp^2 orbitals parallel to the axis between the atoms and with the electrons in it antiparallel, as before. The two unhybridized p orbitals, each with wave function

$$\psi_0 = \vec{u}_0 \cdot \vec{\phi} = u_1^0\phi_{p_x} + u_2^0\phi_{p_y} + u_3^0\phi_{p_z}, \tag{42.61}$$

join into a π ($L_z = 1$) orbital. The p orbitals are transverse to the axis between the carbons, and the wave function is normal to the sp^2 hybridized orbitals.

Next we consider the C_2H_2 (acetylene) molecule, $HC{\equiv}CH$. In it, for each carbon atom, the s orbital and one p orbital hybridize into two sp orbitals, while the remaining two p's remain untouched. One sp orbital joins with a hydrogen, while the other joins with the orbital of the other carbon and forms a σ bond. The two remaining untouched p orbitals of each carbon join with the one from the other atom, and form two π bonds. In all, we have one σ and two π bonds between the carbons. See Fig. 42.3f for the structure, orbitals, and bonds of C_2H_4 and C_2H_2.

Finally, we consider the C_6H_6 (benzene) molecule, which corresponds to a complex bond. The carbons sit at the vertices of a regular hexagon, and bond with one hydrogen each, and have a σ bond with each of the two immediate neighbors, leaving one electron unbonded. Therefore, one s orbital and two p orbitals hybridize into three sp^2 orbitals, one joining with the hydrogen and the other two joining with the immediate neighbors to make two σ bonds. The remaining electrons from each carbon pool together to make a circular motion around the hexagon that is Bohr quantized (like in a hydrogenoid atom) on the circle, meaning we have delocalized bound electrons. We represent these electrons by a circle inside the hexagon. In Fig. 42.3g we represent the naive structure and the actual structure (with the aromatic bond) of C_6H_6, and in Fig. 42.3h we represent the creation of the aromatic bond out of p orbitals.

Important Concepts to Remember

- In a hydrogenoid atom, the states are described by (n, l, m, m_s) or (because of the spin–orbit interaction) as (n, l, j, m_j), and the spectroscopic notation nl^d.
- In shell models, we fill atomic orbitals, extensions of the hydrogenoid orbitals; the Hund rules state that electrons avoid being in the same orbitals if other orbitals of same l are available, and that electrons in equivalent, but different orbitals, have parallel spins in the ground state.
- In LS coupling (normal, or Russell–Saunders) first $\vec{l}_i$ couples into $\vec{L}$ and $\vec{s}_i$ into $\vec{S}$, then $\vec{L}$ and $\vec{S}$ into $\vec{J}$, while in jj, or spin–orbit, coupling, first $\vec{l}_i$ couples and $\vec{s}_i$ into $\vec{j}_i$, then $\vec{j}_i$ into $\vec{J}$; most atoms are in between these two extremes. Note that the coupling doesn't change the number of energy levels, only the energy gaps.
- In the method of atomic orbitals, we use a finite number of unperturbed eigenfunctions for the total Hamiltonian, and minimize.
- Chemical bond types are electrovalent or ionic, metallic (electrons delocalized within the crystalline structure), and covalent, between neutral atoms.
- Energy scales in molecules are: $E_{\text{rot}} \ll E_{\text{vibr}} \ll E_e$, with usually $E_{\text{rot}} \sim 25$ meV, $E_{\text{vibr}} \sim 0.5$ eV and $E_e \sim$ a few eV.
- A molecular orbital is a product of individual electronic orbitals,

$$\phi_r(\vec{r}_1, \ldots, \vec{r}_{ne}) = \phi_{a_1}(\vec{r}_1) \cdots \phi_{a_{ne}}(\vec{r}_{ne}),$$

  and in the method of molecular orbitals one uses linear combinations of these, and the Ritz variational method for them.
- In the LCAO method, one defines a molecular orbital as linear combinations of atomic orbitals (for a single electron with respect to each of the nuclei) and uses a Ritz-like variational method for the linear coefficients.
- In molecules, atomic orbitals hybridize when the difference in energy between them is smaller than the binding energy of the molecule: e.g., sp^3 in CH_4 and C_2H_6, sp^2 in C_2H_4, sp in C_2H_2.
- Hybridized orbitals can join to form bonds: σ ($L_z = 0$), for two parallel such orbitals (sp^3, sp^2, or sp) that are parallel to the common axis and of opposite spin, and $2p$ orbitals can form 2π ($L_z = 1$) bonds, etc. One particular example is the aromatic bond of C_6H_6 with the electron delocalized and quantized on a circle.

Further Reading

See [2].

Exercises

(1) Write down explicitly the $4s$ and $4p$ atomic orbital wave functions.

(2) Use the shell model and the Hund rules to show how the orbitals are populated with electrons in the elements C, O, and Mg.

(3) Consider the element O (oxygen). Write down the LS and jj couplings for the electrons, and show explicitly that the splitting of energy levels is the same for each type of coupling.

(4) Consider the element O. Write down explicitly the method of atomic orbitals for it (for electron–electron interactions), using the basis of only the filled orbitals.

(5) Write down explicitly the method of molecular orbitals for the O_2 molecule.

(6) Write down explicitly the LCAO method for the O_2 molecule, applied to each of the atomic orbitals in O.

(7) Write down the Bohr quantization for the common electron in benzene, and find the energy levels as a function of the "radius" of the benzene molecule (the distance between the center and a C atom).

43 Nuclear Liquid Droplet and Shell Models

In this chapter, we study nuclear models and approximations, the counterpart to the analysis of atoms in the previous chapter.

43.1 Nuclear Data and Droplet Model

In the nucleus we have Z protons (where Z is electric charge of the nucleus), and N neutrons. Both of them are nucleons; the total number of nucleons is known as the atomic number A, so $A = N + Z$.

The proton and neutron mass are approximately the same, $m_p \simeq m_n$, together called m_N, and the atomic number A is then approximately the total nucleus mass divided by m_N. More precisely,

$$M_{\text{nucleus}} \simeq Am_p(1 - 8/100), \tag{43.1}$$

which means that the binding energy per nucleon is very small (about 8%) and is constant as A increases.

Moreover, for most nuclei, we have the radius R versus A law:

$$R = r_0 A^{1/3}, \tag{43.2}$$

where $r_0 \simeq 1.3 \times 10^{-15}$ m $= 1.3$ fm (1 fm $= 10^{-15}$ m is a femtometer, or one fermi). This law and the previous relation ($M \propto A$) imply that both the energy and the volume of the nucleus are proportional to A, which is the property of a liquid. We say then that we have "nuclear matter", forming something like a droplet of liquid, hence we have a "liquid droplet model".

In the ground state, we have a static, spherically symmetric fluid droplet. In an excited state, we can have waves propagating in the fluid, deformations of the droplet that make it nonspherical, etc. Inside this fluid, the individual nucleons are like particles of matter, and their motion is either thermal or brownian.

However, the law $E \propto A$ is only approximate (we mentioned it is valid for *most* nuclei; in fact, there are fluctuations around it for some cases). In fact, there is some structure besides linearity: there are nuclei that have better stability, meaning higher binding energy. The nuclei for which this happens have *either* N or Z as one of the "magic numbers",

$$2, 8, 20, 28, 50, 82, 126. \tag{43.3}$$

For instance, tin (chemically Sn, for *Stannum* in Latin) has $Z = 50$, and it has 10 *stable* isotopes, the most of any element.

Also, if both N and Z are magic numbers, a situation called "double magic", we have elements that are even more stable. One example of this is helium-4, He^4, with $N = Z = 2$; others are oxygen-16, O^{16}, with $N = Z = 8$, and lead-208, Pb^{208}, with $Z = 82$ and $N = 126$.

The situation above, with magic numbers for Z or/and N, is similar to the stability of atomic inert gases, which have full atomic shells, in the shell model of the atom. In fact, more quantitatively, for electrons in a hydrogenoid atom, with a Coulomb potential, the degeneracy of a shell of given n, for electrons characterized by (n, l, m, m_s), is (m_s takes two values and m takes $2l + 1$ values, while l goes from 0 to $n - 1$)

$$d_n = \sum_{l=0}^{n-1} 2(2l + 1) = 2n^2, \tag{43.4}$$

giving, for $n = 1, 2, 3, 4, 5, 6$,

$$2, 8, 18, 32, 50, 72. \tag{43.5}$$

However, the magic number is supposed to be the *total* number of electrons in all the full shells, so

$$\sum_{m=1}^{n} d_m = \sum_{m=1}^{n} 2m^2 = \frac{n(n+1)(2n+1)}{3}, \tag{43.6}$$

giving, for $n = 1, 2, 3, 4, 5, 6$,

$$2, 10, 28, 60, 110, 182, \ldots \tag{43.7}$$

43.2 Shell Models 1: Single-Particle Shell Models

From the previous analysis it follows that one possibility for a nuclear model is to consider a shell model similar to that for an atom, but with another central potential $V(r)$ rather than the Coulomb potential, and where the shell corresponds to a given energy (principal quantum number n).

The wave function for a central potential is, as we saw in Chapter 19,

$$\psi_{nlm}(\vec{r}) = \frac{\chi_{nl}(r)}{r} Y_{lm}(\theta, \phi). \tag{43.8}$$

The form of the reduced radial wave function $\chi_{nl}(r)$ is found from the Schrödinger equation, for which we need the potential $V(r)$. The true potential is something like a rounded-out finite square well, with a depth of V_0 and width R (the maximum radius for the well) and 0 at infinity, as in Fig. 43.1a.

Instead of the true potential, we must use an analytic form that we can solve. A first approximation is a spherical square well, of depth V_0 and radius R. The potential is $V = -V_0$ for $0 \leq r \leq R$ and, in order to be closer to reality, we would also need $V = 0$ for $r > R$; see Fig. 43.1b. A second possible approximation is a spherical (three-dimensional) harmonic oscillator, starting at $V(r = 0) = -V_0$, and reaching $V = 0$ at $r = R$. To be closer to reality, again we would need to put $V = 0$ for $r > R$; see Fig. 43.1c.

Since the potential is central, for $r \in (0, \infty)$, and not one-dimensional (for $x \in (-\infty, +\infty)$), the conditions that one needs to impose are finiteness and integrability at 0 and ∞, as we saw in Chapter 19.

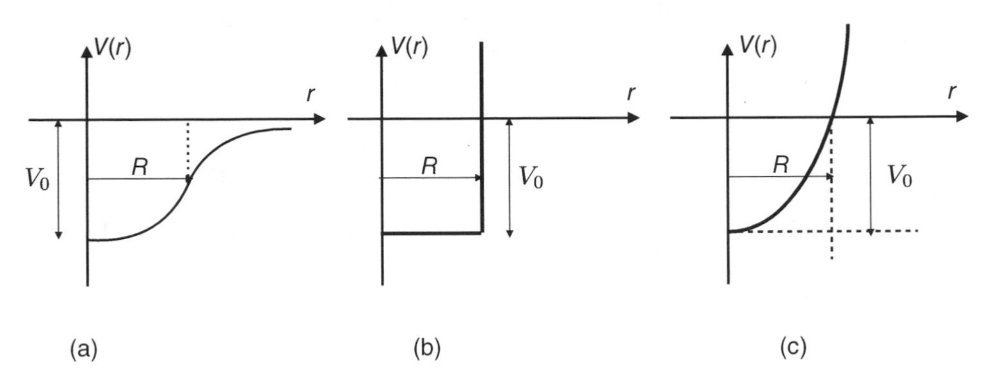

Figure 43.1 (a) Real nuclear potential. (b) Approximation 1: spherical square well. (c) Approximation 2: three-dimensional harmonic oscillator.

Approximation 1 (Spherical Square Well)

The spherical square well has been analyzed in Chapter 19 already, where we found that, for $r < R \equiv r_0$,

$$\frac{\chi_{nl}(r)}{r} = N_{nl} j_l(kr), \tag{43.9}$$

where

$$k = \sqrt{\frac{2m(E+V_0)}{\hbar^2}}. \tag{43.10}$$

For $r > R$, the correct thing to do would be to put $V = 0$, but the simpler thing to do is to consider $V = \infty$, for an infinitely deep square well. If we are interested only in the low-lying energy states (for $V_0 - |V| \ll V_0$), which we are since there are seven magic numbers only so presumably we go up to at most $n = 7$, the difference in the calculation required is minimal.

Then the wave function should vanish for $r \geq R$, meaning that we need to have the quantization condition

$$j_l(kR) = 0 \Rightarrow k_n R = j^0_{n,l}, \tag{43.11}$$

where $j^0_{n,l}$ is the nth zero of j_l. It then also follows that

$$N^2_{nl} = \frac{2}{R^3}\frac{1}{j^2_{l+1}(j^0_{n,l})}. \tag{43.12}$$

If we consider, as is more appropriate, $V = 0$ for $r > R$, the analysis of the boundary conditions, giving the quantization condition and normalization, is more complicated.

Approximation 2 (Spherical Harmonic Oscillator)

A better approximation is, however, the spherical harmonic oscillator, starting at $V = -V_0$, in which case the potential for the nucleus is

$$V_N(r) = -V_0\left[1 - \left(\frac{r}{R}\right)^2\right] = -V_0 + \frac{1}{2}m\omega^2 r^2, \tag{43.13}$$

where we have defined

$$\frac{V_0}{R^2} \equiv \frac{1}{2} m\omega^2, \tag{43.14}$$

and m is the nucleon mass, $m = m_N \simeq m_n \simeq m_p$.

Again, really we should stop the increase of V at $r = R$, when $V = 0$, and put $V = 0$ for $r > R$ instead. However, if we are interested only in the low lying states (and we saw that we have only seven magic numbers, so we should stop at most at $n = 7$), the difference is minimal.

The solution of the radial Schrödinger equation is

$$\frac{\chi_{nl}(r)}{r} = N_{n,l} e^{-\alpha^2 r^2/2} (\alpha r)^l \, {}_1F_1(-n_r, l + 3/2; (\alpha r)^2), \tag{43.15}$$

where

$${}_1F_1(-n_r, l + 3/2; (\alpha r)^2) = L_{n_r-1}^{l+1/2}(\alpha^2 r^2), \tag{43.16}$$

are Laguerre polynomials,

$$\alpha = \sqrt{\frac{m\omega}{\hbar}}, \tag{43.17}$$

the normalization constant is

$$N_{n,l} = \frac{\alpha\sqrt{2\alpha}}{\Gamma(l + 3/2)} \left(\frac{n_r!}{\Gamma(n_r + l + 3/2)} \right)^{1/2}, \tag{43.18}$$

and $n_r = 0, 1, 2, \ldots$ is the number of nodes of the radial wave function (less the node at infinity). The energy (principal) quantum number is, as we saw,

$$n = 2n_r + l = 0, 1, 2, \ldots \tag{43.19}$$

From the quantization condition, we found the eigenenergies

$$E_n = \hbar\omega \left(n + \frac{3}{2} \right). \tag{43.20}$$

The degeneracy of the above energy levels comes from the two values of the spin projection m_s and sum over l from 0 to n,

$$d_n = 2 \sum_{l=0}^{n} (l + 1) = 2 \frac{(n + 1)(n + 2)}{2} = (n + 1)(n + 2). \tag{43.21}$$

But the number in which we are interested, to be tested against the magic numbers, is the total number of states in the fully filled shells, i.e.,

$$\begin{aligned} \sum_{m=0}^{n} d_m = \sum_{m=0}^{n} (m^2 + 3m + 2) &= \frac{n(n + 1)(2n + 1)}{6} + 3\frac{n(n + 1)}{2} + 2(n + 1) \\ &= \frac{(n + 1)(n + 2)(n + 3)}{3}, \end{aligned} \tag{43.22}$$

which gives, for $n = 0, 1, 2, 3, 4, 5, 6, 7$,

$$2, 8, 20, 40, 70, 112, 168, 240, \ldots \tag{43.23}$$

Out of these, the first three numbers match the magic numbers, although the others do not.

For completeness, note that

$$d_n = (n+1)(n+2) = 2, 6, 12, 20, 30, 42, 56, 72, \ldots \tag{43.24}$$

Since only the first three numbers match, we need a better theory but it is clear what it should be, since the same thing happens for the atomic energy levels. At high Z, as for the atomic levels, the magnetic spin–orbit coupling becomes large (whereas the electrostatic energy responsible for energy levels does not increase).

43.3 Spin–Orbit Interaction Correction

In order to understand the nuclear spin–orbit interaction, we review the atomic spin–orbit interaction first. The interaction term in the Hamiltonian comes from the Dirac equation, in the expansion around the nonrelativistic result (see Chapter 54), and one finds

$$H_{\text{spin–orbit}} = \frac{e}{2m_0^2c^2}\frac{1}{r}\frac{dA_0}{dr}\vec{s}\cdot\vec{l}. \tag{43.25}$$

But then

$$eA_0 = -V_e(r) \tag{43.26}$$

is the Coulomb potential, so the spin–orbit interaction Hamiltonian is rewritten as

$$H_{\text{spin–orbit}} = -\text{const}\times\frac{1}{r}\frac{dV_e(r)}{dr}(2\vec{s}\cdot\vec{l}). \tag{43.27}$$

We will assume that the same formula still holds for the nucleus, except that instead of the Coulomb potential $V_e(r)$ we use the nuclear potential (the potential for the nucleus states) $V_N(r)$.

We will use the second approximation, the spherically symmetric harmonic oscillator, which gives better results, and obtain that

$$\frac{1}{r}\frac{dV_N(r)}{dr} = m\omega^2 = \frac{2V_0}{R^2} = \text{const}. \tag{43.28}$$

We also find that $V_0 \simeq E_{\text{binding}} \propto A$ increases for larger nuclei.

Moreover,

$$\begin{aligned}(2\vec{s}\cdot\vec{l}) &= (\vec{l}+\vec{s})^2 - \vec{l}^2 - \vec{s}^2 = \vec{j}^2 - \vec{l}^2 - \vec{s}^2\\ &= \hbar^2\left[j(j+1) - l(l+1) - s(s+1)\right].\end{aligned} \tag{43.29}$$

Since all the nucleons are spin 1/2 fermions, just like electrons, $s = 1/2$ implies that $j = l \pm 1/2$. In this case, for the two values we find

$$\begin{aligned}(2\vec{s}\cdot\vec{l}) &= (l+1/2)(l+3/2) - l(l+1) - 3/4 = l, \quad \text{for } j = l+1/2\\ &= (l-1/2)(l+1/2) - l(l+1) - 3/4 = -l-1, \quad \text{for } j = l-1/2.\end{aligned} \tag{43.30}$$

Then the spin–orbit interaction Hamiltonian is

$$H_{\text{spin–orbit}} = -(\text{const}')\times\begin{pmatrix} l \\ -l-1\end{pmatrix}, \quad \text{which is } \begin{pmatrix} <0 \\ >0\end{pmatrix} \text{ for } \begin{pmatrix} j = l+1/2 \\ j = l-1/2\end{pmatrix}. \tag{43.31}$$

This means that the $j = l - 1/2$ states are shifted upwards in energy, and are shifted more as A increases (so as either N or Z increases), since the constant is roughly proportional to A. Similarly, the $j = l + 1/2$ states are shifted downwards in energy, and shifted more as A increases. Moreover, the shift is also proportional to l for the first case, and to $l + 1$ in the second, so the shift is smaller for small l and larger for large l.

To describe the nuclear states, we use a variant of the spectroscopic notation, with

$$n_r + 1 = n'_r = \frac{n + 1 - l}{2} + 1, \tag{43.32}$$

the number of nodes, including infinity, appearing before the notation for l, so that

$$(n_r + 1)\, l_j = n'_r\, l_j. \tag{43.33}$$

For the first three shells, which as we saw already match the magic numbers, without considering the spin–orbit shifts, we have the sub-shells

$$\begin{aligned} n &= 0:\ 1s_{1/2} \\ n &= 1:\ 1p_{3/2},\ 1p_{1/2} \\ n &= 2:\ 1d_{5/2},\ 2s_{1/2},\ 1d_{3/2}, \end{aligned} \tag{43.34}$$

in the exact order of increasing energy written above, that is

$$(1s_{1/2})_{n=0},\ (1p_{3/2},\ 1p_{1/2})_{n=1},\ (1d_{5/2},\ 2s_{1/2},\ 1d_{3/2})_{n=2}. \tag{43.35}$$

For the next shells, we have the following sub-shells over n_r, l:

$$\begin{aligned} n &= 3:\ 1f,\ 2p \\ n &= 4:\ 1g,\ 2d,\ 3s \\ n &= 5:\ 1h,\ 2f,\ 3p \\ n &= 6:\ 1i,\ 2g,\ 3d,\ 4s, \end{aligned} \tag{43.36}$$

where the $n = 3$ shell splits over j into

$$1f_{7/2},\ 1f_{5/2},\ 2p_{3/2},\ 2p_{1/2}, \tag{43.37}$$

but the first sub-shell, $1f_{7/2}$, for $n = 3, l = 3$, has a spin–orbit interaction term that is sufficiently large to separate it downwards in energy from the $n = 3$ shell, but not enough for it to join the $n = 2$ shell, meaning that it acts as a separate shell all by itself. Its degeneracy is $2j + 1 = 8$.

Next, for $n = 4$, the shell splits into

$$1g_{9/2},\ 1g_{7/2},\ 2d_{5/2},\ 2d_{3/2},\ 2s_{1/2}. \tag{43.38}$$

But now $1g_{9/2}$ has $n = l = 4$ and the spin–orbit interaction is large enough to take it down into the $n = 3$ shell.

Finally, for $n = 5$, the shell splits into

$$1h_{11/2},\ 1h_{9/2},\ 2f_{7/2},\ 2f_{5/2},\ 3p_{3/2},\ 3p_{1/2}. \tag{43.39}$$

The sub-shell $1h_{11/2}$ has $n = l = 5$ and the spin–orbit interaction is large enough to take it down into the $n = 4$ shell. Moreover, from the $n = 6$ shell, the element with $n = l = 6$, namely $1i_{13/2}$, also has a spin–orbit interaction large enough to take it down into the $n = 5$ shell. Then, finally, the sub-shells, ordered by increasing energy into shells (except the $n = 0, 1, 2$ shells, which are already matched), are as follows (where we also give the degeneracy d_n of the shell and the magic number $k_n = \sum_{m=1}^{n} d_m$):

$$\begin{pmatrix} n'_r l_j \\ d_n \\ k_n \end{pmatrix} = \begin{pmatrix} (1f_{7/2})_{n=3} \\ 8 \\ 28 \end{pmatrix}, \begin{pmatrix} (2p_{3/2} \quad 1f_{5/2} \quad 2p_{1/2})_{n=3} \quad (1g_{9/2})_{n=4} \\ 22 \\ 50 \end{pmatrix},$$
$$\begin{pmatrix} (2d_{5/2} \quad 1g_{7/2} \quad 2d_{3/2})_{n=4} \quad (1h_{11/2})_{n=5} \quad (3s_{1/2})_{n=4} \\ 32 \\ 82 \end{pmatrix}, \tag{43.40}$$
$$\begin{pmatrix} (2f_{7/2} \quad 1h_{9/2} \quad 3f_{5/2} \quad 2p_{3/2})_{n=5} \quad (1i_{13/2})_{n=6} \quad (3p_{1/2})_{n=5} \\ 44 \\ 126 \end{pmatrix}.$$

We see that we have managed to reproduce *all* the magic numbers.

43.4 Many-Particle Shell Models

Until now we have considered only single-particle shell models, where the contribution from the other nucleons is assumed to be included in the one-particle potential $V_N(r)$, but in reality we must consider the interactions of the nucleons, at least for those outside the fully filled shells, called (as in the atomic case) *valence particles*. As for atoms, the fully filled shells correspond to an inert core (spherically symmetric, without multipole interactions), contributing only through a "screening" of the nuclear charge and thus modifying V_0 only, but not creating a multiparticle potential.

The simplest correction to the above single-particle shell models is to consider a two-particle potential for the valence particles. Denoting by $\vec{r}_1, \vec{r}_2$ the positions of the two valence particles relative to the nuclear center, and θ_{12} the angle between them, the potential is

$$V = V(\vec{r}_1, \vec{r}_2) = V(r_1, r_2, \cos\theta_{12}). \tag{43.41}$$

Moreover, we can expand it in terms of Legendre polynomials for the angular dependence,

$$V = \sum_k f_k(r_1, r_2) P_k(\cos\theta_{12}). \tag{43.42}$$

Conversely (using the orthonormality of the Legendre polynomials), we have

$$f_k = \frac{2k+1}{2} \int_{-1}^{1} P_k(\cos\theta_{12}) V(r_1, r_2, \cos\theta_{12}) d\cos\theta_{12}. \tag{43.43}$$

The simplest model for the two-particle potential is a contact, delta function, potential,

$$V = -g\delta(\vec{r}_1 - \vec{r}_2) = -g\delta(1 - \cos\theta_{12}) \frac{\delta(r_1 - r_2)}{\pi r_1 r_2}, \tag{43.44}$$

where $\pi r_1 r_2$ is the Jacobian for the transformation between the Cartesian coordinates and the radial coordinates. In this case, we obtain

$$f_k(r_1, r_2) = -\frac{g}{4\pi}(2k+1) \frac{\delta(r_1 - r_2)}{r_1 r_2}. \tag{43.45}$$

We consider states that are tensor products of the single-particle states (in zeroth-order perturbation theory for the interaction potential V, in which we have only single-particle potentials and shells), defined in terms of the quantum numbers $(nljm)$ for each particle (the spin–orbit interaction for each valence nucleon produces the coupling $\vec{j}_i = \vec{l}_i + \vec{s}_i$ which, as for atomic jj coupling, happens at large Z, for several fully filled shells since then the spin–orbit interaction is larger), so

$$|\phi_{n_1 l_1 j_1 m_1}\rangle \otimes |\phi_{n_2 l_2 j_2 m_2}\rangle. \tag{43.46}$$

We want to calculate the interaction energy coming out of this two-particle interaction potential using first-order time-independent perturbation theory, so we will evaluate the potential $\hat{V}$ for the zeroth-order states.

The total angular momenta couples the two individual ones, $\vec{j}_1 + \vec{j}_2 = \vec{J}$, so we replace $|j_1 m_1 j_2 m_2\rangle$ by $|j_1 j_2 JM\rangle$ via Clebsch–Gordan coefficients. We could continue with a full description, but we will just show a simple application instead.

Consider the case when $j_1 = j_2 = j$, so that there are two identical particles in the same shell. Then these particles can "pair up" into parallel but opposite ("antiparallel") angular momenta, $\uparrow\downarrow$, i.e., $J = 0 = M$. In this case, the matrix element in the correct states $|j_1 j_2 JM\rangle = |jj00\rangle$ can be trivially related to the matrix elements for the $|\uparrow\rangle \otimes |\downarrow\rangle$ states (products of one-particle states), and then the sum over m_j gives the degeneracy $d_j = 2j + 1$, so

$$\langle jj00|\hat{V}|jj00\rangle = (2j+1)\int |\phi_{n_1 l_1}(r_1)|^2 |\phi_{n_2 l_2}(r_2)|^2 f_k(r_1, r_2) dr_1 dr_2. \tag{43.47}$$

Substituting (43.45), we obtain

$$\langle jj00|\hat{V}|jj00\rangle = -(2j+1)\frac{g}{8\pi}\int |\phi_{n_1 l_1}(r)|^2 |\phi_{n_2 l_2}(r)|^2 \frac{dr}{r^2}. \tag{43.48}$$

This is the *pairing energy* in first-order time-independent perturbation theory for the two-particle shell model.

The next logical step would be to continue on to multiparticle shell models. But this leads to a self-consistent solution, "Hartree–Fock approximation", and will be treated in Chapter 57, in Part II$_e$, concerning multiparticle calculations.

Important Concepts to Remember

- A nucleus, defined by N and Z (and $A = N + Z$), can be roughly approximated as a liquid droplet of nuclear matter, since its radius is $R = r_0 A^{1/3}$ and $M_{\text{nucleus}}/(m_p A) \simeq$ const and < 1.
- However, we have magic numbers of stability in N or Z, 2,8,20,28,50,82,126, explained by analogy with the atomic shell model as being due to having a full nuclear shell (so the magic number is $\sum_m d_m$), for the quantum numbers arising from quantization in the nuclear potential.
- The nuclear potential can be approximated either as a spherical square well or as a spherical (three dimensional) harmonic oscillator starting at $-V_0$, naturally with $V = 0$ for $r > R$, but in practice $V = \infty$ for $r > R$ gives good results for low quantum numbers.
- For the spherical harmonic oscillator with $V = 0$ for $r > R$, we find the first three magic numbers, but the rest do not match. For them, we need to take into account the spin–orbit interaction correction, and make some assumptions about its parameters.

- To obtain a better result for the energy levels, one has to consider many-particle interactions. A two-particle delta function interaction gives the pairing energy, but better corrections are obtained via the self-consistent Hartree–Fock approximation.

Further Reading

See more about nuclear models in the book [27].

Exercises

(1) If a nucleus is spinning around a given axis, assuming the liquid droplet model how would you modify the law $R = r_0 A^{1/3}$ from the static case?

(2) Calculate the eccentricity, $\Delta R/R$, of the spinning nucleus in exercise 1, in a classical physics approximation.

(3) If the (effective) central potential is replaced with an (effective) azimuthal potential, $V(r, z)$, depending independently on the polar radius r in a plane and on the height z along the axis transverse to the plane, is it possible to have a shell model? Why? If so, consider the harmonic oscillator potential with different constants in the direction z and in the polar radial direction r, and find the (first few) magic numbers.

(4) In the case of a spherical square well with $V = 0$ for $r \geq R$, write down the solutions in the regions $r \leq R$ and $r \geq R$, the gluing (continuity) conditions at $r = R$, and the resulting quantization condition.

(5) Consider the nuclear central potential ($R_2 \gg R_1$, but not by too large a factor)

$$V_N(r) = -V_0 \left(1 - \frac{r^2}{R_1^2}\right) e^{-r/R_2}. \tag{43.49}$$

What can you (qualitatively) infer about the modification of the energy levels with respect to the levels of the spherical harmonic oscillator approximation in the text?

(6) In the case in exercise 5, calculate the spin–orbit interaction correction. What can you (qualitatively) infer about the modification of the various states from the uncorrected case?

(7) Calculate the pairing energy for the two-particle delta function potential, for two nucleons in the ground state, $1s_{1/2}$.

44 Interaction of Atoms with Electromagnetic Radiation: Transitions and Lasers

In this chapter, we first study the exact time-dependent solution for a two-level system in a harmonic potential, after which we study the general first-order perturbation theory in a harmonic potential for interaction with an external electromagnetic field. Finally, we consider a quantized electromagnetic field, and the interaction with the resulting photons, and derive the Planck formula that started quantum mechanics.

44.1 Two-Level System for Time-Dependent Transitions

We start with the generic two-level system considered in Chapter 4, but now with a time-dependent potential for transition, namely

$$i\hbar\frac{d}{dt}\begin{pmatrix}c_1(t)\\ c_2(t)\end{pmatrix} = \hat{H}\begin{pmatrix}c_1(t)\\ c_2(t)\end{pmatrix} = \begin{pmatrix}H_{11} & H_{12}\\ H_{21} & H_{22}\end{pmatrix}\begin{pmatrix}c_1(t)\\ c_2(t)\end{pmatrix}, \tag{44.1}$$

where $H_{12} = V(t)$ and $H_{21} = H_{12}^* = V^*(t)$, since the Hamiltonian is Hermitian. Moreover, as before,

$$E = \frac{H_{11}+H_{22}}{2}, \quad \Delta = \frac{H_{22}-H_{11}}{2}, \tag{44.2}$$

where the nonperturbed energies (at $V(t) = 0$) are

$$\begin{aligned} H_{11} &= E_1^{(0)} = E - \Delta \\ H_{22} &= E_2^{(0)} = E + \Delta > E_1^{(0)}. \end{aligned} \tag{44.3}$$

A useful formalism for describing the evolution of the states is in terms of the density matrix,

$$\rho = |\psi(t)\rangle\langle\psi(t)| = \begin{pmatrix}c_1(t)\\ c_2(t)\end{pmatrix}\begin{pmatrix}c_1^*(t) & c_2^*(t)\end{pmatrix} = \begin{pmatrix}|c_1(t)|^2 & c_1(t)c_2^*(t)\\ c_1^*(t)c_2(t) & |c_2(t)|^2\end{pmatrix}. \tag{44.4}$$

Define first the difference in occupation numbers (probabilities) in the two states,

$$N(t) = \rho_{22} - \rho_{11} = |c_2(t)|^2 - |c_1(t)|^2 = 2|c_2(t)|^2 - 1, \tag{44.5}$$

where we have used that $|c_1(t)|^2 + |c_2(t)|^2 = 1$, since we only have these two states, so the probability for the system to be in either state is one.

Next, we note that $\rho_{21} = \rho_{12}^*$, and then define

$$\rho_{21} = c_2(t)c_1^*(t) = \frac{Q(t) + iP(t)}{2}, \tag{44.6}$$

so that

$$Q = \rho_{21} + \rho_{21}^* = \rho_{21} + \rho_{12}, \quad P = \frac{\rho_{21} - \rho_{21}^*}{i} = \frac{\rho_{21} - \rho_{12}}{i}. \tag{44.7}$$

Then the Schrödinger equation is equivalent to the *Bloch equations*

$$\begin{aligned}
\frac{d}{dt}N(t) &= -\frac{1}{i\hbar}\left[Q(t)(V(t)-V^*(t))+iP(t)(V(t)+V^*(t))\right]\\
\frac{d}{dt}Q(t) &= \omega_0 P(t)+\frac{1}{i\hbar}N(t)(V(t)-V^*(t))\\
\frac{d}{dt}P(t) &= -\omega_0 Q(t)+\frac{1}{\hbar}N(t)(V(t)+V^*(t)),
\end{aligned} \tag{44.8}$$

where we have defined

$$\omega_0 \equiv \frac{2\Delta}{\hbar} = \frac{E_2 - E_1}{\hbar}, \tag{44.9}$$

previously called ω_{21}.

The most relevant case, which can be solved exactly, is that of a harmonic potential,

$$V(t) = V_0 e^{i\omega t}. \tag{44.10}$$

We can then check that the solution for $N(t)$ is (the method for finding it is somewhat lengthy)

$$N(t) = N(0) - \frac{2V_0}{\hbar\Omega}P(0)\sin\Omega t + \frac{2V_0}{\hbar\Omega}\left[\frac{2V_0}{\hbar\Omega}N(0) - \frac{\omega_0-\omega}{\Omega}Q(0)\right](\cos\Omega t - 1), \tag{44.11}$$

where

$$\Omega \equiv \sqrt{(\omega-\omega_0)^2 + \left(\frac{2V_0}{\hbar}\right)^2}. \tag{44.12}$$

The solution for $P(t), Q(t)$ is given as

$$\begin{aligned}
Q(t) &= Q'(t)\cos\omega t + P'(t)\sin\omega t\\
P(t) &= P'(t)\cos\omega t - Q'(t)\sin\omega t\\
P'(t) &= P(0)\cos\Omega t + \left[\frac{2V_0}{\hbar\Omega}N(0) - \frac{\omega_0-\omega}{\Omega}Q(0)\right]\sin\Omega t\\
Q'(t) &= Q(0) + \frac{\omega_0-\omega}{\Omega}P(0)\sin\Omega t - \frac{\omega_0-\omega}{\Omega}\left[\frac{2V_0}{\hbar\Omega}N(0) - \frac{\omega_0-\omega}{\Omega}Q(0)\right](\cos\Omega t - 1).
\end{aligned} \tag{44.13}$$

Considering that $N(t) = 2|c_2(t)|^2 - 1$, and that the most relevant case is when all particles are in the ground state initially, at $t = 0$, we have $|c_2(0)|^2 = 0$, which means $N(0) = -1$. Moreover, in this case $c_2(0) = 0$, so $c_2^*(0) = 0$ also, and thus $Q(0) = P(0) = 0$. In this case we obtain

$$\begin{aligned}
Q(t) &= \frac{2(\omega_0-\omega)V_0}{\hbar\Omega^2}(1-\cos\Omega t)\cos\omega t - \frac{2V_0}{\hbar\Omega^2}\sin\Omega t\sin\omega t\\
P(t) &= -\frac{2V_0}{\hbar\Omega}\sin\Omega t\cos\omega t - \frac{2(\omega_0-\omega)V_0}{\hbar\Omega^2}(1-\cos\Omega t)\sin\omega t\\
N(t) &= -1 + \left(\frac{2V_0}{\hbar\Omega}\right)^2(1-\cos\Omega t) = -1 + \left(\frac{2V_0}{\hbar\Omega}\right)^2 2\sin^2\frac{\Omega t}{2}.
\end{aligned} \tag{44.14}$$

Therefore we obtain that the probability for the system to be in the upper energy level is

$$P_2(t) = |c_2(t)|^2 = \frac{1+N(t)}{2} = \left(\frac{2V_0}{\hbar\Omega}\right)^2\sin^2\frac{\Omega t}{2}, \tag{44.15}$$

which is known as the *Rabi formula*. We also have $P_1(t) = 1 - P_2(t)$.

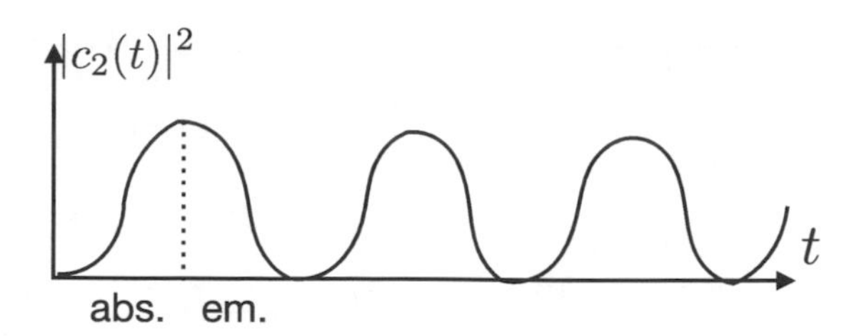

Figure 44.1 Probability $|c_2(t)|^2$ as a function of time t, showing the absorption and emission parts of the cycle.

From the form of Ω, we see that the amplitude is a maximum when Ω is a minimum, which happens when

$$\omega = \omega_0 = \frac{E_2 - E_1}{\hbar}, \tag{44.16}$$

which is known as the *resonance condition*. At resonance,

$$\Omega = \Omega_0 = \frac{V_0}{\hbar}. \tag{44.17}$$

Note also that the amplitude is Lorentzian in shape, since

$$\left(\frac{2V_0}{\hbar\Omega}\right)^2 = \frac{(2V_0)^2}{(2V_0)^2 + [\hbar(\omega - \omega_0)]^2}, \tag{44.18}$$

and it becomes 1 at the resonance. Moreover, when averaging over time, we obtain

$$\langle \sin^2 \Omega t \rangle = \frac{1}{2}, \tag{44.19}$$

so $\langle P_2 \rangle = 1/2$ at resonance, and is smaller otherwise.

At resonance then, we have a cycling between the two energy levels. From $t = 0$ until $\Omega t/2 = \pi/2$, the system absorbs energy from the potential $V(t)$, since $E_2 > E_1$, and the probability $P_2(t) = |c_2(t)|^2$ increases from 0 to a maximum during that time. Then, from $\pi/2$ to π, $P_2(t)$ decreases, so the system is losing energy, by emission of radiation. The cycles of absorption and emission, each for a time $\Omega t/2$ equal to $\pi/2$, repeat ad infinitum; see Fig. 44.1.

The above two-level system in a harmonic potential is relevant for the case of the laser, which actually is the acronym for "light amplification by stimulated emission of radiation" (LASER). In this case, there is radiation in a resonant cavity, with walls made of the atom with two levels. Then the field gives energy to the atom through absorption, followed by stimulated emission, after which the radiation gets reflected and comes back and gets absorbed, etc. The light is amplified coherently and is then led out of the cavity as a laser.

44.2 First-Order Perturbation Theory for Harmonic Potential

The above case was exact, but the two-level system itself is an approximation. In reality, the atom is a multi-level system and interacts with the electromagnetic field, oscillating at some frequency, which generates transitions between the states of the system (so the interaction with the electromagnetic field couples the states indirectly). Thus we can consider the generic interaction operator as the sum of two terms (so that it is Hermitian),

$$\hat{H}_1(t) = \hat{V}(t) = \hat{V}_0 e^{i\omega t} + \hat{V}_0^\dagger e^{-i\omega t}, \tag{44.20}$$

for $t > 0$ (so, in a sudden approximation). For example, we found in Chapter 39 that the interaction Hamiltonian with an electromagnetic field is

$$\hat{H}_1 = \int d^3x \vec{A} \cdot \vec{j} = A_0 \int d^3x (\vec{\epsilon}(\vec{k}) \cdot \vec{j}) \left[e^{i(\vec{k}\cdot\vec{x} - \omega t)} + e^{-i(\vec{k}\cdot\vec{x} - \omega t)} \right], \tag{44.21}$$

and we took the matter current $\vec{j}$ to have matrix elements.

For comparison, for absorption from a constant field, we have the interaction Hamiltonian matrix elements

$$H_{1,fi} = iqA_0\omega_0 \int d^3r \psi_f^*(\vec{r})(\vec{\epsilon}(\vec{k}) \cdot \vec{r})\psi_i(\vec{r}). \tag{44.22}$$

Then, using the first-order time-dependent perturbation theory formula from Chapter 37, integrated over time, we have

$$c_n^{(1)}(t) = \frac{1}{i\hbar} \int_0^t dt' H_{1,ni} e^{i\omega_{ni} t'}. \tag{44.23}$$

Substituting $H_1(t)$, we find

$$\begin{aligned} c_n^{(1)}(t) &= \frac{1}{i\hbar} \int_0^t dt' \left(V_{0,ni} e^{i\omega t'} + V_{0,ni}^\dagger e^{-i\omega t'} \right) e^{i\omega_{ni} t'} \\ &= \frac{1}{\hbar} \left[\frac{1 - e^{i(\omega + \omega_{ni})t}}{\omega + \omega_{ni}} V_{0,ni} + \frac{1 - e^{i(\omega_{ni} - \omega)t}}{-\omega + \omega_{ni}} V_{0,ni}^\dagger \right]. \end{aligned} \tag{44.24}$$

Then, as in Chapter 37, we continue on to find the Fermi golden rule, but now with the replacement

$$\hbar\omega_{ni} = E_n - E_i \to \hbar\omega_{ni} \pm \hbar\omega = E_n - E_i \pm \hbar\omega, \tag{44.25}$$

where the $\pm$ refers to stimulated emission or absorption.

The modified Fermi golden rule is then

$$\frac{dP_{ni}}{dt} = \frac{2\pi}{\hbar} \begin{pmatrix} |V_{0,ni}|^2 \\ |V_{0,ni}^\dagger|^2 \end{pmatrix} \int dE \int \cdots \sum \rho(E_n, \ldots)\delta(E_n - E_i \pm \hbar\omega). \tag{44.26}$$

This formula could have been obtained intuitively, for the stimulated emission or absorption of a photon with energy $\hbar\omega$.

We note that the Hamiltonian $\hat{H}$ is Hermitian, so $|V_{0,ni}^\dagger|^2 = |V_{0,ni}|^2$, which comes from $\langle i|V_0^\dagger|n\rangle = \langle n|V|i\rangle^*$.

44.3 The Case of Quantized Radiation

To improve upon the previous analysis, we must also quantize the electromagnetic field. Strictly speaking, this should be done in the relativistically invariant quantum electrodynamics (QED) formalism, which is beyond the scope of this book, but we can make some shortcuts and "guess" the result.

Briefly, radiation is made up of quanta of a given ω, just as for a harmonic oscillator, where we saw that $E = (N + 1/2)\hbar\omega$ is built up from the vacuum by adding N quanta of energy $\hbar\omega$. Now, the vector potential $\vec{A}$ contains a polarization vector $\vec{\epsilon}_j(\vec{k})$ (where j is the index for the two polarizations transverse to the momentum), a wave factor $e^{i(\vec{k}\cdot\vec{r}-\omega t)}$, and a normalization constant K, plus a Hermitian conjugate term to ensure the reality of the potential (all of which are present in the classical theory), but now the wave factor is multiplied, as for a harmonic oscillator, by a creation or annihilation operator for respectively creating or annihilating a photon. All in all,

$$\vec{A}(\vec{r},t) = K\sum_{\vec{k},j}\vec{\epsilon}_j(\vec{k})\left[a_j(\vec{k})e^{i(\vec{k}\cdot\vec{r}-\omega t)} + a_j^\dagger(\vec{k})e^{-i(\vec{k}\cdot\vec{r}-\omega t)}\right]. \tag{44.27}$$

As we saw when discussing the harmonic oscillator, the only nonzero matrix elements of the creation and annihilation operators are

$$\begin{aligned}\langle N-1|a|N\rangle &= \sqrt{N}\\ \langle N+1|a^\dagger|N\rangle &= \sqrt{N+1}.\end{aligned} \tag{44.28}$$

We see that, with respect to the classical analysis of the electromagnetic field given above, we replace A_0 with $Ka_j(\vec{k})$ in one term and with $Ka_j^\dagger(\vec{k})$ in the other, which are operators in both cases. This means that now we need to consider states not only for the atom but for the radiation itself, and now the total Hilbert space is a product $\mathcal{H}_{\text{total}} = \mathcal{H}_{\text{atom}} \otimes \mathcal{H}_{\text{radiation}}$. Then the initial and final states become

$$|I\rangle = |i\rangle \otimes |\tilde{i}\rangle, \quad |F\rangle = |f\rangle \otimes |\tilde{f}\rangle, \tag{44.29}$$

where $|i\rangle, |f\rangle$ are atomic states and $|\tilde{i}\rangle, |\tilde{f}\rangle$ are radiation states.

Therefore now the matrix elements of the interaction potential are

$$V_{0,FI}^\dagger = K(a_j(\vec{k}))_{\tilde{f}\tilde{i}}M_{fi}, \tag{44.30}$$

where

$$M_{fi} = \sum_{\vec{k},j}\int d^3r\,\vec{\epsilon}_j(\vec{k})\cdot(\vec{j})_{fi}e^{i\vec{k}\cdot\vec{r}}. \tag{44.31}$$

But, as we just saw, the only nonzero matrix element $(a_j(\vec{k}))_{\tilde{f}\tilde{i}}$ is for $|\tilde{i}\rangle = |N\rangle$ and $|\tilde{f}\rangle = |N-1\rangle$, in which case the matrix element is $\sqrt{N_j(\vec{k})}$. As we can see, in this situation one photon is *absorbed*, so the term with $V_{0,FI}^\dagger$ is for *absorption*, as in the classical radiation case.

Similarly,

$$V_{0,FI} = K(a_j^\dagger(\vec{k}))_{\tilde{f}\tilde{i}}M_{fi}^* \tag{44.32}$$

is only nonzero if $|\tilde{i}\rangle = |N\rangle$ and $|\tilde{f}\rangle = |N+1\rangle$, in which case the matrix element is $\sqrt{N_j(\vec{k})+1}$. This is then the situation of *emission* of a photon into the radiation field, so the $V_{0,FI}$ term corresponds to *emission*, again as in the classical radiation case.

Thus, the transition rate in this quantum radiation case is (from the Fermi golden rule)

$$\frac{dP(I\to F)}{dt} = \begin{cases} K^2|M_{fi}|^2\sum_{\vec{k},j}N_j(\vec{k})\int dE\cdots\sum\delta(E_f-E_i-\hbar\omega) & \text{abs.}\\ K^2|M_{fi}|^2\sum_{\vec{k},j}[N_j(\vec{k})+1]\int dE.\cdots\sum\delta(E_f-E_i-\hbar\omega) & \text{em.}\end{cases} \tag{44.33}$$

for the absorption and emission cases, respectively.

We note that the transition rate in emission case is proportional to $N_j(\vec{k}) + 1$, so it can be split into two terms, a term proportional to $N_j(\vec{k})$, which is only nonzero when there are already photons in the radiation field, so this is a *stimulated emission* term (emission stimulated by photons), whereas the term equal to 1 is independent of the existence of photons, so it is a *spontaneous emission* term.

We also note that the absorption rate equals the stimulated emission rate, as in the previous (classical) case. Only the spontaneous emission is different. In the resulting spontaneous emission rate, the right-hand side is called the "Einstein coefficient".

We can turn the sum over $\vec{k}$ into an integral,

$$\int d^3k = \int k^2 dk\, d\Omega = \int \frac{\omega^2 d\omega}{c^3} d\Omega. \tag{44.34}$$

Moreover, the normalization constant K is proportional to $1/\sqrt{\omega}$, so we define $K = K'/\sqrt{\omega}$. Then we finally obtain, for the emission rate in the cases of stimulated or spontaneous emission,

$$\frac{dP_{ni}}{dt} = \frac{2\pi}{\hbar c^3}|M_{ni}|^2 K'^2 \int \omega^2\, d\omega\, d\Omega \begin{pmatrix} N_j(\vec{k}) \\ 1 \end{pmatrix} \delta(E_f - E_i - \hbar\omega). \tag{44.35}$$

44.4 Planck Formula

We can use the above formalism to calculate the Planck formula for thermal radiation, which was what started quantum theory (of course, Planck did not use this derivation, but a much simplified one with only a single assumption, that energy comes in quanta $h\nu$, but here we present the correct proof in quantum mechanics).

We start from the observation that, for a given momentum k (thus angular frequency ω for the radiation), we have the following ratio of the absorption to spontaneous emission rates:

$$\frac{dP_{\text{abs.}}(k)}{dP_{\text{sp.em.}}(k)} = N(\vec{k}). \tag{44.36}$$

Next, we calculate the energy absorbed in an interval $d\omega$ of angular frequency and in a solid angle $d\Omega$. Since we are absorbing photons, we have a factor $\hbar\omega$ for the energy of the photon times a factor $N(\vec{k})$ for how many photons there are in the field. Further, in the continuum, we have

$$\frac{1}{V}\sum_{\vec{k}} \to \frac{1}{(2\pi)^3}\int d^3k = \frac{1}{(2\pi)^3}\int \frac{\omega^2 d\omega\, d\Omega}{c^3}, \tag{44.37}$$

where the correspondence comes from $k_x = 2\pi n_x/L_x$, so $dk_x = 2\pi/L_x$, and a product over dk_x, dk_y, dk_z. Then we get

$$dE(\vec{k}) = \hbar\omega \frac{N(\vec{k})}{(2\pi)^3}\frac{\omega^2 d\omega\, d\Omega}{c^3}. \tag{44.38}$$

This allows us to rewrite

$$\frac{dP_{\text{abs}}(k)}{dP_{\text{sp.em.}}(k)} = N(\vec{k}) = \frac{dE(\omega)}{d\omega\, d\Omega}\frac{(2\pi c)^3}{\hbar\omega^3}. \tag{44.39}$$

Consider now a two-level system, with lower level E_1 and upper level E_2, in equilibrium with the radiation. The probability that the system in E_2 changes by dP_2 via absorption, taking the atom from the E_1 state to the E_2 state, with rate $dP_{\text{abs.}}$, is equal to $dP_{\text{abs.}}$ times the probability P_1 of finding the atom in state E_1,

$$dP_2 = P_1 dP_{\text{abs.}} \tag{44.40}$$

On the other hand, the probability of finding the system in E_1 increases owing to emission from the state E_2, with rate dP_{em} times the probability P_2 of finding the atom in state E_2, so

$$dP_1 = P_2 dP_{\text{em.}} \tag{44.41}$$

But the emission is either stimulated or spontaneous, and the stimulated rate equals the absorption rate, so

$$dP_{\text{em}} = dP_{\text{st.em.}} + dP_{\text{sp.em.}} = dP_{\text{abs.}} + dP_{\text{sp.em.}}, \tag{44.42}$$

so

$$dP_1 = P_2(dP_{\text{abs}} + dP_{\text{sp.em.}}). \tag{44.43}$$

But, at equilibrium (and since we only have two levels for the atom in this picture),

$$dP_2 = dP_1 \Rightarrow P_1 dP_{\text{abs.}} = P_2(dP_{\text{abs.}} + dP_{\text{sp.em.}}). \tag{44.44}$$

Then we obtain

$$\frac{dP_{\text{abs.}}}{dP_{\text{sp.em.}}} = \frac{P_2}{P_1 - P_2} = \frac{1}{P_1/P_2 - 1}. \tag{44.45}$$

However, at equilibrium, for a temperature T, the Boltzmann distribution of the atomic states is

$$P_2 = \frac{1}{Z} e^{-E_2/k_B T}, \quad P_1 = \frac{1}{Z} e^{-E_1/k_B T}, \tag{44.46}$$

so we obtain

$$\frac{dE(\omega)}{d\omega\, d\Omega} \frac{(2\pi c)^3}{\hbar\omega^3} = \frac{dP_{\text{abs.}}}{dP_{\text{sp.em.}}} = \frac{1}{e^{(E_2 - E_1)/k_B T} - 1}, \tag{44.47}$$

Finally, after integrating over $d\Omega$ and getting 4π, we have

$$\frac{dE(\omega)}{d\omega} = 4\pi \frac{\hbar\omega^3}{(2\pi c)^3} \frac{1}{e^{(E_2 - E_1)/k_B T} - 1}. \tag{44.48}$$

Alternatively, using the frequency $\nu = \omega/(2\pi)$, given that $E_2 - E_1 = \hbar\omega$, and having an extra factor of 2 from the two polarizations of light, we obtain the *Planck formula*

$$\frac{dE}{d\nu} = \frac{8\pi\nu^2}{c^3} \frac{h\nu}{\exp\left(\frac{h\nu}{k_B T}\right) - 1}. \tag{44.49}$$

Important Concepts to Remember

- For a two-level system with a time-dependent interaction between states, $H_{12} = V(t)$, specifically for the harmonic $V(t) = V_0 e^{i\omega t}$, we find the Rabi formula, with an oscillating probability of being in the upper state, $|P_2(t)|^2 \propto \sin^2 \Omega t/2$, and a Lorentzian-shaped amplitude with a maximum at resonance, $\omega = \omega_0 = (E_2 - E_1)/\hbar$.

- A laser refers to radiation in a resonant cavity, with atoms of two-level system and coherent amplification at resonance, which is then led out of the cavity as the laser.
- From quantum mechanics the atoms and Fermi's golden rule we obtain stimulated emission and absorption, whereas when quantizing the radiation, the previous results are proportional to the number of photons, but we also obtain the possibility of *spontaneous* emission, which is not.
- The Planck formula is obtained from the energy balance for emission and absorption of photons, and the Boltzmann distribution for the atoms.

Further Reading

Read more about the interaction between matter and electromagnetism in [1] and [2].

Exercises

(1) Show that (44.11) is the solution to the Bloch equations (44.8).

(2) Show that $P(t), Q(t)$ in (44.13) are the corresponding solutions to the Bloch equations.

(3) Consider an exponentially decaying and oscillating potential,

$$\hat{V}(t) = (\hat{V}_0 e^{i\omega t} + \hat{V}_0 e^{-i\omega t})e^{-\gamma t}$$

for the interaction with classical radiation. What is the equivalent of Fermi's golden rule in this case?

(4) Find a limit in which the first-order perturbation theory for a harmonic potential is consistent with the exact, but two-level, calculation of Section 44.1 (giving the same result).

(5) Consider a possible quantum term, of the type $\alpha \vec{A}^2$, for light–light interaction in the case of quantized radiation. Analyze and interpret the resulting terms in the potential $V(t)$ in terms of photons, as was done in the text for the linear term.

(6) If there are no photons (no radiation field), $N_j(\vec{k}) = 0$, how do we interpret the rate of spontaneous emission (what is the physical situation it describes, and what are its limitations)?

(7) In deriving the Planck formula, we used the classical Boltzmann distribution for the atoms. Why is this allowed, given that we are considering quantum interactions?

PART II_d

SCATTERING THEORY

45 One-Dimensional Scattering, Transfer and S-Matrices

In Part II$_d$, we give an *introduction* to the rather large field of scattering theory. In this chapter, we revisit one-dimensional scattering in a potential, restating it in a form that can be generalized to three dimensions and introducing various useful quantities along the way such as transfer and S-matrices. The one-dimensional case is interesting in several ways; an application we will do at the end of the chapter is for discrete one-dimensional systems, or "spin chains" (where the continuous line is replaced by a series of sites or points).

We start with a short review of the set-up for one-dimensional systems from Chapter 7. The one-dimensional Schrödinger equation is

$$-\frac{\hbar^2}{2m}\frac{d^2\psi}{dx^2} + V(x)\psi(x) = E\psi(x), \tag{45.1}$$

and can be rewritten as

$$\psi''(x) = -(\epsilon - U(x))\psi(x), \tag{45.2}$$

where

$$\begin{aligned} U(x) &\equiv \frac{2mV(x)}{\hbar^2} \\ \epsilon &\equiv \frac{2mE}{\hbar^2} \equiv k^2. \end{aligned} \tag{45.3}$$

In Chapter 7, we presented a more general *Wronskian theorem*, but, for applications here, we consider two solutions $\psi_1(x), \psi_2(x)$ of the same Schrödinger equation (meaning, with the same potential *and* energy). Then their Wronskian,

$$W(y_1, y_2) = y_1 y_2' - y_2 y_1', \tag{45.4}$$

is independent of x,

$$\frac{dW}{dx} = 0. \tag{45.5}$$

Moreover, $W \neq 0$ implies that the two solutions are linearly independent, in which case an arbitrary solution can be written as a linear combination of the two,

$$\psi(x) = a\psi_1(x) + b\psi_2(x). \tag{45.6}$$

If we give the initial condition data at x_0, namely $\psi(x_0), \psi'(x_0)$, then the determinant of the linear equations for a and b,

$$\begin{aligned} \psi(x_0) &= a\psi_1(x_0) + b\psi_2(x_0) \\ \psi'(x_0) &= a\psi_1'(x_0) + b\psi_2'(x_0), \end{aligned} \tag{45.7}$$

is W.

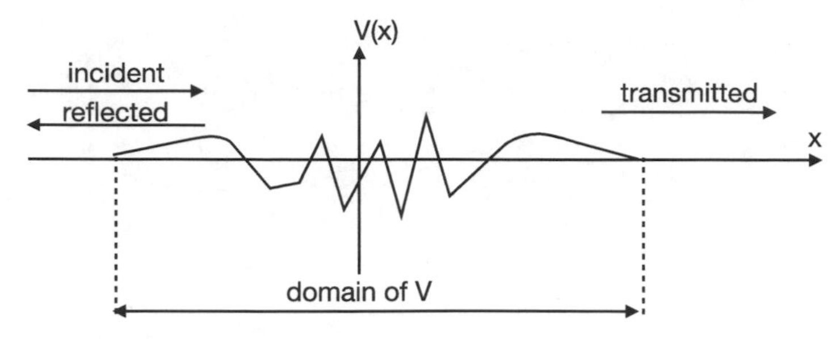

Figure 45.1 Generic scattering situation.

45.1 Asymptotics and Integral Equations

Having in mind applications to 3-dimensional systems, we assume that the potential $V(x)$ is bounded in space (is "local", or has a finite domain), as in Fig. 45.1. More precisely, we impose

$$\int_{-\infty}^{+\infty} dx|V(x)| < \infty. \tag{45.8}$$

Under this assumption, it follows that at $x \to \pm\infty$ we have a free particle solution, meaning a linear combination of $e^{\pm ikx}$, which defines this as a solution with momentum k. Then we must have

$$\begin{aligned}\psi(x \to \pm\infty) &= a_\pm e^{ikx} + b_\pm e^{-ikx} \Rightarrow \\ \psi'(x \to \pm\infty) &= a_\pm ike^{ikx} - b_\pm ike^{-ikx}.\end{aligned} \tag{45.9}$$

Generalizing this behavior to the domain of the potential (the "inside") means that a and b must become functions,

$$\begin{aligned}\psi(x) &= a(x)e^{ikx} + b(x)e^{-ikx} \\ \psi'(x) &= a(x)ike^{ikx} - b(x)ike^{-ikx}.\end{aligned} \tag{45.10}$$

This ansatz is by definition good, since we are writing two arbitrary functions $\psi(x)$ and $\psi'(x)$ in terms of two other arbitrary functions $a(x)$ and $b(x)$. The Schrödinger equation (a second-order differential equation) is equivalent to two first-order differential equations,

$$\begin{aligned}\frac{d}{dx}\psi(x) &= \psi'(x) \\ \frac{d}{dx}\psi'(x) &= (U(x) - k^2)\psi(x).\end{aligned} \tag{45.11}$$

Substituting the ansatz for $\psi(x)$ and $\psi'(x)$ into the first equation above, we obtain

$$a'(x)e^{ikx} + b'(x)e^{-ikx} = 0, \tag{45.12}$$

while substituting it into the second equation, we obtain

$$a'(x)ike^{ikx} - b'(x)ike^{-ikx} = U(x)\psi(x). \tag{45.13}$$

In this way we have obtained two equations for the unknowns $a'(x)$ and $b'(x)$. Adding the first equation to the second divided by ik, we obtain

$$a'(x) = \frac{U(x)\psi(x)}{2ik}e^{-ikx} = \frac{U(x)}{2ik}\left(a(x) + b(x)e^{-2ikx}\right). \tag{45.14}$$

Replacing the sum with the difference, we obtain

$$b'(x) = -\frac{U(x)\psi(x)}{2ik}e^{ikx} = -\frac{U(x)}{2ik}\left(b(x) + a(x)e^{2ikx}\right). \tag{45.15}$$

We define $\psi_1(x)$ as the solution that has only the term e^{ikx} at $-\infty$, so $a(-\infty) = 1, b(-\infty) = 0$. Then integrating the equations for $a'(x), b'(x)$ from $-\infty$ to x, we obtain

$$\begin{aligned} a(x) &= 1 + \frac{1}{2ik}\int_{-\infty}^{x} dx' e^{-ikx'}U(x')\psi_1(x') \\ b(x) &= -\frac{1}{2ik}\int_{-\infty}^{x} dx' e^{ikx'}U(x')\psi_1(x'). \end{aligned} \tag{45.16}$$

Substituting these into (45.10), we obtain

$$\begin{aligned} \psi_1(x) &= e^{ikx} + \frac{1}{2ik}\int_{-\infty}^{x} dx'\left(e^{ik(x-x')} - e^{-ik(x-x')}\right)U(x')\psi_1(x') \\ &= e^{ikx} + \frac{1}{k}\int_{-\infty}^{x} dx' \sin[k(x-x')]U(x')\psi_1(x') \\ \psi_1'(x) &= ike^{ikx} + \frac{1}{2}\int_{-\infty}^{x} dx'\left(e^{ik(x-x')} + e^{-ik(x-x')}\right)U(x')\psi_1(x') \\ &= ike^{ikx} + \int_{-\infty}^{x} dx' \cos[k(x-x')]U(x')\psi_1(x'). \end{aligned} \tag{45.17}$$

We have obtained an integral equation for $\psi_1(x)$. It can be solved iteratively, as an infinite perturbative sum,

$$\psi_1(x) = \sum_{n=0}^{\infty} \psi_1^{(n)}(x), \tag{45.18}$$

where the zeroth-order term is a free wave,

$$\psi_1^{(0)}(x) = e^{ikx}. \tag{45.19}$$

Then, we substitute it into the right-hand side of the equation, thus defining the first-order term, $\psi_1^{(1)}(x)$, then repeat the procedure to find $\psi_1^{(2)}$, etc. In general, then,

$$\psi_1^{(n+1)}(x) = \frac{1}{k}\int_{-\infty}^{x} dx' \sin[k(x-x')]U(x')\psi_1^{(n)}(x). \tag{45.20}$$

One can show that the series is convergent uniformly (though we will not give the proof here) if

$$\frac{1}{k}\int_{-\infty}^{+\infty} dx' V(x') < 1. \tag{45.21}$$

But since the integral is finite by assumption, the relation is true for a sufficiently high k, $k > k_0$, with k_0 the integral in (45.21). Conversely, though, it is *not* true for sufficiently small k, $k < k_0$.

We can make a similar analysis for the solution $\psi_2(x)$ that is pure e^{ikx} at $+\infty$, so that $a(+\infty) = 1, b(+\infty) = 0$. In this case, integrating the differential equations for $a'(x)$ and $b'(x)$, we obtain

$$\begin{aligned} a(x) &= 1 - \frac{1}{2ik}\int_{x}^{+\infty} dx' e^{-ikx'}U(x')\psi_2(x') \\ b(x) &= +\frac{1}{2ik}\int_{x}^{+\infty} dx' e^{ikx'}U(x')\psi_2(x'). \end{aligned} \tag{45.22}$$

Note the change of sign in front of the integral, since now x is the lower limit instead of the upper limit. Substituting these expressions for $a(x), b(x)$ into (45.10), we obtain

$$
\begin{aligned}
\psi_2(x) &= e^{ikx} - \frac{1}{2ik}\int_x^{+\infty} dx' \left(e^{ik(x-x')} - e^{-ik(x-x')}\right) U(x')\psi_2(x') \\
&= e^{ikx} - \frac{1}{k}\int_x^{+\infty} dx' \sin[k(x-x')]\, U(x')\psi_2(x') \\
\psi_2'(x) &= ike^{ikx} - \frac{1}{2}\int_x^{+\infty} dx' \left(e^{ik(x-x')} + e^{-ik(x-x')}\right) U(x')\psi_2(x') \\
&= ike^{ikx} - \int_x^{+\infty} dx' \cos[k(x-x')]\, U(x')\psi_2(x').
\end{aligned}
\tag{45.23}
$$

45.2 Green's Functions

We can rewrite the above integral equations as

$$
\begin{aligned}
\psi_1(x) &= e^{ikx} + \int_{-\infty}^{+\infty} dx' G_1(x,x')U(x')\psi_1(x') \\
\psi_2(x) &= e^{ikx} + \int_{-\infty}^{+\infty} dx' G_2(x,x')U(x')\psi_2(x'),
\end{aligned}
\tag{45.24}
$$

where we have defined the Green's functions

$$
\begin{aligned}
G_1(x,x') &= \frac{\sin[k(x-x')]}{k}\theta(x-x') \\
G_2(x,x') &= \frac{\sin[k(x-x')]}{k}\theta(x'-x),
\end{aligned}
\tag{45.25}
$$

where $\theta(x)$ is the Heaviside function. As a reminder, $\theta(x) = 1$ for $x \geq 0$ and $\theta(x) = 0$ for $x < 0$. Also, we remember that $(d/dx)\theta(x) = \delta(x)$ as a distribution.

These Green's functions have a number of properties:

- G_1, G_2 are continuous and equal to zero at $x = x'$, since there the prefactor of the Heaviside function vanishes, $\sin[k(x-x')] = 0$.
- The derivatives of G_1, G_2 are discontinuous, however. Indeed,

$$
\begin{aligned}
\partial_x G_1(x,x') &= \cos[k(x-x')]\theta(x-x') + \frac{\sin[k(x-x')]}{k}\theta'(x-x') \\
&= \cos[k(x-x')]\theta(x-x').
\end{aligned}
\tag{45.26}
$$

In the second equality above we have used that $\theta'(x-x') = \delta(x-x')$, which is zero everywhere except at $x = x'$; however, there the prefactor vanishes, so the whole second term does too (we explained this in Chapter 2). Then we also find

$$
\partial_x G_2(x,x') = \cos[k(x-x')]\theta(x'-x). \tag{45.27}
$$

- These G_1, G_2 are indeed Green's functions for the free Schrödinger equation. Indeed, calculating the second derivative, we obtain

$$\begin{aligned}\partial_x^2 G_1(x,x') &= -k\sin[k(x-x')]\theta(x-x') + \cos[k(x-x')]\theta'(x-x') \\ &= -k^2 G_1(x,x') + \delta(x-x').\end{aligned} \tag{45.28}$$

Here we have used the fact that we can replace the function multiplying $\theta'(x-x') = \delta(x-x')$ with its value at $x = x'$, which is 1 in this case. Then it follows that $G_1(x,x')$ solves the Green's function equation for the Schrödinger operator in the free case,

$$(\partial_x^2 + k^2)G_1(x,x') = \delta(x,x'). \tag{45.29}$$

We can rewrite the full Schrödinger equation as

$$\left(\frac{d^2}{dx^2} + k^2\right)\psi(x) = U(x)\psi(x) \equiv f(x), \tag{45.30}$$

and if we consider $f(x)$ as a given function (which, in fact, it is not, as we want to solve for $\psi(x)$), we can formally solve for $\psi(x)$ by convoluting the Green's function with f. Indeed, multiplying the equation with $f(x)$ and integrating between $-\infty$ and x, we obtain

$$\int_{-\infty}^{x} dx' f(x')\left(\frac{d^2}{dx^2} + k^2\right)G_1(x,x') = \int_{-\infty}^{x} dx'\delta(x-x')f(x') = f(x), \tag{45.31}$$

or more generally

$$\left(\frac{d^2}{dx^2} + k^2\right)\left[\int_{-\infty}^{x} dx' G(x,x')f(x') + \psi_0(x)\right] = f(x), \tag{45.32}$$

where $\psi_0(x)$ is a general solution of the free (homogenous) equation, $(d^2/dx^2 + k^2)\psi_0(x) = 0$.
Then it follows that the formal solution is

$$\psi(x) = \psi_0(x) + \int_{-\infty}^{x} dx' G_1(x,x')f(x'). \tag{45.33}$$

However, as we noted, in reality $f(x)$ is not given but is equal to $U(x)\psi(x)$. Replacing it in the formal solution above, we actually obtain the integral equations.

We note also that a general Green's function is a sum of the above special Green's function $G_1(x,x')$ and a general solution $g(x,x')$ of the homogenous equation

$$\left(\frac{d^2}{dx^2} + k^2\right)g(x,x') = 0, \tag{45.34}$$

so that

$$\tilde{G}_1(x,x') = G_1(x,x') + g(x,x'). \tag{45.35}$$

45.3 Relations between Abstract States and Lippmann–Schwinger Equation

Until now, we have worked in the x (space) representation. But we can use an abstract approach as well, without choosing a representation. In fact, the formal solution above can be written as

$$|\psi\rangle = |\psi_0\rangle + \hat{G}_0|f\rangle, \tag{45.36}$$

where $\hat{G}_0$ stands in for $G_1(x, x')$ without introducing a given representation, and the index 0 is there to show that the Green's operator is for the free Schrödinger operator $\hat{D}_0$, equal to $d^2/dx^2 + k^2$ in the x representation. In other words, we are solving

$$\hat{D}_0|\psi\rangle = |f\rangle = \hat{U}|\psi\rangle. \tag{45.37}$$

Equivalently, by multiplying with $\hbar^2/(2m)$, we can solve

$$(-\hat{H}_0 + E)|\psi\rangle = \hat{V}|\psi\rangle, \tag{45.38}$$

where $\hat{H}_0$ is the Hamiltonian of the free particle. Then the Green's function equation is

$$(E - \hat{H}_0)\hat{G}_0 = \frac{\hbar^2}{2m}\mathbb{1}, \tag{45.39}$$

where $\mathbb{1}$ is the identity operator, so that $\langle x|\mathbb{1}|x'\rangle = \delta(x - x')$. Finally, this means that we can solve the Green's function equation formally as

$$\hat{G}_0 = \frac{\hbar^2}{2m}\frac{1}{E - \hat{H}_0}. \tag{45.40}$$

However, we see that this operator is not well defined when acting on eigenvectors $|k\rangle$ of the free Hamiltonian, since $(E - \hat{H}_0)|k\rangle = 0$. To make it well defined, we must make the energy slightly complex, $E \to E + i\epsilon$ (we can also add $-i\epsilon$ instead, but that gives another Green's function, not so useful), so that we obtain

$$\hat{G}_0 = \lim_{\epsilon\to 0}\frac{\hbar^2}{2m}\frac{1}{E + i\epsilon - \hat{H}_0}. \tag{45.41}$$

The integral equation is written formally as

$$|\psi\rangle = |k\rangle + \hat{G}_0\hat{V}|\psi\rangle = |k\rangle + \frac{1}{E + i\epsilon - \hat{H}_0}\hat{V}|\psi\rangle, \tag{45.42}$$

and is known as the *Lippmann–Schwinger equation.* To see that this solves the Schrödinger equation formally, we multiply by $E - \hat{H}_0$, obtaining (neglecting $i\epsilon$ for simplicity)

$$\begin{aligned}&(E - \hat{H}_0)|\psi\rangle = (E - \hat{H}_0)|k\rangle + \hat{V}|\psi\rangle = \hat{V}|\psi\rangle \Rightarrow\\&(\hat{H}_0 + \hat{V})|\psi\rangle = E|\psi\rangle.\end{aligned} \tag{45.43}$$

An iterative solution is obtained as before, by substituting the nth-order solution into the right-hand side of the Lippmann–Schwinger equation, to obtain the $(n+1)$th-order solution,

$$\begin{aligned}|\psi\rangle &= |k\rangle + \hat{G}_0\hat{V}|k\rangle + (\hat{G}_0\hat{V})^2|k\rangle + \cdots + (\hat{G}_0\hat{V})^n|k\rangle + \cdots\\&= |\psi^{(0)}\rangle + |\psi^{(1)}\rangle + |\psi^{(2)}\rangle + \cdots + |\psi^{(n)}\rangle + \cdots.\end{aligned} \tag{45.44}$$

We can obtain this same perturbative solution in a different way. Substitute $|\psi\rangle = |k\rangle + |\phi\rangle$ into the full Schrödinger equation $(E - \hat{H}_0 - \hat{V})|\psi\rangle = 0$. Using the free Schrödinger equation, we obtain

$$(E - \hat{H}_0 - \hat{V})|\phi\rangle = \hat{V}|k\rangle. \tag{45.45}$$

This is solved by

$$\begin{aligned}|\phi\rangle &= \frac{1}{E - \hat{H}_0 - \hat{V}}\hat{V}|k\rangle\\&= (\hat{G}_0 + \hat{G}_0\hat{V}\hat{G}_0 + \hat{G}_0\hat{V}\hat{G}_0\hat{V}\hat{G}_0 + \cdots)\hat{V}|k\rangle,\end{aligned} \tag{45.46}$$

where we have defined

$$\hat{G} = \frac{1}{E - \hat{H}_0 - \hat{V}} = \frac{1}{\hat{G}_0^{-1} - \hat{V}}, \tag{45.47}$$

and used the following relation for two arbitrary operators $\hat{A}$ and $\hat{B}$,

$$\frac{1}{\hat{A} - \hat{B}} = \frac{1}{\hat{A}} + \frac{1}{\hat{A}}\hat{B}\frac{1}{\hat{A}} + \frac{1}{\hat{A}}\hat{B}\frac{1}{\hat{A}}\hat{B}\frac{1}{\hat{A}} + \cdots, \tag{45.48}$$

which we leave as a simple exercise to prove.

We note that, indeed, we have obtained the same perturbative solution of the full Schrödinger equation. We can now define the operator $\hat{T}$ by factorizing $\hat{G}_0$ on the right-hand side of the perturbative solution, obtaining

$$\hat{G}_0(\hat{V} + \hat{V}\hat{G}_0\hat{V} + \hat{V}\hat{G}_0\hat{V}\hat{G}_0\hat{V} + \cdots) \equiv \hat{G}_0\hat{T}. \tag{45.49}$$

But then we notice that we can again obtain the $\hat{T}$ operator in its definition, as

$$\hat{T} = \hat{V} + \hat{V}\hat{G}_0(\hat{V} + \hat{V}\hat{G}_0\hat{V} + \cdots) = \hat{V} + \hat{V}\hat{G}_0\hat{T}. \tag{45.50}$$

This equation for $\hat{T}$ is in fact an equivalent form of the Lippmann–Schwinger equation.

45.4 Physical Interpretation of Scattering Solution

We have defined $\psi_1(x)$ as constituting a pure solution at $x \to -\infty$,

$$\psi_1(x) \sim e^{ikx} \Rightarrow \psi_1^*(x) \sim e^{-ikx}, \tag{45.51}$$

while $\psi_2(x)$ has a pure solution at $x \to +\infty$,

$$\psi_2(x) \sim e^{ikx} \Rightarrow \psi_2^*(x) \sim e^{-ikx}. \tag{45.52}$$

Thus at $x \to -\infty$ the two solutions $\psi_1(x)$ and $\psi_1^*(x)$ are linearly independent, with Wronskian

$$W_1(\psi_1, \psi_1^*) = \psi_1^*\psi_1' - \psi_1\psi_1^{*\prime} = 2ik \neq 0, \tag{45.53}$$

so that $(\psi_1(x), \psi_1^*(x))$ form a basis in the space of eigenfunctions near $-\infty$. Similarly, at $x \to +\infty$, the two solutions $\psi_2(x)$ and $\psi_2^*(x)$ are linearly independent, with Wronskian

$$W_2(\psi_2, \psi_2^*) = \psi_2^*\psi_2' - \psi_2\psi_2^{*\prime} = 2ik \neq 0, \tag{45.54}$$

meaning that $(\psi_2(x), \psi_2^*(x))$ is a basis in the space of eigenfunctions near $+\infty$.

Therefore we can expand the basis elements near $-\infty$ in terms of the basis elements near $+\infty$,

$$\begin{aligned} \psi_1(x) &= \alpha(k)\psi_2(x) + \beta(k)\psi_2^*(x) \Rightarrow \\ \psi_1^*(x) &= \beta^*(k)\psi_2(x) + \alpha^*(k)\psi_2^*(x). \end{aligned} \tag{45.55}$$

Applying this relation near $+\infty$, where $\psi_2(x) \sim e^{ikx}$, we obtain

$$\begin{aligned} \psi_1(x \to +\infty) &\to \alpha(k)e^{ikx} + \beta(k)e^{-ikx} \\ \psi_1^*(x \to +\infty) &\to \beta^*(k)e^{ikx} + \alpha^*(k)e^{-ikx}. \end{aligned} \tag{45.56}$$

Substituting the relation between basis elements into the Wronskian W_1, we obtain

$$W_1 = \psi_1^*\psi_1' - \psi_1\psi_1^{*\prime} = [|\alpha(k)|^2 - |\beta(k)|^2](\psi_2^*\psi_2' - \psi_2\psi_2^{*\prime}) = [|\alpha(k)|^2 - |\beta(k)|^2]W_2. \quad (45.57)$$

However since W_1 and W_2 are independent of x, and we have calculated that $W_1 = 2ik = W_2$, it follows that

$$|\alpha(k)|^2 - |\beta(k)|^2 = 1. \quad (45.58)$$

For a generic eigenfunction, expanded in the basis at $-\infty$,

$$\psi(x) = a\psi_1(x) + b\psi_1^*(x), \quad (45.59)$$

which means that near $x \to -\infty$, when

$$\psi(x) \to ae^{ikx} + be^{-ikx}, \quad (45.60)$$

we can use the transformation law between bases, and apply it near $x \to +\infty$, to obtain

$$\psi(x) \to (a\alpha(k) + b\beta^*(k))e^{+ikx} + (a\beta(k) + b\alpha^*(k))e^{-ikx} \equiv a'e^{+ikx} + b'e^{-ikx}. \quad (45.61)$$

It follows that in effect we are acting with a matrix on the state (a, b), namely

$$\begin{pmatrix} a \\ b \end{pmatrix} \to \begin{pmatrix} a' \\ b' \end{pmatrix} = K \begin{pmatrix} a \\ b \end{pmatrix}, \quad (45.62)$$

where we define the *transfer matrix* K:

$$K \equiv \begin{pmatrix} \alpha(k) & \beta^*(k) \\ \beta(k) & \alpha^*(k) \end{pmatrix}. \quad (45.63)$$

Outside the domain of the potential, the time-dependent solution of the Schrödinger equation is

$$\psi_k(x,t) = \psi_k(x)e^{-iEt}, \quad (45.64)$$

and it corresponds to a propagating wave. For $\psi_k(x) \sim e^{ikx}$, we have a wave propagating to the right, while for $\psi_k(x) \sim e^{-ikx}$, we have a wave propagating to the left.

We can now connect with the analysis in Chapter 7. Indeed, there is a solution

$$\psi_+(x) = \psi_1(x) - \frac{\beta}{\alpha^*}\psi_1^*(x), \quad (45.65)$$

that behaves as

$$\psi_+(x) \sim \frac{1}{\alpha^*}e^{ikx} \quad (45.66)$$

at $x \to +\infty$, and as

$$\psi_+(x) \sim e^{ikx} - \frac{\beta}{\alpha^*}e^{-ikx} \quad (45.67)$$

at $x \to -\infty$. This is a wave propagating from the left ($\propto e^{ikx}$), and after encountering the potential domain, having a reflected part $\propto e^{-ikx}$ and a transmitted part $\propto e^{ikx}$.

Similarly, there is a solution

$$\psi_-(x) = \frac{1}{\alpha^*}\psi_1^* \quad (45.68)$$

that behaves as

$$\psi_-(x) \sim \frac{1}{\alpha^*} e^{-ikx} \tag{45.69}$$

at $x \to -\infty$, and as

$$\psi_-(x) \sim \frac{\beta^*}{\alpha^*} e^{ikx} + e^{-ikx} \tag{45.70}$$

at $x \to +\infty$. This is a wave propagating from the right ($\propto e^{-ikx}$), after encountering the potential domain, having a reflected part $\propto e^{ikx}$, and a transmitted part $\propto e^{-ikx}$.

We define a wave as being an "in" wave if it is going towards the potential, and as an "out" wave if it is going away from the potential.

45.5 S-Matrix and T-Matrix

Define abstract states $|k\pm\rangle$ by

$$e^{ikx} = \langle x|k+\rangle, \quad e^{-ikx} = \langle x|k-\rangle. \tag{45.71}$$

Then the transmitted (t) and reflected (r) coefficients in the waves are

$$\begin{aligned} t_+ &= \frac{1}{\alpha^*} \quad r_+ = -\frac{\beta}{\alpha^*} \\ t_- &= \frac{1}{\alpha^*} \quad r_- = \frac{\beta^*}{\alpha^*}, \end{aligned} \tag{45.72}$$

which are normalized correctly,

$$|t_+|^2 + |r_+|^2 = |t_-|^2 + |r_-|^2 = 1. \tag{45.73}$$

Then the free "in" states turn into a combination of free "out" states, as follows:

$$\begin{aligned} |k+\rangle_{\text{in}} &\to t_+|k+\rangle_{\text{out}} + r_+|k-\rangle_{\text{out}} \\ |k-\rangle_{\text{in}} &\to t_-|k-\rangle_{\text{out}} + r_-|k-\rangle_{\text{out}}. \end{aligned} \tag{45.74}$$

This can be written as a matrix action,

$$\begin{pmatrix} |k+\rangle_{\text{in}} \\ |k-\rangle_{\text{in}} \end{pmatrix} \to \hat{S} \begin{pmatrix} |k+\rangle_{\text{out}} \\ |k-\rangle_{\text{out}} \end{pmatrix} = \begin{pmatrix} t_+ & r_+ \\ t_- & r_- \end{pmatrix} \begin{pmatrix} |k+\rangle_{\text{out}} \\ |k_-\rangle_{\text{out}} \end{pmatrix}, \tag{45.75}$$

where we have defined the matrix

$$\hat{S} = \begin{pmatrix} \langle k+|\hat{S}|k+\rangle & \langle k-|\hat{S}|k+\rangle \\ \langle k+|\hat{S}|k-\rangle & \langle k-|\hat{S}|k-\rangle \end{pmatrix} = \begin{pmatrix} t_+ & r_+ \\ t_- & r_- \end{pmatrix} = \begin{pmatrix} 1/\alpha^* & \beta^*/\alpha^* \\ -\beta/\alpha^* & 1/\alpha^* \end{pmatrix}, \tag{45.76}$$

called the *S-matrix*. It is a unitary matrix, since, as we can easily check,

$$\hat{S}\hat{S}^\dagger = \mathbb{1}. \tag{45.77}$$

Thus we can write it in terms of a Hermitian ("real") matrix $\hat{\Delta}$, as

$$\hat{S} = e^{i\hat{\Delta}}. \tag{45.78}$$

Equivalently, we can define an S-operator as the evolution operator between $-\infty$ and $+\infty$,

$$\hat{S} = \hat{U}(+\infty, -\infty) = \lim_{t_2 \to +\infty, t_1 \to -\infty} U(t_2, t_1), \tag{45.79}$$

and the S-matrix as the matrix whose elements are the matrix elements of the $\hat{S}$ operator between the free states $|k\pm\rangle$. This definition also gives a transition between in states (at $t \to -\infty$) and out states (at $t \to +\infty$).

We can define the T-matrix from the previously defined T-operator $\hat{T}$. The perturbative solution of the Schrödinger equation was written as

$$|\psi\rangle = |k\rangle + |\phi\rangle = |k\rangle + \hat{G}_0 \hat{T}|k\rangle = (\mathbb{1} + \hat{G}_0 \hat{T})|k\rangle. \tag{45.80}$$

Acting with $\hat{V}$ from the left, we obtain

$$\hat{V}|\psi\rangle = (\hat{V} + \hat{V}\hat{G}_0\hat{T})|k\rangle = \hat{T}|k\rangle, \tag{45.81}$$

where in the last equality we have used the Lippmann–Schwinger equation. This then gives an alternative form of the Lippmann–Schwinger equation,

$$|\psi\rangle = |k\rangle + \hat{G}_0 \hat{V}|\psi\rangle. \tag{45.82}$$

Moreover, considering the two $|k\pm\rangle$ free-wave solutions, we find

$$|\psi\pm\rangle = |k\pm\rangle + \hat{G}_0\hat{V}|\psi\pm\rangle = |k\pm\rangle + \hat{G}_0\hat{T}|k\pm\rangle. \tag{45.83}$$

This equation also defines an S-matrix, though one in which we have separated an identity term $\mathbb{1}$, giving the first, free-wave, part of the state.

45.6 Application: Spin Chains

As an important application of the one-dimensional scattering formalism, consider a useful one-dimensional system: on a discrete line, made up of a set of points ("sites") $i = 1, 2, \ldots, L$, each site has a "spin" variable. In the simplest case, of $s = 1/2$, the spin variable can have two states, $|\uparrow\rangle$ and $|\downarrow\rangle$. We can define the state $|x_1, x_2, \ldots, x_k\rangle$ as the state with $|\uparrow\rangle$ at sites $x_1, \ldots, x_k \in \{1, 2, \ldots, L\}$ and with $|\downarrow\rangle$ at the rest of the sites; see Fig. 45.2.

Further, we can define a state with one excitation, one spin up moving along a sea of spins down (called a "magnon") with momentum p,

$$|\psi(p)\rangle = \sum_{x=1}^{L} e^{ipx}|x\rangle \equiv \sum_{x=1}^{L} \psi_p(x)|x\rangle. \tag{45.84}$$

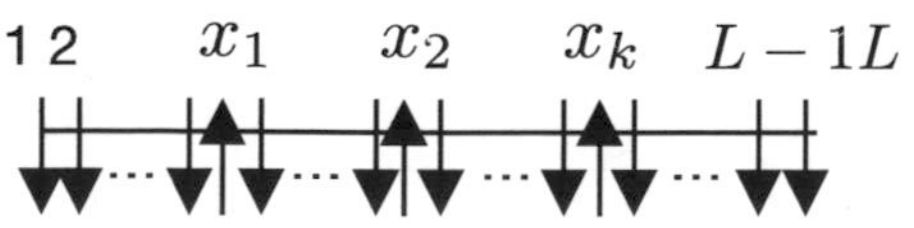

1 2 x_1 x_2 x_k $L-1\,L$

Figure 45.2 Spin chain.

Similarly, the state with two excitations (two magnons) is defined in terms of a two-particle wave function as

$$|\psi(p_1,p_2)\rangle = \sum_{1\leq x_1<x_2\leq L} \psi(x_1,x_2)|x_1,x_2\rangle. \tag{45.85}$$

Next, we can define an ansatz for this wave function $\psi(x_1,x_2)$, known as the Bethe ansatz (as defined by Hans Bethe in 1932),

$$\begin{aligned}\psi(x_1,x_2) &= \psi_{p_1}(x_1)\psi_{p_2}(x_2) + S(p_2,p_1)\psi_{p_2}(x_1)\psi_{p_1}(x_2)\\ &= e^{i(p_1x_1+p_2x_2)} + S(p_2,p_1)e^{i(p_2x_1+p_1x_2)},\end{aligned} \tag{45.86}$$

where the first term is the free, "in", part, and the second term is the scattered, "out", part, while $S(p_2,p_1)$ is an S-matrix.

We note that $\psi(x_1,x_2)$ is defined only for $x_1 < x_2$. However, we can define $\psi(x_2,x_1)$ by periodicity, as (since $x_2 < x_1 + L$)

$$\psi(x_2, x_1 + L) = \psi(x_1,x_2). \tag{45.87}$$

That gives the condition

$$e^{ip_1L} = S(p_1,p_2), \tag{45.88}$$

which in turn gives

$$e^{ip_2L} = S(p_2,p_1), \tag{45.89}$$

meaning the S-matrix is a phase (a 1×1 unitary matrix). These two equations are called the *Bethe ansatz equations*, and from them, and their generalizations in the case of several magnons, we obtain the sets of possible p's (in general complex), or *Bethe roots* (strictly speaking, these are a given function of the p's).

From these two equations, we obtain

$$\begin{aligned}S(p_1,p_2)S(p_2,p_1) &= 1\\ &= e^{i(p_1+p_2)L},\end{aligned} \tag{45.90}$$

which gives the quantization condition

$$p_1 + p_2 = \frac{2\pi}{L}n \mod(2\pi/L). \tag{45.91}$$

In general, we can define the S-matrix between two momenta as the phase

$$S(p_i,p_j) = e^{i\delta_{ij}}. \tag{45.92}$$

In terms of it, we can define the Bethe ansatz for M magnons,

$$\psi(x_1,\dots,x_M) = \sum_{P\in\text{Perm.}(M)} \exp\left[i\sum_{i=1}^{M} p_{P(i)}x_i + \frac{i}{2}\sum_{i<j}\delta_{P(i)P(j)}\right]. \tag{45.93}$$

The Hamiltonian on the spin chain is an operator acting on the sites. The standard example, for which actually Bethe invented his ansatz, is the Heisenberg spin 1/2 Hamiltonian,

$$H = 2J\sum_{j=1}^{L}(P_{j,j+1} - 1), \tag{45.94}$$

where J is a constant, and $P_{j,j+1}$ is the operator that permutes the states at sites j and $j+1$. One can show that in general, when acting on the spin 1/2 Hilbert spaces,

$$P_{ij} = \frac{1}{2} + \frac{1}{2}\vec{\sigma}_i \cdot \vec{\sigma}_j, \tag{45.95}$$

where $\vec{\sigma}_i$ are the Pauli matrices at site i. Then the Hamiltonian can be rewritten as

$$H = J\sum_{j=1}^{L}\left(\vec{\sigma}_j \cdot \vec{\sigma}_{j+1} - 1\right). \tag{45.96}$$

Since we have

$$\begin{aligned}(P_{j,j+1} - 1)|j\rangle &= |j+1\rangle - |j\rangle \\ (P_{j,j+1} - 1)|j+1\rangle &= |j\rangle - |j+1\rangle,\end{aligned} \tag{45.97}$$

the Heisenberg Hamiltonian acting on the one-magnon state gives

$$\begin{aligned}\hat{H}|\psi(p)\rangle &= J\sum_{j=1}^{L} 2e^{ipx}\left(|x-1\rangle + |x+1\rangle - 2|x\rangle\right) \\ &= 2J\left(e^{ip} + e^{-ip} - 2\right)\sum_{x=1}^{L} e^{ipx}|x\rangle \\ &= -8J\sin^2\frac{p}{2}|\psi(p)\rangle \equiv E|\psi(p)\rangle.\end{aligned} \tag{45.98}$$

As can be seen, the one-magnon state is automatically an eigenstate of the Heisenberg Hamiltonian.

We could solve for eigenstates and eigenenergies in the case of several magnons, but we will not do it here.

Important Concepts to Remember

- Defining $\psi_1(x)$ as the solution that has only the term e^{ikx} at $x = -\infty$ and $\psi_2(x)$ as the solution that has only the term e^{ikx} at $x = +\infty$, we have the equations $\psi_{1,2}(x) = e^{ikx} + \int dx' G_{1,2}(x,x')U(x')\psi_1(x')$, with $G_{1,2}$ Green's functions and U the rescaled potential.
- In the abstract case, we have the Lippmann–Schwinger equation $|\psi\rangle = |k\rangle + \hat{G}_0\hat{V}|\psi\rangle$, where $\hat{G}_0 = 1/(E + i\epsilon - \hat{H}_0)$ is the free Green's function, and $|k\rangle$ is a free-wave state, which can be solved by iteration.
- Alternatively, we have $|\psi\rangle = |k\rangle + \hat{G}\hat{V}|k\rangle$, where $\hat{G} = 1/(E + i\epsilon - \hat{H})$, where $\hat{H} = \hat{H}_0 + \hat{V}$. Also, $\hat{G}\hat{V} \equiv \hat{G}_0\hat{T}$, implying $\hat{T} = \hat{V} + \hat{V}\hat{G}_0\hat{T}$.
- Expanding ψ_1 in ψ_2, ψ_2^* as $\psi_1 = \alpha\psi_2 + \beta\psi_2^*$, a generic wave function is $\psi = a\psi_1 + b\psi_1^*$ and which near $x = -\infty$ is $\sim ae^{ikx} + be^{-ikx}$ and near $x = +\infty$ is $\sim a'e^{ikx} + b'e^{-ikx}$, with (a', b') related to (a, b) by the transfer matrix K.
- There exist wave functions that correspond to a wave coming from the left and being reflected and transmitted, or to a wave coming from the right and being reflected and transmitted. Also, we have "in" waves (moving towards the potential domain) and "out" waves (moving away from the potential domain).

- The S-matrix relates out and in waves, is unitary, and corresponds to the matrix elements in free $|k\rangle$ states of the S-operator, $\hat{S} = \hat{U}(+\infty, -\infty)$.
- Spin chains have spin variables at sites along a one-dimensional chain, and have magnon excitations.
- The Bethe ansatz for two magnons is $\psi(x_1, x_2) = e^{ip_1x_1+ip_2x_2} + S(p_2, p_1)e^{ip_2x_1+ip_1x_2}$, where $S(p_2, p_1)$ is the S-matrix, and the standard example for it is the Heisenberg spin 1/2 Hamiltonian $H = 2J\sum_i[P_{i,i+1} - 1]$.

Further Reading

See [2] for one-dimensional scattering, [2],[1] for Lippman–Schwinger formalism, and [6] for spin chains.

Exercises

(1) Consider the potential $V(x) = V_0 e^{-\alpha x^2}$. Calculate perturbatively the first two terms in the solutions $\psi_1(x)$ ($\sim e^{ikx}$ at $x = -\infty$, incoming) and $\psi_2(x)$ ($\sim e^{ikx}$ at $x = +\infty$, outgoing), if $E \leq V_0$.

(2) Show that $G_2(x, x')$ is also a Green's function for the Schrödinger operator, as is $G_1(x, x')$, and that their difference is a solution of the homogenous Schrödinger operator.

(3) Write down the Lippmann–Schwinger equation in the momentum representation, and the corresponding first few terms in its perturbative solution.

(4) Consider a barrier potential, $V = V_0$ for $|x| \leq L/2$ and $V = 0$ for $|x| > L/2$, and an energy $E \leq V_0$. Calculate $\psi_1(x), \psi_2(x)$, and the corresponding transfer matrix K.

(5) In the case in exercise 4, calculate the S-matrix and the Hermitian matrix $\hat{\Delta}$, with $\hat{S} = e^{i\hat{\Delta}}$.

(6) Substitute the two-magnon ansatz into the Schrödinger equation for the Heisenberg spin 1/2 spin chain, and find that the energy of the two-magnon state is just the sum of the energies of the single magnons, and that

$$S(p_1, p_2) = \frac{\phi(p_1) - \phi(p_2) + i}{\phi(p_1) - \phi(p_2) - i}, \quad \phi(p) \equiv \frac{1}{2}\cot\frac{p}{2}. \tag{45.99}$$

(7) Solve the Bethe ansatz equations for two magnons for the case $n = 0$ in the quantization condition (45.91), and write down the corresponding explicit wave functions.

46 Three-Dimensional Lippmann–Schwinger Equation, Scattering Amplitudes and Cross Sections

In this chapter, we start the analysis of three-dimensional scattering. We will consider (except in the last chapter of this part) only elastic scattering, meaning the colliding particles do not change their states during the collision. As we have already seen, this can be reduced to scattering off a potential, where the scattering is in terms of the relative motion.

We begin with a quick review of motion in a potential.

46.1 Potential for Relative Motion

Since we are considering elastic scatterings, it is enough to have only two colliding particles, A and B. In that case, the Hamiltonian for the system is

$$\hat{H} = \hat{H}_A + \hat{H}_B + \hat{V} = -\frac{\hbar^2}{2m_A}\Delta_A - \frac{\hbar^2}{2m_B}\Delta_B - \hat{V}(\vec{r}_A - \vec{r}_B). \tag{46.1}$$

Defining the relative position $\vec{r} = \vec{r}_A - \vec{r}_B$, the center of mass position $\vec{R}$, the relative mass $m = m_A m_B/(m_A + m_B)$, and the total mass $M = m_A + m_B$, we rewrite the above equation as

$$\hat{H} = \hat{H}_{\text{CM}} + \hat{H}_{\text{rel}} = -\frac{\hbar^2}{2M}\Delta_{\vec{R}} - \frac{\hbar^2}{2m}\Delta_{\vec{r}} + V(\vec{r}). \tag{46.2}$$

Then the center of mass and relative variables separate in the solution. The center of mass Schrödinger equation is trivial, and the relative-motion Schrödinger equation is

$$\left(-\frac{\hbar^2}{2m}\Delta_{\vec{r}} + \hat{V}(\vec{r})\right)\psi = E\psi. \tag{46.3}$$

In the absence of a potential (for instance, when we are far from the domain of the potential), we have a free-wave solution,

$$\langle \vec{r}|\vec{p}\rangle = \phi_{\vec{p}}(\vec{r}) = \frac{1}{(2\pi\hbar)^{3/2}} e^{i\vec{p}\cdot\vec{r}/\hbar}. \tag{46.4}$$

Defining the wave vector $\vec{k} = \vec{p}/\hbar$, we can expand the exponential in terms of spherical harmonics $Y_{lm}(\theta, \phi)$,

$$e^{i\vec{k}\cdot\vec{r}} = \sum_{l,m} A_{lm}(\vec{n}_k) j_l(kr) Y_{lm}(\vec{n}_r), \tag{46.5}$$

where the unit vector in the momentum direction is $\vec{n}_k = \vec{k}/k$ and in the position direction is $\vec{n}_r = \vec{r}/r$, and the coefficients are

$$A_{lm}(\vec{n}_k) = 4\pi i^l Y^*_{lm}(\vec{n}_k). \tag{46.6}$$

The position wave function at given l, m, is then

$$u_{lm}(k; \vec{n}_r) = j_l(kr) Y_{lm}(\vec{n}_r). \tag{46.7}$$

46.2 Behavior at Infinity

To describe scattering in a potential, we need to understand first the boundary conditions, just as we did in the one-dimensional case, and in the case of eigenfunctions in a central potential, in Chapters 18 and 19. We saw there that the boundary conditions at $r = \infty$ and $r = 0$ restrict the wave function, in a similar way to that discussed in Chapter 7, in the one-dimensional case. If we have negative energy, $E < 0$, we saw that we obtain bound states, and that leads to the quantization of the energy levels. Indeed, we have two normalizability conditions, one at $r = 0$ and one at $r = \infty$, for the two independent solutions of the second-order differential equation. On the other hand, if we have positive energy, $E > 0$, we have asymptotic states (states that "escape" to infinity), and there are no quantization conditions. At $r \to \infty$, both these states, with behavior $rR(r) = \chi(r) \sim e^{\pm ikr}$ (as we saw) are well behaved. As we will see, these two behaviors lead to outgoing and incoming waves, respectively, for scattering states $|\psi\pm\rangle$ (as opposed to bound states, like those in which we were interested in Chapters 18 and 19). In conclusion, then, the boundary conditions at infinity imply the scattering solution that we want.

For the time being we will consider a potential with a finite range, i.e., a potential $V(r)$ (for a spherically symmetric case) that decays faster than $1/r$ at $r \to \infty$.

In that case, at infinity, the free particle (which experiences no potential) is a good approximation. The Schrödinger equation there (with $E = \hbar^2 k^2/(2m)$) is

$$\frac{\partial^2 \psi}{\partial r^2} + \frac{2}{r}\frac{\partial \psi}{\partial r} + \frac{1}{r^2}\Delta_{\theta,\phi}\psi + k^2\psi = 0. \tag{46.8}$$

It can be rewritten as

$$\frac{\partial^2}{\partial r^2}(r\psi) + \frac{1}{r}\Delta_{\theta,\phi}\psi + k^2(r\psi) = 0. \tag{46.9}$$

But, since we are at $r \to \infty$, the middle term is $\mathcal{O}(1/r)$, so it goes to zero when compared with the other two terms and can be neglected. Then the equation at infinity is

$$\frac{\partial^2}{\partial r^2}(r\psi) + k^2(r\psi) = 0, \tag{46.10}$$

which has the solutions

$$r\psi = Ae^{\pm ikr}. \tag{46.11}$$

To be more precise, the prefactors A have also some dependence,

$$r\psi = A(k, \pm\vec{n}_r)e^{\pm ikr}, \tag{46.12}$$

which means that we have an "out" (outgoing) solution

$$\psi_1 = A(k, \vec{n}_r)\frac{e^{ikr}}{r} \tag{46.13}$$

and an "in" (incoming) solution

$$\psi_2 = B(k, -\vec{n}_r)\frac{e^{-ikr}}{r}, \tag{46.14}$$

where the minus sign in the dependence of B is conventional, to remind us that the wave goes towards $r = 0$.

The most general solution then is a combination of both: $\psi = \psi_1 + \psi_2$. In particular, we show that the free wave is also of this form, at $r \to \infty$, with given A and B,

$$e^{i\vec{k}\cdot\vec{r}} \simeq \frac{2\pi}{ik}\delta(\vec{n}_k - \vec{n}_r)\frac{e^{-ikr}}{r} - \frac{2\pi}{ik}\delta(\vec{n}_k + \vec{n}_r)\frac{e^{ikr}}{r}, \tag{46.15}$$

so that

$$A(k, \vec{n}_r) = \frac{2\pi}{ik}\delta(\vec{n}_k - \vec{n}_r), \quad B(k, -\vec{n}_r) = -\frac{2\pi}{ik}\delta(\vec{n}_k + \vec{n}_r). \tag{46.16}$$

Proof To prove this relation, we calculate the integral of the wave, on a sphere multiplied by an arbitrary function,

$$I = \int d^2\vec{n}_r f(\vec{n}_r) e^{i\vec{k}\cdot\vec{r}} = \int_0^\pi \sin\theta d\theta \int_0^{2\pi} d\phi e^{ikr\cos\theta} f(\theta, \phi). \tag{46.17}$$

Defining $x = \cos\theta$, we calculate

$$\begin{aligned} I &= \int_0^{2\pi} d\phi \int_{-1}^{1} e^{ikrx} f(x, \phi) = \int_0^{2\pi} \frac{d\phi}{ikr} \int_{-1}^{1} f(x, \phi) d\left(e^{ikrx}\right) \\ &= \int_0^{2\pi} \frac{d\phi}{ikr}\left[f(1, \phi)e^{ikr} - f(-1, \phi)e^{-ikr} - \int_{-1}^{1} e^{ikrx}\frac{df(x, \phi)}{dx}dx\right], \end{aligned} \tag{46.18}$$

where in the second line we have integrated by parts. Since at $\cos\theta = 1$ we have $\vec{n}_r = \vec{n}_k$, and at $\cos\theta = -1$ we have $\vec{n}_r = -\vec{n}_k$, it follows that $f(\pm 1, \phi) = f(\pm\vec{n}_k)$. Moreover, at $kr \gg 1$, $\int_{-1}^{1} e^{ikrx}$ is highly oscillatory, and therefore the integral is close to zero, or rather it is much less than the first two terms. These remaining terms are independent of ϕ, since $\vec{n}_k$ is a fixed direction in terms of $\vec{r}$, which means that we can trivially obtain $\int_0^{2\pi} d\phi = 2\pi$. Then finally, we have

$$I \simeq \frac{2\pi}{ikr}\left[f(\vec{n}_k)\frac{e^{ikr}}{r} + f(-\vec{n}_k)\frac{e^{-ikr}}{r}\right] \tag{46.19}$$

when $kr \gg 1$. We note then that this integral I gives the same result as when replacing $e^{i\vec{k}\vec{r}}$ with

$$\frac{2\pi}{ikr}\left[\delta(\vec{n}_k - \vec{n}_r)\frac{e^{ikr}}{r} - \delta(\vec{n}_k + \vec{n}_r)\frac{e^{-ikr}}{r}\right] \tag{46.20}$$

in the original integral, thus proving the relation. *q.e.d.*

46.3 Scattering Solution, Cross Sections, and S-Matrix

After seeing the behavior at infinity of the solution of the Schrödinger equation for a finite-range potential, we need to construct a physical set-up that will describe the scattering of particles, from the stationary (scattered wave) point of view.

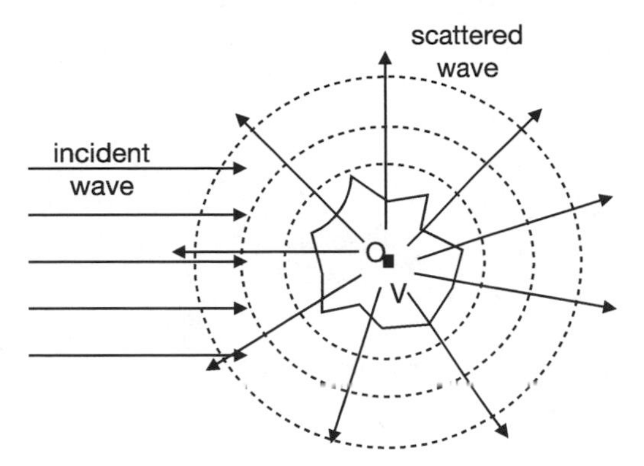

Figure 46.1 Generic scattering situation in three-dimensions with a potential V (the arbitrary shape with solid lines).

Certainly at $r \to \infty$ (away from the potential), we want to have an incident free particle wave, $e^{i\vec{k}\cdot\vec{r}}$. We have already seen that at infinity this wave decomposes into an in part and an out part, as for any wave. Yet, since we are in the stationary-wave point of view (with the time-independent Schrödinger equation), the wave should have a scattered component but that component must only be an outgoing wave (a diverging spherical wave), since it is generated at the moment of scattering; see Fig. 46.1. Then we have the following ansatz for the stationary-wave function at $r \to \infty$:

$$u^+_{\vec{k}}(\vec{r}) \simeq e^{i\vec{k}\cdot\vec{r}} + f_{\vec{k}}(\vec{n}_r)\frac{e^{ikr}}{r} = u_{\rm inc}(\vec{r}) + u_{\rm scatt}(\vec{r}). \tag{46.21}$$

We saw earlier that the probability current is

$$\vec{j} = \frac{\hbar}{2im}(\psi^*\vec{\nabla}\psi - \psi\vec{\nabla}\psi^*). \tag{46.22}$$

When substituting the above wave function, we obtain an incident current, a scattered current, and an interference term,

$$\vec{j}^{\,+}_{\vec{k}} \simeq \vec{j}_{\rm inc} + \vec{j}_{\rm scatt} + \vec{j}_{\rm interf} \tag{46.23}$$

when $r \to \infty$. Specifically, we find

$$\begin{aligned}\vec{j}_{\rm inc} &= \frac{\hbar\vec{k}}{m}\\ \vec{j}_{\rm scatt} &= \frac{\hbar k}{m}\vec{n}_r\frac{|f|^2}{r^2},\end{aligned} \tag{46.24}$$

where $\vec{n}_r = \vec{r}/r$. We note that $|u_{\rm scatt}|^2 = |f|^2/r^2$ is the probability density. Then the interference current term is

$$\vec{j}_{\rm interf} = -\frac{4\pi\hbar}{mr^2}\frac{\vec{k}}{k}({\rm Im} f)\delta(\vec{n}_k - \vec{n}_r), \tag{46.25}$$

which means that we can neglect it except in the forward direction.

We saw in Chapter 37 that the definition of the differential cross section is

$$d\sigma = \frac{dP_{\rm scatt}(d\Omega, dt)/dt}{j_{\rm inc}}, \tag{46.26}$$

and the incident current is $j_{\rm inc} = \hbar k/m$, while the probability current (flow rate) through the surface $dS = r^2 d\Omega$ is

$$dP_{\rm scatt}(d\Omega, dt)/dt = j_{\rm scatt} dS = j_{\rm scatt} r^2 d\Omega = \frac{\hbar k}{m} |f|^2 d\Omega. \tag{46.27}$$

Thus we obtain

$$\frac{d\sigma}{d\Omega} = |f|^2. \tag{46.28}$$

The conservation of probability equation is

$$\vec{\nabla} \cdot \vec{j} + \frac{\partial |\psi|^2}{\partial t} = 0. \tag{46.29}$$

In the stationary (time-independent) case we have $\partial|\psi|^2/\partial t = 0$, so after integrating over a volume V bounded by Σ_∞, we obtain

$$0 = \int_V d^3 r \vec{\nabla} \cdot \vec{j} = \int_{\Sigma_\infty} d\vec{S} \cdot \vec{j} = \int_{\Sigma_\infty} dS j_r = \frac{\hbar}{2im} \int_{\Sigma_\infty} r^2 d\Omega (\psi^* \partial_r \psi - \psi \partial_r \psi^*). \tag{46.30}$$

We can choose the whole of the Σ_∞ boundary to be within the $r \to \infty$ limit, in which case we can replace ψ with its general asymptotical form in terms of $e^{\pm ikr}/r$ with coefficients A and B. Then the conservation of probability becomes

$$\begin{aligned} \frac{\hbar}{2im} \int d\Omega r^2 \frac{2ik}{r^2} [A^* A - B^* B] = 0 \\ \Rightarrow \int d\Omega |A|^2 = \int d\Omega |B|^2. \end{aligned} \tag{46.31}$$

This means that knowing one of the coefficients A, B implies a value for the average of the other, and thus both coefficients must be nonzero. Moreover, it means that there exists a linear relation between them. Indeed, if we have several waves expanded in this way,

$$\psi_n \simeq A_n \frac{e^{ikr}}{r} + B_n \frac{e^{-ikr}}{r}, \tag{46.32}$$

then linear combination of the waves means linear combinations for the coefficients,

$$\sum_n C_n \psi_n \simeq \left(\sum_n C_n A_n \right) \frac{e^{ikr}}{r} + \left(\sum_n C_n B_n \right) \frac{e^{-ikr}}{r}, \tag{46.33}$$

which in turn implies that

$$\int d\Omega \left| \sum_n C_n A_n \right|^2 = \int d\Omega \left| \sum_n C_n B_n \right|^2. \tag{46.34}$$

This is only possible if the out coefficient is an operator times the in coefficient,

$$A = \hat{S} \cdot B, \tag{46.35}$$

which more precisely leads to

$$A(\vec{n}') = \int d^2 \vec{n}\, S(\vec{n}', \vec{n}) B(\vec{n}), \tag{46.36}$$

where we have generalized from $\vec{n}' = \vec{n}_r, \vec{n} = -\vec{n}_r$ to general $\vec{n}, \vec{n}'$.

However, to satisfy the law of conservation of probability $\int d\Omega |A|^2 = \int d\Omega |B|^2$, we also require that $\hat{S}$ is a unitary operator, $\hat{S}^\dagger = \hat{S}^{-1}$ (obtained by substituting $A = \hat{S} \cdot B$ into the conservation law). This $\hat{S}$ refers to the S-operator, or S-matrix.

We note that this is a generalization of the one-dimensional case, where there are only two states $|k\pm\rangle$, to the case of an infinity of states, called "channels", for the in and out states.

46.4 S-Matrix and Optical Theorem

We now expand the free wave $e^{i\vec{k}\cdot\vec{r}}$ in the scattering stationary wave (46.21) into in and out components, changing the notation as follows, $f_{\vec{k}}(\vec{n}_r) = f_k(\vec{n}_r, \vec{n}_k)$ and $u^+_{\vec{k}}(\vec{r}) = u^+_k(\vec{n}_r, \vec{n}_k; r)$, in order to emphasize the dependence on angles. We obtain

$$u^+_k(\vec{n}_r, \vec{n}_k; r) \simeq \left[\frac{2\pi}{ik}\delta(\vec{n}_r - \vec{n}_k) + f_k(\vec{n}_r, \vec{n}_k)\right]\frac{e^{ikr}}{r} + \frac{2\pi}{ik}\delta(-\vec{n}_r - \vec{n}_k)\frac{e^{-ikr}}{r}. \tag{46.37}$$

We generalize $\vec{n}_r$ to $\vec{n}'$ and $-\vec{n}_r$ to $\vec{n}$ as before, leading to

$$\begin{aligned} A(\vec{n}') &= \frac{2\pi}{ik}\delta(\vec{n}' - \vec{n}_k) + f_k(\vec{n}', \vec{n}_k) \\ B(\vec{n}) &= \frac{2\pi}{ik}\delta(\vec{n} - \vec{n}_k). \end{aligned} \tag{46.38}$$

Then the S-matrix relation $A = \hat{S} \cdot B$ becomes

$$\begin{aligned} A(\vec{n}') &= \frac{2\pi}{ik}\delta(\vec{n}' - \vec{n}_k) + f_k(\vec{n}', \vec{n}_k) \\ &= \frac{2\pi}{ik}\int d^2\vec{n} S(\vec{n}', \vec{n})\delta(\vec{n} - \vec{n}_k) = \frac{2\pi}{ik}S(\vec{n}', \vec{n}). \end{aligned} \tag{46.39}$$

Therefore the S-matrix contains a trivial (identity) part, plus a nontrivial part,

$$S(\vec{n}_r, \vec{n}_k) = \delta(\vec{n}_r - \vec{n}_k) + \frac{ik}{2\pi}f_k(\vec{n}_r, \vec{n}_k), \tag{46.40}$$

or, in operator terms,

$$\hat{S} = \mathbb{1} + \frac{ik}{2\pi}\hat{f}. \tag{46.41}$$

Here S and f are *amplitudes for scattering*, with S containing also the trivial (not the interaction) part. We will see that $\hat{f}$ is related to the T-operator defined in the previous chapter.

Since $\hat{S}$ is a unitary operator, as we saw before, meaning that it conserves probability as it relates the in and out states, we find

$$\begin{aligned} \mathbb{1} = \hat{S}\hat{S}^\dagger &= \left(\mathbb{1} + \frac{ik}{2\pi}\hat{f}\right)\left(\mathbb{1} - \frac{ik}{2\pi}\hat{f}^\dagger\right) \\ &= \mathbb{1} + \frac{ik}{2\pi}(\hat{f} - \hat{f}^\dagger) + \left(\frac{k}{2\pi}\right)^2 \hat{f}\hat{f}^\dagger. \end{aligned} \tag{46.42}$$

Finally, we obtain

$$\frac{\hat{f} - \hat{f}^\dagger}{2i} = \frac{k}{4\pi}\hat{f}\hat{f}^\dagger. \tag{46.43}$$

Changing from operators to matrices,

$$\hat{f} = f(\vec{n}', \vec{n}_k) \Rightarrow \hat{f}^\dagger = f^*(\vec{n}_k, \vec{n}'), \tag{46.44}$$

where $\vec{n}' = \vec{n}_r$, and considering forward scattering, so that $\vec{n}_r = \vec{n}_k$, we obtain

$$\mathrm{Im} f(\vec{n}_k, \vec{n}_k) = \frac{k}{4\pi} \int d^2\vec{n}_r |f(\vec{n}_r, \vec{n}_k)|^2 = \frac{k}{4\pi}\sigma_{\text{tot}}, \tag{46.45}$$

where in the last equality we have used the fact that $d^2\vec{n}_r = d\Omega$ and $d\sigma/d\Omega = |f|^2$.

The above form is the most common (though not the most general) form of the *optical theorem* for scattering.

46.5 Green's Functions and Lippmann–Schwinger Equation

We now turn to equations for the wave functions, specifically the generalization to three dimensions of the Lippmann–Schwinger equation. Now the relations we will prove are valid even for the case of a potential with an infinite range, as in the Coulomb case.

We trivially generalize the one-dimensional case from the previous chapter to define the Green's functions in an abstract form

$$(E - \hat{H}_0)\hat{G}_0 = \frac{\hbar^2}{2m}\mathbb{1}. \tag{46.46}$$

However, the coordinate representation is also trivially generalized, since

$$\begin{aligned} \langle \vec{r}|\hat{G}_0|\vec{r}'\rangle &= G_0(\vec{r}, \vec{r}') \\ \langle \vec{r}|\mathbb{1}|\vec{r}'\rangle &= \delta^3(\vec{r} - \vec{r}'). \end{aligned} \tag{46.47}$$

Continuing to use the abstract form, we find

$$\hat{G}_0 = \frac{\hbar^2}{2m}\frac{\mathbb{1}}{E - \hat{H}_0}, \tag{46.48}$$

or, more precisely, since we need to avoid the singularities appearing in the solutions of the (free) Schrödinger equation, $(E - \hat{H}_0)|\psi_{\vec{k}}\rangle = 0$, we need to add $\pm i\epsilon$ to the energy, leading to two Green's functions,

$$\hat{G}_0^{(\pm)} = \lim_{\epsilon\to 0}\frac{\hbar^2}{2m}\frac{1}{E \pm i\epsilon - \hat{H}_0}. \tag{46.49}$$

Also as in the one-dimensional case (trivially generalized in the abstract form) we find the Lippmann–Schwinger equation,

$$|\psi\pm\rangle = |\vec{k}\rangle + \frac{2m}{\hbar^2}\hat{G}_0^{(\pm)}\hat{V}|\psi\pm\rangle, \tag{46.50}$$

where $|\vec{k}\rangle$ is the free particle state of wave vector $\vec{k}$, $(E - \hat{H}_0)|\vec{k}\rangle = 0$. Note here that we have defined $|\psi\pm\rangle$ as the state corresponding to $\hat{G}_0^{(\pm)}$, though we will soon see that it also corresponds to having an outgoing or incoming scattered wave, meaning that $\psi_1 = \psi_+$, $\psi_2 = \psi_-$ are the solutions at infinity.

Proof To prove the Lippmann–Schwinger equation, we multiply it by $(E \pm i\epsilon - \hat{H}_0)$, obtaining

$$\begin{aligned}&(E \pm i\epsilon - \hat{H}_0)|\psi\pm\rangle = (E \pm i\epsilon - \hat{H}_0)|\vec{k}\rangle + (E \pm i\epsilon - \hat{H}_0)\hat{G}_0\frac{2m}{\hbar^2}\hat{V}|\psi\pm\rangle \\ \Rightarrow\quad &(E \pm i\epsilon - \hat{H}_0 - \hat{V})|\psi\pm\rangle = 0,\end{aligned} \tag{46.51}$$

where we have used the fact that the first term in the first line vanishes, by the free particle Schrödinger equation, and in the second, we have used the definition of $\hat{G}_0$ to write it as $\hat{V}|\psi\pm\rangle$.
q.e.d.

Note that we can define $\hat{G}_0$ fully on the complex plane, not just an infinitesimal distance away from the real line,

$$\hat{G}_0(z) = \frac{\hbar^2}{2m}\frac{1}{z - \hat{H}_0}, \tag{46.52}$$

and then the Lippmann–Schwinger equation in this more general case is

$$|\psi(z)\rangle = |\vec{k}_z\rangle + \frac{2m}{\hbar^2}\hat{G}_0(z)\hat{V}|\psi(z)\rangle. \tag{46.53}$$

Then in coordinate space, i.e., multiplying with $\langle\vec{r}|$, we get

$$\psi_z(\vec{r}) = \phi_z(\vec{r}) + \frac{2m}{\hbar^2}\int d^3\vec{r}\,'G_0(z;\vec{r},\vec{r}\,')V(\vec{r}\,')\psi_z(\vec{r}\,'). \tag{46.54}$$

We go back now and specialize to $z = E \pm i\epsilon$, which means that

$$\psi_\pm(\vec{r}) = \phi_E(\vec{r}) + \frac{2m}{\hbar^2}\int d^3\vec{r}\,'G_0^{(\pm)}(\vec{r},\vec{r}\,')V(\vec{r}\,')\psi_\pm(\vec{r}\,'). \tag{46.55}$$

The (free) Green's function in coordinate space,

$$\frac{2m}{\hbar^2}G_0^{(\pm)}(\vec{r},\vec{r}\,') = \left\langle\vec{r}\left|\frac{1}{E \pm i\epsilon - \hat{H}_0}\right|\vec{r}\,'\right\rangle \tag{46.56}$$

is not diagonal but in momentum space it is diagonal, since

$$\begin{aligned}\frac{2m}{\hbar^2}G_0^{(\pm)}(\vec{p},\vec{p}\,') &= \left\langle\vec{p}\left|\frac{1}{E \pm i\epsilon - \hat{H}_0}\right|\vec{p}\,'\right\rangle \\ &= \frac{\langle\vec{p}|\vec{p}\,'\rangle}{E \pm i\epsilon - p^2/2m} = \frac{1}{E \pm i\epsilon - p^2/2m}\delta^3(\vec{p} - \vec{p}\,').\end{aligned} \tag{46.57}$$

This means that we can obtain a simpler form for the Lippmann–Schwinger equation in momentum space,

$$\begin{aligned}\psi_\pm(\vec{p}) &= \phi_E(\vec{p}) + \frac{2m}{\hbar^2}\int d^3\vec{p}\,'G_0^{(\pm)}(\vec{p},\vec{p}\,')V(\vec{p}\,',\psi\pm) \\ &= \phi_E(\vec{p}) + \frac{1}{E \pm i\epsilon - \frac{p^2}{2m}}V(\vec{p}\,',\psi\pm),\end{aligned} \tag{46.58}$$

where

$$V(\vec{p}\,',\psi\pm) \equiv \langle\vec{p}\,'|\hat{V}|\psi\pm\rangle. \tag{46.59}$$

To obtain an explicit form for the Lippmann–Schwinger equation in coordinate space, we must first find the (free) Green's function in coordinate space. We insert momentum space completeness relations in order to relate it to the momentum space Green's function, obtaining

$$\begin{aligned}G_0^{(\pm)}(\vec r,\vec r\,') &= \int d^3p\int d^3p'\langle\vec r|\vec p\rangle\langle\vec p|\hat G_0^{(\pm)}|\vec p\,'\rangle\langle\vec p\,'|\vec r\,'\rangle\\ &= \frac{\hbar^2}{2m}\int d^3p\int d^3p'\frac{e^{i\vec p\cdot\vec r/\hbar}}{(2\pi\hbar)^{3/2}}\frac{\delta^3(\vec p-\vec p\,')}{E\pm i\epsilon-(p^2/2m)}\frac{e^{-i\vec p\,'\cdot\vec r\,'/\hbar}}{(2\pi\hbar)^{3/2}}\\ &= \frac{\hbar^2}{2m}\int\frac{d^3p}{(2\pi\hbar)^3}\frac{e^{i\vec p\cdot(\vec r-\vec r\,')/\hbar}}{E\pm i\epsilon-p^2/2m}\\ &= \frac{1}{(2\pi)^3}\int_0^\infty q^2dq\int_0^{2\pi}d\phi\int_0^\pi d\theta\sin\theta\frac{e^{iq|\vec r-\vec r\,'|\cos\theta}}{k^2-q^2\pm i\epsilon},\end{aligned}\tag{46.60}$$

where the θ,ϕ angles refer to the $\vec r-\vec r\,'$ vector relative to the $\vec p$ vector, $E=\hbar^2k^2/(2m)$, and $p=\hbar q$. Integrating using $\int_0^{2\pi}d\phi=2\pi$ and $\int_0^\pi d\theta\sin\theta=\int_{-1}^1 d(\cos\theta)$, we find

$$\begin{aligned}G_0^{(\pm)}(\vec r,\vec r\,') &= \frac{1}{4\pi^2}\int_0^\infty q^2dq\frac{e^{iq|\vec r-\vec r\,'|}-e^{-iq|\vec r-\vec r\,'|}}{iq|\vec r-\vec r\,'|(k^2-q^2\pm i\epsilon)}\\ &= -\frac{1}{8\pi^2 i|\vec r-\vec r\,'|}\int_{-\infty}^{+\infty}qdq\frac{e^{iq|\vec r-\vec r\,'|}-e^{-iq|\vec r-\vec r\,'|}}{(q^2-k^2\mp i\epsilon)},\end{aligned}\tag{46.61}$$

where in the last equality we have used $\frac{1}{2}\int_{-\infty}^{+\infty}qdq=\int_0^\infty qdq$. The integral has poles at

$$q=\pm\sqrt{k^2\pm i\epsilon}=\pm k(1\pm i\epsilon').\tag{46.62}$$

To calculate the above integral, we use the standard residue theorem on the complex plane. We first extend the integral over the complex plane, and then close the contour of integration over the real line with a semicircle at infinity in the upper half-plane for the term with $e^{iq|\vec r-\vec r\,'|}$, since if $|q|\to\infty$ in the upper half-plane, it becomes $\sim e^{-(\mathrm{Im}\,q)|\vec r-\vec r\,'|}\to 0$. We close the contour with a semicircle at infinity in the lower half plane for the term with $e^{-iq|\vec r-\vec r\,'|}$, since if $|q|\to\infty$ in the lower half-plane, the term becomes $\sim e^{+(\mathrm{Im}\,q)|\vec r-\vec r\,'|}\to 0$. This means that addition of the semicircles does not change the final result; however, now that the contour is closed we can use the residue theorem, and find that the integral is equal to $2\pi i$ times the residue(s) inside the contour, with a plus sign for counterclockwise contour integration and a minus for clockwise contour integration.

That means that the integral over the real line times $\frac{1}{2\pi i}$ gives, in the cases of $G_0^{(\pm)}$, the following results:

$$\begin{aligned}G_0^{(+)} &\to \frac{k}{2k}e^{ik|\vec r-\vec r\,'|}-(-)\frac{(-k)}{2(-k)}e^{-i(-k)|\vec r-\vec r\,'|}=e^{ik|\vec r-\vec r\,'|}\\ G_0^{(-)} &\to \frac{-k}{-2k}e^{i(-k)|\vec r-\vec r\,'|}-(-)\frac{k}{2k}e^{-ik|\vec r-\vec r\,'|}=e^{-ik|\vec r-\vec r\,'|},\end{aligned}\tag{46.63}$$

where in the first line the first term comes from the $+k+i\epsilon$ residue and has a counterclockwise contour and the second term comes from the $-k-i\epsilon$ residue and has a clockwise contour; in the second line the first term comes from the $-k+i\epsilon$ residue and has a counterclockwise contour and the second term comes from the $+k-i\epsilon$ residue and has a clockwise contour, as in Fig. 46.2.

Then we find the coordinate space Green's function

$$G_0^{(\pm)}(\vec r,\vec r\,')=-\frac{1}{4\pi|\vec r-\vec r\,'|}e^{\pm ik|\vec r-\vec r\,'|}.\tag{46.64}$$

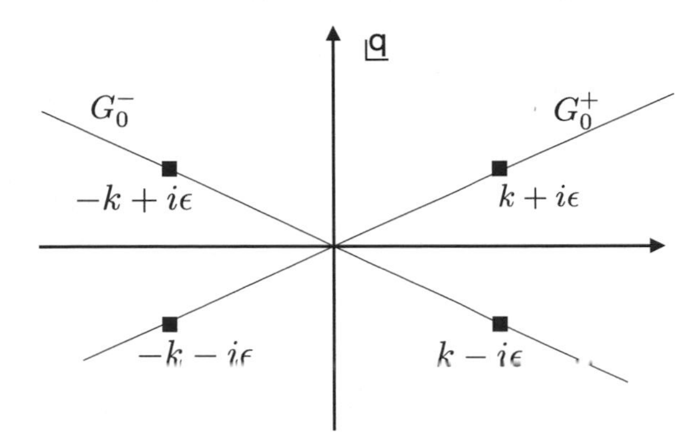

Figure 46.2 Complex poles relevant for G_0^+ and G_0^-.

We note that this is the same Green's function as for the equation

$$(\Delta + k^2)G = \delta^3(\vec{r} - \vec{r}\,'). \tag{46.65}$$

Now we are ready to find an explicit form for the coordinate-space Lippmann–Schwinger equation. Substituting $G_0^{(\pm)}(\vec{r}, \vec{r}\,')$ into (46.55) and substituting $\phi_E(\vec{r}) = e^{i\vec{k}\cdot\vec{r}}$, we find

$$\psi_\pm(\vec{r}) = e^{i\vec{k}\cdot\vec{r}} - \frac{2m}{4\pi\hbar^2}\int d^3\vec{r}\,' \frac{e^{\pm ik|\vec{r}-\vec{r}\,'|}}{|\vec{r}-\vec{r}\,'|} V(\vec{r}\,')\psi_\pm(\vec{r}\,'). \tag{46.66}$$

Specializing to a potential of finite range, so that $\vec{r}\,'$ belongs to the finite volume V that is the domain of the potential, it follows that for $r \to \infty$ we have

$$|\vec{r} - \vec{r}\,'| \simeq \sqrt{r^2 - 2rr'\cos\alpha} \simeq r\left(1 - \frac{r'}{r}\cos\alpha\right) = r - \vec{n}_r \cdot \vec{r}\,', \tag{46.67}$$

where α is the angle between $\vec{r}$ and $\vec{r}\,'$, and $\vec{n}_r = \vec{r}/r$. Then at $r \to \infty$ we keep in the exponential both the infinite phase and the finite phase (the first two terms in the expansion) since the latter is integrated over,

$$e^{\pm ik|\vec{r}-\vec{r}\,'|} \simeq e^{\pm ikr} e^{\mp ik\vec{n}_r\cdot\vec{r}\,'}, \tag{46.68}$$

and in the denominator we can replace $|\vec{r} - \vec{r}\,'|$ with the leading term r, since we only are interested in the leading behavior of the integral.

Then we obtain

$$\psi_\pm(\vec{r}) \simeq e^{i\vec{k}\cdot\vec{r}} - \frac{e^{\pm ikr}}{r}\frac{2m}{4\pi\hbar^2}\int d^3\vec{r}\,' e^{\mp i\vec{k}\,'\cdot\vec{r}\,'} V(\vec{r}\,')\psi_\pm(\vec{r}\,'), \tag{46.69}$$

where we have defined

$$\vec{k}\,' \equiv k\vec{n}_r. \tag{46.70}$$

We note that we have reached the general scattering wave function in the stationary point of view if we make the identification

$$f_k^\pm(\vec{n}_r, \vec{n}_k) \equiv f^{(\pm)}(\vec{k}\,', \vec{k}) \equiv -\frac{2m}{4\pi\hbar^2}\int d^3\vec{r}\,' e^{\mp i\vec{k}\,'\cdot\vec{r}\,'} V(\vec{r}\,')\psi_\pm(\vec{r}\,'), \tag{46.71}$$

corresponding to an outgoing or incoming wave $\psi_\pm$: that is, the state $|\psi\pm\rangle$, previously defined by the term $\pm i\epsilon$ in the Green's function, actually corresponds to the scattered wave function term

$e^{\pm ikr} f^{(\pm)}/r$. Then it follows that, since in most cases we are interested only in the outgoing case $|\psi+\rangle$, we are more interested in the Green's function $G_0^{(+)}$.

Important Concepts to Remember

- For three-dimensional scattering, at $r \to \infty$ we have two asymptotic solutions, outgoing or "out", $\psi_1 = A(k, \vec{n}_r)e^{ikr}/r$, and incoming or "in", $\psi_2 = B(k, -\vec{n}_r)e^{-ikr}/r$.
- A free wave $e^{i\vec{k}\cdot\vec{r}}$ is expanded at infinity in the in and out waves, with $A(k, \vec{n}_r) = (2\pi/ik)\delta(\vec{n}_k - \vec{n}_r)$ and $B(k, -\vec{n}_r) = -(2\pi/ik)\delta(\vec{n}_k + \vec{n}_r)$.
- In the case of scattering in a potential, we have at infinity the stationary wave function ansatz $u^{\pm}_{\vec{k}}(\vec{r}) = e^{i\vec{k}\cdot\vec{r}} + f_{\vec{k}}(\vec{n}_r)e^{ikr}/r = u_{\rm inc}(\vec{r}) + u_{\rm scatt}(\vec{r})$, and differential cross section $d\sigma/d\Omega = |f|^2$.
- For a generic in plus out wave, $\psi = Ae^{ikr}/r + Be^{-ikr}/r$, we have $A = \hat{S} \cdot B$, or $A(\vec{n}) = \int d\vec{n}' S(\vec{n}, \vec{n}')B(\vec{n}')$, where $\hat{S}$ is the unitary S-matrix or S-operator.
- Then we have $\hat{S} = \mathbb{1} + (ik/2\pi)\hat{f}$, or $S(\vec{n}_r, \vec{n}_k) = \delta(\vec{n}_r - \vec{n}_k) + (ik/2\pi)f(\vec{n}_r, \vec{n}_k)$, and the optical theorem, ${\rm Im} f(\vec{n}_r, \vec{n}_k) = k/4\pi \int d\vec{n}_r |f|^2 = (k/4\pi)\sigma_{\rm tot}$.
- The Green's functions and Lippmann–Schwinger equations in the abstract form are trivially generalized from one dimension, and we can further generalize to the complex plane,

$$\hat{G}_0(z) = \frac{\hbar^2}{2m}\frac{1}{z - \hat{H}_0}.$$

- In the coordinate representation, the Lippmann–Schwinger equation becomes

$$\psi_\pm(\vec{r}) = \phi_E(\vec{r}) + \frac{2m}{\hbar^2}\int d^3\vec{r}' G_0^\pm(\vec{r}, \vec{r}')V(\vec{r}')\psi_\pm(\vec{r}'),$$

with

$$G_0^\pm(\vec{r}, \vec{r}') = -\frac{1}{4\pi|\vec{r} - \vec{r}'|}e^{\pm ik|\vec{r}-\vec{r}'|}.$$

Further Reading

See [2] and [1].

Exercises

(1) Consider a central potential of the spherical-step type, $V = V_0 > 0$ for $r \leq R$ and $V = 0$ for $r > R$, and a solution with energy $E > V_0$ and given angular momentum $l > 0$. If the solution is assumed to be square integrable at $r = 0$, calculate the coefficients A and B at infinity.

(2) The decomposition of the free wave at infinity, (46.15), seems counterintuitive since the left-hand side is certainly nonzero (in fact, naively, it is of order 1!) if $\vec{k}$ is other than parallel or antiparallel to $\vec{r}$. In what sense should we understand this relation, then?

(3) If the $f_{\vec{k}}(\vec{n}_r)$ in (46.21) were independent of $\vec{n}_r$, would that contradict unitarity or not?

(4) For the case in exercise 1, calculate the total cross section and the S-matrix.

(5) Consider the wave $\psi = Ae^{ikr/r} + Be^{-ikr/r}$, with $A, B = (2\pi/ik)\delta(\vec{n}_k \pm \vec{n}_r) + a, b$, where a, b are real constants. Can it be understood as a scattering solution? If so, calculate the total cross section. What if a, b are imaginary constants?

(6) Calculate the generalization of $G_0^{\pm}(\vec{r}, \vec{r}\,')$ on the complex plane for energy, $G_0(z; \vec{r}, \vec{r}\,')$.

(7) Calculate the equivalent of (46.69) for the generalization $G_0(z; \vec{r}, \vec{r}\,')$ in exercise 6.

47 Born Approximation and Series, S-Matrix and T-Matrix

In this chapter, we define the Born approximation and series of higher-order approximations, and connect with the time-dependent point of view for scattering, where we define the S- and T-matrices.

47.1 Born Approximation and Series

To solve the Lippmann–Schwinger integral equation, we use the same procedure as in the one-dimensional case. We define the zeroth-order solution, namely the free (noninteracting) wave,

$$\psi_{\pm}^{(0)}(\vec{r}) = \phi_E(\vec{r}) = e^{i\vec{k}\cdot\vec{r}}. \tag{47.1}$$

Then we substitute this solution into the right-hand side of the Lippmann–Schwinger equation to find the first order interacting solution,

$$f^{(1,\pm)}(\vec{k}',\vec{k}) = -\frac{2m}{4\pi\hbar^2}\int d^3\vec{r}\,' e^{i\vec{r}\,'\cdot(\vec{k}\mp\vec{k}')}V(\vec{r}\,') = -\frac{2m}{4\pi\hbar^2}V(\vec{k}'\mp\vec{k}), \tag{47.2}$$

where we have defined the Fourier-transformed potential $V(\vec{q})$, where $\vec{q} = \vec{k}' - \vec{k}$ is the momentum transfer. That means that the first-order term in the wave function solution is

$$\begin{aligned}\psi_{\pm}^{(1)}(\vec{r}) &= -\frac{2m}{4\pi\hbar^2}\int d^3\vec{r}\,'\frac{e^{\pm ikr}}{r}e^{i\vec{r}\,'\cdot(\vec{k}\mp\vec{k}')}V(\vec{r}\,')\\ &= \int d^3\vec{r}\,' G_0^{(\pm)}(\vec{r},\vec{r}\,')e^{i\vec{k}\cdot\vec{r}\,'}V(\vec{r}\,').\end{aligned} \tag{47.3}$$

We can then put this into the right-hand side of the Lippmann–Schwinger equation to find the second-order term in the wave function solution, etc.

Generally, though restricting to the physical + solution (and dropping the index when doing so), we find the recursion relation for the $(n+1)$th-order term,

$$\psi^{(n+1)}(\vec{r}) = \int d^3\vec{r}\,'\frac{2m}{\hbar^2}G_0(\vec{r},\vec{r}\,')\psi^{(n)}(\vec{r}\,')V(\vec{r}\,'). \tag{47.4}$$

This recursion relation defines the *Born series*. The first-order term is the *Born approximation*.

We apply the Born approximation to a spherically symmetric potential, $V(\vec{r}) = V(r)$. Since $|\vec{k}'| = |\vec{k}| = k$, and defining θ as the angle between $\vec{n}_r$ and $\vec{n}_k$, we have (see Fig. 47.1)

$$|\vec{k}' - \vec{k}| = 2k\sin\frac{\theta}{2} = q. \tag{47.5}$$

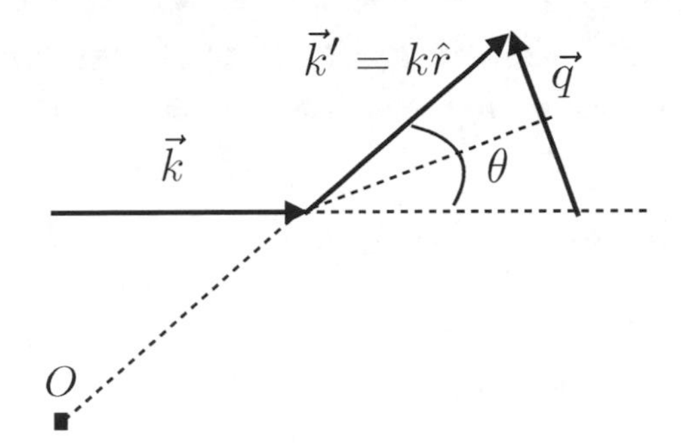

Figure 47.1 Geometry of scattering.

The first-order solution is then

$$\begin{aligned} f^{(1)} &= -\frac{2m}{4\pi\hbar^2}\int_0^\infty r'^2 dr' \int_0^{2\pi} d\phi \int_{-1}^{1} d\cos\theta\, e^{iqr'\cos\theta} V(r') \\ &= -\frac{2m}{\hbar^2}\int_0^\infty r'^2 dr' V(r') \frac{e^{iqr'} - e^{-iqr'}}{2iqr'} \\ &= -\frac{2m}{4\pi\hbar^2 q} 4\pi \int_0^\infty r' dr' V(r') \sin(qr') \\ &= -\frac{2m}{4\pi\hbar^2} V(q), \end{aligned} \tag{47.6}$$

where in the last line we have defined $V(q)$, which matches the first-order term (47.2), specialized to the spherically symmetric case, as well as the same formula in the derivation of the Born approximation through the Fermi golden rule, in Chapter 37. However, note that there we had a different normalization, with a $1/(2\pi\hbar)^{3/2}$ factor.

We will take an example, the Yukawa potential

$$V = V_0 \frac{e^{-\mu r}}{r}, \tag{47.7}$$

as in Chapter 37. There, we left as an exercise the proof of

$$V(q) = \frac{4\pi V_0}{\mu^2 + q^2}, \tag{47.8}$$

which leads to the Born approximation for the wave function,

$$f^{(1)} = -\left(\frac{2mV_0}{\hbar^2}\right) \frac{1}{\mu^2 + q^2}. \tag{47.9}$$

Then the Born approximation for the differential cross section,

$$\frac{d\sigma}{d\Omega} = |f|^2 = \left(\frac{m}{2\pi\hbar^2}\right)^2 |V(q)|^2, \tag{47.10}$$

matches the Chapter 37 calculation using Fermi's golden rule.

47.2 Time-Dependent Scattering Point of View

From the Fermi's golden rule calculation of the Born approximation, which used first-order time-dependent perturbation theory, we realize that we can use a time-dependent point of view for scattering.

We review this calculation (though within our new formalism) for completeness. We define the S-matrix as in the one-dimensional case in Chapter 45,

$$\hat{S} = \hat{U}_I(+\infty, -\infty), \tag{47.11}$$

which however is still the same matrix as that defined in the previous chapter, $A_{\text{out}} = \hat{S} \cdot B_{\text{in}}$, and then the probability of transition is

$$\text{Probab.}(\vec{p}_{\text{in}} \to d\Omega \text{ around } \vec{p}_f) = d\Omega |\langle \vec{p}_f | \hat{S} | \vec{p}_i \rangle|^2. \tag{47.12}$$

However in time-dependent perturbation theory, at first order,

$$\hat{U}_I(+\infty, -\infty) = \mathbb{1} - \frac{i}{\hbar} \int_{-\infty}^{+\infty} dt V_I(t), \tag{47.13}$$

as we saw in Chapter 38. Then we have

$$\frac{dP}{dt} = \frac{2\pi}{\hbar} |\langle \vec{p}_f | \hat{V} | \vec{p}_i \rangle|^2 \rho(E_i, \vec{n}) d\Omega, \tag{47.14}$$

and the analysis in Chapter 37 follows.

In the case of scattering in a Coulomb potential, obtained as the $\mu \to 0$ limit of the Yukawa potential, so that

$$V(q) = \frac{4\pi V_0}{q^2}, \tag{47.15}$$

we obtain the differential cross section

$$\frac{d\sigma}{d\Omega} = \left(\frac{m}{2\pi\hbar^2}\right)^2 \left(\frac{4\pi V_0}{q^2}\right)^2 = \frac{4m^2 V_0^2}{\hbar^4 q^4} = \frac{m^2 V_0^2}{\hbar^4 4k^4 \sin^4 \theta/2}, \tag{47.16}$$

which is the Rutherford formula for $V_0 = Zze_0^2$, derived *classically* by Rutherford (so the quantum mechanics approach to the first-order Born approximation doesn't improve upon it).

47.3 Higher-Order Terms and Abstract States, S- and T-Matrices

We saw in the previous chapter that the differential cross section is

$$\frac{d\sigma}{d\Omega} = |f^+|^2 \tag{47.17}$$

and, from the Lippmann–Schwinger equation,

$$f^+(\vec{k}', \vec{k}) = -\frac{2m}{4\pi\hbar^2} \int d^3\vec{r}' e^{-i\vec{k}' \cdot \vec{r}'} V(\vec{r}') \psi_+(\vec{r}'). \tag{47.18}$$

In terms of abstract operators and states, we have

$$f^+(\vec{k}', \vec{k}) = \frac{m}{2\pi\hbar^2}\langle \vec{k}'|\hat{V}|\psi+\rangle, \tag{47.19}$$

so

$$\frac{d\sigma}{d\Omega} = \left(\frac{m}{2\pi\hbar^2}\right)^2 \left|\langle \vec{k}'|\hat{V}|\psi+\rangle\right|^2, \tag{47.20}$$

where $\langle \vec{k}'| = \langle \vec{p}_f|$ is a final momentum state, so the formula above generalizes the Born approximation formula above (47.10), which can be written as

$$\frac{d\sigma}{d\Omega} = \left(\frac{m}{2\pi\hbar^2}\right)^2 \left|\langle \vec{p}_f|\hat{V}|\vec{p}_i\rangle\right|^2. \tag{47.21}$$

However, $\langle \vec{k}'|\hat{V}|\psi+\rangle$ is not a matrix element in the Hilbert space basis, since we have two different kinds of states on the left and on the right. In order to have a matrix element in the Hilbert space basis, we need to define, as in the one-dimensional case in Chapter 45, the operator T:

$$\hat{V}|\psi+\rangle \equiv \hat{T}|\vec{k}\rangle, \tag{47.22}$$

i.e., we need to replace the action on an interacting state by an action on the free state $|\vec{k}\rangle$. Then

$$\langle \vec{k}'|\hat{V}|\psi+\rangle = \langle \vec{k}'|\hat{T}|\vec{k}\rangle \tag{47.23}$$

is a matrix element in the Hilbert space basis.

But since the density of states is, as we saw in Chapter 37,

$$\rho(E) = \frac{m\hbar k}{(2\pi\hbar)^3}, \tag{47.24}$$

and since the velocity is $v = \hbar k/m$, we can rewrite the differential cross section in terms of the transition matrix, or T-matrix (the matrix element of the T-operator),

$$\frac{d\sigma}{d\Omega} = \frac{2\pi}{\hbar v}\rho(E)|\langle \vec{k}'|\hat{V}|\psi+\rangle|^2 = \frac{2\pi}{\hbar v}\rho(E)|\langle \vec{k}'|\hat{T}|\vec{k}\rangle|^2. \tag{47.25}$$

Then we can identify the amplitude for scattering, f^+, with the T-matrix, up to a constant,

$$f^+(\vec{k}', \vec{k}) = \frac{m}{2\pi\hbar^2}\langle \vec{k}'|\hat{T}|\vec{k}\rangle, \tag{47.26}$$

or, for operators,

$$\hat{f} = -\frac{m}{2\pi\hbar^2}\hat{T}. \tag{47.27}$$

Moreover, from (46.41), we can relate the S-operator to the T-operator (and the S-matrix to the T-matrix), since

$$\hat{S} = \mathbb{1} + \frac{ik}{2\pi}\hat{f} = \mathbb{1} - \frac{imk}{(2\pi\hbar)^2}\hat{T}. \tag{47.28}$$

As in the one-dimensional case (a trivial generalization), the abstract Lippmann–Schwinger equation, and its "solution" through the T-operator, is

$$\begin{aligned} |\psi\pm\rangle &= |\vec{k}\rangle + \left(\frac{2m}{\hbar^2}\hat{G}_0^{(\pm)}\right)\hat{V}|\psi\pm\rangle \\ &= |\vec{k}\rangle + \left(\frac{2m}{\hbar^2}\hat{G}_0^{(\pm)}\right)\hat{T}|\vec{k}\rangle. \end{aligned} \tag{47.29}$$

Also as in the one-dimensional case, the full Born series is defined by the recursion relation

$$|\psi_\pm^{(n+1)}\rangle = \left(\frac{2m}{\hbar^2}\hat{G}_0^{(\pm)}\right)\hat{V}|\psi_\pm^{(n)}\rangle. \tag{47.30}$$

Absorbing $2m/\hbar^2$ into $\hat{G}_0^{(\pm)}$ for simplicity, the full Born series becomes

$$\begin{aligned}|\psi\pm\rangle &= |\vec{k}\rangle + \hat{G}_0^{(\pm)}\hat{V}|\vec{k}\rangle + (\hat{G}_0^{(\pm)}\hat{V})^2|\vec{k}\rangle + \cdots \\ &= |\psi_\pm^{(0)}\rangle + |\psi_\pm^{(1)}\rangle + |\psi_\pm^{(2)}\rangle + \cdots \\ &= |\vec{k}\rangle + \hat{G}_0^{(\pm)}(\hat{V} + \hat{V}\hat{G}_0^{(\pm)}\hat{V} + \cdots)|\vec{k}\rangle.\end{aligned} \tag{47.31}$$

Comparing with the Lippmann–Schwinger equation in terms of $\hat{T}$, we obtain

$$\hat{T} = \hat{V} + \hat{V}\hat{G}_0\hat{V} + \hat{V}\hat{G}_0\hat{V}\hat{G}_0\hat{V} + \cdots = \hat{V} + \hat{V}\hat{G}_0\hat{T}. \tag{47.32}$$

The physical interpretation of the full Born series is obtained once we go to coordinate space, since then the nth-order term is

$$\int d^3\vec{r}_1 \ldots \int d^3\vec{r}_n G_0(\vec{r},\vec{r}_1)V(\vec{r}_1)G_0(\vec{r}_1,\vec{r}_2)V(\vec{r}_2)\cdots G_0(\vec{r}_{n-1},\vec{r}_n)V(\vec{r}_n)\phi(\vec{r}_n). \tag{47.33}$$

Then we see that the incoming wave hits the potential at $\vec{r}_n$ (the last integration variable, the variable of the free-wave function $\phi(\vec{r}_n)$), interacts with the potential, and then propagates with the propagator (Green's function) G_0 until $\vec{r}_{n-1}$, where it interacts again with $V(\vec{r}_n)$, etc., until after the last propagation we hit the point where we are "measuring" the wave function, $\vec{r}$.

We can also define the full Green's function,

$$\begin{aligned}\hat{G} &\equiv \frac{1}{E - \hat{H}_0 - \hat{V}} = \frac{1}{\hat{G}_0^{-1} - \hat{V}} \\ &= \hat{G}_0 + \hat{G}_0\hat{V}\hat{G}_0 + \hat{G}_0\hat{V}\hat{G}_0\hat{V}\hat{G}_0 + \cdots \\ &= \hat{G}_0(\mathbb{1} + \hat{T}\hat{G}_0).\end{aligned} \tag{47.34}$$

Acting with $\hat{G}^{-1}$ on the difference between the full state and the free state, we get

$$(E - \hat{H}_0 - \hat{V})(|\psi\rangle - |\vec{k}\rangle) = \hat{V}|\vec{k}\rangle, \tag{47.35}$$

where we have used the full Schrödinger equation for $|\psi\rangle$ and the free Schrödinger equation for $|\vec{k}\rangle$. Then we have

$$\begin{aligned}|\psi\rangle - |\vec{k}\rangle &= \frac{1}{E - \hat{H}_0 - \hat{V}}\hat{V}|\vec{k}\rangle = \hat{G}\hat{V}|\vec{k}\rangle \\ &= \hat{G}_0(\mathbb{1} + \hat{T}\hat{G}_0)\hat{V}|\vec{k}\rangle,\end{aligned} \tag{47.36}$$

which defines the full Born series once again.

47.4 Validity of the Born Approximation

Until now we have assumed that the Born approximation is valid, but we have not stated a condition for this to be true.

We require that, for $\vec{r} = 0$, meaning at the maximum of the interaction (where the potential is centered), the extra term is small with respect to the free term, which at $\vec{r}$ is 1, thus

$$|\psi_{\pm}^{(1)}(0)| = \left| \int d^3\vec{r}' G_0(0,\vec{r}') e^{i\vec{k}'\cdot\vec{r}'} V(\vec{r}') \right| \\ = \frac{2m}{4\pi\hbar^2} \left| \int d^3 r' \frac{e^{ikr'}}{r'} V(\vec{r}') e^{i\vec{k}\cdot\vec{r}'} \right| \ll 1. \tag{47.37}$$

Note that in the second line we have an exact equality, by just substituting $G_0(0,\vec{r}')$ into the first line.

For the particular case of the Yukawa potential,

$$V(r') = V_0 \frac{e^{-\mu r'}}{r'}, \tag{47.38}$$

and at low energy, $kr' \ll 1$, we obtain the condition

$$\frac{2m}{4\pi\hbar^2} V_0 \int d^3 r' \frac{e^{-\mu r'}}{r'^2} = \frac{2mV_0}{\mu\hbar^2} \ll 1, \tag{47.39}$$

where we have used $d^3r'/r'^2 = r'd\Omega$, integrated over r′, $\int_0^\infty dr' e^{-\mu r'} = 1/\mu$, and $\int d\Omega = 4\pi$.

On the other hand, for a potential where $V \sim V_0$ for a range $r \leq r_0$, and at low energy $kr' \ll 1$, we have

$$\int d^3 r' \frac{V(r')}{r'} \simeq 4\pi \frac{V_0}{2} r_0^2, \tag{47.40}$$

which means that the Born approximation validity condition is

$$\frac{mV_0}{\hbar^2} r_0^2 \ll 1. \tag{47.41}$$

One can make a more precise analysis of the validity condition, in particular one that interpolates to high energy, but we will not do it here. Here we just note that the region of validity of the Born approximation is not a low-energy one in general, and also is different than that of the WKB approximation.

Important Concepts to Remember

- The Born approximation is the first-order approximation to the Lippmann–Schwinger equation, where we put the zeroth-order solution $e^{i\vec{k}\cdot\vec{x}}$ on the right-hand side, and the Born series is obtained as successive terms in the approximation.
- The Born approximation gives the same result as a reinterpretation of the scattering as a time-dependent process, with Fermi's golden rule in first order, giving $dP/dt = (2\pi/\hbar)|\langle p_f|V|p_i\rangle|^2 \rho \, d\Omega$.
- The Lippman–Schwinger equation implies

$$\frac{d\sigma}{d\Omega} = \left(\frac{m}{2\pi\hbar^2}\right)^2 |\langle \vec{k}'|\hat{V}|\psi+\rangle|^2.$$

  with Born approximation

$$\frac{d\sigma}{d\Omega} = \left(\frac{m}{2\pi\hbar^2}\right)^2 |\langle \vec{p}_f|\hat{V}|\vec{p}_i\rangle|^2.$$

- One can define the T-matrix by $\hat{V}|\psi+\rangle = \hat{T}|\vec{k}\rangle$, so that the action on the interacting state is that of the T-matrix on a free state, and then
$$\hat{f} = -\frac{m}{2\pi\hbar^2}\hat{T}.$$
- The Born approximation is valid in a region different from that of the WKB approximation, nor is it the low-energy region. At low energy, for a Yukawa potential $V_0 e^{-\mu r}/r$, we have $mV_0/(\mu\hbar^2) \ll 1$ and for a constant potential, $V \sim V_0$, in a range r_0, we have $mV_0 r_0^2/\hbar^2 \ll 1$.

Further Reading

See [2] and [1].

Exercises

(1) Calculate the first two terms in the Born series for a delta function potential, $V(r) = -V_0\delta^3(\vec{r})$.

(2) Calculate the differential cross section for scattering in the Born approximation for a potential $V(r) = A/r^2$.

(3) Describe physically how it is possible that the Born approximation to quantum mechanical scattering in a Coulomb potential gives the classical-scattering Rutherford formula.

(4) Calculate the first two terms in the Born series for a potential $V(r) = A/(r^2 + a^2)$, with $a =$ constant.

(5) Write down explicitly the Lippmann–Schwinger equation for the T-matrix in the Yukawa case of $V(r) = V_0 e^{-\mu r}/r$.

(6) Write down and solve the Lippmann–Schwinger equation for the T-matrix in the case of the delta function potential $V = -V_0\delta^3(\vec{r})$.

(7) Is there a domain of validity of the Born approximation in the case of the Coulomb potential?

48 Partial Wave Expansion, Phase Shift Method, and Scattering Length

In this chapter we will consider spherically symmetric potentials, in which case we need to describe states and wave functions of a given l. To do that, we define an expansion in angular momentum l, the partial wave expansion, and the associated notions and methods of phase shift and scattering length.

48.1 The Partial Wave Expansion

In the spherically symmetric case, the complete set of mutually commuting variables is $\{H, \vec{L}^2, L_z\}$, so we have basis states that are eigenstates of those operators, $|Elm\rangle$, and which are orthonormal (including on the continuous energy variable),

$$\langle E'l'm'|Elm\rangle = \delta_{ll'}\delta_{mm'}\delta(E - E'). \tag{48.1}$$

In coordinate space, the wave function of a basis state is

$$\langle \vec{r}|Elm\rangle = u_{Elm}(\vec{r}) = R_{El}(r)Y_{lm}(\vec{n}_r). \tag{48.2}$$

We saw in Chapters 18 and 19 that, for a free particle, the solution for the radial wave function is

$$R_{El}(r) = C_l j_l(kr), \tag{48.3}$$

where j_l is the spherical Bessel function. Moreover, the *plane wave* solution, $e^{i\vec{k}\cdot\vec{r}}$, where $\vec{k} = \vec{p}/\hbar$, is expanded in the above basis as

$$\begin{aligned} e^{i\vec{k}\cdot\vec{r}} = e^{ikr\cos\theta} &= \sum_{l=0}^{\infty}(2l+1)i^l j_l(kr)P_l(\cos\theta) \\ &= \sum_{l=0}^{\infty}\sum_{m=-l}^{l} a_{lm} j_l(kr)Y_{lm}(\theta,\phi), \end{aligned} \tag{48.4}$$

where $(2l + 1)i^l \equiv a_l$, the first line is expanded in terms of only θ, since the expanded function is a function only of $\cos\theta$, whereas the second line is the general expansion of a wave function in the $\langle\vec{r}|Elm\rangle$ basis. The relation between the two expansions is provided by the equality (from Chapter 17)

$$P_l(\cos\theta) = \sqrt{\frac{4\pi}{2l+1}}Y_{l,0}(\theta,\phi), \tag{48.5}$$

which means that

$$a_{lm} = \delta_{m,0}a_l\sqrt{\frac{4\pi}{2l+1}}. \tag{48.6}$$

The expansion in terms of θ (on the first line) follows from the relation

$$j_l(kr) = \frac{1}{2i^l}\int_{-1}^{+1} d(\cos\theta) e^{ikr\cos\theta} P_l(\cos\theta). \tag{48.7}$$

Since at large values of the argument the spherical Bessel function becomes

$$j_l(kr) \simeq \frac{e^{ikr} i^{-l} - e^{-ikr} i^l}{2ikr}, \tag{48.8}$$

this means that, at $r \to \infty$, the free plane wave becomes

$$e^{i\vec{k}\cdot\vec{r}} \simeq \frac{e^{ikr}}{r}\frac{1}{2ik}\sum_{l=0}^{\infty}(2l+1)P_l(\cos\theta) - \frac{e^{-ikr}}{r}\frac{1}{2ik}\sum_{l=0}^{\infty}(2l+1)i^{2l}P_l(\cos\theta). \tag{48.9}$$

Then, in terms of the general expansion at infinity, the coefficients of the free plane wave are

$$\begin{aligned} A(k,\vec{n}_r) &= \frac{1}{2ik}\sum_{l=0}^{\infty}(2l+1)P_l(\cos\theta) \\ &= \frac{2\pi}{ik}\delta(\vec{n}_k - \vec{n}_r) \\ B(k,-\vec{n}_r) &= \frac{1}{2ik}\sum_{l=0}^{\infty}(2l+1)i^{2l}P_l(\cos\theta) \\ &= \frac{2\pi}{ik}\delta(\vec{n}_k + \vec{n}_r). \end{aligned} \tag{48.10}$$

Then, when we have scattering in a central (spherically symmetric) potential, we saw that we can make the replacement

$$\begin{aligned} A(k,\vec{n}_r) = A_k(\vec{n}') = \frac{2\pi}{ik}\delta(\vec{n}' - \vec{n}_k) &= \sum_{l=0}^{\infty}(2l+1)\frac{1}{2ik}P_l(\cos\theta) \\ &\to \frac{2\pi}{ik}\delta(\vec{n}' - \vec{n}_k) + f_k(\vec{n}', \vec{n}_k), \end{aligned} \tag{48.11}$$

where $\vec{k}' = k\vec{n}'$ generalizes $k\vec{n}_r$.

However, given the expansion in $\cos\theta$ of the coefficient A before the replacement, it means that the added term should also have an expansion, where we replace $1/(2ik)$ with a general coefficient $a_l(k)$, i.e., we define

$$f_k(\vec{n}', \vec{n}_k) = f(\theta) = \sum_{l=0}^{\infty}(2l+1)a_l(k)P_l(\cos\theta), \tag{48.12}$$

where θ is the angle between $\vec{n}'$ and $\vec{n}_k$.

Here $a_l(k)$ is called the *lth partial wave amplitude*. Given the above expansion, the scattering solution for a potential with a finite range, i.e., a free plane wave and a diverging spherical wave, expands into partial waves as follows:

$$\begin{aligned} \psi^+_{\vec{k}}(\vec{r}) &= e^{i\vec{k}\cdot\vec{r}} + f_k(\vec{n}_r, \vec{n}_k)\frac{e^{ikr}}{r} \\ &\simeq \frac{1}{2ik}\left\{\sum_{l=0}^{\infty}(2l+1)P_l(\cos\theta)\left[\frac{e^{ikr}}{r}(1 + 2ika_l(k)) - \frac{e^{-ikr}i^{2l}}{r}\right]\right\}. \end{aligned} \tag{48.13}$$

48.2 Phase Shifts

The boundary conditions for the wave function in the spherically symmetric scattering case are imposed on $R_{El}(r)$, and define the abstract state $|Elm\rangle$ in coordinate space. But the physical situation we have now, unlike the previous derivations (in Chapters 18 and 19) of bound states in the case of say, the hydrogen atom, is that of a scattering solution, which means an incoming plane wave plus an outgoing (or incoming) spherical wave giving the state $|\psi\pm\rangle$. So, really, it is more useful to write $|Elm\pm\rangle$ for the abstract state and $R^{\pm}_{El}(r)$ to refer to the two outgoing or incoming solutions. If we do not use the $\pm$ index, it means that we are considering a general linear combination of the two solutions.

Then the general solution of the Schrödinger equation $|\psi\rangle$ is expanded in terms of some basis $|Elm\rangle$ (a linear combination of the $|Elm\pm\rangle$ states) as

$$\psi_{\vec{k}}(\vec{r}) = \sum_{l=0}^{\infty}\sum_{m=-l}^{l} K_{lm}R_{El}(r)Y_{lm}(\theta,\phi) = \sum_{l=0}^{\infty}\sum_{m=-l}^{l} K_{lm}u_{Elm}(\vec{r}), \tag{48.14}$$

where $R_{El}(r)$ is a linear combination of $R^{\pm}_{El}(r)$.

At $r\to\infty$, we have the Helmholtz equation

$$(\Delta + k^2)R = 0, \tag{48.15}$$

so the general solution of it is either a linear combination of the spherical Bessel function $j_l(kr)$ and the spherical Neumann function $n_l(kr)$ or $y_l(kr)$, or a linear combination of the spherical Hankel functions of first and second degrees, $h_l^{(1)}(kr)$ and $h_l^{(2)}(kr)$,

$$\begin{aligned} R_{El}(r) &\sim C_l j_l(kr) + D_l n_l(kr) \\ &= A_l h_l^{(1)}(kr) + B_l h_l^{(2)}(kr), \end{aligned} \tag{48.16}$$

where, since $h_l^{(1,2)} = j_l \pm i n_l$, we have

$$C_l = A_l + B_l, \quad D_l = iA_l - iB_l. \tag{48.17}$$

But, at large values of the argument, the spherical Hankel functions behave as

$$\begin{aligned} h_l^{(1)}(kr) &\sim \frac{e^{ikr} i^{-l}}{ikr} \\ h_l^{(2)}(kr) &\sim -\frac{e^{-ikr} i^{l}}{ikr}, \end{aligned} \tag{48.18}$$

which means that if at $r\to\infty$ $B_l = A_l^*$ then $R_{El}(r)$ is real (since $h_l^{(1)*} = h_l^{(2)}$). Moreover, if also A_l is real, the solution has only a $j_l(kr)$ component, as for the free wave, but that is a very special case.

Then the real solution for the radial wave function is

$$R_{El}(r) \sim \frac{A_l e^{i(kr - l\pi/2)} - A_l^* e^{-i(kr - l\pi/2)}}{ikr}, \tag{48.19}$$

and is therefore the sum $R_{El}(r) = R^{(+)}_{El}(r) + R^{(-)}_{El}(r)$, where the first term contains e^{ikr} and the second e^{-ikr}.

We define the phase of A_l as $e^{i\delta_l}$, i.e.,

$$A_l = A_l^0 e^{i\delta_l} \;\Rightarrow\; A_l^* = A_l^0 e^{-i\delta_l}. \tag{48.20}$$

Then δ_l is called the *phase shift*, and the real solution for the wave function becomes

$$R_{El}(r) \sim \frac{2A_l^0}{kr} \sin\left(kr - \frac{l\pi}{2} + \delta_l\right). \tag{48.21}$$

Since

$$\begin{aligned} C_l &= A_l + A_l^* = 2A_l^0 \cos\delta_l \\ D_l &= iA_l - iA_l^* = -2A_l^0 \sin\delta_l, \end{aligned} \tag{48.22}$$

it follows that we obtain the phase shift from the expansion in terms of j_l and n_l,

$$\tan\delta_l = -\frac{D_l}{C_l}. \tag{48.23}$$

We then define the wave function solution of the Schrödinger equation,

$$\psi_k(r,\theta) = \sum_{l=0}^{\infty} K_l R_{El}(r) P_l(\cos\theta), \tag{48.24}$$

which is a *particular* linear combination of the $|Elm\rangle$ solutions, and is also of the scattering-solution type (since the partial wave expansion (48.12) is of the same type). The linear combination of solutions, just like the *plane wave* free solution (48.4), is a particular linear combination of $j_l(kr)Y_{lm}(\theta,\phi)$, with $a_{lm} \propto \delta_{m,0}$. Here $\sum_{m=-l}^{l} K_{lm}Y_{lm}(\theta,\phi)$ reduces to $K_l P_l(\cos\theta)$ when $K_{lm} \propto \delta_{m,0}$.

We can define outgoing and incoming solutions $|\psi\pm\rangle$ from the decomposition of the radial wave functions (48.19), as

$$\begin{aligned} |\psi+\rangle \equiv |\vec{k}+\rangle &= \sum_{l=0}^{\infty}\sum_{m=-l}^{l} A_{lm}^0 e^{i\delta_l}|Elm+\rangle = \sum_{l=0}^{\infty}\sum_{m=-l}^{l} A_l^0 K_{lm} e^{i\delta_l}|Elm+\rangle \\ |\psi-\rangle \equiv |\vec{k}-\rangle &= \sum_{l=0}^{\infty}\sum_{m=-l}^{l} A_{lm}^0 e^{-i\delta_l}|Elm-\rangle = \sum_{l=0}^{\infty}\sum_{m=-l}^{l} A_l^0 K_{lm} e^{-i\delta_l}|Elm-\rangle, \end{aligned} \tag{48.25}$$

where $A_{lm} \equiv A_l^0 K_{lm}$, and the states $|Elm\pm\rangle$ are defined with real coefficients (times $e^{\pm ikr}$) at infinity.

Then the asymptotics at $r \to \infty$ of the linear combination solution defined above is

$$\psi_k(\vec{r}) \simeq \sum_{l=0}^{\infty} A_l^0 P_l(\cos\theta) \frac{e^{i(kr - l\pi/2 + \delta_l)} - e^{-i(kr - l\pi/2 + \delta_l)}}{2ikr}, \tag{48.26}$$

where we have absorbed K_l into A_l^0. We note that the above becomes

$$\langle \vec{r}| \left(|\vec{k}+\rangle - |\vec{k}-\rangle\right). \tag{48.27}$$

The $|\vec{k}\pm\rangle$ are orthonormal,

$$\langle \vec{k}' + |\vec{k}+\rangle = \delta^3(\vec{k}' - \vec{k}) = \langle \vec{k}' - |\vec{k}-\rangle, \tag{48.28}$$

as are the spherical states $|Elm\rangle$,

$$\langle E'l'm'|Elm\rangle = \delta_{ll'}\delta_{mm'}\delta(E - E'). \tag{48.29}$$

However, the product of the in and out states defined in (46.69) is nontrivial,

$$\begin{aligned}\langle \vec{k}' - |\vec{k}+\rangle = \int d^3\vec{r}\, \psi_{\vec{k}'}^{(-)*}(\vec{r})\psi_{\vec{k}}^{(+)}(\vec{r}) &= \delta(\vec{k}' - \vec{k}) + \frac{ik}{2\pi}\delta(k' - k) f_k(\vec{n}', \vec{n}_k) \\ &= \delta(k' - k)\langle \psi - |\psi+\rangle = \langle \vec{k}'|\hat{S}|\vec{k}\rangle \\ &= \delta(k' - k) S(\vec{n}', \vec{n}_k),\end{aligned} \tag{48.30}$$

where, compared with the in and out states defined before in (48.25), we have used states with an extra k modulus tensored in, $|\vec{k}\pm\rangle = |\psi\pm\rangle \otimes |k\rangle$, so that in the product we have an extra $\langle k'|k\rangle = \delta(k' - k)$.

Now we need to put the solution (48.24) with asymptotics (48.26) into a scattering form. In that form, the incoming wave part ($\propto e^{-ikr}/r$) is only from the plane wave ($e^{i\vec{k}\cdot\vec{r}}$) part, so we can identify A_l^0 by comparing with the plane wave asymptotics in (48.9), obtaining

$$B(k, \vec{n}_r) = \sum_{l=0}^{\infty} (2l+1) P_l(\cos\theta) \frac{i^{2l}}{2ik} = \sum_{l=0}^{\infty} A_l^0 P_l(\cos\theta) \frac{i^l e^{-i\delta_l(k)}}{2ik}, \tag{48.31}$$

which means that A_l^0 is given by

$$A_l^0 = (2l+1) e^{i(\delta_l + l\pi/2)}. \tag{48.32}$$

Substituting back into the full $\psi_k(\vec{r})$ in (48.24), we obtain first

$$\psi_k(r, \theta) = \sum_{l=0}^{\infty} K_l R_{El}(r) P_l(\cos\theta). \tag{48.33}$$

But K_l was absorbed into A_l^0 so, up to an irrelevant normalization, we can identify them, implying

$$\psi_k(r, \theta) = \sum_{l=0}^{\infty} (2l+1) e^{i(\delta_l + l\pi/2)} R_{El}(r) P_l(\cos\theta). \tag{48.34}$$

Then, substituting A_l^0 also in asymptotic ($r \to \infty$) form into (48.26), we obtain

$$\begin{aligned}\psi_k(\vec{r}) &\simeq \frac{1}{2ikr} \sum_{l=0}^{\infty} (2l+1) P_l(\cos\theta) [e^{ikr} e^{2i\delta_l} - e^{-i(kr - l\pi)}] \\ &= e^{i\vec{k}\cdot\vec{r}} + \frac{e^{ikr}}{r} \left[\sum_{l=0}^{\infty} (2l+1) P_l(\cos\theta) \left(\frac{e^{2i\delta_l} - 1}{2ik} \right) \right],\end{aligned} \tag{48.35}$$

where in the last line we have recreated $e^{i\vec{k}\cdot\vec{r}}$ from the $\cos\theta$ expansion. Then we find

$$f_k(\vec{n}', \vec{n}_k) = \left[\sum_{l=0}^{\infty} (2l+1) P_l(\cos\theta) \left(\frac{e^{2i\delta_l} - 1}{2ik} \right) \right], \tag{48.36}$$

and, now that we have the asymptotic scattering form, we can identify $a_l(k)$ as

$$a_l(k) = \frac{e^{2i\delta_l(k)} - 1}{2ik} = \frac{e^{i\delta_l(k)} \sin\delta_l(k)}{k} = \frac{1}{k\cot\delta_l(k) - ik}. \tag{48.37}$$

Moreover, with $\vec{n}' = \vec{n}_r$, and $\vec{n}_r \cdot \vec{n}_k = \cos\theta$, we have also an expansion in spherical harmonics,

$$f_k(\vec{n}_r, \vec{n}_k) = \sum_{l=0}^{\infty} (2l+1) P_l(\vec{n}_r \cdot \vec{n}_k) a_l(k) = \sum_{l=0}^{\infty} \sum_{m=-l}^{l} 4\pi a_l(k) Y_{lm}^*(\vec{n}_k) Y_{lm}(\vec{n}_r). \tag{48.38}$$

Considering the scattering solution in (48.13), and comparing with the free case in (48.9), we see that the diverging wave ($\propto e^{ikr}/r$) is just multiplied by the factor

$$1 + 2ika_l(k) = e^{2i\delta_l(k)} \equiv S_l(k), \tag{48.39}$$

called the *lth partial wave S-matrix element.*

Indeed, since the S-operator is related to the f-operator by

$$\hat{S} = \mathbb{1} + \frac{ik}{2\pi}\hat{f}, \tag{48.40}$$

so that, by multiplication with $\langle\vec{n}_r|$ from the left and with $|\vec{n}_k\rangle$ from the right; we have

$$S(\vec{n}_r, \vec{n}_k) = \delta^2(\vec{n}_r - \vec{n}_k) + \frac{ik}{2\pi} f_k(\vec{n}_r, \vec{n}_k), \tag{48.41}$$

which expands into

$$S(\vec{n}_r, \vec{n}_k) = \frac{1}{4\pi}\sum_{l=0}^{\infty}(2l+1)P_l(\cos\theta) + \frac{1}{4\pi}\sum_{l=0}^{\infty}(2l+1)P_l(\cos\theta)2ika_l(k), \tag{48.42}$$

it follows that

$$S_l = 1 + 2ika_l(k) = e^{2i\delta_l(k)}. \tag{48.43}$$

48.3 T-Matrix Element

The f-operator is related to the T-operator by

$$\hat{f} = -\frac{m}{2\pi\hbar^2}\hat{T}, \tag{48.44}$$

which means that their matrix elements are related too, by

$$f(\vec{k}', \vec{k}) = \langle\vec{k}'|\hat{f}|\vec{k}\rangle = -\frac{m}{2\pi\hbar^2}\langle\vec{k}'|\hat{T}|\vec{k}\rangle. \tag{48.45}$$

Then the T-matrix element is

$$\begin{aligned}\langle\vec{k}'|\hat{T}|\vec{k}\rangle &= \langle\vec{k}'|\hat{V}|\vec{k}+\rangle\\ &= \frac{2\pi\hbar^2}{m}\delta(k-k')\sum_{l=0}^{\infty}(2l+1)a_l(k)P_l(\vec{n}'\cdot\vec{n}_k)\\ &\equiv \delta(k-k')T_k(\vec{n}', \vec{n}_k),\end{aligned} \tag{48.46}$$

where again the $\delta(k-k')$ factor appears because we have modified the states, $|\vec{k}+\rangle$ from $|\psi+\rangle$.

Taking in consideration (48.30), and the above T-matrix, we obtain for the S-matrix

$$\begin{aligned}\langle\vec{k}'-|\vec{k}+\rangle = \langle\vec{k}'|\hat{S}|\vec{k}\rangle &= \left\langle\vec{k}'\right|\left(\mathbb{1} - \frac{imk}{(2\pi\hbar)^2}\hat{T}\right)\left|\vec{k}\right\rangle\\ &= \delta^3(\vec{k}-\vec{k}') - \frac{imk}{(2\pi\hbar)^2}\delta(k'-k)T_k(\vec{n}', \vec{k}).\end{aligned} \tag{48.47}$$

Then the S- and T-operators are related by

$$\hat{S} = \mathbb{1} - \frac{imk}{(2\pi\hbar)^2}\delta(k'-k)\hat{T}_k = \mathbb{1} - \frac{ik^2}{(2\pi)^2}\delta(E'-E)\hat{T}_k. \tag{48.48}$$

The S-matrix elements in the spherical, $|Elm\rangle$, basis, are given by

$$\langle E'l'm'|\hat{S}|Elm\rangle = \delta_{ll'}\delta_{mm'}\delta(E-E')S_l, \tag{48.49}$$

because of the $\cos\theta$ expansion of the S-matrix $S(\vec{n}_r, \vec{n}_k)$ in (48.42).

But then the T-matrix element in the same basis is

$$\langle E'l'm'|\hat{T}|Elm\rangle = \delta_{ll'}\delta_{mm'}\delta(E-E')T_l, \tag{48.50}$$

or, taking out the $\delta(E-E')$ factor,

$$\langle E'l'm'|\hat{T}_k|Elm\rangle = \delta_{ll'}\delta_{mm'}T_l. \tag{48.51}$$

Now multiplying (48.48) by $\langle E'l'm'|$ from the left and by $|Elm\rangle$ from the right, we find[1]

$$S_l = 1 - 2\pi i T_l, \tag{48.52}$$

which means that T_l is related to a_l by

$$T_l(k) = -\frac{k}{\pi}a_l(k). \tag{48.53}$$

The matrix element relation generalizes to an operatorial relation,

$$\hat{S} = \mathbb{1} - 2\pi i\delta(E-E')\hat{T}. \tag{48.54}$$

48.4 Scattering Length

At low energies, $k \to 0$, we will see later that, for a finite range, $\delta_l(k) \propto k^{2l+1} \to 0$, and moreover only $l = 0$ contributes, so $\delta_0(k) \to 0$, and it dominates the other $\delta_l(k)$.

But then, the $a_l(k)$ are given by

$$a_l(k) \simeq \frac{\sin\delta_l(k)}{k} \simeq \frac{\tan\delta_l(k)}{k} \simeq \frac{\delta_l(k)}{k}. \tag{48.55}$$

The leading contribution is then

$$a_0(k) \simeq \frac{\delta_0(k)}{k}, \tag{48.56}$$

but it is actually negative, so we define the finite and positive quantity

$$a \equiv -\lim_{k\to 0} a_0(k) = -\lim_{k\to 0}\frac{\delta_0(k)}{k} \tag{48.57}$$

called the scattering length.

[1] There is an extra $(2\pi)^3$, coming from the, now different, normalization of states (with a $1/(2\pi)^{3/2}$ factor), and there is also an extra $1/k^2$ factor, which appears because of the states being defined as $\langle k\vec{n}_k|$ rather than as $\langle\vec{n}_k|$, giving a factor of $1/k$ in the normalization. We take into account this change in normalization because the relation between S and T is usually defined as in the following.

48.5 Jost Functions, Wronskians, and the Levinson Theorem

We have seen before that, for a finite-range potential, i.e., for a potential decaying faster than the Coulomb potential,

$$\lim_{r\to\infty} rV(r) \to 0, \tag{48.58}$$

the radial wave function solution

$$\frac{\chi_{kl}(r)}{r} = R_{kl}(r) \sim \frac{2A_l^0}{kr}\sin\left(kr - \frac{l\pi}{2} + \delta_l\right) \tag{48.59}$$

is a real solution, meaning it contains both an outgoing and an incoming part: $R^+_{kl}(r) + R^-_{kl}(r)$.

If we also impose

$$\lim_{r\to 0} r^2 V(r) = 0, \tag{48.60}$$

it means that there is a discrete spectrum for $E < 0$ (since we are now imposing two normalizability conditions, at infinity and at zero, on two independent solutions of the Schrödinger equation), but the spectrum is still continuous for $E \geq 0$.

The behavior of $\chi_{kl}(r)$ at $r \to 0$ is (as we saw in Chapter 19) $\sim r^{-l}$ or $\sim r^{l+1}$, so that $R_{kl}(r) \sim r^{-l-1}$ or $\sim r^l$.

We then define the *regular (normalizable-at-zero) solution* $\chi_l = \phi_l$, which means the solution that has the behavior $\sim r^{l+1}$. Moreover, we choose the normalization constant such that

$$\lim_{r\to 0} r^{-l-1}\phi_l = 1. \tag{48.61}$$

Note that this result is valid whether the energy is positive (a continuous-spectrum, scattering solution) or negative (a discrete-spectrum, bound-state solution). The physically normalized solution is multiplied by a constant N_{kl},

$$\chi_{kl} = N_{kl}\phi_l(k;r). \tag{48.62}$$

We then extend $\phi_l(k;r)$ to the full complex plane for k using analytic continuation, which means that ϕ_l must be an analytical function of k. Its properties are as follows.

(1) $\phi_l(-k;r) = \phi_l(k;r)$, which is part of the definition of the function: on the real axis, $k < 0$ is defined from the physical case, $k > 0$.
(2) $\phi_l(k;r) = \phi_l^*(k^*;r)$, which is a result of the analyticity imposed on ϕ_l.
(3) If $k \in \mathbb{R}$ then ϕ_l is an eigenfunction of the Hamiltonian, but if k is not real (so that $E^* \neq E$), then ϕ_l is not an eigenfunction of H.

Jost Functions

For k real, we define *uniquely* the Schrödinger equation solutions $f_l^\pm$ that behave at $r \to \infty$ as incoming or outgoing, respectively, i.e.,

$$\lim_{r\to\infty} e^{\pm ikr} f_l^\pm(k,r) = 1, \tag{48.63}$$

where again we choose the normalization constant such that we have 1 on the right-hand side. Therefore

$$\chi_l = f_l^\pm \simeq e^{\mp ikr} \quad \Rightarrow \quad \psi \propto \frac{e^{\mp ikr}}{r}. \tag{48.64}$$

Then, at $r \to 0$, the solutions $f_l^\pm$ are a linear combination of regular ($\sim r^{l+1}$) and irregular, or non-normalizable ($\sim r^{-l}$), solutions, with the irregular solution dominating. This means that, at $r \to 0$,

$$f_l^\pm \sim C_l^\pm r^{-l}, \tag{48.65}$$

where $C_l^\pm$ is well defined (the subleading, regular, component is $\sim D_l^\pm r^{l+1}$).

When $k \in \mathbb{C}$, we have that:

f_l^+ is well defined for Im $k < 0$.

f_l^- is well defined for Im $k > 0$.

This implies that, at $r \to \infty$,

$$|f_l^\pm(k;r)| \propto e^{\pm(\mathrm{Im}\,k)r} \to 0. \tag{48.66}$$

In one dimension, and therefore also in three dimensions but for the radial direction, we defined the Wronskian theorem. We need only to apply it for the same potential $V(r)$, and the same energy, which means that the Wronskian of two solutions ψ_1 and ψ_2,

$$W(\psi_1, \psi_2) = \psi_1\psi_2' - \psi_2\psi_1', \tag{48.67}$$

is constant as a function of r, so

$$\frac{dW(\psi_1, \psi_2)}{dr} = 0. \tag{48.68}$$

Then in particular we can calculate it at $r \to \infty$. Choosing the two solutions to be $f_l^\pm$, in which case, $f_l^\pm \simeq e^{\mp ikr}$, we find

$$W(f_l^+, f_l^-) = 2ik. \tag{48.69}$$

Thus, if for all k there are three solutions, ϕ_l, f_l^+, f_l^-, this means that they are linearly dependent, so ϕ_l is a linear combination of f_l^+ and f_l^-:

$$\phi_l = C_1 f_l^+ + C_2 f_l^-. \tag{48.70}$$

In this case, we define the *Jost functions* $\mathcal{F}^\pm(k)$ as

$$\mathcal{F}^\pm(k) \equiv W(f_l^\pm, \phi_l). \tag{48.71}$$

We can calculate the Wronskian (which is r-independent again, as we have the same potential and the same energy) either at infinity or at zero. We find (from the Wronskian at infinity)

$$\phi_l = \frac{1}{2ik}\left[-\mathcal{F}_l^-(k) f_l^+(k;r) + \mathcal{F}_l^+(k) f_l^-(k;r)\right], \tag{48.72}$$

where (from the Wronskian at zero)

$$\mathcal{F}_l^\pm(k) = (2l+1)C_l^\pm = (2l+1)\lim_{r\to 0} r^l f_l^\pm(k;r), \tag{48.73}$$

so, at $r \to 0$,

$$f_l^\pm(k;r) \simeq \frac{\mathcal{F}_l^\pm(k)}{(2l+1)r^l}. \tag{48.74}$$

The properties of $f_l^\pm$ and the Jost functions $\mathcal{F}_l^\pm$ are:

(1) $f_l^-(k;r) = f_l^+(-k;r)$, since at infinity we have $f_l^\pm \simeq e^{\mp ikr}$.
(2) $[f_l^\pm(-k^*;r)]^* = f_l^\pm(k;r)$, again proven from the behavior at infinity and analyticity.
(3) From the two previous properties, we also obtain $f_l^\pm(k;r) = [f_l^\mp(k^*;r)]^*$.
In particular, if $k \in \mathbb{R}$, then $f_l^- = (f_l^+)^*$, so

$$\mathcal{F}_l^\pm(-k) = \mathcal{F}_l^\mp(k). \tag{48.75}$$

(4) This is generalized through analyticity to the complex plane relation

$$[\mathcal{F}_l^\pm(-k^*)]^* = \mathcal{F}_l^\pm(k). \tag{48.76}$$

But if we go back to $k \in \mathbb{R}$, this reduces to a more general relation than the previous one,

$$\mathcal{F}_l^+(-k) = (\mathcal{F}_l^-(k))^*. \tag{48.77}$$

As ϕ_l is χ_l, from the expansion of ϕ_l into $f_l^\pm$, we obtain

$$S_l(k) = e^{2i\delta_l(k)} = \frac{\mathcal{F}_l^+(k)}{\mathcal{F}_l^-(k)}e^{il\pi} = \frac{\mathcal{F}_l^+(k)}{\mathcal{F}_l^+(-k)}e^{il\pi}. \tag{48.78}$$

Substituting this into the behavior at $r \to \infty$ of (48.72), we find

$$\phi_l(k;r) \sim \frac{\mathcal{F}_l^-(k)}{k}e^{i\delta_l(k)}e^{-l\pi/2}\sin\left(kr - \frac{l\pi}{2} + \delta_l\right). \tag{48.79}$$

If $k \in \mathbb{R}$, then

$$\begin{aligned}\mathcal{F}_l^+(k) &= |\mathcal{F}_l^+(k)|e^{i\alpha_l}\\ \mathcal{F}_l^-(k) &= |\mathcal{F}_l^+(k)|e^{-i\delta_l}.\end{aligned} \tag{48.80}$$

This, together with the relation $e^{2i\delta_l} = \mathcal{F}_l^+(k)e^{il\pi}/\mathcal{F}_l^-(k)$, implies that

$$\begin{aligned}\delta_l &= \alpha_l + l\pi (\text{mod } 2\pi)\\ \delta_l(-k) &= -\delta_l(k).\end{aligned} \tag{48.81}$$

In this analysis of Jost functions, we have assumed that $k \in \mathbb{C}$, though without much justification. We will come back to this in Chapter 50, where we will define more thoroughly the analysis for complex k.

Levinson Theorem

This theorem was proven in 1949 by Levinson, but here we will just present a statement of it, without a proof. We will consider a not very rigorous proof later on.

(a) The first statement regarding $\delta_l(k)$, specifically for the difference between low energy and high energy, is:

$$\delta_0(0) - \delta_0(\infty) = \begin{cases} n_b^0\pi & \text{if } \mathcal{F}_0^+(0) \neq 0\\ (n_b^0 + 1/2)\pi & \text{if } \mathcal{F}_0^+(0) = 0.\end{cases} \tag{48.82}$$

Note that $\mathcal{F}_0^+(0) \neq 0$ means that ϕ_l contains *both* f_l^+ and f_l^- components, whereas the $\mathcal{F}_0^+(0) = 0$ condition means that $\phi_l \propto f_l^+$, giving a quantization condition (but not two, as in the case of obtaining a discrete level, i.e., a bound state), meaning an extra 1/2 term is added to n_b^0. Here n_b^0 is the number of energy levels (number of bound states) of given angular momentum $l = 0$, so it is simple to generalize to n_b^l for angular momentum l. The assumption of this statement is that if $\mathcal{F}_0^+(0) = 0$ then $\mathcal{F}_0^+(k) \sim ak$ as $k \to 0$.

(b) The second statement is:

$$\delta_l(0) - \delta_l(\infty) = n_b^l \pi \qquad \text{if } \mathcal{F}_l^+(0) = 0 \text{ and } \mathcal{F}_l^+(k) \sim Ak^2. \tag{48.83}$$

In both cases, an extra assumption is that the potential has not just finite range but also a faster vanishing of the potential at infinity,

$$\lim_{r\to 0} r^3 V(r) = 0. \tag{48.84}$$

Important Concepts to Remember

- The partial wave expansion is $f_k(\vec{n}', \vec{n}_k) = \sum_{l=0}^{\infty}(2l+1)a_l(k)P_l(\cos\theta)$, in terms of the lth partial wave amplitude $a_l(k)$ and follows the same expansion as that of the other term, $(2\pi/ik)\delta(\vec{n}' - \vec{n}_k)$, in $A_k(\vec{n}')$, which has $1/(2ik)$ instead of $a_l(k)$.
- For a central potential going to 0 at infinity, the real radial wave function at infinity is $R_{El}(r) = [A_l e^{i(kr - l\pi/2)} - A_l^* e^{-i(kr-l\pi/2)}]/(ikr)$, with $A_l = A_l^0 e^{i\delta_l}$, or $R_{El}(r) = (2A_l^0/kr)\sin(kr - l\pi/2 + \delta_l)$, with δ_l the phase shift.
- The partial wave amplitude is related to the phase shift by $a_l(k) = (e^{2i\delta_l(k)} - 1)/(2ik) = [e^{i\delta_l(k)}\sin\delta_l(k)]/k = [k\cot\delta_l(k) - ik]^{-1}$, and to the lth partial wave S-matrix element by $S_l(k) = e^{i\delta_l(k)} = 1 + 2ika_l(k)$, for the same partial wave expansion of $S(\vec{n}_r, \vec{n}_k)$.
- The S-matrix and T-matrix are related by $S_l = 1 - 2\pi i T_l$ and $\hat{S} = \mathbb{1} - 2\pi i\delta(E - E')\hat{T}$.
- At low energies, $k \to 0$, $a_l(k) \simeq \delta_l(k)/k$, and $a \equiv -\lim_{k\to 0} a_0(k) > 0$ is the scattering length.
- We can define radial wave solutions $R_{kl}(r)$ analytically continued to the complex k plane. Defining, for real k, ϕ_l via $\lim_{r\to 0} r^{-l-1}\phi_l(r) = 1$ and $f_l^\pm(k, r)$ via $\lim_{r\to\infty} e^{\pm ikr} f_l^\pm(k; r) = 1$, we define the Jost functions as the Wronskians $\mathcal{F}^\pm(k) = W(f_l^\pm, \phi_l)$.
- The Levinson theorem states that $\delta_0(0) - \delta_0(\infty) = n_b^0\pi$ if $\mathcal{F}_0^+(0) \neq 0$, where n_b^l is the number of bound states with l, and $\delta_0(0) - \delta_0(\infty) = (n_b^0 + 1/2)\pi$ if $\mathcal{F}_0^+(0) = 0$, while $\delta_l(0) - \delta_l(\infty) = n_b^l\pi$ if $\mathcal{F}_l^+(0) = 0$ and $\mathcal{F}_l^+(k) \sim Ak^2$.

Further Reading

For partial waves, phase shifts, and scattering length, see [1] and [2]. For the Levinson theorem, see Levinson's paper [28].

Exercises

(1) Consider scattering onto a delta function potential, $V = -V_0\delta^3(\vec{r})$, in the Born approximation. Calculate the partial wave amplitudes $a_l(k)$.

(2) Consider a spherical well potential $V = -V_0$ for $r \leq R$, and $V = 0$ for $r > R$, in the Born approximation, and waves with $E > 0$. Calculate the phase shifts $\delta_l(k)$ for scattering.
(3) In the case in exercise 2, calculate $S_l(k)$ and the differential cross section.
(4) If $\delta_l(k)$ is real, what do you deduce about $T_l(k)$?
(5) Calculate the scattering length for the case in exercise 1.
(6) Is relation (48.72) well defined for complex k? Why?
(7) If $\lim_{k\to 0} \mathcal{F}_l^+(k)/k^2$ is constant for l even, find the $k \to 0, r \to \infty$ behavior with k of $\phi_l(k;r)$ for even l.

49 Unitarity, Optics, and the Optical Theorem

In this chapter, we will revisit the issue of unitarity and the optical theorem for scattering. An example of the partial wave formalism, for a hard sphere, puts us on the track of quantum mechanical scattering as optics.

49.1 Unitarity: Review and Reanalysis

To understand the application of unitarity, we first remember what unitarity means: in quantum mechanics, the conservation of probability implies that the time evolution is unitary, namely that the evolution operator is unitary, $\hat{U}^\dagger = \hat{U}^{-1}$.

But the S-matrix is related to the evolution operator by $\hat{S} = \hat{U}_I(+\infty, -\infty)$, so $\hat{S}^\dagger = \hat{S}^{-1}$. Then unitarity of the S-matrix is equivalent to the conservation of probability.

On the other hand, in the $|Elm\rangle$ basis, which means in the partial wave formalism, the S-operator $\hat{S}$ is (as we saw) represented by $S_l = e^{2i\delta_l}$, and so

$$\langle E'l'm'|\hat{S}|Elm\rangle = \delta_{ll'}\delta_{mm'}\delta(E - E')S_l. \tag{49.1}$$

The diagonalization comes from the fact that the time evolution operator $\hat{S} = \hat{U}_I(+\infty, -\infty)$ has common eigenfunctions with $\hat{H}$ (with eigenvalue E), $\widehat{\vec{L}^2}$ (with eigenvalue related to l), and $\hat{L}_z$ (with eigenvalue m). This follows from the fact that the time evolution leaves invariant the energy E, angular momentum $\vec{L}^2$, and angular momentum projection L_z. Classically, this means that

$$\partial_t H = \partial_t \vec{L}^2 = \partial_t L_z = 0, \tag{49.2}$$

while in quantum mechanics, invariance under time translation means that the commutator with $\hat{H}$ vanishes, $\partial_t \cdots = 0 \to [\hat{H}, \ldots] = 0$ (the generator of time translation is $\hat{H}$). However, the time evolution invariance is

$$[\hat{H}, \hat{H}] = [\hat{H}, \widehat{\vec{L}^2}] = [\hat{H}, \hat{L}_z] = 0, \tag{49.3}$$

which is true in a spherically symmetric system.

But then, unitarity, $\hat{S}^\dagger = \hat{S}^{-1}$, implies $S_l^* = S_l^{-1}$ and, since $S_l = e^{2i\delta_l(k)}$, this means that $\delta_l^*(k) = \delta_l(k)$, so that the phase shift is real. But this was what we (implicitly) assumed when defining the phase shift.

Equivalently, the unitarity condition is $|S_l(k)| = 1$, which means that the only change in the outgoing wave with respect to the incoming one is a phase, not an amortization ($e^{-|\text{Im}\,\delta_l|}$), which would lead to probability decay (in a physical case, probability decay could only mean that the system is not complete, and the probability "leaks" somewhere else, into another system coupled to the one we are analyzing).

Since

$$ka_l = \frac{S_l - 1}{2i} = \frac{i}{2} - \frac{ie^{2i\delta_l}}{2} = \frac{i}{2} + \frac{e^{-i\pi/2+2i\delta_l}}{2}, \tag{49.4}$$

this means that ka_l maps a circle in the complex plane, centered on $i/2$, of radius 1/2, called the "unitarity circle".

49.2 Application to Cross Sections

To see the effect of unitarity analysis on physical measurements, we need to look at cross sections. As we saw, in the partial wave formalism,

$$\frac{d\sigma}{d\Omega} = |f_k(\theta)|^2. \tag{49.5}$$

But since (see (48.12))

$$\begin{aligned} f_k(\theta) &= \sum_{l=0}^{\infty}(2l+1)P_l(\cos\theta)a_l(k) \\ a_l(k) &= \frac{e^{2i\delta_l} - 1}{2ik} = \frac{S_l - 1}{2ik}, \end{aligned} \tag{49.6}$$

it follows that

$$\frac{d\sigma}{d\Omega} = \frac{1}{4k^2}\sum_{l=0}^{\infty}\sum_{l'=0}^{\infty}(2l+1)(2l'+1)(e^{2i\delta_l(k)} - 1)(e^{-2i\delta_{l'}(k)} - 1)P_l(\cos\theta)P_{l'}(\cos\theta), \tag{49.7}$$

where we have used the fact that $P_l(\cos\theta)$ is real. Then the total cross section is found by integrating over $d\Omega$,

$$\begin{aligned} \sigma_{\text{tot}}(k) &= \int \frac{d\sigma}{d\Omega} d\Omega \\ &= \frac{1}{4k^2}\int_0^{2\pi} d\phi \sum_{l=0}^{\infty}\sum_{l'=0}^{\infty}\int_{-1}^{+1} d(\cos\theta)P_l(\cos\theta)P_{l'}(\cos\theta) \\ &\quad \times (2l+1)(2l'+1)(e^{2i\delta_l(k)} - 1)(e^{-2i\delta_{l'}(k)} - 1). \end{aligned} \tag{49.8}$$

Now we use the orthogonality condition of the Legendre polynomials (see (17.38)),

$$\int_{-1}^{+1} d(\cos\theta)P_l(\cos\theta)P_{l'}(\cos\theta) = \frac{2\delta_{ll'}}{2l+1}, \tag{49.9}$$

so that we obtain

$$\sigma_{\text{tot}}(k) = \frac{\pi}{k^2}\sum_{l=0}^{\infty}(2l+1)\left|e^{2i\delta_l(k)} - 1\right|^2 = \frac{4\pi}{k^2}\sum_{l=0}^{\infty}(2l+1)\sin^2\delta_l(k). \tag{49.10}$$

This means we can define a cross section σ_l through

$$\sigma_{\text{tot}}(k) = \sum_{l=0}^{\infty}\sigma_l(k), \tag{49.11}$$

and, identifying the two expansions, we obtain

$$\sigma_l(k) = \frac{4\pi}{k^2}(2l+1)\sin^2\delta_l(k). \tag{49.12}$$

A more general formula applies also to the case when unitarity is "violated", meaning that $\delta_l(k)$ is complex (which implies that the system under analysis interacts with other systems, having a "probability leak", so the scattering is "inelastic"). This formula is

$$\sigma_l(k) = \frac{\pi}{k^2}(2l+1)\left|e^{2i\delta_l(k)} - 1\right|^2. \tag{49.13}$$

In the case of unitary evolution ($\delta_l(k) \in \mathbb{R}$), since $\sin^2\delta_l(k) \leq 1$ we have a "unitarity bound",

$$\sigma_l(k) \leq \frac{4\pi}{k^2}(2l+1) \equiv \sigma_l^{\max}. \tag{49.14}$$

We will come back to this bound later but, for the moment, we just note that the saturation of the bound is at $\delta_l = \pi/2$, which means that when $\delta_l \simeq \pi/2$, the lth partial wave has a maximal effect.

However, we need to calculate $\delta_l(k)$. To do that, we note that $a_l(k)$ is found from (49.6) by integration with $P_{l'}(\cos\theta)d\cos\theta$, so

$$\int_{-1}^{1} d(\cos\theta)P_{l'}(\cos\theta)f_k(\theta) = 2a_l(k) = \frac{e^{2i\delta_l(k)} - 1}{ik} = 2e^{i\delta(k)}\frac{\sin\delta_l}{k}. \tag{49.15}$$

But the Lippmann–Schwinger equation implies (see (46.71))

$$f_k(\theta) = -\frac{2m}{4\pi\hbar^2}\int d^3\vec{r}\,' e^{\mp i\vec{k}'\cdot\vec{r}\,'}V(\vec{r}\,')\psi_\pm(\vec{k},\vec{r}\,'), \tag{49.16}$$

where we have substituted $G_0^{(+)}(\vec{r},\vec{r}\,')$, and in the spherically symmetric case the l-expansion of the wave function is

$$\psi_+(k;r,\theta) = \sum_{l=0}^{\infty}(2l+1)e^{i(\delta_l + l\pi/2)}R_l(k;r)P_l(\cos\theta). \tag{49.17}$$

Instead of continuing like this, we can use a spherical expansion of the Green's function,

$$G_0^{(+)}(\vec{r},\vec{r}\,') = -ik\sum_{l,m}Y_{lm}^*(\vec{n}_r)Y_{lm}(\vec{n}_r')j_l(kr_<)h_l^{(1)}(kr_>), \tag{49.18}$$

which implies, via the l-expansion of the wave function, the Lippmann–Schwinger equation for the radial wave function,

$$R_l(k;r) = j_l(kr) - \frac{2mik}{\hbar^2}\int_0^\infty dr' r'^2 j_l(kr_<)h_l^{(1)}(kr_>)V(r')R_l(k;r'). \tag{49.19}$$

Since the first term is the expansion of the plane wave, the second term is the expansion of the $f_k(\theta)$, which gives $a_l(k)$, so that

$$a_l(k) = \frac{e^{i\delta_l(k)}\sin\delta_l(k)}{k} = -\frac{2m}{\hbar^2}\int_0^\infty r^2 dr V(r)j_l(kr)R_l(k;r). \tag{49.20}$$

49.3 Radial Green's Functions

We can also expand the Green's functions and the Lippmann–Schwinger equation using a partial wave expansion.

The Green's function partial wave expansion is

$$\begin{aligned} G(z;\vec{r},\vec{r}') = G(z;r,r',\theta) &= \sum_{l=0}^{\infty} \frac{2l+1}{4\pi} \tilde{g}_l(z;r,r') R_l(\vec{n}_r,\vec{n}_k) \\ &= \sum_{l=0}^{\infty} \sum_{m=-l}^{l} \tilde{g}_l(z;r,r') Y^*_{lm}(\vec{n}_r) Y_{lm}(\vec{n}_{r'}), \end{aligned} \tag{49.21}$$

where $\tilde{g}_l(z;r,r')$ is known as the radial Green's function. It satisfies

$$\left\{ z + \frac{\hbar^2}{2m}\left[\frac{1}{r^2}\frac{d}{dr}\left(r^2\frac{d}{dr}\right) - \frac{l(l+1)}{r^2}\right] - V(r)\right\} \tilde{g}_l(z;r,r') = \frac{\delta(r-r')}{r^2}, \tag{49.22}$$

and has the spectral decomposition

$$\tilde{g}_l(z;r,r') = \sum_n \frac{R_{nl}(r)R^*_{nl}(r)}{z-E_n}, \tag{49.23}$$

representing in radial space the general formula

$$\hat{G}(z) = \sum_n \frac{|n\rangle\langle n|}{z-E_n}. \tag{49.24}$$

In the free particle case, the radial Green's function is

$$\tilde{g}_l^{(+)0}(E;r,r') = -\frac{2mi}{\hbar^2} q^2 j_l(qr_<) h_l^{(1)}(qr_>), \tag{49.25}$$

where

$$E = \frac{\hbar^2 q^2}{2m} \tag{49.26}$$

and $r_< = \min(r,r')$, $r_> = \max(r,r')$. It appears in the radial Lippmann–Schwinger equation,

$$R_l(k;r) = j_l(kr) + \int dr' \tilde{g}_l^{(+)0}(E;r,r') V(r') R_l(k;r') r'^2, \tag{49.27}$$

which leads to the equation (49.19). It also leads to the Lippmann–Schwinger equation for the full Green's function $\tilde{g}_l$,

$$\tilde{g}_l(z;r,r') = \tilde{g}_l^{(+)0}(z;r,r') + \int dr'' r''^2 \tilde{g}_l^{(+)0}(z;r,r'') V(r'') \tilde{g}_l(z;r'',r'). \tag{49.28}$$

As an example, consider the radial delta function potential,

$$V(r) = -\lambda\delta(r-a). \tag{49.29}$$

Then the Lippmann–Schwinger equation for the full radial Green's function has the solution

$$\tilde{g}_l(z;r,r') = \tilde{g}_l^{(+)0}(z;r,r') - \lambda a^2 \frac{\tilde{g}_l^{(+)0}(z;r,a)\tilde{g}_l^{(+)0}(z;a,r')}{1+\lambda a^2 \tilde{g}_l^{(+)0}(z;a,a)}. \tag{49.30}$$

This means that the radial Green's function gives

$$R_l(k;r) = j_l(kr) - \lambda a^2 \frac{\tilde{g}_l^{(+)0}(E;r,a) j_l(ka)}{1 + \lambda a^2 \tilde{g}_l^{(+)0}(E;a,a)}. \tag{49.31}$$

The Jost solutions are

$$f_l^{\mp}(k;r) = \pm ikr h_l^{(1,2)}(kr). \tag{49.32}$$

The details are left as an exercise.

49.4 Optical Theorem

We have already proven a version of the optical theorem in Chapter 46, but here we will revisit it.

First, we review the proof given in Chapter 46. Unitarity means $\hat{S}\hat{S}^\dagger = \mathbb{1}$, but since

$$\hat{S} = \mathbb{1} + \frac{ik}{2\pi}\hat{f}, \tag{49.33}$$

it follows that

$$\frac{\hat{f} - \hat{f}^\dagger}{2i} = \frac{k}{4\pi}\hat{f}\hat{f}^\dagger. \tag{49.34}$$

When we represent this abstract statement in a basis, and take $\vec{n}' = \vec{n}_k$, we obtain the total cross section, since

$$\text{Im} f(\vec{n}_k, \vec{n}_k) = \frac{k}{4\pi}\int d^2\vec{n}_r |f(\vec{n}_r, \vec{n}_k)|^2 = \frac{k}{4\pi}\sigma_{\text{tot}}, \tag{49.35}$$

and on the left-hand side we have the forward direction,

$$f(\theta = 0) = f(\vec{n}_k, \vec{n}_k). \tag{49.36}$$

But we want to describe this proof in physical terms, which means in terms of the interaction with a potential V. To do that, we relate the above calculation to the T-matrix, by

$$\hat{f} = -\frac{m}{2\pi\hbar^2}\hat{T}. \tag{49.37}$$

Then, when taking the matrix element in the $\langle\vec{k}|$ basis, we obtain

$$f(\theta = 0) = f(\vec{n}_k, \vec{n}_k) = -\frac{m}{2\pi\hbar^2}\langle\vec{k}|\hat{T}|\vec{k}\rangle. \tag{49.38}$$

But the T-matrix element is related to the potential $\hat{V}$ in between two states, so

$$\text{Im}\langle\vec{k}|\hat{T}|\vec{k}\rangle = \text{Im}\langle\vec{k}|\hat{V}|\psi+\rangle. \tag{49.39}$$

From the Lippmann–Schwinger equation, we obtain

$$\langle\vec{k}'|\hat{V}|\psi+\rangle = \langle\psi+|\hat{V}|\psi+\rangle - \left\langle\psi+\left|\hat{V}\frac{1}{E - \hat{H}_0 - i\epsilon}\hat{V}\right|\psi+\right\rangle. \tag{49.40}$$

On the other hand, from simple complex analysis formula extended to operators,

$$\hat{G}_0^{(+)} = \frac{1}{E - \hat{H}_0 - i\epsilon} = \mathcal{P}\left(\frac{1}{E - \hat{H}_0}\right) + i\pi\delta(E - \hat{H}_0), \tag{49.41}$$

where $\mathcal{P}$ refers to the principal part. But when taking the imaginary part, we obtain a zero if we have a diagonal matrix element of a Hermitian ("real") operator, so

$$\begin{aligned} &\mathrm{Im}\langle\psi+|\hat{V}|\psi+\rangle = 0 \\ &\mathrm{Im}\left\langle\psi+\left|\hat{V}\mathcal{P}\left(\frac{1}{E-\hat{H}_0}\right)\hat{V}\right|\psi+\right\rangle = 0. \end{aligned} \tag{49.42}$$

Then out of the three terms in the imaginary part of the T-matrix element, only one contributes (and is actually purely imaginary), and we obtain

$$\begin{aligned} \mathrm{Im}\langle\vec{k}|\hat{V}|\psi+\rangle &= -\pi\langle\psi+|\hat{V}\delta(E-\hat{H}_0)\hat{V}|\psi+\rangle \\ &= -\pi\langle\vec{k}|\hat{T}\delta(E-\hat{H}_0)\hat{T}|\vec{k}\rangle \\ &= -\pi\int d^3\vec{\tilde{k}}'\langle\vec{k}|\hat{T}|\vec{\tilde{k}}'\rangle\delta\left(E-\frac{\hbar^2\tilde{k}'^2}{2m}\right)\langle\vec{\tilde{k}}'|\hat{T}|\vec{k}\rangle \\ &= -\frac{\pi m}{\hbar^2 k}\int d\Omega k^2\left|\langle\vec{k}|\hat{T}|\vec{k}'\rangle\right|^2, \end{aligned} \tag{49.43}$$

where in the last step we have used

$$\delta\left(E-\frac{\hbar^2\tilde{k}'^2}{2m}\right) = \frac{\delta(k-\tilde{k}')}{\hbar^2\tilde{k}'/m}. \tag{49.44}$$

Then also (since $d^2\vec{n}' = d^2\vec{n}_k = d\Omega$)

$$\int d^3\vec{\tilde{k}}'\delta(k-\tilde{k}') = \int d^2\vec{n}'k^2 \times (|\vec{\tilde{k}}'\rangle \to |\vec{k}'\rangle) = \int d\Omega\, k^2 \times (|\vec{\tilde{k}}'\rangle \to |\vec{k}'\rangle). \tag{49.45}$$

Finally, then, we obtain

$$\begin{aligned} \mathrm{Im} f(\theta=0) &= -\frac{m}{2\pi\hbar^2}\mathrm{Im}\langle\vec{k}|\hat{V}|\psi+\rangle \\ &= \frac{m^2 k}{2\hbar^4}\int d\Omega\left|\langle\vec{k}|\hat{T}|\vec{k}'\rangle\right|^2 \\ &= \frac{k}{4\pi}\int d\Omega\left|-\frac{m}{2\pi\hbar^2}\langle\vec{k}|\hat{T}|\vec{k}'\rangle\right|^2 \\ &= \frac{k}{4\pi}\int d\Omega|f(\theta)|^2 = \frac{k}{4\pi}\sigma_{\mathrm{tot}}. \end{aligned} \tag{49.46}$$

This completes the physical proof of the optical theorem. While the result is the same, and we have taken a longer route to prove it, we still have gained something. The use of the Lippmann–Schwinger equation, which can be solved perturbatively through the Born series, means that we can define the perturbative series on both sides of the equation for the optical theorem, and consider a term-by-term analysis (though we will not do so here). Moreover, in the previous proof, we *assumed* that the S-matrix is unitary, but that means assuming that the quantum mechanical construction of scattering is fully consistent (and we have not proved this until now), whereas here the proof is explicitly written in terms of the potential.

49.5 Scattering on a Potential with a Finite Range

We now give an example of how to calculate δ_l and solve the partial wave problem. Consider the case of a potential that is *strictly* vanishing outside a spherical shell,

$$V(r) = 0 \quad \text{for} \quad r \geq r_0. \tag{49.47}$$

In such a case (the case of a potential with a strict finite range), outside the spherical shell the solution of the Schrödinger equation is a free spherical wave.

Throughout the space (both outside and inside the spherical shell), we found (see (48.34)) that the wave function solution for a spherically symmetric potential is

$$\begin{aligned}\psi_k(r,\theta) &= \sum_{l=0}^{\infty}(2l+1)e^{i(\delta_l+l\pi/2)}\tilde{R}_{El}(r)P_l(\cos\theta)\\ &= \sum_{l=0}^{\infty} R_{El}(r)P_l(\cos\theta).\end{aligned} \tag{49.48}$$

Furthermore, we found an *asymptotic* solution, (48.16), for which the potential was vanishing (the interaction was turning off). But in our case, the potential is really zero, so the solution is exact,

$$R_{El}(r) = C_l j_l(kr) + D_l n_l(kr), \tag{49.49}$$

where

$$\begin{aligned}C_l &= A_l + A_l^* = A_l^0 2\cos\delta_l\\ D_l &= iA_l - iA_l^* = A_l^0(-2\sin\delta_l)\end{aligned} \tag{49.50}$$

and

$$A_l^0 = \frac{R_{El}(r)}{\tilde{R}_{El}(r)} = (2l+1)e^{i(\delta_l+l\pi/2)}. \tag{49.51}$$

In order to find the wave function, as we saw in the one-dimensional case, at the place where the potential jumps, we need to impose continuity of the logarithmic derivative of the wave function. Calculating it at $r_0+\epsilon$, so we can use the above formulas, we obtain

$$q_{0l} \equiv \left.\frac{r dR_{El}(r)}{R_{El}(r)dr}\right|_{r=r_0} = kr_0\frac{j_l'(kr_0)\cos\delta_l - n_l'(kr_0)\sin\delta_l}{j_l(kr_0)\cos\delta_l - n_l(kr_0)\sin\delta_l}. \tag{49.52}$$

Then we calculate q_{0l} at $r_0-\epsilon$, in terms of the wave function solution at $r \leq r_0$; equating it to the above gives the full solution, in terms of the calculated δ_l. Indeed, inverting the formula, we find

$$\tan\delta_l = \frac{kr_0 j_l'(kr_0) - q_{0l}j_l(kr_0)}{kr_0 n_l'(kr_0) - q_{0l}n_l(kr_0)}. \tag{49.53}$$

To find the solution inside the shell, we must impose regularity (normalizability) of the solution at $r=0$, which, as we saw in Chapter 19, means that $\chi_{El}(r=0) = rR_{El}(r)|_{r=0} = 0$.

49.6 Hard Sphere Scattering

The simplest example of the calculation inside radius $r = r_0$, is for a "hard sphere", with infinite potential, $V = \infty$, for $r \leq r_0$.

Then the simplest boundary condition at $r = r_0$ is that the radial wave function vanishes, $R_{El}(r_0) = 0$, which means

$$j_l(kr_0)\cos\delta_l - n_l(kr_0)\sin\delta_l = 0, \tag{49.54}$$

leading to

$$\tan\delta_l = \frac{j_l(kr_0)}{n_l(kr_0)}. \tag{49.55}$$

Equivalently, we can take the limit $q_{0l} \to \infty$ (since the denominator in the definition of q_{0l} has $R_{El}(r_0) = 0$, as the wave function does when it hits an infinite wall) in the general formula for $V(r < r_0)$, and we obtain the same result.

For $l = 0$, and at $r \geq r_0$, we find

$$\tan\delta_0 = \frac{j_0(kr_0)}{n_0(kr_0)} = \frac{\sin(kr_0)/(kr_0)}{-\cos(kr_0)/(kr_0)} = -\tan(kr_0) = \tan(-kr_0). \tag{49.56}$$

Then we identify the arguments of the tangent, since both of them are bounded,

$$\delta_0 = -kr_0. \tag{49.57}$$

The radial wave function (for $r \geq r_0$) is

$$R_{El}(r) = 2A_l^0\left(j_l(kr)\cos\delta_l - n_l(kr)\sin\delta_l\right), \tag{49.58}$$

which means that for $l = 0$ we obtain

$$\begin{aligned} R_{E0}(r) &= 2A_0^0\left(\frac{\sin kr_0}{kr_0}\cos\delta_0 + \frac{\cos kr_0}{kr_0}\sin\delta_0\right) \\ &= \frac{2A_0^0}{kr_0}\sin(kr_0 + \delta_0) = \frac{2A_0^0}{kr_0}\sin[k(r - r_0)], \end{aligned} \tag{49.59}$$

a sinusoid shifted by r_0, i.e., starting at r_0 instead of zero.

Then, we calculate the scattering length,

$$a = -\lim_{r\to 0}\frac{\delta_0(k)}{k} = r_0. \tag{49.60}$$

Note that now we do not need to take the low-energy limit.

49.7 Low-Energy Limit

We now consider the low-energy limit for the hard sphere potential. For Bessel functions at small argument, $x = kr_0 \ll 1$,

$$\begin{aligned} j_l(kr_0) &\simeq \frac{(kr_0)^l}{(2l+1)!!} \\ n_l(kr_0) &\simeq -\frac{(2l-1)!!}{(kr_0)^{l+1}}. \end{aligned} \tag{49.61}$$

In the $k \to 0$ limit, we saw that $\delta_l \simeq \tan \delta_l$, so

$$\delta_l(k) = \tan \delta_l(k) = \frac{(kr_0)^{2l+1}}{(2l+1)!!(2l-1)!!}. \tag{49.62}$$

For the partial wave amplitude, in the low-energy limit, we obtain

$$a_l(k) \simeq \frac{\delta_l(k)}{k} \propto k^{2l}. \tag{49.63}$$

This means that δ_0 dominates strongly, giving an isotropic ($l = 0$) result. The lth partial wave cross section is

$$\sigma_l = 4\pi(2l+1)\frac{\sin^2 \delta_l(k)}{k^2} \propto k^{4l}, \tag{49.64}$$

which means all these cross sections go to zero, except σ_0, which stays constant. Then

$$\begin{aligned} \sigma_0 &= 4\pi \lim_{k\to 0} \frac{\sin^2 \delta_0(k)}{k^2} = 4\pi a^2 \\ a &= -\lim_{k\to 0} \frac{\delta_0(k)}{k} = r_0, \end{aligned} \tag{49.65}$$

so the total cross section is

$$\sigma_{\text{tot}} \simeq \sigma_0 \simeq 4\pi a^2 = 4\pi r_0^2, \tag{49.66}$$

which is four times the geometrical cross section for radius r_0, πr_0^2.

Moreover, the differential cross section is also constant,

$$\frac{d\sigma}{d\Omega} = |f|^2 \simeq |a_0(k)P_0(\cos\theta)|^2 = \left|\lim_{k\to 0} \frac{\delta_0(k)}{k}\right|^2 = a^2 = r_0^2, \tag{49.67}$$

where we have used $P_0(\cos\theta) = 1$. This is consistent with the previous result, as the integration over the solid angle just gives a factor equal to the total angle, 4π.

Since we have four times the geometrical cross section (which would be the classical result), we have to ask, why do we have a larger result? One answer is that at low energy the wavelengths of the particles are larger than the range of the potential, leading to an extreme quantum regime, the opposite of the classical limit.

On the other hand, at *high energy*, $kr_0 \gg 1$, we still find a result higher than the classical one, even though now the wavelengths are going to zero. The result is (the derivation can be found in [1])

$$\sigma_{\text{tot}} = 2\pi r_0^2 = \sigma_{\text{reflection}} + \sigma_{\text{shadow}}, \tag{49.68}$$

where $\sigma_{\text{reflection}} = \pi r_0^2$ is the classical result, and σ_{shadow} is the Fraunhofer diffraction result in optics. This is consistent with the geometric optics approximation of the quantum mechanical wave function (semiclassical), which is a short-wavelength approximation.

Important Concepts to Remember

- Unitarity means that $S_l^* = S_l^{-1}$, so that δ_l is real, $|S_l| = 1$, and $ka_l = \frac{S_l - 1}{2i}$, meaning that ka_l maps a circle on the complex plane called the unitarity circle.
- The total cross section expands as $\sigma_{\text{tot}} = \sum_l \sigma_l$, with $\sigma_l = 4\pi/k^2(2l+1)\sin^2\delta_l$, so $\sigma_l \leq 4\pi/k^2(2l+1)$, called the unitarity bound.

- We can write a Lippmann–Schwinger equation for the radial wave function $R_l(k;r)$, $R_l(k;r) = j_l(kr) - \frac{2mik}{\hbar^2}\int_0^\infty dr'r'^2 j_l(kr_<)h_l^{(1)}(kr_>)V(r')R_l(k;r')$, and the Green's function has also a decomposition in spherical harmonics $Y^*_{lm}(\vec{n}_r)Y_{lm}(\vec{n}'_r)$, with the coefficients of the radial Green's function

$$g_l(k;r,r') = \sum_n \frac{R_{nl}(r)R_{nl}(r')}{z - E_n},$$

satisfying a similar L–S equation.
- The optical theorem can be proven physically, without assuming $\hat{S}\hat{S}^\dagger = \mathbb{1}$, and in the proof we use the Lippmann–Schwinger equation, which can be solved perturbatively via the Born series so that we can reduce it to a term-by-term equation.
- When scattering on a hard sphere of radius r_0, the scattering length (without actually employing a low-energy limit) is $a = r_0$, and the low-energy differential cross section is constant, $d\sigma/d\Omega = r_0^2$, so we get four times the geometric cross section, owing to a Fraunhofer diffraction contribution.

Further Reading

See [2] and [1] for more details.

Exercises

(1) Use the Lippmann–Schwinger equation for ψ to write an expression for σ_l in terms of integrals involving the potential $V(r)$ and the wave function.

(2) Show the details of going from (49.19) to (49.20).

(3) In the case of the radial delta function potential, prove that the radial wave function $R_l(k;r)$ is (49.31) and the Jost solutions are (49.32).

(4) Expand the optical theorem first in angular momentum l and then in the Born series, for both $f(\theta = 0)$ and $\langle\vec{k}|\hat{T}|\vec{k}'\rangle$.

(5) For a potential with finite range, consider $V = V_0 > 0$ inside $r < r_0$, and find the solution in this region, in the case $E > V_0$, imposing normalizability at $r = 0$.

(6) For the hard sphere, calculate σ_l at general k.

(7) Calculate the high-energy limit ($k \to \infty$) for the partial wave cross section σ_l for a hard sphere, and their relative weight in σ_{tot}.

50 Low-Energy and Bound States, Analytical Properties of Scattering Amplitudes

In this chapter we will consider the relation between low-energy scattering (meaning $E > 0$) and bound states (with $E < 0$), and we will find that, on considering complex values of k (or E), the same result can be obtained by analysis on the complex plane. This leads us to consider more general analytical properties of the scattering amplitudes.

50.1 Low-Energy Scattering

We have seen that the phase shift δ_l is calculated from the potential via

$$ka_l(k) = e^{i\delta_l}\sin\delta_l = -\frac{2mk}{\hbar^2}\int_0^\infty dr\, r^2 V(r) j_l(kr) R_l(k;r). \tag{50.1}$$

Moreover, at low energy, $k \to 0$ implies $\delta_l \to 0$, and we have

$$ka_l(k) \simeq \delta_l \propto k^{2l+1} \to 0, \tag{50.2}$$

called *threshold behavior*, which means that the $l = 0$ mode dominates the phase shifts.

Physically, we can understand the dominance of $l = 0$ from the effective potential

$$V_{\text{eff}}(r) = V(r) + \frac{\hbar^2}{2m}\frac{l(l+1)}{r^2}. \tag{50.3}$$

At low energy, since the second term in V_{eff} blows up for $r \to 0$ if $l \neq 0$, we will never come close to zero, where $l \neq 0$ dominates the effective potential. This means that only the $l = 0$ term contributes, for which $V_{\text{eff}} = V(r)$.

Step Potential

We next consider the simplest finite-range potential, the one that is constant inside its range,

$$V = \begin{cases} V_0, & r < r_0 \\ 0, & r \geq r_0. \end{cases} \tag{50.4}$$

If $V_0 > 0$ we have a repulsive potential, and if $V_0 < 0$ we have an attractive one.

At low energy, $kr_0 \ll 1$, $l = 0$ dominates and the equation (48.34) implies

$$\psi_k(r,\theta) \simeq e^{i\delta_0} R_{E,l=0} = e^{i\delta_0}(j_0(kr)\cos\delta_0 - n_0(kr)\sin\delta_0) \simeq \frac{e^{i\delta_0}\sin(kr+\delta_0)}{kr}, \tag{50.5}$$

where in the first equality we have used the dominance of $l = 0$, in the second $P_0(\cos\theta) = 1$, and in the third $r \to \infty$.

For the reduced radial wave function, we obtain (for $r \geq r_0$)

$$\chi = rR_{E,l=0}(r) = \text{const} \times \sin(kr_0 + \delta_0), \tag{50.6}$$

where we have chosen a different normalization, such that the constant is 1.

Within the potential range, we impose regularity at $r \to 0$, $\chi(0) = 0$. Since $V = V_0$ is constant over the range, the wave function is a free particle one, but with energy shifted by V_0,

$$E - V_0 = \frac{\hbar^2 \tilde{k}^2}{2m}, \tag{50.7}$$

so the real solution for $l = 0$ (the spherically symmetric case) or at large r is

$$\chi(r) = N \frac{e^{i\tilde{k}r} - e^{-i\tilde{k}r}}{2i} = N \sin \tilde{k}r, \tag{50.8}$$

where N is a normalization constant. Of course, this is only true if $E > V_0$. If $E < V_0$, we write $\tilde{k} = i\kappa$, so

$$V_0 - E = \frac{\hbar^2 \kappa^2}{2m}, \tag{50.9}$$

leading to the wave function

$$\chi(r) = \frac{N}{i} \frac{e^{\kappa r} - e^{-\kappa r}}{2} = \tilde{N} \sinh \kappa r. \tag{50.10}$$

We need to join the inside and outside solutions at $r = r_0$.

We saw in the previous chapter that among the lth partial wave cross sections

$$\sigma_l = \frac{4\pi}{k^2}(2l + 1) \sin^2 \delta_l, \tag{50.11}$$

at low energy the $l = 0$ term dominates,

$$\sigma_0 = 4\pi \frac{\sin^2 \delta_0(k)}{k^2} \to 4\pi a^2. \tag{50.12}$$

However, the dominance of σ_0 also happens at $\sin^2 \delta_0 \to 0$ and $k \to 0$, not only at $\delta_0 \to 0$. It can also happen at $\delta_0 = \pi$.

If $-V_0 = |V_0|$ is large then $\tilde{k}$ is large (even though k is small). Thus, eventually we reach $\tilde{k}r_0 = \pi$, so

$$\sin \tilde{k}r = \sin \pi = 0, \tag{50.13}$$

even though $kr_0 = 0$. But the continuity of the inside and outside solutions means that

$$\sin(kr_0 + \delta_0) = N \sin(\tilde{k}r_0) = 0, \tag{50.14}$$

so $\delta_0 = \pi$, since we have increased δ_0 as $|V_0|$ is increased from 0, which correspond to the free particle (when $\delta_0 = 0$).

50.2 Relation to Bound States

The outside ($r \geq r_0$) wave function solution for $k \to 0$ can be rewritten as

$$\chi(r) = \sin(kr + \delta_0) = \sin\left[k\left(r + \frac{\delta_0}{k}\right)\right] \simeq k\left(r + \frac{\delta_0}{k}\right). \tag{50.15}$$

However, since we are in the regime $E \simeq 0$, $l = 0$, $r > r_0$ so that $V = 0$, the Schrödinger equation is

$$\frac{d^2(rR)}{dr^2} = \frac{d^2\chi}{dr^2} = 0, \tag{50.16}$$

with solution

$$\chi \simeq C(r-a), \tag{50.17}$$

where C, a are constants. But matching with the first form in the $k \to 0$ limit, we obtain that

$$a = -\lim_{k\to 0}\frac{\delta_0}{k}, \tag{50.18}$$

meaning a is the scattering length.

But the condition of matching the inside and outside solutions means the logarithmic derivative must be continuous at r_0. We define it slightly differently, in terms of $\chi(r)$, as

$$\tilde{q}_0 \equiv \frac{\chi'(r_0)}{\chi(r_0)}. \tag{50.19}$$

In terms of the outside function, its value at arbitrary r is

$$\tilde{q}_0 = \frac{k\cos(kr+\delta_0)}{\sin(kr+\delta_0)} = k\cot\left[k\left(r+\frac{\delta_0}{k}\right)\right] \simeq k\frac{1}{k\,(r+\delta_0/k)} = \frac{1}{r-a}. \tag{50.20}$$

If we choose $r = 0$, even though this is outside the domain of validity of the function ($r \geq r_0$), we obtain

$$\tilde{q}_{\text{outside}}(r\to 0) = k\cot\delta_0 = -\frac{1}{a}, \tag{50.21}$$

consistent with the definition of the scattering length. Its significance follows from the fact that $\tilde{q}$ is continuous at r_0: we have

$$\chi_{\text{outside}}(r) \simeq \sin[k(r-a)] = 0 \tag{50.22}$$

for $r = a$, regardless of whether $r = a \geq r_0$ or not, which means that the scattering length $r = a$ is the intercept of the outside wave function; see Fig. 50.1.

If $\delta_0 < 0$, coming from the repulsive potential $V_0 > 0$, then $a > 0$. If $\delta > 0$, coming from the attractive potential $V_0 < 0$, then $a < 0$. However, in the attractive-potential case $V_0 < 0$, if $|V_0|$ is large, meaning large $\tilde{k}$, then we have lower periodicity of the wave function, so the wave function drops faster at $r > r_0$ and so its first zero after r_0 corresponds to a different scattering length $a' > 0$.

Indeed, the low-energy ($E \simeq 0$) outside solution $r > r_0$ is

$$\chi \simeq C(r-a). \tag{50.23}$$

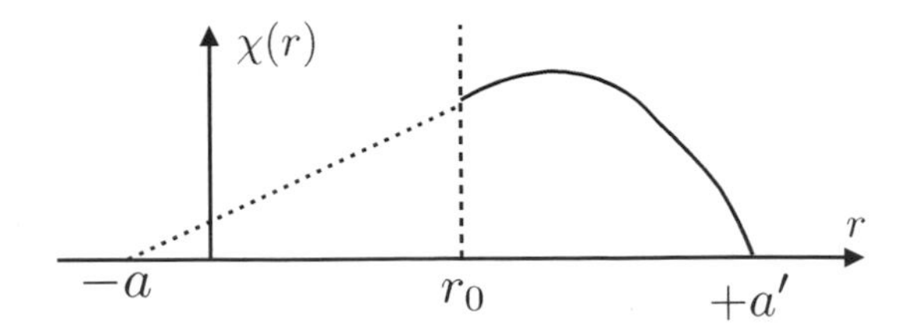

Figure 50.1 Wave function and scattering length.

This is a scattering solution for $E > 0$ (so $E = 0+$), though it can be written also as

$$\chi \simeq Be^{-\kappa r}, \tag{50.24}$$

where $\kappa \simeq 0$ (very small) and

$$B = -Ca, \quad -B\kappa = C < 0. \tag{50.25}$$

After this rewriting, the wave function looks also like a bound state ($E < 0$), though very slightly so, $E = 0-$. In both the scattering ($E = 0+$) and bound state ($E = 0-$) solutions,

$$\frac{\hbar^2 \tilde{\kappa}^2}{2m} = E - V_0 \simeq |V_0|. \tag{50.26}$$

Therefore in both cases the inside wave function is

$$\chi \simeq \sin \tilde{\kappa} r. \tag{50.27}$$

Then the logarithmic derivative inside is (in both cases)

$$\tilde{q}_0^{\text{inside}} = \frac{-\kappa e^{-\kappa r_0}}{e^{-\kappa r_0}} = -\kappa = \tilde{\kappa} \cot \tilde{\kappa} r_0, \tag{50.28}$$

to be equated to the outside value,

$$\tilde{q}_0^{\text{outside}} = \frac{1}{r_0 - a} \simeq -\frac{1}{a}, \tag{50.29}$$

in this larger-scattering-length case, $r_0 \ll a$. Thus the binding energy of the bound state is obtained by equating the cases $E = 0+$ and $E = 0-$,

$$I = -E_{\text{bound state}} = \frac{\hbar^2 \kappa^2}{2m} = \frac{\hbar^2}{2ma^2}. \tag{50.30}$$

This is a relation between the binding energy I of a mode close to zero and the (large value of the) scattering length a for scattering on an attractive potential of large depth.

50.3 Bound State from Complex Poles of $S_l(k)$

The partial wave S-matrix $S_l = e^{2i\delta_l(k)}$ appears in the ratio of the e^{ikr}/r and e^{-ikr}/r modes of the real wave function solution. From (48.19), the radial wave function is

$$\begin{aligned} R_{El}(r) &\simeq A_l^0 \left[e^{i\delta_l} \frac{e^{i(kr - l\pi/2)}}{r} - e^{-i\delta_l} \frac{e^{-i(kr - l\pi/2)}}{r} \right] \\ &= \tilde{A}_l^0 \left[\frac{e^{i(kr - l\pi/2)}}{r} - \frac{1}{e^{2i\delta_l(k)}} \frac{e^{-i(kr - l\pi/2)}}{r} \right], \end{aligned} \tag{50.31}$$

where $e^{2i\delta_l(k)} = S_l(k)$ appears in the denominator of the incoming spherical wave, and we have redefined A_l^0 by a factor $e^{i\delta_l(k)}$.

Dropping A_l^0, putting $l = 0$ as it is the most relevant case, *yet the calculation is actually valid for any l*, from (48.34) we obtain

$$\begin{aligned}\psi_k(r,\theta) &\simeq e^{i\delta_0} R_{E,l=0} = A_0^0 \left[S_0(k)\frac{e^{ikr}}{r} + \frac{e^{-ikr}}{r}\right] \\ &= \tilde{A}_0^0 \left[\frac{e^{ikr}}{r} - \frac{1}{S_0(k)}\frac{e^{-ikr}}{r}\right].\end{aligned} \quad (50.32)$$

At this point we generalize to complex k, in order to connect to the description of bound states. Writing $k = i\kappa$, now with $\kappa > 0$, the scattering solution transforms into

$$\psi_k(r,\theta) = \tilde{A}_0^0 \left[\frac{e^{-\kappa r}}{r} - \frac{1}{S_0(i\kappa)}\frac{e^{+\kappa r}}{r}\right], \quad (50.33)$$

which matches the bound state wave function solution provided that the second term vanishes, which means $S_0(i\kappa) \to \infty$ for $\kappa \in \mathbb{R}_+$. Therefore the function $S_0(k)$ has a pole on the imaginary line in the upper half-plane $k = i\kappa$.

This means that, starting from scattering with $k \in \mathbb{R}$, we generalize to the complex plane, $k \in \mathbb{C}$, and then the poles $k = i\kappa$ correspond to bound states.

If there are no singularities other than a single bound state $k = i\kappa$, with $\kappa > 0$, the s wave S-matrix $S_0(k)$ can be constructed. This implies that there are no singularities even at infinity; otherwise, we need to consider k not too large (if this is not true, then $S_0(k) \to 0$ at infinity for a physical case).

Besides the pole condition for $S_0(k)$, we have also the following conditions: (1) $|S_0(k)| = 1$ for $k \in \mathbb{R}_+$, derived from unitarity, and (2) $S_0 = 1$ for $k = 0$, since then $S_l = e^{2i\delta_l} \simeq e^{iCk^{2l+1}} \to 1$.

The simplest solution for the conditions on $S_0(k)$ is

$$S_0(k) = -\frac{k+i\kappa}{k-i\kappa} = e^{2i\delta_0(k)}. \quad (50.34)$$

This means that the partial wave amplitude is

$$a_{l=0}(k) = \frac{S_{l=0}-1}{2ik} = -\frac{1}{i(k-i\kappa)} = \frac{1}{-\kappa-ik}. \quad (50.35)$$

Equating it with the definition

$$a_{l=0}(k) = \frac{1}{k\cot\delta_{l=0}(k) - ik} \quad (50.36)$$

gives

$$k\cot\delta_{l=0}(k) = -\kappa \;\Rightarrow\; -\frac{1}{a} = \lim_{k\to 0} k\cot\delta_{l=0}(k) = -\kappa, \quad (50.37)$$

the same relation between bound states and scattering length as we had before.

Thus extending $\delta_l(k)$ and $S_l(k)$ to the complex-k plane is useful for obtaining relations between different physical regimes.

50.4 Analytical Properties of the Scattering Amplitudes

Therefore, we will set up the analytical properties of the functions $\delta_l(k)$ and $S_l(k)$, extended over the complex-k plane. If $k \in \mathbb{C}$, the energy is also complex, $E \in \mathbb{C}$. The exceptions are $k \in \mathbb{R}$, leading to $E \in \mathbb{R}_+$ and $k \in i\mathbb{R}$ (imaginary k), leading to $E \in \mathbb{R}_-$. In general, since

$$E = \frac{\hbar^2}{2m} k^2, \tag{50.38}$$

we obtain

$$\begin{aligned} \text{Re}\, E &= \frac{\hbar^2}{2m}[(\text{Re}\, k)^2 - (\text{Im}\, k)^2] \\ \text{Im}\, E &= \frac{\hbar^2}{2m} 2\, \text{Re}\, k\ \text{Im}\, k. \end{aligned} \tag{50.39}$$

Thus if we consider k over the upper half-plane ($\text{Im}\, k > 0$), it sweeps *all* of the complex E plane (both the real and imaginary parts reach arbitrarily large positive and large negative values). However, this implies that this complex E plane is not a regular plane (over which functions are single valued), but rather a *Riemann sheet* of the domain of E, called the physical sheet.

Indeed, when $r \to \infty$, the reduced radial wave function in the scattering regime, with $k \in \mathbb{R}$, is

$$\chi(r) \simeq A(E) e^{ikr} + B(E) e^{-ikr}, \tag{50.40}$$

where

$$-\frac{A(E)}{B(E)} = e^{2i\delta_0(E)} = S_0(E). \tag{50.41}$$

But if $k = i\kappa$ instead, meaning

$$E = -\frac{\hbar^2}{2m} \kappa^2 < 0, \tag{50.42}$$

then a real $\chi(r)$ for $E < 0$ means real $A(E), B(E)$.

The relation between $E > 0$ and $E < 0$ is by analytical continuation through the upper half-plane in k ($\text{Im}\, k > 0$), so $k = i\kappa$. But at an arbitrary point on the upper half of the k plane, we have

$$A(E^*) = A^*(E), \quad B(E^*) = B^*(E). \tag{50.43}$$

Indeed, $\text{Im}\, E$ does change sign, but, since $\text{Im}\, k > 0$, in fact only $\text{Re}\, k$ actually changes sign, and

$$e^{ikr} = e^{i \text{Re}\, k\, r} e^{-\text{Im}\, k\, r} \leftrightarrow e^{-i\text{Re}\, k\, r} e^{-\text{Im}\, k\, r} = e^{-i\text{Re}\, k\, r} e^{-\text{Re}\, \kappa\, r}, \tag{50.44}$$

the outgoing wave function e^{ikr} becomes the bound state wave function $e^{-\kappa r}$, and the coefficient is its complex conjugate, so

$$A(E^*) = A^*(E), \tag{50.45}$$

and similarly for $B(E^*) = B^*(E)$.

Then the physical Riemann sheet for E is defined by

$$\text{Im}\, k = \text{Re}\, \kappa > 0, \tag{50.46}$$

and we need to consider a cut for $E \in \mathbb{R}_+$ (from $E = 0$ to infinity), taking us through to a different Riemann sheet. The analyticity conditions $A(E^*) = A^*(E)$ and $B(E^*) = B^*(E)$ also define the E plane, or physical Riemann sheet.

50.5 Jost Functions Revisited

To continue the analysis of analyticity properties, we go back to the Jost functions

$$\mathcal{F}_l^{\pm}(k) = W(f_l^{\pm}, \phi_l), \tag{50.47}$$

where the regular solution is

$$\phi_l = \frac{1}{2ik}\left[-\mathcal{F}_l^{-}(k) f_l^{+}(k;r) + \mathcal{F}_l^{+}(k) f_l^{-}(k;r)\right], \tag{50.48}$$

and the Jost solutions are defined by

$$\lim_{r\to\infty} e^{\pm ikr} f_l^{\pm}(k;r) = 1. \tag{50.49}$$

But then the coefficients B and $-A$ are matched to $\mathcal{F}_l^{\pm}(k)$,

$$\begin{aligned} \lim_{r\to\infty} f_l^{+}(k;r) = e^{-ikr} \quad &\Rightarrow \quad \mathcal{F}_l^{+}(k) = B(E) \\ -\lim_{r\to\infty} f_l^{-}(k;r) = e^{+ikr} \quad &\Rightarrow \quad -\mathcal{F}_l^{-}(k) = A(E). \end{aligned} \tag{50.50}$$

Then, since $k \to -k^*$ means that Re k changes sign but Im k does not, this corresponds to E^*, so from the analyticity of A and B we obtain

$$\mathcal{F}_l^{\pm}(-k^*) = \pm B(E^*)/A(E^*) = \pm[B(E)/A(E)]^* = \left[\mathcal{F}_l^{\pm}(k)\right]^*, \tag{50.51}$$

which is indeed one of the Jost function's properties. The other relevant one is

$$\mathcal{F}_l^{\pm}(-k) = \mathcal{F}^{\mp}(k), \tag{50.52}$$

implying also

$$\left[\mathcal{F}_l^{\pm}(k^*)\right]^* = \mathcal{F}_l^{\mp}(k). \tag{50.53}$$

Then the lth partial wave S-matrix is written in terms of the Jost functions as

$$S_l(k) = e^{2i\delta_l(k)} = \frac{\mathcal{F}_l^{+}(k)}{\mathcal{F}_l^{-}(k)} e^{il\pi} = \frac{\mathcal{F}_l^{+}(k)}{\mathcal{F}_l^{+}(-k)} e^{il\pi}. \tag{50.54}$$

This means that the analytical properties of $S_l(k)$ can be derived from the analytical properties of the Jost functions $\mathcal{F}_l^{\pm}(k)$.

Indeed, we have a *theorem*. The zeroes of the Jost functions $\mathcal{F}_l^{+}(k)$ in the lower half-plane (meaning Im $k < 0$) lie on the imaginary line $i\mathbb{R}$, correspond to bound states, and are simple zeroes (with multiplicity one).

To prove this, we see that $\mathcal{F}_l^{+}(-k) = 0$ implies $S_l(k) \to \infty$, meaning a pole of $S_l(k)$. But we have already seen that a pole of $S_0(k)$ is a bound state $k = i\kappa_n$. Moreover, $S_l(k) \to \infty \Leftrightarrow \mathcal{F}_l^{+}(-k) = 0$, so $\mathcal{F}_l^{+}(k)$ has zeroes at $k = -i\kappa_n$ corresponding to bound states. *q.e.d.*

We then also obtain easily that $S_l(k)$ has zeroes at $k = -i\kappa_n$ and that $\mathcal{F}_l(k)$ has poles at $k = +i\kappa_n$; see Fig. 51.1 in the next chapter.

For bound states of $\mathcal{F}_l^{-}(k)$, the same analysis follows, except now we have zeroes in the upper half-plane (Im $k > 0$), since

$$\mathcal{F}_l^{-}(k) = \mathcal{F}_l^{+}(-k). \tag{50.55}$$

Because $\mathcal{F}_l^+(k) = W(f_l^+, \phi_l)$, the reduced radial wave function for physical bound state is just the analytically continued regular solution with a normalization constant in front,

$$\chi_l(k_n = -i\kappa_n; r) = N_{nl}\phi_l(k_n = -i\kappa_n; r), \tag{50.56}$$

where

$$N_{nl}^2 = \frac{-4\kappa_n^2}{\mathcal{F}_l^-(-i\kappa_n)\, d\mathcal{F}_l^+(-i\kappa)/d\kappa\big|_{\kappa=\kappa_n}}. \tag{50.57}$$

The proof of the normalization constant is left as an exercise.

Since

$$S_l(k) = e^{2i\delta_l(k)} = \frac{\mathcal{F}_l^+(k)}{\mathcal{F}_l^-(k)}e^{il\pi}, \tag{50.58}$$

the analytical property $[\mathcal{F}_l^\pm(k^*)] = [\mathcal{F}_l^\mp(k)]^*$ implies that

$$\begin{aligned} S_l(k^*) &= \frac{\mathcal{F}_l^+(k^*)}{\mathcal{F}_l^-(k^*)}e^{il\pi} = \frac{[\mathcal{F}_l^-(k)]^*}{[\mathcal{F}_l^+(k)]^*}e^{il\pi} \Rightarrow \\ S_l^*(k^*) &= \frac{\mathcal{F}_l^-(k)}{\mathcal{F}_l^+(k)}e^{-il\pi} = S_l^{-1}(k), \end{aligned} \tag{50.59}$$

so $S_l(k)$ satisfies

$$S_l(k)S_l^*(k^*) = 1. \tag{50.60}$$

On the other hand, the analytical property $\mathcal{F}_l^\pm(-k) = \mathcal{F}_l^\mp(k)$ implies that

$$S_l(-k) = \frac{\mathcal{F}_l^+(-k)}{\mathcal{F}_l^-(-k)}e^{il\pi} = \frac{\mathcal{F}_l^-(k)}{\mathcal{F}_l^+(k)}e^{il\pi}, \tag{50.61}$$

so $S_l(k)$ satisfies

$$S_l(-k)S_l(k) = e^{2il\pi}, \tag{50.62}$$

where the right-hand side equals 1 if l is integer.

Important Concepts to Remember

- For scattering at low energy, a_0 and σ_0 dominate; this is known as threshold behavior.
- For a finite-range negative step potential, $V = V_0 < 0$ for $r \leq r_0$, we can relate scattering ($E = 0+$) and bound state ($E = 0-$) solutions by

$$-E_{\text{boundstate}} \equiv \frac{\hbar^2\kappa^2}{2m} = \frac{\hbar^2}{2ma^2}$$

  (so $\kappa = 1/a$), with a the (large) scattering length.
- Generalizing $S_l(k)$ to the complex k plane, the poles $k = i\kappa$ of $S_0(k)$ correspond to bound states. For a single pole, $S_0(k) = (k + i\kappa)/(k - i\kappa)$, giving $a = 1/\kappa$.
- Expanding the reduced radial wave function at infinity, $\chi(r) \simeq A(E)e^{ikr} + B(E)e^{-ikr}$, and extending to $k \in \mathbb{C}$, the coefficients are Jost functions, $B(E) = \mathcal{F}_l^+(E)$ and $A(E) = -\mathcal{F}_l^-(E)$, and $S_l(k) = \mathcal{F}_l^+(k)/\mathcal{F}_l^-(k)e^{il\pi}$, with $S_l(-k)S_l(k) = e^{il\pi}$.

Further Reading

For more on the analytical properties of wave functions, see [4].

Exercises

(1) For a step potential, with energy $E < V_0$, connect the solution at $r < r_0$ with the solution at $r > r_0$ (in the general case).

(2) Find the first correction to the approximate relation (50.30) between the binding energy I of the bound state close to zero and the large scattering length a.

(3) Find the equivalent of the relation replacing (50.30) if we still have $r_0 \ll a$, but $E = V_0/2 > 0$ and very small ($kr_0 \ll 1$).

(4) Extend the analysis in complex space leading to $S_l(k), a_l(k), \delta_l(k)$ and coming from the existence of a bound state, from the similar analysis for $l = 0$.

(5) In the complex k plane do we still have real $\delta_l(k)$?

(6) Write down the Jost functions $\mathcal{F}_l^+(k)$ and $\mathcal{F}_l^-(k)$ in the case of a single bound state.

(7) Derive the normalization constant (50.57).

51 Resonances and Scattering, Complex k and l

In this chapter we consider resonances in scattering and their interpretation and analytical properties in the complex k plane, where k is the wave number. We end with a few remarks about continuing l into the complex plane.

51.1 Poles and Zeroes in the Partial Wave S-Matrix $S_l(k)$

We saw that $k_n = -i\kappa_n$ is a zero for $\mathcal{F}_l^+(k) = 0$, corresponding to a bound state. Then $k_n = +i\kappa_n$ is a pole for $S_l(k)$.

But then the question arises: can one not have poles *outside* the imaginary axis in the complex k plane? If there are extra poles, they cannot be on the real axis, since for $k \in \mathbb{R}$, $|S_0(k)| = 1$, whereas we want $S_0(k) = \infty$. Thus, we require

$$k_{\text{pole}} = k_1 + ik_2. \tag{51.1}$$

Since $S_0(k) = \infty$, from the previous chapter we know that this corresponds to having only the "out" wave at $r \to \infty$,

$$\chi(r) \sim e^{ikr} \Rightarrow \chi^* \sim e^{-ik^*r}. \tag{51.2}$$

Then the Schrödinger equations for χ and for χ^* are

$$\begin{aligned}\frac{d^2}{dr^2}\chi(r) + k^2\chi(r) - V(r)\chi(r) = 0 \Rightarrow \\ \frac{d^2}{dr^2}\chi^*(r) + k^{*2}\chi^*(r) - V(r)\chi^*(r) = 0.\end{aligned} \tag{51.3}$$

Multiplying the first equation by χ^* and subtracting the second, multiplied by χ, we obtain

$$\frac{d}{dr}\left(\chi^*\frac{d}{dr}\chi - \chi\frac{d}{dr}\chi^*\right) + (k^2 - k^{*2})\chi^*\chi = 0. \tag{51.4}$$

Integrating $\int_0^r dr'$, we obtain

$$\left(\chi^*\frac{d}{dr}\chi - \chi\frac{d}{dr}\chi^*\right)\Bigg|_0^r + (k^2 - k^{*2})\int_0^r dr'|\chi(r')|^2 = 0. \tag{51.5}$$

Regularity of the solution at $r = 0$ means that $\chi(0) = \chi^*(0) = 0$. Since $r \to \infty$, we can use $\chi(r) \sim e^{ikr}, \chi^*(r) \sim e^{-ik^*r}$, obtaining

$$(k + k^*)\left[ie^{-2r\,\text{Im}\,k} + 2i\,\text{Im}\,k\int_0^r dr'|\chi(r')|^2\right] = 0. \tag{51.6}$$

Then we can have either $k + k^* = 0$, which means that the poles lie on the imaginary line (the bound states), or

$$\mathrm{Im}\, k = -\frac{e^{-2r\,\mathrm{Im}\, k}}{2\int_0^r dr'|\chi(r')|^2} < 0. \tag{51.7}$$

This would mean that there are poles in the lower half-plane, $k = k_1 - ik_2$, where $k_2 \in \mathbb{R}_+$.

We have analyzed the poles for $S_0(k)$ because we started with (50.32), but that was actually valid for any l, so all the analysis above can be generalized to $S_l(k)$ for any l. Since

$$S_l(k) = \frac{\mathcal{F}_l^+(k)}{\mathcal{F}_l^-(k)} e^{il\pi} = \frac{\mathcal{F}_l^+(k)}{\mathcal{F}_l^+(-k)} e^{il\pi}, \tag{51.8}$$

the $k = k_1 - ik_2$ pole for $S_l(k) \leftrightarrow$ a pole for $\mathcal{F}_l^+(k)$ or a zero of $\mathcal{F}_l^+(-k) = \mathcal{F}_l^-(k) \Leftrightarrow$ a zero for $\mathcal{F}_l^{*-}(k) = \mathcal{F}_l^+(k^*) \Leftrightarrow$ the $k = k_1 + ik_2$ zero for $S_l(k)$,

$$k_{\mathrm{zero}} = k_1 + ik_2. \tag{51.9}$$

Therefore for every pole in the lower half-plane, we have a corresponding zero in the upper half-plane at the complex conjugate point.

However, we also have $\mathcal{F}_l^\pm(-k^*) = [\mathcal{F}_l^\pm(k)]^*$. Thus if $k = -k_1 + ik_2$ is a zero for $\mathcal{F}_l^+(k) \Leftrightarrow$ $k = k_1 - k_2$ is a zero for $\mathcal{F}_l^-(k)$ ($\Leftrightarrow$ pole for $S_l(k)$) $\Leftrightarrow$ also $-k_1 - ik_2$ is a zero for $\mathcal{F}_l^- \Leftrightarrow k_1 + ik_2$ is a zero of $\mathcal{F}_l^+(k)$.

To summarize, we have (see Fig. 51.1)

$$\begin{aligned} &\text{zero for } \mathcal{F}_l^-(k)\colon k = \pm k_1 - ik_2 \leftrightarrow \text{pole for } S_l(k) \\ &\text{zero for } \mathcal{F}_l^+(k)\colon k = \pm k_1 + ik_2 \leftrightarrow \text{zero for } S_l(k). \end{aligned} \tag{51.10}$$

However, in the physical case, k is real. If the poles and zeroes are *close* to the real line, i.e., if $k_2 \ll k_1$, we can approximate the form of $S_l(k)$. Then

$$S_l(k) = e^{2i\delta_l(k)} = \frac{k - k_{\mathrm{zero}}}{k - k_{\mathrm{pole}}} \tilde{S}_l(k) \left(= \frac{E - E_{\mathrm{zero}}}{E - E_{\mathrm{pole}}} \tilde{\tilde{S}}_l(E) \right), \tag{51.11}$$

where $\tilde{S}_l(k)$ varies very little. For physical values of k, we have two cases. For bound states, k is on the imaginary axis, and if also $l = 0$ then $\tilde{S}_0(k) = -1$. But for the poles or zeroes near the

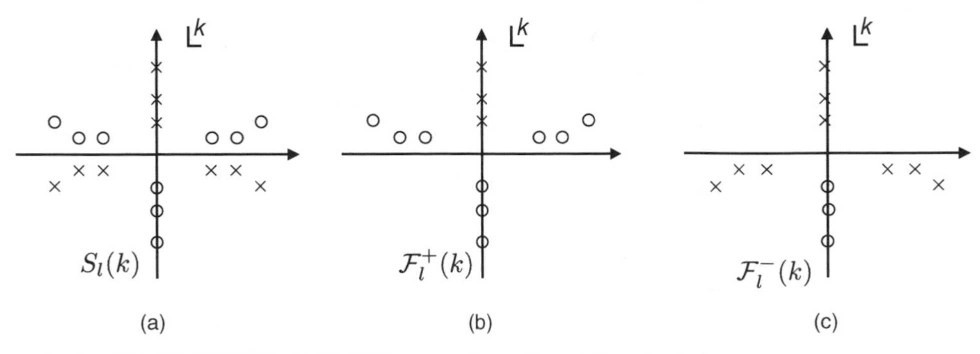

Figure 51.1 Zeroes and poles of (a) $S_l(k)$; (b) $\mathcal{F}_l^+(k)$; (c) $\mathcal{F}_l^-(k)$. Zeroes are denoted by a circle, and poles by a cross.

real axis, $\tilde{S}_l(k) \simeq 1$ since this corresponds to scattering, and if k is far from the pole or zero, then $S_l(k) \simeq \tilde{S}_l(k) \simeq 1$ is close to the unperturbed case. Then

$$S_l(k) \simeq \frac{k - k_1 - ik_2}{k - k_1 + ik_2}. \tag{51.12}$$

This $S_l(k)$ is valid close to the pole or zero.

51.2 Breit–Wigner Resonance

From the formula $S_l(k) = e^{2i\delta_l(k)}$ above, we can calculate the partial wave amplitude by substituting it into the general formula,

$$\begin{aligned} a_l(k) &= \frac{S_l(k) - 1}{2ik} \simeq \frac{1}{k} \frac{-k_2}{(k - k_1) + ik_2} = \frac{1}{k\frac{k-k_1}{(-k_2)} - ik} \\ &= \frac{e^{i\delta_l(k)} \sin \delta_l(k)}{k} = \frac{1}{k \cot \delta_l(k) - ik}. \end{aligned} \tag{51.13}$$

Identifying the term with $\cot \delta_l$ with the similar term in the first line, we obtain

$$\cot \delta_l(k) = \frac{k - k_1}{-k_2}. \tag{51.14}$$

Moreover, the lth partial wave cross section is

$$\sigma_l(k) = \frac{4\pi}{k^2}(2l+1) \sin^2 \delta_l(k) = 4\pi(2l+1)|a_l(k)|^2 \simeq \frac{4\pi}{k^2}(2l+1)\frac{k_2^2}{(k-k_1)^2 + k_2^2}. \tag{51.15}$$

But this formula has a maximum at $k = k_1$, which is called a *resonance*.

We have seen that the zero of $S_l(k)$ is at $k_{\text{zero}} = k_1 + ik_2$, and correspondingly there is a pole at $k_{\text{pole}} = k_1 - ik_2$. Then

$$E_{\text{zero/pole}} = \frac{\hbar^2}{2m} k^2_{\text{zero/pole}} = \frac{\hbar^2}{2m}(k_1^2 - k_2^2 \pm 2ik_1k_2) = E_{\text{res}} \pm i\frac{\Gamma}{2}, \tag{51.16}$$

where we have defined the resonance energy E_{res} and the *width* Γ by

$$\begin{aligned} E_{\text{res}} &= \frac{\hbar^2}{2m}(k_1^2 - k_2^2) \\ \frac{\Gamma}{2} &= \frac{\hbar^2}{2m} 2k_1k_2. \end{aligned} \tag{51.17}$$

If k is close to k_1 ($k \simeq k_1$), and correspondingly the real energy E is close to the real energy E_1, related by

$$E = \frac{\hbar^2}{2m}k^2 \in \mathbb{R}, \quad E_1 = \frac{\hbar^2}{2m}k_1^2 \in \mathbb{R}, \tag{51.18}$$

we obtain

$$E - E_1 = \frac{\hbar^2}{2m}(k + k_1)(k - k_1) \simeq \frac{\hbar^2}{2m} 2k_1(k - k_1), \tag{51.19}$$

as well as (since $k_1 \gg k_2$)

$$E_{\text{res}} \simeq E_1. \tag{51.20}$$

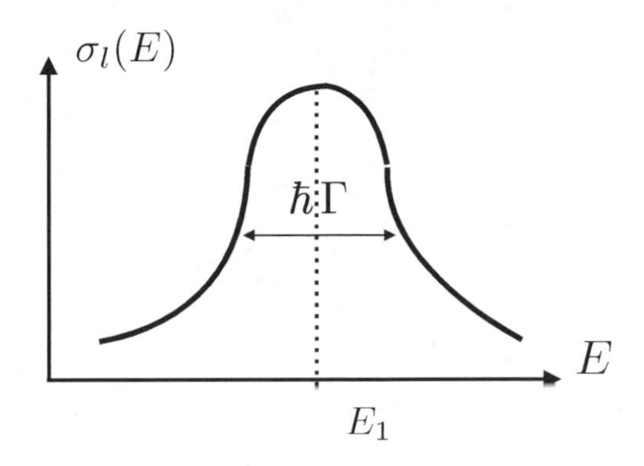

Figure 51.2 Breit–Wigner resonance.

Then from (51.14) we have

$$\tan\delta_l(k) = \frac{-k_2}{k - k_1} = \frac{-\Gamma/2}{E - E_1}, \tag{51.21}$$

which in turn means that

$$S_l(k) = e^{2i\delta_l(k)} = \frac{1 + i\tan\delta_l(k)}{1 - i\tan\delta_l(k)} = \frac{E - E_1 - i\Gamma/2}{E - E_1 + i\Gamma/2}. \tag{51.22}$$

Moreover, now we have

$$(k - k_1)^2 = \frac{(E - E_1)^2}{\left(\frac{\hbar^2}{2m}2k_1\right)^2}, \quad k_2^2 = \frac{(\Gamma/2)^2}{\left(\frac{\hbar^2}{2m}2k_1\right)^2}, \tag{51.23}$$

which we substitute in $\sigma_l(k)$ to obtain

$$\sigma_l(k) \simeq \frac{4\pi}{k^2}(2l + 1)\frac{(\Gamma/2)^2}{(E - E_1)^2 + (\Gamma/2)^2}. \tag{51.24}$$

This is the Breit–Wigner formula for resonance in energy, giving a bell-like shape for the cross section; see Fig. 51.2. The maximum of $\sigma_l(k)$ is at $E = E_1 \simeq E_{\text{res}}$, which is therefore the resonance energy. On the other hand Γ is the *width at half maximum value* for σ_l. Indeed, if $E - E_1 = \pm\Gamma/2$, $\sigma_l(E)/\sigma(E_1) = 1/2$.

We can write a slightly more general formula. Until now, we have assumed that we can directly invert (51.14), so that

$$\delta_l(k) = \tan^{-1}\left(\frac{-\Gamma/2}{E - E_1}\right), \tag{51.25}$$

and replace it in $S_l(k) = e^{2i\delta_l(k)}$ to find the Breit–Wigner formula for $\sigma_l(k)$. But it could be that there is a "background" value (a constant) for δ_l, called ξ_l, i.e.,

$$\delta_l(k) \to \tilde{\delta}_l(k) = \delta_l(k) + \xi_l = \tan^{-1}\left(\frac{-\Gamma/2}{E - E_1}\right) + \xi_l = \tan^{-1}\left(\frac{-\Gamma/2}{E - E_{\text{res}}}\right) + \xi_l. \tag{51.26}$$

If $\xi_l = n\pi$, that does not change the value of of $\tan\delta_l(k)$, but otherwise it does. Then, defining

$$\begin{aligned} \tan\delta_l &= \frac{-\Gamma/2}{E - E_1} \equiv -\frac{1}{\epsilon} \\ \tan\xi_l &\equiv -\frac{1}{q}, \end{aligned} \tag{51.27}$$

from the partial wave S-matrix

$$S_l = e^{2i\delta_l} \to e^{2i\tilde{\delta}_l} = e^{2i\delta_l} e^{2i\xi_l}, \tag{51.28}$$

we obtain the partial wave cross section

$$\sigma_l = \frac{4\pi}{k^2}(2l+1)\frac{(q+\epsilon)^2}{(1+q^2)(1+\epsilon^2)}. \tag{51.29}$$

In the normal case $\xi_l = 0$, implying $q \to \infty$, we obtain

$$\sigma_l = \frac{4\pi}{k^2}(2l+1)\frac{1}{1+\epsilon^2} = \frac{4\pi}{k^2}(2l+1)\frac{(\Gamma/2)^2}{(E-E_{\text{res}})^2 + (\Gamma/2)^2}. \tag{51.30}$$

In the extreme non-normal case, $q = 0 \leftrightarrow \xi_l = \pi/2$, we obtain a new formula,

$$\tilde{\sigma}_l = \epsilon^2 \sigma_l = \frac{4\pi}{k^2}(2l+1)\frac{(E-E_{\text{res}})^2}{(E-E_{\text{res}})^2 + (\Gamma/2)^2}. \tag{51.31}$$

A final observation is that

$$\cot\delta_l \simeq \frac{E-E_{\text{res}}}{-\Gamma/2} \tag{51.32}$$

means that at resonance $\cot\delta_l = 0$, and a little away from it, the above formula gives the first term in the Taylor expansion.

51.3 Physical Interpretation and Its Proof

Now that we have defined the poles of $S_l(k)$ and the related Breit–Wigner resonance formulas analytically, we need to understand the physical interpretation of this mathematical result. The first observation is that the poles are *near* the real axis, and they influence the behavior *on* the real axis, where physical scattering is located. But how are these poles created, or equivalently, what do the poles represent? The poles that we considered previously were bound states. But in bound states the system stays bound for an infinite amount of time, i.e., there is an infinite lifetime. The infinite lifetime refers to the particle being inside a potential well, with the energy of the bound state satisfying $E_n < V$ on both sides of the potential well, and with $E_n < 0$, so that there are no asymptotic states (the particle cannot escape to infinity).

Bound states are also stationary, which means that their time dependence is just a phase,

$$\psi(r,t) \propto e^{-iE_n t/\hbar}, \tag{51.33}$$

where $E_n = -|E_n|$. Since this corresponds to an imaginary k pole, $k_n = i\kappa_n$, the "out" wave turns into a decaying exponential, $e^{ik_n r} = e^{-\kappa_n r}$, but under the transition the energy stays real, just changing sign,

$$E_n = E(k_n) = \frac{\hbar^2 k_n^2}{2m} = -\frac{\hbar^2\kappa_n^2}{2m} = -|E_n|. \tag{51.34}$$

Turning to the resonance scattering, $k \simeq k_{\text{pole}} = k_1 - ik_2$, we are still close to the real line, so actually $k \sim k_1$, meaning the "out" wave stays out, $e^{ikr} \simeq e^{ik_1 r}$. On the other hand, the energy now

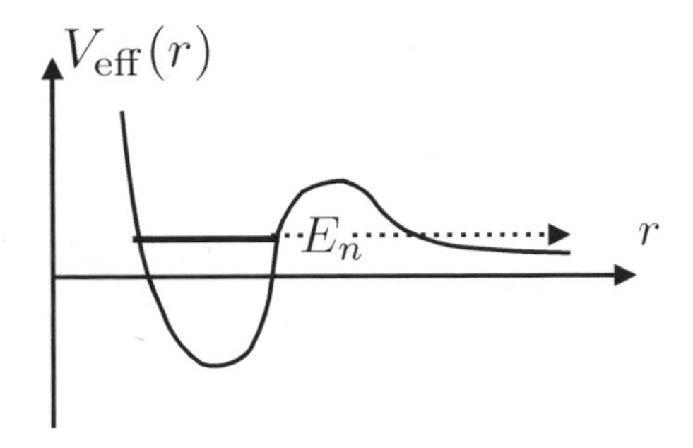

Figure 51.3 Scattering by entering, and then tunneling out of, a potential barrier around a quasi-bound state.

becomes complex, $E \simeq E_{\rm pole} = E_{\rm res} - i\Gamma/2$, leading to a time dependence of the wave function that is not just a phase, but also a decaying exponential,

$$\psi(r,t) \propto e^{-iEt/\hbar} \simeq e^{-iE_{\rm res}t/\hbar} e^{-\Gamma t/2\hbar}. \tag{51.35}$$

Then the probability density decays in time,

$$|\psi(r,t)|^2 \propto e^{-\Gamma t/\hbar}, \tag{51.36}$$

with a finite *lifetime*

$$\tau = \frac{\hbar}{\Gamma} \tag{51.37}$$

for the state. This means we now have a *quasi-bound state*. With respect to the (true) bound state, we still have $E_n < V$ on both sides of the potential well, but now $E_n > 0$, which means that there *are* asymptotics at infinity. This means that the well has a potential barrier on the side at large r, and we can tunnel out of it, see Fig. 51.3.

In conclusion, the physical picture is the following: we have a particle coming in from infinity, it gets captured into the quasi-bound state, and then it tunnels out of the potential well, or "leaks out".

To prove this picture, we consider the radial wave function extended on the complex k plane,

$$\begin{aligned} R_l(k;r) &= C\left(S_l(k)\frac{e^{ikr}}{r} - \frac{e^{-ikr}}{r}\right) \\ &\simeq C\left(\frac{k-k_1-ik_2}{k-k_1+ik_2}\frac{e^{ikr}}{r} - \frac{e^{-ikr}}{r}\right). \end{aligned} \tag{51.38}$$

Choosing the constant

$$C = \frac{1}{k-k_1-ik_2}, \tag{51.39}$$

the reduced radial wave function becomes

$$\chi_l(k;r) \simeq \frac{e^{ikr}}{k-k_1+ik_2} - \frac{e^{-ikr}}{k-k_1-ik_2}. \tag{51.40}$$

From it, we construct a wave packet by integrating χ_l, multiplied by the stationary phase, near the pole,

$$\psi(r,t) = \int dk\, \chi_l(k;r) e^{-iEt/\hbar}. \tag{51.41}$$

Since we are integrating near the pole, we write $k = k_1 + \delta k$ and integrate over δk, with $dk = d\delta k$. In this case, the energy is

$$E = \frac{\hbar^2 k^2}{2m} \simeq \frac{\hbar^2}{2m} k_1^2 + \frac{\hbar^2}{m} k_1 \delta k = E_1 + \hbar v_1 \delta k, \tag{51.42}$$

where we have defined the velocity of the pole,

$$v_1 \equiv \frac{\hbar k_1}{m}. \tag{51.43}$$

Then the reduced radial wave function is

$$\chi_l(k;r) \simeq \frac{e^{ik_1 r} e^{i\delta k\, r}}{\delta k + ik_2} - \frac{e^{-ik_1 r} e^{-i\delta k\, r}}{\delta k - ik_2} \tag{51.44}$$

and the wave packet is

$$\psi(r,t) = e^{i(k_1 r - E_1 t/\hbar)} \int d\delta k \frac{e^{i\delta k(r - v_1 t)}}{\delta k + ik_2} - e^{-i(k_1 r - E_1 t/\hbar)} \int d\delta k \frac{e^{-\delta k(r + v_1 t)}}{\delta k - ik_2}. \tag{51.45}$$

Since most of the integral comes from the region near the pole ($\delta k \sim 0$), we can extend the integration region to the whole real line, $\int_{-\infty}^{+\infty} d\delta k$ with minimal change. To calculate this integral over the real line, we must close the contour of integration in the complex plane, with a half-circle at infinity.

In the first integral:

If $r - v_1 t > 0$ then in order for the contribution of the half-circle to vanish, we need to have Im $\delta k > 0$ (so that the exponential is decaying), meaning the integration is in the upper half-plane. Then we have a closed contour C with a counterclockwise integration, meaning $\frac{1}{2\pi i} \oint_C$ gives the sum of residues inside C, i.e., the residues in the upper half-plane, of which there are none.

If $r - v_1 t < 0$ then for the contribution of the half-circle to vanish, we need to have Im $\delta k < 0$, i.e., integration in the lower half-plane. But now we have a closed contour C with clockwise integration, so $-\frac{1}{2\pi i} \oint_C$ gives the sum of residues in the lower half-plane. That means the residue at $\delta k = -ik_2$, giving finally

$$\int d\delta k \frac{e^{i\delta k(r - v_1 t)}}{\delta k + ik_2} = -2\pi i \,\mathrm{Res}(\delta k = -ik_2)\theta(v_1 t - r) = -2\pi i e^{k_2(r - v_1 t)}\theta(v_1 t - r). \tag{51.46}$$

In the second integral:

If $r + v_1 t > 0$ for the contribution of the half-circle to vanish, we need Im $\delta k < 0$, which means the integration is in lower half-plane. This gives a clockwise integration, and $-\frac{1}{2\pi i} \oint_C$ gives the residues in the lower half-plane, of which there are none.

If $r + v_1 t < 0$ then we need Im $\delta k > 0$, so integration is in the upper half-plane. This gives a counterclockwise integration, with $\frac{1}{2\pi i} \oint_C$ giving the sum of the residues in the upper half-plane, meaning $\delta k = ik_2$, giving finally

$$\int d\delta k \frac{e^{i\delta k(r + v_1 t)}}{\delta k - ik_2} = 2\pi i \,\mathrm{Res}(\delta k = ik_2)\theta(-v_1 t - r) = 2\pi i e^{k_2(r + v_1 t)}\theta(-v_1 t - r). \tag{51.47}$$

Then the wave packet time dependence splits into three regions:

If $t < -r/v_1$, we have the in wave. This time region corresponds to the coming wave propagating towards the center, since at $r \to \infty$ we have

$$\psi(r,t) \simeq 2\pi i e^{-i(k_1 r - E_1 t/\hbar)} e^{k_2(r + v_1 t)}, \tag{51.48}$$

which means that the wave comes in from $r = \infty$ with velocity $-v_1$.

If $-r/v_1 < t < r/v_1$, in the asymptotic region $r \to \infty$ we have

$$\psi(r,t) \simeq 0, \tag{51.49}$$

which means the wave is concentrated near $r = 0$. Thus the particle has been captured in a metastable or quasi-bound state.

If $t > r/v_1$, we have the out wave. This region corresponds to the outgoing wave propagating from the center, since at $r \to \infty$ we have

$$\psi(r,t) \simeq -2\pi i e^{i(k_1 r - E_1 t/\hbar)} e^{k_2(r - v_1 t)}, \tag{51.50}$$

where the last exponential is decaying, since $r - v_1 t < 0$. Since

$$\frac{\Gamma}{2} = \frac{\hbar^2}{2m} 2k_1 k_2 = \hbar v_1 k_2, \tag{51.51}$$

we obtain that the probability density at fixed $r \to \infty$ in this region is

$$|\psi(r,t)|^2 \propto e^{-2k_2 v_1 t} = e^{-\Gamma t/\hbar} = e^{-t/\tau}, \tag{51.52}$$

which is what we wanted to prove. Indeed then, at large times, the wave function implies that there is a probability of tunneling out of the metastable state.

51.4 Review of Levinson Theorem

Now that we have understood how to obtain $\delta_l(k)$, we will come back to the Levinson theorem, to present a simple proof. The theorem can be stated as

$$\delta_l(0) - \delta_l(\infty) = n_b^l \pi + \phi, \tag{51.53}$$

where $\phi = 0$ or $\pi/2$.

We start with a small potential, namely potential that is small with respect to the energy associated with a de Broglie wavelength of the order of the range of the potential,

$$|V(r)| \ll \frac{\hbar^2}{2mr_0^2}. \tag{51.54}$$

In this case, the Born approximation is valid at all energies, including $E = \infty$, so $\delta_l(E) \ll 1$, and $\delta_l(\infty) \simeq 0$. On the other hand, $\delta_l(0) = \phi$ (which is also zero, except for a background value). Since the potential is shallow, we have no bound states, $n_b^l = 0$. Then consider $E = \epsilon$, a small and fixed energy, and $\delta_l(\epsilon) - \delta_l(\infty)$ as $V(r)$ deepens. As each new bound state of energy ϵ appears as $V(r)$ deepens, and its radial wave function contains a new node with respect to the previous bound state with energy ϵ (since the bound state with energy ϵ has the largest energy among bound states, i.e., it is $E_{n_{\max}}$, for which we have $n_{\max} + 1$ total nodes). But since $E = \epsilon$ is small, this last node is at large r (since the periodicity is inversely proportional to k), for which the wave function is

$$R_{El}(r) \sim \sin\left(kr - \frac{l\pi}{2} + \delta_l\right), \tag{51.55}$$

which means that $kr - l\pi/2 + \delta_l$ changes by π at the new node's appearance, so $\delta_l(\epsilon)$ changes by π when each new node appears at fixed r. That means that

$$\delta_l(\epsilon \to 0) - \delta_l(\infty) = n_b^l \pi + \phi, \tag{51.56}$$

and after we take the limit $\epsilon \to 0$, we obtain the Levinson theorem. *q.e.d.*

51.5 Complex Angular Momentum

Until now, we have learned new information from considering k in the complex plane (and so also the energy E in the complex plane), but we have kept the angular momentum as a natural number, $l \in \mathbb{N}$.

However, we can also learn from an extension to a complex angular momentum, $l \in \mathbb{C}$, where now the energy is kept real, $E \in \mathbb{R}$. The result is called *Regge theory*.

The reduced radial wave function in the bound state case is

$$\chi_{El}(r) = rR_{El}(r) = A(l,E)e^{-\kappa r} + B(l,E)e^{+\kappa r}, \tag{51.57}$$

where $\kappa = \sqrt{-2mE}/\hbar$. We can relate it to the scattering solution through analytical continuation through the *lower half-plane* (different from the upper half-plane relevant to complex k). The scattering solution is (the proof is just a modification of the continuation of the energy dependence through the upper half-plane)

$$\chi_{El}(r) = A^*(l,E)e^{-ikr} + B^*(l,E)e^{ikr}. \tag{51.58}$$

This means that

$$A^*(l,E) = -\mathcal{F}_l^-(k) = -(\mathcal{F}_l^+(k))^*, \quad B^*(l,E) = \mathcal{F}_l^+(k) = (\mathcal{F}_l^-(k))^*. \tag{51.59}$$

If $l \in \mathbb{R}$ and $E > 0$, then, identifying the scattering χ_{El} with the one analytically continued from the bound state, we obtain an analyticity constraint of $A(l,E)$, namely

$$A(l,E) = B^*(l,E). \tag{51.60}$$

But analytically continuing to the complex plane, $l \in \mathbb{C}$, while still keeping $E \in \mathbb{R}_+$, we obtain

$$A(l^*,E) = B^*(l,E). \tag{51.61}$$

The general solution of the Schrödinger equation, even in the case of complex l, has the following behavior near $r \to 0$,

$$R_{El}(r) \sim \alpha r^l + \beta r^{-l-1}. \tag{51.62}$$

In the case of $l \in \mathbb{N}$, we need to put $\beta = 0$ so that we have only $R_{El} \sim r^l$. If $l \in \mathbb{C}$, we have an alternative: we can define R_{El} to be valid for general r^{-l-1}. Namely, we require $r^l \gg r^{-l-1}$, which means $\mathrm{Re}\, l \geq \mathrm{Re}(-l-1)$, so

$$\mathrm{Re}(2l + 1/2) > 0. \tag{51.63}$$

Since

$$S(l,E) = e^{2i\delta(l,E)} = e^{i\pi l}\frac{A(l,E)}{B(l,E)}, \tag{51.64}$$

we can generalize to

$$\frac{A(E)}{B(E)} = \frac{\mathcal{F}_l^+(k)}{\mathcal{F}_l^-(k)} \to \frac{\mathcal{F}^+(l,k)}{\mathcal{F}^-(l,k)} = \frac{A(l,E)}{B(l,E)}. \tag{51.65}$$

Then we obtain

$$
\begin{aligned}
S^*(l,E) &= e^{-i\pi l^*}\frac{A^*(l,E)}{B^*(l,E)}\\
S(l^*,E) &= e^{+i\pi l^*}\frac{A(l^*,E)}{B(l^*,E)},
\end{aligned}
\tag{51.66}
$$

leading to the analyticity condition

$$S^*(l,E)S(l^*,E) = 1. \tag{51.67}$$

The poles of $S(l,E)$ are then the zeroes of $B(l,E)$ in the complex l plane, called *Regge poles*. These line up as points on curves

$$l = \alpha_i(E) \tag{51.68}$$

called *Regge trajectories*.

We will not continue with the analysis of Regge theory, since it is beyond the scope of this book.

Important Concepts to Remember

- For $S_l(k)$ we have poles in the lower half-plane, $k_{\text{pole}} = k_1 - ik_2$, and for each such pole a corresponding zero in the upper half-plane, $k_{\text{zero}} = k_1 + ik_2$.
- For poles close to $\mathbb{R}$, $k_1 \gg k_2$, near the pole

$$S_l(k) \simeq \frac{k - k_1 - ik_2}{k - k_1 + ik_2},$$

  and we have a resonance,

$$\sigma_l = \frac{4\pi}{k^2}(2l+1)\frac{k_2^2}{(k-k_1)^2 + k_2^2},$$

  with $E_{\text{zero/pole}} = E_{\text{res}} \pm i\frac{\Gamma}{2}$, and with $\tan\delta_l = (-\Gamma/2)/(E - E_1)$.
- The resonance satisfies the Breit–Wigner formula,

$$\sigma_l = \frac{4\pi}{k^2}(2l+1)\frac{(\Gamma/2)^2}{(E-E_1)^2 + (\Gamma/2)^2},$$

  though with a background value for δ_l, $\delta_l \to \delta_l + \xi_l$, we can change it.
- The resonance has the physical interpretation that there is a quasi-bound state, with a potential barrier but with $E > 0$, so we have an in wave for a particle going towards the center of the potential, then a metastable (quasi-bound) state, then an out wave for a particle tunneling out (decaying) with probability $\propto e^{-t/\tau}$, $\tau = \hbar/\Gamma$.
- We can also consider complex angular momentum l, with real energy, leading to Regge theory, in which case $S(l,E)$ has poles at points $l = \alpha_i(E)$ forming curves known as Regge trajectories.

Further Reading

For more on the analytical properties of wave functions, see [4]. See also [2] and [1].

Exercises

(1) Write down the approximate value for the partial wave S-matrix $S_l(k)$ for the case of two resonances and two bound states.

(2) Consider the case where among the partial wave cross sections σ_l only σ_1 is at (or very near) a resonance. What can you say about $\sigma_{\rm tot}$?

(3) Consider the case where $S_l(k)$ has a single pole (resonance), but k_2 is comparable with k_1. Calculate $\tan\delta_l$ and the equivalent of the Breit–Wigner formula in this more general case, $\sigma_l(E)$.

(4) If we have two resonances (complex poles) for σ_l close to each other, and we are at resonance on the real k line, write down the wave function in terms of the physical interpretation of time-dependent scattering with in and out waves, as in the text.

(5) In the case in exercise 4, calculate the asymptotic wave function in the metastable region of time, and in the out region.

(6) In the case of the Levinson theorem, for n_b^l, do we need to count only the poles of $S_l(k)$ that are *exactly* on the imaginary line, or can they be slightly away from the imaginary line?

(7) Consider that $S_l(k)$ for all $l \in \mathbb{N}$ has a single resonance pole, very close to the real line. If we consider now complex momentum instead, what can we learn about the Regge trajectories?

52 The Semiclassical WKB and Eikonal Approximations for Scattering

In this chapter we study semiclassical approximations for scattering. We start with the WKB approximation, extended to three dimensions but then reduced to effectively one dimension. Then we give an alternative approach, the eikonal approximation, where the geometrical optics approximation is further refined to a straight line. We end with a special application, the Coulomb potential scattering, where the semiclassical approximation is actually exact.

52.1 WKB Review for One-Dimensional Systems

Before the extension of the WKB analysis to three dimensions, we review the one-dimensional case.

Constructing the semiclassical wave function starts with the redefinition of the time-dependent but stationary wave function,

$$\psi(\vec{r},t) = e^{iS(\vec{r},t)/\hbar} = \psi(\vec{r})e^{-iEt/\hbar}, \tag{52.1}$$

where the exponent separates the time dependence as

$$S(\vec{r},t) = s(\vec{r}) - Et. \tag{52.2}$$

Then the time-independent wave function is

$$\psi(\vec{r}) = e^{is(\vec{r})/\hbar}. \tag{52.3}$$

This already shows that we can apply this formalism to scattering, if $\psi(\vec{r})$ corresponds to the scattering wave function $\psi^{(+)}(\vec{r})$. Note that we applied the WKB method before to *bound* states, leading to Bohr–Sommerfeld quantization.

The expansion of the function $s(\vec{r})$ in $\hbar$,

$$s(\vec{r}) = s(\vec{r}) + \hbar s_1(\vec{r}) + \hbar^2 s_2(\vec{r}) + \cdots \tag{52.4}$$

leads to the semiclassical approximation if we keep only the first two terms, s_0 (classical) and s_1 (semiclassical).

The zeroth-order term is the classical on-shell action between the initial condition $\vec{r}_0(t_0)$ and the final point $\vec{r}(t)$,

$$s_0(\vec{r}) = S_0[\vec{r}_0(t_0) \to \vec{r}(t)]. \tag{52.5}$$

This arises as the extremum of the path integral for the propagator,

$$U(t,t_0) = \int \mathcal{D}\vec{r}(t)\, \exp\left\{\frac{i}{\hbar}S[\vec{r}_0(t_0) \to \vec{r}(t)]\right\}. \tag{52.6}$$

One Dimension

Specializing to one dimension, the Schrödinger equation implies the equation for $s(x)$,

$$\left(\frac{ds(x)}{dx}\right)^2 - 2m(E - V(x)) + \frac{\hbar}{i}\frac{d^2}{dx^2}s(x) = 0. \tag{52.7}$$

It is the quantum-corrected version of the Hamilton–Jacobi equation for the classical action.

The solution of the equation to the first order in $\hbar$ (i.e., $s_0 + \hbar s_1$) gives, for the wave function in the classically allowed region $E > V(x)$,

$$\psi(x) = \frac{1}{[2m(E - V(x))]^{1/4}} \exp\left[\pm\frac{i}{\hbar}\int_{x_0}^{x} dx'\sqrt{2m(E - V(x'))}\right], \tag{52.8}$$

where the exponent is $\frac{i}{\hbar}s_0(x)$, where $s_0(x)$ is the on-shell action (the classical action on the classical trajectory).

We use connection formulas to transition from a solution in a classically allowed region to a solution in a forbidden one (and vice versa), through the classical turning point of the trajectory.

The transition formula from the classically allowed region $x > x_1$ to the classically forbidden region $x < x_1$ is

$$\begin{aligned}&\frac{\sqrt{2}}{[2m(E - V(x))]^{1/4}} \sin\left[\frac{1}{\hbar}\int_{x_1}^{x} dx'\sqrt{2m(E - V(x'))} + \frac{\pi}{4}\right], \quad x > x_1\\ \rightarrow &\frac{1}{[2m(V(x) - E)]^{1/4}} \exp\left[-\frac{1}{\hbar}\int_{x}^{x_1} dx'\sqrt{2m(V(x') - E)}\right], \quad x < x_1.\end{aligned} \tag{52.9}$$

For the case where the classically allowed region is to the left, $x < x_2$, and the forbidden region is to the right, $x > x_2$, we have

$$\begin{aligned}&\frac{1}{[2m(V(x) - E)]^{1/4}} \exp\left[-\frac{1}{\hbar}\int_{x_2}^{x} dx'\sqrt{2m(V(x') - E)}\right], \quad x > x_2\\ \rightarrow &\frac{\sqrt{2}}{[2m(E - V(x))]^{1/4}} \sin\left[\frac{1}{\hbar}\int_{x}^{x_2} dx'\sqrt{2m(E - V(x'))} - \frac{\pi}{4}\right], \quad x < x_2.\end{aligned} \tag{52.10}$$

52.2 Three-Dimensional Scattering in the WKB Approximation

We now apply the formalism to three-dimensional scattering. Then the quantum-corrected Hamilton–Jacobi equation arises from the Schrödinger equation, in terms of $s(\vec{r})$,

$$\frac{(\vec{\nabla}s)^2}{2m} + V(\vec{r}) - E - \frac{i\hbar}{2m}\Delta s = 0. \tag{52.11}$$

But we are interested in the $\hbar$-expansion of $s(\vec{r})$, whose zeroth-order term is $s_0(\vec{r})$. To obtain the equation for s_0 we drop the last term in (52.11), the only one depending on $\hbar$, and obtain the classical Hamilton–Jacobi equation,

$$\frac{(\vec{\nabla}s_0)^2}{2m} + V(\vec{r}) = E \equiv \frac{\hbar^2 k^2}{2m}, \tag{52.12}$$

which defines k.

Then we reduce the system in the spherically symmetric case to one dimension, namely to the radial direction. We first reduce to the radial wave function by writing

$$\psi(\vec{r}) = R(r)Y_{lm}(\vec{n}_r), \tag{52.13}$$

and then to the reduced radial wave function $\chi(r) = rR(r)$. But the radial variable has a domain equal to half the real line, $r \in (0, \infty)$, so we have to redefine the variable to $x = \ln(kr)$ so that $x \in (-\infty, +\infty)$.

The equation for $\chi(r)$ is (19.7), namely

$$\frac{d^2}{dr^2}\chi(r) + \frac{2m}{\hbar^2}[E - V_{\text{eff}}(r)]\chi(r) = 0, \tag{52.14}$$

where the effective potential has the centrifugal term added,

$$V_{\text{eff}}(r) = V(r) + \frac{\hbar^2 l(l+1)}{2mr^2}. \tag{52.15}$$

In terms of $x = \ln(kr)$, meaning $dr = e^x dx/k$, we obtain

$$\frac{d^2}{dx^2}\chi(x) - \frac{d}{dx}\chi(x) + \frac{2m}{\hbar}\frac{e^{2x}}{k^2}[E - V_{\text{eff}}(x)]\chi(x) = 0. \tag{52.16}$$

But to get rid of the term with the first derivative, we need to further set

$$\chi(x) = e^{x/2}W(x). \tag{52.17}$$

Then the equation of motion becomes

$$\frac{d^2}{dx^2}W(x) + Q^2(x)W(x) = 0, \tag{52.18}$$

where

$$\begin{aligned} Q^2(x) &\equiv \frac{2m}{\hbar^2}r^2[E - V_{\text{eff}}(r)] - \frac{1}{4} \\ &= r^2\left[\frac{2m}{\hbar^2}(E - V(r)) - \frac{(l+1/2)^2}{r^2}\right], \end{aligned} \tag{52.19}$$

so that Q^2 takes the place of $E - V(x)$ in the one-dimensional case.

We can easily see that if $E > 0$ (meaning, for scattering solutions) and $V(r) \to 0$ for $r \to \infty$, then $Q^2(x) \to +\infty$ at $r \to \infty$ ($x \to +\infty$). Then, if $V(r)$ doesn't blow up faster than $1/r^2$ at $r \to 0$, $Q^2 < 0$ at $r \to 0$ ($x \to -\infty$), so there is a single turning point x_0, at which $Q^2(x_0) = 0$. Since Q^2 represents $E - V(x)$ in the one-dimensional case, we consider $-Q^2$ as $V(x) - E$. At infinity, $-Q^2 \to -\infty$ and at $r = 0$, $-Q^2 > 0$, with a classical turning point, $E - V(x) = 0$, in the middle.

The effective one-dimensional quantum-corrected Hamilton–Jacobi equation is obtained by further redefining

$$W(x) = e^{i\tilde{s}(x)/\hbar}, \tag{52.20}$$

which means

$$\psi(\vec{r}) = e^{is(\vec{r})/\hbar} = \sqrt{\frac{k}{r}}e^{i\tilde{s}(x)/\hbar}Y_{lm}(\vec{n}_r). \tag{52.21}$$

The quantum-corrected Hamilton–Jacobi equation is

$$\left(\frac{d\tilde{s}}{dx}\right)^2 - \hbar^2 Q^2(x) = i\hbar\frac{d^2\tilde{s}(x)}{dx^2}. \tag{52.22}$$

Expanding in $\hbar$,

$$\tilde{s}(x) = \tilde{s}_0(x) + \hbar\tilde{s}_1(x) + \hbar^2\tilde{s}_2(x) + \cdots \tag{52.23}$$

the equation of the leading term $\tilde{s}_0(x)$ is obtained by dropping the right-hand side, which is proportional to $\hbar$, leaving just the classical Hamilton–Jacobi equation, with solution

$$\tilde{s}_0(x) = \pm\int_{x_0}^{x} dx'\hbar Q(x'). \tag{52.24}$$

The semiclassical solution correspnds to $\tilde{s}_1(x)$, as usual:

$$\tilde{s}_1(x) = A + \frac{i}{2}\ln|Q(x)|, \tag{52.25}$$

where A is a constant, so we have the usual one-dimensional WKB solution in terms of $W(x)$,

$$W_{\text{WKB}}(x) = \frac{1}{\sqrt{k(x)}}\left[C_1\exp\left(i\int_{x_0}^{x} dx'k(x')\right) + C_2\exp\left(-i\int_{x_0}^{x} dx'k(x')\right)\right], \tag{52.26}$$

where $Q^2(x) \equiv k^2(x) > 0$, so it is in the classically allowed region, including the asymptotic region $r \to \infty$ $(x \to +\infty)$.

For the classically forbidden region $Q^2(x) \equiv -\kappa^2(x) < 0$, inside the range of the potential, we find

$$W_{\text{WKB}}(x) = \frac{1}{\sqrt{\kappa(x)}}\left[D_1\exp\left(-\int_{x_0}^{x} dx'\kappa(x')\right) + D_2\exp\left(+\int_{x_0}^{x} dx'\kappa(x')\right)\right]. \tag{52.27}$$

At the effective one-dimensional turning point of the classical trajectory, $Q^2(x) = 0$, we have the continuity formula, transitioning from the inside region (inside the range of the potential) to the outside region (including the asymptotic region),

$$W_{\text{WKB}}(x) = \frac{1}{\sqrt{\kappa(x)}}\exp\left[-\int_{x}^{x_0} dx'\kappa(x')\right] \to \frac{\sqrt{2}}{\sqrt{k(x)}}\sin\left(\frac{\pi}{4} + \int_{x_0}^{x} dx'k(x')\right). \tag{52.28}$$

To describe scattering, we need to consider physical solutions in the $r \to \infty$ limit, which are in the classically allowed region. For the reduced radial wave function $\chi(r) = \sqrt{kr}W$, we go back to the r dependence, $dx = dr/r$, implying $Q(x) = k(x) \to k(r)/r = Q(r)/r$, obtaining

$$\begin{aligned}\chi_{\text{WKB}}(r) &= \left[1 - \frac{2mV(r)}{\hbar^2k^2} - \frac{(l+1/2)^2}{r^2k^2}\right]^{-1/4} \\ &\times \sin\left[\frac{\pi}{4} + \int_{r_0}^{r} dr'\sqrt{k^2 - \frac{2m}{\hbar^2}V(r') - \frac{(l+1/2)^2}{r'^2}}\right].\end{aligned} \tag{52.29}$$

But at $r \to \infty$, which means that the integral in the exponent is extended to $\int_{r_0}^{\infty}$, the argument of the sine is $kr - l\pi/2 + \delta_l^{\text{WKB}}$, leading to a *phase shift in the WKB approximation*,

$$\delta_l^{\text{WKB}}(k) = \frac{\pi}{4} + \frac{l\pi}{2} - kr_0 + \int_{r_0}^{\infty} dr'\left[\sqrt{k^2 - \frac{2m}{\hbar^2}V(r') - \frac{(l+1/2)^2}{r'^2}} - k\right]. \tag{52.30}$$

52.3 The Eikonal Approximation

We now consider another version of the WKB approximation called the eikonal approximation. The region of the validity is a subset of the region of the validity of the WKB approximation, namely the region where $V(x)$ varies little over the de Broglie wavelength λ. This implies the WKB condition,

$$\frac{\delta_\lambda \lambda}{\lambda} \ll 1. \tag{52.31}$$

We further refine the domain of the approximation by saying that $E \gg |V|$ (the actual value of the potential, not just the variation of it, is small with respect to the energy). This means we are at *high energy*, which is different from the case for the Born approximation, which is valid for both low and high energies.

However, as a WKB approximation, we are within *geometrical optics*. That means that a wave is replaced by a light ray path, i.e., the integral over all paths (the path integral) is restricted to just the classical path. Indeed, that restricts $s(\vec{r})$ to the classical on-shell action $s_0(\vec{r})$, where $x(t)$ is on the classical trajectory, which solves the classical Hamilton–Jacobi equation.

Instead of reducing the three- dimensional problem to a one-dimensional problem as in the WKB method above, we now make a further approximation: that the classical trajectory is approximately a straight path, which is true at high energies, since then there is only a small deflection. This makes the problem calculable, since to actually *find* the classical trajectory as a solution of the Hamilton–Jacobi equation is potentially very difficult.

The approximation defined above is called the *eikonal approximation*, coming from the Greek "eikon", meaning icon or image, since in the geometric optics case the light rays remaining straight through the interaction means that we obtain an undistorted "image" of an emitting object, after the interaction.

Since the classical trajectory is a straight line, we can define it as the Oz axis, where the origin O is the projection onto the axis of the central point of the spherically symmetric potential; see Fig. 52.1. The distance of the projection from the central point to the origin O is the *impact parameter* b. Since $r = |\vec{r}| = \sqrt{b^2 + z^2}$, the Hamilton–Jacobi equation reduces to a one-dimensional equation for the parameter z of the classical trajectory, giving

$$\left(\frac{d}{dz}\frac{s_0}{\hbar}\right)^2 = k^2 - \frac{2m}{\hbar^2}V(r) = k^2 - \frac{2m}{\hbar^2}V(\sqrt{b^2 + z^2}). \tag{52.32}$$

It integrates to

$$\frac{s_0}{\hbar} = \int_{-\infty}^{z} dz' \sqrt{k^2 - \frac{2m}{\hbar^2}V(\sqrt{b^2 + z'^2})} + C, \tag{52.33}$$

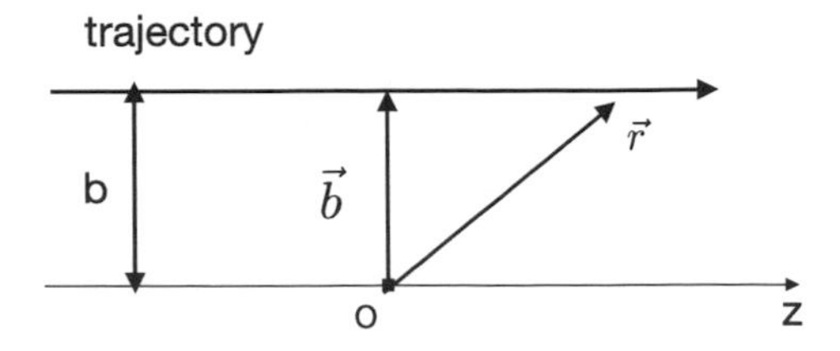

Figure 52.1 Classical trajectory in the eikonal approximation, and the impact parameter.

where C is a constant, which we can choose such that $s_0/\hbar \to 0$ when we turn off the potential, $V \to 0$. Then, since

$$\sqrt{k^2 - \frac{2mV}{\hbar^2}} = k\sqrt{1 - \frac{V}{E}} \simeq k\left(1 - \frac{V}{2E}\right) \simeq k - \frac{mV}{\hbar^2 k}, \tag{52.34}$$

we find

$$\begin{aligned} \frac{s_0}{\hbar} &= kz + \int_{-\infty}^{z} dz' \left[\sqrt{k^2 - \frac{2m}{\hbar^2}V(\sqrt{b^2 + z'^2})} - k\right] \\ &\simeq kz - \frac{m}{\hbar^2 k}\int_{-\infty}^{z} dz' V(\sqrt{b^2 + z'^2}). \end{aligned} \tag{52.35}$$

Then we obtain a scattering wave function for the argument $\vec{r} = \vec{b} + z\vec{e}_z$ in the eikonal approximation,

$$\psi_+(\vec{r}) = \psi(\vec{b} + z\vec{e}_z) = e^{ikz}\exp\left[-\frac{im}{\hbar^2 k}\int_{-\infty}^{z} dz' V(\sqrt{b^2 + z'^2})\right], \tag{52.36}$$

where $e^{ikz} = e^{i\vec{k}\cdot\vec{r}}$.

To obtain the scattering amplitude f we substitute the above ψ_+ into the right-hand side of (46.71), obtained by substituting the free Green's function $G_0^{(+)}(\vec{r}, \vec{r}\,')$ in the Lippmann–Schwinger equation, namely

$$\begin{aligned} f_{\vec{k}}^{+}(\vec{n}_r, \vec{n}_k) = f^{(+)}(\vec{k}', \vec{k}) &= -\frac{2m}{4\pi\hbar^2}\int d^3\vec{r}\,' e^{-i\vec{k}'\cdot\vec{r}\,'} V(\vec{r}\,')\psi_+(\vec{r}\,') \\ &= -\frac{2m}{4\pi\hbar^2}\int d^3\vec{r}\,' e^{-i\vec{k}'\cdot\vec{r}\,'} V(\sqrt{b^2 + z'^2}) e^{i\vec{k}\cdot\vec{r}\,'} \exp\left[-\frac{im}{\hbar^2 k}\int_{-\infty}^{z'} d\tilde{z}\, V(\sqrt{b^2 + \tilde{z}^2})\right], \end{aligned} \tag{52.37}$$

where if we replace the last exponential $\exp[\dots]$ with 1, we have the Born amplitude $f^{(1)}(\vec{k}', \vec{k})$. This means that the eikonal approximation resums some interactions (each having potential V), replacing $V \times 1$ with $V \times \exp[\dots \int V] \sim V \times \sum_n (\dots \int V)^n$.

To do the integral, we integrate over cylindrical coordinates z, b and the angle ϕ that rotates the trajectory around the center of the potential, so $d^3\vec{r}\,' = \int dz \int db \int b d\phi$. The deflection turns $\vec{k}$ into $\vec{k}'$, but the deflection angle θ is very small, and the modulus of $\vec{k}$ is unchanged, which means that $\vec{k} - \vec{k}'$ is perpendicular to $\vec{k}$, which is proportional to $\vec{e}_z$. Thus, since $(\vec{k} - \vec{k}') \cdot \vec{e}_z = 0$, we find

$$(\vec{k} - \vec{k}') \cdot \vec{r}\,' = (\vec{k} - \vec{k}') \cdot (\vec{b} + z'\vec{e}_z) \simeq (\vec{k} - \vec{k}) \cdot \vec{b}. \tag{52.38}$$

However, $\vec{k} - \vec{k}'$ has modulus $k\theta$ (since $\theta \ll 1$). And its direction makes an angle ϕ with the impact parameter vector $\vec{b}$ (see Fig. 52.2), so finally

$$(\vec{k} - \vec{k}') \cdot \vec{r}\,' \simeq k\theta(b\cos\phi). \tag{52.39}$$

Then the scattering amplitude is

$$f^{(+)}(\vec{k}', \vec{k}) = -\frac{2m}{4\pi\hbar^2}\int_0^{\infty} db\, b \int_0^{2\pi} d\phi e^{-ikb\theta\cos\phi} \int_{-\infty}^{+\infty} dz\, V \exp\left[-\frac{im}{\hbar^2 k}\int_{-\infty}^{z} dz' V\right]. \tag{52.40}$$

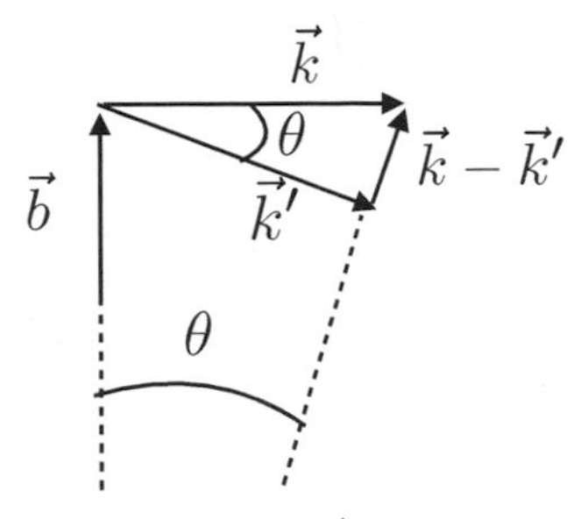

Figure 52.2 Geometry of scattering by a small angle θ, with impact parameter $\vec{b}$.

Using the formulas

$$\int_0^{2\pi} d\phi e^{-ikb\theta\cos\phi} = 2\pi J_0(kb\theta)$$

$$\int_{-\infty}^{+\infty} dz\, V \exp\left[-\frac{im}{\hbar^2 k}\int_{-\infty}^{z} dz' V\right] = \int_{-\infty}^{+\infty} dz \frac{d}{dz}\left\{\frac{i\hbar^2 k}{m}\exp\left[-\frac{im}{\hbar^2 k}\int_{-\infty}^{z} dz' V\right]\right\}, \tag{52.41}$$

we obtain

$$f^{(+)}(\vec{k}',\vec{k}) = -ik\int_0^\infty db\, bJ_0(kb\theta)\left[e^{2i\Delta(b)} - 1\right], \tag{52.42}$$

where

$$\Delta(b) \equiv -\frac{m}{2\hbar^2 k}\int_{-\infty}^{+\infty} dz\, V(\sqrt{b^2+z^2}). \tag{52.43}$$

If we have a finite-range potential, with range r_0, if $b > r_0$ then $\Delta(b) = 0$, so the integrand is zero, leading to

$$f^{(+)}(\theta) = -ik\int_0^{r_0} db\, bJ_0(kb\theta)\left[e^{2i\Delta(b)} - 1\right]. \tag{52.44}$$

This amplitude corresponds to the optical theorem in an interesting way. To see this, we take the imaginary part of the forward amplitude, obtaining

$$\text{Im} f^{(+)}(\theta = 0) = -k\text{Re}\int_0^{r_0} db\, bJ_0(0)\left[e^{2i\Delta(b)} - 1\right] = 2k\int_0^{r_0} db\, b\sin^2\Delta(b). \tag{52.45}$$

For the optical theorem to be true, the total cross section must be

$$\sigma_{\text{tot}} = \int_0^{r_0} db(4\pi b)2\sin^2\Delta(b). \tag{52.46}$$

This can be obtained from the partial wave expansion.

To see that, we first consider the fact that the eikonal approximation is a high-energy approximation, therefore the maximum angular momentum is large. But the angular momentum is $l\hbar = bp$, and since $p = \hbar k$, we find $l = bk$. Then $l_{\max} = kb_{\max} = kr_0 \gg 1$, and thus the sum over l becomes an integral over kb,

$$\sum_{l=0}^{l_{\max}} \leftrightarrow \int_0^{r_0} kdb, \tag{52.47}$$

and the phase shift turns into $\Delta(b)$,

$$\delta_l(k) \leftrightarrow \Delta(b)|_{b=l/k} . \tag{52.48}$$

Thus if $b > b_{\max} = r_0$, $\delta_l = 0 \leftrightarrow \Delta(b) = 0$. Since also $l \gg 1$ and $\theta \ll 1$, we have

$$P_l(\cos\theta) \simeq J_0(l\theta) = J_0(kb\theta), \tag{52.49}$$

and also $2l + 1 \simeq 2l \leftrightarrow 2kb$. Then the amplitude, from the partial wave expansion formula, is, since $a_l(k) = (e^{2i\delta_l(k)} - 1)/(2ik)$,

$$\begin{aligned} f(\theta) &\leftrightarrow k \int_0^{r_0} db \frac{2kb}{2ik} \left[e^{2i\Delta(b)} - 1\right] J_0(kb\theta) \\ &= -ik \int_0^{r_0} db\, bJ_0(kb\theta) \left[e^{2i\Delta(b)} - 1\right], \end{aligned} \tag{52.50}$$

which is the eikonal approximation formula, obtained from the partial wave expansion.

The total cross section is now obtained similarly,

$$\begin{aligned} \sigma_{\text{tot}} &= \frac{4\pi}{k^2} \sum_{l=0}^{l_{\max}} (2l+1) \sin^2 \delta_l(k) \\ &\leftrightarrow \frac{4\pi}{k^2} \int_0^{r_0} kdb\, 2bk\, \sin^2 \Delta(b), \end{aligned} \tag{52.51}$$

and this formula is consistent with the optical theorem.

52.4 Coulomb Scattering and the Semiclassical Approximation

We will try to apply the formalism of the semiclassical WKB approximation to the case of Coulomb scattering.

Since

$$\sqrt{k^2 - \frac{2mV(r)}{\hbar^2} - \frac{(l+1/2)^2}{r^2}} \simeq k \left[1 - \frac{V(r)}{2E} - \frac{\hbar^2 (l+1/2)^2}{4mEr^2}\right], \tag{52.52}$$

we find

$$\int_{r_0}^{\infty} dr' \sqrt{k^2 - \frac{2mV(r)}{\hbar^2} - \frac{(l+1/2)^2}{r^2}} \simeq k \int_{r_0}^{r} dr' - \frac{k}{2E} \int_{r_0}^{r} dr' V(r') - \frac{\hbar^2 (l+1/2)^2}{4mE} \int_{r_0}^{r} \frac{dr}{r^2}. \tag{52.53}$$

But $r \to \infty$ does not make sense in the Coulomb case $V(r) = A/r$, but only if $V = A/r^\alpha$ with $\alpha > 1$.

Instead of using the WKB approximation for the radial reduction, for the Coulomb potential $V = e_0^2/r$ (for the interaction of two *electrons*, i.e., negative charges, unlike in the hydrogenoid atom), we start directly from the Schrödinger equation,

$$\Delta\psi + \left(k^2 - \frac{2m}{\hbar^2} \frac{e_0^2}{r}\right) \psi = 0. \tag{52.54}$$

Making the change of variables

$$\psi(\vec{r}) = e^{ikr} F(\vec{r}), \tag{52.55}$$

the Schrödinger equation becomes

$$\Delta F + 2ik\frac{\vec{r}}{r}\cdot\vec{\nabla}F + \left(2ik - \frac{2me_0^2}{\hbar^2}\right)\frac{1}{r}F = 0. \tag{52.56}$$

However, instead of reducing the problem to the radial coordinate, we will use Cartesian coordinates (x, y, z) to reduce the problem to the variable $u = k(r - z)$. Therefore our ansatz for the scattering solution of the Schrödinger equation is $F = F(u)$. Then we find

$$\begin{aligned} \Delta F &= \frac{2ku}{r}\frac{d^2F}{du^2} + \frac{2u}{r}\frac{dF}{du} \\ \frac{\vec{r}}{r}\cdot\vec{\nabla}F &= k\left(1 - \frac{z}{r}\right)\frac{dF}{du} = \frac{u}{r}\frac{dF}{du}. \end{aligned} \tag{52.57}$$

This reduces the Schrödinger equation to

$$\frac{2k}{r}u\frac{d^2F}{du^2} + \frac{2k}{r}(1 + iu)\frac{dF}{du} + \frac{2k}{r}\left(i - \frac{me_0^2}{k\hbar^2}\right)F = 0. \tag{52.58}$$

Defining $v = -iu$, the equation becomes

$$v\frac{dF}{dv^2} + (1 - v)\frac{dF}{dv} - \left(1 + i\frac{me_0^2}{k\hbar^2}\right)F = 0, \tag{52.59}$$

which is the equation for the confluent hypergeometric function ${}_1F_1(a, b; v)$ (see (18.41)),

$$v\frac{d^2F}{dv^2} + (b - v)\frac{dF}{dv} - aF = 0; \tag{52.60}$$

this means that the solution has $b = 1, a = 1 + i\alpha$, where

$$\alpha = \frac{me_0^2}{k\hbar^2}. \tag{52.61}$$

But the confluent hypergeometric function of an imaginary variable has the asymptotics

$${}_1F_1(a, b; -iu) \simeq \frac{\Gamma(b)}{\Gamma(a)}\frac{e^{i\pi(b-a)/2}}{u^{b-a}}e^{-iu} + e^{-i\pi a/2}\frac{\Gamma(a)}{\Gamma(b-a)}\frac{1}{u^a}. \tag{52.62}$$

In our case, we have

$$\begin{aligned} {}_1F_1(1 + i\alpha, 1; -iu) &\simeq \frac{1}{\Gamma(1 + i\alpha)}\frac{e^{\pi\alpha/2}}{u^{-i\alpha}}e^{-iu} + \frac{1}{\Gamma(-i\alpha)}\frac{e^{-i\pi(1+i\alpha)/2}}{u^{1+i\alpha}} \\ &= \frac{e^{\pi\alpha/2}}{\Gamma(1 + i\alpha)}\left[e^{-iu + i\alpha\ln u} + e^{-i\pi/2}\frac{\Gamma(1 + i\alpha)}{\Gamma(-i\alpha)}\frac{e^{-i\alpha\ln u}}{u}\right], \end{aligned} \tag{52.63}$$

where we have used $1/u^{-i\alpha} = e^{+i\alpha\ln u}$ and $1/u^{1+i\alpha} = u^{-1}e^{-i\alpha\ln u}$.

Therefore the full wave function is (since $\psi = e^{ikr}{}_1F_1(\dots)$)

$$\psi = \frac{e^{\pi\alpha/2}}{\Gamma(1 + i\alpha)}\left[e^{ikz - i\alpha\ln[k(r-z)]} - i\frac{\Gamma(1 + i\alpha)}{\Gamma(-i\alpha)}\frac{e^{ikr - i\alpha\ln[k(r-z)]}}{k(r - z)}\right]. \tag{52.64}$$

In the first term, $e^{ikz} = e^{i\vec{k}\cdot\vec{r}}$ is the incoming plane wave, and there is an extra slowly varying phase.

In the second term, using $\Gamma(1+i\alpha) = i\alpha\Gamma(i\alpha)$ and $r - z = r(1-\cos\theta) = 2r\sin^2\theta/2$ and factorizing e^{ikr}/r, we obtain the scattering amplitude

$$f(\theta) = \frac{\Gamma(i\alpha)}{\Gamma(-i\alpha)} \frac{\alpha}{2k\sin^2\theta/2} e^{-i\alpha\ln[2kr\sin^2\theta/2]}. \quad (52.65)$$

However, since $[\Gamma(i\alpha)]^* = \Gamma(-i\alpha)$, the first factor is a phase and the differential cross section is

$$\begin{aligned}\frac{d^2\sigma}{d\Omega} &= |f(\theta)|^2 = \left(\frac{\alpha}{2k\sin^2\theta/2}\right)^2 = \left(\frac{e_0^2 m}{2\hbar^2 k^2 \sin^2\theta/2}\right)^2 \\ &= \frac{e_0^2}{4m^2v^4}\frac{1}{\sin^4\theta/2}.\end{aligned} \quad (52.66)$$

This is nothing other than the classical Rutherford formula, even though we have obtained an *exact* formula for (nonrelativistic) quantum scattering, since we haven't used any approximation. We have seen before that in the Born approximation we can find the classical result. So any semiclassical corrections to the classical result vanish, and in fact all quantum corrections vanish. Thus, Coulomb scattering is a very special case.

Important Concepts to Remember

- We can use the WKB approximation also for three-dimensional scattering, using as before the transition formulas from the classically allowed region to the classically forbidden region, and reducing the three-dimensional Schrödinger equation to a one-dimensional Schrödinger equation *with domain* $\mathbb{R}$, by writing $x = \ln(kr)$, $\chi(x) = e^{x/2}W(x)$, and the equation $\frac{d^2}{dx^2}W(x) + Q^2(x)W(x) = 0$.
- The WKB approximation for $W(x) = \exp\left[\frac{i}{\hbar}\tilde{s}(x)\right]$ is obtained from the quantum-corrected Hamilton–Jacobi equation for $\tilde{s}(x)$, leading to an expression for $\chi_{\rm WKB}(r)$ that at infinity goes like $\sin(kr - l\pi/2 + \delta_l^{\rm WKB})$, giving a formula for the phase shift in the WKB approximation.
- The WKB approximation is a geometrical optics approximation, of classical ray paths, but one can add to it an eikonal approximation, of straight ray paths, since the classical paths (solutions of the Hamilton-Jacobi equation) are difficult to obtain.
- The eikonal approximation, a high-energy approximation, resums some Born series terms, and gives, for a finite-range potential with range r_0, $f^+ = ik\int_0^\infty db\, bJ_0(kb\theta)[e^{2i\Delta(b)} - 1]$, with $\Delta(b) = -(m/2\hbar^2 k)\int dx V(\sqrt{b^2+z^2})$ and $\sigma_{\rm tot} = \int_0^{r_0} db(4\pi b)2\sin^2\Delta(b)$.
- The *exact* Coulomb scattering calculation reproduces the Rutherford formula, so there are no WKB approximation corrections, in fact no quantum corrections of any kind, in this *nonrelativistic* case.

Further Reading

See [2] and [1].

Exercises

(1) Consider a one-dimensional system with potential $V(x) = V_0/(x^2 + a^2)$. Write the WKB approximation for the wave function for scattering, for energy $0 < E < V_0/a^2$.

(2) Consider a three-dimensional system with central potential $V(r) = V_0/(r^2 + a^2)$. Write the WKB approximation for the wave function for scattering of given angular momentum l, for energy $0 < E < V_0/a^2$.

(3) For the case in exercise 2, calculate the cross section as a formal sum over l.

(4) For the same potential as above, $V(r) = V_0/(r^2 + a^2)$, calculate $\Delta(b)$ for the eikonal approximation. If the potential turns off ($V(r) = 0$) for $r \geq r_0$, calculate $\Delta(b)$ and, if r_0 is large, calculate an approximate value for the scattering amplitude in the eikonal approximation.

(5) For a Yukawa potential, $V(r) = V_0 e^{-\mu r}/r$, write an approximate value for the cross section in the eikonal approximation.

(6) If we have a Coulomb potential that turns off ($V(r) = 0$) for a large $r \geq r_0$, calculate the total (approximate) cross section in the eikonal approximation, and compare with the exact value from the Rutherford formula.

(7) Are there any resonances for Coulomb scattering? Check that the analytical properties of $S_l(k)$ in this case are satisfied.

53 Inelastic Scattering

In this chapter we study inelastic scattering, which means there are energy, momentum, and maybe even particles leaking out of the system into another system (or "sink"). A standard example is the scattering of a fundamental particle (such as an electron) off a compound particle (such as an atom), which can absorb energy and get excited to a different state. After analyzing this from the point of view of an unknown other system, and that of a known other system, we generalize to scattering in the multi-channel case, and also the scattering of identical particles.

53.1 Generalizing Elastic Scattering from Unitarity Loss

In the first generalization we abandon unitarity, which means that there is a probability leak into an unknown other system. We previewed this in Chapter 49. Then, everything follows via the partial wave expansion, but the S-matrix is not unitary, $\hat{S}^\dagger \hat{S} \neq \mathbb{1}$. In terms of partial waves,

$$S_l^* S_l \neq 1, \tag{53.1}$$

which means that S_l is not a phase factor. Or, defining in the same way $S_l = e^{2i\delta_l}$, it means that $\delta^* \neq \delta$, i.e., δ has a nonzero imaginary part.

Since S_l is the ratio of the "out" wave to the "in" wave,

$$\psi \sim S_l(k)\frac{e^{ikr}}{r} + \frac{e^{-ikr}}{r}, \tag{53.2}$$

and the probability leaks into something else between the in and the out wave functions, which means that $|S_l(k)| < 1$, or

$$\text{Im } \delta_l \geq 0 \Rightarrow |S_l| = e^{-2\text{Im } \delta_l(k)} \leq 1. \tag{53.3}$$

In the eikonal approximation,

$$\begin{aligned} \delta_l(k) &\leftrightarrow \Delta(b)|_{b=l/k} \\ S_l(k) = e^{2i\delta_l(k)} &\leftrightarrow e^{2i\Delta(b)}, \end{aligned} \tag{53.4}$$

where $\text{Im } \Delta(b) \geq 0$.

The total cross section expands into partial waves,

$$\sigma_{\text{tot}} = \sum_{l=0}^{l_{\max}} \sigma_l, \tag{53.5}$$

and for σ_l we have a more general formula, previewed in (49.13),

$$\sigma_l = \frac{\pi}{k^2}(2l+1)|e^{2i\delta_l(k)} - 1|^2 = \frac{\pi}{k^2}(2l+1)|S_l(k) - 1|^2, \tag{53.6}$$

corresponding in the eikonal approximation to

$$\frac{\pi}{k}2b|e^{2i\Delta(b)}-1|^2. \tag{53.7}$$

The sum over l turns into an integral, $\sum_{l=0}^{l_{\max}} \to \int_0^{r_0} kdb$, so

$$\sigma_{\rm tot} = 2\pi \int_0^{r_0} db\ b|e^{2i\Delta(b)}-1|^2. \tag{53.8}$$

We can define then a maximally inelastic scattering, the "black disk" eikonal, which means that inside a disk of radius r_0 we have complete absorption, $S_l = 0$, i.e., a "black" region. This is achieved by setting

$$\text{Im}\,\Delta(b) = +\infty, \quad \text{for} \quad b < r_0, \tag{53.9}$$

and, as always, $S_l = 1$ (there is no interaction or deflection), or $\Delta(b) = 0$, for $b > r_0$.

Then the total cross section of the black disk eikonal is

$$\sigma_{\rm tot} = \pi r_0^2, \tag{53.10}$$

which is just the classical, geometrical, cross section. Note that this is the total *inelastic* cross section.

On the other hand, we can construct the amplitude as we did in the previous chapter. Defining $\vec{q} = \vec{k}' - \vec{k}$, with modulus $q = |\vec{q}| \simeq k\theta$, we write $-i(k\theta)b\cos\phi$ as $-i\vec{q}\cdot\vec{b}$, where the vector parameter $\vec{b}$ is in the two-dimensional plane transverse to the initial direction $\vec{k}$. Then the eikonal amplitude becomes

$$f(\theta, r_0) = -\frac{ik}{2\pi}\int_0^{r_0} db\ b \int_0^{2\pi} d\phi\ e^{-i\vec{q}\cdot\vec{b}}(e^{2i\Delta(b)}-1), \tag{53.11}$$

where

$$\int_0^{2\pi} d\phi e^{-i\vec{q}\cdot\vec{b}} = 2\pi J_0(qb) = 2\pi J_0(k\theta b). \tag{53.12}$$

Taking the imaginary part of the forward ($\theta = 0$) amplitude for the black disk eikonal ($e^{2i\Delta(b\leq r_0)} = 0$), we obtain

$$\text{Im}\, f(0, r_0) = k\int_0^{r_0} db\ b = \frac{kr_0^2}{2}. \tag{53.13}$$

If the optical theorem is to be satisfied, the total cross section must be

$$\sigma_{\rm tot} = \frac{4\pi}{k}\text{Im}\, f = 2\pi r_0^2. \tag{53.14}$$

However, if we consider only inelastic scattering then unitarity is violated, as we said, since $|S_l| < 1$, and thus the optical theorem (which follows from unitarity) is also violated. But the total cross section from the optical theorem is larger. In fact, the eikonal approximation is at high energy, where we have already seen that the result is $2\pi r_0^2$, comprising the geometrical cross section and the shadow one,

$$\sigma_{\rm tot} = 2\pi r_0^2 = \pi r_0^2 + \pi r_0^2 = \sigma_{\rm geometric} + \sigma_{\rm shadow}. \tag{53.15}$$

The inelastic cross section is the geometric part of the above.

The eikonal amplitude can be further rewritten as

$$f(\theta, r_0) = -\frac{ik}{2\pi}\int_0^{r_0} db\, b \int_0^{2\pi} d\phi e^{-i\vec{q}\cdot\vec{b}}(e^{2i\Delta(b)} - 1)$$
$$= -\frac{ik}{2\pi}\int d^2\vec{b}\, e^{-i\vec{q}\cdot\vec{b}}(e^{2i\Delta(b)} - 1). \quad (53.16)$$

Calculating the *black disk* ($e^{2i\Delta(b\leq r_0)} = 0$) eikonal amplitude at arbitrary θ, we obtain

$$f(\theta, r_0) = \frac{ik}{2\pi}\int_0^{r_0} db\, b 2\pi J_0(qb)$$
$$= \frac{ik}{q^2}\int_0^{qr_0} dx\, x J_0(x) = \frac{ikr_0}{q} J_1(qr_0), \quad (53.17)$$

where we have used $\int_0^a dx\, x J_0(x) = a J_1(a)$. Taking the limit $q \to 0$ by using the limit $J_1(x) \simeq x/2$ for $x \to 0$, we obtain

$$f(\theta \to 0, r_0) = \frac{ikr_0^2}{2}. \quad (53.18)$$

In order to obtain the correct *inelastic* cross section, we use the optical theorem but *divide by an extra 2*, since the optical theorem gives the total cross section, which is in equal parts inelastic and shadow, as explained above.

$$\sigma_{\text{tot,inelastic}} = \frac{1}{2}\frac{4\pi}{k}\text{Im}\, f(0, r_0) = \pi r_0^2. \quad (53.19)$$

53.2 Inelastic Scattering Due to Target Structure

One possible way to have inelastic scattering is for the target to have structure, leading to excited states, that can be reached by absorbing energy from the projectile.

That is, the projectile is particle A, and the target is system B, with internal structure. The standard example is the scattering of an electron (particle A) off an atom (system B) with excited internal states. Then the state of the combined system AB (electron plus atom) is the product of the momentum state of the particle A and the state of the target B,

$$|\vec{k}n_0\rangle = |\vec{k}\rangle_A \otimes |n_0\rangle_B, \quad (53.20)$$

with the wave function factorizing into a free particle wave function for A times a multiparticle wave function for the N components of B,

$$\psi \sim e^{i\vec{k}\cdot\vec{x}}\psi_{n_0}(\vec{r}_1, \ldots, \vec{r}_N). \quad (53.21)$$

In the case of the atom we have $N = Z$, the electric charge number, which is the same as the number of electrons in the atom.

The transition between the initial state and the final state is

$$|\vec{k}, n_0\rangle = |\vec{k}\rangle_A \otimes |n_0\rangle_B \to |\vec{k}'n\rangle = |\vec{k}'\rangle \otimes |n\rangle_B. \quad (53.22)$$

The final state wave function is

$$\psi \sim e^{i\vec{k}'\cdot\vec{x}}\psi_n(\vec{r}_1,\ldots,\vec{r}_N). \tag{53.23}$$

The elastic case corresponds to $n = n_0$, otherwise we have inelastic scattering.
In Chapter 46, we found the differential cross section as

$$\frac{d\sigma}{d\Omega} = \frac{r^2|\vec{j}_{\text{scatt}}|}{|\vec{j}_{inc}|} = \frac{r^2(\hbar k/m)(|f|/r^2)}{(\hbar k/m)} = |f|^2. \tag{53.24}$$

We want to generalize the final formula by calculating the change in the initial and scattered currents. In the scattered current the energy, and therefore the wave vector k, of the projectile is changed by exciting the internal state of the target, so $k' \neq k$. But the projectile itself does not change, so m stays fixed. The more general formula is then

$$\frac{d\sigma}{d\Omega}(|n_0\rangle \to |n\rangle) = \frac{r^2 j_{\text{scatt}}}{j_{\text{inc}}} = \frac{k'}{k}|f|^2. \tag{53.25}$$

If we restrict to the Born approximation, namely the first-order time-dependent perturbation theory, Fermi's golden rule, calculated in Chapter 47 as

$$f^{(1)}(\vec{k}',n;\vec{k},n_0) = \frac{2m}{4\pi\hbar^2}\langle\vec{k}',n|\hat{V}|\vec{k},n_0\rangle, \tag{53.26}$$

leads to the first-order differential cross section,

$$\frac{d\sigma^{(1)}}{d\Omega} = \frac{k'}{k}\left|\frac{2m}{4\pi\hbar^2}\langle\vec{k}',n|\hat{V}|\vec{k},n_0\rangle\right|^2. \tag{53.27}$$

In general (outside the Born approximation), the amplitude is written as

$$f(\vec{k}',n;\vec{k},n_0) = \frac{2m}{4\pi\hbar^2}\langle\vec{k}',n|\hat{V}|\vec{k},n_0+\rangle, \tag{53.28}$$

and the first-order differential cross section is

$$\frac{d\sigma}{d\Omega} = \frac{k'}{k}\left|\frac{2m}{4\pi\hbar^2}\langle\vec{k}',n|\hat{V}|\vec{k},n_0+\rangle\right|^2. \tag{53.29}$$

The potential contains the interaction of the projectile A with all the components of the target B. In the particular case of the atom, we have the interaction of the incoming electron with the nucleus of charge $+Ze$ situated at $\vec{r} = 0$, and the $N = Z$ electrons, situated at $\vec{r}_i$,

$$V = -\frac{Ze_0^2}{r} + \sum_{i=1}^{N}\frac{e_0^2}{|\vec{r} - \vec{r}_i|}. \tag{53.30}$$

In the Born approximation, the amplitude for the elastic scattering becomes explicitly

$$f^{(1,+)}(\vec{k}',\vec{k}) = \frac{2m}{4\pi\hbar^2}\int d^3\vec{r}'e^{i\vec{r}'\cdot(\vec{k}-\vec{k}')}V(\vec{r}'), \tag{53.31}$$

where $\vec{k} - \vec{k}' = -\vec{q}$. In the inelastic generalization, this becomes

$$
\begin{aligned}
f^{(1,+)}(\vec{k}', n; \vec{k}, n_0) &= \frac{2m}{4\pi\hbar^2}\int d^3\vec{r}' e^{-i\vec{q}\cdot\vec{r}'}\langle n|V(\vec{r}')|n_0\rangle \\
&= \frac{2m}{4\pi\hbar^2}\int d^3\vec{r}' e^{-i\vec{q}\cdot\vec{r}'} \\
&\times \int \prod_{i=1}^{N} d^3\vec{r}_i\, \psi_n^*(\vec{r}_1,\ldots,\vec{r}_N)V(\vec{r}',\vec{r}_1,\ldots,\vec{r}_N)\psi_{n_0}(\vec{r}_1,\ldots,\vec{r}_N).
\end{aligned}
\tag{53.32}
$$

But we have that

$$
\begin{aligned}
\sum_{i=1}^{N}\int d^3\vec{r}'\frac{e^{-i\vec{q}\cdot\vec{r}'}}{|\vec{r}'-\vec{r}_i|} &= \sum_{i=1}^{N}\int d^3\vec{r}\frac{e^{-i\vec{q}\cdot(\vec{r}+\vec{r}_i)}}{r} \\
&= \sum_{i=1}^{N}\frac{4\pi}{q^2}e^{-i\vec{q}\cdot\vec{r}_i} = \frac{4\pi}{q^2}\int d^3\vec{r}\,\rho(\vec{r})e^{-i\vec{q}\cdot\vec{r}},
\end{aligned}
\tag{53.33}
$$

where the target's electron density is

$$
\rho = \sum_{i=1}^{N}\delta^3(\vec{r}-\vec{r}_i).
\tag{53.34}
$$

We define the *form factors* (in momentum space)

$$
F_{n,n_0}(\vec{q}) = \frac{1}{Z}\langle n|\sum_{i=1}^{N}e^{-i\vec{q}\cdot\vec{r}_i}|n_0\rangle.
\tag{53.35}
$$

Then the integral in the amplitude is

$$
\int d^3\vec{r}\, e^{-i\vec{q}\cdot\vec{r}}\left\langle n\left|\left(-\frac{Ze_0^2}{r}+\sum_{i=1}^{N}\frac{e_0^2}{|\vec{r}-\vec{r}_i|}\right)\right|n_0\right\rangle = Ze_0^2\frac{4\pi}{q^2}\left[-\delta_{n,n_0}+F_{n,n_0}(\vec{q})\right],
\tag{53.36}
$$

leading to the differential cross section

$$
\begin{aligned}
\frac{d\sigma}{d\Omega} &= \frac{k'}{k}\frac{4m_e^2}{\hbar^4}\frac{Z^2e_0^4}{q^4}\left|-\delta_{n,n_0}+F_{n,n_0}(\vec{q})\right|^2 \\
&= \frac{k'}{k}\frac{4Z^2}{a_0^2q^4}\left|-\delta_{n,n_0}+F_{n,n_0}(\vec{q})\right|^2.
\end{aligned}
\tag{53.37}
$$

53.3 General Theory of Collisions: Inelastic Case and Multi-Channel Scattering

After considering a simplified treatment of inelastic scattering, from the point of view of a fundamental projectile, we consider the general theory of collisions. In this general case, both projectile and target can have *structure*, as for instance in atom-on-atom collision, or nucleus-on-nucleus collision, as in the RHIC (relativistic high-energy ion collider) and LHC (large hadron collider) experiments.

In the most general case, then, consider N *fundamental* particles (particles that cannot be separated in sub-parts, at least not at the available energies), like for instance electrons and nucleons. They can form different *fragments*, as for instance nuclei or atoms, in different ways, if there are bound states (if there is a nucleus, or an atom, in the given conditions). Examples include nuclear reactions through scattering, in which we collide two nuclei, but after the scattering interaction other nuclei appear.

A division of all the N fundamental particles into fragments is called a channel. We note that changing the state or the quantum numbers (as for instance $|n_0\rangle$ to $|n\rangle$ for an atom) of a fragment remains within the channel (though we can formally consider it as being in a different channel, if we want); the term channel is reserved for different combinations into fragments. For scattering, we will have both an "in" channel (before the collision) and an "out" channel (after the collision).

Also, if conservation of energy, momentum, angular momentum, etc., impedes the appearance in the final state of a fragment, related to a given channel, then we say that the "channel is closed".

The energy E_i of fragment i is given by the kinetic energy of the center of mass motion of the fragment, ϵ_i, subtracting from it the binding energy w_i,

$$E_i = \epsilon_i - w_i. \tag{53.38}$$

The simplest nontrivial example of channels corresponds to a system of $N = 3$ fundamental particles, called, say, a, b, c. Then the three two-particle bound states correspond to fragments (ab), (bc), and (ac). To admit the bound states of the fragments, we need to have the binding energies $w_{ab}, w_{bc}, w_{ac} > 0$.

Then there are four channels, with the total energy divided among the various fragments,

$$\begin{aligned}
&\text{I}, \quad a + (bc) \colon E = \epsilon_a + \epsilon_{bc} - w_{bc}\\
&\text{II}, \quad b + (ac) \colon E = \epsilon_b + \epsilon_{ac} - w_{ac}\\
&\text{III}, \quad c + (ab) \colon E = \epsilon_c + \epsilon_{ab} - w_{ab}\\
&\text{IV}, \quad a + b + c \colon E = \epsilon_a + \epsilon_b + \epsilon_c.
\end{aligned} \tag{53.39}$$

Putting the fragments in the order of their binding energies, $w_{bc} > w_{ac} > w_{ab} > 0$, and having the middle channel, II, as the in channel, we will analyze the case of the out channel that is the least bound, III, i.e., $c + (ab)$.

Energy conservation for the collision, between the in and the out channels, gives

$$\epsilon_b + \epsilon_{ac} - w_{ac} = \epsilon'_c + \epsilon'_{ab} - w_{ab}. \tag{53.40}$$

Then we obtain

$$\epsilon_b + \epsilon_{ac} = w_{ac} - w_{ab} + \epsilon'_c + \epsilon'_{ab} > w_{ac} - w_{ab} > 0. \tag{53.41}$$

This is the condition for the $c + (ab)$ channel to be open. If the total incoming kinetic energy, $\epsilon_b + \epsilon_{ac}$, is $< w_{ac} - w_{ab}$, we have a closed channel.

53.4 General Theory of Collisions

Before considering channels further, we describe a general, abstract-state, model of scattering.

Suppose that we have orthonormal states $|E\alpha\rangle$,

$$\langle E\alpha|E'\alpha'\rangle = \delta(E - E')\delta_{\alpha\alpha'}, \tag{53.42}$$

of eigenvalue energy E of some unperturbed Hamiltonian $\hat{H}_0$,

$$\hat{H}_0|E\alpha\rangle = E|E\alpha\rangle. \tag{53.43}$$

The scattering states are

$$|E\alpha\pm\rangle = |E\alpha\rangle + \hat{G}^{\pm}(E)\hat{V}|E\alpha\rangle, \tag{53.44}$$

where $\hat{G}^{\pm}(E)$ are the full Green's functions. The states have the same norm as the free states,

$$\langle E\alpha\pm|E'\alpha'\pm\rangle = \langle E\alpha|E'\alpha'\rangle, \tag{53.45}$$

and they obey the Lippmann–Schwinger equation

$$|E\alpha\pm\rangle = |E\alpha\rangle + \hat{G}_0^{\pm}(E)\hat{V}|E\alpha\pm\rangle. \tag{53.46}$$

The full Green's function also satisfies the Lippmann–Schwinger equation:

$$\hat{G}^{\pm} = \hat{G}_0^{\pm} + \hat{G}_0^{\pm}\hat{V}\hat{G}^{\pm}. \tag{53.47}$$

The S-matrix is defined by

$$\langle E\alpha-|E'\alpha+\rangle = \langle E\alpha|\hat{S}|E'\alpha'\rangle = \delta(E-E')\delta_{\alpha\alpha'} - 2\pi i\delta(E-E')\mathcal{T}_{\alpha\alpha'}(E), \tag{53.48}$$

where the T-matrix is

$$\mathcal{T}_{\alpha\alpha'}(E) = \langle E\alpha-|\hat{V}|E'\alpha'\rangle_{E=E'} = \langle E\alpha|\hat{V}|E'\alpha'+\rangle_{E=E'}. \tag{53.49}$$

In a spherical basis, this S-matrix reduces to the partial wave S-matrix,

$$S_{Elm,E'l'm'} = \tilde{S}_l(E)\delta(E-E')\delta_{ll'}\delta_{mm'}, \tag{53.50}$$

leading to

$$S(\vec{k}_1,\vec{k}_2) = e^{2i\delta_l}\frac{\pi}{2k_1^2}\delta(k_1-k_2)\delta_{l_1l_2}\delta_{m_1m_2}. \tag{53.51}$$

Define the operators $\Omega^{(\pm)}$ that create the scattering states from the free states,

$$\Omega^{(\pm)}|E\alpha\rangle = |E\alpha\pm\rangle, \tag{53.52}$$

specifically

$$\Omega^{(\pm)} = \lim_{t\to\mp\infty} U_I(0,t), \tag{53.53}$$

and giving the S-operator as

$$\hat{S} = \Omega^{(-)\dagger}\Omega^{(+)}. \tag{53.54}$$

If we specialize to a single channel, yet with target structure (electron–atom scattering), the initial (free) state is

$$|E\alpha\rangle = |E_A,\vec{n}_A\rangle \otimes |\epsilon_{B_0},\beta_0\rangle, \tag{53.55}$$

and the final state is

$$|E_a,\vec{n}'_A\rangle \otimes |\epsilon_B,\beta\rangle. \tag{53.56}$$

Conservation of energy leads to

$$E = E_A + \epsilon_{B_0} = E_a + \epsilon_B. \tag{53.57}$$

The Lippmann–Schwinger equation is

$$|E\alpha+\rangle = |E;\vec{n}_A,\beta_0+\rangle = |E_A,\vec{n}_A\rangle \otimes |\epsilon_{B_0},\beta_0\rangle + \hat{G}_0^+(E)\hat{V}|E;\vec{n}_A,\beta_0+\rangle. \tag{53.58}$$

The corresponding scattering wave function for projectile position $\vec{r}$ and center of mass target position $\vec{R}$, as $r \to \infty$, is

$$\left(\langle\vec{r}| \otimes \langle\vec{R}|\right)|E\alpha+\rangle = v^+(\vec{r},\vec{R}) \sim e^{i\vec{k}\cdot\vec{r}}u_{n_0}(\vec{R}) + \sum_n \frac{e^{ik'r}}{r} f_{nn_0}(\vec{k}',\vec{k})u_n(\vec{R}). \tag{53.59}$$

We expand the scattering solution in terms of the target states,

$$v^+(\vec{r},\vec{R}) = \sum_n F_n^+(\vec{r})u_n(\vec{R}). \tag{53.60}$$

Then at infinity for the projectile, $r \to \infty$, we find

$$F_n^+(\vec{r}) \sim e^{i\vec{k}\cdot\vec{r}}\delta_{nn_0} + \frac{e^{ik'r}}{r} f_{nn_0}(\vec{k}',\vec{k}) \tag{53.61}$$

for an open channel, and

$$F_n^+(\vec{r}) \sim 0 \tag{53.62}$$

for a closed channel.

Then the differential cross section becomes

$$\frac{d\sigma^{\epsilon_{B_0},\beta_0\to\epsilon_B\beta}}{d\Omega_a} = |f|^2\frac{p_a}{p_A}, \tag{53.63}$$

which splits between the elastic ($n = n_0$) and inelastic ($n \neq n_0$) cases,

$$\begin{aligned}\frac{d\sigma_{n_0}^{\text{elastic}}}{d\Omega} &= |f_{n_0n_0}|^2\\ \frac{d\sigma_n^{\text{inelastic}}}{d\Omega} &= |f_{nn_0}|^2\frac{k'}{k}.\end{aligned} \tag{53.64}$$

53.5 Multi-Channel Analysis

We now generalize to the multi-channel case. As we said, the presence of different channels means dividing the fundamental particles and their interaction into fragments. Denoting a channel by Γ, we have

$$\hat{H} = \hat{H}_0^\Gamma + \hat{V}^\Gamma. \tag{53.65}$$

The free states $|E\alpha\Gamma\rangle$ correspond to eigenvalues of the free Hamiltonian of the channel,

$$\hat{H}_0^\Gamma|E\alpha;\Gamma\rangle = E|E\alpha;\Gamma\rangle. \tag{53.66}$$

These states are orthonormal within the channel,

$$\langle E\alpha;\Gamma|E'\alpha';\Gamma\rangle = \delta(E-E')\delta_{\alpha\alpha'}, \tag{53.67}$$

but there is a nonzero overlap between channels,

$$\langle E\alpha;\Gamma|E'\alpha';\Gamma'\rangle \neq 0 \tag{53.68}$$

for $\Gamma \neq \Gamma'$.

The scattering states are defined as

$$|E\alpha;\Gamma\pm\rangle \equiv |E\alpha;\Gamma\rangle + \hat{G}(E)\hat{V}^{\Gamma}|E\alpha;\Gamma\rangle \tag{53.69}$$

and satisfy the Lippmann–Schwinger equation,

$$|E\alpha;\Gamma\pm\rangle = |E\alpha;\Gamma\rangle + \hat{G}_0(E)\hat{V}^{\Gamma}|E\alpha;\Gamma\pm\rangle. \tag{53.70}$$

The orthogonality conditions of the states are as follows:

(1) The scattering states in different channels *are* orthogonal,

$$\langle E\alpha\Gamma\pm|E\alpha\Gamma'\pm\rangle = 0 \tag{53.71}$$

for $\Gamma \neq \Gamma'$.

(2) The free states within a single channel are not complete,

$$\sum_\alpha \int_0^\infty dE|E\alpha\Gamma\rangle\langle E\alpha\Gamma| \neq \mathbb{1}, \tag{53.72}$$

while the scattering states of a single channel give just the projector to the scattering states of the channel,

$$\sum_\alpha \int_0^\infty dE|E\alpha\Gamma\pm\rangle\langle E\alpha\Gamma\pm| \equiv P^{\Gamma\pm}. \tag{53.73}$$

(3) The sum of the projectors onto different channels, added to the projector Λ_b onto bound states, is complete,

$$\sum_\Gamma P^{\Gamma\pm} + \Lambda_b = \mathbb{1}. \tag{53.74}$$

The operators turning free states into scattering states are defined for each channel, by projecting onto the (free) states of the channel with Λ_Γ:

$$\Omega^{(\pm)\Gamma} = \hat{U}^{\Gamma}(0,+\infty)\Lambda_\Gamma. \tag{53.75}$$

Within the channel these operators are unitary,

$$[\Omega_\Gamma^{(\pm)}]^\dagger\Omega_\Gamma^{(\pm)} = \Lambda_\Gamma. \tag{53.76}$$

The S-operator is defined for arbitrary in and out channels,

$$S^{\Gamma'\Gamma} = \Omega_{\Gamma'}^{(-)\dagger}\Omega_\Gamma^{(+)}. \tag{53.77}$$

The S-matrix is

$$\begin{aligned}\langle E'\alpha'\Gamma'|S^{\Gamma'\Gamma}|E\alpha\Gamma\rangle &= \langle E'\alpha'\Gamma'-|E\alpha\Gamma+\rangle \\ &= \delta(E-E')\delta_{\alpha\alpha'}\delta_{\Gamma\Gamma'} - 2\pi i\delta(E-E')T_{\alpha'\alpha}^{\Gamma'\Gamma}(E),\end{aligned} \tag{53.78}$$

where the T-matrix is given by

$$\begin{aligned}T_{\alpha'\alpha}^{\Gamma'\Gamma}(E) &= \langle E\alpha'\Gamma'|\hat{V}_\Gamma|E\alpha\Gamma+\rangle \\ &= \langle E\alpha'\Gamma'-|\hat{V}_{\Gamma'}|E\alpha\Gamma\rangle.\end{aligned} \tag{53.79}$$

If both the in and the out channels have two fragments, small one, A (the projectile), transformed into A', and large one, B (the target), transformed into B', the inelastic differential cross section has an extra factor coming from the mass of the projectile,

$$\frac{d\sigma^{\text{inelastic}}}{d\Omega} = |f_{\alpha',\Gamma';\alpha,\Gamma}|^2 \frac{k'}{k} \frac{m_A}{m_{A'}}. \tag{53.80}$$

53.6 Scattering of Identical Particles

The relevant case of electron scattering off an atom, which itself has N electrons, reveals a new characteristic, namely when identical particles scatter (the scattering electron versus the atom's electrons). But identical particles need to be symmetrized, as we have seen (their wave functions are either totally symmetric (S), or totally antisymmetric (A)).

We assume that the target (the atom, containing electrons) is already symmetrized,

$$\hat{H}_0[\hat{P}|E\alpha\rangle] = E\hat{P}|E\alpha\rangle, \tag{53.81}$$

where P is the symmetrization operator.

But then we add to the system the projectile A, which is also an electron, so we need to consider states that are symmetric or antisymmetric in all the $N+1$ electrons. That means we need to symmetrize $j = 1$, corresponding to the projectile, with $j = 2, \ldots, N+1$, for the electrons of the atom.

The symmetrized system is obtained by the action of the operator Λ, symmetrizing $j = 1$ with all the other j's,

$$\Lambda|E\alpha\rangle = \frac{1}{N+1}\left[|E\alpha\rangle + \sum_{j=2}^{N+1} P_{1j}|E\alpha\rangle\right]. \tag{53.82}$$

The correctly normalized symmetric/antisymmetric states are then

$$\begin{aligned} |E\alpha\rangle_{S/A} &= \sqrt{N+1}\Lambda|E\alpha\rangle \\ &= \frac{1}{\sqrt{N+1}}\left[|E\alpha\pm\rangle \pm \sum_{j=2}^{N+1} P_{1j}|E\alpha\pm\rangle\right]. \end{aligned} \tag{53.83}$$

The S-matrix is

$$\begin{aligned} S_{E'\alpha',E\alpha} &= \langle E'\alpha' - |E\alpha+\rangle \pm N\langle E'\alpha' - |P_{12}|E\alpha+\rangle \\ &= \delta(E-E')\delta_{\alpha\alpha'} - 2\pi i\delta(E-E')\mathcal{T}_{\alpha\alpha'}(E), \end{aligned} \tag{53.84}$$

where the T-matrix is given by

$$\begin{aligned} \mathcal{T}_{\alpha\alpha'}(E) &= \langle E'\alpha' - |\hat{V}|E\alpha\rangle \pm N\langle E'\alpha' - |P_{12}\hat{V}|E\alpha\rangle \\ &= \langle E'\alpha'|\hat{V}|E\alpha+\rangle \pm N\langle E'\alpha'|P_{12}\hat{V}|E\alpha+\rangle. \end{aligned} \tag{53.85}$$

It splits into a direct term (index d) and an exchange term (index exch),

$$\mathcal{T}_{\alpha\alpha'}(E) = \mathcal{T}^{d}_{\alpha\alpha'}(E) \pm N\mathcal{T}^{\text{exch}}_{\alpha\alpha'}(E). \tag{53.86}$$

Applying this to the simplest system, electron scattering off a hydrogen atom ($N = 1$), we replace

$$\begin{aligned} \mathcal{T}^d_{\alpha\alpha'}(E) &\to t^d_{\vec{p}'n'l'm_l';\vec{p},n,l,m_l}\delta_{m_1'm_1}\delta_{m_2'm_2} \\ \mathcal{T}^{\text{exch}}_{\alpha\alpha'}(E) &\to t^{\text{exch}}_{\vec{p}'n'l'm_l';\vec{p},n,l,m_l}\delta_{m_1'm_2}\delta_{m_2'm_1}. \end{aligned} \tag{53.87}$$

Then the differential cross section is

$$\frac{d\sigma_{m_1'm_2';m_1,m_2}}{d\Omega} = (\dots)|t^d\delta_{m_1'm_1}\delta_{m_2'm_2} \pm t^{\text{exch}}\delta_{m_1'm_2}\delta_{m_2'm_1}|^2. \tag{53.88}$$

Considering $m_i = \pm$ (the states of the spin projection), we obtain

$$\begin{aligned} \frac{d\sigma_{++,++}}{d\Omega} &= \frac{d\sigma_{--,--}}{d\Omega} = |t^d - t^{\text{exch}}|^2 \\ \frac{d\sigma_{+-,+-}}{d\Omega} &= \frac{d\sigma_{-+,-+}}{d\Omega} = |t^d|^2 \\ \frac{d\sigma_{+-,-+}}{d\Omega} &= \frac{d\sigma_{-+,+-}}{d\Omega} = |t^{\text{exch}}|^2, \end{aligned} \tag{53.89}$$

and the rest of the scattering cross sections vanish.

That means that the unpolarized differential cross section is

$$\left(\frac{d\sigma}{d\Omega}\right)^{\text{unpolarized}} = \frac{3}{4}|t^d - t^{\text{exch}}|^2 + \frac{1}{4}|t^d + t^{\text{exch}}|^2. \tag{53.90}$$

To calculate the transition amplitude, we use the Born approximation for $|E\alpha\pm\rangle \to |E\alpha\rangle$, and $\hat{V} = \hat{V}_1 + \hat{V}_{12}$, giving

$$\mathcal{T}_{\alpha\alpha'}(E) = \langle E'\alpha'|(\hat{V}_1 + \hat{V}_{12})|E\alpha\rangle \pm N\langle E'\alpha'|P_{12}(\hat{V}_1 + \hat{V}_{12})|E\alpha\rangle. \tag{53.91}$$

Important Concepts to Remember

- Inelastic scattering means scattering when energy, momentum, and/or particles can leak into a "sink", namely the case in which one or both of the particles (projectile and target) have internal structure and can become excited.
- In inelastic scattering $\hat{S}$ is not unitary, so S_l is not a phase, i.e., δ_l is not real: $|S_l| < 1$, or Im $\delta_l > 0$.
- For a "black disk" eikonal, Im $\Delta(b) = +\infty$ for $b < r_0$ and $\Delta = 0$ for $b > r_0$, so the total inelastic cross section is $\sigma_{\text{tot}} = \sigma_{\text{geometrical}} = \pi r_0^2$, while the total cross section is $2\pi r_0^2$.
- In inelastic scattering with a target structure (so that the target can absorb energy and momentum),

  $$\frac{d\sigma}{d\Omega}(|n_0\rangle \to |n\rangle) = \frac{k'}{k}|f|^2.$$

  For an atom with N electrons,

  $$\frac{d\sigma}{d\Omega} = \frac{k'}{k}\frac{4Z^2}{a_0^2q^4}|-\delta_{n,n_0} + F_{n,n_0}(\vec{q})|^2,$$

  with $F_{n,n_0}(\vec{q})$ the form factors.
- In general, both projectile and target can have structure, and that structure can change during the collision (as in nuclear reactions occurring through collision), so we have different divisions, named channels, of the total number of fundamental particles into fragments.

- In the case of several channels Γ (different ones for initial and final states), we have

$$\frac{d\sigma}{d\Omega} = \frac{k'}{k}\frac{m_A}{m_{A'}}|f_{\alpha',\Gamma';\alpha,\Gamma}|^2.$$

- When scattering involves identical fundamental particles, organized into fragments, as for instance an electron scattering off an atom with electrons, the total state needs to be (anti)symmetrized, leading to a direct term and an exchange term in the amplitude, so $d\sigma/d\Omega \sim |t^d\delta\cdots + t^{\text{exch}}\delta\ldots|^2$.

Further Reading

See [2] and [1].

Exercises

(1) Consider the mapping of the black disk eikonal into the general, $\delta_l(k)$, representation. Calculate the total inelastic cross section in this representation.

(2) Calculate the differential cross section $d\sigma(\theta)/d\Omega$ for the black disk eikonal, integrate it, and compare with the total inelastic cross section.

(3) Consider the scattering of a projectile A (nonidentical to the target components) off a hydrogenoid atom in the Born approximation, involving a jump of the electron from the ground state to the first excited state. Calculate the differential cross section.

(4) Consider a system of $N = 4$ fundamental particles. Write down the channels for the scattering of the various fragments. Taking the in channel as one where each of the two fragments have two particles, find the condition for the out channel to be one in which the fragments have respectively one and three particles.

(5) Describe the general theory of scattering (as in the text) for the case when the target is a single fundamental particle (and the projectile is composite), so both in and out channels have a single-particle target.

(6) How do you write the inelastic differential cross section for the case where the in channel has two parallel projectiles, and the out channel has a single outgoing projectile (fragment) of general composition (neither of the initial projectiles, nor their sum)?

(7) Specialize the scattering of the identical particle formalism for the case of an electron scattering off a helium atom to describe the differential cross sections of various spin projections in terms of direct and exchange T-matrices.

PART II_e

MANY PARTICLES

54 The Dirac Equation

In this chapter we consider relativistic corrections to the electron (fermion) theory, in the form of the Dirac equation. The correct treatment of the Dirac equation is in quantum field theory, but here we will deal with only the first relativistic corrections, in which case we can ignore all quantum field corrections. In some treatments in the literature, one talks about building a relativistic quantum mechanics. However, there is no relativistic quantum mechanics; joining relativity with quantum mechanics leads to quantum field theory, as we will show.

54.1 Naive Treatment

In basic nonrelativistic quantum mechanics we have the Schrödinger equation, which in the free case is

$$-i\hbar\partial_t\psi - \frac{\hbar^2}{2m}\Delta\psi = 0, \tag{54.1}$$

and is invariant under the Galilei transformations,

$$\vec{x}' = \vec{x} - \vec{v}t. \tag{54.2}$$

In the momentum representation, the free Schrödinger equation is

$$i\hbar\partial_t\psi = H\psi = \frac{\hbar^2}{2m}\vec{p}^{\,2}\psi = 0, \tag{54.3}$$

which is the nonrelativistic energy relation.

We would like to build a replacement for the Schrödinger equation that is invariant under the Lorentz transformation. We will replace the energy relation with a relativistic one. Acting twice with the same operator on the wave function we obtain

$$-\hbar^2\partial_t^2\psi = H^2\psi, \tag{54.4}$$

where

$$E^2 \to H^2 = \vec{p}^{\,2}c^2 + m^2c^4. \tag{54.5}$$

Then we have a replacement of the Schrödinger equation defined by

$$\begin{aligned}
&-\hbar^2\partial_t^2\psi = (\vec{p}^{\,2}c^2 + m^2c^4)\psi \;\Rightarrow\\
&[-\hbar^2\partial_t^2 + \hbar^2c^2\vec{\nabla}^2 - m^2c^4]\psi = 0 \;\Rightarrow\\
&\left[\vec{\nabla}^2 - \frac{1}{c^2}\frac{\partial^2}{\partial t^2} - \left(\frac{mc}{\hbar}\right)^2\right]\psi = 0.
\end{aligned} \tag{54.6}$$

This is the Klein–Gordon equation, but it is not what we wanted. The equation acts on a wave function ψ that is in a scalar representation of the Lorentz group. In fact, we need to find an equation acting on the spinor representation (spin 1/2, for the electron) of the Lorentz group.

Moreover, we need to have a first-order equation in ∂_t, like the usual Schrödinger equation, and correspondingly (because of Lorentz invariance) also in $\vec{p} = \frac{\hbar}{i}\vec{\nabla}$.

Putting together these two requirements, we find that we need to take the "square root" of the Klein–Gordon equation, in terms of matrices (the coefficients are matrices). Then we have

$$c^2\vec{p}^2 + (mc^2)^2 = (c\vec{\alpha}\cdot\vec{p} + \beta mc^2)^2 \equiv H^2, \tag{54.7}$$

meaning that the Hamiltonian is

$$H = c\vec{\alpha}\cdot\vec{p} + \beta mc^2, \tag{54.8}$$

with matrix coefficients $\vec{\alpha}$ and β, satisfying (from the defining equation, by equating terms with $p_i p_j$, or p_i, or no p_i at all, on both sides of it),

$$\begin{aligned}(\alpha_i)^2 &= \beta^2 = 1\\ \alpha_i\alpha_j + \alpha_j\alpha_i &= \{\alpha_i,\alpha_j\} = 0, \quad i \neq j\\ \alpha_i\beta + \beta\alpha_i &= \{\alpha_i,\beta\} = 0.\end{aligned} \tag{54.9}$$

Moreover, since the Hamiltonian is Hermitian, it means that the coefficient matrices are also Hermitian, so $\alpha_i^\dagger = \alpha_i$ and $\beta^\dagger = \beta$. They are also traceless and have eigenvalues ± 1 (since $\alpha_i^2 = \beta^2 = 1$). The matrices must be 4×4, which we can prove as follows. The dimension must be even, since we know that the functions on which we act with them must eventually depend on spin, which takes two values ($\pm 1/2$). And for 2×2 matrices, the complete set of matrices is $(\mathbb{1}, \vec{\sigma})$, so the full set of independent matrices that anticommute comprises the three Pauli matrices σ_i; yet we need four, for the α_i and β. That leaves the next even dimension up, namely 4×4 matrices.

The matrices are also not unique; after a unitary transformation, with matrix S ($S^\dagger = S^{-1}$), we have another solution for the matrices:

$$\vec{\alpha} \to S^{-1}\vec{\alpha}S, \;\; S^{-1}\beta S. \tag{54.10}$$

The *Dirac equation*, replacing the Schrödinger equation, is then

$$i\hbar\frac{\partial}{\partial t}\psi = (c\vec{\alpha}\cdot\vec{p} + \beta mc^2)\psi = \left(\frac{\hbar c}{i}\vec{\alpha}\cdot\vec{\nabla} + \beta mc^2\right)\psi. \tag{54.11}$$

One useful choice for the matrices α, β (reachable from another by a unitary transformation) is

$$\vec{\alpha} = \begin{pmatrix}\mathbf{0} & \vec{\sigma}\\ \vec{\sigma} & \mathbf{0}\end{pmatrix}, \quad \beta = \begin{pmatrix}\mathbb{1} & \mathbf{0}\\ \mathbf{0} & -\mathbb{1}\end{pmatrix}. \tag{54.12}$$

Since the $\vec{\alpha}, \beta$ coefficients are 4×4 matrices, they act on four-dimensional column vectors ψ, not just on a single wave function. However, we have already encountered a two-component vector as the spin 1/2 wave function ($m_s = +1/2$ and $m_s = -1/2$ components). Here we have two two-component vectors, which actually correspond to both the electron e^- and its antiparticle e^+ (though the separation of the two is related to some specific choice for $\vec{\alpha}, \beta$).

The correct treatment for producing the Dirac equation and its solutions is in quantum field theory. In the literature, there are statements about the existence of a relativistic quantum mechanics, but that is misleading; the correct relativistic treatment, joining relativity with quantum mechanics, is quantum field theory. The fact that a strictly relativistic quantum mechanics does not exist can be understood from three points of view:

(1) For any particle, there is an antiparticle. For spin 1/2 particles, the two are different (for real fields, such as real scalars, the particles are their own antiparticles). Thus if the energy is $E > m_p c^2 + m_{\bar{p}} c^2$, we can create a particle and antiparticle pair, which means that the number of particles is not conserved in quantum field theory, which describes real situations. But in usual quantum mechanics, particles follow quantum paths that never end or begin, so the number of particles is conserved.

(2) Even if $E < m_p c^2 + m_{\bar{p}} c^2$, we can create a particle–antiparticle pair for a short time. Indeed, the Heisenberg uncertainty principle, regarding E and t implies that $\Delta E \cdot \Delta t \sim \hbar$, so for low enough Δt, we can have $E + \Delta E > m_p c^2 + m_{\bar{p}} c^2$. Therefore we can always create a *virtual* pair of particles. There are no asymptotic particles (at large space and time separation) owing to energy and momentum conservation, but we have quantum fluctuations. Thus, even for low energies, we can always have quantum paths of particles that begin and end, violating the conservation of particle number.

(3) In the usual quantum mechanics, we also violate causality, even if we have Lorentz invariance (a relativistic theory), with $E = \sqrt{\vec{p}^{\,2} c^2 + m^2 c^4}$. We will show that under a time evolution that violates causality we still have a nontrivial result, contrary to what we should have in a good theory.

The propagator matrix element between $\vec{x}_0$ at time zero and $\vec{x}$ at time t is

$$\begin{aligned} U(t) &= \langle \vec{x} | e^{-i\hat{H}t} | \vec{x}_0 \rangle = \int d^3\vec{p} \int d^3\vec{p}\,' \langle \vec{x} | \vec{p} \rangle \langle \vec{p} | e^{-i\hat{H}t/\hbar} | \vec{p}\,' \rangle \langle \vec{p}\,' | \vec{x}_0 \rangle \\ &= \frac{1}{(2\pi)^3} \int d^3\vec{p} \, \exp\left(-\frac{i}{\hbar} t \sqrt{\vec{p}^{\,2} c^2 + m^2 c^4} \right) e^{i\vec{p}\cdot(\vec{x}-\vec{x}_0)/\hbar}. \end{aligned} \tag{54.13}$$

But

$$\begin{aligned} \int d^3\vec{p}\, e^{i\vec{p}\cdot\vec{x}/\hbar} &= \int_0^\infty p^2 dp\, 2\pi \int_{-1}^{+1} d(\cos\theta) e^{ipx\cos\theta/\hbar} = \int_0^\infty p^2 dp \left[\frac{2\pi\hbar}{ipx} \left(e^{ipx/\hbar} - e^{-ipx/\hbar} \right) \right] \\ &= \int p^2 dp \frac{4\pi\hbar}{px} \sin\left(\frac{p}{\hbar} x \right). \end{aligned} \tag{54.14}$$

Then the propagator matrix element is

$$U(t) = \frac{\hbar}{2\pi^2 |\vec{x} - \vec{x}_0|} \int_0^\infty p\, dp \sin\left(\frac{p}{\hbar} |\vec{x} - \vec{x}_0| \right) \exp\left(-\frac{i}{\hbar} t \sqrt{p^2 c^2 + m^2 c^4} \right). \tag{54.15}$$

To approximate it, we use the saddle point approximation of an integral, by Taylor expanding around the extremum in an exponent,

$$I = \int dx e^{f(x)} \simeq e^{f(x_0)} \int d\delta x e^{f''(x_0)\delta x^2/2} = e^{f(x_0)} \sqrt{\frac{2\pi}{f''(x_0)}}. \tag{54.16}$$

If we consider a separation much outside the causal cone, $x^2 \gg c^2 t^2$, the saddle point (extremum) condition implies

$$\begin{aligned} &\frac{1}{\hbar} \frac{d}{dp} (ipx - it\sqrt{p^2 c^2 + m^2 c^4}) = 0 \;\Rightarrow\; x = \frac{tpc^2}{\sqrt{p^2 c^2 + m^2 c^4}} \\ &\Rightarrow p = p_0 = \frac{imcx}{\sqrt{x^2 - c^2 t^2}}. \end{aligned} \tag{54.17}$$

Thus the saddle point evaluation of the propagator matrix element is

$$U(t) \propto \exp\left(\frac{i}{\hbar}(p_0 x - t\sqrt{p_0^2 c^2 + m^2 c^4})\right) \sim \exp\left(-\frac{mc}{\hbar}\sqrt{x^2 - c^2 t^2}\right) \neq 0 \tag{54.18}$$

This means that there is a breakdown of causality even well outside the causal cone, albeit with an exponentially small value. In quantum field theory causality is recovered, but we will not explain how, since it requires information beyond the scope of this book.

54.2 Relativistic Dirac Equation

Our previous treatment of the Dirac equation was based on the idea of a relativistic version of quantum mechanics. But, as we have just seen, that is not consistent so we need a derivation based on quantum field theory.

We need to consider a field associated with spin 1/2 particles. These are "spinor" fields, encompassing the Hilbert space acted upon by so-called *gamma matrices* γ_μ. These are objects satisfying the "Clifford algebra", objects which when squared give the identity, and which anticommute among each other, just like $\vec{\alpha}$ and β. Together, these conditions make the Clifford algebra

$$\{\gamma^\mu, \gamma^\nu\} = 2g^{\mu\nu}\mathbb{1}, \tag{54.19}$$

where $g^{\mu\nu}$ is the metric, in our case the Minkowski metric $g^{\mu\nu} = \eta^{\mu\nu} = \mathrm{diag}(-1, +1, +1, +1)$. If we replace γ^0 with $i\gamma^0$, the Clifford algebra will have the Euclidean metric $\tilde{g}^{\mu\nu} = \delta^{\mu\nu}$. In three dimensions, the Pauli matrices σ_i satisfy this Euclidean Clifford algebra.

The Hilbert space for the Clifford algebra has a basis that is changed by a unitary transformation, $\psi \to S\psi$, which amounts to a transformation on the gamma matrices, $\gamma^\mu \to S\gamma^\mu S^{-1}$, meaning the gamma matrices are not unique, just like $\vec{\alpha}$ and β.

A choice of γ^μ corresponds to a representation of the Clifford algebra. A useful representation is the Weyl representation,

$$\gamma^0 = -i\begin{pmatrix} \mathbf{0} & \mathbb{1} \\ \mathbb{1} & \mathbf{0} \end{pmatrix}, \quad \gamma^i = -i\begin{pmatrix} \mathbf{0} & \sigma^i \\ -\sigma^i & \mathbf{0} \end{pmatrix}, \tag{54.20}$$

satisfying $(\gamma^0)^2 = -\mathbb{1}, (\gamma^i)^2 = \mathbb{1}$. Together, the gamma matrices in the Weyl representation are written as

$$\gamma^\mu = -i\begin{pmatrix} \mathbf{0} & \sigma^\mu \\ \bar{\sigma}^\mu & \mathbf{0} \end{pmatrix}, \tag{54.21}$$

where

$$\sigma^\mu = (\mathbb{1}, \sigma^i), \quad \bar{\sigma}^\mu = (\mathbb{1}, -\sigma^i). \tag{54.22}$$

Then the Dirac equation is the Lorentz-invariant 4×4 matrix equation that is linear in ∂_t and has rest energy mc^2. That uniquely gives, in the "theorist's units" with $c = \hbar = 1$, the equation

$$(\gamma^\mu \partial_\mu + m)\psi = 0. \tag{54.23}$$

Reinstating $\hbar$ and c, we find

$$\gamma^0 \frac{\hbar}{c}\frac{\partial}{\partial t}\psi + \hbar\gamma^i \partial_i \psi + mc\psi = 0. \tag{54.24}$$

Multiplying by $i\gamma^0 c$, we obtain

$$i\hbar\frac{\partial}{\partial t}\psi = i\hbar c\gamma^0\gamma^i\partial_i\psi + imc^2\gamma^0\psi, \tag{54.25}$$

which means that we have

$$\vec{\alpha} = -\gamma^0\vec{\gamma}, \;\; \beta = i\gamma^0. \tag{54.26}$$

54.3 Interaction with Electromagnetic Field

To describe the interaction of the spin 1/2 fermions (electrons) with the electromagnetic field, we use minimal coupling, replacing $\vec{p}$ with $\vec{p} - q\vec{A}$, and H with $H + q\phi$.

Then the Hamiltonian with the interaction with the electromagnetic field is

$$H = c\vec{\alpha}\cdot\left(\vec{p} - \vec{A}\right) + \beta mc^2 + q\phi. \tag{54.27}$$

Substituting into the Schrödinger equation

$$i\hbar\frac{\partial}{\partial t}\psi = \hat{H}\psi, \tag{54.28}$$

we obtain the Dirac equation including the interaction with the electromagnetic field,

$$\left[\frac{\hbar}{i}\frac{1}{c}\frac{\partial}{\partial t} + \frac{q}{c}\phi + \vec{\alpha}\cdot\left(\frac{\hbar}{i}\vec{\nabla} - q\vec{A}\right) + \beta mc\right]\psi = 0. \tag{54.29}$$

In the explicitly relativistic invariant notation, we have

$$\left[\gamma^\mu\left(\frac{\hbar}{i}\partial_\mu - qA_\mu\right) + mc\right]\psi = 0. \tag{54.30}$$

54.4 Weakly Relativistic Limit

The main reason to consider in this book on quantum mechanics the Dirac equation, which is really part of a quantum field theory treatment, is to calculate the first nontrivial relativistic corrections to the nonrelativistic result. This is usually called the weakly relativistic limit.

Interaction with a Magnetic Field

The first application is to the case of $\phi = 0$, leading to interaction with just a magnetic field. In the stationary case, $\psi(t) = \psi e^{-iEt/\hbar}$, the Dirac equation for interaction with $\vec{A}$ reduces to

$$E\psi = \hat{\psi} = (c\vec{\alpha}\cdot\vec{p}_{\text{kin}} + \beta mc^2)\psi, \tag{54.31}$$

where the kinetic momentum is

$$\vec{p}_{\text{kin}} = \vec{p} - q\vec{A}. \tag{54.32}$$

Dividing the 4-vector ψ into two two-component vectors χ and ϕ, we find the matrix equation (using $\vec{\alpha}$ and β in (54.12))

$$\begin{pmatrix} E - mc^2 & -c\vec{\sigma} \cdot \vec{p}_{\rm kin} \\ -c\vec{\sigma} \cdot \vec{p}_{\rm kin} & E + mc^2 \end{pmatrix} \begin{pmatrix} \chi \\ \phi \end{pmatrix} = 0, \tag{54.33}$$

which divides into two equations,

$$\begin{aligned} (E - mc^2)\chi &= c\vec{\sigma} \cdot \vec{p}_{\rm kin}\phi \\ (E + mc^2)\phi &= c\vec{\sigma} \cdot \vec{p}_{\rm kin}\chi. \end{aligned} \tag{54.34}$$

In the nonrelativistic limit, $\vec{\sigma} \cdot \vec{p} \sim mv$, and $E + mc^2 \simeq 2mc^2$, meaning that (from the second relation)

$$\frac{\phi}{\chi} \sim \frac{mvc}{2mc^2} = \frac{1}{2}\frac{v}{c} \ll 1. \tag{54.35}$$

Thus the two-component vector ϕ is small, while χ is large.

Substituting

$$\phi \simeq \frac{\vec{\sigma} \cdot \vec{p}_{\rm kin}}{2mc}\chi, \tag{54.36}$$

obtained from the second Dirac equation, into the first equation, we obtain

$$(E - mc^2)\chi = c\vec{\sigma} \cdot \vec{p}_{\rm kin}\phi = \frac{(\vec{\sigma} \cdot \vec{p}_{\rm kin})^2}{2m}\chi, \tag{54.37}$$

which is called the Pauli equation, for the large two-component vector χ.

Using the relation between Pauli matrices,

$$\sigma_i \sigma_j = \delta_{ij} + i\epsilon_{ijk}\sigma_k, \tag{54.38}$$

we find

$$\begin{aligned} \frac{1}{2m}\sigma_i\sigma_j (p_i - qA_i)\left(p_j - qA_j\right) &= \frac{\left(\vec{p} - q\vec{A}\right)^2}{2m} + \frac{i}{2m}\epsilon_{ijk}\left(\frac{\hbar}{i}\partial_i - qA_i\right)\left(\frac{\hbar}{i}\partial_j - qA_j\right)\sigma_k \\ &= \frac{\left(\vec{p} - q\vec{A}\right)^2}{2m} - \frac{\hbar q}{2m}\vec{\sigma} \cdot \vec{B}, \end{aligned} \tag{54.39}$$

where we have used $\epsilon_{ijk}(-\partial_i A_j + \partial_j A_i) = -B_i$.

Then the Dirac equation for χ including the interaction with $\vec{A}$ (the Pauli equation) is

$$(E - mc^2)\chi = \left[\frac{\left(\vec{p} - q\vec{A}\right)^2}{2m} - \frac{\hbar q}{2m}\vec{\sigma} \cdot \vec{B}\right]\chi. \tag{54.40}$$

Here $E - mc^2$ is the energy appearing in the nonrelativistic Schrödinger equation, and the $\vec{\sigma} \cdot \vec{B}$ term is the spin interaction with magnetic field, with the Landé factor $g = 2$, as we have seen before.

Interaction with an Electric Field

Interaction with a central electric potential $\phi(r)$,

$$V(r) = e\phi(r), \tag{54.41}$$

gives two coupled Dirac equations for χ and ϕ with an extra term proportional to the identity,

$$\begin{aligned}(E - V - mc^2)\chi &= c\vec{\sigma}\cdot\vec{p}_{\text{kin}}\phi \\ (E - V + mc^2)\phi &= c\vec{\sigma}\cdot\vec{p}_{\text{kin}}\chi.\end{aligned} \tag{54.42}$$

Solving the second equation as before,

$$\phi = \frac{c\vec{\sigma}\cdot\vec{p}_{\text{kin}}}{E - V + mc^2}\chi, \tag{54.43}$$

and substituting into the first equation, we get an equation for the large two-component vector χ,

$$(E - mc^2 - V)\chi = c^2(\vec{\sigma}\cdot\vec{p}_{\text{kin}})\frac{1}{E - V + mc^2}(\vec{\sigma}\cdot\vec{p}_{\text{kin}})\chi. \tag{54.44}$$

This equation is now expanded in the small parameter

$$\frac{E - mc^2 - V}{2mc^2} \ll 1. \tag{54.45}$$

The zeroth-order term is

$$(E - mc^2 - V)\chi \simeq c^2\frac{(\vec{\sigma}\cdot\vec{p}_{\text{kin}})^2}{2mc^2}\chi, \tag{54.46}$$

which becomes (moving V to the right-hand side) the usual interaction with a magnetic field and an electric potential (the Pauli equation with V in it)

$$(E - mc^2)\chi = \left(\frac{(\vec{\sigma}\cdot\vec{p}_{\text{kin}})^2}{2m} + V\right)\chi. \tag{54.47}$$

Next, we consider the first-order corrections in the expansion parameter,

$$\begin{aligned}(E - mc^2 - V)\chi &= \left(c^2\frac{(\vec{\sigma}\cdot\vec{p}_{\text{kin}})}{2mc^2}\left(1 + \frac{E - mc^2 - V}{2mc^2}\right)^{-1}(\vec{\sigma}\cdot\vec{p}_{\text{kin}})\right) \\ &\simeq \left[\frac{(\vec{\sigma}\cdot\vec{p}_{\text{kin}})^2}{2m} - (\vec{\sigma}\cdot\vec{p}_{\text{kin}})\frac{E - mc^2 - V}{4m^2c^2}(\vec{\sigma}\cdot\vec{p}_{\text{kin}})\right]\chi.\end{aligned} \tag{54.48}$$

At this point, we want to consider the pure electric case, with $\vec{A} = 0$, and $\vec{p}_{\text{kin}} = \vec{p}$. Then, using

$$(E - mc^2 - V)\vec{\sigma}\cdot\vec{p}\chi = \vec{\sigma}\cdot\vec{p}(E - mc^2 - V)\chi + \vec{\sigma}\cdot[E - mc^2 - V, \vec{p}]\chi, \tag{54.49}$$

we find

$$(E - mc^2)\chi = \left\{\frac{\vec{p}^2}{2m} + V - \frac{p^4}{8m^3c^2} - \frac{(\vec{\sigma}\cdot\vec{p})(\vec{\sigma}\cdot[\vec{p}, V])}{4m^2c^2}\right\}\chi. \tag{54.50}$$

The first two terms on the right-hand side are the nonrelativistic result, while the third term is the relativistic correction to the energy coming from the expansion of $E = p^2c^2 \vec{+} m^2c^4$,

$$E \simeq mc^2 + \frac{p^2}{2m} - \frac{p^4}{8m^3c^2}. \tag{54.51}$$

Decomposing $\sigma_i \sigma_j = \delta_{ij} + i\epsilon_{ijk}\sigma_k$, we obtain the corrected Schrödinger equation

$$(E - mc^2)\chi = \left(\frac{\vec{p}^2}{2m} + V - \frac{\vec{p}^4}{8m^3c^2} - i\frac{\vec{\sigma} \cdot (\vec{p} \times [\vec{p}, V])}{4m^2c^2} - \frac{\vec{p} \cdot [\vec{p}, V]}{4m^2c^2} \right) \chi, \tag{54.52}$$

where there are three relativistic corrections to the Hamiltonian. But the second is

$$-i\frac{\vec{\sigma} \cdot (\vec{p} \times [\vec{p}, V])}{4m^2c^2} = -i\frac{\vec{\sigma} \cdot (\vec{p} \times [-i\hbar\vec{\nabla}, V])}{4m^2c^2}, \tag{54.53}$$

where

$$[-i\hbar\vec{\nabla}, V] = -i\hbar(\vec{\nabla}V) = -i\hbar\frac{\vec{r}}{r}\frac{dV}{dr}. \tag{54.54}$$

However, $-\vec{p} \times \vec{r} = \vec{L}$ is the angular momentum and $\hbar\vec{\sigma}/2 = \vec{S}$ is the spin, so the second relativistic correction is

$$+\frac{1}{2m^2c^2}(\vec{S} \cdot \vec{L})\frac{1}{r}\frac{dV}{dr}, \tag{54.55}$$

which is the spin–orbit interaction, giving Thomas precession.

Finally however, the last term gives the so-called Darwin term. But the term is not Hermitian,

$$(\vec{p} \cdot [\vec{p}, V])^\dagger = [V, \vec{p}] \cdot \vec{p} = -\vec{p} \cdot [\vec{p}, V]. \tag{54.56}$$

This means that the conservation of probability is broken (the conservation of probability implies unitary evolution, through a Hermitian Hamiltonian). We need an extra term in the Hamiltonian to compensate for the probability loss and restore Hermiticity.

The probability loss is due to the fact that we dropped the small two-component vector ϕ. Indeed, the normalization of probability is given by

$$\langle\psi|\psi\rangle = \langle\chi|\chi\rangle + \langle\phi|\phi\rangle = 1. \tag{54.57}$$

Therefore we must replace the resulting norm of ϕ with a term involving only χ. Using (54.43), approximated as

$$|\phi\rangle \simeq \frac{\vec{\sigma} \cdot \vec{p}}{2mc^2}|\chi\rangle, \tag{54.58}$$

to relate the two vectors, we obtain

$$\langle\phi|\phi\rangle = \left\langle \chi \middle| \left(\frac{\vec{\sigma} \cdot \vec{p}}{2mc} \right)^2 \middle| \chi \right\rangle \tag{54.59}$$

to be added to $\langle\chi|\chi\rangle$, i.e., to replace it by

$$\left\langle \chi \middle| \left[1 + \left(\frac{\vec{\sigma} \cdot \vec{p}}{2mc} \right)^2 \right] \middle| \chi \right\rangle. \tag{54.60}$$

Thus we are replacing $|\chi\rangle$ by

$$|\tilde{\chi}\rangle = \sqrt{1 + \left(\frac{\vec{\sigma} \cdot \vec{p}}{2mc} \right)^2} |\chi\rangle \simeq \left[1 + \frac{(\vec{\sigma} \cdot \vec{p})^2}{8m^2c^2} \right] |\chi\rangle, \tag{54.61}$$

inverted as

$$|\chi\rangle = \left(1 - \frac{(\vec{\sigma} \cdot \vec{p})^2}{8m^2c^2} \right) |\tilde{\chi}\rangle. \tag{54.62}$$

Evaluating the energy in the Schrödinger equation $E - mc^2$ in the $|\tilde{\chi}\rangle$ state, we obtain (since $(\vec{\sigma}\cdot\vec{p})^2 = \vec{p}^2$)

$$\begin{aligned}(E - mc^2)|\tilde{\chi}\rangle &= \left(1 + \frac{\vec{p}\cdot\vec{p}}{8m^2c^2}\right)(E - mc^2)|\chi\rangle \\ &= \left(1 + \frac{\vec{p}\cdot\vec{p}}{8m^2c^2}\right)\hat{H}\left(1 - \frac{\vec{p}\cdot\vec{p}}{8m^2c^2}\right)|\tilde{\chi}\rangle \\ &= \left(\hat{H} + \frac{[\vec{p}\cdot\vec{p},\hat{H}]}{8m^2c^2}\right)|\tilde{\chi}\rangle,\end{aligned} \tag{54.63}$$

where in the second equality we have used $(E-mc^2)|\chi\rangle = \hat{H}|\chi\rangle$, and replaced $|\chi\rangle$ with its expression in terms of $|\tilde{\chi}\rangle$. The only nontrivial commutator with $\vec{p}\cdot\vec{p}$ is V, the rest of the terms in $\hat{H}$ commute trivially. Therefore we have an extra term in the Hamiltonian:

$$(E - mc^2)|\tilde{\chi}\rangle = \left(\hat{H} + \frac{[\vec{p}\cdot\vec{p},V]}{8m^2c^2}\right)|\tilde{\chi}\rangle. \tag{54.64}$$

Then the Darwin term, including the last relativistic correction and the term coming from the normalization, is

$$H_D = \frac{1}{8m^2c^2}\left(-2\vec{p}\cdot[\vec{p},V] + [\vec{p}\cdot\vec{p},V]\right) = -\frac{1}{8m^2c^2}[\vec{p},\cdot[\vec{p},V]] = +\frac{\hbar^2}{8m^2c^2}\Delta V. \tag{54.65}$$

54.5 Correction to the Energy of Hydrogenoid Atoms

The relativistic correction to the Schrödinger equation is given by

$$H' = -\frac{(\vec{p}^2)^2}{8m^3c^2} + \frac{1}{2m^2c^2}\vec{L}\cdot\vec{S}\frac{1}{r}\frac{dV}{dr} + \frac{\hbar^2}{8m^2c^2}\Delta V = H_1 + H_2 + H_3, \tag{54.66}$$

where H_1 is the relativistic correction to the energy of a particle, H_2 is the spin–orbit coupling, and H_3 is the Darwin term. We will apply this to the hydrogenoid atom, with

$$V = -\frac{Ze_0^2}{r}. \tag{54.67}$$

Since we have spin–orbit coupling, it is convenient to add the orbital and spin angular momenta together to give the total angular momentum, $\vec{L} + \vec{S} = \vec{J}$. The eigenstates in the central potential are then eigenstates $|nljm_j\rangle$ of the complete set $(\hat{H}, \vec{L}^2, \vec{J}^2, J_z)$. In the coordinate representation,

$$\langle\vec{r}|nljm_j\rangle = R_{nl}(r)\langle\theta,\phi|ljm_j\rangle = R_{nl}(r)\left(C_{ljm_j}Y_{lm_j-1/2}(\theta,\phi)|\xi\rangle + D_{ljm_j}Y_{lm_j+1/2}(\theta,\phi)|\eta\rangle\right), \tag{54.68}$$

where $|\xi\rangle = |\uparrow\rangle$ and $|\eta\rangle = |\downarrow\rangle$.

The first-order time-independent perturbation theory for the energy is given by

$$W^{(1)} = \langle H'\rangle_{nljm_j}, \tag{54.69}$$

where the matrix element is diagonal,

$$\langle nljm_j|H'|nl'j'm_j'\rangle = \delta_{ll'}\delta_{jj'}\delta_{m_jm_j'}\langle H'\rangle_{nljm_j}. \tag{54.70}$$

The corrected energy is

$$W_{nj} = E_n + W_{nj}^{(1)}. \tag{54.71}$$

Using the matrix elements of powers of momentum and radius, one finds for the first term in (54.66),

$$\langle H_1 \rangle_{nljm_j} = -\frac{1}{2}mc^2 \frac{(\alpha Z)^4}{n^3}\left(\frac{1}{l+1/2} - \frac{3}{4n}\right), \tag{54.72}$$

for the spin–orbit coupling

$$\langle H_2 \rangle_{nljm_j} = \begin{cases} \pm\frac{1}{4}mc^2 \frac{(\alpha Z)^4}{n^3} \frac{1}{(l+1)(l+1/2)}, & \text{for } j = l \pm 1/2, \;\; l \neq 0 \\ 0, & \text{for } l = 0, \end{cases} \tag{54.73}$$

and for the Darwin term,

$$\langle H_3 \rangle_{nljm_j} = \frac{1}{2}mc^2 \frac{(\alpha Z)^4}{n^3}\delta_{l0}. \tag{54.74}$$

Since at $l \ll 1$, we have

$$-\frac{1}{l+1/2 \pm 1/2} \simeq -\frac{1}{l+1/2}\left[1 \mp \frac{1}{2(l+1/2)}\right], \tag{54.75}$$

this means that at large l or j, we find the total relativistic correction as

$$W_{nj}^{(1)} = -\frac{1}{2}mc^2 \frac{(\alpha Z)^4}{n^3}\left(\frac{1}{j+1/2} - \frac{3}{4n}\right). \tag{54.76}$$

Important Concepts to Remember

- The Dirac equation comes from the quantum field theory treatment of the electron, though it is often mistakenly described as a relativistic version of quantum mechanics.
- From the Hamiltonian becoming relativistic energy, we get the Klein–Gordon equation, $\left[\vec{\nabla}^2 - c^{-2}\partial_t^2 - (mc/\hbar)^2\right]\psi = 0$, and the Dirac equation is a sort of matrix square root of it, $i\hbar\partial_t\psi = (-i\hbar c\vec{\alpha}\cdot\vec{\nabla} + \beta mc^2)\psi$.
- There is no relativistic quantum mechanics since: the particle number is not conserved, owing at least to particle–antiparticle annihilation (and more); for a short time we can have virtual particles that can also change the particle number; in quantum mechanics we can violate causality.
- The relativistic form of the Dirac equation is $(\gamma^\mu\partial_\mu + mc)\psi = 0$ (where ∂_t has a factor $\hbar/c$ in front), with γ^μ the gamma matrices satisfying the Clifford algebra, $\{\gamma^\mu, \gamma^\nu\} = 2g^{\mu\nu}$, ψ a spinor field and $\alpha^i = -\gamma^0\gamma^i$ and $\beta = i\gamma^0$.
- The coupling to electromagnetism is via the minimal coupling, $\vec{p} \to \vec{p} - q\vec{A}$, relativistically $\frac{\hbar}{i}\partial_\mu \to \frac{\hbar}{i}\partial_\mu - qA_\mu$.
- The resulting coupling to a magnetic field is $[(\vec{p} - q\vec{A})^2/2m - (\hbar q/2m)\vec{\sigma}\cdot\vec{B}]$, and the coupling to an electric field is the spin–orbit interaction (giving Thomas precession),

$$\frac{\hbar^2}{2m^2c^2}(\vec{S}\cdot\vec{L})\frac{1}{r}\frac{dV}{dr},$$

and the Darwin term,

$$\frac{\hbar^2}{2m^2c^2}\Delta V.$$

- The relativistic corrections to the hydrogenoid atom come from: the first relativistic correction to the energy, $-p^4/(8m^3c^2)$, the spin–orbit interaction, and the Darwin term, leading to $-\frac{mc^2}{2}\frac{(\alpha Z)^4}{n^3}(1/(j+1/2)-3/(4n))$.

Further Reading

See [2], [1] and [3].

Exercises

(1) Does a wave function satisfy the Klein–Gordon equation? What can you deduce about the sign of the rest energy mc^2 in the Dirac equation?

(2) Find the Dirac equation in 1+1 dimensions (one space dimension) and find a representation for the matrices involved.

(3) Show that if we consider $\gamma_5 = i\gamma^0\gamma^1\gamma^2\gamma^3$ together with γ^0 and γ^i, they form a Clifford algebra in five dimensions.

(4) Write down the Klein–Gordon equation with coupling to electromagnetism.

(5) Consider a shell model for a nucleus, with the potential for a nucleon being approximated by the Yukawa potential. Calculate the first relativistic correction to the Hamiltonian for the nucleon.

(6) Check the missing steps in the relativistic corrections to the hydrogenoid atom.

(7) Calculate the relativistic corrections to a hydrogenoid atom in a magnetic field.

55 Multiparticle States in Atoms and Condensed Matter: Schrödinger versus Occupation Number

In this chapter we return to multiparticle states, from the point of view of atomic physics and condensed matter physics, with a large number of identical particles. After reviewing the Schrödinger representation and the approximations needed, we lay the foundations of the occupation number picture, which will be continued in the next chapter in terms of Fock states and second quantization.

55.1 Schrödinger Picture Multiparticle Review

The abstract-state Schrödinger equations, in the time-dependent and time-independent varieties, are

$$i\hbar\partial_t|\psi\rangle = \hat{H}|\psi\rangle \quad\Rightarrow\quad \hat{H}|\psi\rangle = E|\psi\rangle. \tag{55.1}$$

This equation applies to both single-particle and multiparticle states. Applying it to N *identical* particles, the coordinate basis is $|\vec{r}_1\dots\vec{r}_N\rangle$, and the basis has exchange symmetry under exchange of $\vec{r}_i$ with $\vec{r}_j$ (particle i with particle j),

$$|\vec{r}_1\dots\vec{r}_i\dots\vec{r}_j\dots\vec{r}_N\rangle = \pm|\vec{r}_1\dots\vec{r}_j\dots\vec{r}_i\dots\vec{r}_N\rangle, \tag{55.2}$$

where the $\pm$ correspond to Bose–Einstein and Fermi–Dirac statistics, respectively. If the particles are all different, we have no symmetry. If only some of them are identical, the exchange symmetry is valid for them only.

The completeness relation on the N-particle Hilbert space is

$$\int d^3\vec{r}_1\cdots\int d^3\vec{r}_N|\vec{r}_1\dots\vec{r}_N\rangle\langle\vec{r}_1\dots\vec{r}_N| = \mathbb{1}. \tag{55.3}$$

We can use, alternatively, the unsymmetrized basis $|\vec{r}_1\rangle\otimes\cdots\otimes|\vec{r}_N\rangle$ in the identical case also (since in the case of all different particles it is a basis by definition), with completeness relation

$$\int d^3\vec{r}_1\cdots\int d^3\vec{r}_N|\vec{r}_1\rangle\langle\vec{r}_1|\otimes\cdots\otimes|\vec{r}_N\rangle\langle\vec{r}_N| = \mathbb{1}. \tag{55.4}$$

Inserting the symmetrized basis, we find the Schrödinger equation acting on wave functions,

$$i\hbar\partial_t\langle\vec{r}_1\dots\vec{r}_N|\psi\rangle = \int d^3\vec{r}'_1\cdots\int d^3\vec{r}'_N\langle\vec{r}_1\dots\vec{r}|\hat{H}|\vec{r}'_1\dots\vec{r}'_N\rangle\langle\vec{r}'_1\dots\vec{r}'_N|\psi\rangle, \tag{55.5}$$

or

$$\begin{aligned} &i\hbar\partial_t\psi_N(\vec{r}_1,\dots,\vec{r}_N)\\ &= H_{\vec{r}_1,\dots,\vec{r}_N}\psi_N(\vec{r}_1,\dots,\vec{r}_N) \;\Rightarrow\; H_{\vec{r}_1,\dots,\vec{r}_N}\psi_N(\vec{r}_1,\dots,\vec{r}_N) = E\psi_N(\vec{r}_1,\dots,\vec{r}_N). \end{aligned} \tag{55.6}$$

The symmetry properties of N independent but identical particles imply the general invariant states

$$|12\ldots N\rangle = \frac{1}{\sqrt{N_{\text{perms}}}} \sum_{\text{perms } P} \left(sgn(P)\hat{P}|1\rangle \otimes |2\rangle \otimes \cdots \otimes |N\rangle \right), \tag{55.7}$$

where in the Bose–Einstein case $sgn(P)$ is replaced by 1, leading to the wave function $\psi_{S/A}(\vec{r}_1, \ldots, \vec{r}_N)$.

In the Fermi–Dirac (antisymmetric) case, we have

$$\psi_A(\vec{r}_1, \ldots, \vec{r}_N) = \left| \begin{pmatrix} \psi_{a_1}(\vec{r}_1) & \cdots & \psi_{a_N}(\vec{r}_1) \\ \vdots & & \vdots \\ \psi_{a_1}(\vec{r}_N) & \cdots & \psi_{a_N}(\vec{r}_N) \end{pmatrix} \right|, \tag{55.8}$$

known as the Slater determinant.

The symmetrized wave function satisfies

$$\psi_{N,S/A}(\vec{r}_1, \ldots, \vec{r}_i, \ldots, \vec{r}_j, \ldots, \vec{r}_N) = \pm\psi_{N,S/A}(\vec{r}_1, \ldots, \vec{r}_j, \ldots, \vec{r}_i, \ldots, \vec{r}_N). \tag{55.9}$$

We consider a Hamiltonian containing a free part that acts on a single particle, namely the kinetic part for each particle, and a potential that acts on several particles,

$$\begin{aligned} \hat{H} &= \hat{H}_0 + \hat{V}(\vec{r}_1, \ldots, \vec{r}_N) \\ \hat{H}_0 &= \hat{T} = \sum_{i=1}^{N} \frac{\vec{p}_i^{\,2}}{2m_i} \\ \hat{V}(\vec{r}_1, \ldots, \vec{r}_N) &= \sum_{i=1}^{N} \hat{V}(\vec{r}_i) + \sum_{i,j=1}^{N} \hat{V}(\vec{r}_i, \vec{r}_j) + \sum_{i,j,k=1}^{N} \hat{V}(\vec{r}_i, \vec{r}_j, \vec{r}_k) + \cdots \end{aligned} \tag{55.10}$$

Here the one-particle potential $\sum_{i=1}^{N} \hat{V}(\vec{r}_i)$ can be a Coulomb potential from a fixed source, in which case it could be added to the free one-particle Hamiltonian $\hat{H}_0$, while the three-particle potential $\hat{V}(\vec{r}_i, \vec{r}_j, \vec{r}_k)$ and any higher-N N-particle potentials come from quantum field theory corrections to the classical potential.

The only relevant term in the potential is the two-particle potential $\hat{V}(\vec{r}_i, \vec{r}_j) = \hat{V}(|\vec{r}_i - \vec{r}_j|)$ (the Coulomb potential between the particles, for instance).

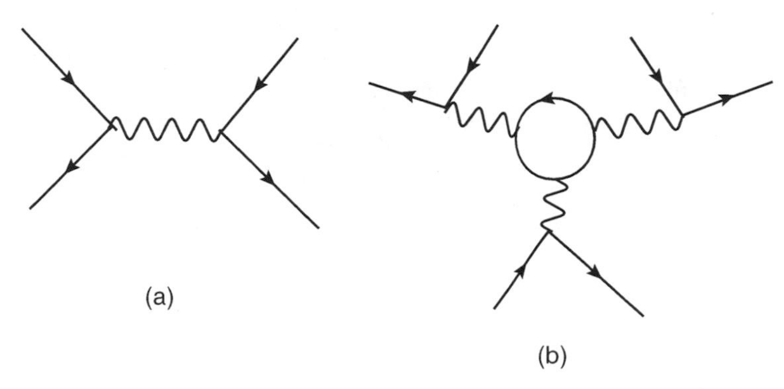

Figure 55.1 (a) The 2-point potential can come from a quantum mechanical Feynman diagram, but without loops (since it is classical). (b) The three-point potential can only come from loop corrections, and so from quantum field theory.

55.2 Approximation Methods

The multiparticle Schrödinger equation is hard to solve exactly, so in the relevant cases for atomic or molecular physics and condensed matter physics, one can use approximation methods to get a workable wave function.

A useful example for this chapter is the LCAO (linear combination of atomic orbitals) approximation to the multi-electron wave function for the atom or molecule. Consider the basis of factorized wave functions

$$\phi_r(\vec{r}_1, \dots, \vec{r}_N) \equiv \phi_{a_1}(\vec{r}_1) \cdots \phi_{a_N}(\vec{r}_N), \tag{55.11}$$

where r is an index encompassing the relevant subset of all the $a_1, \dots, a_N$, or its symmetrized alternative, in the fermionic case a Slater determinant.

In the case of the molecular wave function, we must also add a dependence on the positions $\vec{R}_\alpha$ of the nuclei in the molecule, giving $\phi_r(\{\vec{r}_i\}, \{\vec{R}_\alpha\})$.

Then the wave function is expanded in a finite subset of the total set of basis states,

$$\psi(\{\vec{r}_i\}, \{\vec{R}_\alpha\}) = \sum_r C_r \phi_i(\{\vec{r}_i\}, \{\vec{R}_\alpha\}), \tag{55.12}$$

and we minimize over C_r.

55.3 Transition to Occupation Number

The approximation method amounts to using a subset of the complete set of states, but if we keep *all* the states, we will get an exact statement. The basis wave functions are denoted

$$\phi_{a_1 \dots a_N}(\vec{r}_1, \dots, \vec{r}_N) = \phi_{a_1}(\vec{r}_1)\phi_{a_2}(\vec{r}_2) \cdots \phi(\vec{r}_N). \tag{55.13}$$

The exact expansion in this basis is with coefficients in all the basis states, i.e.,

$$\psi(\vec{r}_1, \dots, \vec{r}_N; t) = \sum C_{a_1 \dots a_N}(t) \phi_{a_1 \dots a_N}(\vec{r}_1, \dots, \vec{r}_N). \tag{55.14}$$

In the LCAO approximation, one takes a subset of states and minimizes the Schrödinger equation over the coefficients.

Alternatively, to obtain the same expansion, we can use successive expansions in each particle. That is, first expand in wave functions of $\vec{r}_1$,

$$\psi(\vec{r}_1, \dots, \vec{r}_N; t) = \sum_{a_1} C_{a_1}(\vec{r}_2, \dots, \vec{r}_N; t) \phi_{a_1}(\vec{r}_1), \tag{55.15}$$

and then the coefficients are expanded in the wave functions for $\vec{r}_2$,

$$C_{a_1}(\vec{r}_2, \dots, \vec{r}_N; t) = \sum_{a_2} C_{a_1 a_2}(\vec{r}_3, \dots, \vec{r}_N; t) \phi_{a_2}(\vec{r}_2), \tag{55.16}$$

etc., until we are back to

$$\psi(\vec{r}_1, \dots, \vec{r}_N; t) = \sum_{a_1, \dots, a_N} C_{a_1 \cdots a_N}(t) \phi_{a_1}(\vec{r}_1) \cdots \phi_{a_N}(\vec{r}_N). \tag{55.17}$$

The symmetry properties of the wave function are transmitted to symmetry properties of the coefficients,

$$C_{a_1\ldots a_i\ldots a_j\ldots a_N}(t) = \pm C_{a_1\ldots a_j\ldots a_i\ldots a_N}(t). \tag{55.18}$$

A relevant case occurs when the indices a_i are indexing energies, so that $a_i \to E_i$. The energies can even be approximately continuous, though not exactly continuous. In this case, we will not consider degeneracies: the energy indexes the available states. From now on we will replace a_i with E_i everywhere.

Note then that E_i takes value in the set $\{1, 2, 3, \ldots, \infty\}$, ordered by increasing energy.

In the following we will mostly follow Bose–Einstein (boson) statistics, while Fermi–Dirac (fermion) statistics will be treated later.

Then, if index 1 (the lowest-energy state) occurs n_1 times among the row of indexes $(E_1, \ldots, E_N)$, index 2 occurs n_2 in the same row of indexes, etc., until ∞ occurs n_∞ times in the row, we have that the total number of occurences of any value equals N,

$$\sum_{i=1}^{\infty} n_i = N. \tag{55.19}$$

In that case, the coefficients can have the dependence on $(E_1, \ldots, E_N)$ replaced with dependence on the *occupation numbers* $n_1, n_2, \ldots, n_\infty$. The numerical equality is then related to a different functional dependence, $\tilde{C}$ instead of C,

$$C_{E_1\ldots E_N}(t) = \tilde{C}_{n_1\ldots n_\infty}(t). \tag{55.20}$$

Now we can translate the normalization condition in the indices E_i,

$$\sum_{E_1\ldots E_N} |C_{E_1\ldots E_N}(t)|^2 = 1, \tag{55.21}$$

into a normalization condition in the indices n_i. But the sum over E_i's equals the sum over n_i, with a combinatorial factor for choosing n_1 objects of type 1, n_2 of type 2, etc. out of a total of N,

$$C^N_{n_1\ldots n_\infty} = \frac{N!}{n_1! \cdots n_\infty!}. \tag{55.22}$$

The normalization condition in $\tilde{C}$ is now

$$1 = \sum_{n_1,\ldots,n_\infty} |\tilde{C}_{n_1\ldots n_\infty}(t)|^2 C^N_{n_1\ldots n_\infty} = \sum_{n_1,\ldots,n_\infty} \left| \sqrt{\frac{N!}{n_1!\ldots n_\infty!}} \tilde{C}_{n_1\cdots n_\infty}(t) \right|^2. \tag{55.23}$$

This then allows for the definition of new coefficients, appropriately normalized:

$$\sqrt{\frac{N!}{n_1! \cdots n_\infty!}} \tilde{C}_{n_1\ldots n_\infty}(t) = C'_{n_1\ldots n_\infty}(t). \tag{55.24}$$

55.4 Schrödinger Equation in Occupation Number Space

We now consider the Schrödinger equation with a Hamiltonian that contains a kinetic term that is a one-particle operator, perhaps including also the one-particle potential $V(\vec{r}_i)$, and a potential that is a two-particle operator,

$$\hat{H} = \sum_{i=1}^{N} T(\vec{r}_i) + \sum_{i,j=1}^{N} \hat{V}(\vec{r}_i, \vec{r}_j) + \cdots \tag{55.25}$$

We set to zero the three-particle and higher parts of the potential (since they are small, and come from quantum field theory corrections to the classical potential).

We will transition from the Schrödinger equation in the coordinate representation to a representation in terms of occupation numbers.

One-Particle Operator Acting on Wave Functions

We start with the action of the (generalized) kinetic operator, meaning the one-particle part of the Hamiltonian. That means we calculate

$$\sum_{i=1}^{N} T(\vec{r}_i)\psi(\vec{r}_1, \ldots, \vec{r}_N; t) = \sum_{i=1}^{N} T(\vec{r}_i) \sum_{E'_1 \ldots E'_N} C_{E'_1 \ldots E'_N}(t)\phi_{E'_1}(\vec{r}_1) \cdots \phi_{E'_N}(\vec{r}_N). \tag{55.26}$$

More precisely, we want to peel off the existence of extra wave functions ($N - 1$ of them) on the right-hand side. To do that, we calculate

$$T_{(1)} \equiv \int d^3\vec{r}_1 \cdots \int d^3\vec{r}_N \phi^*_{E_1}(\vec{r}_1) \cdots \phi^*_{EN}(\vec{r}_N) \sum_{i=1}^{N} T(\vec{r}_i)\psi(\vec{r}_1, \ldots, \vec{r}_N; t). \tag{55.27}$$

Using the orthonormality of the wave functions,

$$\int d^3\vec{r}_j \psi^*_{E_j}(\vec{r}_j)\psi_{E'_j}(\vec{r}_j) = \delta_{E_j E'_j}, \tag{55.28}$$

for all the wave functions except for $\psi_{E_i}(\vec{r}_i)$, on which we have $T(\vec{r}_i)$ acting, we obtain

$$\begin{aligned} T_{(1)} &= \sum_{i=1}^{N} \sum_{E'_i} \int d^3\vec{r}_i \phi^*_{E_i}(\vec{r}_i) T(\vec{r}_i) \phi_{E'_i}(\vec{r}_i) C_{E_1 \ldots E_{i-1} E'_i E_{i+1} \ldots E_N}(t) \\ &= \sum_{i=1}^{N} \sum_{E'_i} \langle \phi_{E_i} | \hat{T}_i | \phi_{E'_i} \rangle C_{E_1 \ldots E_{i-1} E'_i E_{i+1} \ldots E_N}(t). \end{aligned} \tag{55.29}$$

This means that in the coefficients, we replace E_i with E'_i, and sum over it.

Next we change to the occupation number indexes:

$$\begin{aligned} C_{E_1 \ldots E_{i-1} E'_i E_{i+1} \ldots E_N}(t) &= \tilde{C}_{n_1 \ldots n_{E_i}-1, \ldots n_{E'_i}+1, \ldots n_\infty} \\ &= \sqrt{\frac{n_1! \cdots n_\infty!}{N!}} C'_{n_1 \ldots n_{E_i}-1, \ldots n_{E'_i}+1, \ldots n_\infty}. \end{aligned} \tag{55.30}$$

In the first equality we note that we have removed one occupied state from the value E_i and added it at E'_i. In the second equality, we have just redefined the coefficients in terms of correctly normalized ones.

When going to occupation number space, the sum over particles translates into a sum over states, the multiplicity being the occupation numbers,

$$\sum_{i=1}^{N} \langle \phi_{E_i} | = \sum_{E_i} n_{E_i} \langle \phi_{E_i} |. \tag{55.31}$$

At this point, we can simplify the notation, and replace $|\phi_{E_i}\rangle$ with $|i\rangle$ and n_{E_i} with n_i everywhere. Then we get the simple one-particle operator action

$$T_{(1)} = \sum_{i,i'=1}^{N} \langle i|\hat{T}|i'\rangle \sqrt{\frac{n_1! \cdots n_\infty!}{N!}} C'_{n_1 \dots n_i - 1, \dots, n_{i'}+1, \dots, n_\infty}(t). \tag{55.32}$$

Two-Particle Operator Acting on Wave Functions

Next, we consider the action of the two-particle operator, i.e., the two-particle potential

$$\sum_{i>j=1}^{N} V(\vec{r}_i, \vec{r}_j) = \frac{1}{2} \sum_{i,j=1, i\neq j}^{N} V(\vec{r}_i, \vec{r}_j), \tag{55.33}$$

and define the same kind of coefficient coming from it as for $T_{(1)}$,

$$V_{(2)} = \int d^3\vec{r}_1 \cdots \int d^3\vec{r}_N \phi^*_{E_1}(\vec{r}_1) \cdots \phi^*_{E_N}(\vec{r}_N) \frac{1}{2} \sum_{i,j=1, i\neq j}^{N} V(\vec{r}_i, \vec{r}_j). \tag{55.34}$$

Again using the orthonormality of the one-particle wave functions, except for those involving $\vec{r}_i, \vec{r}_j$, we find

$$\begin{aligned} V_{(2)} &= \frac{1}{2} \sum_{i,j=1, i\neq j} \sum_{E'_i} \sum_{E'_j} \int d^3\vec{r}_i \int d^3\vec{r}_j \phi^*_{E_i}(\vec{r}_i) \phi^*_{E_j}(\vec{r}_j) V(\vec{r}_i, \vec{r}_j) \psi_{E'_i}(\vec{r}_i) \phi_{E'_j}(\vec{r}_j) \\ &\quad \times C_{E_1 \dots E_{i-1} E'_i E_{i+1} \dots E_{j-1} E'_j E_{j+1} \dots E_N}(t) \\ &= \frac{1}{2} \sum_{i,j=1, i\neq j} \sum_{E'_i} \sum_{E'_j} \langle \phi_{E_i} \phi_{E_j} | \hat{V} | \phi_{E'_i} \phi_{E'_j} \rangle C_{E_1 \dots E_{i-1} E'_i E_{i+1} \dots E_{j-1} E'_j E_{j+1} \dots E_N}(t). \end{aligned} \tag{55.35}$$

We then change to the occupation number representation, now with two changes in the indices,

$$\begin{aligned} C_{E_1 \dots E_{i-1} E'_i E_{i+1} \dots E_{j-1} E'_j E_{j+1} \dots E_N}(t) &= \tilde{C}_{n_1 \dots n_{E_i} - 1, \dots, n_{E'_i} + 1, \dots, n_{E_j} - 1, \dots, n_{E'_j} + 1, \dots, n_\infty} \\ &= \sqrt{\frac{n_1! \cdots n_\infty!}{N!}} C'_{n_1 \dots n_{E_i} - 1, \dots n_{E'_i} + 1, \dots, n_{E_j} - 1, \dots, n_{E_j} + 1, \dots, n_\infty}, \end{aligned} \tag{55.36}$$

where the factorials in the square root have the same occupation numbers as the coefficients C'.

Now, besides replacing the single sum over particles replaced with a sum over particles that includes the multiplicity of occupation numbers,

$$\sum_{i=1}^{N} \langle \phi_{E_i} | = \sum_{E_i} n_{E_i} \langle \phi_{E_i} |, \tag{55.37}$$

we have to account for double sum, in which, if two energies are equal, the second sum has one state less,

$$\sum_{i=1}^{N} \sum_{j=1}^{N} \langle \phi_{E_i} \phi_{E_j} | = \sum_{E_j \neq E'_i} n_{E_i} n_{E_j} \langle \phi_{E_i} \phi_{E_j} | + \sum_{E_i} n_{E_i} (n_{E_i} - 1) \langle \phi_{E_i} \phi_{E_i} |. \tag{55.38}$$

Relabeling $|\phi_{E_i}\rangle$ as $|i\rangle$ and $n_{E_i} = n_i$ as before, we find the two-particle potential operator:

$$\begin{aligned} V_{(2)} &= \sum_{E_i E_j}\sum_{E'_i E'_j} \frac{1}{2} n_{E_i}(n_{E_j} - \delta_{E_i E_j})\langle \phi_{E_i}\phi_{E_j}|\hat{V}_{(2)}|\phi_{E'_i}\phi_{E'_j}\rangle \\ &\quad \times \sqrt{\frac{n_1!\cdots n_\infty!}{N!}} C'_{n_1\ldots n_{E_i}-1,\ldots n_{E'_i}+1,\ldots n_{E_j}-1,\ldots n_{E'_j}+1,\ldots n_\infty}(t) \\ &= \sum_{i,j,i',j'} \frac{1}{2} n_i(n_j - \delta_{ij})\langle ij|\hat{V}_{(2)}|i'j'\rangle \\ &\quad \times \sqrt{\frac{n_1!\cdots n_\infty!}{N!}} C'_{n_1\ldots n_i-1,\ldots n_{i'}+1,\ldots n_j-1,\ldots n_{j'}+1,\ldots n_\infty}(t), \end{aligned} \tag{55.39}$$

where again the factorials inside the square root have the same occupation numbers as the coefficients C'.

Then, the Schrödinger equation, multiplied by the wave functions and integrated over position, gives

$$\int d^3\vec{r}_1 \int d^3\vec{r}_2 \cdots \int d^3\vec{r}_N \phi^*_{E_1}(\vec{r}_1)\cdots\phi^*_{E_N}(\vec{r}_N)\times i\hbar\partial_t\psi(\vec{r}_1,\ldots,\vec{r}_N;t) = T_{(1)} + V_{(2)}. \tag{55.40}$$

Since the left-hand side gives $i\hbar\partial_t C_{E_1\ldots E_N}(t)$, we obtain Schrödinger equation for the coefficients, first in terms of the original $C_{E_1\ldots E_N}(t)$ coefficients,

$$\begin{aligned} i\hbar\partial_t C_{E_1\ldots E_N}(t) &= \sum_{i=1}^{N}\sum_{E'_i}\langle\phi_{E_i}|\hat{T}_i|\phi_{E'_i}\rangle C_{E_1\ldots E_{i-1}E'_i E_{i+1}\ldots E_N}(t) \\ &\quad + \frac{1}{2}\sum_{i,j=1,i\neq j}\sum_{E'_i}\sum_{E'_j}\langle\phi_{E_i}\phi_{E_j}|\hat{V}|\phi_{E'_i}\phi_{E'_j}\rangle C_{E_1\ldots E_{i-1}E'_i E_{i+1}\ldots E_{j-1}E'_j E_{j+1}\ldots E_N}(t), \end{aligned} \tag{55.41}$$

and finally in terms of the $C'_{n_1\ldots n_\infty}(t)$ coefficients,

$$\begin{aligned} i\hbar\sqrt{\frac{n_1!\cdots n_\infty!}{N}}\partial_t C'_{n_1\ldots n_\infty}(t) &= \sum_{i,i'=1}^{N}\langle i|\hat{T}|i'\rangle\sqrt{\frac{n_1!\cdots(n_i-1)!\cdots(n_{i'}+1)!\cdots n_\infty!}{N}} \\ &\quad \times C'_{n_1\ldots n_i-1,\ldots n_{i'}+1,\ldots n_\infty}(t) + \sum_{i\neq j,i',j'}\frac{1}{2}n_i n_j\langle ij|\hat{V}_{(2)}|i'j'\rangle \\ &\quad \times\sqrt{\frac{n_1!\cdots(n_i-1)!\cdots(n_{i'}+1)!\cdots(n_j-1)!\cdots(n_{j'}+1)!\cdots n_\infty!}{N!}} \\ &\quad \times C'_{n_1\ldots n_i-1,\ldots n_{i'}+1,\ldots n_j-1,\ldots n_{j'}+1,\ldots n_\infty}(t) \\ &\quad + \sum_{i=j,i',j'}\frac{1}{2}n_i(n_i-1)\langle ij|\hat{V}_{(2)}|i'j'\rangle \\ &\quad \times\sqrt{\frac{n_1!\cdots(n_i-2)!\cdots(n_{i'}+1)!\cdots(n_{j'}+1)!\cdots n_\infty!}{N!}} \\ &\quad \times C'_{n_1\ldots n_i-2,\ldots n_{i'}+1,\ldots n_{j'}+1,\ldots n_\infty}(t). \end{aligned} \tag{55.42}$$

55.5 Analysis for Fermions

In the analysis before, we were considering bosonic particles. To analyze the case of fermions, we must replace the factorized basis (products of the one-particle wave functions) with Slater determinant, with combinatorial coefficient

$$[C^N_{n_1\ldots n_\infty}]^{-1/2}\left|\begin{pmatrix}\phi_{E_1}(\vec r_1) & \cdots & \phi_{E_N}(\vec r_1)\\ \vdots & & \vdots\\ \phi_{E_1}(\vec r_N) & \cdots & \phi_{E_N}(\vec r_N)\end{pmatrix}\right|. \quad (55.43)$$

But this combinatorial coefficient is the same as before,

$$[C^N_{n_1\ldots n_\infty}]^{-1/2}=\sqrt{\frac{n_1!\cdots n_\infty!}{N!}}=\frac{1}{\sqrt{N!}}, \quad (55.44)$$

since the occupation numbers for fermions n_i are either 0 or 1, in both cases obtaining $n_i! = 1$.

The one-particle and two-particle operators act on particles with the same coefficients $C_{E_1\ldots E_N}(t)$, where all the E_i are different (since for fermions we have at most one particle per state). Then the coefficients can be rewritten in terms of $\tilde C_{n_1\ldots n_\infty}(t)$ and $C'_{n_1\ldots n_\infty}(t)$,

$$C_{E_1\ldots E_N}(t)=\tilde C_{n_1\ldots n_\infty}(t)=\frac{1}{\sqrt{N!}}C'_{n_1\ldots n_\infty}(t). \quad (55.45)$$

The one-particle operator acting on the Slater determinant gives

$$\langle\phi_{E_i}|\hat T_i|\phi_{E'_i}\rangle C_{E_1\ldots E_{i-1}E'_iE_{i+1}\ldots E_N}(t), \quad (55.46)$$

where the term with coefficient $C_{E_1\ldots E_{i-1}E'_iE_{i+1}\ldots E_N}(t)$ acts by replacing the Slater determinant with

$$\left|\begin{pmatrix}\phi_{E_1}(\vec r_1) & \cdots & \phi_{E'_i}(\vec r_1) & \cdots & \phi_{E_N}(\vec r_1)\\ \vdots & & & & \vdots\\ \phi_{E_1}(\vec r_N) & \cdots & \phi_{E'_i}(\vec r_N) & \cdots & \phi_{E_N}(\vec r_N)\end{pmatrix}\right|. \quad (55.47)$$

Important Concepts to Remember

- The multiparticle Schrödinger equation has one-particle, two-body, three-body, etc., potentials, where only the one-particle and two-body potentials have a classical counterpart, the others coming from quantum field theory.
- We can expand the multiparticle wave function in a basis of factorized wave functions, $\phi_r(\vec r_1,\ldots,\vec r_N)=\phi_{a_1}(\vec r_1)\cdots\phi_{a_N}(\vec r_N)$, and in the LCAO approximation we keep only a finite number of these basis functions.
- The coefficients $C_{a_1\ldots a_N}(t)$ have the correct symmetry properties (for bosons or fermions). It is useful to consider $a_i\equiv E_i$ (energy eigenstates), and transition to occupation number states for *all* the energy eigenstates, $C_{E_1\ldots E_N}(t)\equiv\tilde C_{n_1\ldots n_\infty}(t)$.

- The one-particle (kinetic and potential) operator in occupation number is $\sum_i \sum_{E'_i} \langle \phi_{E_i} | \hat{T}_i | \phi_{E'_i} \rangle C_{E_1 \dots E_{i-1} E'_i E_{i+1} \dots E_N}(t)$, and the two-particle operator is $\frac{1}{2} \sum_{i \neq j} \sum_{E'_i, E'_j} \langle \phi_{E_i} \phi_{E_j} | \hat{V} | \phi_{E'_i} \phi_{E'_j} \rangle C_{E_1 \dots E_{i-1} E'_i E_{i+1} \dots E_{j-1} E'_j E_{j+1} \dots E_N}(t)$.
- We can write a Schrödinger equation for the coefficients $C_{E_1 \dots E_N}(t) \equiv \tilde{C}_{n_1 \dots n_\infty}(t)$.
- For fermions the coefficients multiply changed Slater determinants.

Further Reading

More details are given in the book [27].

Exercises

(1) How does translational invariance and rotational symmetry constrain the three-body potential coming from quantum field theory?

(2) Consider the LCAO approximation for an atom with three electrons. Write the expansion, restricting yourself to four basis elements.

(3) Write the coefficients of the expansion in exercise 2 as coefficients in terms of energy, $C_{E_1 \dots}$, show explicitly their symmetry properties, and then rewrite them as coefficients in the occupation number picture.

(4) Consider five bosons with a one-body potential that can be approximated as harmonic oscillator. Write the wave function expansion in terms of harmonic oscillator states with $n \leq 3$.

(5) For the case in exercise 4, write the one-particle (kinetic) operator in the Hamiltonian, in the occupation number basis.

(6) For the case in exercise 4, consider a two-particle potential of the Coulomb ($\propto 1/r$) type. Write explicitly the first six nontrivial terms (those having an occupation number change in the two-particle operator in the Hamiltonian), in the occupation number basis.

(7) Consider three fermions with a one-body potential that can be approximated as a harmonic oscillator potential. Write the first four one-particle terms in the Hamiltonian that acts on the Slater determinant (for a wave function with arbitrary coefficients).

56 Fock Space Calculation with Field Operators

In this chapter we continue with the occupation number picture, and build the Fock space corresponding to it. Then we build field operators, and operators acting on them. This is the beginning of the quantum field theory formalism, though in its nonrelativistic version, with $v \ll c$, and no antiparticles. This is a whole field of study, of many-body physics, of which we describe only the foundation.

56.1 Creation and Annihilation Operators

Assume, as in the previous chapter, that there are one-particle states, which means that there are (quasi-)particles. The states are indexed as $(1, 2, \ldots, \infty)$, and each multiparticle state has occupation numbers $(n_1, n_2, \ldots, n_\infty)$ for the one-particle states. Consider basis vectors for the multiparticle Hilbert space that are tensor products of each single-particle state,

$$|n_1 n_2 \ldots n_\infty\rangle = |n_1\rangle \otimes |n_2\rangle \otimes \cdots \otimes |n_\infty\rangle. \tag{56.1}$$

In the coordinate representation for the n-particle Hilbert space, we have

$$\begin{aligned} \langle \vec{r}_1, \ldots, \vec{r}_{n_1} | n_1 \rangle &= \phi_1(\vec{r}_1) \cdots \phi_1(\vec{r}_{n_1}) \\ \langle \vec{r}_{n_1+1}, \ldots, \vec{r}_{n_1+n_2} | n_2 \rangle &= \phi_2(\vec{r}_{n_1+1}) \cdots \phi_2(\vec{r}_{n_1+n_2}), \text{ etc.} \end{aligned} \tag{56.2}$$

The occupation number basis is complete in the multiparticle Hilbert space, so

$$\sum_{n_1 \ldots n_\infty} |n_1 \ldots n_\infty\rangle\langle n_1 \ldots n_\infty| = \mathbb{1}. \tag{56.3}$$

We now define creation and annihilation operators, first in the bosonic case, $b_i^\dagger, b_i$. These operators act in the same way as for the harmonic oscillator, just on the occupation number states. This means that on the single-particle state i, with occupation number n_i, the actions of b_i and $b_i^\dagger$ are

$$\begin{aligned} b_i |n_i\rangle &= \sqrt{n_i}\, |n_i - 1\rangle \\ b_i^\dagger |n_i\rangle &= \sqrt{n_i + 1}\, |n_i + 1\rangle. \end{aligned} \tag{56.4}$$

We can also define the number operator in the state i,

$$N_i = b_i^\dagger b_i \rightarrow N_i |n_i\rangle = b_i^\dagger b_i |n_i\rangle = n_i |n_i\rangle. \tag{56.5}$$

For the states with $n_i = 0$ or $n_i = 1$, we have, in particular,

$$\begin{aligned} b|0\rangle &= 0, \qquad & b|1\rangle &= |1\rangle \\ b^\dagger|0\rangle &= |1\rangle, \qquad & b^\dagger|1\rangle &= \sqrt{2}|2\rangle. \end{aligned} \tag{56.6}$$

The creation and annihilation operators satisfy the independent harmonic oscillator commutation rules,

$$[b_i, b_j^\dagger] = \delta_{ij}$$
$$[b_i, b_j] = [b_i^\dagger, b_j^\dagger] = 0. \tag{56.7}$$

We can then finally extend, trivially, the actions of b_i and $b_i^\dagger$ to the full multiparticle Hilbert space, with its basis $|n_1 \ldots n_i \ldots n_\infty\rangle$.

56.2 Occupation Number Representation for Fermions

We can define similarly the occupation number states (and representation) for fermions. The fermionic creation and annihilation operators $a_i^\dagger, a_i$ for fermionic harmonic oscillators were defined in Chapters 20 and 28 via their anticommutation rules, owing to the antisymmetric statistics. The anticommutators are

$$\{a_i, a_j^\dagger\} = \delta_{ij}$$
$$\{a_i, a_j\} = \{a_i^\dagger, a_j^\dagger\} = 0. \tag{56.8}$$

We can import the anticommutators from the fermionic harmonic oscillator to the single-particle states i.

Since we have fermions, the single-particle states can be either empty (unoccupied) or full (occupied), so the states are $|0\rangle$ or $|1\rangle$. Then we must solve for the anticommutation rules for the actions of the operators a_i and $a_i^\dagger$ on the states $|0\rangle$ and $|1\rangle$. The solution is

$$0|0\rangle = 0, \qquad a|1\rangle = |1\rangle$$
$$a^\dagger|0\rangle = |1\rangle, \qquad a^\dagger|1\rangle = 0. \tag{56.9}$$

With respect to the bosonic case, we only need to change the last action, since there is no $|2\rangle$ state so it must be replaced with 0: $|2\rangle \equiv 0$.

In terms of the occupation numbers $n_i = 0, 1$, the relations above can be summarized as

$$a_i|n_i\rangle = \sqrt{n_i}|n_i - 1\rangle$$
$$a_i^\dagger|n_i\rangle = \sqrt{1 - n_i}|n_i + 1\rangle. \tag{56.10}$$

Equivalently, we can just use the same definition as for the bosonic case but with the identification $|2\rangle = 0$.

We can also define fermionic number operators

$$N_i = a_i^\dagger a_i \;\Rightarrow\; N_i|n_i\rangle = a_i^\dagger a_i|n_i\rangle = n_i|n_i\rangle, \tag{56.11}$$

for $n_i = 0, 1$.

Finally, the extension of the actions of $(a_i^\dagger, a_i)$ on the basis $|n_1 \ldots n_\infty\rangle$ of the full multiparticle Hilbert space will be explained later, in the definition of Fock space.

56.3 Schrödinger Equation on Occupation Number States

Define the abstract state

$$|\psi(t)\rangle = \sum_{n_1\dots n_\infty} C_{n_1\dots n_\infty}(t)|n_1 n_2 \dots n_\infty\rangle, \tag{56.12}$$

with a general wave function, i.e., coefficient $C_{n_1\dots n_\infty}$.

Going back to the Schrödinger equation for the coefficients of occupation number states (55.42), rewriting it by dividing by the square root of the factorials on the left-hand side and isolating the $i = i'$ kinetic term, in which $n_{i'} = n_i - 1$, so that $\sqrt{n_i(n_{i'}+1)} = n_i$, we obtain

$$\begin{aligned}
i\hbar\partial_t C'_{n_1\dots n_\infty}(t) &= \sum_i \langle i|\hat{T}|i\rangle n_i C'_{n_1\dots n_i\dots n_\infty}(t)\\
&+ \sum_{i,i'} \langle i|\hat{T}|i'\rangle \sqrt{n_i}\sqrt{n_{i'}+1}C'_{n_1\dots n_i-1,\dots n_{i'}+1,\dots n_\infty}(t)\\
&+ \sum_{i\neq j,i',j'} \langle ij|\hat{V}_{(2)}|i'j'\rangle \frac{1}{2}\sqrt{n_i}\sqrt{n_j}\sqrt{n_{i'}+1}\sqrt{n_{j'}+1}\\
&\times C'_{n_1\dots n_i-1,\dots n_{i'}+1,\dots n_j-1,\dots n_{j'}+1,\dots n_\infty}(t)\\
&+ \sum_{i=j,i'j'} \langle ii|\hat{V}_{(2)}|i'j'\rangle \frac{1}{2}\sqrt{n_i(n_i-1)}\sqrt{n_{i'}+1}\sqrt{n_{j'}+1}\\
&\times C'_{n_1\dots n_i-2,\dots n_{i'}+1,\dots n_{j'}+1,\dots n_\infty}(t).
\end{aligned} \tag{56.13}$$

Multiplying by the occupation number states, we obtain the Schrödinger equation for general states,

$$\begin{aligned}
i\hbar\partial_t|\psi(t)\rangle &= \sum_{n_1\dots n_\infty;\sum_i n_i=N} i\hbar\partial_t C'_{n_1\dots n_\infty}(t)|n_1 n_2\dots n_\infty\rangle\\
&= \sum_{n_1\dots n_\infty;\sum_i n_i=N} \left\{\sum_i \langle i|\hat{T}|i\rangle C'_{n_1\dots n_i\dots n_\infty}(t) n_i |n_1\dots n_i\dots n_\infty\rangle\right.\\
&+ \sum_{i\neq i'} \langle i|\hat{T}|i'\rangle C'_{n_1\dots n_i-1,\dots n_{i'}+1,\dots n_\infty}(t)\sqrt{n_i}\sqrt{n_{i'}+1}|n_1\dots n_i,\dots n_{i'},\dots n_\infty\rangle\\
&+ \sum_{i\neq j,i',j'} \langle ij|\hat{V}_{(2)}|i'j'\rangle \frac{1}{2} C'_{n_1\dots n_i-1,\dots n_{i'}+1,\dots n_j-1,\dots n_{j'}+1,\dots n_\infty}(t)\\
&\times \sqrt{n_i}\sqrt{n_j}\sqrt{n_{i'}+1}\sqrt{n_{j'}+1}|n_1\dots n_i,\dots n_{i'},\dots n_j,\dots n_{j'},\dots n_\infty\rangle\\
&+ \sum_{i=j,i',j'} \langle ii|\hat{V}_{(2)}|i'j'\rangle \frac{1}{2} C'_{n_1\dots n_i-2,\dots n_{i'}+1,\dots n_{j'}+1,\dots n_\infty}(t)\\
&\left.\times \sqrt{n_i(n_i-1)}\sqrt{n_{i'}+1}\sqrt{n_{j'}+1}|n_1\dots n_i,\dots n_{i'},\dots n_{j'},\dots n_\infty\rangle\right\}.
\end{aligned} \tag{56.14}$$

Redefine in the second term (in the first term no redefinition is needed)

$$n_i - 1 = \tilde{n}_i, \quad n_{i'} + 1 = \tilde{n}_{i'}. \tag{56.15}$$

In the third term, we redefine

$$\begin{aligned} n_i - 1 &= \tilde{n}_i \quad n_j - 1 = \tilde{n}_j \\ n_{i'} + 1 &= \tilde{n}_{i'} \quad n_{j'} + 1 = \tilde{n}_{j'}, \end{aligned} \tag{56.16}$$

and in the fourth term, we redefine

$$n_i - 2 = \tilde{n}_i, \quad n_{i'} + 1 = \tilde{n}_{i'}, \quad n_{j'} + 1 = \tilde{n}_{j'}. \tag{56.17}$$

Then the states and prefactors multiplying the matrix elements and coefficients C' become, under the redefinitions,

$$\begin{aligned} n_i|n_1 \ldots n_i \ldots n_\infty\rangle &= b_i^\dagger b_i |n_1 \ldots n_i \ldots n_\infty\rangle, \\ \sqrt{\tilde{n}_i + 1}\sqrt{\tilde{n}_{i'}}|n_1 \ldots \tilde{n}_i + 1, \ldots \tilde{n}_{i'} - 1, \ldots n_\infty\rangle &= b_i^\dagger b_{i'}|n_1 \ldots \tilde{n}_i \ldots \tilde{n}_{i'} \ldots n_\infty\rangle, \\ \sqrt{\tilde{n}_i + 1}\sqrt{\tilde{n}_{i'}}\sqrt{\tilde{n}_j + 1}\sqrt{\tilde{n}_{j'}}|n_1 \ldots \tilde{n}_i + 1, \ldots \tilde{n}_{i'} - 1, \ldots \tilde{n}_j + 1, \ldots \tilde{n}_{j'} - 1, \ldots n_\infty\rangle & \\ = b_i^\dagger b_{i'} b_j^\dagger b_{j'}|n_1 \ldots \tilde{n}_i \ldots \tilde{n}_{i'} \ldots \tilde{n}_j \ldots \tilde{n}_{j'} \ldots n_\infty\rangle, & \\ \sqrt{(\tilde{n}_i + 1)(\tilde{n}_i + 2)}\sqrt{\tilde{n}_{i'}}\sqrt{\tilde{n}_{j'}}|n_1 \ldots \tilde{n}_i + 2, \ldots \tilde{n}_{i'} - 1, \ldots \tilde{n}_{j'} - 1, \ldots n_\infty\rangle & \\ = b_i^\dagger b_i^\dagger b_{i'} b_{j'}|n_1 \ldots \tilde{n}_i \ldots \tilde{n}_{i'} \ldots \tilde{n}_{j'} \ldots n_\infty\rangle, & \end{aligned} \tag{56.18}$$

and we can replace the sum over n_i with a sum over $\tilde{n}_i$, etc., since the sum is unchanged as the prefactor is zero for the extra terms in the sum. Then, in all the terms, the original basis vectors are reobtained, and from them the full $|\psi(t)\rangle$.

Finally then, the Schrödinger equation for general states is

$$\begin{aligned} i\hbar\partial_t|\psi(t)\rangle = &\sum_i \langle i|\hat{T}|i\rangle b_i^\dagger b_i|\psi(t)\rangle \\ &+ \sum_{i,i'} \langle i|\hat{T}|i'\rangle b_i^\dagger b_{i'}|\psi(t)\rangle \\ &+ \sum_{i\neq j,i',j'} \langle ij|\hat{V}_{(2)}|i'j'\rangle \frac{1}{2} b_i^\dagger b_j^\dagger b_{i'} b_{j'}|\psi(t)\rangle \\ &+ \sum_{i=j,i'j'} \langle ii|\hat{V}_{(2)}|i'j'\rangle \frac{1}{2} b_i^\dagger b_i^\dagger b_{i'} b_{j'}|\psi(t)\rangle. \end{aligned} \tag{56.19}$$

The right-hand side of the equation can be identified with $\hat{H}|\psi(t)\rangle$ in occupation number states, which means that the Hamiltonian acting on the occupation number states is

$$\hat{H} = \sum_{i,i'} b_i^\dagger \langle i|\hat{T}|i'\rangle b_{i'} + \sum_{i,j,i',j'} b_i^\dagger b_j^\dagger \langle ij|\hat{V}_{(2)}|i'j'\rangle b_{i'} b_{j'}. \tag{56.20}$$

56.4 Fock Space

We can define the total number operator,

$$\hat{N} = \sum_i \hat{N}_i = \sum_i b_i^\dagger b_i, \tag{56.21}$$

which has as eigenvalue the total number of particles/occupied states

$$N = \sum_i n_i. \tag{56.22}$$

We could consider the space to be of given total number N, since in (nonrelativistic) quantum mechanics the number of particles does not change. However, as the individual operators $b_i, b_i^\dagger$ change the total number to $N - 1$ and $N + 1$, respectively, in order to define their action we must consider the total Hilbert space as the space of *all* values of N, called *Fock space*.

In this space, the normalized basis states are

$$|n_1 \dots n_\infty\rangle = \frac{(b_1^\dagger)^{n_1}}{\sqrt{n_1!}} \cdots \frac{(b_\infty^\dagger)^{n_\infty}}{\sqrt{n_\infty!}}|0\rangle. \tag{56.23}$$

Indeed, the normalization constant C for the state

$$|n\rangle = C(b^\dagger)^n|0\rangle \tag{56.24}$$

comes from the normalization condition

$$1 = \langle n|n\rangle = |C|^2\langle 0|b^n(b^\dagger)^n|0\rangle = |C|^2 n\langle 0|b^{n-1}(b^\dagger)^{n-1}|0\rangle = \cdots = |C|^2 n!\,\langle 0|0\rangle = |C|^2 n!\,, \tag{56.25}$$

where in the iteration step we have commuted the last b past n $b^\dagger$s, leading to $n(b^\dagger)^{n-1}$.

56.5 Fock Space for Fermions

For fermions, the definition of the occupied states $|n\rangle$ is

$$|n\rangle = (a^\dagger)^n|0\rangle, \tag{56.26}$$

where

$$a|n\rangle = \sqrt{n}|n-1\rangle, \quad a^\dagger|n\rangle = \sqrt{n+1}|n+1\rangle, \tag{56.27}$$

and where $n = 0$ or 1 only and $|2\rangle \equiv 0$. Note that then we also have $\sqrt{n!} = 1$.

The multiparticle states are defined in the same way,

$$|n_1 \dots n_\infty\rangle = (a_1^\dagger)^{n_1} \cdots (a_i^\dagger)^{n_i} \cdots (a_\infty^\dagger)^{n_\infty}|0\rangle. \tag{56.28}$$

However, because of the anticommutating operators for $i \neq j$,

$$\{a_i, a_j^\dagger\} = 0, \tag{56.29}$$

the action of the annihilation operator a_i on the Fock space states is different from its action just on a single state, by a sign: before reaching $(a_i^\dagger)^{n_i}$ in this string, we have to anticommute past the rest of the creation operator, creating a sign,

$$a_i|n_1 \dots n_i \dots n_\infty\rangle = (-1)^S\sqrt{n_i}|n_1 \dots n_i - 1, \dots n_\infty\rangle, \tag{56.30}$$

where

$$S = n_1 + n_2 + \cdots + n_{i-1}. \tag{56.31}$$

Similarly, the creation operator acts on the Fock space to produce the same sign,

$$a_i^\dagger |n_1 \dots n_i \dots n_\infty\rangle = (-1)^S \sqrt{n_i + 1} |n_1 \dots n_i + 1, \dots n_\infty\rangle. \tag{56.32}$$

Finally, the number operator acts without a sign,

$$N_i |n_1 \dots n_i \dots n_\infty\rangle = a_i^\dagger a_i |n_1 \dots n_i \dots n_\infty\rangle = n_i |n_1 \dots n_i \dots n_\infty\rangle. \tag{56.33}$$

56.6 Schrödinger Equations for Fermions, and Generalizations

The analysis for the Schrödinger equation for fermions follows that for bosons, since we have the same definition of the action of creation and annihilation operators, but with the addition of the condition $|2\rangle = 0$ and with the sign $(-1)^S$ in intermediate equations. But at the end, when $|\psi(t)\rangle$ is re-formed, the sign disappears (we leave the details as an exercise).

Then the Schrödinger equation for fermions has the same form as for bosons,

$$i\hbar\partial_t |\psi(t)\rangle = \hat{H} |\psi(t)\rangle, \tag{56.34}$$

where the Hamiltonian in the occupation number space is

$$\hat{H} = \sum_{i,i'} a_i^\dagger \langle i|\hat{T}|i'\rangle a_i + \frac{1}{2} \sum_{i,j,i',j'} a_i^\dagger a_j^\dagger \langle ij|\hat{V}_{(2)}|i'j'\rangle a_{i'} a_{j'}. \tag{56.35}$$

We can then use a formalism where we treat both bosons and fermions together, with annihilation operator $c_i = (b_i, a_i)$ according to whether we have bosons or fermions, and similarly for the creation operators.

Then the more general one-particle operator

$$\hat{A} = \sum_{i=1}^{N} \hat{A}^{(i)}, \tag{56.36}$$

becomes in the occupation number picture

$$\hat{A} = \sum_{i,i'} c_i^\dagger \langle i|\hat{A}|i'\rangle c_{i'}. \tag{56.37}$$

Similarly the more general two-particle operator

$$\hat{B} = \frac{1}{2} \sum_{i,j=1}^{N} \hat{B}^{(ij)}, \tag{56.38}$$

becomes in the occupation number picture

$$\hat{B} = \frac{1}{2} \sum_{i,j,i',j'=1}^{N} c_i^\dagger c_j^\dagger \langle ij|\hat{B}|i'j'\rangle c_{i'} c_{j'}. \tag{56.39}$$

56.7 Field Operators

The occupation number representation leads to so-called second quantization, which is regarded as the quantization of the wave function, where instead of complex numbers they take values in operators (numbers are replaced with operators). However, that is actually a misnomer, since the correct procedure is to build quantum field theory, where the wave function is replaced by a quantum field. But here we are dealing with nonrelativistic quantum mechanics, so $v \ll c$, and there are no antiparticles in the theory, only particles. This means that we have annihilation and creation operators only for particles, c_i and $c_i^\dagger$, but not the corresponding operators for antiparticles, as in the relativistic quantum field theory.

We build the *field operators* as the product of the annihilation or creation operators with the single-particle wave functions $\phi_i(\vec{r})$, i.e.,

$$\begin{aligned} \hat{\psi}(\vec{r}) &= \sum_i \phi_i(\vec{r}) c_i \Rightarrow \\ \hat{\psi}^\dagger(\vec{r}) &= \sum_i \phi_i^*(\vec{r}) c_i^\dagger. \end{aligned} \tag{56.40}$$

More precisely, the above definitions are only for the boson case, $c_i = b_i$.

In the case of fermions, there is an extra index for spin $\sigma = \pm 1/2$ or 1, 2,

$$\begin{aligned} \hat{\psi}_\sigma(\vec{r}) &= \sum_i \phi_{i\sigma}(\vec{r}) a_{i,\sigma} \Rightarrow \\ \hat{\psi}_\sigma^\dagger(\vec{r}) &= \sum_i \phi_{i\sigma}^*(\vec{r}) a_{i,\sigma}^\dagger. \end{aligned} \tag{56.41}$$

We can also write field operators involving spin wave functions,

$$\begin{aligned} \hat{\psi}(\vec{r}, s) &= \sum_{\sigma=1,2} \hat{\psi}_\sigma(\vec{r}) \chi_\sigma(s) \\ \hat{\psi}^\dagger(\vec{r}, s) &= \sum_{\sigma=1,2} \hat{\psi}_\sigma^\dagger(\vec{r}) \chi_\sigma(s). \end{aligned} \tag{56.42}$$

The field operators are operators in the occupation number Hilbert space, through $c_i, c_i^\dagger$, and, as such, they obey commutation and anticommutation relations derived from those for c_i and $c_i^\dagger$,

$$\begin{aligned} [\hat{\psi}_\sigma(\vec{r}), \hat{\psi}_{\sigma'}^\dagger(\vec{r}\,')]_\mp &= \delta_{\sigma\sigma'} \delta^3(\vec{r} - \vec{r}\,') \\ [\hat{\psi}_\sigma(\vec{r}), \hat{\psi}_{\sigma'}(\vec{r}\,')]_\mp &= [\hat{\psi}_\sigma^\dagger(\vec{r}), \hat{\psi}_{\sigma'}^\dagger(\vec{r}\,')]_\mp = 0. \end{aligned} \tag{56.43}$$

In the above, we have used the same indices σ as for fermions. For bosons, we just drop all the σ indices.

Using the orthonormality and completeness of the single-particle wave functions $\phi_{i\sigma}(\vec{r})$, we can rewrite the result of acting with one-particle operators in the occupation number picture as an integral with the field operators,

$$\begin{aligned} \hat{A} &= \sum_{i,i'} c_i^\dagger \langle i|\hat{A}|i'\rangle c_{i'} \\ &= \int d^3\vec{r} \sum_\sigma \hat{\psi}_\sigma^\dagger(\vec{r}) A(\vec{r}) \hat{\psi}_\sigma(\vec{r}). \end{aligned} \tag{56.44}$$

Similarly, the two-particle operators in the occupation number picture also become an integral with field operators,

$$\begin{aligned} \hat{B} &= \frac{1}{2} \sum_{i,j,i',j'} c_i^\dagger c_j^\dagger \langle ij|\hat{B}|i'j'\rangle c_{i'} c_{j'} \\ &= \frac{1}{2} \int d^3\vec{r} \int d^3\vec{r}\,' \sum_{\sigma,\sigma'} \hat{\psi}_\sigma^\dagger(\vec{r}) \hat{\psi}_{\sigma'}^\dagger(\vec{r}\,') B(\vec{r}, \vec{r}\,') \hat{\psi}_{\sigma'}(\vec{r}\,') \hat{\psi}_\sigma(\vec{r}). \end{aligned} \tag{56.45}$$

We could further generalize to operators that are nontrivial in their spin action, $\hat{A}_{\sigma\sigma'}$, $\hat{B}_{\sigma_1,\sigma_2,\sigma_1',\sigma_2'}$, so that the relation becomes a matrix relation in σ space.

Remaining in the case of trivial dependence on spin, the Hamiltonian becomes

$$\hat{H} = \int d^3\vec{r} \sum_\sigma \psi_\sigma^\dagger(\vec{r}) T(\vec{r}) \psi_\sigma(\vec{r}) + \frac{1}{2} \int d^3\vec{r} \int d^2\vec{r}\,' \sum_{\sigma,\sigma'} \hat{\psi}_\sigma^\dagger(\vec{r}) \hat{\psi}_{\sigma'}^\dagger(\vec{r}\,') V(\vec{r},\vec{r}\,') \hat{\psi}_{\sigma'}(\vec{r}\,') \hat{\psi}_\sigma(\vec{r}). \tag{56.46}$$

56.8 Example of Interaction: Coulomb Potential, as a Limit of the Yukawa Potential, for Spin 1/2 Fermions

Consider the example of the Yukawa interaction,

$$V(\vec{r},\vec{r}\,') = \frac{e_0^2 e^{-\mu|\vec{r}-\vec{r}\,'|}}{|\vec{r}-\vec{r}\,'|}, \tag{56.47}$$

which leads to the Coulomb potential in the limit $\mu \to 0$.

Consider then the state labeled by $\vec{k} = \vec{p}/\hbar$ and spin σ, with spin wave functions

$$\chi_\sigma = \begin{pmatrix} 1 \\ 0 \end{pmatrix}, \quad \text{or} \quad \begin{pmatrix} 0 \\ 1 \end{pmatrix}. \tag{56.48}$$

Then the kinetic energy in these states in momentum space is

$$\langle \vec{k}_1 \sigma_1 | T | \vec{k}_1' \sigma_1' \rangle = \frac{\hbar^2 \vec{k}_1^2}{2m} \delta_{\sigma_1 \sigma_1'} \delta_{\vec{k}_1, \vec{k}_1'}. \tag{56.49}$$

The potential in the same states is

$$\langle \vec{k}_1 \sigma_1, \vec{k}_2, \sigma_2 | \hat{V} | \vec{k}_1' \sigma_1', \vec{k}_2', \sigma_2' \rangle = \frac{e_0^2}{V} \delta_{\sigma_1 \sigma_1'} \delta_{\sigma_2 \sigma_2'} \delta_{\vec{k}_1+\vec{k}_2, \vec{k}_1'+\vec{k}_2'} \frac{4\pi}{(\vec{k}_1 - \vec{k}_2)^2 + \mu^2}. \tag{56.50}$$

In terms of the matrix elements, the Hamiltonian is

$$\begin{aligned} \hat{H} = & \sum_{\vec{k}_1 \sigma_1} \sum_{\vec{k}_1' \sigma_1'} \langle \vec{k}_1 \sigma_1 | T | \vec{k}_1' \sigma_1' \rangle a^\dagger_{\vec{k}_1 \sigma_1} a_{\vec{k}_1' \sigma_1'} \\ & + \sum_{\vec{k}_1 \sigma_1, \vec{k}_2 \sigma_2, \vec{k}_1' \sigma_1', \vec{k}_2' \sigma_2'} a^\dagger_{\vec{k}_1 \sigma_1} a^\dagger_{\vec{k}_2, \sigma_2} \langle \vec{k}_1 \sigma_1, \vec{k}_2, \sigma_2 | \hat{V} | \vec{k}_1' \sigma_1', \vec{k}_2' \sigma_2' \rangle a_{\vec{k}_1' \sigma_1'} a_{\vec{k}_2' \sigma_2'}. \end{aligned} \tag{56.51}$$

Taking the limit $\mu \to 0$ and substituting the matrix element, with the replacements $\vec{k}_1' \to \vec{k}, \vec{k}_2' \to \vec{k}'$, and $\vec{k}_1 - \vec{k}_2 \to \vec{q}$, we get

$$H = \sum_{\vec{k},\sigma} \frac{\hbar^2 \vec{k}^2}{2m} a^\dagger_{\vec{k}\sigma} a_{\vec{k}\sigma} + \frac{e_0^2}{V} \sum_{\vec{k},\vec{k}',\vec{q}} \sum_{\sigma,\sigma'} \frac{4\pi}{\vec{q}^2} a^\dagger_{\vec{k}+\vec{q},\sigma} a^\dagger_{\vec{k}'-\vec{q},\sigma'} a_{\vec{k}',\sigma'} a_{\vec{k},\sigma}. \tag{56.52}$$

Important Concepts to Remember

- If we have one-particle states in a multiparticle system, in the occupation number picture we can define creation and annihilation operators, $b_i^\dagger, b_i$ that act as for a harmonic oscillator.
- Similarly, for fermions, in the occupation number (equal to 0 or 1) picture, we can define creation and annihilation operators $a_i^\dagger, a_i$ that act as for a fermionic harmonic oscillator.
- In terms of $b_i, b_i^\dagger$, the occupation number picture Hamiltonian is $\hat{H} = \sum_{i,i'} b_i^\dagger \langle i|\hat{T}|i'\rangle b_{i'} + \sum_{i,j,i',j'} b_i^\dagger b_j^\dagger \langle ij|\hat{V}_{(2)}|i'j'\rangle b_{i'} b_{j'}$
- Fock space is the Hilbert space in the occupation number picture, for all possible values of the total number N.
- In the fermionic Fock space, $a_i|n_1 \dots n_\infty\rangle = (-1)^S |n_1 \dots n_{i-1}, n_i - 1, n_{i+1} \dots n_\infty\rangle$, with $S = n_1 + \cdots n_{i-1}$ and the same sign for $a_i^\dagger$.
- By analogy with (relativistic) quantum field theory, in fact a nonrelativistic version of the same, we can construct field operators, as $\hat{\psi}(\vec{r}) = \sum_i \phi_i(\vec{r})\hat{c}_i$, $c_i = (b_i, a_i)$, where for $c_i = a_i$ we actually have also a spin index, so that $\hat{\psi}_\sigma(\vec{r}) = \sum_i \phi_{i,\sigma}(\vec{r}) a_i$ and we can write $\hat{\psi}(\vec{r}; s) = \sum_s \hat{\psi}_\sigma(\vec{r})\chi_s(s)$.
- For the Coulomb potential, as a limit of the Yukawa potential, we find

$$\hat{H} = \sum_{\vec{k},\sigma} \frac{\hbar^2 \vec{k}^2}{2m} \hat{a}^\dagger_{\vec{k},\sigma} \hat{a}_{\vec{k},\sigma} + \frac{e_0^2}{V} \sum_{\vec{k},\vec{k}'\vec{q}} \sum_{\sigma,\sigma'} \hat{a}^\dagger_{\vec{k}+\vec{q},\sigma} \hat{a}^\dagger_{\vec{k}'-\vec{q},\sigma'} a_{\vec{k}',\sigma'} a_{\vec{k},\sigma}.$$

Further Reading

More details can be found in the book [27].

Exercises

(1) Find the eigenvalue of the operator e^{ab} on single-particle states.

(2) If the Hamiltonian acting on the occupation number states of a system can be written as $\hat{H} = \sum_i t_i b_i^\dagger b_i + V \sum_{i<j} b_i^\dagger b_j^\dagger b_{i+1} b_{j-1}$, how would you interpret this physically?

(3) Consider a system with seven one-particle states, and three bosons in the system. How many states are there in the Fock space? How many multiparticle states are possible?

(4) Find the eigenvalue of the fermionic operator $\exp[(\sum_i a_i)\alpha]$ in the fermionic Fock space.

(5) Show the details of the proof of the Schrödinger equation for the occupation number states for fermions, and show the resulting absence of a sign.

(6) Calculate the occupation number Hamiltonian for a delta function two-body interaction $V(\vec{r}_i - \vec{r}_j) = V_0 \delta(\vec{r}_i - \vec{r}_j)$.

(7) Explain physically the Coulomb occupation number Hamiltonian (56.52) in terms of Feynman diagrams.

57 The Hartree–Fock Approximation and Other Occupation Number Approximations

In this chapter, we will consider a self-consistent approximation method for multiparticle systems, the Hartree–Fock approximation. It arises from the multiparticle picture, either the original one or, better, the occupation number picture. We will make connections with the original derivation of Hartree and Fock. Then we sketch the general ideas of the other approximations in the occupation number picture: the usual perturbation theory in the style of nonrelativistic quantum field theory, and the self-consistent Bethe–Salpeter equation.

57.1 Set-Up of the Problem

We would like to solve perturbatively for the energies of multiparticle states. The most relevant case is that of multi-electron states, meaning we have fermionic one-particle states.

The perturbative solution is via a self-consistent equation, where the Hamiltonian contains field operators for the rest of the fermions (electrons) other than the one whose interaction we are analyzing.

We will start directly with the occupation number Hamiltonian derived in the previous chapter,

$$\begin{aligned}
\hat{H} &= \hat{H}_0 + \hat{H}_1 \\
\hat{H}_0 &= \sum_{i,i'} \langle i|\hat{T}_{(1)}|i'\rangle a_i^\dagger a_{i'} \\
\hat{H}_1 &= -\frac{1}{2} \sum_{i,j,i',j'} a_i^\dagger a_j^\dagger \langle ij|\hat{V}_{(2)}|i'j'\rangle a_{i'} a_{j'} .
\end{aligned} \tag{57.1}$$

Note that the minus sign in front of the expression for $\hat{H}_1$ is there because the particles are fermions (the general formula for bosons or fermions with the above ordering has $\pm$ in front).

Because of the Fermi–Dirac antisymmetric statistics, we can rewrite $\hat{H}_1$ equivalently by reordering $a_{i'}a_{j'}$ as $-a_{j'}a_{i'}$, followed by redefinition of the summation variables, i' as j' and j' as i'. Averaging over the two equivalent forms, we substitute in $\hat{H}_1$

$$\langle ij|\hat{V}_{(2)}|i'j'\rangle \to \frac{1}{2}\left(\langle ij|\hat{V}_{(2)}|i'j'\rangle - \langle ij|\hat{V}_{(2)}|j'i'\rangle\right). \tag{57.2}$$

Equivalently, consider in the original formulation the basis N-particle state

$$|\psi\rangle = |E_1\rangle_1 \otimes |E_2\rangle_2 \otimes \cdots \otimes |E_N\rangle_N, \tag{57.3}$$

and antisymmetrize it to obtain the Slater determinant,

$$|\bar{\psi}\rangle = \frac{1}{\sqrt{N!}} \sum_{\text{perms. } P} P|\psi\rangle. \tag{57.4}$$

We consider the average of $\hat{H}_1$ for the Slater determinant state $|\bar{\psi}\rangle$. Then we define $\hat{V}^{(ij)}$ as the two-particle interaction potential, acting on the left on the ith single-particle Hilbert space, and on the right on the jth single-particle Hilbert space, so that

$$\langle\bar{\psi}|\hat{H}_1|\bar{\psi}\rangle = \frac{1}{2}\sum_{i,j}\langle\psi|\hat{V}^{(ij)}|\bar{\psi}\rangle. \tag{57.5}$$

Then, for $|\bar{\psi}\rangle$ containing states with E_i and E_j, this results (after using the orthonormality of the rest of the other one-particle states in $|\psi\rangle$ and $|\bar{\psi}\rangle$, $\langle i|j\rangle = \delta_{ij}$) in a matrix element for two-particle states,

$$\langle\bar{\psi}|\hat{H}_1|\bar{\psi}\rangle = \frac{1}{2}\sum_{i,j}\left(\langle E_iE_j|\hat{V}^{(ij)}|E_iE_j\rangle - \langle E_iE_j|\hat{V}^{(ij)}|E_jE_i\rangle\right). \tag{57.6}$$

This is the same result as that from the occupation number method above. On the other hand, transitioning from the above original representation result, by writing

$$|\psi\rangle = \ldots a_i^\dagger \ldots a_j^\dagger \ldots |0\rangle, \tag{57.7}$$

means that we can relate the N-particle matrix element to the two-particle one in the two-particle operator,

$$\langle\bar{\psi}|\hat{V}^{(ij)}|\bar{\psi}\rangle = \frac{1}{2}\left(\langle ij|\hat{V}_{(2)}|ij\rangle - \langle ij|\hat{V}_{(2)}|ji\rangle\right). \tag{57.8}$$

57.2 Derivation of the Hartree–Fock Equation

We want to solve the time-independent Schrödinger equation, $\hat{H}|\psi\rangle = E|\psi\rangle$, which means that we need to consider

$$\langle\psi|(\hat{H} - E)|\psi\rangle = 0. \tag{57.9}$$

Varying the matrix equation, we obtain

$$\langle\delta\psi|(\hat{H} - E)|\psi\rangle. \tag{57.10}$$

If the state is

$$|\psi\rangle = a_{i_1}^\dagger a_{i_2}^\dagger \cdots a_{i_N}^\dagger |0\rangle, \tag{57.11}$$

then its variation is

$$|\delta\psi\rangle = \delta a_{i_1}^\dagger a_{i_2}^\dagger \cdots a_{i_N}^\dagger|0\rangle + a_{i_1}^\dagger \delta a_{i_2}^\dagger \cdots a_{i_N}^\dagger|0\rangle + \cdots + a_{i_1}^\dagger \cdots a_{i_{N-1}}^\dagger \delta a_{i_N}^\dagger|0\rangle. \tag{57.12}$$

The variation occurs through a unitary transformation of the single-particle state by $\hat{U}$, with $\hat{U}\hat{U}^\dagger = \mathbb{1}$, such that

$$\hat{U} \simeq \mathbb{1} + i\epsilon\hat{K} \Rightarrow \hat{K}^\dagger = \hat{K}. \tag{57.13}$$

The transformation $\hat{U}$ acts on states, so on the annihilation and creation operators, mixing the various i's, so that

$$a_i \to U_{ij}a_j, \quad a_i^\dagger \to U_{ij}^\dagger a_j^\dagger. \tag{57.14}$$

For an infinitesimal transformation,

$$\delta a_i^\dagger = -i\epsilon \sum_j K_{ji} a_j^\dagger. \tag{57.15}$$

We can have two types of transformation:

(a) If $\delta a_i^\dagger \propto a_i$ only, it means that $|\delta\psi\rangle \propto |\psi\rangle$, so (57.10) becomes just the usual Schrödinger equation,

$$\langle\psi|\hat{H}|\psi\rangle = E\langle\psi|\psi\rangle. \tag{57.16}$$

(b) If $\delta a_i^\dagger$ contains other $a_j^\dagger$, meaning $K_{ji} \neq 0$ for $i \neq j$, then $\langle\delta\psi|\psi\rangle = 0$ and then we obtain

$$\langle\delta\psi|\hat{H}|\psi\rangle = 0. \tag{57.17}$$

The interpretation of this result is that, in the sector of $|\psi\rangle$'s, see (57.11), where there is *only one* creation operator $a_i^\dagger$ that changes from i to j (in all the terms in $|\delta\psi\rangle$, there is also only one $\delta a_i^\dagger$ among the other ones that is untouched, and thus $\delta a_i^\dagger \propto \sum_j K_{ji} a_j^\dagger$) and the matrix element of $\hat{H}$ vanishes.

In particular, for relevant combinations of $(ij, i'j')$,

$$\langle\delta\psi|\hat{H}_{(2)}|\psi\rangle = \langle 0|\dots a_i \dots a_j \dots |\hat{H}_{(2)}|\dots a_{i'}^\dagger \dots a_{j'}^\dagger \dots |0\rangle = \langle ij|\hat{H}_{(2)}|i'j'\rangle. \tag{57.18}$$

This means that the terms in the occupation number Hamiltonian $\hat{H}$ that have only one $a_i^\dagger$-change, or otherwise (by the previous equation) if $\langle ij|\hat{H}_{(2)}|i'j'\rangle$ has an index on the left equal to one on the right, must vanish. These terms are

$$\begin{aligned}
&-\frac{1}{2}\sum_{i,j,j'}\langle ij|\hat{V}_{(2)}|ij'\rangle a_i^\dagger a_j^\dagger a_i a_{j'} - \frac{1}{2}\sum_{i,j,i'}\langle ij|\hat{V}_{(2)}|i'i\rangle a_i^\dagger a_j^\dagger a_{i'} a_i \\
&-\frac{1}{2}\sum_{i,j,j'}\langle ij|\hat{V}_{(2)}|jj'\rangle a_i^\dagger a_j^\dagger a_j a_{j'} - \frac{1}{2}\sum_{i,j,i'}\langle ij|\hat{V}_{(2)}|i'j\rangle a_i^\dagger a_j^\dagger a_{i'} a_j \\
&= -\sum_{k,i,i'}\Big[-\frac{1}{2}\langle ki|\hat{V}_{(2)}|ki'\rangle a_i^\dagger a_k^\dagger a_k a_{i'} + \frac{1}{2}\langle ki|\hat{V}_{(2)}|i'k\rangle a_i^\dagger a_k^\dagger a_k a_{i'} \\
&\qquad + \frac{1}{2}\langle ik|\hat{V}_{(2)}|ki'\rangle a_i^\dagger a_k^\dagger a_k a_{i'} - \frac{1}{2}\langle ik|\hat{V}_{(2)}|i'k\rangle a_i^\dagger a_k^\dagger a_k a_{i'}\Big] \\
&= -\sum_{k,i,i'}\Big[-\frac{1}{2}\langle ik|\hat{V}_{(2)}|i'k\rangle a_i^\dagger a_k^\dagger a_k a_{i'} + \frac{1}{2}\langle ik|\hat{V}_{(2)}|ki'\rangle a_i^\dagger a_k^\dagger a_k a_{i'} \\
&\qquad + \frac{1}{2}\langle ik|\hat{V}_{(2)}|ki'\rangle a_i^\dagger a_k^\dagger a_k a_{i'} - \frac{1}{2}\langle ik|\hat{V}_{(2)}|i'k\rangle a_i^\dagger a_k^\dagger a_k a_{i'}\Big],
\end{aligned} \tag{57.19}$$

where in the first equality we have commuted $a^\dagger$ past $a^\dagger$ or a past a (with different indices), and then relabeled the summation indices as the common set (kii'), and in the second equality we have used $\langle ij|\hat{V}_{(2)}|kl\rangle = \langle ji|\hat{V}_{(2)}|lk\rangle$ in the first two terms.

Then the relevant terms terms in the Hamiltonian are

$$\sum_{ii'} a_i^\dagger \left[\sum_k a_k^\dagger a_k \langle ik|\hat{V}_{(2)}|i'k\rangle - \sum_k a_k^\dagger a_k \langle ik|\hat{V}_{(2)}|ki'\rangle\right] a_{i'}, \tag{57.20}$$

and since $a_k^\dagger a_k = N_k$ is the occupation number in mode k, which is 1 for an occupied state and 0 for an unoccupied state, it means that we can replace it with changing the sum over all k to a sum over occupied states k,

$$\begin{aligned}&\sum_{ii'} a_i^\dagger \left[\sum_{k \text{ occ.}} \langle ik|\hat{V}_{(2)}|i'k\rangle - \sum_{k \text{ occ.}} \langle ik|\hat{V}_{(2)}|ki'\rangle\right] a_{i'}\\ &\equiv \sum_{ii'} a_i^\dagger \langle i|\hat{V}_{H-F}|i'\rangle a_{i'},\end{aligned} \tag{57.21}$$

where we have defined the matrix element of the *Hartree–Fock potential* $\hat{V}_{H-F}$,

$$\langle i|\hat{V}_{H-F}|i'\rangle = \sum_{k \text{ occ.}} \langle ik|\hat{V}_{(2)}|i'k\rangle - \sum_{k \text{ occ.}} \langle ik|\hat{V}_{(2)}|ki'\rangle. \tag{57.22}$$

Thus the condition for the terms in the Hamiltonian with only one transition of a_i to vanish is

$$\langle i|\hat{T}_{(1)}|j\rangle + \langle i|\hat{V}_{H-F}|j\rangle = 0, \quad \text{for} \quad i \neq j. \tag{57.23}$$

Therefore $\hat{T}_{(1)} + \hat{V}_{H-F}$ is diagonal, or in other words the one-particle states are eigenstates for it,

$$(\hat{T}_{(1)} + \hat{V}_{H-F})|i\rangle = E_i|i\rangle. \tag{57.24}$$

This is the *Hartree–Fock equation* for the single-particle Hamiltonian ($\hat{V}_{H-F}$ is a single-particle operator), which is to be solved iteratively for the eigenstates $|i\rangle$ and the eigenenergies E_i.

So the new Hamiltonian in occupation number space is

$$\hat{H}_{H-F} = \sum_{i,i'} \langle i|\hat{T}_{(1)}|i'\rangle a_i^\dagger a_{i'} + \sum_{i,i'} \langle i|\hat{V}_{H-F}|i'\rangle a_i^\dagger a_{i'}. \tag{57.25}$$

57.3 Hartree and Fock Terms

Writing the Schrödinger equation for wave functions, we obtain

$$\left[-\frac{\hbar^2}{2m}\Delta_{\vec{r}} + V(\vec{r})\right]\phi_{i'}(\vec{r}) + V_{H-F}^{(1)}(\vec{r})\phi_{i'}(\vec{r}) - \int d^3\vec{r}\,' V_{H-F}^{(2,\text{exch})}(\vec{r},\vec{r}\,')\phi_i(\vec{r}\,') = 0. \tag{57.26}$$

In this equation, Hartree considered the first term in V_{H-F}, and Fock added the second term.

The equation is obtained as follows.

The first term,

$$\sum_{k \text{ occ}} \langle ik|\hat{V}_{(2)}|i'k\rangle = \langle i|\hat{V}_{H-F}^{(1)}|i'\rangle \tag{57.27}$$

becomes, in wave functions, on introducing the completeness relation for $|\vec{r}\rangle$,

$$\int d^3\vec{r}\,\phi_i^*(\vec{r})V_{H-F}^{(1)}\phi_{i'}(\vec{r}) = \int d^3\vec{r}\,\phi_i^*(\vec{r}) \sum_{k \text{ occ.}} \int d^3\vec{r}\,'\phi_k^*(\vec{r}\,')V_{(2)}(\vec{r},\vec{r}\,')\phi_k(\vec{r}\,')\phi_{i'}(\vec{r}), \tag{57.28}$$

which defines the Hartree potential,

$$V_H(\vec{r}) = V_{H-F}^{(1)}(\vec{r}) = \int d^3\vec{r}\,' \sum_{k,\text{occ.}} |\phi_k(\vec{r}\,')|^2 V_{(2)}(\vec{r},\vec{r}\,') = \int d^3\vec{r}\,'\rho(\vec{r}\,')V_{(2)}(\vec{r},\vec{r}\,'). \tag{57.29}$$

We have defined the probability density $\rho(\vec{r}) = \sum_{k,\text{occ.}} |\phi_k(\vec{r})|^2$ of the electrons. By multiplying with e, to give $\rho_e = e\rho$, we obtain the electron density, which means that this first term has a classical interpretation.

The second term is called the *exchange term*, since it is a purely quantum term, without a classical analog. In it,

$$\sum_{k,\text{occ.}} \langle ik|\hat{V}_{(2)}|ki'\rangle = \langle i|\hat{V}^{(2)}_{H-F}|i'\rangle \tag{57.30}$$

becomes, in wave functions,

$$\int d^3\vec{r}\, \phi_i^*(\vec{r}) V^{(2)}_{H-F} \phi_{i'}(\vec{r}) = \int d^3\vec{r}\, \phi_i^*(\vec{r}) \sum_{k\ \text{occ.}} \int d^3\vec{r}\,' \phi_k^*(\vec{r}\,') V_{(2)}(\vec{r},\vec{r}\,') \phi_{i'}(\vec{r}\,') \phi_k(\vec{r}), \tag{57.31}$$

which defines the Fock "exchange" potential acting on a wave function,

$$\begin{aligned} V_F \phi_{i'}(\vec{r}) = V^{(2)}_{H-F} \phi_{i'}(\vec{r}) &= \int d^3\vec{r}\,' \left(\sum_{k,\text{occ}} \phi_k^*(\vec{r}\,') \phi_k(\vec{r}) \right) V_{(2)}(\vec{r},\vec{r}\,') \phi_{i'}(\vec{r}\,') \\ &\equiv \int d^3\vec{r}\,' V^{(2,\text{exch})}_{H-F}(\vec{r},\vec{r}\,') \phi_{i'}(\vec{r}\,'). \end{aligned} \tag{57.32}$$

Here we have defined the two-particle exchange term of the potential,

$$V^{(2,\text{exch})}_{H-F}(\vec{r},\vec{r}\,') = \sum_{k\ \text{occ.}} \phi_k^*(\vec{r}\,') \phi_k(\vec{r}) V^{(2)}(\vec{r},\vec{r}\,'). \tag{57.33}$$

This completes our definition of the Hartree–Fock equation for wave functions.

We now specialize to the case of the Coulomb interaction,

$$V = \frac{e_0^2}{|\vec{r} - \vec{r}\,'|}. \tag{57.34}$$

Then keeping only the first of the Hartree–Fock terms, we obtain the Hartree equation,

$$\left[-\frac{\hbar^2}{2m} \Delta + V(\vec{r}) + V_H(\vec{r}) \right] \phi_i(\vec{r}) = E_i \phi_i(\vec{r}). \tag{57.35}$$

This has an interpretation as a semiclassical potential, in which the Hartree term V_H is a classical potential with a quantum charge density,

$$V_H(\vec{r}) = \int d^3\vec{r}\,' \rho(\vec{r}\,') V(\vec{r} - \vec{r}\,') = \int d^3\vec{r}\,' \rho_e(\vec{r}\,') \frac{e_0}{|\vec{r} - \vec{r}\,'|}, \tag{57.36}$$

where $\rho_e(\vec{r}) = e_0 \rho(\vec{r})$ is the electric charge density.

On the other hand, the second term, the Fock term, is due to the Coulomb exchange potential

$$V^{(2,\text{exch})}_{H-F}(\vec{r},\vec{r}\,') = \sum_{k\ \text{occ.}} e_0 \phi_k^*(\vec{r}\,') \phi_k(\vec{r}) \frac{e_0}{|\vec{r} - \vec{r}\,'|}, \tag{57.37}$$

leading to

$$V_F \phi_i(\vec{r}) = \int d^3\vec{r}\,' V^{(2,\text{exch})}_{H-F}(\vec{r},\vec{r}\,') \phi_i(\vec{r}\,'). \tag{57.38}$$

57.4 Other Approximations in the Occupation Number Picture

Besides the Hartree–Fock approximation, solving iteratively the Hartree–Fock equations, we can also employ perturbation theory, but not for single-particle wave functions. Instead, we use a perturbation theory in terms of the field operators for the occupation number. It is in fact the nonrelativistic version of the formalism of quantum field theory. Here we will give just a rough sketch, an outline, of the method.

In order to define perturbation theory, we use the Dirac picture, with respect to the free Hamiltonian, in the occupation number version,

$$\hat{H} = \sum_i \hbar\omega_i \hat{c}_i^\dagger \hat{c}_i, \tag{57.39}$$

where

$$\hbar\omega_i \delta_{ii'} = \langle i|\hat{T}_{(1)}|i'\rangle, \tag{57.40}$$

and $\hat{T}_{(1)}$ contains the kinetic energy and the one-particle potential.

The operators in the Dirac (interaction) picture are (in terms of the Schrödinger picture operator $\hat{A}_S$)

$$\hat{A}_I(t) = e^{i\hat{H}_0 t/\hbar}\hat{A}_S e^{-i\hat{H}_0 t/\hbar}, \tag{57.41}$$

while the operators in the Heisenberg picture are

$$\hat{A}_H(t) = e^{i\hat{H}t/\hbar}\hat{A}_S e^{-i\hat{H}t/\hbar}. \tag{57.42}$$

Then we can define the creation and annihilation operators in the Dirac picture, $\hat{c}_{i,I}(t), \hat{c}_{i,I}^\dagger(t)$.

A crucial object in the theory is the Green's function, defined as

$$G_{\sigma\sigma'}(\vec{r},t;\vec{r}',t') \equiv -i\frac{\langle\psi_0|T\{\hat{\psi}_{H,\sigma}(\vec{r},t)\hat{\psi}^\dagger_{H,\sigma'}(\vec{r}',t')\}|\psi_0\rangle}{\langle\psi_0|\psi_0\rangle}. \tag{57.43}$$

This will have an expansion similar to the single-particle Green's function, generalizing

$$\hat{G}(z) \sim \sum_i \frac{|i\rangle\langle i|}{z - E_i}, \tag{57.44}$$

implying that the Green's function has poles at the energies of the bound states E_i.

Moreover, the Green's function is part of a more general correlation function of generic operators $\hat{A}, \hat{B}$ in the Heisenberg picture; for the ground state wave function $|\psi_0\rangle$ we have

$$\frac{\langle\psi_0|T\{\hat{A}_H(t)\hat{B}_H(t')\}|\psi_0\rangle}{\langle\psi_0|\psi_0\rangle}. \tag{57.45}$$

The "Feynman theorem" will relate these correlation functions to those of the interaction picture operator $\hat{A}_I(t), \hat{B}_I(t)$ and $|\phi_0\rangle$, the ground state of $\hat{H}_0$.

Then the Green's functions are related to observables. For one-particle operators,

$$\hat{A}_{\sigma\sigma'} = \int d^3\vec{r}\, \hat{a}_{\sigma\sigma'}(\vec{r}), \tag{57.46}$$

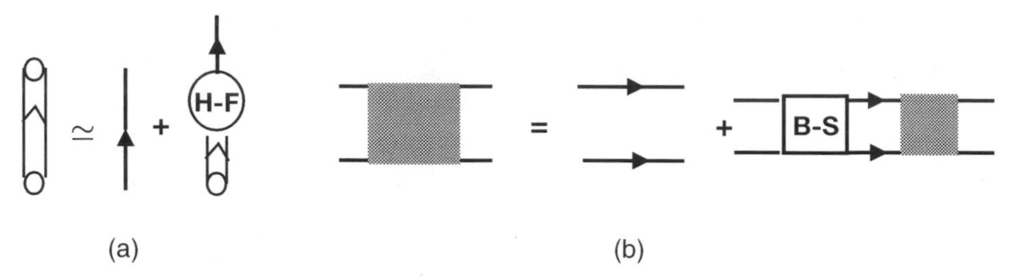

Figure 57.1 (a) Hartree–Fock equation for Green's functions. (b) Bethe–Salpeter equation for four-point Green's functions.

the average in the ground state is

$$\begin{aligned}\langle a(\vec{r})\rangle &\equiv \langle\psi_0|a_{\sigma\sigma}|\psi_0\rangle = -i \lim_{\vec{r}\to\vec{r}',t'\to t}\sum_{\sigma,\sigma'} A_{\sigma,\sigma'}G_{\sigma',\sigma}(\vec{r},t;\vec{r}',t')\\ &= -i\,\mathrm{Tr}[AG].\end{aligned} \tag{57.47}$$

Similarly, the average in the ground state of the two-particle operator is

$$\langle\hat{B}\rangle = -\frac{1}{2}\int d^3\vec{r}'\int d^3\vec{r}\sum_{\sigma_1\sigma_2\sigma_1'\sigma_2'} B_{\sigma_1\sigma_1'\sigma_2\sigma_2'}(\vec{r},\vec{r}')\lim_{t'\to t}G_{\sigma_2\sigma_2',\sigma_1\sigma_1'}(\vec{r},t;\vec{r}',t'). \tag{57.48}$$

Hartree–Fock versus Bethe–Salpeter Equation

We can consider a self-consistent equation for the standard perturbation theory approximation, the Hartree–Fock equation *for the Green's functions*,

$$G^{(2)} \simeq G_0^{(2)} + G_0^{(2)}\cdot V_{H-F}^{(2)}\cdot G^{(2)}, \tag{57.49}$$

which depicted pictorially is in Fig. 57.1a. As can be seen, the equation is for two-point Green's functions, i.e., propagators.

Alternatively, we can write a Bethe–Salpeter equation, or Dyson equation, for the four-point Green's functions $G^{(4)}$. This equation for $G^{(4)}$ is

$$G^{(4)} = G_{0,1}^{(2)}G_{0,2}^{(2)} + K_{B-S}^{(4)}\cdot G_{0,1}^{(2)}G_{0,2}^{(2)}\cdot G^{(4)}, \tag{57.50}$$

depicted pictorially in Fig. 57.1b.

However, as we saw, the Green's function has poles at the corrected bound states E_i, and near the pole at i, the Green's function becomes

$$\hat{G}(z) \simeq \frac{|i\rangle\langle i|}{z-E_i}, \tag{57.51}$$

which can be generalized to the case of several particles, in particular to the propagation of two particles, described by the four-point Green's functions $\hat{G}^{(4)}$. Near a pole it has a similar form,

$$\hat{G}^{(4)}(E) \sim \frac{|(ij)\rangle\langle(ij)|}{E-E_i}, \tag{57.52}$$

which means that the Bethe–Salpeter equation for the two-particle wave function $|(ij)\rangle$ is

$$|(ij)\rangle \simeq K_{B-S}^{(4)}\cdot G_{0,1}^{(2)}G_{0,2}^{(2)}|(ij)\rangle. \tag{57.53}$$

Important Concepts to Remember

- The Hartree–Fock approximation is an approximation for the energy of an electron in a multi-electron state, using the occupation number picture, via a self-consistent equation with field operators for the *other* electrons.
- The Hartree–Fock potential $\hat{V}_{H-F}$ is defined by $\langle i|\hat{V}_{H-F}|i'\rangle = \sum_{k\text{ occ.}} \langle ik|\hat{V}^{(2)}|i'k\rangle - \sum_{k\text{ occ.}} \langle ik|\hat{V}^{(2)}|ki'\rangle$.
- The Hartree–Fock equation states that then the Hamiltonian is diagonal, so $(\hat{T}^{(1)} + \hat{V}_{H-F})|i\rangle = E_i|i\rangle$, and the mean-field Hartree–Fock Hamiltonian in occupation number space is $\hat{H}_{H-F} = \sum_{i,i'} \langle i|\hat{T}^{(1)}|i'\rangle a_i^\dagger a_{i'} + \sum_{i,i'} \langle i|\hat{V}_{H-F}|i'\rangle a_i^\dagger a_{i'}$.
- On wave functions, from $\hat{V}_{H-F}$ we obtain two terms, the direct term or Hartree potential, $V^{(1)}_{H-F}\phi_{i'}(\vec{r})$, with $V_H(\vec{r}) = V^{(1)}_{H-F}(\vec{r}) = \int d^3\vec{r}\,' \sum_{k\text{ occ.}} |\phi_k(\vec{r}\,')|^2 V^{(2)}(\vec{r},\vec{r}\,') = \int d^3\vec{r}\,' \rho(\vec{r}\,') V^{(2)}(\vec{r},\vec{r}\,')$, and the exchange or Fock term, $V_F\phi_{i'}(\vec{r}) = V^{(2)}_{H-F}\phi_{i'}(\vec{r}) = \int d^3\vec{r}\,' V^{(2,\text{exch.})}_{H-F}(\vec{r},\vec{r}\,')\phi_{i'}(\vec{r}\,')$, with $V^{(2,\text{exch.})}_{H-F}(\vec{r},\vec{r}\,') = \sum_{k\text{ occ.}} \phi_k(\vec{r})\phi_k(\vec{r}\,') V^{(2)}(\vec{r},\vec{r}\,')$.
- The Hartree equation contains only the Hartree potential, as a semiclassical contribution from the charge of the other electrons, $\int d^3\vec{r}\,'\rho(\vec{r}\,')V(\vec{r}-\vec{r}\,')$, while the Fock term contains the purely quantum exchange term.
- In the occupation number picture, we can also introduce a nonrelativistic version of quantum field theory, via field operators. Interaction picture operators, $\hat{A}_I(t) = e^{i\hat{H}_0/\hbar}\hat{A}_S e^{-i\hat{H}_0/\hbar}$ are used for perturbation theory, and Heisenberg picture operators $\hat{A}_H(t) = e^{i\hat{H}/\hbar} A_S e^{-i\hat{H}/\hbar}$ are used for definitions.
- Green's functions

$$G_{\sigma\sigma'}(\vec{r},t;\vec{r}\,',t') \equiv -i\frac{\langle\psi_0|T\{\hat{\psi}_{H,\sigma}(\vec{r},t)\hat{\psi}^\dagger_{H,\sigma'}(\vec{r}\,',t')\}|\psi_0\rangle}{\langle\psi_0|\psi_0\rangle}$$

and more general correlation functions

$$\frac{\langle\psi_0|T\{\hat{A}_H(t)\hat{B}_H(t')\}|\psi_0\rangle}{\langle\psi_0|\psi_0\rangle}$$

can be expanded, via the Feynman theorem, using the perturbation theory of interaction picture operators; one-particle operators $\hat{A} = \int d^3\vec{r}\, a_{\sigma\sigma'}(\vec{r})$ have an average in the ground state $\langle a\rangle = -i\,\text{Tr}[AG]$ and similar formulas hold for operators for higher numbers of particles.
- One can write a Hartree–Fock equation for (two-point) Green's functions, $G^{(2)} \simeq G_0^{(2)} + G_0^{(2)} \cdot V^{(2)}_{H-F} \cdot G^{(2)}$, or alternatively a Dyson equation, or version of the Bethe–Salpeter equation, for the four-point Green's function, $G^{(4)} = G^{(2)}_{0,1}G^{(2)}_{0,2} + K^{(4)}_{B-S} \cdot G^{(2)}_{0,1}G^{(2)}_{0,2} \cdot G^{(4)}$.

Further Reading

More details are given in the books [2] and [27].

Exercises

(1) Consider bosons instead of fermions, and set up the corresponding problem for a self-consistent approximation.

(2) Continue with this method to find the equivalent of the Hartree–Fock potential $\hat{V}_{H-F}$ for bosons.

(3) Given the Hartree–Fock equation, rewrite the Hartree–Fock Hamiltonian.

(4) Calculate the Hartree potential for the helium atom in its ground state.

(5) Calculate the two-particle exchange term in the Hartree–Fock approximation for the lithium atom in its ground state.

(6) Write explicitly the Hartree–Fock equation for the 2-point Green's function, in the coordinate ($\vec{r}$ and spin σ) representation.

(7) Write explicitly, in the coordinate ($\vec{r}$ and spin σ) representation, the Dyson or Bethe–Salpeter equation.

58 Nonstandard Statistics: Anyons and Nonabelions

In this chapter we consider statistics other than Bose–Einstein and Fermi–Dirac, namely, statistics involving multiplication with an abelian phase $e^{i\alpha}$ (different from ± 1), called *anyonic statistics*, or with a nonabelian factor U, called nonabelian anyonic statistics. The corresponding particles, anyons and nonabelions, live in 2+1 dimensions and are associated with the abelian and nonabelian Berry phase and connection. Therefore, we will review these concepts first. The implementation in physical materials of the new statistics is associated with the Chern–Simons action, which we will also introduce.

58.1 Review of Statistics and Berry Phase

In Chapter 20 we considered the permutation of particles within a quantum state,

$$|\psi\rangle = |x_1 x_2\rangle \rightarrow \hat{P}_{12}|\psi\rangle = |x_2 x_1\rangle. \tag{58.1}$$

We found that $\hat{P}_{12}$ has eigenvalues C_{12}. Then we imposed that the application of the interchange operator twice returns the situation to the initial state, $\hat{P}_{12}^2 = \mathbb{1}$, implying that the eigenvalue is $C_{12} = \pm 1$. This behavior was associated with Bose–Einstein (bosons) and Fermi–Dirac (fermions) statistics. But it is not necessary to have $\hat{P}_{12}^2 = \mathbb{1}$. In fact, we can "interchange" particles continuously, through a path around each other, leading to a Berry phase, as considered in Chapter 20.

If the Hamiltonian is time dependent, $\hat{H} = \hat{H}(t)$, via the *adiabatic* time dependence of parameters $K = K(t)$, the time dependence of the state that at $t = 0$ is $|n\rangle$ and has energy E_n has an extra phase, called the Berry phase γ_n, so that

$$|\psi(t)\rangle = \exp\left[-\frac{i}{\hbar}\int_0^t E_n(t')dt'\right] e^{i\gamma_n(t)}|n(K(t))\rangle. \tag{58.2}$$

Here the *Berry connection* is

$$\vec{A}_n(K) = i\langle n(K(t))|\vec{\nabla}_K|n(K(t))\rangle. \tag{58.3}$$

If K corresponds to a rescaled position $\vec{R}/(\hbar c)$, then the Berry phase is

$$\gamma_n = \int_C \frac{\vec{A}_n(\vec{R})}{\hbar c}\cdot d\vec{R}, \tag{58.4}$$

and is an Aharanov–Bohm-like phase.

We saw that the Berry phase on a closed contour C, modulo $2\pi m$, is invariant and so cannot be removed by a gauge transformation since, under it,

$$\gamma_n'(C) = \gamma_n(C) + 2\pi m. \tag{58.5}$$

We also defined a nonabelian generalization, in the case when the state with E_n has degeneracy N with index $|a\rangle$, $a = 1, \dots, N$. Then the adiabatic Hamiltonian $\hat{H}(K(t))$ has states $|n, K\rangle$ depending on time through the parameters. In this case, the nonabelian Berry connection is

$$\vec{A}^{ab}(K) = i\langle a, K(t)|\vec{\nabla}_K|b, K(t)\rangle, \tag{58.6}$$

with values in in the adjoint representation of the group $U(N)$ of unitary transformations. The nonabelian Berry phase is

$$\gamma^{(ab)} = \oint_C \vec{A}^{(ab)}(\vec{K}) \cdot d\vec{K}. \tag{58.7}$$

The Berry phase factor is path ordered,

$$\mathcal{P}e^{i\gamma^{(ab)}}, \tag{58.8}$$

implying that the expansion of the exponential in powers is ordered by the value of $\vec{K}$ along C: the later in the path, the further to the right.

58.2 Anyons in 2+1 Dimensions: Explicit Construction

We now construct explicitly an anyon system, where the particles are neither bosons or fermions and under the permutation of two particles we obtain a phase $e^{i\alpha}$, different from ± 1. The possibility of an abelian phase $e^{i\alpha}$ is restricted to 2+1 dimensions, which means that the system has an infinitesimal extent in a third spatial dimension.

Consider the following construction of an anyon system. We start with a system of N particles, for instance electrons, interacting with an electromagnetic field $A_\mu = (A_0, \vec{A})$ since they are charged. The Hamiltonian is

$$\hat{H} = \sum_{i=1}^{N} \frac{|\vec{p}_i - q\vec{A}(\vec{r}_i)|^2}{2m} + \sum_{i<j} v(\vec{r}_i - \vec{r}_j) + \sum_{i=1}^{N} qA_0(\vec{r}_i), \tag{58.9}$$

where $\vec{r}_i$ are the positions of the particles, and $v(\vec{r}_i - \vec{r}_j)$ is a two-particle potential.

The multiparticle wave function $\psi(\vec{r}_1, \dots, \vec{r}_N)$ obeys the Schrödinger equation,

$$\hat{H}\psi(\vec{r}_1, \dots, \vec{r}_N) = E\psi(\vec{r}_1, \dots, \vec{r}_N). \tag{58.10}$$

Now we make a canonical transformation (a unitary transformation) of the wave function from ψ to $\phi = U\psi$ as follows:

$$\phi(\vec{r}_1, \dots, \vec{r}_N) = U\psi(\vec{r}_1, \dots, \vec{r}_N) = \left\{\prod_{i<j} \exp\left[-i\frac{\theta}{\pi}\alpha(\vec{r}_i - \vec{r}_j)\right]\right\} \psi(\vec{r}_1, \dots, \vec{r}_N), \tag{58.11}$$

where $\alpha(\vec{r}_i - \vec{r}_j)$ is the angle made by the relative distance between the particles $\vec{r}_i - \vec{r}_j$ with respect to a fixed axis. Under U the Hamiltonian transforms as $\hat{H} \to U^{-1}\hat{H}U$. We can check easily that

$$U^{-1}(\vec{p}_i - q\vec{A}(\vec{r}_i))U = \vec{p}_i - q\vec{A}(\vec{r}_i) - q\vec{a}(\vec{r}_i), \tag{58.12}$$

where we have defined an emergent (or statistical) gauge field $\vec{a}$ by

$$q\vec{a}(\vec{r}_i) = \frac{\theta}{\pi}\sum_{j\neq i}\vec{\nabla}_i\alpha(\vec{r}_i - \vec{r}_j) = \frac{\theta}{\pi}\sum_{j\neq i}\frac{\hat{z}\times(\vec{r}_i - \vec{r}_j)}{|\vec{r}_i - \vec{r}_j|^2}. \tag{58.13}$$

Here $\hat{z}$ is the unit vector in the direction perpendicular to the two-dimensional material (the third spatial direction).

Note that, since $\alpha(\vec{r}_i - \vec{r}_j)$ is the angle of the relative distance with the fixed axis,

$$\alpha(\vec{r}_i - \vec{r}_j) = \alpha(\vec{r}_j - \vec{r}_i) + \pi. \tag{58.14}$$

Substituting this in the definition of $\phi = U\psi$, when we exchange the two particles i and j, after the canonical transformation by U we obtain an extra $e^{i\theta}$, besides the original ± 1 valid for ψ:

$$\phi(\ldots,\vec{r}_j,\ldots,\vec{r}_i,\ldots) = \pm e^{i\theta}(\ldots,\vec{r}_i,\ldots,\vec{r}_j,\ldots). \tag{58.15}$$

Then, if $\theta = (2k+1)\pi$, the statistics is changed from Bose–Einstein to Fermi–Dirac and from Fermi–Dirac to Bose–Einstein and, for more general θ, we change to fractional, or anyonic, statistics. The field a_μ is called the *statistical gauge field*, since it changes the statistics.

In the new representation, the Hamiltonian is

$$\hat{H} = \sum_{i=1}^{N}\frac{|\vec{p}_i - q\vec{A}(\vec{r}_i) - q\vec{a}(\vec{r}_i)|^2}{2m} + \sum_{i<j} v(\vec{r}_i - \vec{r}_j) + \sum_{i=1}^{N} qA_0(\vec{r}_i), \tag{58.16}$$

where $\vec{a}$ is added to $\vec{A}$, and its field strength (due to the magnetic field) is

$$f_{12}(\vec{r}_i) \equiv b(\vec{r}_i) = \vec{\nabla}\times\vec{a}(\vec{r}_i) = \frac{2\theta}{q}\sum_{j\neq i}\delta(\vec{r}_i - \vec{r}_j) = \frac{2\theta}{q^2}\rho_{\rm charge}(\vec{r}_i), \tag{58.17}$$

and where $\rho_{\rm charge}(\vec{r}_i) = q\sum_{j\neq i}\delta(\vec{r}_i - \vec{r}_j)$ is the charge density of the anyon system. Note that in 2+1 dimensions the magnetic field is a scalar B, not a vector, and the magnetic flux associated with it is just $\Phi = \int BdS$.

Since $\vec{A}\to\vec{A}+\vec{a}$ after the canonical (unitary) transformation, if $B = F_{12}(\vec{r}) = 0$ before the transformation then the total magnetic field after it is

$$\tilde{B} = \tilde{F}_{12} = \vec{\nabla}\times(\vec{A}+\vec{a})(\vec{r}_i) = \frac{2\theta}{q}\sum_{j\neq i}\delta(\vec{r}_i - \vec{r}_j). \tag{58.18}$$

This means that we have a delta function magnetic flux associated with the particle (at its position),

$$\Phi = \int_S \tilde{F}_{12}dS = \int_S f_{12}dS = \frac{2\theta}{q}, \tag{58.19}$$

where S is a small surface around the particle. Therefore, we have attached a magnetic flux at the position of the particle, turning it into an *anyon*.

The explicit construction above was through the canonical (unitary) transformation, leading to magnetic field B for the statistical gauge field, but we can define B as a delta function for any gauge field, including an electromagnetic field itself, and still obtain an anyon behavior.

Indeed, the Aharonov–Bohm phase obtained when one of the anyons is interchanged with another, by continuously moving it around the other, is

$$\exp\left(iq\oint_C \vec{A}\cdot d\vec{r}\right) = \exp\left(iq\Phi[C]\right) = e^{2i\theta}. \tag{58.20}$$

However, rotating one anyon around another returns it to the original position, therefore corresponds to two interchanges. That means that the factor for a single interchange is

$$C_{12} = e^{i\theta}. \tag{58.21}$$

Moreover, the same argument can be used for the more general Berry phase γ, associated with a Berry connection $\vec{A}$. Then the Berry phase for one particle going around the other in parameter space gives

$$e^{2i\theta} = e^{i\gamma} \Rightarrow \theta = \frac{\gamma}{2}. \tag{58.22}$$

58.3 Chern–Simons Action

We have obtained the relation between the magnetic field (here the total magnetic field) and the electric charge density,

$$F_{12} = \frac{2\theta}{q}\rho_{\text{charge}} \equiv \frac{2\theta}{q}J_0, \tag{58.23}$$

but ρ_{charge} is the zeroth component of the relativistic current J_μ (in 2+1 dimensions), $\rho_{\text{charge}} = J_0$.

The above equation is the (12) component of a relativistically covariant equation,

$$F_{\mu\nu} = \frac{2\pi}{k}\epsilon_{\mu\nu\rho}J^\rho, \tag{58.24}$$

with a "Chern-Simons quantization level" k, where

$$k = \pi\frac{q}{\theta}, \tag{58.25}$$

and $\epsilon^{\mu\nu\rho}$ is the totally antisymmetric Levi–Civita tensor, with $\epsilon^{012} = +1$.

The action from which this relativistically covariant equation of motion comes is

$$\begin{aligned} S_{CS} &= \frac{k}{4\pi}\int_M d^{2+1}x\, \epsilon^{\mu\nu\rho}A_\mu\partial_\nu A_\rho \\ &= \frac{k}{4\pi}\int_M d^{2+1}x\Big[-A_1\dot{A}_2 + A_2\dot{A}_1 \\ &\qquad + A_0\partial_1A_2 - A_2\partial_1A_0 - A_0\partial_2A_1 + A_1\partial_2A_0\Big] \\ &= \frac{k}{2\pi}\int_M d^{2+1}x[A_2\dot{A}_1 + A_0(\partial_1A_2 - \partial_2A_1)] \\ &= \frac{k}{2\pi}\int_M d^{2+1}x[A_2\dot{A}_1 + A_0F_{12}], \end{aligned} \tag{58.26}$$

where we have used partial integration in the third equality, assuming that the boundary term at infinity vanishes.

Adding a source term,

$$S_{\text{source}} = \int d^{2+1}x\, J^\mu A_\mu = \int d^{2+1}x[\vec{J}\cdot\vec{A} + J^0A_0], \tag{58.27}$$

the total action,

$$S = S_{CS} + S_{\rm source} = \int d^{2+1}x \left[\frac{k}{4\pi} \epsilon^{\mu\nu\rho} A_\mu \partial_\nu A_\rho + J^\mu A_\mu \right], \quad (58.28)$$

has the equations of motion

$$\frac{k}{4\pi} \epsilon^{\mu\nu\rho} F_{\nu\rho} = J^\mu \quad \Rightarrow \qquad F_{\mu\nu} = \frac{2\pi}{k} \epsilon_{\mu\nu\rho} J^\rho, \quad (58.29)$$

as required.

Alternatively, the equation of motion for A_0 (considering $\vec{A}$ as fixed) is just the relation (without the relativistically covariant generalization)

$$F_{12} = \frac{2\theta}{q} J^0. \quad (58.30)$$

Moreover, the Chern–Simons level k is quantized. Its quantization is related to Dirac quantization, as we will now show.

The gauge transformation is in general

$$\delta \vec{A} = \vec{\nabla} \lambda, \quad \delta A_0 = \dot{\lambda}, \quad (58.31)$$

but we restrict to transformations that depend only on time, $\lambda = \lambda(t)$. Then the Chern–Simons action transforms as

$$\delta S_{CS} = \frac{k}{2\pi} \int_M d^{2+1}x \dot{\lambda} F_{12} = \frac{k}{2\pi} \int_{t_1}^{t_2} dt \int_M dS F_{12} \dot{\lambda} = \left[\frac{\int_M BdS}{2\pi} \right] k \left(\lambda(t_2) - \lambda(t_1) \right). \quad (58.32)$$

We now consider periodicity in time, which is related to finite temperature as we saw in Chapter 28 (the relation to finite temperature appearing after the Wick rotation to imaginary, or Euclidean, time), $t_2 - t_1$ being the periodicity. Since λ is the gauge parameter, under the gauge transformation the wave function of the anyon changes by (see Dirac quantization in Chapter 27)

$$\psi \to e^{iq\lambda/\hbar} \psi, \quad (58.33)$$

and in order for the gauge transformation to be well defined as we go around the periodicity cycle, we need to have the periodicity

$$\lambda(t_2) - \lambda(t_1) = \frac{2\pi\hbar}{q} m, \quad m \in \mathbb{Z}. \quad (58.34)$$

Substituting this in the variation of the Chern–Simons action δS_{CS}, together with the fact that the magnetic flux is quantized in units of $\Phi_0 = h/q$, as seen in Chapter 24 concerning the Hall effect,

$$\frac{\int BdS}{2\pi} = \frac{\Phi_p}{2\pi} = \frac{p}{2\pi} \frac{h}{q} = p \frac{\hbar}{q}, \quad p \in \mathbb{Z}, \quad (58.35)$$

and defining the product of the integers m and p as an integer n, we obtain

$$\delta S_{CS} = kn \frac{2\pi\hbar}{q} \frac{\hbar}{q}. \quad (58.36)$$

Since the action appears in the path integral for quantum calculations in the form $e^{iS/\hbar}$, and therefore the variation under the gauge transformation of $e^{iS/\hbar}$ must be trivial, $e^{i\delta S/\hbar} = 1$, we obtain

$$\delta S_{\rm CS} = 2\pi\hbar N \quad \Rightarrow \quad k\frac{\hbar}{q^2} \in \mathbb{Z}. \tag{58.37}$$

It is usual to rescale the gauge field by the charge, $A_\mu \to A_\mu/q$, such that the electromagnetic kinetic term has $1/q^2$ in front,

$$S_{\rm em.kin} = -\frac{1}{4q^2}\int F_{\mu\nu}F^{\mu\nu}; \tag{58.38}$$

then the quantization condition is simply that k is an integer in units of $1/\hbar$, or $k\hbar \in \mathbb{Z}$.

Then the Chern–Simons action $S_{\rm CS}$ generates anyons, and

$$\theta = \frac{\pi}{k}, \quad k \in \mathbb{Z}. \tag{58.39}$$

58.4 Example: Fractional Quantum Hall Effect (FQHE)

The Hall effect refers to the current density $\vec{j}\,(j_y)$ induced in a direction perpendicular to the applied electric field $\vec{E}\,(E_x)$ and a constant magnetic field $\vec{B}\,(B_z)$, with $j_y = \sigma_H E_x$. Then the integer quantum Hall effect (IQHE) refers to a quantized value of the Hall conductivity, $\sigma_H = n\sigma_0$, whereas the fractional quantum Hall effect (FQHE) refers to fractional values of the same, $\sigma_H = (q/r)\sigma_0$, with $q, r \in \mathbb{Z}$.

We will not consider further the theory of the FQHE, first encountered in Chapter 24, but will just say that for the particular case where

$$\sigma = \frac{1}{r}\sigma_0, \tag{58.40}$$

the action that describes both the Hall effect on the conductivity, and the particles responsible for this effect, is

$$S_{\rm eff} = \int_M d^{2+1}x\left[\frac{1}{2\pi}\epsilon^{\mu\nu\rho}A_\mu\partial_\nu a_\rho - \frac{r}{4\pi}\epsilon^{\mu\nu\rho}a_\mu\partial_\nu a_\rho\right]. \tag{58.41}$$

The classical equations of motion of the action (for a_μ) are

$$f_{\mu\nu} = \partial_\mu a_\nu - \partial_\nu a_\mu = \frac{1}{r}F_{\mu\nu} = \frac{1}{r}(\partial_\mu A_\nu - \partial_\nu A_\mu) \quad \Rightarrow \quad a_\mu = \frac{1}{r}A_\mu. \tag{58.42}$$

Then the on-shell action (solved for a_μ) in terms of the electromagnetic field A_μ is

$$S'_{\rm eff} = \frac{1}{r}\frac{1}{4\pi}\int_M d^{2+1}x\,\epsilon^{\mu\nu\rho}A_\mu\partial_\nu A_\rho, \tag{58.43}$$

and from it we obtain the fractional Hall conductivity $\sigma_H = \sigma_0/r$ (though we will not consider it further here).

But if we do not solve for a_μ, we obtain the kinetic term for the statistical field interacting with anyons. Indeed, the source term for electromagnetism, in the form of charged particles, $\int eJ^\mu A_\mu$, charge e, implies that there should also be a source term for the statistical gauge field a_μ, in the form of "quasiparticles" of charge q under a_μ,

$$\int d^{2+1}x\, qJ^\mu a_\mu = q\int dt\, a_0(\vec{x},t). \tag{58.44}$$

Adding this source term to $S_{\rm eff}$, the equation of motion for a_μ becomes

$$\frac{F_{12}}{2\pi} - \frac{rf_{12}}{2\pi} + q\delta(\vec{x} - \vec{x}(t)) = 0. \tag{58.45}$$

This implies that we have a delta function flux either in F_{12} (the electromagnetic field) or in f_{12} (the statistical field). But having a delta function flux in the electromagnetic field is an idealization that does not sit well with reality, since the flux lines going up through the third direction should return back down somewhere else. Instead, we can have a delta function flux in the statistical gauge field, that is,

$$\frac{f_{12}}{2\pi} = \frac{q}{r}\delta(\vec{x} - \vec{x}(t)). \tag{58.46}$$

Comparing this with the relation between the magnetic field and the charge density for anyons, we obtain that the anyon statistics parameter θ is

$$\theta = \frac{\pi}{r}q^2, \tag{58.47}$$

and, as we said, the factor q^2 can be absorbed into a redefinition of the gauge fields.

58.5 Nonabelian Statistics

Until now we have considered an abelian phase statistics, but we can have also a nonabelian statistics. Moore and Read have shown that, even in the FQHE, we can have nonabelian statistics.

That means that under the interchange of particles, the possibilities for the change in the wave function do not belong to the permutation group $\mathbb{Z}_N$ but rather to the braid group $\mathcal{B}_N$. We can think of the anyons as abelian representations of the braid group $\mathcal{B}_N$. But there are nonabelian representations of the braid group, and the particles charged under these representations are called nonabelian anyons, or nonabelions.

Under the interchange (braid), the wave function changes as

$$\begin{aligned}&\psi_{p:\{i_1,\dots,i_r,\dots,i_s,\dots,i_N\}}(z_1,\dots,z_{i_r},\dots,z_{i_s},\dots,z_N)\\ &= \sum_q B_{pq}[i_1,\dots,i_N]\psi_{q:\{i_1,\dots,i_N\}}(z_1,\dots,z_N),\end{aligned} \tag{58.48}$$

where p is a shorthand for the set of indices $\{i_1,\dots,i_N\}$, and $z\in\mathbb{C}$ is the two-dimensional spatial coordinate.

The Moore–Read wave function for N "electrons without spin" is

$$\psi_{Pf}(z_1,\dots,z_N) = \text{Pfaff}\left(\frac{1}{z_i - z_j}\right)\prod_{i<j}(z_i - z_j)^m \exp\left[-\frac{1}{4}|z_j|^2\right], \tag{58.49}$$

where z_i is the position of a "vortex" and the Pfaffian is a kind of "square root" of the determinant, when N is even (there is no space to explain this further). It is rather hard to show that a bound

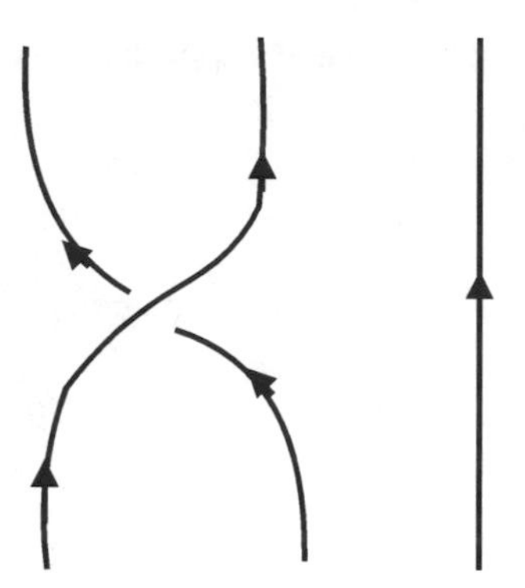

Figure 58.1 A basic braiding move.

state of four excitations obeys nonabelian statistics. Equivalently, and also hard to show, we have a nonabelian Berry connection, leading to a nonabelian Berry phase, also with nonabelian statistics.

An example of the action of the nonabelian braid on bound states of excitations is as follows. Consider oscillators with annihilation and creation operators c_i and $c_i^\dagger$, with commutation relations

$$\{c_i, c_j^\dagger\} = \delta_{ij}, \quad \{c_i, c_j\} = \{c_i^\dagger, c_j^\dagger\} = 0. \tag{58.50}$$

Then define the Majorana fermion zero modes (modes of zero energy) γ_i as

$$\gamma_i = c_i + c_i^\dagger, \tag{58.51}$$

satisfying the Clifford algebra

$$\{\gamma_i, \gamma_j\} = \delta_{ij}, \quad i = 1, \ldots, 2n. \tag{58.52}$$

Then the complex fermion made up from the Majorana fermions is

$$\psi_k = \frac{\gamma_{2k+1} + i\gamma_{2k}}{2}, \quad k = 1, \ldots, n, \tag{58.53}$$

and has a 2^n-dimensional Hilbert space. For each vortex, we have one Majorana zero mode, but the relevant excitations forming the nonabelion are nonlocal (they come from several vortices).

An example of braiding, in fact the one that generates the braid group, is where the ith vortex is braided with the $(i+1)$th vortex, in an anticlockwise direction:

$$R : \{\gamma_i \to \gamma_{i+1}, \quad \gamma_{i+1} \to -\gamma_i, \quad \gamma_j \to \gamma_j, \quad j \neq i, i+1\}. \tag{58.54}$$

A graphical representation of a basic braiding move is given in Fig. 58.1.

Important Concepts to Remember

- Instead of the Bose–Einstein and Fermi–Dirac statistics, in 2+1 dimensions we can also have particles whose statistics are defined by interchange by a phase $e^{i\theta}$, called anyons, or by a nonabelian matrix, called nonabelions.
- We can construct anyons with $e^{i\theta}$ by a unitary transformation from the wave function of a system of ordinary particles (for instance electrons), obtaining an extra coupling to a statistical gauge field qa_μ, $q\vec{a} = \frac{\theta}{\pi} \sum_{j\neq i} \vec{\nabla}_i \alpha(\vec{r}_i - \vec{r}_j)$.

- To construct the anyon we attach a flux at the position of the particle, $\tilde{B} = \tilde{F}_{12} = \vec{\nabla} \times (\vec{A} + \vec{a})(\vec{r}_i) = \frac{2\theta}{q} \sum_{j \neq i} \delta(\vec{r}_i - \vec{r}_j)$.
- The constructed anyon satisfies the equation of motion for the Chern–Simons action with a source, $\int d^{2+1}x \left[\frac{k}{4\pi} \epsilon^{\mu\nu\rho} A_\mu \partial_\nu A_\rho + J^\mu A_\mu \right]$, with quantized level k, quantized via Dirac quantization.
- Anyons and the associated Chern–Simons action appear also in the theoretical description of the fractional quantum Hall effect (FQHE), for $\sigma_r = \sigma_0 / r$.
- A nonabelian statistics is also possible, with interchange under the braiding group, exemplified on Majorana fermion zero modes, as found from the Moore–Read wave function.

Further Reading

More details about the FQHE and nonabelian statistics can be found in the review by David Tong [29], and the Moore–Read paper [30].

Exercises

(1) The nonabelian Berry phase, by its very definition, changes the wave function, even though we return to the initial state in terms of $K(t)$ (in the abelian case, the wave function changes by a phase, so $|\psi|^2$ doesn't change). Can you give an example where this is consistent?

(2) Show the steps in the proof of the relation (58.12), for constructing an anyon from a statistical gauge field.

(3) Consider the (01) component of the equation of motion $F_{\mu\nu} = \frac{2\pi}{k} \epsilon_{\mu\nu\rho} J^\rho$, coming from the relativistic Chern–Simons action with a source. What physical property does it describe? Note that this is now related to the existence of anyons.

(4) Show that the Chern–Simons action (without a source) is equal to a (3 + 1)-dimensional action of a gauge invariant, topological (metric independent, and with discrete values) type.

(5) When describing the FQHE in the text, we gave the option of solving for the statistical gauge field a_μ and obtaining a Chern–Simons-like action for the electromagnetic field A_μ, or of adding a source term and obtaining an anyon. But why did we not consider the equation of motion for A_μ, either in the first or in the second case?

(6) When discussing the FQHE anyon as a statistical gauge delta function flux added to the quasiparticles, we argued that for the electromagnetic field $F_{\mu\nu}$ to be a delta function is unphysical because of flux conservation. But why then is it OK for the statistical $f_{\mu\nu}$ to be a delta function?

(7) Show that the factor $(z_i - z_j)^m$ in the Moore–Read wave function (also present in the phenomenological "Laughlin wave function" for a theoretical description of the FQHE) implies there are m "vortices" at each position. A vortex ansatz is $f(r)e^{i\alpha}$, where r, α are the radius and the polar angle in the plane, respectively, and one can impose a consistency condition on the ansatz (and so find it).

References

[1] J.J. Sakurai and San Fu Tuan, *Modern Quantum Mechanics*, revised edition, Addison–Wesley, 1994.

[2] Albert Messiah, *Quantum Mechanics*, Dover Publications, 2014.

[3] R. Shankar, *Principles of Quantum Mechanics*, second edition, Springer, 1994.

[4] L.D. Landau and L.M. Lifshitz, *Quantum Mechanics (Non-Relativistic Theory)*, Butterworth-Heinemann, 1981.

[5] H. Nastase, *Introduction to Quantum Field Theory*, Cambridge University Press, 2019.

[6] H. Nastase, *String Theory Methods for Condensed Matter Physics*, Cambridge University Press, 2017.

[7] H. Nastase, *Classical Field Theory*, Cambridge University Press, 2019.

[8] Howard Georgi, *Lie Algebras in Particle Physics: From Isospin to Unified Theories*, Westview Press, 1999.

[9] Michael V. Berry, "Quantal phase factors accompanying adiabatic changes", *Proc. Roy. Soc. Lond. A* **392** (1984) 45.

[10] R. Rajaraman, *Solitons and Instantons: An Introduction to Solitons and Instantons in Quantum Field Theory*, first edition, North-Holland Personal Library, Vol. 15, 1987.

[11] Paul A.M. Dirac, *Lectures in Quantum Mechanics*, Dover Publications, 2001.

[12] John Preskill, *Caltech Lecture Notes on Quantum Information*, http://theory.caltech.edu/ preskill/ph219/.

[13] E. Knill, R. Laflamme, H. Barnum, *et al.*, "Introduction to quantum information processing", arXiv:quant-ph/0207171 [quant-ph].

[14] Richard Jozsa, "An introduction to measurement-based quantum computation", arXiv:quant-ph/0508124 [quant-ph].

[15] Daniel Gottesman, "An introduction to quantum error correction and fault-tolerant quantum computation", arXiv:0904.2557 .[quant-ph].

[16] J.S. Bell, "On the Einstein–Podolsky–Rosen paradox", *Physics* **1** (3) (1964) 195.

[17] E.P. Wigner, *American Journal of Physics* **38** (1970) 1005.

[18] J. Maldacena, S.H. Shenker, and D. Stanford, "A bound on chaos", *J. High-Energy Phys.* **08** (2016) 106, arXiv:1503.01409 [hep-th].

[19] R. Jefferson and R.C. Myers, "Circuit complexity in quantum field theory", *J. High-Energy Phys.* **10** (2017) 107, arXiv:1707.08570 [hep-th].

[20] M.A. Nielsen, "A geometric approach to quantum circuit lower bounds", *Quantum Inform. Comput.* **6** (2006) 213 [quant-ph/0502070].

[21] A.R. Brown and L. Susskind, "Second law of quantum complexity", *Phys. Rev. D* **97** (8) (2018) 086015, arXiv:1701.01107 [hep-th].

[22] S. Xu and B. Swingle, "Accessing scrambling using matrix product operators", *Nature Phys.* **16** (2) (2019) 199, arXiv:1802.00801 [quant-ph].

[23] M. Srednicki, "Chaos and quantum thermalization", cond-mat/9403051.

[24] R. Nandkishore and D. A. Huse, "Many-body localization and thermalization in quantum statistical mechanics", *Ann. Rev. Cond. Matter Phys.* **6** (2015) 15.

[25] W.H. Zurek, "Decoherence, Einselection, and the quantum origins of the classical", *Rev. Mod. Phys.* **75** (2003) 715, arXiv:quant-ph/0105127 [quant-ph].

[26] Marcos Rigol, Vanja Dunjko, and Maxim Olshanii, "Thermalization and its mechanism for generic isolated quantum systems", *Nature* **452** (2008) 854, arXiv:0708.1324 [cond-mat].

[27] A.L. Fetter and J.D. Walecka, *Quantum Theory of Many-Particle Systems*, Dover, 2003.

[28] N. Levinson, "Determination of the potential from the asymptotic phase", *Phys. Rev.* **75** (1949) 1445.

[29] David Tong, "Lectures on the quantum Hall effect", arXiv:1606.06687 [hep-th].

[30] G.W. Moore and N. Read, "Nonabelions in the fractional quantum Hall effect", *Nucl. Phys. B* **360** (1991) 362.

Index

Printed in the United States
by Baker & Taylor Publisher Services